ECOLOGY AND EVOLUTION OF ACOUSTIC COMMUNICATION IN BIRDS

ECOLOGY AND EVOLUTION OF ACOUSTIC COMMUNICATION IN BIRDS

Edited by **Donald E. Kroodsma**

Department of Biology
University of Massachusetts
Amherst, Massachusetts

and **Edward H. Miller**

Department of Biology
Memorial University of Newfoundland
St. John's, Newfoundland

Comstock Publishing Associates a division of

Cornell University Press | Ithaca and London

First published 1996 by Cornell University Press.

Printed in the United States of America

♾ The paper in this book meets the minimum requirements of the American National Standard for Information Sciences—Permanence of Paper for Printed Library Materials, ANSI Z39.48-1984.

Library of Congress Cataloging-in-Publication Data
Ecology and evolution of acoustic communication in birds / edited by
Donald E. Kroodsma and Edward H. Miller.
p. cm.
Includes bibliographical references (p.) and index.
ISBN 0-8014-3049-6 (alk. paper). — ISBN 0-8014-8221-6 (pbk. : alk. paper)
1. Birdsongs. 2. Animal communication. 3. Birds—Vocalization. I. Kroodsma, Donald E.
II. Miller, Edward H.
QL698.5.E36 1996
598.259—dc20 95-47777

We dedicate this book to the memory of Theodore A. Parker III

Ted was renowned for his knowledge of bird sounds, especially in the Neotropics, and could identify nearly 4000 species by their vocalizations alone. An expert on Neotropical biodiversity in general, not just birds, Ted was committed to applying his knowledge of bird sounds to conservation issues. Eager to begin his work, he bypassed a formal graduate education (for a personal glimpse of Ted in action, see Stap 1990). As the leader of Conservation International's Rapid Assessment Team, Ted used his extensive knowledge of bird sounds to help chart and preserve the biodiversity of Latin America. His thousands of tape recordings (about 20,000 recordings of more than 1600 species, housed at Cornell University's Library of Natural Sounds) and his field efforts are the most extensive in the Neotropics to date, and they alone could provide the basis for theses and research projects for decades. More important, Ted was an inspiration, both personally and by reputation, in every country he visited—for conservationists, for politicians, and for young ornithologists interested in understanding bird sounds or using them to promote conservation. Ted bridged disciplines and boundaries: academic and popular ornithology, ornithology and conservation, and the north temperate scientific establishment with the emerging scientific community of Latin America. Those of us who knew Ted remember him best for his boundless generosity in sharing his knowledge with others, for his sheer enjoyment of all our planet's riches, and for doing all he could to help preserve that biodiversity (see Remsen and Schulenberg in press).

Ted was killed in action, in a plane crash on 3 August 1993, while surveying a coastal forest 350 miles southwest of Quito, Ecuador (see Kaufmann 1993, Myers 1993, O'Neill 1993, K. J. Zimmer 1993, Collar 1995, Schulenberg 1995, and Remsen et al. in press). By dedicating this book to Ted, we salute both his extraordinary efforts and his success in a life that was all too short. His painstaking efforts to record Neotropical birds, species by species, enabled him and others to locate and identify taxa far more efficiently. In his explorations, he uncovered a wealth of exciting biological issues of the type featured in Chapter 15 of this volume. Perhaps most important, his knowledge of bird sounds and his subsequent surveys allowed the avifauna and related biodiversity of the Neotropics to be inventoried in a more rapid and systematic manner (Parker 1991). This volume celebrates the diversity of birds and their styles of vocal communication in a way that we think Ted would have appreciated. We hope that the book will also help readers to appreciate the magnitude of our planet's riches and the responsibility we all share to preserve that great wealth.

CONTENTS

PREFACE

In seeking contributors to this volume, we set out to recruit active and original workers on acoustic communication in birds. In June 1993 we wrote to 39 colleagues whose research and thinking we respected, and we asked: What ideas excite you most now? Please tell us in a relatively brief chapter. Keep your approach broad and integrative, with an ecological and evolutionary focus. Identify the status of current research in your field and the important ideas that you feel will dominate research in the future.

The response was swift; a few prospective authors declined because of other commitments, but most responded with great enthusiasm. After surveying the topics on which the participants chose to write, we expanded the coverage of topics by supplementing the original list of authors with other invitations. During the following year, we lost some authors and gained a few, and some were asked to be coauthors. The book took shape and in December 1994, only 18 months after we had made our first contacts, the completed manuscript was sent to Cornell University Press for review.

The 26 chapters in this book cover a broad spectrum. They are organized into five parts plus an appendix. Each part is preceded by an introduction that explains the relationships among the parts, the rationale for grouping particular chapters, and relationships among chapters within parts.

Part I is on development, because it seems logical to begin at the beginning. Individual birds must become competent at communicating, and that competence begins early in life. The focus in this section is on acknowledging and then trying to understand the diversity of developmental modes among birds; three chapters cover chosen species (two songbirds and a parrot). Display repertoires develop in individuals, so Part II focuses on vocal repertoires, both how they are classified (by us and the birds) and how birds use them. Part III covers vocal variation in time and space. How vocal displays develop dictates, to a considerable extent, the extent of temporal and spatial variation that occurs within and among populations. The five chapters in this part explore diverse topics: cultural evolution, systematics, speciation, and differences in communication systems among tropical and temperate species. To be useful in communication, all the vocal displays used by birds must be controlled by the brain and perceived appropriately by the sensory system—topics addressed in Part IV. The chapters in this part discuss the neuroendocrine control system, morphological and physiological constraints on sound production, display detection, sex differences in display recognition, and individual recognition.

Individuals develop (Part I) repertoires of displays (Part II) that vary in space and time (Part III); the great variety of displays must be produced and perceived (Part IV) during acts of communication (Part V). The five chapters in Part V are concerned with the use of interactive playback to study dynamic exchanges between individuals, communication networks in populations, the dawn chorus and other diel patterns of displaying, and the evolution of female choice based on vocal displays. The book concludes with an appendix that describes where and how to archive recordings and pleads for cooperation among bird recordists in preserving the precious tape recordings on which our knowledge is based. For some species and for some natural areas, these recordings will be the only ones ever made, so recordists have an overriding responsibility to document their recordings and to archive them where they can be curated properly.

We hope that our efforts to produce this volume will be rewarded in several ways. First, we want the book to be a showcase for the exciting research being done on acoustic communication in birds. Second, we hope that this book will suggest rich ideas for future work. We especially hope to excite young investigators and to direct them to intriguing problems in avian bioacoustics. The wealth of challenging research problems suggested by the chapter authors should occupy researchers for decades, and there is a pressing need for work on certain poorly known groups and threatened areas. Last, and perhaps most important, we are concerned for the survival of the birds themselves. The rate of habitat loss everywhere is high and accelerating, and we hope that fundamental knowledge of bird behavior will help to arrest these losses. Some studies will have direct benefits in conservation; others will contribute indirectly, by enhancing awareness and appreciation. In the end, humanity will save only those parts of the natural world that it has come to understand and love, and we hope this book will make a significant contribution to that appreciation.

We are especially grateful to the contributors. We asked for "not just another book chapter," but for their best effort. They accepted our challenge and produced remarkable chapters that capture the excitement and broad scope of modern research on bird sounds. We appreciate the authors' tolerance when we requested just a little bit more—sometimes clarification of a phrase, sometimes modification of letter sizes in figures, sometimes substantial reorganization or rewriting. Finally, we are grateful to the authors for contributing to the greater endeavor of understanding how and why birds vocalize; without this group of scientists, our world would be far duller, and humanity would know far less about birds.

Toby Gaunt, Ohio State University, provided a critical and insightful review of the entire book manuscript. We appreciate his thoughtful suggestions and sense of humor. Penny Jaques was instrumental in converting alien formats to our chosen word-processing package.

We thank Robb Reavill and especially Helene Maddux of Cornell University Press. They have been enthusiastic supporters of this project and have helped us realize our hope for a useful and interesting book published in a timely manner, at a price affordable for students. We also thank Mindy Conner for her attention to fine detail in the copyediting.

We are grateful to mentors who made a significant impact on our respective careers. Don Kroodsma thanks Sewall Pettingill, University of Michigan Biological Station, John Wiens, Oregon State University, and Peter Marler, Rockefeller University. Ted Miller thanks Stu Macdonald and Earl Godfrey, both at the National Museum of Canada, Dave Boag, University of Alberta, Bill Threlfall, Memorial University of Newfoundland, and Ian McLaren, Dalhousie University.

DONALD E. KROODSMA
EDWARD H. MILLER

Amherst, Massachusetts
St. John's, Newfoundland

CONTRIBUTORS

Luis F. Baptista, Department of Ornithology, California Academy of Sciences, Golden Gate Park, San Francisco, California 94118, USA

Michael D. Beecher, Animal Behavior Program, Departments of Psychology and Zoology, University of Washington, Box 351525, Seattle, Washington 98195, USA

Eliot A. Brenowitz, Departments of Psychology and Zoology NI-25, University of Washington, Seattle, Washington 98195, USA

Vincent Bretagnolle, Centre d'Etudes Biologiques de Chizé, Centre National de la Recherche Scientifique, 79360, Beauvoir sur Niort, France

Gregory F. Budney, Library of Natural Sounds, Laboratory of Ornithology, Cornell University, 159 Sapsucker Woods Road, Ithaca, New York 14850, USA

Torben Dabelsteen, Department of Population Biology and Centre for Sound Communication, Zoological Institute, University of Copenhagen, Tagensvej 16, DK-2200 Copenhagen N, Denmark

Robert J. Dooling, Department of Psychology, University of Maryland, College Park, Maryland 20742, USA

J. Bruce Falls, Department of Zoology, University of Toronto, Toronto, Ontario M5S 1A1, Canada

Susan M. Farabaugh, Department of Anatomy, School of Medicine, University of Auckland, Private Bag 92019, Auckland, New Zealand

Millicent Sigler Ficken, Department of Biological Sciences, University of Wisconsin, Milwaukee, Wisconsin 53201, USA

Sandra L. L. Gaunt, Borror Laboratory of Bioacoustics, Department of Zoology, Ohio State University, 1735 Neil Avenue, Columbus, Ohio 43210, USA

Robert W. Grotke, Library of Natural Sounds, Laboratory of Ornithology, Cornell University, 159 Sapsucker Woods Road, Ithaca, New York 14850, USA

Jack P. Hailman, Department of Zoology, Birge Hall, 430 Lincoln Drive, University of Wisconsin, Madison, Wisconsin 53706, USA

Andrew G. Horn, Department of Biology, Dalhousie University, Halifax, Nova Scotia B3H 4J1, Canada

Henrike Hultsch, Institut für Verhaltensbiologie, Frei Universität Berlin, Haderslebener Strasse 9, D-12163 Berlin, Germany

Andrew King, Departments of Psychology and Biology, Indiana University, Bloomington, Indiana 47405, USA

Georg M. Klump, Institut für Zoologie, Technische Universität Munchen, Lichtenbergstrasse 4, D-85748 Garching, Germany

Donald E. Kroodsma, Department of Biology, University of Massachusetts, Amherst, Massachusetts 01003, USA, and Visiting Fellow, Laboratory of Ornithology, Cornell University, Ithaca, New York 14850, USA

Marcel M. Lambrechts, CNRS-CEFE, B.P.-5051, F-34033 Montpellier cedex 1, France

Alejandro Lynch, Department of Biology, University of Papua New Guinea, Box 320, University P.O., Papua New Guinea

Jochen Martens, Institut für Zoologie, Johannes Gutenberg–Universität, Saarstrasse 21, D-55099 Mainz, Germany

Peter K. McGregor, Behaviour and Ecology Research Group, Department of Life Science, University of Nottingham, University Park, Nottingham NG7 2RD, United Kingdom

Edward H. Miller, Department of Biology, Memorial University of Newfoundland, St. John's, Newfoundland A1B 3X9, Canada

Eugene S. Morton, Department of Zoological Research, National Zoological Park, Smithsonian Institution, Washington, D.C. 20008, USA

Ken Otter, Department of Biology, Queen's University, Kingston, Ontario K7L 3N6, Canada

Robert B. Payne, Museum of Zoology, University of Michigan, Ann Arbor, Michigan 48109, USA

Richard Ranft, Wildlife Section, British Library National Sound Archive, 29 Exhibition Road, London SW7 2AS, United Kingdom

Laurene Ratcliffe, Department of Biology, Queen's University, Kingston, Ontario K7L 3N6, Canada

William A. Searcy, Department of Biology, University of Miami, Coral Gables, Florida 33124, USA

W. John Smith, Department of Biology, University of Pennsylvania, Philadelphia, Pennsylvania 19104, USA

David A. Spector, Department of Zoological Research, National Zoological Park, Smithsonian Institution, Washington, D.C. 20008, USA. Current address: Department of Biological Sciences, Central Connecticut State University, New Britain, Connecticut 06050, USA

Cynthia A. Staicer, Department of Biology, Dalhousie University, Halifax, Nova Scotia, B3H 4J1, Canada

F. Gary Stiles, Instituto de Ciencias Naturales, Universidad Nacional de Colombia, Apartado 7495, Bogotá, Colombia

Philip Kraft Stoddard, Department of Biological Sciences, Florida International University, Miami, Florida 33199, USA

Dietmar Todt, Institut für Verhaltensbiologie, Frei Universität Berlin, Haderslebener Strasse 9, D-12163 Berlin, Germany

Olga D. Veprintseva, Phonotheka of Animal Voices, Institute of Biophysics, Academy of Sciences, 142292 Puschino-on-Oka, Moscow Region, Russia

Jacques M. E. Vielliard, Instituto de Biología, Universidade Estadual de Campinas, CP 6109, 13083–970 Campinas, São Paulo, Brazil

Meredith West, Departments of Psychology and Biology, Indiana University, Bloomington, Indiana 47405, USA

Ken Yasukawa, Department of Biology, Beloit College, Beloit, Wisconsin 53511, USA

NOMENCLATURE

We followed C. G. Sibley and Monroe (1990, 1993) and Monroe and Sibley (1993) for English and scientific names of bird species in all but two chapters. Vincent Bretagnolle (Chapter 9) chose to follow Warham (1990) for most scientific and common names; we note his few disagreements with Warham in the index. In Chapter 12, Jochen Martens provides new systematic information on some species groups, so, after ensuring that most names in his contribution were consistent with C. G. Sibley and Monroe (1990, 1993) (e.g., *Phylloscopus borealoides* and *lorenzii;* exceptions are noted in the index), we deferred to his professional judgment. Elsewhere, subspecies names of birds, plus names of mammals, anurans, and other animal groups, were left to the discretion of the authors.

Latin names of subgenera are italicized throughout the text, with one minor exception. In Chapter 8 the names of subgenera are not italicized to avoid possible confusion between genera and subgenera with identical names.

Some common names that correspond to higher-level classification are used frequently in this book. We have standardized them as follows. Members of the Passeriformes are referred to as passerines; that term also appears as an adjective (e.g., passerine song). Similarly, the passerine suborder Suboscines is used both as a noun (suboscine) and as an adjective (e.g., suboscine species). Members of the passerine suborder Oscines are referred to as oscines or songbirds; again, oscine appears as both a noun and an adjective.

PART I DEVELOPMENT

Introduction

How do vocalizations develop in individuals? From the earliest beginnings in the egg, how does a young bird come to know what to vocalize and the appropriate circumstances in which to vocalize? How many kinds of vocalizations develop? How similar are the vocalizations of neighbors? How important is imitation in the developmental process, and to what extent can displays develop without auditory experience?

Parallels between the development of oscine song and human speech provide one focus for such questions (Marler and Peters 1981). Songbirds learn to sing just as humans learn to speak. For example, young songbirds "subsing," and human babies babble; hearing is necessary for normal development because auditory feedback is required during the learning process; and dialects and large repertoire sizes are consequences of learning in both species. Given these parallels and the extensive research they have fostered, it is not surprising that the chapters on development in this book emphasize songbirds.

The first chapter in this section asks, Why do developmental modes vary so much in passerines? By shifting the focus in our laboratory studies from the *how* to the *why* of development, Don Kroodsma advocates a comparative approach that seeks to understand the ecological significance of natural variation in development. The extent to which songbirds imitate or improvise songs might be related, for example, to site fidelity or residency of populations, as explored in an example with *Cistothorus* wrens. How vocalizations develop (whether birds imitate or improvise) may also be related to the size of vocal repertoires and how those displays vary geographically. Kroodsma argues that a fundamental knowledge of life histories coupled with a comparative framework for asking questions about ontogeny can help to explain the ecology and evolution of song development.

The chapter by Meredith West and Drew King also concentrates on variation, but primarily on the great developmental variation that can occur among individuals of a single species. West and King advocate using nonlinear dynamics and chaos theory to organize the diverse knowledge now available about the ontogeny of birdsong. Viewing communication as an aggregate product of ecology, genes, and actions of organisms will also help to reveal how those factors interact throughout development, they argue. The systems approach advocated by West and King stresses that intraspecific variation should not be treated as unwelcome experimental "noise," but rather as a pervasive and important phenomenon in behavioral development that needs to be studied in its own right.

Luis Baptista evaluates the roles of the environment and the genome in shaping vocalizations. He teases apart the relative roles played by nature and nurture in affecting the size of vocal repertoires and the characteristics of vocal displays (e.g., duration, rhythm, frequency, and tonal quality) that a young bird learns. Baptista encourages a careful search for evidence of vocal learning in groups where it has been thought to be absent and identifies two exciting areas for future research: the diverse vocal ontogenies among hummingbirds, and sex differences in vocal development and production.

Mike Beecher shows in his chapter that laboratory-based developmental studies provide an incomplete and sometimes erroneous picture of how birds learn under natural conditions. Beecher discusses four song-learning rules that seem to guide young Song Sparrows (*Melospiza melodia*) in nature so that the number of songs shared with neighbors is maximized, and he shows how laboratory studies of Song Sparrows failed to reveal these fundamental learning strategies. Beecher advocates looking first at song development under natural conditions and then using the laboratory as a place to simulate critical conditions that have been identified in the field. Such a field-first approach will yield a more balanced appreciation of song development strategies among birds.

Dietmar Todt and Henrike Hultsch explore song development in a species with a beautiful and complex song: the Common Nightingale (*Luscinia megarhynchos*). Examining how males learn and control their repertoires of roughly 200 different song types reveals the high potential for complex behavior that is enabled by vocal learning. Todt and Hultsch suggest some "rules" that govern how a nightingale learns his songs and manages his singing performances. The hierarchy of decisions that nightingales make both during vocal development and when singing, and how the brain controls those decisions, is an exciting area for future work.

Susan Farabaugh and Bob Dooling integrate current knowledge of the ecology, the development of learned vocalizations, social interaction in adult vocal learning, and auditory perceptual abilities to address vocal learning in the Budgerigar (*Melopsittacus undulatus*). Vocal behavior of Budgerigars is compared with that of other parrots and of songbirds. With their exemplary integration of laboratory and field approaches, the authors pose many questions for future research on vocal communication in parrots.

1 Ecology of Passerine Song Development

Donald E. Kroodsma

Song development in birds "shows a diversity of patterns and a variety of consequences" (Slater 1989, p. 40). Among the suboscines, for example, normal songs develop in individuals that are isolated or even deafened at an early age (Kroodsma and Konishi 1991). In contrast, among oscines, individuals typically need an external model and intact hearing for normal song development to occur (Konishi 1989). The need for auditory feedback may be the only common denominator among songbirds, because styles of song learning vary enormously among species (Slater 1983a, 1989, Kroodsma 1988a, Marler 1991a). The males of some species can learn from tape-recorded songs, for example, but others need live tutors. Some learn early in life, some throughout life. Some imitate precisely, others improvise generally. Some learn from fathers, some prefer other social tutors. The consequences of these diverse ontogenies are equally diverse (see, e.g., Krebs and Kroodsma 1980). Vocal signals among songbird species differ in usage patterns, overall complexity, variety (i.e., vocabulary and repertoire size), the extent to which signals of one individual are similar to those of nearby or distant individuals, and much more.

But why is each species the way it is? If every signal is used to "manage" other individuals for selfish gain (Smith, this volume), why, then, do species differ so enormously in their styles of vocal development and the consequences (the adult vocalizations) that are needed in this management process? Signals (and their development) must coevolve with other life history parameters (Kroodsma 1983a). What aspects of social systems and population biologies determine the developmental styles and signals that work best in particular ecological circumstances?

I believe that the primary reason we have not realized greater progress in answering these questions is simple. Most developmental studies have focused on how birds develop songs (i.e., on mechanisms), not on why the songs develop the way they do (i.e., function, evolution). Without a comparative, ecological framework, the *how* of development learned from numerous isolated studies provides little help in understanding the *why*. Inevitably, the myriad facts (e.g., Slater 1989) remain largely unconnected, bits of a grand evolutionary picture. I believe we can attain an evolutionary perspective for development if we choose our questions, methods, and subjects more carefully. We can learn the *how* as we study the *why* of development, but not necessarily vice versa.

Ecology and Evolution in the Laboratory?

Exactly what can laboratory studies (e.g., Rothstein et al. 1989) tell us about how birds behave in nature? What can they say about, for example, the absolute timing of learning, the timing relative to dispersal, the extent of imitation and improvisation, or what is learned and from whom? I am increasingly convinced that laboratory studies can at most show only what a bird is capable of doing in an environment never before encountered in the species' evolutionary history. We must therefore be highly cautious about inferring from laboratory results what birds do in nature (Kroodsma 1985b). In this section I provide six examples that illustrate both the kind of caution I believe is necessary and the difficulty of extrapolating from studies of mechanisms to an understanding of evolution.

Six Examples

We can be most confident of knowing how songs develop in nature for those species that develop normal songs under a variety of artificial conditions in the laboratory. *Empidonax* flycatchers, for example, seem imperturbable (Kroodsma 1984, 1985a, 1989b), and Eastern Phoebes (*Sayornis phoebe*) produce normal songs even if they can't hear themselves sing (Kroodsma and Konishi 1991). Hand-reared male Grey Catbirds (*Dumetella carolinensis*) that are untutored in the laboratory appear to produce normal-sized repertoires of song elements to which males in nature respond aggressively (Kroodsma et al. unpubl. ms.). From these results, we can infer that the flycatchers, and perhaps the catbirds, can produce appropriate songs in nature, too, without exposure to normal song. In our simple laboratory microcosms we discover subsets of our subject's potentials. Thus, if we can demonstrate a capability in the laboratory, a setting typically devoid of the rich social milieu found in nature, the bird undoubtedly has that capability in nature, too (but see West and King, this volume). We must be aware, however, that a bird may also have other abilities. Demonstrating that a flycatcher or a catbird can develop normal song without imitation doesn't mean that, under the right circumstances, it would not or could not imitate.

More controversial is what we know about the ecological aspects of the timing of vocal learning (M. C. Baker and Cunningham 1985). Suppose that all hand-reared White-crowned Sparrows (*Zonotrichia leucophrys*) imitate tape-tutored songs before 50 days of age but not after, and that dispersal in nature never occurs before day 51. I believe that the most we can conclude is that real birds in nature are undoubtedly also capable of learning before day 50. Other capabilities of White-crowned Sparrows can be revealed by altering the laboratory environment in dramatic or more subtle ways. Even if we demonstrate that all hand-reared juveniles with social tutors learn after day 50, however, we have demonstrated only a potential for learning, not what birds in fact do in nature. Combining fieldwork (such as Baptista and Morton 1982, 1988, DeWolfe et al. 1989) with carefully devised laboratory studies of ontogenetic differences between White-crowned Sparrow populations (D. A. Nelson et al. 1995) might explain both the *how* and the *why* of vocal learning in this fascinating species.

Even though a laboratory experiment can alter song development by manipulating ecologically relevant variables, extrapolating to what birds do in nature remains difficult. For example, male Marsh Wrens (*Cistothorus palustris*) laboratory reared on daylengths simulating an early-season hatching date (June) were more likely to learn additional songs the next year than were males reared on daylengths simulating a late-season hatching date (August; Kroodsma and Pickert 1980). In many marshes, birds that hatch at the end of the breeding season do not hear adult song at all during their hatching year, and they might even disperse farther than birds that hatched earlier in the season (Dhondt and Huble 1968). An extended learning period for these late-hatching wrens makes sense, but the laboratory data simply cannot reveal when and where June-hatched and August-hatched Marsh Wrens learn. The sensitive period for learning can be extended not only by shorter daylengths but also by using socially more realistic tutoring environments (Kroodsma 1978a). Clearly, if we are to determine what birds do in nature, we must don our hip boots and study behavior in the field.

The difficulty of getting from how to why is illustrated especially well by the extensive studies on the Zebra Finch (*Taeniopygia guttata*), one of the true workhorses of song development studies (Slater et al. 1988). These birds conveniently breed in captivity and make great "laboratory preparations" for studying physiological and social processes. Unfortunately, as Slater (1989) lamented, we know so little about Zebra Finches in nature that placing the laboratory knowledge of this species into a functional ecological context is impossible. In spite of knowing so much about which factors influence song development, we know little about why Zebra Finches are the way they are.

West and King's superb series of studies on the Brown-headed Cowbird (*Molothrus ater;* 1988a, b, this volume) also illustrate my point that research on mechanisms, as fascinating as it might be, does not necessarily inform us about ecology and evolution. One confounding factor is that laboratory subjects are good at doing experiments we had not intended (King and West 1987a). As a result, our conclusions may pertain only to a limited, and unknown, combination of conditions in the laboratory. In their earliest efforts King and West (1977, p. 1004) were "tempted to label the cowbird's response to isolate song as idiosyncratic or unrepresentative of other songbirds," but they recognized that the lack of information about other species prevented firm conclusions. Unintended experiments done by the birds themselves hampered understanding of the ecological significance of song development in this species, but the lack of comparative data was far more important.

But even a carefully designed comparative approach involving closely related species, such as the fine series of experiments on Song Sparrows (*Melospiza melodia*) and Swamp Sparrows (*M. georgiana*) by Marler and his colleagues (e.g., Marler 1991b), does not guarantee understanding of why species are the way they are. By rearing individuals of the two species under identical conditions in the laboratory, the researchers were able to attribute many differences in behavioral development to genetic differences between the two species. These differences included mean note duration in the earliest subsong, the size of the song repertoire developed, the propensity to produce songs of natural structure, and the willingness (or ability) to mimic songs or song elements of other species. Intended to examine "species differences in mechanisms of behavioural development" (Marler 1991b, p. 65), these studies were not designed to address why these two sparrow species differ from each other in the ways they do. The two species were chosen for study because they were congeneric and convenient, not because their life histories differed in some identified manner that might influence the evolution of different styles of vocal development. As a result, the experiments did not provide enough information to formulate ecological or evolutionary hypotheses as to why the two species differ in the ways they do.

Slater's extensive review (1989) impresses me with many facts about how a number of species develop songs under a variety of conditions in our laboratories. We know little, however, about how all these facts fit into an ecological or evolutionary framework. No coherent framework exists for relating one study to another, because species are typically chosen for study out of some intrinsic interest in that particular species, or perhaps merely out of convenience. Without a carefully designed comparative framework, each separate study on vocal development provides only information on how species X developed song under laboratory condition Y, and a series of such studies provides few ecological or evolutionary insights or hypotheses about why species X is the way it is.

A Comparative Approach to the Ecology of Vocal Development

I believe that a carefully implemented comparative approach (D. R. Brooks and McLennan 1991, Harvey and Pagel 1991) using rigorous hypothesis testing (Fig. 1.1) can tell us much about the ecology and evolution of vocal development in birds, especially the oscines. This approach begins with the identification of closely related populations or taxa that differ in some fundamental aspect of behavior or population biology. Florida and California Red-winged Blackbirds (*Agelaius phoeniceus*) are more sedentary and exhibit more local variation in song than do the migratory blackbirds of northeastern North America (Kroodsma and James 1994). The *nuttalli* subspecies of the White-crowned Sparrow is largely resident, but *oriantha* migrates (Baptista 1975, Baptista and Morton 1988). White-crowned Sparrows of the *nuttalli* subspecies also have relatively small

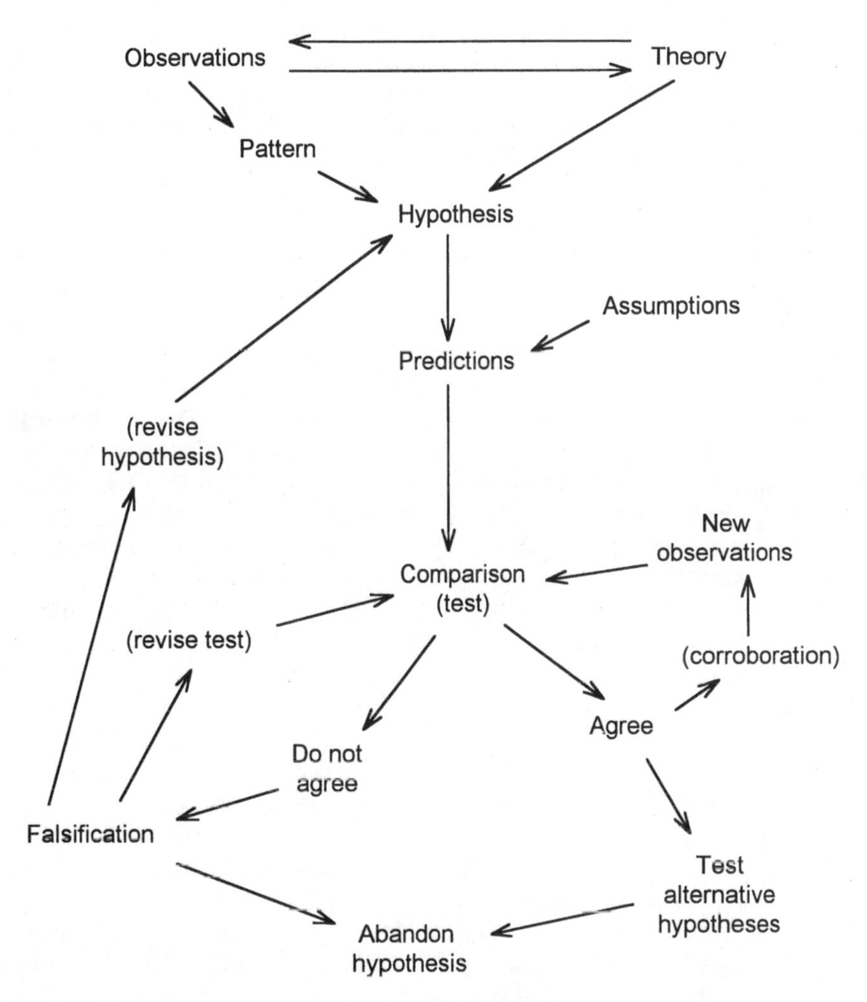

Fig. 1.1. Progress in understanding the relationships between patterns of geographic variation (observations, patterns) and the processes of song development will come only through rigorous tests of hypotheses. A search for general explanations demands several ingredients in this research program: clearly stated hypotheses, predictions, tests, and retests; equally critical appraisal of tests whether data agree or disagree with the hypothesis; careful evaluation of alternative hypotheses; and use of additional, independent taxonomic groups in successive attempts to falsify (not prove) a hypothesis. After Wiens (1989, p. 23).

dialect areas with sharp boundaries, but the song forms are more widely distributed in some other subspecies (Baptista 1977) and in the closely related White-throated Sparrow (*Z. albicollis;* Lemon and Harris 1974). The congeneric Marsh Wren and Sedge Wren (*Cistothorus platensis*) in North America differ in their degree of site fidelity (Kroodsma and Verner 1978). Brown-headed Cowbirds are brood parasites (Rothstein 1990), but not all cowbird species use that reproductive strategy (S. M. Lanyon 1992). Zebra Finches in Australia (*T. g.*

castanotis) are highly nomadic (Keast 1958), but perhaps the finches of the Lesser Sunda Islands in Indonesia (*T. g. guttata*) are not, and differences in lifestyles might have led to related differences in song development (see Clayton 1990a). If subsets of a song repertoire function differently, as they do in many warbler species (Staicer 1989, Spector 1992), how does development of the repertoire subsets relate to function, and what are the geographic consequences of those differences (Byers and Kroodsma 1992)?

After identifying what seem to be patterns in the observations, we formulate hypotheses, make predictions, and openly assess the assumptions (Fig. 1.1) about how song development should proceed in closely related taxa. A test of the hypothesis follows. If the data are consistent with the hypothesis, further attempts to falsify the hypothesis are in order. New taxa, perhaps even entire phylogenetic groups (see, e.g., Irwin 1988, S. M. Lanyon 1992), are selected to determine whether the hypothesis can be repeatedly corroborated among independent evolutionary groups. Or, alternative hypotheses, for which the data might also be consistent, might need testing. If the data are inconsistent with the hypothesis, the stated hypothesis has been falsified, but either the test or the hypothesis could then be revised. Only by this form of rigorous hypothesis testing can true progress be made in understanding the relationship between the process of vocal development and the pattern of vocal variation produced (see also Wiens 1989).

An Example of the Comparative Approach: *Cistothorus* Wrens

To illustrate the process by which I believe we can study the ecology of vocal development, I provide an example from my research on *Cistothorus* wrens. The American Ornithologists' Union (1983) recognizes two species in North America, the Sedge Wren and the Marsh Wren. These species are similar in many ways: both occur at high population densities in communities of low avifaunal diversity, both can be highly polygynous, and males of both species sing large repertoires of song types (from 50 to 200, depending on location; Kroodsma 1977b). One aspect of their population biology differs, however. Marsh Wrens of many populations are resident, and those that migrate often return to the same site (Welter 1935, Kale 1965, Verner and Engelson 1970, Verner 1971, Kroodsma unpubl. data for New York and Saskatchewan). In contrast, Sedge Wrens appear to be somewhat nomadic, and the occurrence of individuals at a given site over one season or between one season and the next is less predictable (Kroodsma and Verner 1978, Burns 1982). The relative degree of site fidelity among these wren populations has probably coevolved with habitat stability; that is, the wet meadows required by the Sedge Wrens are less predictable than the more stable marshes used by the Marsh Wrens.

Song development in these two species seems to nicely match their respective population biologies (stated here as a hypothesis, which was developed after a descriptive study of song development in the two species). Male Marsh Wrens develop songs via imitation, as has been repeatedly demonstrated in the laboratory (see, e.g., Kroodsma and Pickert 1980). In nature, territorial males often countersing with matching song types (Verner 1976), a considerable feat given the large repertoires of 100 or more song types from which the males can choose. Marsh Wrens thus form relatively stable communities of territorial males in which songs learned from one another can be countersung in ritualistic exchanges (Verner 1976, Kroodsma 1979). Unlike Marsh Wrens, male Sedge Wrens in a laboratory experiment imitated no songs from a training tape. Although these Sedge Wrens did imitate some songs from each other, the overwhelming mode of vocal development was improvisation; males developed species-typical songs with no identifiable model. Males from an Illinois population shared the same low percentage (about 5%) of songs with each other as they did with males from distant populations (Kroodsma and Verner 1978). Sedge Wrens thus seem to form relatively unstable communities of males that develop their songs largely via improvisation, perhaps according to species-wide rules that minimize geographic variation by maximizing individual variation. Their species-typical songs are not restricted to relatively local populations, as in Marsh Wrens, and may enable communication among any subset of birds that become neighbors (or mates).

This hypothesis, formed in the late 1970s, was not tested further until 1994. I then reasoned that if the proposed relationship between song ontogeny and ecology were true of North American *Cistothorus* wrens, then sedentary populations of Neotropical Sedge Wrens, if sufficiently isolated from North American populations, should behave more like North American Marsh Wrens than like North American Sedge Wrens. This expectation was largely confirmed at Brasilia National Park, Brazil, where Sedge Wren males countersinging at close range matched each other in the same kind of striking exchanges that can often be heard among sedentary Marsh Wrens of western North America (e.g., Verner 1976, D. E. Kroodsma pers. obs.). In one series of 20 songs, for example, 8 out of 10 exchanges between two males were matches, with the songs of one male following and unmistakably matching the songs of his neighbor. Brazilian Sedge Wrens thus clearly imitate each other, just as Marsh Wrens do throughout North America. I believe that such matched countersinging using large song repertoires cannot develop unless the birds are highly sedentary or site faithful, but banding studies are needed at Brasilia to document the population processes there. Other *Cistothorus* populations are also available for study. *Cistothorus platensis* occurs in Central America, throughout South America, and on the Falkland Islands, and two related species (*C. apolinari* and *C. meridae*) occur in the Andes (Ridgely and Tudor 1994).

Although the data from the laboratory and the field seem consistent with my proposed hypothesis, an alternative hypothesis must also be considered. Perhaps Sedge Wrens can develop their large repertoires via imitation, just as Marsh Wrens do, but Sedge Wrens need live tutors. Dispersal from the learning site could explain the different patterns of geographic variation: in North American populations the imitated songs are dispersed over the geographic range of the species, but at Brasilia National Park the songs (and birds, of course) are resident. Although this hypothesis seems less likely (North American Sedge Wrens improvise far more normal songs in the laboratory than Marsh Wrens do), it must be tested with more learning opportunities for Sedge Wrens under naturalistic laboratory conditions.

Other taxa must also be considered if we are to determine whether the relationship between site fidelity and song ontogeny is a general one. Rufous-sided Towhees (*Pipilo erythrophthalmus*) may provide a good test species. In Florida, they appear to be resident, and males share most of their repertoires with immediate neighbors (Ewert and Kroodsma 1994). In contrast, New England towhees are migratory and share few songs with their neighbors. I predict that under identical (laboratory) conditions, New England towhees will improvise song elements more than will Florida towhees.

I find the *Cistothorus* complex particularly exciting because the available (though limited) data suggest relationships between ecology, style of song development, and adult vocal behaviors. Furthermore, this example offers hope that a comparative approach, judiciously applied, will provide satisfying answers to some of our questions about the ecology and evolution of communication systems.

Evolution of Vocal Learning

Questions about the diversity of developmental strategies among oscine songbirds are usually rooted in a comparison with the suboscines. Songbirds have specialized forebrain song nuclei (Konishi 1989) and, through auditory feedback (Konishi 1965), imitate songs of adults (Kroodsma and Baylis 1982, Slater 1989, Marler 1991a), much as humans learn their spoken language from adults. Suboscines differ strikingly. Less well known overall, they are a mostly New World suborder with more than 1000 largely Neotropical species (including the cotingas, tyrant flycatchers, antbirds, ovenbirds, woodcreepers, and the like). The songs of suboscines are relatively simple, much like the nonlearned call notes of songbirds; their repertoires are small; and geographic variation is minimal (Kroodsma 1988a). All available evidence suggests that, unlike the songbirds, suboscines have no imitative or feedback process in their song development.

The evidence for these conclusions about suboscines is drawn from studies of several species. Laboratory experiments have now shown that (1) individuals of three flycatcher species did not need exposure to adult songs after 8–10 days of age (Kroodsma 1984, 1985a), (2) songs of hand-reared Eastern Phoebes were not influenced by either tape or live tutoring (Kroodsma 1989b), (3) the phoebes did not require auditory feedback to develop normal songs (Kroodsma and Konishi 1991), and (4) the forebrains of suboscines lack the neural control centers found among all songbirds studied to date (Nottebohm 1980b, Kroodsma and Konishi 1991, Gahr et al. 1993, Brenowitz and Kroodsma, this volume). Furthermore, fledglings of some suboscine species use their "song" as a type of contact call immediately after leaving the nest (Kroodsma 1984). At that age, songbirds are just beginning to memorize sounds in their environment; recognizable production of those sounds occurs a month or more later, and only after extensive practice, or subsong (Kroodsma 1981a). Although we cannot fully characterize the processes by which suboscines develop their songs, the ontogenetic process and its consequences are extraordinarily different from those found among the imitating songbirds.

The big question, then, prompted by these differences between suboscines and songbirds, is, Why have songbirds evolved the ability to learn their songs? (see Nottebohm 1972). The most parsimonious evolutionary scenario is that vocal learning and the associated neural control system arose only once in the Passeriformes, in the ancestor of modern songbirds but not in the ancestral suboscine (see Raikow 1982 and C. G. Sibley et al. 1988 for comments on monophyly of the Passeriformes). But what circumstances encountered by ancestral songbirds led to vocal learning? Although we can say that the two most obvious consequences of vocal learning are larger (sometimes enormous) song repertoires and learned song dialects (Krebs and Kroodsma 1980), we cannot infer that selection for these characteristics necessarily led to song learning. The ability to learn could have originated for a different reason, such as for controlling inner car damage during production of loud vocalizations (Nottebohm 1991). Some authors argued that song learning by songbirds has promoted speciation (e.g., M. C. Baker and Cunningham 1985, cf. Baptista and Trail 1992), but, again, consequences need to be distinguished from causes. Exactly how vocal learning came to be used by ancestral songbirds to manage their social environment remains an evolutionary puzzle.

Perhaps our best hope for understanding the origins of vocal learning lies in scouring the relatively unknown suboscines for clues. Suboscines whose songs are relatively elaborate (e.g., Great Crested Flycatcher, *Myiarchus crinitus:* Smith, this volume) or (micro)geographically variable would be prime subjects, because such characters would suggest a reduced genetic control of song behavior. Applying the comparative approach on a fine scale to the largely unknown Neotropical suboscines, and studying life histories and vocal development of closely related species or groups, might provide some answers.

Ontogeny, Repertoires, and Geographic Variation among Songbirds

Unlike flycatchers, songbirds can use vocal learning to generate large song repertoires and microgeographic song variation. I am intrigued by the interplay between vocal ontogeny, repertoire size, and geographic variation, especially because not all songbirds exploit the potential that vocal learning provides. Some songbirds, in fact, forgo both large song repertoires and microgeographic song variation, acting in effect more like suboscines than typical oscines. In this section

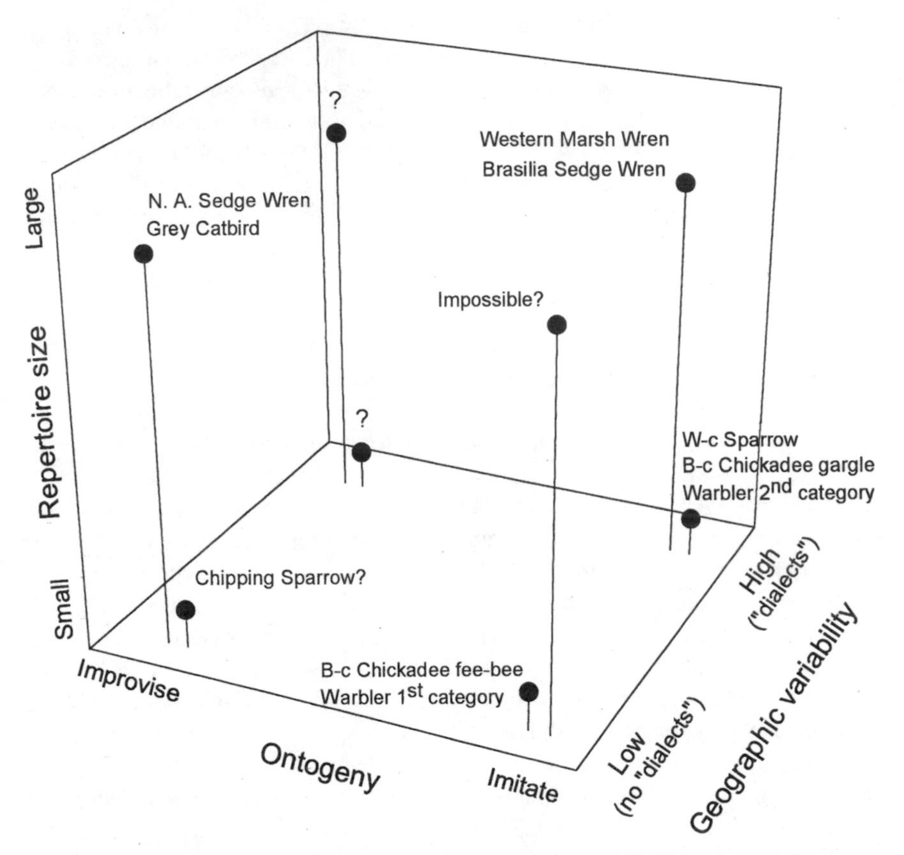

Fig. 1.2. A graphical approach can be used to explore relationships among life history parameters and song ontogeny (imitation vs. improvisation), song repertoire size (small to large), and variation of vocalizations across geographic space (low to high). First, place closely related species or different song forms for the same species in the three-dimensional space; then identify life history features that could have led to observed patterns of convergence and divergence; next, find additional pairs of species to test initial hypotheses. See text and Fig. 1.1 for elaboration of these ideas.

I will relate song ontogeny to its two obvious consequences, using a model (Fig. 1.2) that invites further exploration of the relationships among these factors.

Vocal learning. The aspect of vocal development on which I focus is whether birds imitate or improvise (*sensu* Marler and Peters 1982b) in developing their species-typical songs. Songbirds are celebrated for their imitative abilities, and numerous studies have shown that birds in the laboratory copy details of songs presented to them by either taped or live models (e.g., Marler 1981, 1991a, R. B. Payne 1981a, Baptista and Petrinovich 1986, Petrinovich and Baptista 1987, Hultsch and Todt 1989b), and even that external nonvocal interactions sometimes guide song learning (King and West 1983, West and King 1988a, b). Juvenile Marsh Wrens are so dependent on song imitation, for example, that their repertoires can be easily manipulated in the laboratory (Brenowitz and Kroodsma, this volume).

Not all songbird species have this flair for imitation, however. Male Grey Catbirds seem to use extensive improvisation and invention (referred to hereafter as simply improvisation). Regardless of whether catbird males experience no, few, or many catbird songs after six to eight days of age, they develop normal-sized song repertoires (up to 300–400 song units) of seemingly normal songs (Kroodsma et al. unpubl. ms). Nestling male Sedge Wrens taken from a population in Michigan also developed repertoires of fairly typical songs, largely by improvising (Kroodsma and Verner 1978). These catbirds and wrens are capable of imitating, as revealed by observations in both the laboratory and the field, but the dominant force in their development seems to be improvisation, not imitation. Improvisation is probably a form of vocal "learning," because the birds most likely learn from themselves via auditory feedback during development. Improvisation and imitation, then, are the two extremes of developmental styles that could be favored in different life histories; "copy errors" (see, e.g., J. M. Williams and Slater 1991) may thus not be errors at all, but rather an adaptive means of generating signal variation.

Repertoire size. One consequence of vocal learning—large song repertoires (Fig. 1.2)—is legendary among songbirds. Common Nightingales (*Luscinia megarhynchos*) sing hundreds of song types (Hultsch and Todt 1989b, Todt and Hultsch, this volume), as do Northern Mockingbirds (*Mimus polyglottos;* Derrickson 1987), Marsh Wrens (Kroodsma 1989a), and other superlative songsters (Krebs and Kroodsma 1980). Brown Thrashers (*Toxostoma rufum*) sing thousands of songs (Kroodsma and Parker 1977, Boughey and Thompson 1981), as may males of other thrasher species (J. Verner pers. comm.).

At the other extreme are songbirds that have a more modest vocabulary. Individual male White-crowned Sparrows (Baptista 1975), Chipping Sparrows (*Spizella pusilla;* Borror 1959), Common Yellowthroats (*Geothlypis trichas;* Borror 1967), Kentucky Warblers (*Oporornis formosus;* E. S. Morton and Young 1986), Indigo Buntings (*Passerina passerina;* R. B. Payne et al. 1988), and so on, have, for the most part, a single commonly used song type in their repertoires. Exactly how many song types an individual sings can be debated; a male Black-capped Chickadee (*Parus atricapillus*), for example, can whistle a single song form on different frequencies (see Ratcliffe and Weisman 1985, E. S. Morton and Young 1986, B. G. Hill and Lein 1987, Horn et al. 1992). Investigators would agree, however, that the range of repertoire sizes among oscines is from one song to hundreds and even thousands.

Geographic variation. The second consequence of vocal learning, and the third dimension in my model (Fig. 1.2), focuses on the extent of geographic variation in song. The young birds of some species imitate the songs of adults and then remain in the general area where the songs were learned, so that neighboring

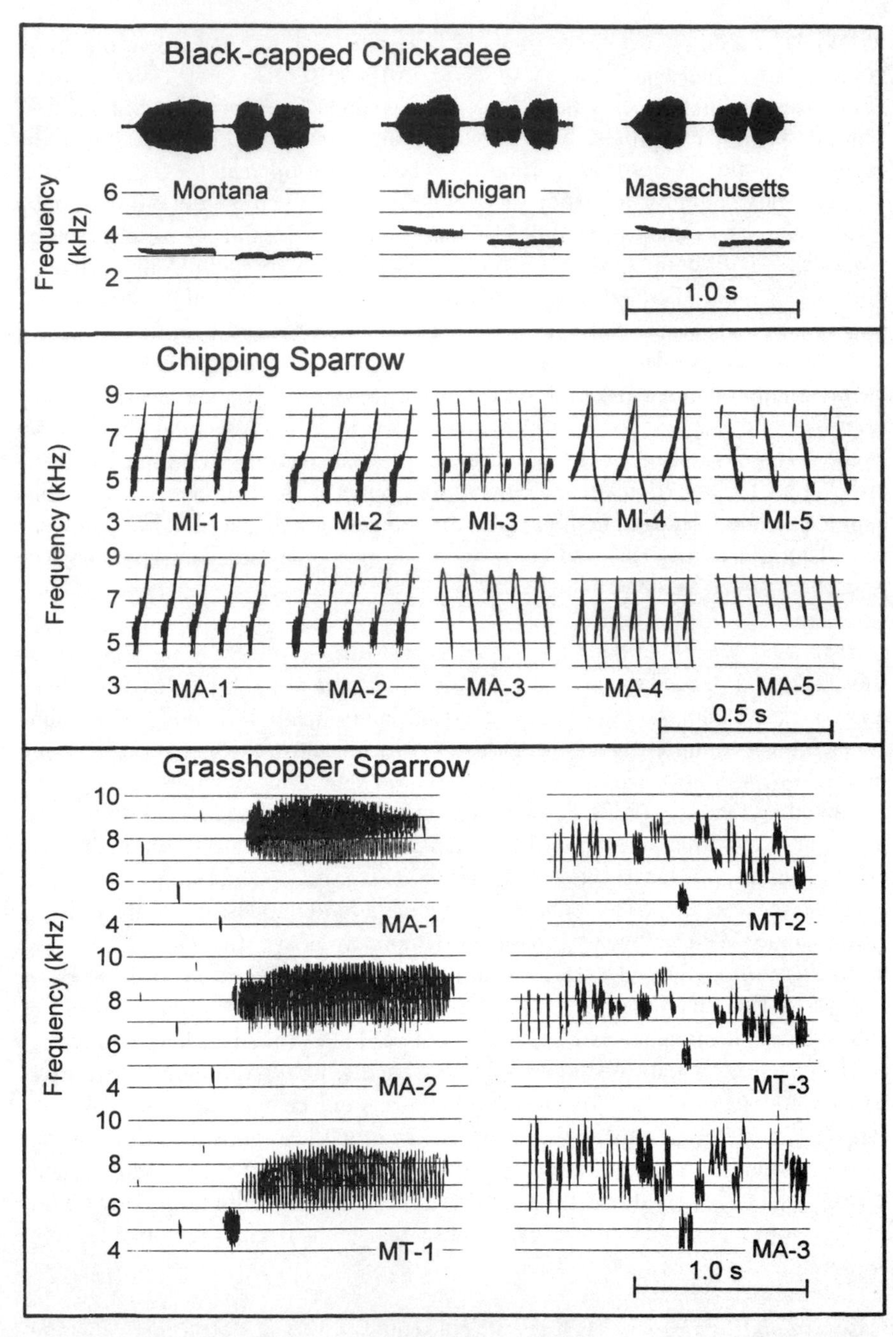

Fig. 1.3. A three-species sampler of how signals can vary across geographic space illustrates the diverse patterns and the challenges involved in understanding "the ecology of song development." (A) Songs of three Black-capped Chickadees, showing in both sonogram and waveform display the uniformity in a single song pattern from Montana (Missoula) to Michigan (Zeeland) to Massachusetts (Amherst). (B) Songs of 10 Chipping Sparrows from Michigan

territorial males have nearly identical songs (Kroodsma 1974, Jenkins 1978, Rothstein and Fleischer 1987, DeWolfe et al. 1989, Beecher et al. 1994b). Sometimes sharp boundaries separate different song neighborhoods (i.e., "dialects"), as among White-crowned Sparrows (Marler 1970, Baptista 1975, M. C. Baker and Cunningham 1985) and Corn Buntings (*Miliaria calandra;* McGregor 1980). Among other species, however, dialect boundaries seem less obvious, though local neighborhoods of song forms are often evident. This microgeographic variation appears to be largely a consequence of song imitation and reduced dispersal from the site of learning.

The interplay of ontogeny, repertoire size, and geographic variation. Ironically, geographic song variation among songbirds, which must "learn" to sing (unlike the suboscine flycatchers), can be reduced by one of two highly different ontogenetic processes: by either minimizing or maximizing individual variation (Fig. 1.3). Individual variation can be minimized if all the birds are somehow guided to learn only a limited range of vocalizations. Such guidance might be provided by a female (West and King 1988a, b) or by a process of neural selection (Marler and Nelson 1992). The consequence of such processes can be a striking uniformity of learned song elements, or entire songs, over considerable geographic expanses. Examples of clearly imitated songs with little geographic variation are the fee-bee of the Black-capped Chickadee (Fig. 1.3; Kroodsma et al. 1995) and first-category songs of certain paruline warblers (Weisman et al. 1990, Kroodsma 1981b, Spector 1992). How a learning process can maintain such stereotypy in one or a few song forms over wide geographic expanses is a mystery (J. M. Williams and Slater 1991, Marler and Nelson 1992); the chickadees are especially puzzling because their potential for developing dialects and repertoires of diverse song forms is fully evident in the laboratory. Deviant song forms arise in wild chickadees, too, but they are lost over successive generations, presumably because pressures to conform maintain the predominant form.

Reduced geographic variation in songs can also be achieved by improvisation, a process that maximizes individual variation (Fig. 1.3). If males use an improvisational program to generate species-typical songs, then each individual will produce a unique repertoire of songs that is equally like (or unlike) the repertoire of both his immediate neighbors and more distant individuals. Neighboring Sedge Wrens, for example, have unique repertoires of 100 or more songs, apparently generated largely via improvisation, and share only a few songs (about 5%) with both immediate neighbors and more distant individuals (Kroodsma and Verner

(Zeeland; MI-1 to MI-5) and Massachusetts (Amherst; MA-1 to MA-5), showing a small portion of the great variety of songs that occur within a given population, and how males within and between populations can share the same song form (compare MI-1 and MA-1, MI-2, and MA-2; see also Borror 1959). (C) Songs of Grasshopper Sparrows from Montana (MT) and Massachusetts (MA) illustrate how the simple song form is relatively invariant over geographic space (compare MA-1, MA-2, and MT-1) and how the complex form varies geographically (the note sequences of MT-2 and MT-3 are essentially identical, but unlike MA-3). Sonograms were produced on a Kay Elemetrics DSP 5500 Sonagraph with an effective bandwidth of 300 Hz; note differences in frequency and time scales.

1978). Heavy reliance on improvisation would seem to guarantee a relative lack of geographic variation, unless the populations were genetically differentiated and males in each used different improvisational programs.

Geographic variation in song is also affected by other song developmental strategies, of course, though I choose not to focus on them here. Variation will be reduced, for example, if males imitate the most common song in a population (J. M. Williams and Slater 1991) or if males learn an "average" song; variants are then eliminated during each generation, leaving only the common form (e.g., Song Sparrow: Beecher et al. 1994b, this volume; Black-capped Chickadee?). In contrast, variation will be increased if an individual imitates songs of only one tutor, regardless of how common his song forms are; some sharing of songs will then occur in a population, but many song forms will co-occur (e.g., Chipping Sparrow?; Fig. 1.3). These individual developmental strategies, coupled with the extent of improvisation in ontogeny, will also strongly influence geographic signal variation.

The variation of vocal signals over geographic space is undoubtedly determined by the audience the signal is used to "manage." Signals that manage males or females from potentially extensive geographic areas must have a general form that is useful for all individuals. Signals used on a more local scale, however, perhaps primarily in countersinging duels with immediate neighbors, could consist of imitated components largely restricted to a small neighborhood (i.e., a local dialect). I find it difficult to consider signal variation over geographic space as a functionless by-product of vocal learning (Slater 1989, J. M. Williams and Slater 1991); rather, I envision selection for a style of vocal development that maximizes the "management potential" of a signal over some "best" geographic area. How a signal develops may thus be a consequence of both the listening audience, which will affect the extent of geographic variation in a song, and the size of the song repertoire needed to socially manage the particular tasks for which the signal is used (Fig. 1.2).

One consequence of such interrelationships is that simultaneous selection for two factors—reduced geographic variation (i.e., wider audience for signal) and large song repertoires—may preclude precise song imitation and promote song improvisation. So far, the best examples of reduced geographic variation in imitated songs come from songbird species with small song repertoires. A single imitated song form predominates throughout the geographic ranges of the Black-capped Chickadee (Kroodsma et al. 1995), the Blue-winged Warbler (*Vermivora pinus;* Kroodsma 1981b), and the Golden-winged Warbler (*Vermivora chrysoptera;* Highsmith 1989). The Chestnut-sided Warbler (*Dendroica pensylvanica*) has only about five song types found seemingly everywhere (Kroodsma 1981b, Byers in press). Maintaining high stereotypy in learned songs becomes increasingly difficult as song repertoires become larger (Kroodsma 1974). If strong selection were to occur simultaneously for both larger song repertoires (Kroodsma 1977b, Catchpole 1980, 1987) *and* signals needed to manage individuals from a wide geographic range, the result would probably be reduced reliance on precise imitation and more reliance on improvisation or invention. I thus predict that no songbird species will be found in which males have (1) large (say, >6–10) song repertoires of (2) precisely imitated songs (3) that remain constant over wide geographic expanses.

Another way exists, however, for birds to develop large repertoires via imitation and still maintain low geographic variation. This process requires considerable dispersal from the site of learning, so that imitated songs are continually dispersed over broad geographic expanses. Geographic variation is reduced, then, not by all males sharing identical repertoires but by neighboring males having highly dissimilar repertoires. Do life history strategies of some species involve selection for this kind of imitation and dispersal? I know of no good examples, but this pattern might be more difficult to detect and is worth further consideration as a developmental strategy (and I will test it with North American Sedge Wrens).

Further exploration of the relationships among ontogeny, repertoire size, and geographic variation requires the study of key comparative situations. The North American Marsh and Sedge Wrens discussed earlier illustrate my point. Selection for large repertoires has occurred in both species, perhaps because of dense populations of highly competitive males in polygynous mating systems (Verner and Engelson 1970, Crawford 1977, Kroodsma 1977b, 1983a), but it seems that differences in habitat stability have promoted imitation among the Marsh Wrens and improvisation among the Sedge Wrens. Exactly what factors favor imitation are unknown, but it may be favored in stable communities because of some intrinsic advantage of song matching (E. S. Morton 1986, Horn and Falls, this volume). The Grey Catbird also deserves further study. Both the extensive improvisation and the relative absence of imitation by males from migratory New England populations used in our laboratory experiments suggest that geographic variation in catbird song will be minimal. Such a prediction is relatively easy to test; it involves comparing song repertoires of neighboring and nonneighboring males from a significant portion of the catbird's geographic range. Such a study can be extended to other populations of either the Grey Catbird or its sister species, especially more sedentary birds from tropical habitats. My working hypothesis predicts that the more sedentary songbird populations would evolve a greater reliance on song imitation.

This comparative approach can be extended to a study of ontogeny, geographic variation, and repertoire size of functionally different songs used by the same individual. Different categories of imitated vocalizations can differ in the extent of geographic variation, with local dialects present in one sound but not another. Males of certain *Vermivora* and *Dendroica* warblers have one to three first-category songs (*sensu* Spector 1992) that do not vary geographically and up to five second-category songs that do (see Nolan 1978, Lemon et al. 1985, 1987, Shackell et al. 1988, Staicer 1989, and Spector 1991 for variations on this warbler phenomenon). The fee-bee of the Black-capped Chickadee is geographically widespread (Fig. 1.3), but the learned gargle call has local dialects (M. S. Ficken et al. 1987, Kroodsma et al. 1995). A similar phenomenon occurs among

Grasshopper Sparrows (*Ammodramus savannarum*); one song form varies little geographically, but a more complex one has striking local dialects (Kroodsma unpubl. data; Fig. 1.3).

The primary audience for each vocalization dictates the style of ontogeny and the resultant degree of geographic variation. Among the warblers, the second-category songs are used in aggressive male-male situations (Lein 1978a, Lemon et al. 1985, Spector 1992), and the immediate audience consists of adult males who typically return faithfully to the same site each year and become increasingly adept at learning the specific songs of the local population (Lemon et al. 1994, D. A. Spector and B. E. Byers unpubl. data; see also Rothstein and Fleischer 1987). The first-category songs are used primarily for attracting and addressing females (Morse 1970, Kroodsma et al. 1989), and thus they vary less geographically. The relative stereotypy of these male-female songs must relate to managing potential mates from an audience that extends beyond the immediate neighborhood—probably from an area as extensive as the geographic distribution of the song itself.

The vocalizations of the Black-capped Chickadee and the Grasshopper Sparrow provide intriguing parallels with the warbler song system. Like the second-category songs of the warblers, the gargle of the chickadees and the complex song of the sparrow also vary geographically and are used in aggressive situations, apparently with audiences of local individuals. The more invariant fee-bee of the chickadee, like the first-category songs of the warblers, increases dramatically in usage when a male loses a mate (Otter and Ratcliffe 1993), suggesting that this geographically invariant signal is used to address a wider audience of potential mates. The simple song of the sparrow is also used in less agonistic situations, with patterns of use much like warbler first-category songs (R. L. Smith 1959).

These hypotheses also predict a relationship between the similarity of songs and the degree of genetic variation among localities: the more similar the songs, the greater the genetic similarity. The extent of geographic variation in the signaling system has likely coevolved with the relative need to manage social interactants from near or distant localities. The geographic distribution of signaling characters thus reveals the pool of potentially interacting and relevant individuals. To the extent that the vocal signals are used in mate choice (R. B. Payne 1983b), the extent of vocal and genetic variation among populations should also covary; thus we should link studies of population genetics, signal variation over space, and signal ontogeny. Chipping Sparrows, for example, have no detectable geographic variation in mtDNA haplotypes (Zink and Dittmann 1993) and no obvious geographic variation in songs (Borror 1959). How they develop their songs is not well known; hand-reared males in the laboratory do not imitate tutor tapes (P. Marler unpubl. data), but they might imitate live tutors. If they do imitate, perhaps birds with learned songs disperse widely from the site of learning, or perhaps yearlings preferentially learn rare songs or learn the songs of individuals without regard to the commonness of a song form in the overall population, thereby perpetuating the great diversity of songs within and between localities. Additional studies of the population genetics, signal variation, and ontogeny of the Chipping Sparrow and other *Spizella* species should help illuminate the interrelationships of these parameters.

Conclusions

Knowledge of life histories coupled with a comparative framework for asking questions about ontogeny will help us to understand the ecology and evolution of song development. The relatively unknown suboscines, confined largely to the Neotropics, merit closer scrutiny because variation within the suboscines may help to explain the evolution of vocal learning among the songbirds. The songbirds themselves hold great untapped potential for study, with far more than 4000 evolutionary experiments (one for each "phylogenetic species": McKitrick and Zink 1988) occurring simultaneously throughout the world. The diversity of life histories in this group is extraordinary, and we could expect vocal signals to arise in each species that best help individuals to socially "manage" other individuals for selfish gain. Convergent and parallel evolution should be expected both in how individuals are managed and in how signals develop and are used in communication. With appropriately chosen species—that is, species selected not for convenience but because of identified life history features—we can continue to study the mechanisms of vocal development and simultaneously begin to understand the relationships among ontogeny, ecology, and evolution. Discovering these relationships and understanding "the ecology of vocal development" is a major challenge but, with appropriate methods, also an attainable goal.

Acknowledgments

I thank Peter Marler for his fine tutelage, from 1972 to 1980, at Rockefeller University, where I formed the basis for my comparative approach to song development. Graduate students at the University of Massachusetts, especially Tim Armstrong, Bruce Byers, Tod Highsmith, Peter Houlihan, David Spector, and Cindy Staicer, have provided both inspiration and feedback since 1980. I especially thank Jacques Vielliard and Maria Luisa da Silva for hosting me in Brazil during January 1994; Jamie Smith for a critical reading of the manuscript; Gene Morton and Don Owings for sharing their manuscript on communication and "management"; the Cornell Library of Natural Sounds, Greg Budney, Jim Rivers, and Peter Vickery for help with Grasshopper Sparrows; and Bruce Byers for help in preparing figures. I thank the National Science Foundation for financial support (IBN-9111666).

2 Eco-gen-actics: A Systems Approach to the Ontogeny of Avian Communication

Meredith West and Andrew King

When we began our studies of vocal learning in the 1970s, we chose to investigate birdsong because it represented a natural example of learning. We believed that focusing on species-typical competencies would afford a more powerful means to explore learning and development than using paradigms in which animals were required to perform arbitrary responses to obtain food, water, or relief from stress. That many closely related species sing also seemed an advantage as it would foster comparative analyses. Moreover, birdsong offered diverse topics for study. The acts of vocalizing, discriminating, attending, imitating, improvising, and memorizing were some of the more obvious attributes. Thus, the whole enterprise was inviting, and made all the more appealing by the beauty and grace of birds.

Although we have not lost our admiration for birds, we now wonder if the aforementioned attributes of song are not too much of a good thing. More time and more research have accumulated more facts but few new principles with which to manage the proliferation of data (Slater et al. 1988, King and West 1990, Kroodsma, this volume). Two decades after our start, we find the study of birdsong more intimidating than ever. What do we do when our subjects generate more exceptions than rules? How do we find order within and across so many levels of analysis? The lack of unifying principles is not a new problem (Marler 1960, Kroodsma 1978a), and this is perhaps the most worrisome fact of all. Recognition and articulation of the problem have not helped to solve it. In our view, the field of birdsong risks losing its power to integrate Tinbergen's four questions (as Hinde optimistically suggested when the field of birdsong was reviewed over a decade ago [see Kroodsma and Miller 1982]). How can we hope to connect *across* levels of function, development, causation, and adaptive value if we cannot make these connections *within* levels?

Here we suggest that a possible solution is to follow the lead of those who have turned to systems theories when phenomena appear to be made up of interdependent parts, to contain much variation, and to involve multiple time scales (Thelen and Ulrich 1991, Timberlake 1993, Cole 1994). The process of communicating possesses these traits. In this chapter, we offer suggestions in the form of four principles, about ways to prepare the study of birdsong for a systems perspective.

Songbirds: Variations on Too Many Themes?

Some of the attributes of avian communication that provide stumbling blocks to integration are illustrated by three species, chosen because they do not exhibit some of the typical impediments to the collection of data. In each case, field and laboratory studies of the species were possible, sufficient subjects were available, and the animals could be studied over long time frames. But each case also reveals the growing pressure on investigators to reevaluate the most basic generalizations about the nature of song learning—such as the primacy of imitation in song learning, the nature of the sensitive periods necessary for vocal maturation, and the species-wide functions of songs. Such evidence suggests a state of "dis-integration" as once-global concepts differentiate into many local forms.

Studies of male White-crowned Sparrows (*Zonotrichia leucophrys*) now suggest the operation of at least two interconnected forms of learning: memory-based processes and action-based processes (Marler 1991c). The latter process involves operant mechanisms and reinforcement, features originally thought unnecessary for vocal learning in this and other species. The data also reveal considerable intraspecific variation in when memories are acquired, which memories are used, and which actions shape vocal behaviors. In particular, ecological events such as the acquisition of a territory appear to have profound effects on the developmental transition from subsong to stereotyped song (DeWolfe et al. 1989). Recent investigations have also introduced a greater emphasis on female song in some populations, a finding of special interest to us because the females appear to be differentially affected by typical experimental procedures such as tape tutoring (Baptista et al. 1993a). Whereas young males can be tutored readily in the laboratory, females from the same population seem refractory to tutoring. Why would sex differences occur in the laboratory but be less evident in the field? Action-based and memory-based principles may have to incorporate ecological and gender differences. Thus, the species once chosen for research because of the simplicity of its song has joined the ranks of other songbirds in terms of generating intraspecific puzzles.

Studies of Indigo Buntings (*Passerina cyanea*) by R. B. Payne and his colleagues (see, e.g., R. B. Payne and Payne 1993b) focus on the principles guiding social learning: What rules do young males use to choose song models? The answer is complicated. Using one of the largest field data sets ever collected, the investigators examined features of adults that may make them attractive tutors to young song learners. No single feature has emerged as crucial, although age and timing of arrival on breeding territories come the closest. These two attributes are complex variables, however, not easily partitioned into acoustic and social attributes. R. B. Payne and Payne (1993b, p. 1063) concluded that potential tutors must meet a "standard of appearance and behaviour" by means of "active behavioural assessment" that is a "continual process both within and between seasons"; thus the weight of evidence suggests no single local copying rule, no single time frame, and no simple concept to explain the pattern of social learning.

Studies of Brown-headed Cowbirds (*Molothrus ater*) raise another issue: there appears to be no simple way to specify the function of vocal signals. For example, the function of species-typical song appears to differ among the subspecies *ater, obscurus,* and *artemisiae.* The song appears to be necessary to trigger copulatory responsiveness in females from North Carolina, but not in females from California (Rothstein et al. 1988, King and West 1990). The California females sometimes respond to just the sight of the male or to another display, the flight whistle (which has its own distinct ontogenetic pattern, different from that of song) (O'Loghlen and Rothstein 1993). That multiple pathways to vocal competence exist is not terribly surprising, although multiple routes are at odds with the way many scientists conceive of ontogeny in a brood parasite (Lehrman 1970, Mayr 1974, Todd and Miller 1993). But, more important, the data suggest that different pathways lead to different endpoints within the same species, a finding difficult to reconcile with most accounts of "species-typical" behavior.

To sum up, studies of White-crowned Sparrows, Indigo Buntings, and Brown-headed Cowbirds have raised many new questions about the ontogeny of birdsong. Moreover, these studies have made it clear that extra-auditory influences act at many levels of song development. Whereas songs can be compared across species on measures of frequency, amplitude, and duration, no common set of metrics conveys how well a male or female sparrow defends a territory, how much aggression a young male sees, how bright a bunting's plumage is, or how attentive a cowbird is to potential consorts. If the aforementioned physical measures of acoustic structure do not contain the fundamental information used by communicating organisms, then the study of birdsong may be perceived as too arbitrary and too artificial to provide a functional base from which to study mechanisms of communication.

Dynamic, or nonlinear, systems offer a way to preserve the functional validity of complex phenomena such as communication because they do not rely on reduction to a common denominator. The dynamic systems approach recognizes that the building blocks of complex behavior are multiple, interacting components formed of qualitatively different elements. Thus, such approaches are inherently more forgiving in the face of diversity. But they are not forgiving in terms of the amount of description required to identify the building blocks, perhaps explaining why they have been recommended for many years but adopted infrequently (see Schneirla 1953, Lehrman 1974). They are not an easy way out. We begin with an example that illustrates the costs and benefits of the dynamic systems approach.

Learning to Walk: Uncovering New Pathways

Until recently, the locomotor development of human infants was typically represented by maturational milestones, depicting how movements such as creeping or crawling ultimately give way to walking. The appearance of an age-related progression through the stages led to the search for a central pattern-generator that controls the pace and form of the changes. Support for such a mechanism came from two sources: the seemingly uniform appearance of motor types in the "average" infant, and the species universality of the outcome. Not all the data could be reconciled, however, including conspicuous cross-cultural differences in the timing of the progression (Bril and Sabatier 1986). The data began to make more sense when the behaviors were studied from a dynamic-systems perspective (Thelen and Ulrich 1991).

For example, very young infants make recognizable "walking" movements when lying down by kicking their feet in the air. They do not have the neuromuscular coordination and head-to-body-weight ratio to make these movements when upright. But put the baby in enough water, thereby reducing the role of gravity, and the bits and pieces of movements assemble in a new way. Supported by the water, the baby can maintain a more vertical posture and use its legs and feet to make stepping movements (Thelen 1984). So too, supporting a baby on a treadmill that moves the feet leads to the increasing efficiency of stepping as the baby gains experience with the apparatus and its own limbs.

These data are not compatible with the idea of a central pattern generator, because such a mechanism cannot explain the differential competence of the same individual at the same age in different contexts. But the data do fit with observed cross-cultural differences, as would be expected if plasticity is part of the system. Differences in early opportunities to move and be moved have demonstrably different consequences, but researchers did not see that until they changed the time periods to be investigated (Bril and Sabatier 1986, Thelen and Smith 1994). Moreover, sources of change in infancy are bidirectional; babies modify the behavior of caregivers as much as the other way around. In the case of locomotor development, as babies gain the ability to support the head, and then the upper body, people hold and place them in different positions, thereby offering new support surfaces affording opportunities to make new movements. No one teaches caregivers to vary the ways they hold infants any more than anyone teaches infants how to walk. Transitions occur because the organisms involved are responsive and responding to the natural rhythms of change (West and Rheingold 1978).

Thelen's approach relies on the power of measurable biomechanical forces to explain changes, eliminating the need to infer programmed maturational processes. The pattern of stepping emerges from the aggregate properties of many cooperating parts. It is not the baby's age that predicts the change in abilities; it is myriad relationships among size and weight of the baby's head, body size, mass, muscle tone, preferred level of activity, and opportunities to move and be moved—that is, the changes in the condition of the organism and context—that are probabilistically associated with the passage of time.

A systems approach to locomotor development required new measures and new approaches to data analysis (Thelen and Ulrich 1991). Thelen used sophisticated optical tracking devices and electromyography to precisely measure changes in the position of limbs and joints. The kinematic images of stepping she generated made static drawings of "typical" infant motor patterns seem like the caricatures they really were. Two observers independently measured muscle mass, limb length, head size, and other indexes of growth to find points of transition—for example, when the weight-bearing potential of the legs was sufficient for infants to alternate steps. This attention to detail illustrates the burden of description that the systems approach places on investigators.

Ecogenactics: Uniting Forces for Change

The study of human infants has a longer empirical history than does the empirical study of songbirds. As such, investigators of human infants may have been in a better position to capitalize first on ideas from systems perspectives, but researchers studying animal behavior also have begun to seize on the power of such approaches. Technological advances, especially the ability to preserve long behavioral sequences on videotape, has made the approach more practical. Among the first animal behaviorists to adopt systems approaches are those who study social insects (Cole 1994). Studying ants or termites one by one is fundamentally different from studying them in groups. The system (i.e., the colony) cannot be generated by studying single individuals. This is not to say that one cannot learn about morphology or anatomy from individuals, but one cannot learn about their sociality. We believe that the same principle applies to the inherently social nature of communication.

Systems approaches also are emerging from less obvious paradigms, for example, the study of operant learning. We said at the outset that we originally sought out birdsong to study learning because we were unsure that the more traditional approaches would be as fruitful. But we were wrong on a critical assumption: animals do not act in arbitrary ways. They do not adhere to the simplicity imposed on them; instead, they create chains of behavior, and new means of reinforcement, where none was expected (Timberlake 1993). A systems perspective views an experimenter's actions as a "modifier of an already-functioning biological system" (Timberlake 1993, p. 125). If the study of animals under such seemingly austere circumstances needs less simplistic models of causality, it seems likely that those studying a complexly determined process such as communication will also profit.

We believe that the way to place the study of acoustic communication into a state of readiness for such approaches is first to frame the quest for ontogenetic principles with broader boundaries, ones that hold more parts of the system. To begin the process, we offer the construct *eco-gen-actics,* a term whose trip across the tongue may be awkward, but at least causes one to think about how to connect each of the components. We intend it as a means to keep visible what often has remained hidden. Let us explain first by analogy. Many of the chapters in this book were probably composed on computer screens in which several operations were going on at once—typing, monitoring mail, automatic filing, or printing. In computerese, the machine was managing multiple windows, only some of which were visible or known to the operator. We seek the same perspective for the study of development: a way to prevent nonobvious processes from being overlooked. Ecogenactics compounds into one word the multiple windows of environments, genes, and actions in relation to the development of behavior. Such an aggregate term reminds us that all these forces act simultaneously and interdependently.

Are reminders necessary? We think so, especially in a time when at least one part of the system—genes—dominates basic biology. We seek a word that puts genes in the middle of the broader network of exogenetic inheritance (Montagu 1959, Oyama 1985, West and King 1987, Gottlieb 1992). Genes inherit ecologies as much as the other way around. So too, it is clear that animals must act to survive, but it is not always made clear that such acts are producers of change, not products of it (Oppenheim 1982). Without added emphasis on the animal's contribution to its own development, we miss critical mechanisms, as learning theorists have come to recognize.

We see the effort to look through multiple windows as a form of horizontal organization for a paradigm that has already proven its worth in guiding vertical integration—from song to neuron (Konishi 1985). But the search for structure-function relationships has been refined as knowledge has increased, so that the actual integration is no longer between brain and behavior but between neural structure and neural function. The idea of a song-nucleus neural system as a series of linearly arranged stations is rapidly being discarded as researchers learn more about how brains actually work. The effects of systems theory also are clearly evident at this level of analysis: Margoliash et al. (1994) suggested that the "richness of neurobiological complexity" will demand concepts more capable of capturing the synergistic nature of neuronal and behavioral activity.

In an instructive overview on communication, Colin Beer conveyed the essence of our agenda: "The general point is that *what developmental study looks for depends on what is thought to be there for it to find,* which depends on how the social communication system is viewed in its syntactic, semantic, and pragmatic aspects" (Beer 1982, p. 300; italics added). The four principles we discuss below are guides to what we should expect to find when we study vocal ontogeny using a systems approach.

Principle 1. The context of communicating, not the sensory channels for communicating, must be the primary unit of analysis. The nature of communication was well described by W. J. Smith in *The Behavior of Communicating* (1977, p. 12): "Interaction is the arena for communication, the sphere within which the behavior of communicating serves those who engage in it. . . . however else communication is analyzed, it must be analyzed from an interactional perspective that seeks to explain the contributions of the behavior of communicating to the task of interacting." In such a view, labeling a communication system by the predominant sensory modality (e.g., acoustic or optical communication) may be misleading, and may bias us towards too narrow a view. This lesson can also be found in systems approaches to the study of the senses themselves, an approach dubbed "ecological" perception (J. J. Gibson 1966, 1979, E. J. Gibson 1969, 1991). Ecological perception reverses the typical train of thought about the role of the sensation, "asking not what's inside your head but what your head is inside of" (W. M. Mace 1977, p. 43). The essential point is that the senses evolved in relation to specific ecologies, and the analysis of those ecologies often reveals behavioral mechanisms that explain perceptual phenomena such as depth perception or size constancy. The role of the perceivers differs—the Gibsons assumed an active perceiver, performing whatever behaviors are needed to go about the business of living. A "performatory" system (J. J. Gibson 1966, p. 46) is at the core—animals do not perceive or communicate for the sake of perceiving or producing a display, but for the sake of managing a social environment.

The Gibsons also argued that the classification of perception is best formulated in relation to the function of behavior, not on the basis of anatomy of the nervous system. Unlike sensory organs, behavior cannot be dissected out of the animal. The Gibsons emphasized that many important behaviors are undedicated, or amodal; that is, they are composed of actions naturally integrated across several senses as animals carry out day-to-day tasks. A simple but powerful example is an animal moving its head—animals do this for many reasons, but the consequences for perception are profound in terms of picking up perceptual information (B. E. Stein and Meredith 1993). Many perceptual illusions rely on restraining head movement in order to achieve their paradoxical effects. The use of comparable restraints on an animal's ability to act reduces the investigator's power to find actual mechanisms by focusing on what an animal *can* do, as opposed to what it naturally *does* do, while developing.

Other support for an integrative view comes from an emphasis on the "merging of the senses" (B. E. Stein and Meredith 1993). In experiments on cats' tracking of potential prey, those investigators found that the sensory sum did not partition according to simple rules; sounds alone or sights alone could activate the cats' receptive fields. The amount of activation achieved by separate presentations, however, was not even half that achieved by joint presentation of sound and object. Moreover, the amount of enhancement of the sensory response depended on the intensity of each sensory source. When intensities were reduced, the merging produced a greater neuronal response. Thus, the cat's system integrates disparate sources and amounts of energy. Returning to songbirds, the relatively greater weighting of acoustic information, as suggested by playback studies, may indicate asymmetrical tuning. Until vocal and visual stimulation (and other sensory sources) are studied in a single organism within a particular time frame, however, we cannot know. The loud song of many passerines, coupled with their ability to be cryptic in many habitats, may explain why they are called "songbirds" and not "sightbirds," but the name also may have lulled us into a false sense of security about the levels of analysis necessary to study birdsong.

If context is important to communication, then environments must be reported and analyzed in greater detail. Our reading of literature on avian communication suggests that standards of reporting and analysis require improvement if a systems approach is to succeed (and if many of the experiments are to be replicated). At the most rudimentary level, we were disappointed in many reports of vocal development in songbirds at what was *not* said about the environments in which the subjects lived, especially in light of the status accorded to social interactions in concepts such as action-based learning (see ten Cate 1989 for a relevant example about sexual imprinting). All too often we find no information on how birds were actually housed, how many were in a cage, how many were of which sex, what else they could see or hear, and whether social interactions between cagemates or neighbors were actually observed and tallied. These data are part of the basic experimental design, not optional information.

Somctimes, too, information is given but its implications are not pursued; for example, authors may note that some birds were given hormone injections to induce singing or to stimulate reproductive responsiveness. Hormones operate on suites of characters (Ketterson and Nolan 1994), but in many reports of song development the extent of the "suite" is surveyed very casually, if at all. Sometimes, other species are present with the target species but their behavior is assumed (apparently) to have no effect. Inattention to such contextual features suggests that many researchers regard them as innocuous or incidental. Such a priori assumptions reinforce the tendency to find only what we expect, and to reduce our understanding of the proximate dynamics of ontogeny.

We offer one example of proximate dynamics to show that attention to such detail can reveal new ways to understand early stages in vocal development. When male cowbirds from North Carolina and Texas were deprived of exposure to adult males and housed with female cowbirds from their own natal areas in sound-attenuating chambers, their song content differed reliably from that of males housed with birds of other species (King and West 1988). The former produced songs with more local geographic markers than did the latter. These acoustic differences were also apparent in playback tests to females: the males housed with conspecific females produced songs that elicited more copulatory responses from independent sets of females, captured from the same site as the males, than the males housed with birds of other species (from four areas covering the three subspecies: North Carolina, Texas, South Dakota, and Indiana) (King and West, in prep.). Thus, even though females may not sing, they do exert social influences, influences sufficient to override any sensory template that presumably should be the same for males within a single local population.

These facts do not tell us anything, however, about the mechanics of social transmission. For one population from North Carolina, part of the explanation may involve the visual responses of females to the songs of males. During the early spring, when captive males repeated many song types in the presence of females, the females responded with flicks of their wings known as wing strokes, which in turn influenced the males' singing: males temporarily ceased cycling through different variations and repeated the song that elicited the wing stroke. The males' experiences in the spring carried over into the breeding season: the songs in the final repertoires for that season contained the song types that had elicited more wing strokes the previous spring. (Wing strokes have never been observed during the breeding season itself [King and West 1990]). Whether the social shaping that resulted in songs of higher potency in North Carolina males is accomplished by the same behaviors in other populations remains to be examined. We also know that males housed so that they could see and hear the interactions of a group of males developed less effective songs than males that only heard the group's songs. What did they see? They saw males attack or threaten one another, especially in settings where females were being courted (West and King 1980). The point is that housing conditions make a difference.

To summarize, a systems view of communication requires that we assume an ecological perspective even when conditions seem quite unecological—that is, in laboratory settings of our design. When we attempted to tutor hand-reared Common Starlings (*Sturnus vulgaris*) with tape recordings (West et al. 1983), we found little evidence of the tutored material in the repertoires. But the starlings apparently perceived the tutoring events differently. They mimicked the sounds of the tape recorder reels turning, the tape hiss, and the coughing of the human "in charge," but the songs on the tape were not copied. Does this mean starlings cannot learn by tape tutoring? No, it means that one must look at other sounds present in the birds' environment that might be more salient. The sounds we assumed were present and the sounds the birds actually heard were quite different.

Principle 2. The integrity of communicative systems is determined by testing the connections between elements in multiple settings. Systems approaches require multiple assessments at points of transition throughout the period of ontogeny. Such assessments reveal which elements form strong associations regardless of context, and which do not. For example, the finding that female sparrows learn from song models in the wild but not in the laboratory suggests that female sensitivity to acoustic stimulation is not a stable feature of her communicative system. A systems approach invites a different attitude toward animals in the laboratory. When we view birds in their natural settings, we know that our proximate power over them is small—we turn our heads, or have them turned by an unexpected event, and the birds are gone. When we take an animal into the laboratory, it is easy to accept the illusion that we know which of the many differences we have imposed are significant, and thus we may feel less obliged to observe the animal's behavior closely. But although the animal may still be there when we look tomorrow, the transition in its behavior may not—the burden of description remains.

From our cowbird studies we have discovered the importance of learning about the development of skills associated with song use in order to assess fully the effects of manipulating the birds' early development. Where we once thought we needed one assay, we now find that we need several. For example, in a long-term study of cowbirds from an ancestral part of the species' range, the Black Hills of South Dakota, we asked the same question we had asked of other populations:

How does vocal development proceed in juvenile males that are deprived of experience with adult males (Freeberg et al. 1995)? The males were housed socially with either female cowbirds or canaries, but not with adult male cowbirds. We used a broader set of measures than those used in previous studies to assess species identification. We suspected that this species-level element would not be seriously disturbed for two reasons. First, the brood parasitic habit of the cowbird has led many to suspect that the species must have a safety net built into the system to compensate for the unusual upbringing of the young (King and West 1977). Second, the males were not naive: they had been caught in flocks of wild cowbirds in South Dakota at least 50 days after fledging. They thus had experience with conspecific birds and had already successfully located their own species.

From late August until May, we housed one group of males with female cowbirds from South Dakota (FH males) and another group with canaries (CH males). In May, we recorded their repertoires and then placed the CH males first in large flight cages and then in large aviaries with unfamiliar female cowbirds and unfamiliar canaries. Our observations in these new settings eradicated any idea of a "safety net." First in the cages and then in the aviaries, we watched the CH males sing and pursue canaries, although female cowbirds were present and socially receptive. The CH males' persistence was remarkable. Even when the canaries said no by retreating and the female cowbirds said yes by approaching and soliciting with copulatory postures, the males were not diverted from their course. Their behavior did not change even after they had witnessed the successful courtship of females by experienced South Dakota males. They showed few signs of species identification. The error seemed to be at so fundamental a level as to undermine any concept of a fail-safe mechanism and to make questions of song effectiveness appear inconsequential.

Based on the behavior of the CH male cowbirds, we predicted that the FH males would be strongly attracted to female cowbirds. The FH males paid no attention to canaries in either the cages or the aviaries, but only when they were confined together in cages did they show signs of attentiveness to female cowbirds (i.e., singing, chasing, and copulating). Once in the aviaries, they allocated most of their singing to one another, even though twice as many females were present.

These data told us that we had another new feature to study in other populations: the development of skills associated with song use. We also had something old to grapple with in a new way: the results of concurrent playback tests of song potency did not predict the FH males' performance in the aviary. Females responded with copulatory postures as often to FH males' songs as they did to songs of wild-caught South Dakota adults (they responded even less to CH males' songs, as we had found for other populations). We knew from previous studies of other populations that song potency is a necessary but not sufficient condition to explain copulatory success (West et al. 1981). But what we had not appreciated was that males might have to *learn* to direct their songs to females of their own species. Perhaps FH males were attentive in the context of a flight cage because the cage provided physical scaffolding to support such behavior. But when males had to approach conspecific females in a less confined situation (the aviary), they were unable to do so.

We followed these birds through their second year, and provided different social conditions to them. Half the males from the combined FH and CH set were randomly assigned to be housed as a group with females from South Dakota. The other half were housed as a group with females and older South Dakota males because we wanted to know if the addition of older males could remediate the experiences of the first year. The following breeding season we again tested social skills by introducing the males to aviaries containing South Dakota females and canaries. We also recorded and played back their final repertoires to South Dakota females. We found no carryover effects from the first year's experience. Among the most successful males were two of the males originally from the CH group who had directed most of their courtship to canaries. All the males turned over their repertoires; only two retained one of the first-year song types. Playback potency in the second year did not correlate with playback potency in the first year. Nor did second-year potency predict success in the aviaries: the males housed with older males developed significantly more consortships than those housed only with females and agemates. Their songs, however, were less potent when played back to females. Thus, first-year experience did not predict second-year performance. These data, like those from the first year, illustrated that assessing song function by playback assays potential, but not actual, competency.

The links between song content and use also may be quite fragile within, and perhaps beyond, a male's first year. In his studies of suboscines, Kroodsma (this volume) found that hand-reared males of some species sang quite normally in the laboratory and concluded that they "undoubtedly have that capability in nature, too." He bases his argument on the premise that the "rich social milieu" in nature would be more likely to support such behavior than the "simple" conditions of the laboratory. Our studies of the South Dakota cowbirds suggest that such stability across contexts cannot be assumed. The best predictor of stability may be the degree of similarity between the conditions in which the animal first learned a certain skill and the subsequent environment in which that skill is tested. Burghardt (1977, p. 79) focused on this distinction when he stated that "the processes involved in the developmental shaping of behavior may have little in common with those subsequently altering the behavior." Systems theorists emphasize this point by stressing the need to understand "initial conditions" (Timberlake 1993, Cole 1994). Which initial conditions are most important for cowbirds? Cowbirds begin life with a host species, overwinter in mixed-species flocks, and interact in natal and breeding habitats that vary greatly in a male's exposure to conspecific adults (O'Loghlen and Rothstein 1993). Thus far, the data suggest that the possibility for new ontogenetic changes reoccurs with each change in context; researchers must thus remain vigilant to detect changes at many developmental points.

Male song potency also illustrates the effects of context. The lack of correlation between playback potency and performance in a reproductive context for the FH males and all the second-year males relates to differences in social context. Previously, we had shown that playback potency of males recorded while living in breeding aviaries was correlated with their copulatory success (West et al. 1981). In that case, the males were singing and living in the same context. Not so for South Dakota males in more recent studies, whose singing was recorded in one housing context and breeding behavior was studied in another. A practical consequence is that the playback potency of songs of acoustically deprived males, a common measure in the study of birdsong, can predict their performance only in equally deprived circumstances, thus rendering the results of questionable utility for developmental studies. The data also emphasize that assessments of normal development based only on acoustic structure may be compromised by the lack of functional tests. Thus, converging measures of behaviors of species-typical receivers are needed to characterize song development.

Studies of other songbirds support the need to look at transition points in birds' lives. In studies of White-crowned Sparrows, investigators were able to target points of ecological transition (e.g., settling into a breeding site and establishing a territory) that appeared to be especially important in understanding the rates of change in vocal ontogeny of males and the frequency of singing in different populations of females (DeWolfe et al. 1989, Baptista et al. 1993a). For Indigo Buntings, the choice of song models is the product of the compound probability of several factors, including the timing of exposure to adults, number of yearlings in the area with brighter plumage, and total number of neighbors. Contrary to what one would have assumed, the best predictor of which birds the young birds copy is not the adult males' breeding success. A systems scenario might suggest that birds that arrive early have an advantage because their songs have a primacy effect on would-be pupils. Some songs may also be easier to perform. Such multiply determined rules allow flexibility because they are formed by looking through a number of windows at how communicative variables are assembled in time and space.

Principle 3. To find the control parameters underlying the assembly and reassembly of behavior, one must identify and manipulate points of social and physiological transition. The time required to develop a vocal repertoire is multiply determined by the opportunities to hear song, produce song, and sing enough to reach some vocal stability. These multiple determiners offer researchers an opportunity to use a technique encouraged by developmental theorists: simulating "microgenesis," that is, "attempts to simulate developmental changes by real-time experimental manipulations" (Thelen and Ulrich 1991, p. 34). The term *microgenesis* is the contribution of L. Vygotsky, a developmental psychologist who studied the development of specific skills to find parameters that might be generalizable to later, more complex or global skill acquisition. The objective is to identify points of ontogenetic vulnerability (when behavior seems most malleable) and to determine the specific conditions that lead to stability and instability. We offer two examples.

Marsh Wrens (*Cistothorus palustris*) in the northeastern United States hatch any time between mid-June and late August. Whether a wren hatches early or late has multiple implications (Kroodsma and Pickert 1980). The early hatchers hear much more song from adults, experience longer photoperiods, and have more days in which to sing. In an experimental simulation of these conditions, the investigators found that this collection of variables affected immediate and subsequent vocal development. Birds living under the simulated conditions of late hatchers (the August group) molted between days 45–71; the early-hatchers (the May group) molted between days 63 and 95. The birds also sang different amounts during their first fall, with the August group producing an average of 11 minutes of singing (out of a possible total of 72 minutes), compared with 54 minutes for the May males. The quality of the singing actions also differed: the songs of the August birds were variable; the May birds progressed to stable forms. Finally, the migratory restlessness of the groups differed, with the August group beginning and ending such behavior sooner. The following spring, the birds were exposed to tutor tapes of 12 new song types. Only August males acquired new types, and those that learned new material were the ones that had heard and produced the fewest songs the previous fall. The May group probably could perceive that their songs already resembled the tutor songs in terms of structural stability. The August group may thus have perceived more of a discrepancy and sang to reduce it. (Marsh Wrens engage in countersinging, and probably use feedback from one another when developing their songs.) Such data identify exogenous and endogenous points of transition. The best candidate for a control parameter predicting repertoire stability might be the male's perception of the match between his performance and that of the surrounding tutors. Neither availability of song tutors, amount of sunlight, temperature, resources within a habitat, nor physiological constraints on energy metabolism are fixed—they are points of change identified only by studying the animal's life cycle.

A second example also suggests that changes in the content of song repertoires may have complex origins. In the experiments on South Dakota cowbirds described earlier in this chapter, the males' first extended opportunities to observe other males came at the same time as their first opportunities to court females. Thus, the first chance for these males to hear and compare sounds occurred quite late, and, for the FH males, this activity took precedence over persistent courtship. These males spent more time singing to one another than to females. That the males could not court and countersing at the same time (as do wild males), even when their only "competition" was males singing to canaries, suggests that the rapid pace of behavioral events surrounding breeding may not support or reinforce the initial conditions needed to learn new behavior. Thus, as with the Marsh Wrens, diagnosing when changes in development can occur requires separating multiple sources of influences. Male cowbirds typically observe and interact with males and females at the same time of year; by doing so, they can compare the differential properties of the sexes: females may deliver wing strokes, but males may deliver physical attacks. Our experimental procedures separated those events and perhaps highlighted the differential contribution of each sex for young learners.

A possible problem in manipulating social context is deciding on the best measures of the outcome. In our experiments with South Dakota cowbirds, for example, we would have reached a different conclusion about the stability of the FH males' behavioral abilities if we had stopped the study when the males emerged from the chambers or when they sang "normally" to females in flight cages. It was the more complex and realistic setting of the aviary that perturbed the males into new states that then remained quite stable during the breeding season. The CH males probably experienced less perturbation throughout the first year because they were always with or near canaries and produced such atypical songs that female cowbirds and other males ceased to seek encounters with them. It may also be that the data reveal an instance of an environmentally defined sensitive phase like the one Kroodsma and Pickert found in their manipulation of wrens. The presence of males in the aviaries may have signaled a new, as yet unexplored, social context. Thus, what to us and to the aviary females was the breeding season may have been the "preseason" for the FH males.

Our studies of North Carolina male cowbirds interacting with females during the early spring, when males and females typically return to their breeding grounds, represented another opportunity to view microgenesis (West and King 1988b). In that study, we provided captive males with more concentrated and undiluted doses of female stimulation than would ever occur in nature. As a result, we moved the males into extreme states of vocal production such that even the merest flicker of response, a wing stroke, led to rapid vocal reassembling. Thus, we compressed events that might normally happen over longer periods and at different scales of intensity. The females thus acted like Thelen and Ulrich's treadmill by providing attentive support for vocal movements, support normally not available or accessible in nature.

The makings of fascinating developmental examples are now emerging from studies of nonoscine passerines (see Kroodsma, this volume). Such studies also suggest that ontogenetic simulations will be a useful tool, but the time to study them may have to be recalibrated to take into account the apparently different growth patterns of these species. The tyrannids studied to date appear to produce recognizable vocalizations much earlier than oscines. Observers have reported the presence of well-formed sounds in some species soon after fledging. Preventing suboscines from hearing the songs of adults from an early age or deafening them at a slightly older age leads to substantially less disruption in vocal production than it does in oscines. Although these data seem to hang together to support the conclusion that these taxa follow a different ontogenetic trajectory, several questions remain. If some species are vocally proficient during the first week after fledging, might their auditory system be "tunable" even before hatching? Perhaps the "experience-expectant" and "experience-dependent" systems operate in different time windows (Greenough et al. 1987). Thus, deafening 35 days after hatching may have different effects in a suboscine than in the "average" songbird. Moreover, housing young suboscines with other young birds may have greater effects in such (possibly) rapid learners. The "initial conditions" thus appear quite different. Taken as a whole, then, the conclusion that song development in suboscines is a closed system seems plausible, but when the pieces are viewed separately, it seems that learning experiences may generally take place much earlier in suboscines than in oscines. Because the vocalizations seem simple and because dialects are often absent in suboscines, investigators may assume that they "ought to find" less flexibility. But the use of simpler themes with fewer variations, coupled with earlier motor competencies in producing sound, may just make the learning happen more rapidly. If the research on oscines has a lesson to teach, it is to be on guard about assuming that we need to manipulate only the most obvious parameters to find plasticity.

Principle 4. The study of communication requires procedures to preserve and analyze variation within and across subjects and settings. More measurement of variation is essential if the goal is to study ontogeny. Studies that have documented interspecific levels of variation in many facets of avian communication have furthered our understanding of evolutionary perspectives (Kroodsma 1988a, Baptista and Gaunt 1994). But interspecific and even subspecific differences are far too large to meet the needs of those who study vocal development. We need to know about individual animals before we can average their differences into populational variables. Reducing the range of variation prematurely may mask developmental processes.

The need to pay attention to intraspecific variation applies to the most routine practices. For example, a sampling of as yet unknown significance (brought to our attention by Christopher Evans) concerns the subjects used for developmental studies. Songbirds tend to come into the laboratory in handfulls. Except for a few domesticated species, an experimenter cannot order subjects from a supply house. This constraint is well known, but it is not always considered as a potential variable. Are we in fact pseudoreplicating ontogenetic results by measuring young who share genetic and environmental lineages because they come from the same nest? The matter is made even more complex by extra-pair paternity in some species. The problem is especially serious when questions are being posed about the innate status of a behavior (see Kroodsma 1989d for an example of attention to this problem). If the songs of young from the same nest are similar, we may find what we want to find but still not know why. Using the multiply determined perspective we advocate here, we see these nestlings as related in many ways: they *may* share genes, but they *do* share parents, an auditory environment, and their pre- and postnatal histories. If the subjects for these studies were mammals from a single litter, the standard statistical procedure would be to consider the litter as one data point. The reason for such a constraint has to do with the variables just mentioned: littermates are epigenetically linked. (One researcher told us that the practice of considering siblings separately seems so obvious that many investigators carry out such procedures without even reporting it; if so, we advocate a change in reporting practices.)

Intraspecific variation has received the most attention in the testing and sampling of song repertoires. How many songs from how many birds are required to estimate a population's typical repertoire size? Acquiring information on vocal variation is both time-consuming and tedious, and researchers too often neglect such mundane "housekeeping" chores. We especially worry that measures established for one part of a species' range may be mismeasures for another. Cowbirds in the West use songs and whistles in a different proximate relationship than do cowbirds in the East. Can repertoire size standards derived from captive eastern birds be used with confidence to represent other populations (O'Loghlen and Rothstein 1993)? Stated more generally, when can data on the function of song in one species or population be used to infer its function in another species or population?

Greater attention to variation among signals is at the heart of ongoing efforts to improve playback designs. Kroodsma (1989b, 1990) put forth systematic arguments calling for redesigns of the playback procedures used to measure the functions of vocal signals. He explicated some ways in which inadequate attention to variation can lead to false conclusions, especially by using pseudoreplication. The implication of these suggestions is that only "some experiments may be possible to do thoroughly, but it is better to know the limitations than it is to collect inappropriate data and delude ourselves" (Kroodsma 1989b, p. 608). This principle applies to other testing paradigms as well. In an extensive review of the playback methods used by many investigators to test females, Searcy (1992a) discussed the growing amount of data on the nature and kinds of perceptual discrimination by female songbirds as assayed by solicitation displays, a procedure he ranked highly in practicality, generality, sensitivity, and interpretability. We recognize that this evaluation was not based on developmental issues—he was dealing with issues of playback techniques in general—but when we look at the use of the female solicitation display from an ontogenetic viewpoint, we come to different conclusions about some of the costs and benefits of the assay.

Most important, we worry that some of the now standard methods used by many researchers may reduce the value of the data for understanding the development or evolution of responsiveness to vocal signals. The solicitation display has now been used to study more than 20 species, and the data indicate that females have the potential to react to many dimensions of song (reviewed by Searcy 1992a). But in all those studies except those originating in our laboratory, estradiol was used to enhance female responsiveness. We refrain from this common practice for several reasons. The shift from living in the wild to living in a laboratory is a complex event. The laboratory is also an environment, and wild-caught animals need time to learn its properties. We routinely wait nine months before using females for playback experiments, allowing them time to settle into the laboratory environment and become accustomed to the routines of human care and human curiosity. This extended period is one reason why we have not found the use of estradiol implants necessary. If females do not all come into breeding condition at the same time (and they typically do not), we can afford to wait a few more days, after waiting nine months. If other researchers cannot wait, then the burden would seem to fall on them to show that the addition of the implant does not distort the female's responsiveness. The use of estradiol to stimulate reproductive behaviors and to facilitate generalizations across species, must be weighed against the potential costs. Although more data can be collected from more individuals in less time, these data may tell us nothing about females' responsiveness under natural conditions.

Further, assessing generality seems premature until the natural range of variation is analyzed. Even with the administration of hormones, not all females respond. And even without hormones, some of our females respond to every playback. We find the differences among females are important in that the range in vocal production among males may be linked to the range in perceptual responsiveness of females (King and West 1989). In one experiment, we housed North Carolina males from August until May with females whose responsiveness to playback had been measured during the previous May and June. Males placed with females that had responded to most of the playback songs developed very small repertoires, but males housed with females whose playback responsiveness was low produced many more song types (King and West 1989). If the females' responsiveness had been made uniform, we might not have been able to make the connection between different parts of the communication system—that males attend to females' responsiveness, and their attention affects song production.

Data obtained in another setting suggest that differences in female responsiveness are also worth pursuing because they offer a possible explanation for another dimension of vocal variation, song sharing. Male cowbirds in local populations often share the same song types (Dufty 1985). When we played back the songs of several groups of males ($N = 25$ over three experiments), we found that females ($N = 24$ different females) responded reliably more often to shared than to unique songs. Data on song sharing and measures of breeding success in captivity indicate that high levels of song sharing are associated with higher rates of reproductive behavior such as copulations (King and West, unpubl. data).

These data suggest a different view of repertoire size in this species: similar may be better. A female may be better able to assess a male if his song resembles a vocal uniform. (Why do athletes not in team competition wear uniforms, if not to focus the judges' attention on the relevant athletic skills?) Since male cowbirds offer no material resources to females, any means by which they can display their physiological quality may be very important.

Thus, ignoring individual variation might have the unwanted effect of masking interrelationships between production and perception. If the emergence of behavioral function is not of primary concern to a researcher, many of the matters discussed here can be evaluated differently. But in the study of development, variation is the link to mechanisms. It is part of the solution, not the problem. If we do not measure naturally occurring variation, we deny ourselves access to the very material from which complex behavior arises, and we rob students of ontogeny of vital resources.

Ecogenactics: The Heuristics of Change

With the term *ecogenactics* we seek to recognize and expand the network of possible influences that are *routinely* considered in the study of vocal ontogeny. We find that the word helps us to focus on the diverse nature of developmental processes. By connecting ecologies, genes, and actions, we can see that species-typical environments and behaviors are as inevitable a source of influence as an animal's genes. A Marsh Wren inherits not only its parents' genes but an ecological legacy of early or late hatching with respect to the species' breeding cycle. A cascade of consequences results from such environmental variation. The outcome of early or late hatching (i.e., development of a species-typical repertoire) is thus multiply constructed by (1) the actions of parents in producing clutches at different times, (2) transitions in the amount of singing produced as the reproductive state of males changes, (3) the influence of physiological changes related to molting, and (4) the availability of adult males the following spring. Thus, we see ecogenactics as a declaration of the natural interdependencies of environmental, genetic, and behavioral processes. Unless we obtain knowledge about each of these processes, we set too low a standard to do justice to the concept of communication.

Advances in sciences are one of the major reasons why new words appear and old ones lose their appeal. In outlining a strategy of "strong inference" to maximize scientific effectiveness, Platt (1964, p. 352) targeted the dangers of complacency: "What I am saying is that, in numerous sciences, we have come to like our habitual ways, and our studies that can be continued indefinitely. . . . And this is not the way to use our minds most effectively." He advocated greater emphasis on the generation of alternative hypotheses at every possible point in the scientific enterprise. A systems approach does just that because it begins with the premise of branching pathways and outcomes.

The four principles described above capture some ways to articulate these alternatives: we need to expand the context in which we measure communication; we must guard against prematurely concluding that developmental processes crystallize in the same way or same time frame within and across species; we must use converging measures to determine the full measure of our manipulations on developmental outcomes; and we must see variation as a valuable resource, not a vexing problem. Said most simply, we need to incorporate more information so that we can falsify or eliminate long-held hypotheses. The method of strong inference (sensu Platt 1964) has served other biological sciences well, and it may serve to maintain the field of birdsong as a scientific enterprise.

To conclude, we advocate a comprehensive attitude towards the study of development. Too often, ontogeny is analyzed by reducing and removing supposedly critical variables or stimulation, then verifying that the observed outcome is abnormal. Studying how and when things go wrong is no substitute for studying how and when things go right. A truly developmental approach also requires construction—the challenge comes in assembling the parts, tinkering with the timing of their interactions, and simulating opportunities for change. Whereas linear processes can withstand study by isolation and separation, interaction among elements is at the very core of nonlinear processes. Fundamental properties disappear if the elements are separated. This is the major lesson we must learn. If we do not use the proper methods to preserve connections, we cannot study the phenomenon of communication. And if we do not capture the nature of communication, we leave future students of animal behavior a legacy of uncertain value.

Acknowledgments

We thank the National Science Foundation for support. We thank K. Cleal, S. Duncan, M. Engle, T. Freeberg, M. Goldstein, R. Titus, and L. Twardy for comments and D. Kroodsma and E. Miller for thoughtful editorial advice.

3 Nature and Its Nurturing in Avian Vocal Development

Luis F. Baptista

The roles of environment and genetics in shaping behaviors are among the central themes of ethological theory (Lorenz 1961, Konishi 1985) and are still vigorously debated (West and King 1987, Barlow 1991). Konishi (1985) lauded the important role of birdsong studies in resolving the debate over instinct versus learning in the ontogeny of behavior. Johnston (1988) disagreed, arguing that the "acquired-innate dichotomy" was an outmoded concept and that the interactionists' viewpoint best explains behavioral development. Bekoff (1988a, p. 631) countered that applying the interactionist perspective to all behaviors was "misleading, myopic, premature and philosophically unsound."

Early studies indicated that vocal acquisition through imitative learning was a characteristic restricted to oscines (Konishi and Nottebohm 1969). Since that time, however, vocal learning has been documented for various nonoscine taxa (Kavanau 1987, Baptista 1993, Groothuis 1993).

Students of avian vocal development usually distinguish between long, complex utterances called songs and briefer, simpler sounds termed calls (Thielcke 1970a). Early experiments on ontogeny suggested that songs are learned in many bird species but that calls develop independently of learning (Thorpe 1958, Thielcke 1961a). This generalization has been undermined by studies that reveal cases of learned calls (Rothstein and Fleischer 1987, Baptista 1990a, 1993, Groth 1993). I treat the development of both classes of vocalizations here.

Johnston (1988) maintained that isolation experiments, so important in studies of vocal development, provide no information about the influence of genetics in behavioral development. Searcy (1988c) responded, and I agree, that these experiments, performed on many species, demonstrate phylogenetic trends in the relative importance of imitation in vocal ontogeny of diverse taxa.

In this essay I review findings indicating that the relative contributions of nature and nurture to avian vocal development fall on a continuum (S. M. Smith 1983). Although many species of birds learn song, for example, one often finds constraints underlying various aspects of the learning process. Experiments indicate that genetically controlled predispositions guide naive juveniles to select song types from the acoustic environment and also control the size of the repertoire produced. Moreover, not all features of song are acquired by imitation. By analyzing song in terms of syllable structure, frequency, rhythm, total duration, syntax, and number of elements, one finds species in which all characters develop independently of learning at one extreme, and species in which only one or a few vocal features develop through imitation at the other. The vocalizations of hybrids between bird species or domestic breeds suggest that different features of vocalizations have different patterns of heritability.

Repertoire Size

Repertoire size can vary with age (Nottebohm and Nottebohm 1978, Loffredo and Borgia 1986), geography (Mulligan 1966, Bitterbaum and Baptista 1979, Kroodsma 1981c), and phylogeny (Baptista and Trail 1992). Some variation is influenced mainly by genetic factors, whereas other variation (most?) results from complex interactions between genes and the environment.

Environmental influences on song development are well illustrated by the House Finch (*Carpodacus mexicanus*), which was introduced into New York City in the 1940s from California. In the 1970s and 1980s, Mundinger (1975, 1982) found that the eastern birds had song repertoires of 2.2 songs per individual, with extensive sharing of themes between individuals. Repertoire size in southern California populations (near the probable source of the eastern birds) was 4.0 songs per individual, and although syllable sharing was extensive, few entire themes were shared (Bitterbaum and Baptista 1979). The number of themes in eastern populations was 2 to 6; the number of themes ranged from 13 to 71 in California populations.

Because eastern House Finches had been introduced only three decades earlier, it is unlikely that the differences Mundinger observed are genetic. The smaller repertoires of eastern birds may be the consequence of widely dispersed individuals in a new environment, with few song models to imitate, and fewer learned song themes as a result. In the dense populations of southern California, juveniles interact with many adult individuals; they acquire many more syllables than do eastern birds, and the syllables are organized into a greater number of themes. Bitterbaum and Baptista (1979) predicted that as eastern population sizes increased, both population density and song variety and complexity would increase.

Repertoire size in some bird species appears to be under strong genetic control. Marler and Sherman (1985) raised Song Sparrows (*Melospiza melodia*) and Swamp Sparrows (*M. georgiana*) in isolation and found that experimental birds sang an average of 5.5 and 1.6 themes, respectively, as adults. The repertoires were about half the normal size (10.3 vs. 3.0), but the species differential was maintained in the laboratory, with isolate Song Sparrows developing about three times as many song types as isolate Swamp Sparrows.

Marsh Wrens (*Cistothorus palustris*) provide another example. Males in California populations sing about 150 themes, whereas males in New York populations sing only about 58 (Canady et al. 1984). Kroodsma and Canady (1985) hand-reared western and eastern wrens and tutored each group with a tape containing 200 song types. Western birds learned more song types than eastern birds (medians of 108 vs. 43.5, respectively), indicating a possible genetic basis for repertoire size.

Stimulus Filtering

Because of the important roles song plays in advertisement, territoriality, and (in some cases) as an isolating mechanism (e.g., estrildids and viduines), a bird must be endowed with the disposition to learn and produce "correct" (species-specific) vocalizations (R. B. Payne 1973a, b, Kroodsma and Byers 1991, Baptista and Trail 1992). Social factors such as social bonds and various forms of social interaction with conspecific birds can ensure that correct vocalizations are acquired (Nicolai 1959, Zann 1990, 1993, Baptista and Gaunt 1994); however, various experiments indicate that even in the absence of social stimuli, a preference to learn species-specific vocalizations directs the learning process, so that correct sounds are selected and incorrect ones are rejected.

The preference to learn conspecific song over alien song was discussed early in the eighteenth century by Baron Ferdinand von Pernau, who observed Chaffinches (*Fringilla coelebs;* Stresemann 1947, Thielcke 1988a). After the sonograph was developed, Thorpe (1958) tutored hand-raised Chaffinches with tapes of a variety of cardueline songs. He found that Chaffinches could learn to sing conspecific as well as heterospecific syllables, but they sang only Chaffinch songs as adults. Additional support for the idea of preferential learning comes from White-crowned Sparrows (*Zonotrichia leucophrys*). Marler (1970) found that White-crowned Sparrows tutored with Song Sparrow songs sang abnormal songs typical of birds that had never heard songs of their own species, but they did not sing syllables from the alien tutors. Konishi (1985) hand-raised White-crowned Sparrows from eggs and subjected naive fledglings to tapes containing songs of their own and other oscine species. In almost all cases, White-crowned Sparrows copied only conspecific song.

The Swamp Sparrow is a persuasive example of a species preference in song learning (Marler and Peters 1977). When synthetic songs containing mixtures of Swamp and Song Sparrow song syllables with varying complexity and in varied temporal patterns were played to hand-raised birds, Swamp Sparrows selected conspecific syllables out of these jumbles and constructed simple single-syllable songs typical of their species.

Based on evidence of prenatal learning in precocial bird species, Johnston (1988) questioned the evidence for song-learning preferences based on tape-tutoring experiments. He argued that embryos may have already heard their father's song before hatching and before being brought into the laboratory. Both the White-crowned Sparrows (Konishi 1985) and the *Melospiza* sparrows (Marler and Sherman 1985) were raised in the laboratory from eggs, however, precluding prenatal experience with conspecific song.

Cross-fostering studies by Eales (1987) also demonstrate a preference to learn conspecific over heterospecific song. When Zebra Finches (*Taeniopygia guttata*) were raised by White-rumped Munias (*Lonchura striata*) in double cages with a screen separating them from a pair of Zebra Finches they could see and hear, all the cross-fostered males learned songs exclusively from their White-rumped Munia foster fathers; they acquired no syllables from the Zebra Finch males in the next cage. In a second experiment, Zebra Finches were raised by White-rumped Munias under the described conditions, but the young Zebra Finches were transferred to the other side of the double cage when they were 35 days old. The young Zebra Finches could then interact physically with conspecific birds. Four of the birds developed songs consisting exclusively of Zebra Finch syllables from the second tutor. A fifth bird sang songs containing White-rumped Munia and Zebra Finch syllables. Members of a third group, consisting of six Zebra Finches, were raised by their own parents and were separated by the screen from a second pair of Zebra Finches. The young were transferred across the screen at 35 days of age and thereafter allowed to interact with the second conspecific pair. Five experimental birds sang the song of the father; the sixth sang a hybrid song containing both the father's and the second male's syllables. Although both the second and third groups could interact with the conspecific unrelated male, only those initially raised by White-rumped Munias tended to learn from the new conspecific tutor. This result suggests that additional learning was likely only when the initial model consisted of alien White-rumped Munia syllables, and was therefore unsuitable.

Zebra Finches raised by White-rumped Munias learned alien songs as accurately as conspecific songs (Clayton 1989). In contrast, White-rumped Munias tutored by Zebra Finches (Clayton 1989) or African Silverbills (*Lonchura cantans*) learned fewer syllables from their alien foster fathers than did individuals raised by their own parents (Dietrich 1980, Clayton 1989, Baptista unpubl. data). This difference indicates that White-rumped Munias learn conspecific song with greater facility than they do heterospecific song, again suggesting a mechanism for filtering out alien sounds as models to be copied.

Mundinger (1988) demonstrated heritability of learning dispositions among domestic canary (*Serinus canaria*) breeds. When he tutored individually isolated Border, Roller, and Border × Roller crosses with tapes containing songs of both canary strains, he found that Borders selected Border syllables and Rollers selected Roller syllables as models for imitation. The hybrids sang songs containing syllables from songs of both breeds. The experiments also show that humans have selected for learning preferences in these canary breeds.

The preference for conspecific songs may be altered by certain experimental conditions. White-crowned Sparrows tutored by live Red Avadavats (*Amandava amandava*), Dark-eyed Juncos (*Junco hyemalis*), and Song Sparrows in cages that allowed them to see and interact with their alien tutors across a screen, and only to hear conspecific singers in the room, imitated their alien tutors (Baptista and

Petrinovich 1984, 1986), supporting Baptista and Morton's (1981) hypothesis that living tutors may cancel out preferences to learn conspecific song or surmount barriers (Marler 1970) to learn heterospecific song.

Some preferences are revealed by the behavior of young birds. White-crowned Sparrows collected as nestlings, hand-reared, and exposed to playback of conspecific and heterospecific songs as fledglings produced more location and begging calls in response to White-crowned Sparrow songs than to songs of alien species. Fledglings thus "possess a perceptual predisposition to respond to conspecific song" during the sensitive phase (D. A. Nelson and Marler 1993, p. 807).

What is the nature of the filtering mechanism? Some authors have suggested that the filtering is on the sensory level (Mulligan 1966, Marler and Mundinger 1972). Playback studies, however, have revealed that birds may store songs without vocalizing them (Slater et al. 1988, Baptista et al. 1993a). Because a number of species will produce heterospecific sounds as young birds (at the stage of producing subsong) but not as adults, Slater et al. (1988) suggested that filtering is effected by a "culling" process. Some species may learn heterospecific songs from live tutors but not from tapes, so social factors may also function as filters (Slater et al. 1988, West and King, this volume). Neuromotor filters may operate in species such as the White-rumped Munia, which produce excellent copies of conspecific songs from live tutors but produce poor copies of heterospecific songs. These data suggest that it is difficult or impossible for White-rumped Munias to mimic certain heterospecific sounds.

Sex Differences in Learning

There have been few extensive studies of female song and its ontogeny (Baptista and Gaunt 1994), but it seems clear that female performance varies from one species to the next. For example, male and female Marsh Tits (*Parus palustris*) produce songs learned from the father (Rost 1987). In contrast, although male Zebra Finches learn both songs and calls from fathers, the females appear to learn neither (Zann 1985, 1990). Playback studies indicate that a female may recognize her father's songs, so she can store songs even though she does not vocalize them (D. M. Miller 1979b).

Sex differences in prerequisites for song learning seem to exist in the White-crowned Sparrow. Females of the sedentary subspecies *nuttalli* and the migratory race *oriantha* sing at certain times of the year (Baptista et al. 1993b). Both natural female song and the songs induced by testosterone injections are similar in their details to male song (Kern and King 1972, Baptista and Morton 1982, 1988, Tomback and Baker 1984). Males readily learn under laboratory conditions, but efforts to teach songs to females with tape recordings and living tutors have met with little success (Cunningham and Baker 1983, Baptista and Petrinovich 1986). Perhaps vocal traditions are transmitted along sexual lines in the White-crowned Sparrow just as they are in Hill Mynas (*Gracula religiosa*) and Common Starlings

(*Sturnus vulgaris*) (Bertram 1970, Hausberger 1993, Henry 1994). Perhaps the female White-crowned Sparrows failed to learn because their tutors were males (Baptista and Petrinovich 1986).

At least one species is polymorphic regarding the female's disposition to sing. White-throated Sparrows (*Zonotrichia albicollis*) exist as white-striped and tan-striped morphs (A. M. Houtman and Falls 1994). White-striped females sing in the wild, but tan-striped individuals tend not to sing. All 33 white-striped females responded to playback with singing, whereas none of the 37 tan-striped ones did so (Lowther and Falls 1968). It would be interesting to know if there is also polymorphism in disposition to learn.

Females of different species clearly differ in how they acquire and produce vocalizations. At one extreme are species that acquire and use songs (Marsh Tits), and at the other are species that store but do not vocalize songs (Zebra Finches). In between are species such as White-crowned Sparrows, which sing only at certain times of the year, and White-throated Sparrows, in which some birds only store songs (tan-striped) and others store and sing them (white-striped).

What Is Learned

So far I have discussed learning preferences, number of song types learned or sung, and sex differences in learning. In this section I focus on the song form itself and treat syllable structure as well as temporal, frequency, and tonal characteristics. The literature reveals that even among taxa that learn vocalizations, one or more of these characteristics may develop normally without reference to an adult song as a model.

Syllables

The conditions needed for normal syllable structure to develop vary widely among birds. Various galliforms, columbiforms, and suboscines produce species-typical sounds when raised in isolation or when deafened and thus deprived of audiosensory feedback (Konishi 1963, Nottebohm and Nottebohm 1971, Baptista and Abs 1983, J. A. Baker and Bailey 1987, Kroodsma and Konishi 1991). Common Quail (*Coturnix coturnix*) raised from eggs in isolation will produce conspecific crows as chicks if treated with testosterone (Schleidt and Schalter 1973). In contrast, Anna's Hummingbirds (*Calypte anna*), White-crowned Sparrows, and a number of oscine species raised in total isolation from other birds (i.e., "isolates") utter syllables that bear little or no resemblance to those in the songs of wild adult conspecific birds (Marler 1970, Kroodsma 1982a, Baptista and Petrinovich 1984, 1986, Petrinovich 1985, Baptista and Schuchmann 1990). Members of these taxa depend on imitation of adult models to produce species-typical syllables.

Although syllables are the building blocks of song, not all oscines acquire syllables through imitation. For example, Short-toed Tree-Creepers (*Certhia brachydactyla*) develop normal begging calls when raised in isolation, and these calls change in structure with increasing age (Thielcke 1965a). The begging calls eventually develop into adult social calls with little or no modification (Thielcke 1971). Adult song consists of calls arranged in a learned, fixed sequence (syntax), each sequence characteristic of a population. Birds raised in isolation sing songs with normal or relatively normal syllabic structure but abnormal syntax, rhythm, and number of syllables (Thielcke 1970b).

The relationships between songs and calls and how song is derived from calls vary among firefinches and waxbills. The Bar-breasted Firefinch (*Lagonosticta rufopicta*) produces a complex metallic-sounding song that is distinct from its social calls (Harrison 1962, R. B. Payne 1973a). In contrast, the related Red-billed Firefinch (*L. senegala*) sings a song consisting of one alarm call (or sometimes several) followed by two to seven consecutive contact notes (R. B. Payne 1990b). These calls develop from begging calls independent of the bird's learning experience (Nicolai 1964, Güttinger and Nicolai 1973, R. B. Payne 1973a). Indeed, cross-fostering experiments have demonstrated that all social calls of African estrildid finches in the genera *Lagonosticta* and *Estrilda* develop independent of imitative learning (Güttinger and Nicolai 1973). In the related African Firefinch (*L. rubricata*), song consists of around six different social calls delivered in varying and apparently random sequence. Alarm calls are, however, absent from their song (Goodwin 1982, p. 132). In Jameson's Firefinch (*L. rhodopareia*) and the Black-throated Firefinch (*L. larvata*), contact calls and song are identical, although each is associated with different postures and different contexts (Goodwin 1964). Among the waxbills are species such as the Common Waxbill (*Estrilda astrild*) and the Orange-cheeked Waxbill (*E. melpoda*), which sing songs distinctly different from social calls (Goodwin 1982, Baptista pers. obs.). In contrast are the sibling species *E. erythronotos* and *E. charmosyna*, which produce songs identical with their two-note contact call (Immelmann et al. 1965, Goodwin 1982, Scheer 1991). Thus, among the firefinches and waxbills, song may be distinct from calls or may consist of one, two, or more social calls, which may or may not be in a fixed syntax acquired through learning. These data lend credence to Thielcke's (1961a) thesis that songs evolved gradually from ritualized calls.

Data from three other songbirds also suggest a close relationship between songs and calls. Northern Cardinals (*Cardinalis cardinalis*) acquire some syllables from adults, but others develop from begging calls, independent of imitative learning (Lemon 1975). Although Swamp Sparrows acquire the number of syllables and their order by imitation, the number of syllable types is limited (Marler and Pickert 1984). Syllable types in this species may be "genetically preordained" (Marler and Pickert 1984, p. 673). Coal Tits (*Parus ater*) reared in isolation produced songs that differed in structure from songs of wild adult males (Thielcke 1973a). One of six isolate song types used in playback studies evoked full responses from wild adults. Close examination revealed that the effective isolate song resembled one of the natural song types and also resembled the alarm calls of both wild adults and isolates. Syllables in this same song type in nature apparently develop from modified alarm calls.

A parallel to the waxbills may be found in the Brown Towhee complex (*Pipilo*). The White-throated (*P. albicollis*) and Canyon Towhees (*P. fuscus*) sing songs distinct from their calls, but songs of California (*P. crissalis*) and Abert's Towhees (*P. aberti*) consist of a "vigorous and accelerating series of the ordinary (=excitement?) calls" (Marshall 1964, p. 347).

Thus, although syllable structure in song appears to be learned in most of the oscines studied, syllables in songs of several Old and New World taxa consist of social calls or partially modified calls arranged in a fixed syntax. The experiments cited above indicate that the development of these calls is often independent of learning experience.

Learning of syllables and other song characteristics is believed to be an evolutionarily derived character (Nottebohm 1972, A. S. Gaunt and Gaunt 1985a); however, Mundinger's (1988) experiments on song development in some domestic canary breeds (Borders and Rollers) demonstrate that reversals from learning to nonlearning are possible. All the syllables in songs of Border canaries raised in isolation were abnormal in structure, indicating that normal development requires imitation. In contrast, some of the flutelike syllables selected for in Roller canaries appeared in songs of isolates and thus developed independently of learning experience. Humans have thus selected for "heritable" syllable types.

Duration

Heritability of duration is evident in songs of the domestic nonoscine avian breeds that have been selected for different song lengths. Crowing in domestic Red Junglefowl (*Gallus gallus*) develops normally even in birds deafened at the time of hatching (Konishi 1963). Breeds of large domestic roosters tend to have long crows with lengthened terminal syllables (Sossinka 1982), whereas bantam breeds tend to have brief crows similar to the ancestral state. The extreme in selection for long songs is exemplified by the Totenko breed, which has crows 15–20 seconds long, compared with 1.5–2.5 seconds in other breeds (Siegel et al. 1965). Selection of crow duration in several breeds was demonstrated by Siegel et al. (1965), who also suggested a polygenic mode of inheritance. Selection for prolonged vocalizations in domesticated Rock Pigeons (*Columba livia*) was first noted by Darwin (1868), who reported that some breeds (Laughers and Trumpeters) were selected for qualities of voice. Some 13 extant "voice breeds" are known collectively as Trumpeters (Levi 1965). A complete bow-coo (functionally equivalent to oscine courtship song) of the Rock Pigeon is 0.26–0.68 seconds long. Laughers and Trumpeters produce bow-coos 5–60 seconds long (Fig. 3.1; Baptista and Abs 1983) because humans have selected for songs with many additional terminal elements. Long songs were inherited in a cross between a Laugher and a Modena, a breed with a wild-type brief song.

Zebra Doves (*Geopelia striata*) exhibit much geographic variation in the dura-

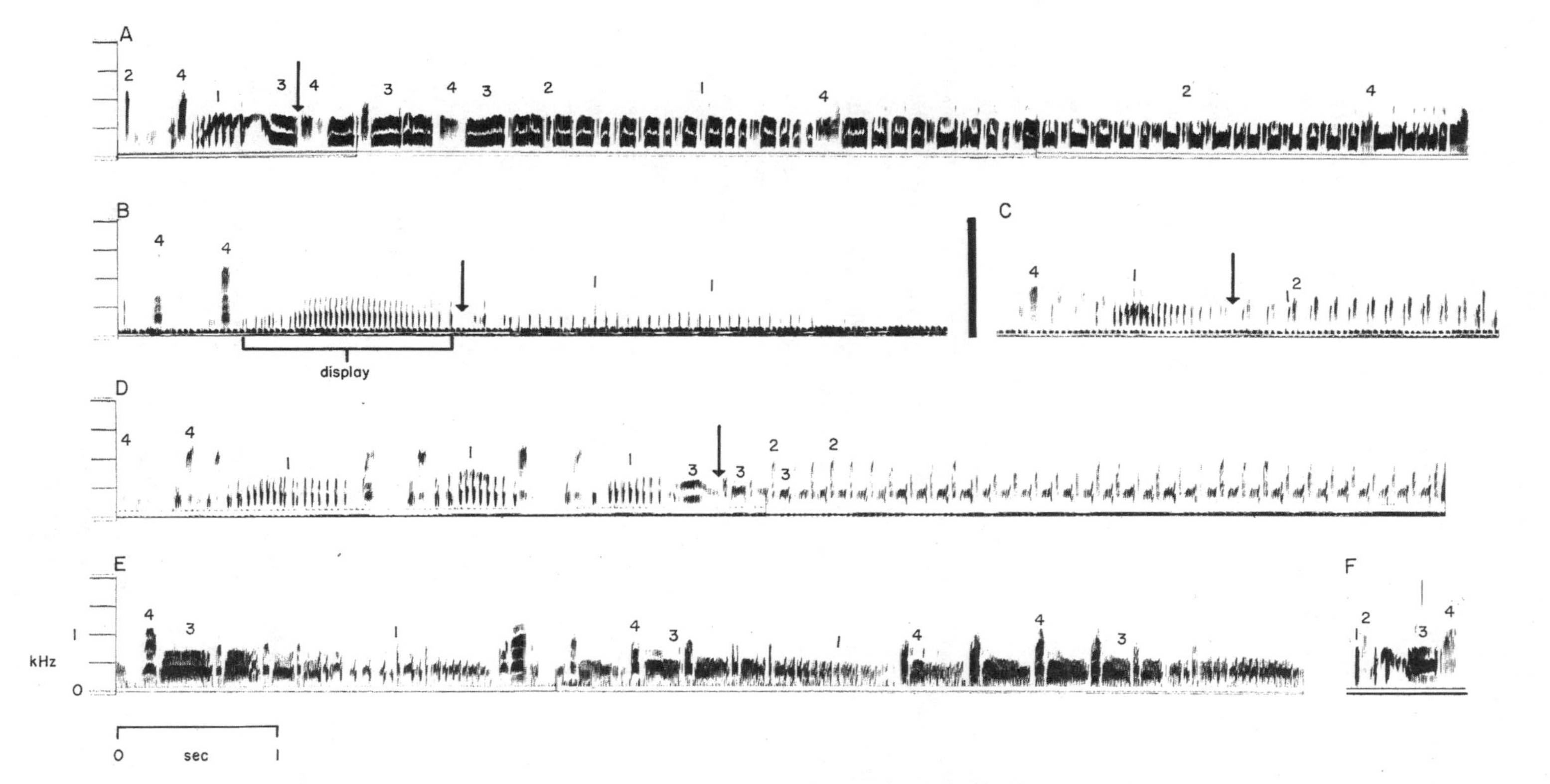

Fig. 3.1. Bow-coos of domestic pigeon breeds (A–E) compared with those of the ancestral Rock Pigeon (F). Numbers used to designate elements in the ancestral call are also used to designate homologous elements in calls of the other breeds. Key to numbers: 1 = *tut*, 2 = *wut*, 3 = *coo* or *who*, 4 = *wok*. (A) Trumpeting (drumming) display of an English Trumpeter male. Note that all four elements of the ancestral vocalization are present in the English Trumpeter's display. (B–D) Trumpeting displays of three Altenburg Trumpeters. Arrows point to where display trills typical of the ancestral form end and the trumpeting coda begins. In B and C, note that the display portion has been modified into a prolonged trill and that the coo portion has been eliminated. A *coo* is present in D. Trills consist of *tut* notes in B, *tut* and *wut* notes in C, and *wut* and *wok* notes in D. Note also the uniform rhythm in trumpeting of this breed compared with the others. (E) Partial trumpeting display of Thailand Laugher. Note the rich harmonic structure of the *woks* and the longer *who* notes. Reproduced with permission from Baptista and Abs 1983.

tion and number of elements in their perch-coo (Harrison 1969), and cross-fostering experiments have demonstrated that these songs are not learned (Layton 1991). Frith (1982) called the geographic variations dialects. In Malaysia, captive birds have been bred for long perch-coos (see below). Birds are named according to song lineages and are entered in open competitions that sometimes attract as many as 2000 entrants (Layton 1991).

Heritability of call duration was demonstrated in hybrids between galliform species. By artificially inseminating female Common Pheasants (*Phasianus colchicus*) with semen from White Leghorn roosters, McGrath et al. (1972) produced hybrids. The duration of the distress calls of male hybrids was similar to that of female hybrids (0.41 and 0.42 seconds, respectively) and pheasants (0.38 seconds) but briefer than that of chickens (0.68 seconds).

The duration of coo displays in doves is species typical, and so is the interval between them (S. J. J. F. Davies 1970, 1974, Baptista et al. 1983); Whitman's (1919) hybridization studies suggest that interval duration is heritable. To assess heritability of both temporal measures, I crossed closely and distantly related dove species. The behavior of F_1 interspecific hybrids either resembled one parental type, was intermediate between the parents, or resembled neither (Franck 1974, R. B. Payne 1980, Baptista and Horblit 1990). Clearly, F_1 behavior is not predictable on simple genetic grounds, although some systematic trends are evident.

Nest-coos are delivered by columbiforms of both sexes at a nest site or potential nest site and, in most species, are accompanied by a wing-flicking display (Goodwin 1983). The nest-coo of the African Collared-Dove (*Streptopelia roseogrisea*) is approximately 1.6 seconds long and is separated from the next coo by approximately 2.3 seconds (Fig. 3.2). Nest-coos of the unrelated Socorro Dove (*Zenaida graysoni*) are approximately 1.9 seconds long (Baptista et al. 1983), with successive calls separated by an interval of approximately 8.9 seconds. The nest-coo of a hybrid between the species was approximately 1.6 seconds long, with intercoo intervals of about 1.6 seconds. The hybrid's nest-coo was thus most similar to that of its *S. roseogrisea* parent.

A dove advertising for a mate produces a sound termed a perch-coo. This vocalization is functionally equivalent to the oscine advertising song. The perch-coo of the Socorro Dove is approximately 5.2 seconds long, with intercoo intervals of approximately 10.3 seconds. The perch-coo of *S. roseogrisea* differs from its nest-coo in minor details of the introductory syllables (Mairy 1976), but perch-coos are separated by much briefer intervals (approximately 0.16 seconds). The hybrid's perch-coo resembled that of the *S. roseogrisea* parent in duration, and successive coos were separated by intervals of approximately 0.76 seconds. Thus the hybrid was similar to its African Collared-Dove parent in this form of coo as well.

I also studied a hybrid between a Namaqua Dove (*Oena capensis*) and the related Tambourine Dove (*Turtur tympanistria*). When courting a female, the male of a Collared or Namaqua Dove performs a bow display accompanied by vocalizations—the bow-coo. Tambourine Doves, Tambourine × Namaqua

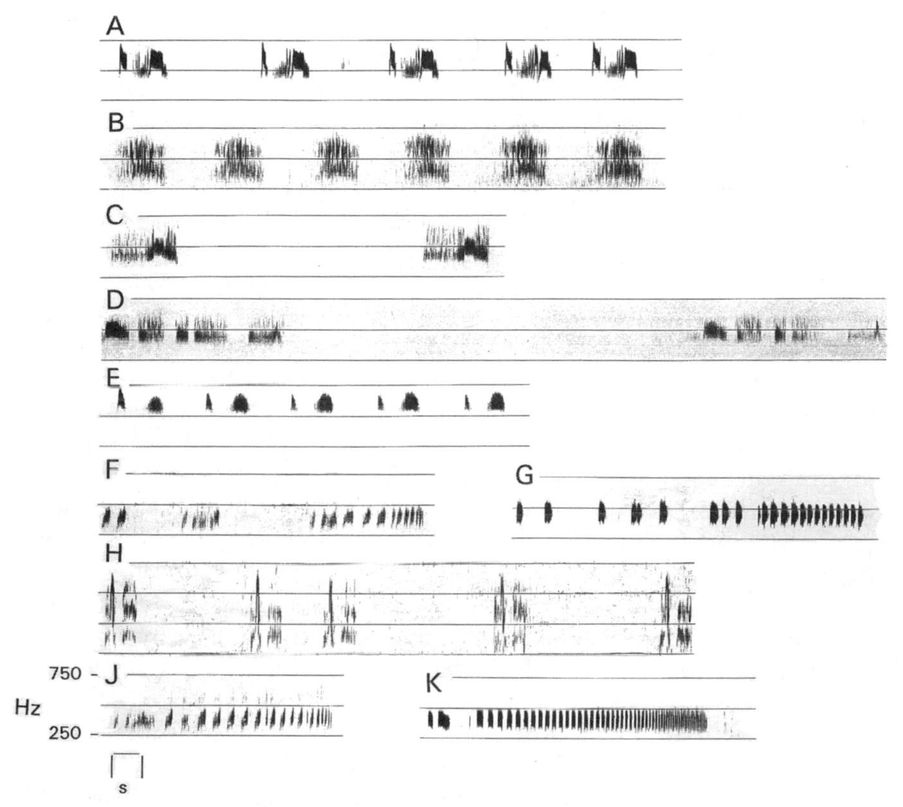

Fig. 3.2. Coos of New and Old World doves and their F_1 hybrids. (A) Nest-coos of African Collared-Dove (*Streptopelia roseogrisea*). (B) Nest-coos of a Socorro Dove (*Zenaida graysoni*) × Collared Dove hybrid. Note the similarities in the duration of coos and the intervals between coos when compared with A. (C) Socorro Dove nest-coos; note the long intervals between coos. (D) Socorro Dove Perch coos; note the long intervals between coos. (E) Series of Namaqua Dove (*Oena capensis*) perch-coos; note the brief intervals bewteen coos. (F) Single perch-coo of a Namaqua Dove × Tambourine Dove (*Turtur tympanistria*) hybrid. (G) Tambourine Dove perch-coo. (H) Bow-coos of Namaqua Dove. (J) Bow-coo of hybrid between Tambourine and Namaqua Doves. (K) Bow-coo of Tambourine Dove.

hybrids, and *Zenaida* doves do not perform bows during courtship. Instead they inflate their crops, posture, and coo, in a static display homologous to the bow-coo of other dove species. Namaqua Dove bow-coos consist of two explosive notes with a total duration of approximately 1.1 seconds; the intercoo intervals are approximately 3.4 seconds long (Fig. 3.2H). This display form lasted approximately 8.7 and 6.3 seconds in the Tambourine Dove and hybrid, respectively, with successive coos separated by intervals of approximately 8.1 seconds and approximately 7.4 seconds in each taxon (Fig. 3.2J, K).

The perch-coos of the Namaqua Dove were brief (approx. 1.5 seconds long) and separated by brief intervals of approximately 1.3 seconds (Fig. 3.2E). The perch-coos of the Tambourine Dove and the hybrid consisted of a series of brief notes, but entire calls lasted approximately 13 (Tambourine Dove) and approximately 12 seconds (hybrid) (Fig. 3.2F, G); perch-coos were separated by long pauses, as were the displays described in the previous paragraph. Thus the duration and intervals of both forms of coo display in the hybrid were most similar to those of its Tambourine Dove parent.

Darwin (1868) saw the process of domestication as analogous to speciation. In the domestication process, humans selectively breed for certain traits in an artificially isolated segment of a gene pool. In speciation, selection operates on individuals in a naturally isolated gene pool (e.g., on montane or oceanic islands). Studies of vocalizations of domestic pigeons can illustrate how natural selection has altered the vocalizations of living avian species. Under domestication, various elements of the ancestral forms of Rock Pigeon calls have become elaborated. In the Laugher and Trumpeter breeds, one form of coo is 30–60 seconds long, compared with the ancestral condition of being less than 1 second long. A parallel may be seen in the brief coos of the Namaqua Dove (Fig. 3.2E, H) and the long coos of the Tambourine Dove (Fig. 3.2G, K). Dispositions to produce long songs are heritable, as evidenced by crosses between trumpeting and nontrumpeting pigeon breeds (Fig. 3.1; Baptista and Abs 1983) or Cape and Tambourine Doves (Fig. 3.2F, J).

Vocalizations also may be simplified during domestication and evolution. For example, the courtship whistles typical of the ancestral Mallard (*Anas platyrhynchos*) have become reduced to a low grunting sound in the domestic Aylesbury breed as a result of selection (Desforges and Wood-Gush 1976). A parallel may be seen in songs of the dove genus *Zenaida.* Whereas the Mourning Dove (*Z. macroura*) and the Socorro Dove produce a loud coo display of five to six notes (Baptista et al. 1983), the Galapagos Dove (*Z. galapagoensis*) utters two scarcely audible notes (Gifford 1931, Nicolai 1969, Baptista unpubl. data). Songs of the mannikin genus *Lonchura* are long and elaborate (M. F. Hall 1962, Güttinger 1970). In the Pictorella Munia (*Heteromunia pectoralis*), song has been simplified to two soft notes (M. F. Hall 1962).

Even among songbirds, which learn their songs, duration may be constrained. Song duration in both Song Sparrows (Mulligan 1963) and White-crowned Sparrows (Baptista 1975) follows a normal frequency distribution. Although much individual and geographic variation in syllable structure and rhythm exists in the songs of both species (Mulligan 1966, Baptista and King 1980), total song duration is probably constrained by heritability. The song durations of isolate-reared and tutored White-crowned Sparrows do not differ (Petrinovich 1985), for example, and Marler and Sherman (1985) found no differences in duration between songs of isolate and wild Song Sparrows. Kroodsma (1977a) found that songs of isolate Song Sparrows averaged slightly longer than those of wild birds but were still within the range of variability of songs recorded in the wild (Mulligan 1966).

Experiments with several oscines indicate that heterospecific songs may be learned, but song duration will be stretched or compressed to fit the species-specific time template. For example, Thorpe (1958) tape-tutored a naive Chaf-

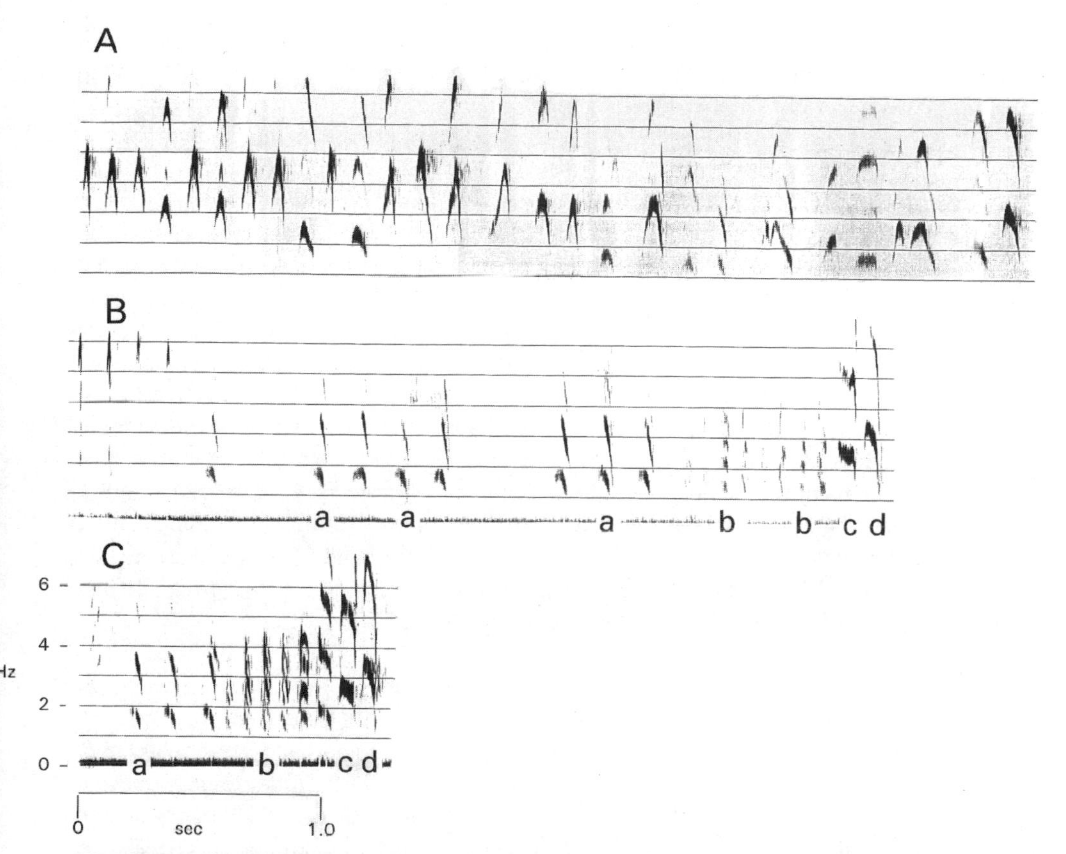

Fig. 3.3. Song characteristics and song learning in the African Silverbill (*Lonchura cantans*). (A) Song phrase. (B) Song phrase of African Silverbill tutored by White-rumped Munia (*L. striata*). (C) Song of the White-rumped Munia tutor of B; note the truncated nature of the tutor's song phrases. Compare syllables a, b, c, and d in tutor song C with pupil song B.

finch with a 3-second-long Tree Pipit (*Anthus trivialis*) song. The Chaffinch learned the heterospecific song but eliminated some of the notes so that the final song produced was typical of the 2.5-second song of its species. A second example concerns White-rumped Munias and Zebra Finches (Clayton 1989). White-rumped Munia song phrases are longer and have more elements per phrase and more repeated elements than do Zebra Finch songs. White-rumped Munias learned syllables from Zebra Finch foster parents but incorporated them into longer phrases typical of their own species. In contrast, Zebra Finches learned White-rumped Munia syllables, but phrases were abbreviated to an intermediate duration. A third example is provided by African Silverbills, whose songs have even longer phrases than those of the White-rumped Munia. A silverbill fostered by a White-rumped Munia acquired the song syllables of that species, but the silverbill's songs averaged longer (Fig. 3.3). Conversely, White-rumped Munias tutored by silverbills learned the silverbills' syllables but reduced them to fit the shorter song phrase typical of White-rumped Munias (Baptista unpubl. data). All three estrildid species thus seem to adjust learned syllables to fit a phrase length close in duration to conspecific song.

Rhythm

Variation in the duration of individual notes and the intervals between notes in complex vocalizations provides rhythms that are characteristic of species. Lade and Thorpe (1964) cross-fostered members of the dove genus *Streptopelia* and found that the rhythms in their coos developed independent of learning. The rhythms in the coos of F_1 hybrids from six intrageneric crosses were intermediate between the parents in some crosses and disorganized in others. I reanalyzed Lade and Thorpe's data and found that inheritance was affected by the degree of relatedness.

Reciprocal crosses between the closely related species *S. roseogrisea* and *S. decaocto* (Eurasian Collared-Dove) produced hybrids with calls intermediate in rhythm between the parents. *Streptopelia roseogrisea* and *S. decaocto* are distant relatives of the Spotted Dove (*S. chinensis*) and are even more distantly related to the Laughing Dove (*S. senegalensis;* Goodwin 1983). Hybrid *chinensis* × *roseogrisea* or *chinensis* × *decaocto* produced slightly rhythmic calls, but the rhythm was not intermediate in character. Hybrids between *S. senegalensis* and the distantly related *S. roseogrisea* produced completely arrhythmic calls. The European Turtle-Dove (*S. turtur*) is more closely related to *S. roseogrisea* than are *S. chinensis* and *S. senegalensis* (Goodwin 1983). Hybrid *turtur* × *roseogrisea* produced calls that were arrhythmic, but their trills had a quality typical of *S. turtur.* Thus, within this group, the degree of relatedness of the parental species affects rhythmic attributes in their hybrid offspring.

Another example of how call rhythm is retained in hybrids between closely related species is provided by doves from a different lineage. Coos of the Namaqua Dove consist of two notes (Fig. 3.2E, H). Coos of the Tambourine Dove and a Tambourine Dove × Namaqua Dove hybrid consisted of series of brief notes of similar morphology uttered in rapid succession (Fig. 3.2F, G, J, K).

The rhythm of vocalizations in intergeneric hybrids between unrelated species is especially disrupted. Display-coos of *S. roseogrisea* begin with a trilled croo followed by a coo that is almost identical to the nest-coo (Fig. 3.2A; Mairy 1976). In *Z. graysoni,* the display-coo consists of five distinct notes (Fig. 3.3D), four of which show sustained frequency with no trills (as in *S. roseogrisea*). The coos of a *Z. graysoni* × *S. roseogrisea* hybrid consisted of only a single whooor note (Fig. 3.2B) and thus did not resemble the coos of either parent.

By selecting for different degrees of complexity in songs of domestic pigeon breeds, humans have also selected for different vocal rhythms. The bow-coo of the Rock Pigeon may be divided into an introductory trill portion and a terminal coo portion, followed by a wok sound or a croo-coo-wok (Baptista and Abs 1983). In the Laugher and English Trumpeter (breeds kept by Darwin), the croo-coo is followed by a long coda consisting mostly of coo notes in the Trumpeter (Fig. 3.1A) and of coo, wok, and croo calls in the Laugher (Fig. 3.1E). In the Altenberger Trumpeter the croo or trill portion has been selected for, and in some individuals the coo portion is missing altogether (Fig. 3.1B).

An index of variation in rhythm is provided by the coefficient of variation (CV, in percentage) of note duration in the lengthened terminal passages of the different breeds. The CV is 11% for the terminal notes of the Altenberger Trumpeter's call, which sounds (even to the unaided ear) very uniform in rhythm (reminiscent of an electric motor); and 51% for notes in the English Trumpeter's call, attributable in part to lengthened coos inserted between series of brief coos (Fig. 3.1A). With CV = 36%, uniformity of rhythm in the coos of the Laugher is intermediate in rhythmicity (Baptista and Abs 1983).

Common Quail calls do not vary structurally throughout Europe (Moreau and Wayre 1968), but rhythm of call delivery does (Guyomarc'h et al. 1984). On returning to France from their winter quarters in Africa, male quail gather in singing assemblies of 4 to 15 individuals. These assemblies may consist of the same individuals for several successive years. The first note of the triplet call is individual specific and identifies each male in an assembly. The silent intervals between the notes in the triplets are shared by all males in a singing group, giving rise to dialects reminiscent of those found in singing assemblies of some hummingbird species (S. L. L. Gaunt et al. 1994). These quail dialects, however, develop independent of learning (Guyomarc'h pers. comm.).

Assembly calls of the Northern Bobwhite (*Colinus virginianus*) also develop independent of learning experience (J. A. Baker and Bailey 1987). Calls from different geographic localities are not distinguishable by ear, but statistical analyses reveal frequency and temporal (rhythmic) differences among populations (Goldstein 1978).

The relationships between vocal development, microgeographic variation, and vocal similarity among birds obviously are imperfectly understood. Common Quail, Northern Bobwhite, and Zebra Doves exhibit geographic variation in vocalizations, yet vocal learning apparently does not occur in those species. If this is so, then vocal similarity between neighbors need not involve learning. In the Common Quail, Zebra Dove, and some *Ptilinopus* fruit doves in New Guinea (Beehler et al. 1986, Baptista unpubl. data), dialects are easily detectable by the unaided ear or by examining sonograms. In the Northern Bobwhite, however, the differences are more subtle.

Rhythm also may be fixed in species that learn song. The Song Sparrow's songs usually begin with a phrase containing notes that accelerate in tempo (Mulligan 1966). This rhythm appears in songs of individuals raised in isolation, indicating that it need not be learned from other individuals (Kroodsma 1977a). Although the European Greenfinch (*Carduelis chloris*) imitates a variety of heterospecific vocalizations in the wild (Güttinger 1974), greenfinches apparently convert their imitations to a more appropriate rhythm (Güttinger 1979). Greenfinch × canary hybrids were successfully tutored with canary song but sang with the greenfinch rhythm, indicating that the hybrids inherited a predisposition to use the greenfinch rhythm (Güttinger 1979).

Frequency

Bird vocalizations are restricted to certain frequency ranges. I reexamined the existing data and found that whereas dispositions to produce sounds of certain frequencies may be under direct genetic control, mean frequencies and frequency envelopes often are correlated with body size.

The frequency attributes of vocalizations in *Coturnix* and some goose species decline in individuals as they grow (Würdinger 1970, Schleidt and Schalter 1973, ten Thoren and Bergmann 1987). Vocal frequency is inversely proportional to mass in adult Northern Bobwhites of different sizes (Goldstein 1978). Goodwin (1965) noted that large breeds of domestic pigeons tend to have lower-pitched coos: the 30–37-ounce King, for example, has a much deeper voice than the 23-ounce Homer (Baptista pers. obs.). Finally, vocal frequency and body size are inversely related across tyrannid species (W. E. Lanyon 1960, 1978). This simple and close relationship makes sense for these species, because there is no vocal learning that could cloud a size-frequency relationship (Kroodsma 1984, 1985a, J. A. Baker and Bailey 1987); however, frequency varies inversely with body size even in different oscine species (Schubert 1976, Bowman 1979, Ryan and Brenowitz 1985).

During the process of domestication of the Red Junglefowl, humans selected for calls of different frequencies (Sossinka 1982). McGrath et al. (1972) studied the heritability of vocal frequency in distress calls of hybrids between Common Pheasants and chickens. The hybrids belonged to two size classes, with males being larger than females (1775 vs. 875 grams). The call frequency of the hybrids was intermediate between the call frequencies of the parents. Although the authors concluded that control of call frequency was polygenic, their data show that frequency is inversely proportional to body size for individuals of both parental forms and the two classes of hybrids. Therefore, frequency differences seem to be determined by body size.

Hummingbirds in the genus *Calypte* perform elaborate aerial displays accompanied by dives and dive sounds to threaten trespassers on their territories. The dive sound of Costa's Hummingbird (*C. costae*) is a very long version of one of the song types it utters while perched (S. Wells et al. 1978). The dive sound of Anna's Hummingbird (*C. anna*) is one of the phrases of its complex song (Baptista and Matsui 1979; but see dissenting view in Stiles 1982) and sometimes is uttered while perched (S. Russell pers. comm., D. Bell pers. comm.). The dive sounds of Costa's Hummingbirds are higher in frequency than those of Anna's Hummingbird, and the sounds produced by intermediate-sized hybrids between the species also are intermediate in frequency (S. Wells et al. 1978).

Clayton's study (1990a) of song development in two subspecies of Zebra Finch (*T. g. guttata* and *T. g. castanotis*) provides a good example of the relationship of vocal frequency to body size. The subspecies differ in body size, with *castanotis* being larger. Clayton cross-fostered eggs reciprocally and found that *guttata* readily learned syllables from foster fathers but transposed them to the higher frequency typical of their own subspecies' songs. Conversely, cross-fostered *castanotis* learned syllables from foster fathers but transposed them to the lower frequency typical of their own subspecies.

The song of the domestic canary breed known as the Roller was selected for trills of low frequency. Although most syllables in songs of isolates do not resemble those found in normal Roller song, the low frequency does, indicating that frequency is independent of learning (Poulson 1959, Mundinger 1988). Rollers are smaller than Borders or Yorkshire canaries, which have higher-frequency songs, so vocal frequency and body size in Rollers are not inversely related as in other domestic breeds. Nevertheless, control of frequency in the Roller's song may be attributable to specializations of the syrinx rather than to direct genetic control. For example, although large species of owls tend to have large syringeal air passages and long tympaniform membranes, and to produce lower-frequency notes (A. H. Miller 1934), there are exceptions to this trend. Male Flammulated Owls (*Otus flammeolus*) weigh only 50–60 grams, compared with 100–120 grams in male Western Screech-Owls (*O. kennicottii*), yet Flammulated Owl calls are five to six halftones lower. The lower-frequency call is a result of the long, thick tympaniform membranes being covered with papillae and rugae, which reduce the rate of vibration of the membranes (Miller 1947). Examination of the syrinx or single-unit analyses in the vocal center of the Roller canary may reveal how low vocal frequency is achieved in this breed. There is thus no good evidence for direct genetic control of vocal frequency in birds.

Differences in frequency modulation, however, cannot be accounted for by differences in body size and likely are under genetic control. Distress calls of chickens differ from those of Common Pheasants in being uniform in frequency, whereas those of pheasants and hybrids begin with a frequency increase and then decrease (McGrath et al. 1972). A Common Black-headed Gull (*Larus ridibundus*) raised with Little Gulls (*L. minutus*) produced calls with a series of syllables and a harmonic structure similar to those of the Little Gull as an adult. This is the first demonstration of vocal learning in gulls; however, the Common Black-headed Gull had the low frequency in terminal elements of its call characteristic of its own species (Groothuis 1993).

Tonal Quality

Tonal quality may be the product of several factors, including (1) noise content of the vocalization; distress calls of chickens, for example, are relatively noise free, whereas those of the related Common Pheasant contain noise and therefore sound harsh (McGrath et al. 1972; noise is sound energy with "no specific energy, but is a random mixture of all audio frequencies": White and White 1980, p. 332); (2) harmonic structure and differential distribution of sound energy at different frequencies (Davis 1964, Marler 1969); and (3) variation in relatively pure frequency calls due to minute modulations of frequency or amplitude (S. Gaunt pers. comm.).

Inheritance of tonal quality appears to be sex linked in Common Quail (Moreau and Wayre 1968). The advertising crow of the Japanese Quail (*Coturnix c. japonica*) consists of three harsh (noisy) notes. The pickerwick call of *C. c. coturnix* consists of three clear (noise-free), sharp notes (Moreau and Wayre 1968). Reciprocal crosses between the two subspecies produced individuals that uttered calls with noise characteristics resembling the female parent's.

Cross-fostering experiments indicate that all call characteristics, including tonal quality, develop independent of learning in columbiforms and mannikins (Lade and Thorpe 1964, Güttinger and Nicolai 1973). The studies indicate that these sounds are under genetic control; but what happens when the genome is altered? The results differ between crosses of closely and distantly related species.

Lade and Thorpe (1964) made six intrageneric crosses between *Streptopelia* doves and found that the hybrid's vocalizations had tonal qualities similar to those of one of the parent species. My results, from an intergeneric cross between *S. roseogrisea* and *Z. graysoni,* differed. The coos of *S. roseogrisea* are pure-toned (Fig. 3.2A), whereas those of *Z. graysoni* consist of whistles with a noisy background, which imparts a hoarse quality (Fig. 3.2C). The hybrid's coo consisted of two superimposed bands of noise (Fig. 3.2B) and thus sounded like neither parent.

Crosses among estrildid finches reveal similar disruption of call structure. The African Silverbill is a member of a separate radiation from "typical" *Lonchura,* and sometimes is placed in a separate genus, *Euodice* (Güttinger 1970, Immelmann et al. 1977). Its contact calls consist of a high-frequency (approx. 5 kHz), pure-toned inverted chevron (Fig. 3.4A). The contact call of the White-rumped Munia consists of a single note that may rise slightly in frequency; the frequency ranges from 3.75 to 5 kHz (Fig. 3.4D). Hybrids between African Silverbills and White-rumped Munias produced calls that resembled neither parent's. These calls had a fundamental frequency peaking at approximately 3.75 kHz and were shaped like a Roman arch, with a harmonic peaking at approximately 7 kHz (Fig. 3.4G). Moreover, whereas the calls of both parents are relatively noise free, the calls of the hybrid were noisy and harsh. A second example involves the Java Sparrow (*Padda* (=*Lonchura*?) *oryzivora:* Güttinger 1976). Its contact call is an inverted chevron at approximately 2.75–4 kHz, and is thus lower in frequency and has a smaller frequency range than calls of the White-rumped Munia (Fig. 3.4B). Hybrids between Java Sparrows and White-rumped Munias produced a call with a fundamental at approximately 1.75–3 kHz, with an overtone peaking at approximately 5.5 kHz (Fig. 3.4H).

The calls of both hybrid types thus differed from those of the parents in two ways: (1) calls of the parental types were relatively pure-toned, whereas those of the hybrids contained an overtone; and (2) the fundamental frequency of the hybrids' calls was lower than calls of parents.

In contrast, calls of intrageneric hybrids between closely related species are

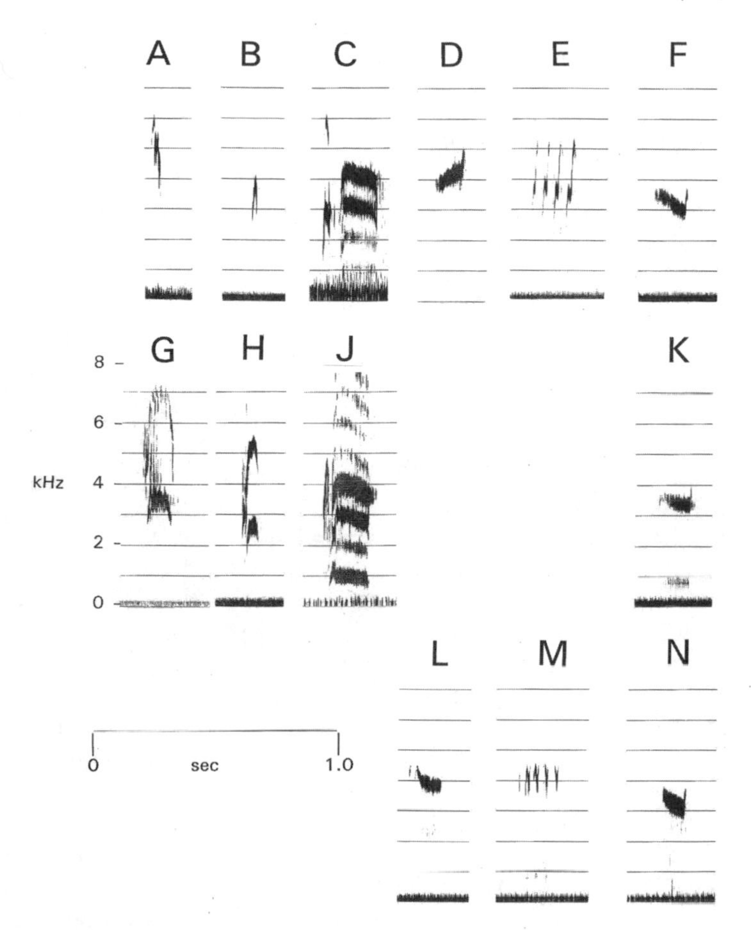

Fig. 3.4. Distance calls of mannikin species and their hybrids. (A) African Silverbill (*Lonchura cantans*). (B) Java Sparrow (*P. oryzivora*). (C) Scaly-breasted Munia (*L. punctulata*). (D) Male White-rumped Munia (*L. striata*). (E) Female White-rumped Munia. (F) Female White-bellied Munia (*L. leucogastra*). (G–K) F_1 hybrids: (G) *cantans* × *striata;* (H) *P. oryzivora* × *L. striata;* (J) *punctulata* × *striata;* (K) *striata* × *leucogastra.* Note the similarities between C and J, and between F and K. (L–N) Backcrosses: (L) Male (*striata* × *leucogastra*) × *striata,* (M) female (*striata* × *leucogastra*) × *striata,* (N) male (*striata* × *leucogastra*) × *leucogastra.* Note the similarities between D and L, E and M, and F and N. Note also that the fundamental frequency in calls of hybrids G and H is much lower than those in calls of the parental types.

often similar to the calls of one of their parents. I provide three examples. The contact calls of the Scaly-breasted Munia (*L. punctulata*) consist of a brief note followed by a longer note with a fundamental at approximately 1 kHz, and several harmonics (Fig. 3.4C). The calls of some hybrids between Scaly-breasted and White-rumped Munias (*L. striata*) lacked the short first note; however, all the hybrids' calls had a long note with a fundamental at approximately 1 kHz, were rich in harmonics, and were very similar to those of the Scaly-breasted Munia (Fig. 3.4J). A second example is the contact calls of the White-bellied Munia (*L.*

leucogastra). In both sexes, these are pure-toned and consist of a single note that drops rapidly from approximately 3.75 kHz to approximately 2.75 kHz, then rises abruptly to approximately 3.75 kHz (Fig. 3.4F). Calls of three male *striata* × *leucogastra* hybrids resembled those of the mother in shape but tended to be sustained in frequency as in *striata* (Fig. 3.4K). A hybrid *striata* × *leucogastra* backcrossed to a *leucogastra* female produced a hybrid with a call similar to its mother's (Fig. 3.4N). The final example concerns the contact calls of *striata.* These differ between the sexes, being a single note in males and a short trill in females (Fig. 3.4D, E; Zann 1985, Yoneda and Okanoya 1991). Backcrosses between *striata* × *leucogastra* males and *striata* females produced female hybrids that used trills, as in *striata* (Fig. 3.4M). Male hybrids produced single notes sustained in frequency and with a mean frequency at approximately 4 kHz, as in *L. striata,* but with a downward-sloping introduction reminiscent of its *L. leucogastra* grandmother. Thus, although the call most resembled *L. striata* in frequency, it still retained some of the shape of the call of *L. leucogastra.*

In sum, the contact calls of hybrids between distantly related estrildid finches bear no resemblance to calls of either parent; but hybrids between closely related species produce calls that do resemble those of one parent. One explanation for this pattern is that control of vocal production is polygenic and similar genes control the behavior in closely related species. Hybrids between distantly related species are products of very different gene complexes, so larger changes have occurred in the genomes of F_1 hybrids, resulting in large disruptions to call structure. The degree of disruption seems to increase as relatedness decreases in parental forms in dove hybrids also (Lade and Thorpe 1964, see above).

It is noteworthy that backcrosses in both doves and mannikins produced vocalizations most similar or identical to those of the species to which they were backcrossed. These results parallel those observed in other kinds of displays in finch backcrosses (Baptista and Horblit 1990). Backcrossing restores a genome relatively close to its original form, giving rise to normal or nearly normal displays and vocalizations typical of the species to which the hybrid was backcrossed.

Tonal quality may be a product of inheritance even in species that learn vocalizations. For example, Konishi (1985, p. 140) noted that "when White-crowned Sparrows copy synthetic sounds that sound mechanical to the human ear, they do not sing in a mechanical voice but in the voice of their normal song." White-crowned Sparrows learn syllables of Red Avadavats but sing them in the tonal quality of their own species (Baptista and Petrinovich 1984). Although White-crowned Sparrows may learn Song Sparrow songs, playback experiments indicate that, based on one response measure, Song Sparrows can distinguish between conspecific and mimicked songs (Baptista and Catchpole 1989), perhaps because the White-crowned Sparrows retain conspecific tonal quality in their mimicked songs.

Constraints may even occur in the tonal quality of sounds produced by consummate mimics such as the Superb Lyrebird (*Menura novaehollandiae*). Some 70%

of the syllables in its songs are mimicked from other species. Power spectra reveal that mimicked and conspecific utterances are similar (Robinson 1975). Lyrebirds thus either select sounds to be copied based on tonal quality or are limited in the range of tones they can mimic.

Conclusions

Temporal, frequency, and tonal characteristics of vocalizations may be genetically determined even in birds whose songs or calls are substantially learned. Biologists have long viewed nature as interacting with nurture to shape vocal development, contrary to Johnston's (1988) contention. Studies of isolates and hybrids reveal that the development of vocalizations may, in some cases, be entirely or in part independent of learning. Thus the interactionist view does not apply to the ontogeny of all behaviors.

Studies of hybrids should continue to be a fertile field, for by carefully noting the taxonomic relationships of the species being hybridized, researchers may also learn which behaviors are potentially controlled by genes and how genes change during evolution. Although we do not understand the pathways between genes and behavior, great disruption of behavior in offspring of more distantly related forms is further evidence for substantial genetic control of behavior. We need many more hybrid studies carried into the F_2 generation and involving backcrosses to clarify this point.

Oscines continue to be the favorite subjects of studies in vocal ontogeny. Groothuis's (1993) findings should urge us to look closely at families that we have traditionally assumed do not exhibit vocal learning, especially when geographic variation in vocalizations is known (e.g., in some dove species). Vocal reproduction of model syllables is an obvious manifestation of learning; to demonstrate learning of subtle temporal or frequency attributes, however, may require multivariate statistical analyses (Groth 1993, S. L. L. Gaunt et al. 1994).

Song learning has been unequivocally demonstrated in a hummingbird species (Baptista and Schuchmann 1990). Song complexity varies greatly among hummingbird species and even between populations within species (S. Wells et al. 1978, Baptista and Gaunt 1994). Like estrildid finches, related hummingbird species may or may not produce songs distinct from calls (S. Wells and Baptista 1979). Both females and males of some hummingbird species sing (S. L. L. Gaunt et al. 1994). What roles do genes and environment play in vocal development in this diverse group of birds? Hummingbirds hold a treasure trove of secrets waiting to be revealed.

Song development in females is also a neglected field. The evidence I have presented here shows that females and males learn in different ways. Moreover, females vary widely in their disposition to learn or produce vocalizations. Vocal development in females merits more attention from both ethologists and neurophysiologists.

Acknowledgments

I thank Sylvia Hope, Sandra Gaunt, Toby Gaunt, Michael Ghiselin, Don Kroodsma, Bruce Lee, Edward Miller, and Robert Payne for constructive criticisms on an earlier version of the manuscript; Andrea Jesse for preparing the figures; Paraskevas Kandianidis and Judy and Joe Passantino for providing some of the birds used in my studies; and the National Science Foundation for funding some of the work on White-crowned Sparrows (DEB-77-12980). Kathleen Berge prepared the manuscript for submission.

4 Birdsong Learning in the Laboratory and Field

Michael D. Beecher

In the study of song learning, longitudinal studies of natural populations in the field have always been shunned in favor of laboratory experiments. The advantage of the laboratory experimental approach has always seemed obvious: it gives researchers the ability to manipulate and control variables. Some workers have been slow to recognize the major disadvantage of the laboratory approach, however, which is that if key variables are controlled out of the experiment, the outcome may be misleading (West and King, this volume). The problem became obvious when Baptista and Petrinovich (1984, 1986) demonstrated very different patterns of song learning in White-crowned Sparrows (*Zonotrichia leucophrys*) exposed to tape recordings of song and sparrows exposed to live, singing birds. Such "live tutor" experiments revealed the importance of social variables, variables that had been controlled out of earlier laboratory studies.

My own realization of the limitations of laboratory studies came when my colleagues and I began making inferences about song learning in a natural population of Song Sparrows (*Melospiza melodia:* Beecher et al. 1994b) and comparing them with extrapolations from laboratory studies of song learning in this and related species (Marler and Peters 1987, 1988, Beecher et al. unpubl. ms.). Not only did the laboratory studies fail to identify critical variables in song learning, but they also showed patterns of learning that differed greatly from those we had observed in the field.

In this chapter I contrast the pictures of song learning provided by longitudinal field studies and laboratory experiments on the Song Sparrow and attempt to draw general conclusions about the roles of the two approaches in the study of song learning. I argue that the key variables in song learning are social and ecological ones, and that the study of song learning in the field should precede attempts to dissect the process in the laboratory (Kroodsma, this volume). My recommendation for the study of song learning is identical to one offered by Kroodsma and Byers (1991, p. 325) for the study of song function: "To experiment first is human, to describe first, divine."

Laboratory Studies of Song Learning

The classical experiments of Marler and his colleagues on song learning in White-crowned Sparrows (e.g., Marler 1970) and subsequent work in this tradition viewed song learning in terms of the interaction of acoustic exposure variables and species-typical (innate) learning programs and perceptual predispositions. This perspective was the rationale for removing from experiments all aspects of the species- and population-typical song-learning context except for song. In particular, the young bird was not exposed to other birds except for the songs he heard over the loudspeaker in his isolation chamber. From this Spartan paradigm have come many important generalizations about song learning. The best known one is that there is a sensitive period for song memorization early in the bird's life (roughly the second month of life in the White-crowned Sparrow and several other species, including the Song Sparrow).

The tape-tutor laboratory experiment also provided insights into species differences in song learning. For example, the song repertoires of male Marsh Wrens (*Cistothorus palustris*) in western North America are two to three times as large as those of wrens in eastern North America (Kroodsma and Verner 1978); the key ecological correlate of this difference may be the higher population densities of western populations (Kroodsma 1983a, 1988a). Kroodsma and Canady (1985) traced the difference in song repertoire size to different innate song-learning programs: when young males of both populations were exposed to identical tape-tutor regimes in the laboratory, western Marsh Wrens learned about two and half times as many song types as did their eastern counterparts. To consider another species difference, a song of a Song Sparrow has a relatively complex syntax consisting of a particular permutation of five or more distinct elements (pure tones, buzzes, note complexes, trills, etc.), whereas the song of the closely related Swamp Sparrow (*Melospiza georgiana*) consists of a single note repeated as a trill. Marler and Peters (1988) presented evidence that Song Sparrows are sensitive to syntactical variables during song learning, but Swamp Sparrows are not: Song Sparrows were induced to copy Swamp Sparrow song elements if they were embedded in Song Sparrow–like syntax, whereas Swamp Sparrows avoided Song Sparrow song elements regardless of the syntactical context.

There is no question that the classical, rigidly controlled tape-tutor experiment is well suited for exposing the interaction of acoustic-exposure variables and species- or population-typical perceptual and learning predispositions. Problems arise only when we suppose that this paradigm gives us a complete description of song learning, one that can be extrapolated fairly directly to song learning in the real world. The problems are of two types: either we extrapolate incorrectly from the lab to the natural context, or we are unable to extrapolate at all. As an example of an extrapolation error, researchers once believed that White-crowned Sparrow males learn their song type (dialect) at or near the natal nest, because males cease learning from tape tutors in the lab at about day 50, and disperse from the natal area in the field on or about day 50. It now seems clear, however, that males of this species typically acquire a song type that best matches those of their postdispersal neighbors (Baptista and Morton 1988, DeWolfe et al. 1989; for an overview of this debate, see Kroodsma et al. 1984).

The second type of problem arises when we cannot extrapolate from laboratory to field at all because we have controlled the variables of interest out of the laboratory study (West and King, this volume). For example, a tape-tutor study does not allow one to answer questions such as the following: Does a young bird learn from one or several tutors? If from several, does a bird mix elements from different tutors? Is he influenced by which songs the tutors share and which they do not? More generally, the tape-tutor protocol precludes identifying social variables that may be critical in song learning. Consider, for example, the common generalization from the classical tape-tutor studies that birds do not learn heterospecific song in nature because it is screened out by an innate perceptual "template." Although White-crowned Sparrows, Song Sparrows, and other songbirds do preferentially learn conspecific song when social interactions are eliminated, they readily learn heterospecific song in the laboratory if they interact socially with heterospecific tutors, and they do so even in the field in unusual cases when they interact with individuals of other species (Emlen et al. 1975, Baptista and Morton 1988).

The first strong indication of the limitations of the highly controlled laboratory approach was the discovery that birds learn better from live tutors than from tape-recorded song. Those cases ranged from species in which birds simply would not learn from taped songs (e.g., Zebra Finches, *Taeniopygia guttata:* Thielcke 1970a, Price 1979) to species in which birds learned more readily from live tutors than from tapes (e.g., Marsh Wrens: Kroodsma 1978a, Kroodsma and Pickert 1984a). Tape-tutored birds often have "poor" songs or smaller than normal song repertoires (Kroodsma and Pickert 1980, 1984a, b). Finally, experiments revealed that certain "rules" of song learning derived from tape-tutor studies are broken, or at least bent, when the song tutors are live birds (Baptista and Petrinovich 1984, 1986). The classical tape-tutor studies and the view of song learning they fostered could not have predicted that exposure to songs of a live heterospecific bird *after* the sensitive period can be more effective than exposure to tape recordings of conspecific birds during the sensitive period.

Introducing live tutors in place of tape recordings is, however, only one step toward a more ecologically valid song-learning context. Further steps should include multiple tutors, neighborhood features (e.g., shared songs among tutors), the ability of the young bird to move freely among the tutors and interact with them, and so on. Only the first of these steps has been tried (with Zebra Finches: Eales 1985, Clayton 1987a, Slater et al. 1991; and Song Sparrows: Beecher et al. unpubl. ms.). Each step sacrifices experimental control—some tutors will sing more than others, shared songs make tutor determination more difficult, free movement means that each subject gets a unique (individually designed) exposure schedule, and so on—but at the same time each step exposes a potential variable of song learning. Once the key variables are identified, researchers can then begin to move toward more experimental control in an effort to describe more precisely and assess the roles of individual variables. For example, perhaps it will prove possible to "simulate" live tutors with loudspeakers placed in different locations or to give young birds the ability to activate a particular tutor's song by flying to a particular perch (Adret 1993).

Field Studies of Song Learning

General Considerations

Song learning can be studied in the field as well as in the laboratory. That alternative has always been available, yet it has been shunned by most students of song learning. Two of the most complete field studies of song learning are Kroodsma's (1974) work on Bewick's Wrens (*Thryomanes bewickii*) and Jenkins's (1978) on Saddlebacks (*Philesturnus carunculatus*). I suspect that these studies might have modified the course of the study of song learning had laboratory studies of the same species been available for comparison, but without discrepancies between laboratory and field results to raise questions, the field approach struck many investigators as just a more difficult, less controlled way to approach song learning. I argue, however, that longitudinal field studies of song learning are quite feasible provided that a few conditions are met, and also that, in conjunction with the opened-up laboratory approach recommended above, they can provide a much more complete understanding of song learning.

To study song learning in the field, the investigator must be able to color-band all or most of the birds in a study population, and song learning must take place within that population. If birds learn their songs before natal dispersal, they must be banded in the nest, but if they learn their songs after dispersal, which is probably the typical case, they can be banded after they have entered the study population. The Song Sparrows we banded in the nest and subsequently recaptured within the population after dispersal (typically a mile or so from the nest) sang the song types of their postdispersal area rather than of their natal area.

Virtually all natural populations in which song learning has been studied to date have been resident populations in which song learning occurs after dispersal (see, e.g., Kroodsma 1974, Jenkins 1978). Migratory populations will present essentially the same picture as resident populations if young birds commence song learning after dispersal from the natal area but before migration (probably the typical case) and return to the postdispersal site the following spring (perhaps also the typical case; see M. L. Morton 1992). Thus longitudinal studies of song learning in migratory populations are feasible if young birds can be banded after natal dispersal but before migration, and can be recaptured the following year. To the best of my knowledge, no one has done such a study. It is worth doing because it could distinguish between two proposed strategies of song learning: do young birds first learn their songs in the first spring following migration, as suggested by R. B. Payne (1981a, 1983b, this volume) for Indigo Buntings (*Passerina cyanea*), or do they memorize all song material their first summer and then select from this pool the songs most similar to those of the adult neighbors they encounter in the first breeding season, as suggested by D. A. Nelson (1992a) for Field Sparrows (*Spizella pusilla*)?

A longitudinal study can proceed, given a sufficiently large study population of individually marked birds of known age, so long as a researcher can obtain reasonably complete census information. Ideally, one would track young birds from the moment they enter the population, but radio transmitters small enough to attach to young passerine birds are not yet available. Young birds probably are always relatively inconspicuous before becoming territorial (Arcese 1987, 1989a), which occurs during the early "sensitive" phase of song learning. Nevertheless, if one records all songs of all birds in a population, and if a young bird has learned his songs from birds in that population, his tutors can be identified, just as in a laboratory experiment, on the basis of similarity of song types between student and potential tutors. In the laboratory one generally knows which songs a bird has heard, but tutor identification may actually be easier in some respects in the field. In our Song Sparrow studies, for example, song copying is more faithful and precise in the field than in the laboratory.

Study Population and Methods of Study

Our study site is an undeveloped 3-km^2 park bordering Puget Sound in Seattle, Washington. The population is resident, and typically about 150 Song Sparrow males are on territories each year. Most birds immigrate into the study population from surrounding areas. First-year males are identified if we (1) banded them in the nest, (2) netted them as juveniles (identifiable by an incompletely pneumatized skull and juvenile plumage), or (3) netted them in winter or spring as young adults in an area where all adult males have been banded. Case 2 is our most common method of identifying young birds.

We record complete song repertoires (all song types) of birds in the field and analyze them on a Kay DSP 5500 Sonograph. A Song Sparrow sings his song types with eventual variety, and in free singing appears to use the different types interchangeably and with approximately equal frequency (but see below). Although a Song Sparrow sings variations on each type, variation within types is small compared with variation among types (Stoddard et al. 1988, Podos et al. 1992, Nowicki et al. 1994); moreover, types are clearly defined by the eventual variety style of singing. The complete repertoire of a Song Sparrow can be obtained in two to five hours of recording. Most birds are recorded on at least two days, and many in two or more years. We have never observed a Song Sparrow to add or delete a song from his repertoire subsequent to early in his first breeding season.

We consider as possible song tutors all older birds in the study population that were on territory in the subject's hatching year. We identify the older bird with the most similar rendition of the type (complete with idiosyncratic features not seen in other renditions of the type) as the young bird's probable tutor for that type. This judgment is rarely difficult; Song Sparrow songs are complex, so similar songs stand out. In some cases two or more older birds sing versions of a song that are highly similar to the young bird's (see below). In addition to the tutor-student classification, we also classify songs as shared (very similar) or unshared. Procedurally, this classification is fairly easy and very much like the tutor-student classification, because highly similar songs stand out on the background of high song diversity.

Song-learning studies rest heavily, and uneasily, on the human observer's ability to judge song similarity. As a general rule, I believe we should not rely on human observers to judge song similarity, but instead should develop methods of having the birds themselves make these judgments (Beecher and Stoddard 1990, Beecher et al. 1994a). I think human judgments are relatively danger free, however, at the high similarity end of the scale, where we use the human as a null-difference instrument. Approximately 95% of our decisions to classify two songs as "tutor" and "student" or as "shared" are highly reliable (high interjudge agreement) and probably valid. The remaining 5% of our decisions are more difficult but should not be allowed to hold the analysis hostage; instead these cases should be reexamined at the end in terms of principles suggested by the analysis. I do so later in this chapter.

Our judgments of song similarity are made on the basis of our extensive field observations of singing in this species and the results of our field playback studies (Stoddard et al. 1988, 1990, 1991, Beecher et al. 1995) and laboratory perceptual experiments (Beecher and Stoddard 1990, Stoddard et al. 1992b, Horning et al. 1993, Beecher et al. 1994a). The following are some important points derived from these studies.

1. Although the close song matches between tutors and students are impressive, we do not expect perfect matches because each Song Sparrow has the ability and proclivity to vary each and every song he sings (Stoddard et al. 1988, Podos et al. 1992, Nowicki et al. 1994). We have shown, moreover, that birds perceptually classify variants on a type as the same type (Stoddard et al. 1992b).

2. The early parts of a song are generally more important than the later parts for classification as to type. We derive this conclusion from the observation that Song Sparrows generally vary the later parts of a song more than the early parts when singing, and from our laboratory perceptual experiments showing that the early parts of a song are more perceptually salient than the later parts (Horning et al. 1993).

3. The validity of our song-similarity judgments is supported by the results of our playback experiments on song matching (Stoddard et al. 1992a, Beecher et al. 1995). When presented with a stimulus song that, according to the human judge,

matches one of his own, the bird replies with the predicted matching type, but when presented with a nonmatching stimulus song, the bird responds with different types on different occasions—that is, apparently randomly.

Although our first analyses of song learning in this population focused on identifying song tutors (Beecher et al. 1994b, Nordby et al. unpubl. ms.), we subsequently compared the young bird's songs not with those of his presumed tutors but with his postdispersal neighbors in his hatching year and his neighbors in his first breeding season (Campbell et al. unpubl. ms.). We designate as neighbors birds on a bordering territory or one territory removed; thus defined, a bird can have 5–15 neighbors. Some of these neighbors may have been identified as tutors, but the majority will not have been. As will be seen, this analysis produces a congruent pattern of results, indicating that our conclusions about song learning in this population are not biased by our method of identifying tutors.

The Song-learning Strategy of the Song Sparrow

I present the results of our field studies in reference to two major issues. First is the field-laboratory contrast: How does the pattern of song learning inferred from the field study contrast with the pattern inferred from laboratory studies? Can we reconcile the differences we find? Second is the question of function: Can we characterize the pattern observed in the field as a *strategy* of song learning? And if so, what advantages might a bird learning songs in this way have over a bird learning according to some other rules? Below, I discuss four important ways in which the results of our field studies differ from results obtained in laboratory studies.

1. Young birds copy whole songs precisely. Our first striking observation in the field was that young Song Sparrows usually developed nearly perfect copies of the songs of their older neighbors (Fig. 4.1). The similarities were remarkable, with differences between tutor and student often being no greater than those one normally observes in repetitions of the same song by one bird. The biggest difference apparent in the example shown in Fig. 4.1 is in the third song, in which the young bird appears to have simplified the song by dropping the high-frequency section near the end.

These field results differ remarkably from the laboratory findings (Marler and Peters 1987, 1988, Beecher et al. unpubl. ms.). I emphasize the comparison to our own lab study in particular because we used four live birds as tutors rather than the tape-recorded songs Marler and Peters used. In the laboratory, birds copy elements of particular tutors, but they often (in our study, usually) combine elements of the different songs of different tutors to form "hybrid" songs—songs made up of parts of different song types. Fig. 4.1 contains a hypothetical example of a hybrid song that could have been produced by that bird in the field had he copied his tutors in the same fashion as do birds in the lab.

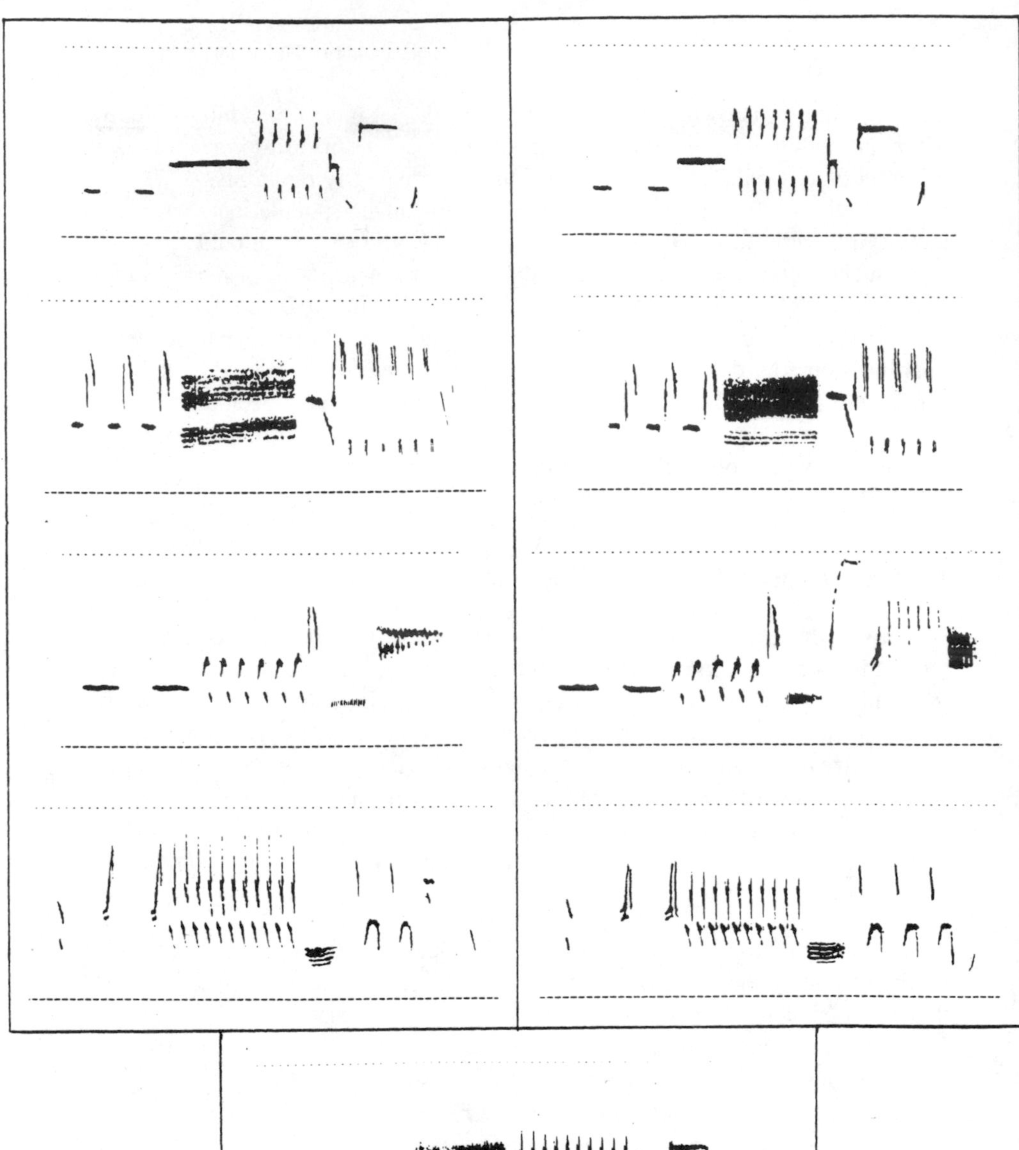

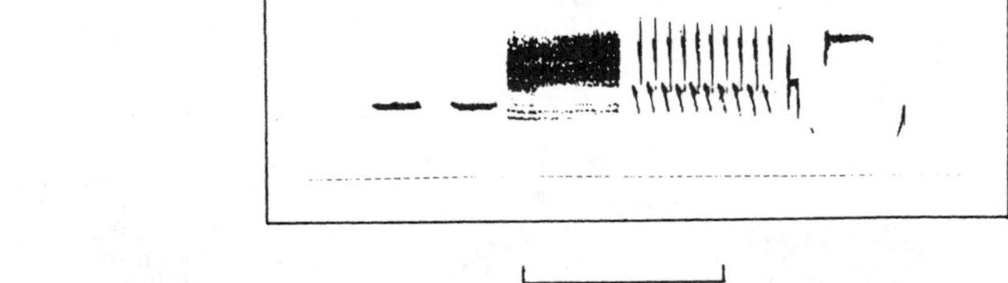

Fig. 4.1. In the field, young Song Sparrows learn nearly perfect copies of their tutor songs. Left panel: four of the nine song types of a young bird. Right panel: corresponding songs of one of his tutors. Bottom panel: the bird might develop a song like this if, instead of learning entire song types, he learned elements and rearranged them into hybrid songs. Note that this hypothetical hybrid song contains one element each from the four tutor songs on the right. Such hybrid songs are common in lab subjects but rare in our field population. Frequency markers at the bottom and top of each sonogram are 0 and 10 kHz, time marker is 1 second, and the analyzing filter bandwidth was 117 Hz.

2. A young bird learns from several birds that were neighbors during his hatching summer, and he establishes his territory in this range. Our second finding was that it usually took three or more tutors to account for a young bird's entire repertoire of eight or nine song types (Fig. 4.2). Invariably, these tutors were neighbors in the young bird's hatching summer. Usually by the following spring (the young bird's first breeding season), some of these tutor-neighbors had died. The young bird ultimately established his territory within the territorial range of his ex-tutors, often replacing a dead tutor (Fig. 4.3).

These data are consistent with Arcese's earlier observations (1987, 1989a) of territory establishment by young Song Sparrows. Arcese showed that young birds have "floater" ranges covering the territories of five or six adult males, and that some time between their first summer and following spring, the floaters ultimately take over one of the territories or insert a new territory within that range.

Laboratory experiments have not predicted, and cannot predict, any aspect of these findings. Tape-tutor studies cannot do so because the social factor is controlled out. Even studies with multiple live tutors cannot predict very much. For example, it is hardly surprising that in our laboratory live-tutor experiments (Beecher et al. unpubl. ms.) birds learned song elements from all four tutors, given that they were exposed to all of them continuously and at close range during the sensitive period. The young birds themselves had no control over this exposure, in contrast with the situation in the field. In particular, they could not concentrate on one of the tutors, get away from others, or interact with more than those four. Introducing live tutors into a laboratory experiment does not make it ipso facto "more natural." It may be more natural in some ways (live vs. tape tutors), and less natural in others (constant, close, unmodulated exposure to multiple adult males). The fundamental point of the field-lab contrast here is that only in the field could one discover that young birds actively acquire their songs from several neighboring tutors. At this point we do not know whether a bird actually visits all of these tutor-neighbors or listens to all their songs from a single location; both are possible, although the evidence (ours and that of Arcese [1987, 1989a]) definitely favors the first hypothesis. But the fact that the bird learns from multiple tutors who are neighbors is surely the centerpiece of the young Song Sparrow's song-learning strategy, and this could not have been discovered in the laboratory.

Our results provide some insight into the timing of song learning. As indicated above, birds identified as tutors were always present in the young bird's hatching year but often were not present the following spring (in our population, survival between breeding seasons is about 60–70% for adult males). New neighbors (except for other one-year-old birds) are uncommon, however, because adult males rarely change territories from one breeding season to the next, and those that do usually just move to an adjacent territory. Thus virtually the only new neighbors a young bird is likely to encounter are birds of his own age. In conclusion, the data implicate the bird's hatching summer as the key time for song learning but do not rule out further learning in the following year (at least up to about April, after which the bird's repertoire appears to be fixed for the rest of his life).

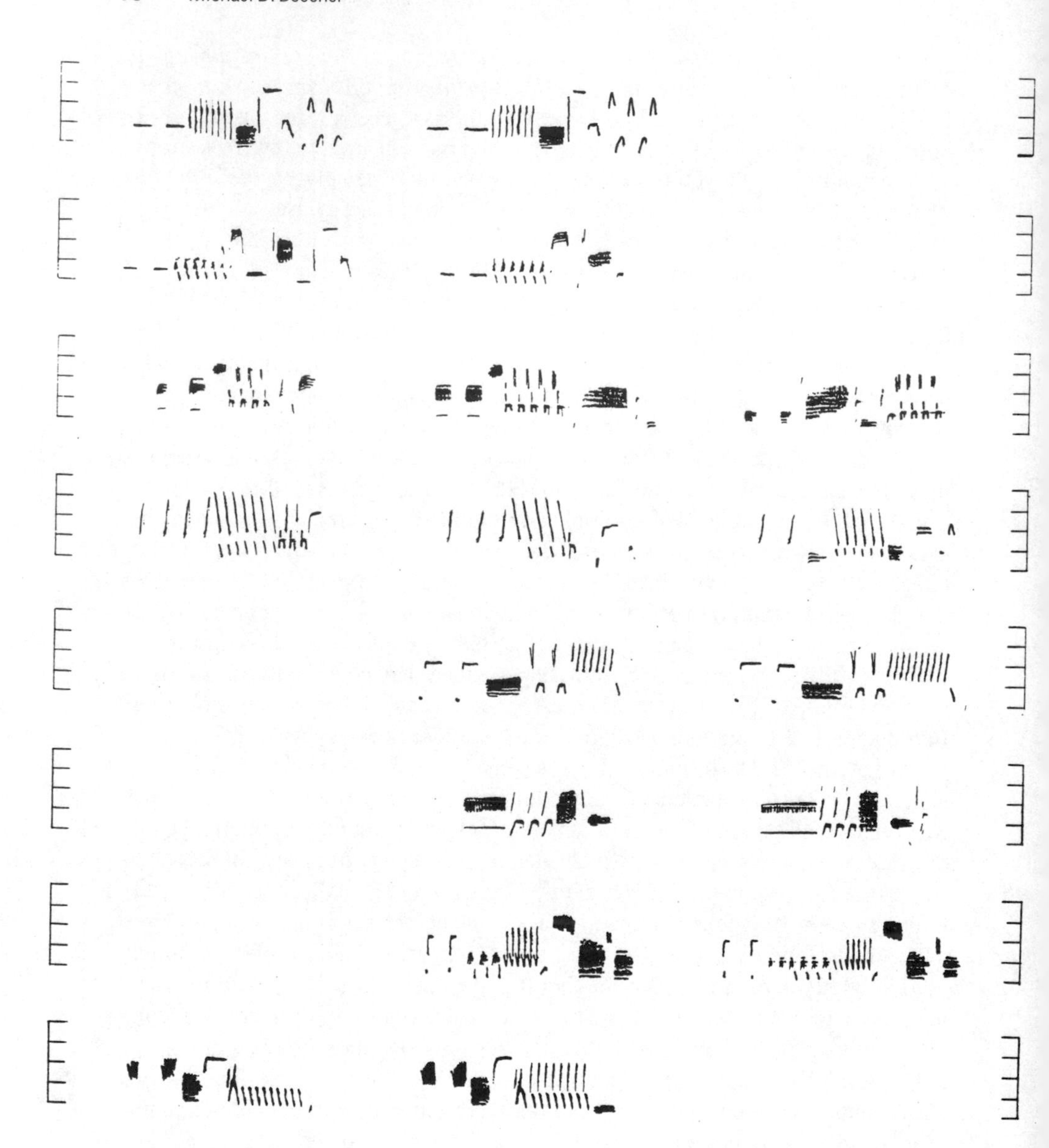

Fig. 4.2. In the field, young male Song Sparrows learn their songs from three or more older males. Shown are 8 of the 11 song types of a young bird (center), and song types of three tutors (one upper left, one lower left, one right). When two tutor songs are shown, the one on the left is clearly the better match. Frequency scale 2–10 kHz in 2-kHz steps, time marker is 1 second, and the analyzing filter bandwidth was 117 Hz.

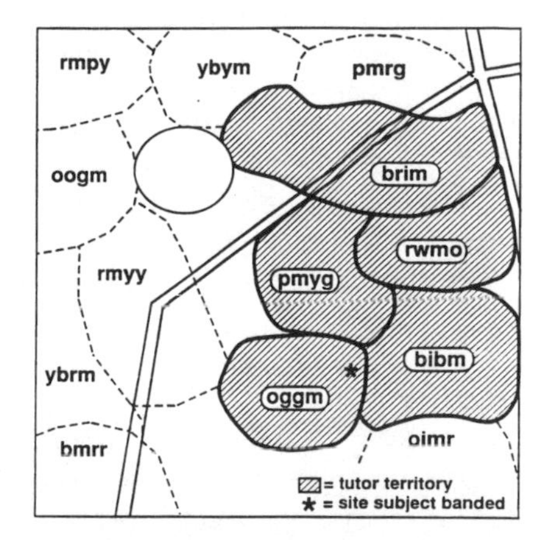

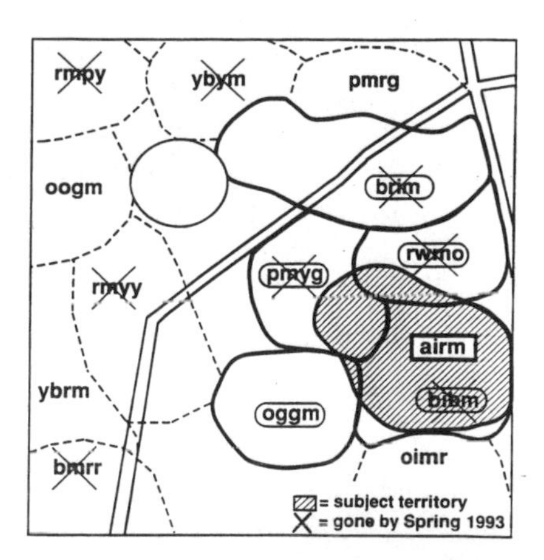

Fig. 4.3. A young bird's tutors are neighbors in the area where he settles after dispersing in his hatching year, but not all tutors survive to the young bird's first breeding season the next spring. Left: map of territories of 13 adult birds where the young male AIRM settled in his hatching year (1992); the star indicates where he was banded. The territories of the five birds that were subsequently identified as AIRM's tutors on the basis of his crystallized repertoire in 1993 are outlined and hatched. Right: the same configuration, overlaid by AIRM's 1993 territory (hatched), and with dead birds crossed out (8 of the 13, including four of the five tutors). Although the actual 1993 territories of birds other than AIRM are not shown, birds OGGM (the sole surviving tutor) and OIMR remained in approximately the same places. Also indicated on the map are a pond (circle) and two intersecting paths; open areas were unoccupied (e.g., steep hills, meadows).

Marler (1990) attempted to reconcile laboratory evidence suggesting that learning is restricted to the bird's first summer with field data suggesting that learning may continue into the bird's first breeding season. He proposed a two-stage theory of song learning: an early sensory or memorization phase in the bird's hatching summer, and a later "action-based" phase early in the following breeding season. In the bird's first summer, he memorizes a pool of songs, as suggested by early laboratory studies of the sensitive period (Marler and Peters 1981, 1982c). At the beginning of the next breeding season, the bird interacts with adult males who may or may not include his original tutors. As a result of these interactions, the bird selects, from the large pool of songs he memorized the previous summer, those songs or song elements that best match the songs of his present neighbors. Field evidence for this kind of learning has been obtained for Indigo Buntings (R. B. Payne 1982), White-crowned Sparrows (Baptista and Morton 1988), and Field Sparrows (D. A. Nelson 1992a), although in none of these cases is it known when, where, or from whom the original songs were learned. All that is known is that the bird appears in his first breeding season with more song types than he will keep, and the song types he keeps tend to be those that best match his neighbors' songs.

If this hypothesis is true, we should find supporting evidence in our field data. In our recent field study, we found that a yearling male learned or retained more songs from his primary tutor (the tutor from whom the subject learned the most songs) if the primary tutor survived past January 1 (Nordby et al. unpubl. ms.).

Furthermore, when we considered just birds with surviving primary tutors, we found that the young bird learned or retained more songs from his primary tutor if the young bird established a territory adjacent to the tutor's than if he settled farther away (which seems to happen only when there are not vacancies, because few or no tutors die or many other young birds move into the same area). Although there are other interpretations, both these results suggest late influence. Specifically, it may be that, of the songs he memorized the previous summer, the young bird is most likely to retain the songs of tutors that are still alive and nearby the following spring.

3. Young birds preserve tutor and type in their songs. In our field population, the exceptions to the "perfect copy" rule actually clarify the rule and suggest an additional principle. The first exception occurs when young birds "blend" two tutors' somewhat different versions of a single song type instead of copying one or the other. These songs are not true hybrids because, although the elements are selected from two different tutors, they are selected from the same or a very similar song type. The second exception, which is rare, occurs when the young bird combines elements from two dissimilar song types of the *same* tutor. The principle followed by the bird, then, seems to be as follows: combine song elements of different songs only if they are different tutors' versions of the same type, or if they are different song types of the same tutor. We summarize this principle as the student "preserving tutor, type, or both" in his songs. We have never found a clear example of a bird hybridizing a song type of one singer with a distinctly dissimilar song type of a different singer. Yet Song Sparrows do this commonly in the laboratory.

4. Young birds preferentially learn shared songs. This brings us to our fourth contrast between laboratory and field results. In the field population, neighbors typically share a portion of their song repertoires. Typically, two neighbors share about four of their eight or nine song types. Our discovery in the field is that young birds preferentially learn (or retain) the tutor-shared types. The young bird whose song types are diagrammed in Fig. 4.4 (along with those of his tutors) retained seven types that were shared by two or more of his tutors, and only two that were unique to one of these tutors, despite the fact that in his tutor group there were 11 shared types and 13 unshared types. In the full sample analyzed by Beecher et al. (1994b), birds learned or retained 84% of tutor-shared types and only 21% of tutor-unique types, even though only 37% of the tutors' song types were shared types.

We can look at this issue in another way by asking how many of the song types in a yearling male's crystallized repertoire are shared with his postdispersal, hatch year neighbors (vs. with his hatch year tutors, as detailed above). In this analysis, we counted all neighbors either adjacent to or one territory removed from a young bird's late-summer or early-fall territory; this group invariably included the birds identified as tutors. Again, this is a "back comparison" because we were compar-

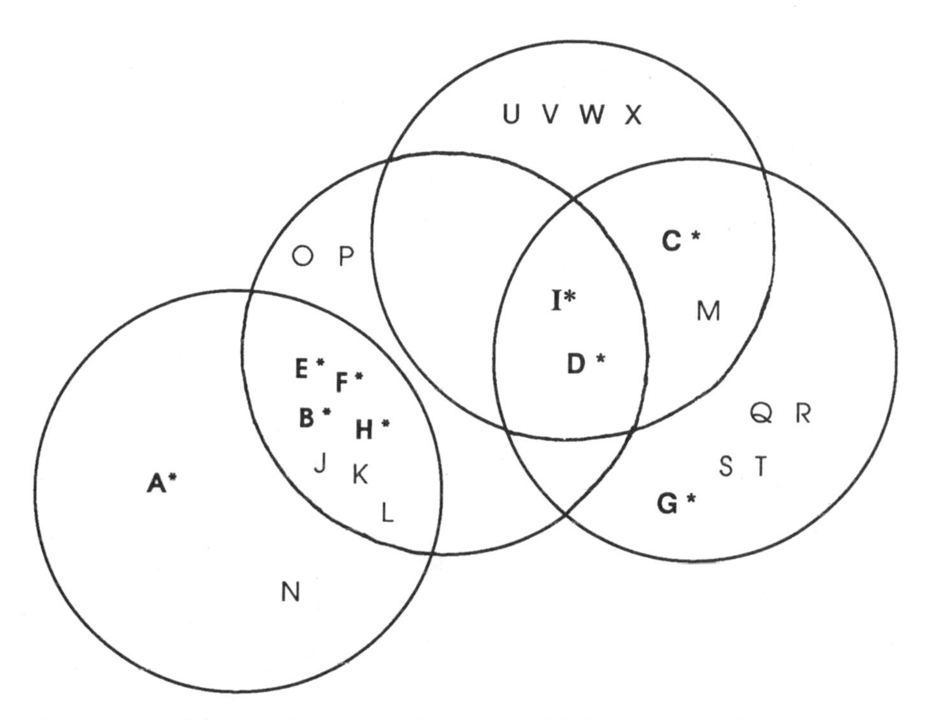

Fig. 4.4. A Venn diagram of the four tutors (four circles) of bird MXPX. Each song type is indicated by a letter, and shared songs of two or more birds are represented by the same letter. Shared types are therefore in the intersections of the tutor circles (e.g., types I and D were shared by three tutors). The nine song types learned by MXPX are indicated in boldface and starred. He learned 7 of the 11 tutor-shared types but only 2 of the 13 tutor-unique types.

ing the bird's crystallized repertoire of the following spring with the songs of the birds he potentially heard during the sensitive period for song memorization the summer before. For both data sets, we found that birds had more shared songs in their repertoires than would be expected if they randomly sampled either among song types or among total songs (Table 4.1).

How much of the learning preference for shared types occurs because shared songs are sung by two or more birds, and how much because they are heard more often? This question cannot be directly analyzed in the field, where the two factors are inevitably confounded. Nor, for that matter, can it be analyzed in the laboratory, at least not with live tutors. I do not think it is a critical point, however, for two reasons. First, from the point of view of evaluating this song-learning strategy, it does not really matter whether a bird preferentially learns shared songs because he hears them from more birds or because he simply hears them more (or for both reasons). Second, the results of laboratory investigations to date uniformly suggest that effects of "dosage" (frequency of hearing) are small or even absent, a common generalization being that frequency of hearing a song, beyond a certain minimum, has little effect on which songs are learned and which are not (Petrinovich 1985, Slater 1989).

Table 4.1. Proportion of a bird's song repertoire that he shared with two or more of his tutors (Beecher et al. 1994b) or two or more of his hatch year neighbors (Campbell et al. unpubl. ms.)

	Reference group	
	Tutors	Neighbors
Mean size of reference group	3.57	10.2
Proportion of subject's songs that are shared songs within the reference group	.70	.62
Proportion of shared song types within the reference group (relative to total number of types)	.37	.35
Proportion of shared songs within the reference group (relative to total number of songs)	.57	.54

Notes: These are independent data sets, but they give parallel results. The focal bird's sharing with his reference group is compared with sharing within the reference group. Two sharing proportions are given for each reference group: first, the proportion of shared song types in the group relative to the total number of types; and, second, the proportion of shared songs relative to the total number of songs. In the first case a song type is counted only once, regardless of how many birds sing it; in the second case a song type is counted once for each bird in the reference group that sings it. For example, the reference group shown in Fig. 4.5 shared 11 of 24 song types (46%), and 24 of 37 songs (65%), a bit higher than both sets of overall figures given here.

The learning preference for shared songs is a key finding, for it provides the first indication of the basis by which a bird selects the songs of his final repertoire from among the many possible tutor songs. It also gives us a start on a functional explanation for repertoire selectivity: shared songs are preferred, suggesting that shared songs may be adaptive in some way.

No laboratory study to date has looked for a possible preference for shared song types. Tape-tutor studies cannot do this because (1) shared types are avoided so as to simplify the identification of which songs were learned, and (2) information as to the identity of singers is absent, except for information in the song itself, and, for Song Sparrows at least, there is little such information (Beecher et al. 1994a). Even the few multiple live-tutor studies performed to date used tutors with unshared types, for the reason just stated. (We are presently carrying out a study with four live tutors who were neighbors in the field and share some song types.)

Conclusions

Our field studies suggest a different set of song-learning "rules" than those derived from laboratory studies. The rules can be summarized as follows: a young Song Sparrow constructs his song repertoire by (1) sampling the repertoires of several (at least three or four) older tutor-neighbors; (2) preserving, within limits, the identity of both song type and song tutor; and (3) preferentially learning types shared among these tutors. If we view this collection of rules as an adaptive learning strategy, then it functions to maximize the number of songs the bird shares with his neighbors—not only his original neighbors but also the younger birds who will eventually replace them, because they will have learned many of the same songs.

These rules provide a mechanism whereby birds come to share large portions of their repertoires. Rule 1 (Learn songs from all males on the floater range) has the obvious advantage that a young bird will necessarily share songs with each of the birds that remains his neighbor. But what is the advantage of rule 3 (Learn shared song types)? If the bird learned unique songs, he would have songs "personalized" for particular tutor-neighbors (i.e., songs shared with those neighbors only). The learning preference for shared types rather than unique types suggests that it is more important to have a shared song, even if (or perhaps because) it is shared with several neighbors, than it is to have a personalized song for each neighbor. Personalized songs are good only until a tutor dies or moves, but a shared song is good until all the birds singing it in the neighborhood have died or moved, and even beyond then because other young birds moving into the area will preferentially learn shared types. This argument assumes that the bird is constrained to memorize in his hatching summer the songs he will need in the future. If this assumption is true, then learning shared songs maximizes the chances of the bird having songs in common with future neighbors in a world in which he is uncertain who those neighbors will be.

We have begun analyzing long-term data to test the idea that preferential learning of shared song types, combined with a relatively low turnover rate of territorial males, maintains high levels of song sharing in the population. We have found that song-sharing levels are similar whether a bird is compared with his neighbors from his hatching (song-learning) year (summarized in Table 4.1), with his actual neighbors of his first breeding season, or with his neighbors of his second breeding season (Campbell et al. unpubl. ms.). Thus, by learning shared types, a bird guarantees that he will have songs in common with his new neighbors despite neighbor turnover caused by death and movement.

An alternative to learning all songs in the first summer and preferentially learning shared songs is to add or delete songs each breeding season. If additions and deletions are made to match the song types of new neighbors, then this strategy would achieve the same end of maintaining song sharing despite neighbor turnover. Such annual adjustments of the repertoire apparently occur in several songbirds (e.g., Great Tits, *Parus major:* McGregor and Krebs 1989; American Redstarts, *Setophaga ruticilla:* Lemon et al. 1994). Why it does not happen in Song Sparrows and other species is unknown.

Connections between social rules and perceptual rules. How do we connect song-learning rules derived from field studies with laboratory studies of song learning? On the timing of learning, lab and field studies agree, at least partly, that there is a sensitive period for song learning in the first summer, but so far only field studies show late learning. On the question of song repertoire construction—how a bird "chooses" the songs he keeps for his final repertoire—the lab studies are essentially silent while the field studies provide a clear picture

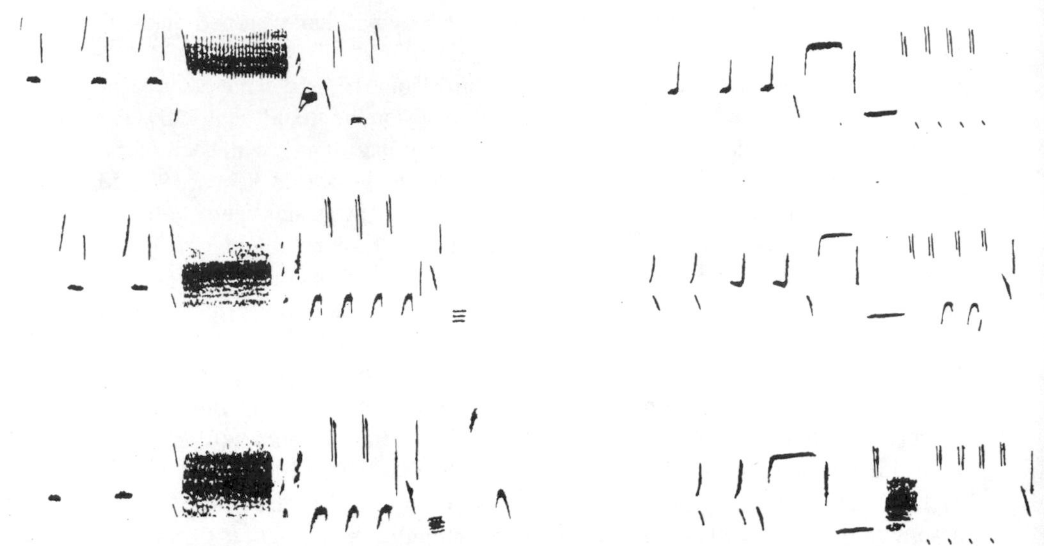

Fig. 4.5. Two examples of a young bird "blending" the songs of two tutors; the two types are similar but not easily classified as shared. In each example, the young bird's song is in the middle, and the two tutor songs are shown above and below it. Left: the bird took the first part of the upper tutor song and the remainder from the lower tutor song. The similarity between the two tutor songs lies not in the microstructure of the elements but in having the same elements in the same order. Specifically, both begin with two or three notes at approximately 4 kHz (birds often vary the number of notes within an introductory phrase) and spaced approximately 40 milliseconds apart. The notes of one tutor song but not the other are embellished with higher "grace" notes. The introductory notes are followed by a buzz about half a second in duration. Note that in both tutor songs the buzz is preceded by a downsweep and is followed by a two-voiced upsweep. The ends of the two songs differ, though not greatly (again, the end is the part of the song the bird most often varies and is least important for type classification). Right: the young bird took the introductory notes of both tutors, the 4-note complex that is identical in both tutor songs (omitting the buzz of the lower tutor), the trill of both tutors (but modified the bottom part of it), and the final note complex of the lower tutor. The two tutor songs are difficult to classify as shared, despite strong similarities in their microstructures, because they differ in their introductory notes and the presence of a buzz in only one of them. Note that if the buzz from the lower tutor song is removed, it is very similar to the upper tutor song except for the introductory notes (the notes at the very end of the song are not diagnostic, as this part of the song is invariably changed from one singing of the song to the next).

of a song-learning strategy. A young bird samples the songs of the territorial males occupying his first-summer floater range and selects songs for his own song repertoire so that he has at least one song from each bird, plus songs shared by the birds. As Table 4.1 indicates, these generalizations apply to the larger neighborhood as well as to the smaller tutor group. Finally, on the question of song construction, laboratory and field studies present conflicting pictures: laboratory results indicate that birds typically improvise new hybrid songs from elements of different tutor songs, but field results indicate that birds avoid doing so, presumably because hybrid songs would violate rules 1 and 2 (above).

I believe that hybrid songs were the rule in our laboratory study mainly because we used four tutors that shared no song types (each tutor had come from a different area). Thus we presented the young birds with an abnormal tutor set. Now, if a young Song Sparrow has a preference for learning shared song types, what will he do when presented with tutors that share no song types? Our theory is that the bird has a learning-perceptual predisposition to key in on song types that are sung by two or more tutors. He must therefore make a judgment about the similarity between songs of different tutors. If the songs are similar enough, he lumps them as the same type, and if these different tutor versions of the type are different in enough ways, he must blend them. The main point is that the more different the songs that are being blended, the more the resulting song will look to the experimenter like a hybrid song, while from the bird's perspective he has produced a version of a tutor-shared song type. A set of dissimilar tutor songs may cause the bird to "stretch" the similarity criterion. We think that this explanation is the most likely one (and certainly the most interesting one) for the hybrid songs produced by Song Sparrows in our laboratory experiment (and perhaps too in Marler and Peters's tape-tutor studies [1987, 1988]). It is a very testable proposition: hybrid songs should be uncommon when there are many shared songs in the tutor song set, but common when there are few tutor-shared songs.

In our study population, hybrid songs are extremely rare, probably because tutor-neighbors always have good matches (95% of the song types are easily classified as shared or unshared). Occasionally we see a tricky case, in which a bird appears to regard two songs of his tutor-neighbors as shared but we, the human observers, might not so classify them. For example, a young bird blended the tutor songs in Fig. 4.5. One can infer that the young bird regarded the pair of tutor songs as the same type (i.e., similar enough to blend). Cases like those shown in Fig. 4.5 are relatively rare in our population, presumably because birds usually have more similar songs from which to choose.

A corollary of this argument is that hybrid songs may be more common in populations in which song sharing among neighbors is weak or absent—as Kramer and Lemon (1983) reported for a migratory Song Sparrow population they studied.

Advantages of song sharing. Our field studies suggest a "strategy" of song learning in the Song Sparrow designed to maximize song sharing with neighbors-to-be. Note that by preferentially learning shared types, Song Sparrows ultimately share more song types with more neighbors than they would if they learned all (or most) of the tutor song types, because unique types would be shared with only one tutor, and then only until he died or moved. We can think of several possible advantages of sharing songs with neighbors. First, song sharing may be advantageous to a young bird in competitive interactions. The advantage may accrue through mimicry (the young bird is confused with older established residents; see, e.g., R. B. Payne 1983b) or through some other mechanism (e.g., shared songs may be better threat signals; see Krebs et al. 1981b). A long-term mimicry advantage in our population is unlikely, however, given our finding that long-term neighbors can recognize one another from a single song type (Stoddard et al. 1990, 1991). Second, song sharing may provide an advantage to a young bird in cooperative interactions with neighbors. Beletsky and Orians (1989) showed that male Red-winged Blackbirds (*Agelaius phoeniceus*) with familiar neighbors have greater breeding success than males with unfamiliar neighbors. The authors pointed out that this effect may favor cooperative behavior in this and other similar species. It may also favor communication signals that encode familiarity, as shared songs do. A third possible advantage, perhaps relating to the second one, is suggested by our recent experiments (Beecher et al. in press). We found that a Song Sparrow will reply to a neighbor song by singing a song type he shares with that neighbor; we call this repertoire matching. These hypothetical advantages all pertain to male-male interactions, but females may also play a role in shaping the song-learning strategy, for example, by preferring the common song types in a neighborhood.

Field studies of the Song Sparrow give a different picture of song learning than do laboratory studies, partly because field studies reveal social variables that are difficult or impossible to manipulate in the lab. But the difference also pertains to variables that are at least partly acoustic or perceptual and that, in theory, can be manipulated in the lab (albeit not easily). For example, with an appropriate selection of tutors, one should be able to demonstrate a preference for tutor-shared types in the lab. Moreover, as suggested above, it should not be difficult to identify in the lab those variables that determine whether a young bird will develop hybrid songs. But I believe our Song Sparrow field research shows the benefits that come from looking first at song learning under natural conditions. The laboratory then becomes a place where researchers attempt to "simulate" the key conditions that have been tentatively identified in the field. The field-first approach deserves to becomes more popular, and it should lead ultimately to a more balanced appreciation of the phenomena of avian song learning.

Acknowledgments

My colleagues in the song-learning studies described here were Elizabeth Campbell, Cully Nordby, and Philip Stoddard. Other colleagues in the Song Sparrow work include John Burt, Patti Loesche, Cindy Horning, Michelle Elekonich, Mary Willis, and Adrian O'Loghlen. I appreciate the generous support of the National Science Foundation. Finally, thanks are due to Don Kroodsma for his yeoman effort to improve the writing in my manuscript.

5 Acquisition and Performance of Song Repertoires: Ways of Coping with Diversity and Versatility

Dietmar Todt and Henrike Hultsch

Diversity and versatility are key features of singing in birds. A seemingly infinite diversity of sounds is produced by vocal modulations of frequency and time (Nowicki and Marler 1988), yet each species can be readily distinguished from others by certain rules. Species differ, too, in how many different songs they produce and how the songs are presented during a typical performance (Hartshorne 1973, Kroodsma 1977c, 1982b, Ince and Slater 1985). Such variations also occur within species because individuals differ in repertoire size and patterns of song delivery.

A singing performance by an individual songbird clearly reveals many hierarchical levels for us to consider. The most obvious unit is the song, or strophe, which is typically two to three seconds in duration, but levels of analysis occur both above and below the level of song. In species with repertoires, any given song is also a rendition of a particular song type, and the specific rules by which different song types are used provide a higher level of organization. The sequence of song types is usually more variable than the sequence of elements within each song.

The Common Nightingale (*Luscinia megarhynchos*) offers researchers a unique opportunity to understand how a bird can acquire, memorize, administer, and use a large variety of songs. Male nightingales are renowned for their vocal virtuosity, and for several reasons the nightingale is an outstanding model animal for research that addresses these issues at several levels of song organization. First, the roughly 200 song types that adult males develop and sing are discrete and readily distinguishable. Second, males sing long bouts of songs, so it is easy to collect a comprehensive database for analysis. Third, males sing in a variety of distinct social contexts (from nocturnal solo singing to typical diurnal singing), so that functions of different behaviors can be more readily determined. Fourth, like many other songbirds, nightingales imitate each song type precisely, even in standard laboratory settings, but the quantity of their imitated songs far surpasses that of most other songbirds that have been studied.

In our studies of nightingales, we first describe the composition of repertoires and the rules of repertoire delivery. We next identify correlations between features of singing and biological or experimental factors (e.g., age of songsters, social context of song usage), and then manipulate the factors to see what effects that has on the singing. Last, we design and test models that attempt to explain how and why birds sing. These models also account for the mechanisms underlying the performance and use of songs.

Performance of Song Repertoires by Wild Nightingales

Songbirds modify their singing according to the season, time of day, ecological features, and, above all, the social context. The influence of social context on singing has been described for a number of species, and repertoire delivery varies according to whether the subject is advertising its territory, countersinging, or addressing a mate (Thompson 1972, Kroodsma 1977c, Todt et al. 1981, Falls and d'Agincourt 1982, Catchpole 1983, Kramer and Lemon 1983, Horn and Falls 1988a). Singing in situations of courtship or close-range male-male interactions, for instance, is typically more complex and versatile than singing used in territorial advertisement. This generalization applies to nightingales as well. Hultsch (1980, 1991a, 1993a) distinguished five categories of singing, three of which are clearly tied to the circadian activity of the songster (nocturnal singing, dawn chorusing, and daytime advertisement).

In this section, we review how wild nightingales use their songs. We begin by describing the simplest condition, night singing, and then report on an experiment and findings that illuminate the rules this species follows in its singing.

Solo Singing and Rules of Sequential Order

The best way to uncover rules of song organization is to study solo singing. This behavior provides a baseline that can be used to examine singing for influences caused by social or environmental factors. Ideal baseline behavior would be shown by songsters not exposed to significant visual or auditory stimuli and not engaged in other activities, such as locomotion or foraging. In addition, it would be best to have birds who sing long enough to provide sufficient material for analyses. These prerequisites are met in the nocturnal singing of the nightingale. Individuals singing in the dark may orchestrate solo performances of more than 2000 songs, in which even rare song types are included several times.

In examining nocturnal solo singing, we seek answers to two questions. First, how does a male choose his next song type? Early studies of song type transitions showed that certain song types are highly associated with one another (Hinde 1958, Isaac and Marler 1963, Todt 1968, 1970a, Lemon and Chatfield 1971, K. Nelson 1973, Dobson and Lemon 1977, Slater 1983b). Second, how soon does a male reuse a given song type? Recurrence of song types can be quantified as either the number of other songs or as the time interval that elapsed between successive occurrences of a given type. Shapes of frequency distributions of both recurrence

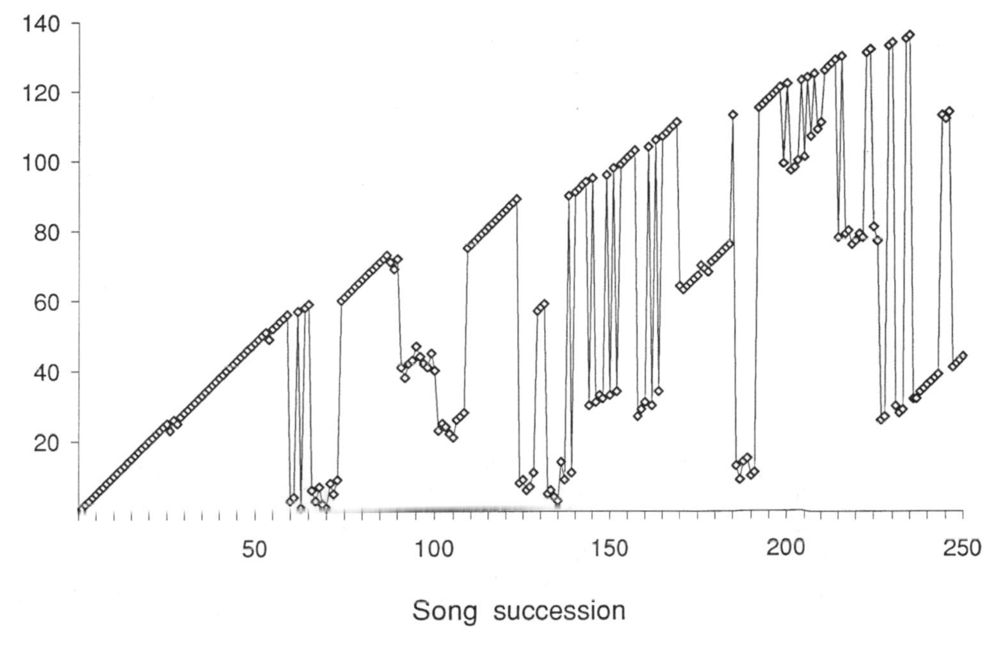

Fig. 5.1. Clustering and regular recurrence of song types during the performance of a 250-song sequence by a nightingale. Each song in the sequence is represented by a rhombus, and successive songs are adjacent to each other or connected by a line. Songs of the same type occur at the same level on the vertical axis. In this sequence, 136 different song types occurred.

numbers and intervals often point to periodic reuse of song types (Todt 1977, Wolffgramm 1980). The relatively few species for which rules of song type recurrence have been determined include the Eurasian Blackbird (*Turdus merula;* Todt 1969, 1975c), the Common Redstart (*Phoenicurus phoenicurus;* Thimm 1973), the Marsh Wren (*Cistothorus palustris;* Verner 1976), the Island Canary (*Serinus canaria;* Wolffgramm 1973), the Common Nightingale (Todt 1971, Hultsch 1980), and the Thrush Nightingale (*Luscinia luscinia;* Naguib et al. 1991, Naguib and Kolb 1992).

Consider first the rules of song type transition for Common Nightingales. The large repertoire of 200 or so song types is subdivided into clusters of 2–12 types. Within clusters, rarely used types typically occur only once, but commonly used types often occur several times (Fig. 5.1). Seldom, however, is a type repeated immediately (e.g., A-A). These clusters of sequentially associated song types are characterized by the directionality of the transition preference within them. In some clusters, the sequence of song types is unidirectional (e.g., B-C-D), in others the sequence is bidirectional (e.g., E-F and F-E), and in still others it is multidirectional (e.g., G-H and H-I, and I-G). Clusters also differ in the consistency of song type sequences. Overall, the size of clusters, the directionality of transition, and the strength of transition vary among individuals (Todt 1971, Hultsch 1980).

These clusters recur, so that on average about 60–80 songs of other types are used before a given song type recurs. The recurrence number of 60–80 is not a mere function of the repertoire size of an individual but varies with the frequency of use of a given song type: rare song types, for example, normally recur after a sequence that is two, three, or even four times as long as the average intersong string (i.e., after 120, 180, or 240 songs). Because the periodic recurrence of a song type is not a consequence of a rigid sequence of song type delivery, the periodic recurrence has to be distinguished as a separate rule of song delivery (Todt 1969, 1977, Hultsch 1980, Wolffgramm 1980).

Rule Testing in a Nightingale: A Playback Experiment

The functional aspects of rules derived from solo singing can be elucidated by exposing individuals whose program of singing has been examined beforehand to a selection of acoustic stimuli. In one series of playbacks, for instance, we wanted to test whether and how rules of clustering and recurrence would be followed during escalated interactions.

We therefore selected a wild nightingale whose singing had the following properties. First, his repertoire was markedly smaller than normal (61 song types). Second, the songs occurred in a highly regular succession and formed four sequential clusters, two of which were special: in one, song type succession was clearly unidirectional, but in the other it was clearly multidirectional. Finally, this male was isolated from conspecific song, as the nearest other male was more than 6 km distant.

In a series of playback experiments conducted during this male's nocturnal singing, we used an interactional playback design (Dabelsteen 1992, Dabelsteen and McGregor, this volume) originally developed for experiments on Eurasian Blackbirds (Todt 1970b, 1971, 1975c, 1981). The nightingale was exposed to playbacks of six different segments taken from his own song sequences. Three segments were from the unidirectional cluster and three were from the multidirectional cluster. We randomized use of these segments during the night and played a given segment only after the bird had produced one of a predetermined set of song types that the experimenter could identify by ear. We recorded both the playback stimuli and the bird's vocalizations (on separate channels), allowing us to determine relationships between the onset of the playback stimuli and the vocalizations of the subject that followed them.

Our experimental results were as follows. First, although the male often matched the playback songs, and thus raised the occurrence frequency of certain song types, the stimulation did not affect song type periodicity. Second, in a significant number of cases, in both the unidirectional and the multidirectional clusters, the bird abandoned his normal singing program and instead switched into the sequence of songs being played to him. Third, in the multidirectional clusters, the male often matched the song that had just been played back. In the unidirectional clusters, however, he responded more often with the next appropriate song type in the rigidly defined sequence. Thus, if X-Y was the rigid sequence, the male often responded with Y after hearing X from the loudspeaker (this "leading" has also been called a "convalent" response; Todt 1971).

Rules derived from solo singing thus help us to understand how males might interact vocally. Like other songbird species (see Todt 1975c, 1981, Krebs and Kroodsma 1980, Horn and Falls, this volume), nightingales will match songs from a playback tape, especially in response to clusters in which sequences of song types are not rigidly determined. In unidirectional clusters, on the other hand, males remain ahead of the playback of a given conspecific bird and become "vocal leaders" (Todt 1971). Vocal leader-follower relationships were also reported by Verner (1976) and Kroodsma (1979) for Marsh Wrens and by D. G. Smith and Norman (1979) for Red-winged Blackbirds (*Agelaius phoeniceus*). Kroodsma (1979) suggested that the leader-follower relationship is linked to social dominance. These formalized interactions, in which males lead a tape or another male, are especially intriguing, though difficult to document, and deserve further study.

Sequential Decision Making

A singing male nightingale must decide which song type to use next. Attempts to model the decision-making process for songbirds began with a study on Chaffinches (*Fringilla coelebs;* Hinde 1958), and over the years a number of models have been proposed, mostly for thrushes (Todt 1975c, Todt and Wolffgramm 1975, Dobson and Lemon 1977, C. L. Whitney 1981). These models typically consider endogenous factors, such as the components that govern (1) the periodicity of song types, reflected by rules of both recurrence number and interval; (2) whether or not a song type can occur in successive utterances; and (3) the association of song types with one another. The influence of endogenous components can be altered by exogenous factors, however, such as auditory stimuli from other birds. Song types compete with one another for expression, and the various rules or components determine the particular song type that has the highest potential for expression at a given instant (Todt 1968, 1969, 1970a, Wolffgramm 1975, Slater 1978, C. L. Whitney 1981).

Time-Specific Responses and Interaction by Song

The coordination of temporal and sequential aspects of song delivery (Thimm et al. 1974, Thimm 1980) is especially apparent when singers are interacting. Male nightingales sometimes engage in dialogues (Todt and Hultsch 1994) in which a bird listens and responds to another bird's songs in a sophisticated manner (Fig. 5.2). Often birds alternate, or take turns, with their songs by adjusting their temporal patterning of singing to that of their neighbors. This turn taking can be remarkably stable, as demonstrated by playbacks in which songs were edited to be longer or shorter than normal. For instance, in response to such playbacks, birds

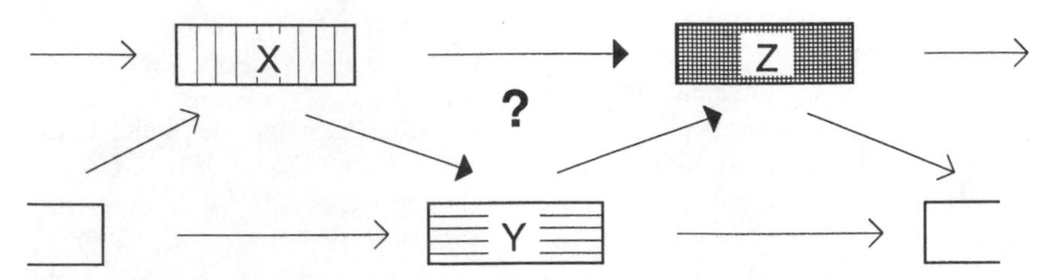

Fig. 5.2. A dialogue-like interaction between two songsters. The bars symbolize their songs, and the question mark stands for the temporal and pattern relationships that we study. The interaction is dialogue-like if the timing of songs is adjusted, indicating a type of turn taking, and if Z is a function of both X and Y but is not determined exclusively by either.

could prolong the normal intersong interval from about 3.5 seconds up to 12 seconds (Hultsch 1980). Nightingales do not always take turns, however; sometimes they overlap the songs of their neighbors. Rather than waiting to respond until about 1 second after the end of a neighbor's song, a male may begin his song about 1 second after his neighbor has begun. Songs are about 3 seconds in duration, so this behavior leads to significant overlap of signals.

Alternating and overlapping seem to be used in different social contexts. Nightingales typically alternate if they hear sequences of song types that are unfamiliar to them. And, as in other songbirds, this kind of alternating can be interpreted as a strategy to avoid overlap of vocalizations (R. W. Ficken et al. 1974, Gochfeld 1978, Hultsch and Todt 1982, Popp 1989). The avoidance of overlap is increasingly important when the signals are similar to one another in frequency range and volume. Overlapping, on the other hand, is an aggressive maneuver that occurs predominantly during territorial settlement, especially when a male is in close proximity to a neighbor. Overlapping may reduce a rival's singing time because the challenged neighbor often interrupts his own singing or prolongs his intersong intervals (Hultsch and Todt 1982).

As is evident during alternating and overlapping signaling, temporal relationships between songs may have a significant impact on the message of exchanged signals. For instance, vocal matching may basically address a particular songster, but superimposed on this address is a message encoded in the temporal pattern of song delivery (Todt 1981, Wolffgramm and Todt 1982, Hultsch and Todt 1986). Alternating song matching may be a form of vocal "greeting," whereas overlapping matching seems to be an agonistic response. For song exchange to be socially effective, the songsters apparently need not perform matching responses repeatedly during a countersinging bout. In fact, we frequently observed matching to occur only sporadically in the dyadic performance. Perhaps such behavior is a strategy to maintain the salience of the signal. Understanding when and why birds match, insert, or overlap will help us to understand how these complex interactions between neighboring birds may have evolved.

One especially puzzling form of dyadic song performance occurs when a male obviously ignores another male. Some nightingales orchestrate their repertoires without paying any detectable attention to the timing or quality of song from neighbors or playbacks. This "autonomous singing" seems to be characteristic of individuals with well-established territories (Hultsch and Todt 1982). Currently, we cannot exclude that such singing might nevertheless convey unique messages.

Future work on nightingales and other songbirds should examine the temporal aspects of interactions more carefully. Perhaps the alternating and overlapping that occur in nightingale songs are gross categories of interaction that could be placed on a finer scale in some continuum of interaction types.

Patterns of interaction require that males share certain aspects of their repertoires. To match types, for example, males must have the same types in their repertoires. If, in addition, sharing extends to unidirectional clusters of song types, males can anticipate the next song type of a neighbor and sing it before he does. We are especially interested in how repertoires and performance rules come to be shared by males. We will therefore look next at song acquisition and development.

How Birds Acquire Song Repertoires

Song learning has been extensively studied in both the field and the laboratory (Kroodsma and Miller 1982, Marler, 1987, 1991a, Slater 1989). Species differ considerably in how they cope with standardized laboratory settings. Some learn from loudspeakers, but others need a social tutor (social selectivity). Some species readily learn heterospecific song patterns, but others do not (signal selectivity). Some have a rigidly defined timing of vocal learning (sensitive phase), whereas others learn throughout life (Kroodsma and Pickert 1980, R. B. Payne 1981a, Clayton 1987b, Marler and Peters 1989, Baptista 1990b, this volume, Böhner 1990, Margoliash et al. 1991, Chaiken et al. 1993, Kroodsma, this volume).

The sensitive phase of song acquisition in nightingales starts around day 15 after hatching and lasts at least through the first three months of life (Hultsch and Kopp 1989). During this phase, an individual can acquire many different song types. Nightingales also learn later in life, so they are well adapted to develop large song repertoires.

Nightingales learn in a socially selective manner. In choice designs (Todt et al. 1979), for example, birds readily acquired master songs (model songs) presented by a live tutor but failed to acquire such songs if they were presented on tape in the absence of a live model. In the field, the role of the live tutor is probably played by a bird's father, who frequently interacts with his offspring (e.g., when feeding them). In the laboratory, tutor acceptance depends on subject-tutor interactions and on the age of subjects when interactions begin. In contrast, the biological properties of the tutor seem to be less important. Like young Eurasian Bullfinches (*Pyrrhula pyrrhula;* Nicolai 1959), nightingales can learn songs from a human caretaker provided that person hand-raised the bird from about day 6 after hatching (Freyschmidt et al. 1984). The social selectivity of nightingales decreases later in life. Experiments with nine-month-old birds showed that males can acquire new songs from tape alone (Hultsch 1991a). This age corresponds roughly to the age when wild birds may be exposed to conspecific songs in their African winter quarters (J. Nicolai pers. comm.). The decrease in selectivity is the result of previous auditory experience with conspecific model songs and is not just a consequence of age (Todt et al. 1979, Todt and Böhner 1994).

Teaching young nightingales. To teach young nightingales, we present them with a number of master songs played back through a loudspeaker while their tutor is present. Normally, the tutoring sessions and the subjects' behaviors (e.g., their movements and vocalizations) are recorded by audiovisual equipment, allowing investigation of the immediate effects of stimuli (Müller-Bröse and Todt 1991). The master songs to which the subjects are exposed are normal song types drawn from performances of wild males. In previous experiments we showed that each of the master songs, now totaling 214 song types, could be acquired by our subjects and never occurred in the repertoire of an experimental bird unless he had been tutored with that type (Fig. 5.3).

In our standard experimental design, we choose at random the song types to be used in training. A given song type occurs in only one string of songs, and each string is therefore different. Intersong intervals typically last four seconds.

After a young nightingale has developed his adult repertoire, we determine which of his song types match the tutor songs. Because the birds heard tutor songs only during a specific set of training sessions (Todt et al. 1979, Hultsch et al. 1984), we can identify the particular circumstances under which each was learned (Kroodsma 1978a, Marler and Peters 1987). Imitations are easily identified because copying of tutor songs is highly accurate (Fig. 5.3); minor additions or deletions of notes occur mainly in the first song sections.

Variables affecting the learning success. As an index of learning success we use the proportion of song types learned from each tutored string. We know that age and social experience affect the learning success, but we explore other experimental variables, too, such as exposure frequency, the number of master songs in a tutored string, presentation rate, serial position, and the distribution of tutoring sessions over time.

We were astounded to discover that males needed only 10–20 exposures to learn a set of 10 different song types. Further work showed that a set of 60 different song types could be acquired, stored, and produced if a male heard that long string only once a day for 20 days (Hultsch and Todt 1989a). When the presentation frequency was reduced to less than 10 exposures, the learning success decreased in terms of both acquisition rate and copy quality of the imitations (Hultsch 1993a).

The number of songs in the master strings also influenced learning success. Contrary to what might have been expected, decreasing the number of song types in a string actually reduced learning success, and males learned poorly if the master string was shorter than five song types. One explanation that may account

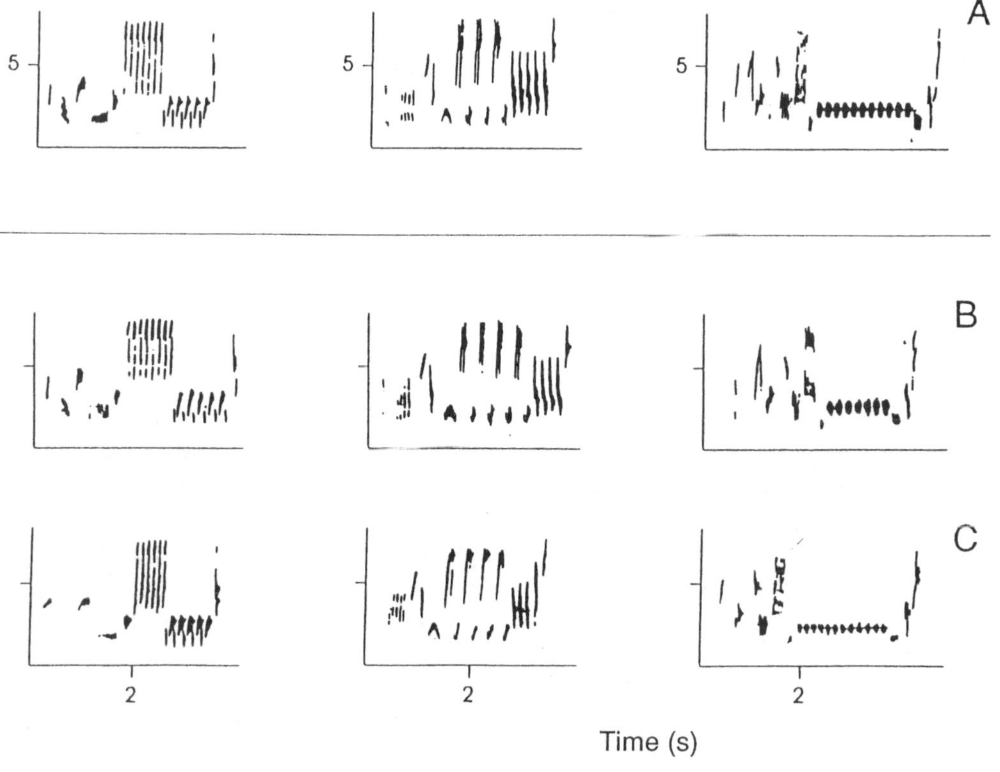

Fig. 5.3. Sonograms of nightingale songs illustrating the birds' normal copy quality. (A) Three song types recorded from a wild nightingale. These songs served as master song types. (B, C) Copies of the master songs by two laboratory-raised nightingales.

for this reduced success is that it reflected a perceptional or motivational bias toward longer and temporally coherent stimulus arrays during the auditory acquisition. But it is not possible to delineate such effects from peculiarities of song storage or retrieval.

The context effect. In nature, nightingales probably both hear and learn their future repertoires of song types from different males. We wanted to determine if and how males would learn from more than one tutor, and how songs from two tutors would be presented in a song performance. We hand-reared a group of nightingales and used two human tutors. One tutor presented the "A" string and the other tutor presented the "B" string, with different master songs in each string. String presentations were separated by one hour, during which the birds were alone, and the schedule of exposure was alternated daily. We discovered that the birds learned the master songs from both tutors and segregated the songs into two distinct clusters, one learned from each tutor (Hultsch and Todt 1989b). This result illuminates previous findings on nightingales in the field, where males often share "subrepertoires" with different males in their song neighborhood, sharing that might be functionally useful in countersinging.

Thus strings of master songs were kept intact if they were presented by different tutors, but strings were also kept intact when presented sequentially by the same tutor. In additional experiments, we showed that master strings were learned intact when separated from each other by more than five minutes. With less than five minutes between master strings, the young nightingales began to mix the clusters (Hultsch and Todt 1992a).

The primer effect. In the learning tasks discussed above, we always presented strings of master songs in the same sequence (e.g., ABCDEF . . .) for logistical reasons. Consequently, a master string formed a kind of sequential gestalt, with A always occurring at the beginning, other songs in the middle, and others at the end. Could it be that the birds formed a concept about that gestalt and learned it as a superunit? In an experiment to explore this issue, we destroyed the regularity of string presentation by playing the song types in 15 different random orders in successive presentations. Because the sequence was randomized, we thought the birds might have difficulty learning the string as a cluster, regardless of whether they learned master strings as superunits or as individual song types. We also expected that birds would not learn a specific sequence from such a program.

We were surprised by the results. First, learning success was just as high as in the rigid, stereotyped sequence of presentation. Second, master strings were learned intact, showing that rigid sequences are not necessary for males to form a coherent representation of the string to which they were exposed. Third, birds delivered imitations not in a random order, as songs had been presented in the training, but rather sang them in a sequentially associated fashion. Perhaps especially surprising, the males used the song types in a sequence that best approximated the first sequence they had heard in their training. And, remarkably, a single exposure was sufficient for the birds to memorize information about song succession in that sequence. Referring to the particular salience of the first exposure in the acquisition process, we call this phenomenon the *primer effect* (Hultsch and Todt in press).

The serial-order effect. In our studies of the primer effect, the nightingales heard relatively few sequences of their tutor songs, and the learned sequences were highly multidirectional. If the amount of exposure was increased, we wondered, could we convert the multidirectional sequences to unidirectional sequences?

It turned out that the unidirectionality of the learned sequence was directly related to the number of presentations. Sequences were especially unidirectional when birds experienced the tutor string 50–100 times. We named this phenomenon the *serial-order effect,* a term borrowed from studies on human serial-item learning (Hultsch and Todt 1992b). In some respects, the serial order effect is not surprising. Species that normally utter their song types in variable succession (e.g., Marsh Wrens) adopt an invariable singing style that clearly reflects the serial order of master songs after hearing them several thousand times (Kroodsma 1979). We expect such huge exposure frequencies to increase nightingales' unidirectionality of sequences, too.

The package effect and its implications. Even after 100 exposures to a tutor sequence, however, certain sequential "noise" persisted within the serial order of song types. The songs involved in this noise, we discovered, were typically sequential neighbors in a given master string. Thus, information from a longer string of master songs is segmented into subsets of sequentially associated items, or packages (Hultsch and Todt 1989c). These packages have a limited size (usually three to five song types) and are mainly bidirectional or multidirectional associations of distinct sets of acquired song types. The packages describe song type associations that are hierarchically inferior to the context groups, so that a top-down structural hierarchy would read as follows: context groups are composed of song type packages, which in turn are composed of songs. Although both context groups and songs reflect organization acquired from the training exposure, package groups are self-induced by the nightingale itself.

Theoretically, nightingales could impose these packages on their behavior at any of three stages: when they acquire, store, or retrieve and produce the songs. We have studied this issue in several ways, and our evidence suggests that packaging occurs when the nightingales acquire their songs (Hultsch 1992). Particularly striking data come from nightingales raised in acoustic isolation, which do not package song types (Wistel-Wozniak and Hultsch 1992).

Properties of memory. We envision the learning accomplishments of nightingales as the coordinated operation of a short-term memory, a recognition memory, and a battery of submemories. The short-term memory segments strings into packages and has two operational features: a capacity buffer that limits the number of units to about four song types, and a time-constrained memory span of about 35 seconds, when the data on perceived song types are retained at a particular level of processing or analysis (Hultsch 1992). The recognition memory sorts incoming information by distinguishing between novel and familiar stimulus patterns and categorizes the familiar patterns by song type, package type, and context group. Acquired song material is further processed, in parallel, in a battery of submemories, each of which stores information about a given string segment (Hultsch and Todt 1989c).

Short-term memory and a recognition mechanism are common postulates in learning models (Kendrick et al. 1986, Martinez and Kesner 1986, Bateson 1987). In the nightingale, these memories could explain the "chunking" of information to form packages. They would also account for the significance of the first exposure in generating these song type associations: when a bird hears a string of master songs a second time, the songs are identified by the recognition memory, and the information is subsequently transmitted to the particular submemory that already holds a memory trace about the same type of pattern. Parallel data processing in a

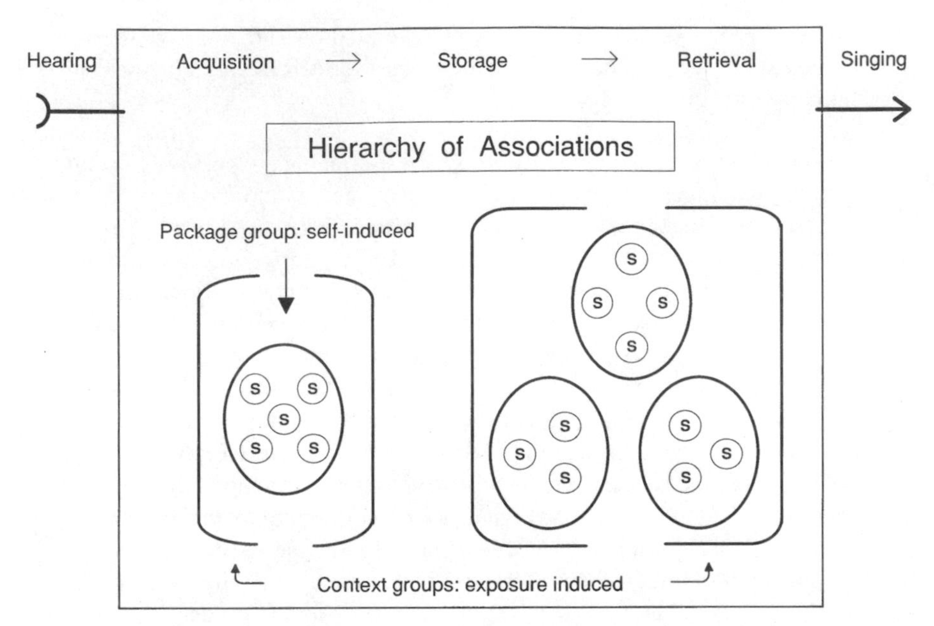

Fig. 5.4. Black box indicating song type acquisition, storage, and retrieval, and also our hypothesis about the format in which acquired song types are represented. Our findings suggest a hierarchical format in which context groups (the two groups in brackets) contain package groups (four oval figures), which in turn contain song types (circles).

battery of submemories, on the other hand, would explain why long master strings are learned as effectively as short ones, even when heard only 10–20 times. Each submemory holds a package of information and processes information on acquired song material separately, and additional information is stored in a way that associates the packages with a given context.

Song types, packages, and context groups reflect a structural hierarchy. These three levels undoubtedly reflect not only a hierarchy of stored information in the long-term memory but also a decisional hierarchy pertaining to the retrieval and performance of song types (Fig. 5.4). This suggestion clearly requires further examination, but we predict that these hierarchical levels express themselves in particular ways and determine how the bird actually decides to perform his repertoire.

The formation of package and context associations by hand-reared nightingales provides a new perspective on the sequential clusters found originally in the singing behavior of wild birds. These clusters could be explained as instruction-based song type associations, which could be memorized by fledglings during the early period of auditory learning. In addition, associations could also be acquired later in life, during periods of male-male interactions. Adult birds use their singing to communicate during such interactions and may profit from establishing shared subrepertoires with a neighbor.

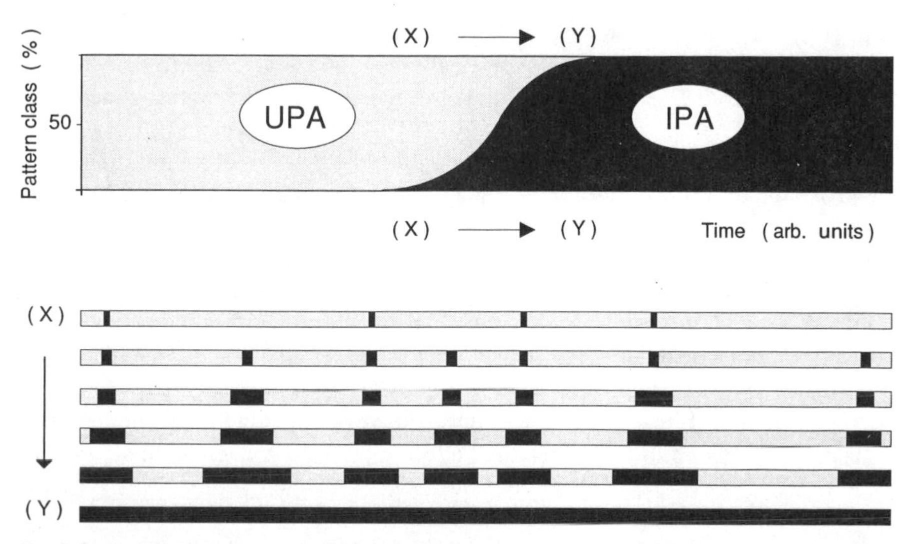

Fig. 5.5. Song development is studied as a change in the proportions of unidentified (UPA) and identified (IPA) patterns of vocalizations between times X and Y. Top: Overall transition from 0% IPA to 100% (black area). Bottom: Six bouts of vocalization, the first recorded near time X, the last recorded near time Y. During ontogeny and within such bouts, phases of UPAs (gray area) alternate with phases of IPAs (black).

Song development: precursors and trajectories. Studies of song learning provide both valuable data and cues for noninvasive research into mechanisms of song memorization. Some crucial questions concern features of song type precursors and the developmental trajectories that describe the transition to the final stage of song crystallization. Below we outline these issues for nightingales.

Nightingales hatch by the end of May, and their early period of auditory learning starts around day 15 after hatching. Typically, birds begin to produce subsong at about six months of age, and after a short winter hiatus, the first precursors of acquired imitations can be discerned in January. Vocal ontogeny then proceeds through several stages of plastic singing, and the adult singing form is complete after about four months. By the end of April, nightingales produce structurally stereotyped songs that are segregated by typical intersong intervals. In the laboratory, the full song of first-year males lasts for approximately four weeks. Thus, the general pattern of song development is similar to that described for other songbirds such as the Swamp Sparrow (*Melospiza georgiana;* Marler and Peters 1982a, Marler 1991b).

Like other songbirds, nightingales are continuous songsters during their early song development period. Embedded in this early coherent stream of vocalizations are the first precursors of song types. Precursors differ from the final version of song types both in the phonology of their syllables, which is not yet stereotyped, and in the serial order of song sections, which may be inverted or incomplete. Nevertheless, such precursors can be reliably identified. In addition to these "identified patterns" (or IPAs), birds produce "unidentified patterns" (UPAs;

Fig. 5.5) that cannot be categorized. In young nightingales, bouts of IPAs alternate with bouts of UPAs in the early, acoustically continuous vocal performances. As the numbers of different IPA types increase, the duration of IPA bouts increases and that of UPAs decreases, until finally all UPAs disappear completely.

We first thought that producing UPAs might serve as a sort of search strategy to improve the retrieval of information acquired during auditory learning, so that a bird randomly invented vocal patterns and reinforced them if they matched stored information. One prediction from this hypothesis was that the first occurrence of song type precursors would be randomly distributed in a stream of UPAs. Distributions of IPAs did not confirm this prediction. Rather, new IPA types occurred within bouts of precursors that had already emerged earlier. The temporal cohesiveness of IPA phases suggests that they reflect episodes of pattern retrieval from the bird's memory. Currently, however, neither the function nor the causes of UPAs and the characteristic pattern of UPA-IPA alternation are understood. We know more about the developmental changes within bouts of song precursors, as we describe below.

Ontogeny of hierarchical levels. Developmental changes within bouts of IPAs show that song types, packages, and context groups all are identifiable during development. Imitations of acquired song types, for example, emerge in packages, and the development of packages is completed before song types and context groups. The number of different packages detected in a bird's performance increases over a period of approximately four weeks, but from the outset, packages developed from the same context are closely associated.

Song intervals during development also inform about how these hierarchies are developed. During the phase of continuous vocal production or song rehearsal, for example, the intervals between imitations acquired from the same master string or context group are significantly shorter than the intervals between imitations of different context groups (Hultsch 1993b). Perhaps this effect indicates that access to the stored representation of song types is quick or delayed, depending on whether retrieval is from within a given context group (switch "on") or from a different context group (switch "not yet on"). Curiously, this temporal effect wanes later in ontogeny and can no longer be assessed by the time of song crystallization. We infer from these interval patterns that context groups are higher-level units of memorization, but the full range of implications of these temporal events warrants further examination.

Ontogenetic changes in repertoire size. About 5% of the imitations we identified during the course of song development were omitted from the bird's final repertoire (Hultsch 1991b). This constriction is much less marked for nightingales than for Swamp Sparrows (Marler and Peters 1982c). We discovered that imitations rejected from the final repertoire had been performed especially infrequently during plastic song and were also relatively poor copies of the models.

Increasing the repertoire size is more common than decreasing it, however, and normally this process is linked with a change in repertoire composition (Freyschmidt et al. 1984). Changes are achieved either by acquiring additional song types or by inventing new ones. Nightingales usually generate such inventions by recombining parts of song types in novel ways, but the recombinations are limited to song types that belong to the same package (Hultsch 1993b). Packages remain closed to further acquisition of song types.

Context groups, in contrast, can be increased by additional songs. When adult nightingales were exposed to a master string with two parts, a familiar portion that the bird learned during his first period of song acquisition and a novel portion, the novel portion was integrated into the same context group as the familiar song type sequence (Hultsch 1991c). This phenomenon suggests that additional learning can lead to sharing of at least parts of repertoires among conspecific neighbors (Hultsch and Todt 1981), and that context groups indeed can be regarded as subrepertoires.

We have many more questions we would like to ask of our nightingales. For example, we are curious about whether nightingales that are exposed to two strings of master song types that share one or more song types would associate their imitations into a single context group. Questions like this one will guide our future explorations.

Performance of Song Repertoires by Hand-raised Nightingales

Our studies of how nightingales acquire songs have given us the ability to explore the control mechanisms of singing in a noninvasive manner. We can compare both the capacities and the constraints of mechanisms, for example, and we can compare features of singing in normal wild birds with those of trained birds.

These comparisons have been very instructive. For example, songs recur periodically in the performances of both wild and laboratory-trained birds, and song types occur in sequential clusters (in either packages or context groups). Also, laboratory-reared birds use not only the matching and leading responses of wild birds, but also a third kind of interactional response: when context groups are shared by two trained birds, a bird often matches context groups by switching to any song within that group. This pattern of interaction further supports the idea of context groups as subrepertoires.

Our data thus reveal that context groups are both units of memorization and units of retrieval from memory. Another observation that supports this idea is that individual repertoire delivery profiles can be best described as preferences to perform particular context groups rather than preferences to perform particular song types. This idea is supported by another finding, which concerns a rule about how many songs acquired from one context group are performed before a bird switches to imitations from another group. The length of such bouts is 5–15 renditions and is independent of the size of the context group. Thus, if a context group contains more than 15 song types, a context bout will contain only a subset of the song type constituents. If, on the other hand, a context group contains fewer than 5 song types, the bout will contain several renditions of particular song types.

This performance rule further supports the hypothesis that context groups are equivalent subassemblies located on the same hierarchical level of representation, and that they are a unit in song type retrieval. In addition, the rule contributes to our understanding of the role of these associations in the vocal communication of birds.

The capacity of nightingales to acquire, organize, and deliver repertoires is truly remarkable. Our conceptual model of the mechanism that presumably underlies the development of song type associations was tested in the laboratory with the aim of explaining how package groups are self-induced and how the context groups are exposure induced. Our study of developmental trajectories in song ontogeny led us to conclude that both song types and context groups are significant units of memorization and retrieval. Although they maintain their integrity well into adulthood, these units nevertheless display marked flexibility, which can be interpreted as an adaptation to the problems and purposes of communication.

Conclusions

Our studies have revealed a number of the mechanisms by which nightingales deal with diversity and versatility. The memorization of large repertoires, for instance, seems to profit from a structural hierarchy (hierarchy of embedment), and the performance of songs appears to be facilitated by a decisional hierarchy (hierarchy of connection). Both strategies are probably important for species that must acquire, manage, and perform large vocal repertoires. The structural hierarchy is evident from our results on sequential associations among song types, formerly described as sequential clusters but now distinguished as packages and context groups. The decision-making hierarchy, on the other hand, can be postulated from our findings on the dynamics of singing processes. It is reasonable to assume that the organizational principles described above are used by other species, too. We therefore think it is important to look for these phenomena among other songbirds so that we can obtain an overall view of how they acquire and orchestrate their vocal repertoires.

Our studies of the singing behavior of nightingales invite further research on how individual songs are organized and whether different types of songs encode different sorts of information in this species. The possibility that particular song types carry unique messages has been documented for several oscines (Lein 1978b, Kroodsma 1988b, Spector et al. 1989, Staicer 1989, Lemon et al. 1994).

Such an issue points to comparative aspects that could include learned vocal accomplishments of other bird species as well, such as Grey Parrots (*Psittacus erithacus;* Todt 1974, Pepperberg 1993).

Finally, we would like to know how songs and song performances are controlled by the brain, especially in birds with large repertoires. Where and how is the phenomenon of song type periodicity generated? What sort of neural network can explain how matching songs are chosen during a performance? Could this network be related to how repetition of certain song types is facilitated or inhibited? Why and when does the brain generate and use the time windows that play a remarkable role in so many areas of song acquisition and performance? And in the central nervous system, how are the rules of hierarchical organization represented? Are the rules represented structurally or physiologically, and if so, where? Given the remarkable recent progress in neurobiological approaches to song learning, it seems reasonable that at least some of these questions can be addressed in the near future (Konishi 1989, Marler 1991c, Doupe 1993, Nottebohm 1993).

Birdsong also merits particular attention from a more interdisciplinary perspective. We are intrigued, for example, by comparisons between development of song in birds and verbal behavior in human beings (Marler and Peters 1981). Such comparisons are heuristically interesting, especially if they are conducted on a formal level and by means of testable paradigms. We addressed this issue by comparing the songs of birds to the sentences of humans (Todt and Hultsch 1994). We found differences, of course, but we also found two intriguing similarities.

First, songs, like sentences, are units of interaction characterized by a mutual change of roles. Also like sentences, songs may end in a way that facilitates turn taking (Osherson and Lasnik 1990). Additionally, both sorts of units seem to be shaped in length by two opposing forces. They should be long enough to transport the intended messages, but not so long that a potential answer or reply is delayed. Perhaps these interactional rules explain why the basic length of both songs and sentences is about three seconds.

Second, songs clearly are units of acquisition, and song type packages probably form because there are constraints on the capacity of short-term memory. This chunking of information is remarkably similar to that observed in human short-term memory. What Bower (1970) termed "divide and conquer" as a strategy of the adult human to manage a larger number of serial data items seems also to hold true in the field of song learning. The short-term memory capacity in humans is constrained to about seven items, and such items form a concrete unit of information (Simon 1974). If it is true that the basic manifestation of such a unit is the sentence, then sentences are clearly similar to songs in this respect.

It is clear, of course, that similarities between songs and sentences should not be interpreted as results of homologous mechanisms. On the other hand, such similarities may well reflect analogous properties of the mechanisms that have evolved through selection to optimally process and use information during communication. Songbirds, which are able to manage large repertoires of signals, provide excellent models for research on the biological basis of cognitive correlates of vocal communication.

Acknowledgments

We are grateful to the many people who joined our field studies, helped with raising and tutoring the birds, or took part in conducting experiments and analyzing data: Fern Duvall, Jörn Freyschmidt, Mathias Fischer, Kurt Hammerschmidt, Dietmar Heike, Marina Hoffmann, Claudia Jörg, Marie-Luise Kopp, Reinhart Lange, Beatrix Lunk, Harriet Plötz, Darja Ribaric, Katharina Riebel, Gisela Schwarz-Mittelstaedt, and Alexandra Wistel-Wozniak. Jörg Böhner, Pat Hawley, Oliver Todt, and Stefanie Wendland provided much-appreciated comments on an earlier draft of the manuscript. Finally, we thank Don Kroodsma and Ted Miller for their invaluable help; without their substantial work on this chapter it would never have achieved the form and style it now has. The research reported here has been supported by several funds of the Deutsche Forschungsgemeinschaft (To 13/15 to To 13/21; Hu 376/3-1, -2).

6 Acoustic Communication in Parrots: Laboratory and Field Studies of Budgerigars, *Melopsittacus undulatus*

Susan M. Farabaugh and Robert J. Dooling

Vocal learning is thought to have evolved independently in at least three orders of birds (Nottebohm 1972, Kroodsma 1982a): Passeriformes (Kroodsma 1982a), Psittaciformes (Gramza 1970, Nottebohm 1970, Rowley and Chapman 1986), and Apodiformes (Trochilidae; Snow 1968, R. H. Wiley 1971, Baptista and Schuchmann 1990). Vocal learning may also have evolved in the Piciformes (Ramphastidae; E. O. Wagner 1944). Most studies of vocal learning in birds have been concerned with the learned song of oscine passerines, a taxonomic group that accounts for nearly half of all bird species. Of the three nonpasserine orders, the Psittaciformes has the most species for which vocal learning has been documented, and it is likely that most of the more than 330 parrot species learn at least part of their vocal repertoire. Further, many parrots have a complex vocal repertoire and are capable of intra- and interspecific vocal imitation throughout their lives. Comparing vocal learning in songbirds and parrots may lead us to a better understanding of its evolution and the basic biological processes that underlie its expression.

In contrast with the vast number of field and laboratory studies on the ecology, behavior, and vocalizations of songbirds, few such studies have been done on parrots, partly because it is so difficult to follow and record individuals that live in large, mobile flocks. In spite of this, some remarkable fieldwork has provided important information on parrot ecology, on vocal behavior such as the structure and function of different vocal signals (Hardy 1963, Brockway 1964a, b, 1969, Zann 1965, Brereton and Pidgeon 1966, Power 1966a, Cameron 1968, Wyndham 1980b, Pidgeon 1981, D. A. Saunders 1983, Rowley 1990, McFarland 1991b), and on individual repertoires, geographic variation (D. A. Saunders 1983), and vocal learning and recognition (Rowley 1980, Rowley and Chapman 1986). The most detailed studies of parrot vocal communication have been performed in the laboratory, including studies of the perception, production, and development of learned vocalizations (e.g., Todt 1975a, b, Dooling et al. 1987a–c, Pepperberg 1990, Farabaugh et al. 1992, 1994). Regrettably, most of the species that have been studied in the field have not been studied in the laboratory, and vice versa. Detailed laboratory and field studies of the same species would provide data for a thorough comparison of vocal learning in parrots with that in songbirds.

At present, the best candidate for a comparison of this type is the Budgerigar (*Melopsittacus undulatus*), a small flock-living parakeet native to central Australia. A popular cage bird throughout the world, it is probably the most studied of all the parrot species. Wyndham (1980a–e, 1981, 1983) conducted an extensive field study of its behavior and ecology, and domesticated Budgerigars have been the subjects of numerous laboratory investigations of their breeding behavior (e.g., Brockway 1964a, b, 1969, Stamps et al. 1985, 1987, 1989, 1990, Baltz and Clark 1994) and their vocal behavior and learning, hearing, and auditory perception (e.g., Brockway 1964a, b, 1969, Dooling and Saunders 1975, Park and Dooling 1985, 1986, Dooling et al. 1987a–c, Okanoya and Dooling 1987, 1990, 1991, S. D. Brown et al. 1988, S. D. Brown and Dooling 1992, 1993, Farabaugh et al. 1992, 1994). In this chapter we describe the life history of the Budgerigar, integrating data on vocal communication with information on the species's biology and ecology. We relate this information to the behavior and ecology of other parrots and compare vocal learning in Budgerigars with that in other parrots and in songbirds.

What Do We Know about the Behavior and Ecology of Wild Budgerigars?

The Budgerigar is the most nomadic of the parrot species inhabiting inland Australia (Rothwell and Amadon 1964, Serventy 1971, Rowley 1974). For this reason, Wyndham did not attempt to band birds and follow known individuals, although his many papers (1980a–e, 1981, 1983) provide a great deal of information about the species. We hope to build a picture of Budgerigar behavior and ecology by gleaning information from Wyndham's papers and by offering equivalent data (when appropriate) from individually marked birds of other Australian parrots. The four species of inland Australian parrots of which detailed studies of marked individuals have been made are Galahs (*Eolophus roseicapillus;* Rowley 1980, 1983a, 1990, Buckland et al. 1983, Rowley and Chapman 1986), Pink Cockatoos (*Cacatua leadbeateri;* Rowley and Chapman 1986, 1991), Slender-billed Black-Cockatoos (*Calyptorhynchus latirostris;* D. A. Saunders 1974, 1982, 1983, 1986), and Western Corellas (*Cacatua pastinator;* G. T. Smith 1991).

How comparable are these species to the Budgerigar? The arid interior of Australia experiences extremely variable productivity. Years with plentiful rain and massive plant growth are interspersed with dry years or even severe drought. Rowley and Chapman (1991) described a range of strategies that Australian parrots use to survive dry years. At one extreme are species that escape to better areas when local conditions deteriorate, and at the other are species that endure by depending on their knowledge of a wide number of specialized foods and watering places. Large cockatoos such as Pink Cockatoos and Western Corellas are examples of enduring specialist species, and smaller seed-eating parrots such as Galahs and, especially, Budgerigars are examples of nomadic species (Wyndham

1980a, 1983, Rowley 1990, Rowley and Chapman 1991, G. T. Smith 1991). Of the four well-studied species of interior Australia, Galahs' ecology is most similar to that of Budgerigars, and we will frequently mention Galahs in the following sections. It is important to be aware, however, that Galahs (especially breeding adults) are much more sedentary than Budgerigars.

Budgerigar Early Development

The nestling period. The developing Budgerigar experiences a complex social environment. Hatching is asynchronous, and clutches are composed of chicks of different ages. In a large clutch, the eldest can be 12 days older than the youngest. Both parents feed the nestlings, but males and females differ in their feeding behavior. Females feed the youngest chick first, but males feed the noisiest chick first (Stamps et al. 1985, 1987, 1989, 1990). As the chicks grow they become more interactive. Nestlings preen each other and their parents, and older nestlings sometimes feed younger ones. Budgerigars gain independence shortly after fledging, and we therefore expected learning and production of adultlike vocalizations to begin early in life in this species. By following the development of Budgerigar vocalizations from hatching through independence (Powell 1993), we found that the food-begging calls of nestling Budgerigars undergo a sequence of dramatic transformations. The young Budgerigar grows from a tiny hatchling that passively produces high-frequency peeps to a fully grown nestling that actively begs for food with loud, structurally complex calls that vary in duration and intensity. In the week before fledging, each chick develops an individually distinctive, stereotyped food-begging call.

Social interactions in the nest are important for both vocal learning and social development. Birds deafened as nestlings or reared in isolation exhibited both abnormal behavior and abnormal vocalizations (Dooling et al. 1987b, Farabaugh et al. 1992, Powell 1993). Budgerigars reared in isolated groups developed normal behavior and vocalizations after joining a flock, but Budgerigars reared alone were never entirely normal and showed evidence of social bonding to inanimate objects (e.g., warbling to the colored plastic base of a water tube) or parts of their own bodies (e.g., courtship feeding their own feet; Farabaugh et al. 1992). Through social interactions in the nest with parents and siblings, young Budgerigars probably learn to associate their own vocalizations with food, social contact, and physical characteristics of the body, especially the adult face.

The importance of early social interactions for learning is evident in field observations of naturally cross-fostered Galahs (Rowley and Chapman 1986, 1991). Pink Cockatoos sometimes displace nesting Galahs from their nest holes after the Galahs have begun laying, and as a result, the cockatoos may raise a young Galah with their own chicks. Galahs raised by Pink Cockatoos produced contact calls like their foster parents' calls and very different from the contact calls of Galahs. The two species differ in appearance: Pink Cockatoos are large and light pink, with white wings and a large yellow-and-red crest; Galahs are smaller with dark gray wings, a rose-colored breast, and a short pink crest. Yet, such naturally cross-fostered birds continue to associate with their foster species for many years (possibly all their lives), even though they encounter their own species daily. The cross-fostered Galahs even adopt the more specialized diet of the Pink Cockatoos (Rowley and Chapman 1986, 1991). In parrots, early social contact involves important visual and vocal associations that guide learning throughout life.

The fledgling period. In Budgerigars and Galahs, fledging represents the beginning of separation from the parents and nestmates. The fledged young immediately join a juvenile flock, or crèche (Rowley 1980, 1983a, Wyndham 1980b). A fledgling Galah's first flight, closely attended by its parents, is from the nest tree to the crèche (Rowley 1980). Because hatching is not completely synchronous and growth rates of nestmates differ in Galahs (Rowley 1990), and because hatching is asynchronous in Budgerigars (Brockway 1964b, Wyndham 1980b), the oldest birds fledge while their younger siblings are still in the nest. Parents of both species continue to feed their nestlings while making intermittent trips to the crèche to feed their fledglings. The parents give contact calls as they fly to the crèche, and their fledglings respond by calling and flying toward them. Budgerigars feed fledglings for no more than two weeks (Brockway 1964a, Wyndham 1980b); Galahs feed for up to six weeks (Rowley 1983a). In contrast, the young of the larger and more sedentary cockatoos remain with their parents for much longer periods, up to six months for Pink Cockatoos (Rowley and Chapman 1991) and as long as a year for Slender-billed Black-Cockatoos (D. A. Saunders 1982).

In the week before fledging, a Budgerigar nestling begins to produce food-begging calls composed of a repeated, individually distinctive, stereotyped pattern of frequency modulation (Powell 1993). At this point a nestling's food-begging calls vary only in the number of repetitions of its stereotyped pattern. This repeating unit becomes the bird's first contact call after fledging (Fig. 6.1) and is used for up to a month. It is significant that young Budgerigars develop individually distinctive vocalizations, which could be used by the parents to recognize their young, just before they fledge and become mobile. Although no field data exist regarding recognition of nestling vocalizations by parent Budgerigars, there is evidence that Galah parents recognize their nestlings. By switching broods of nestling Galahs of various ages between nests and observing the parents' behavior, Rowley (1980) found that parent Galahs could recognize their offspring, but not until their last week in the nest. It is reasonable to suggest that the onset of parental recognition of young Galahs may be based on voice, and that young Galahs may also develop individually distinctive begging calls. This issue could be settled by a careful study of the vocalizations of nestling Galahs, or by experimentally switching broods of Budgerigars. It would also be worthwhile to study early vocal development and parental recognition in other parrot species.

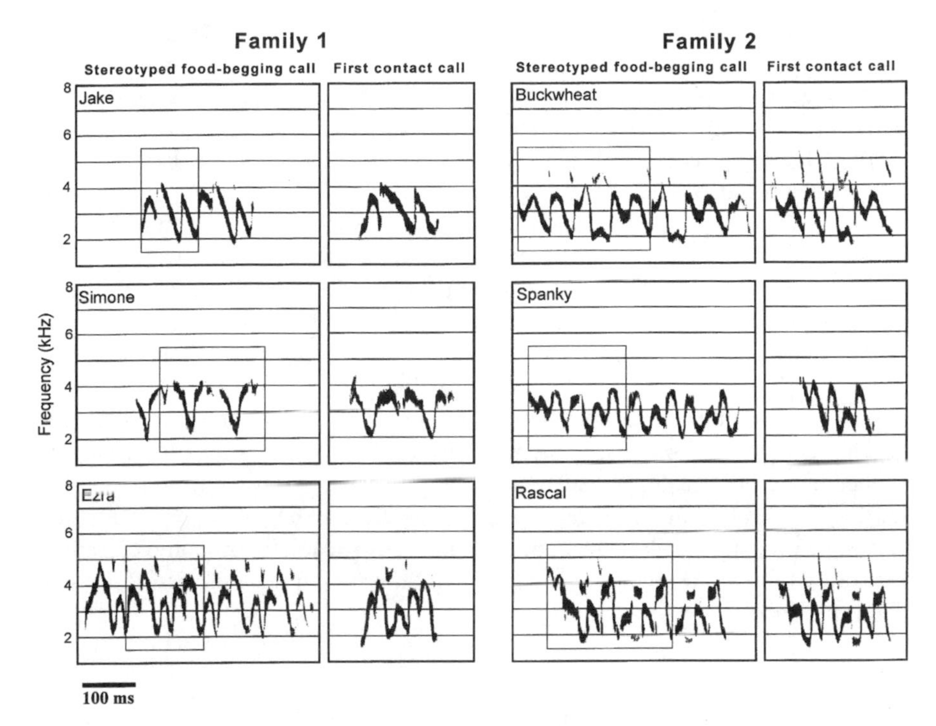

Fig. 6.1. Development of contact calls from food-begging calls in the Budgerigar (*Melopsittacus undulatus*). Sonograms of a stereotyped food-begging call and first contact call from each of six Budgerigars from two different nests are shown. A box has been drawn around the repeating pattern in the stereotyped food-begging call that is later seen in that same bird's first contact call. Also note that each bird has a distinctive pattern, but there are also similarities among the siblings (from Powell 1993). Sonograms in this figure and Figs. 6.2 and 6.3 were made on a Kay Elemetrics 7800 Sonagraph in expanded mode (1.28 seconds/sonogram sheet) with frequency range of 0–8 kHz, analyzing filter bandwidth of 300 Hz.

Independence. Galahs and Budgerigars become independent shortly after fledging. The long-term studies of tagged Galahs by Rowley and his colleagues provide detailed data on the movements of these birds after they reach independence. As the juvenile Galahs in the crèche begin to feed on their own, they join their parents in the foraging flock (Rowley 1980, 1990). Parents return to their nest trees to roost, and the juveniles roost together near the feeding area. Over time the juvenile flock drifts farther away, eventually leaving the natal area. During the rest of the year the juvenile flocks disperse widely. In their second year they settle and join a locally nomadic flock of nonbreeding birds hatched during the last two or three seasons (Buckland et al. 1983, Rowley 1983a). Thus, Galahs do not associate with their own young beyond independence. Independent young Budgerigars also join into flocks and leave the natal area (Wyndham 1980b). Budgerigars are highly nomadic, and even breeding adults join flocks and leave the area once breeding is finished. Flocks contain both juveniles and adults, but they are skewed toward one age class or the other (Wyndham 1980b), perhaps as a result of age differences in departure time from the breeding area. Juveniles depart as they gain independence, but adults depart only after the entire clutch has achieved independence. If adults renest, departure times are even more skewed.

Our laboratory studies (E. F. Powell and S. M. Farabaugh unpubl. data) indicate that the newly independent Budgerigar has a well-developed vocal repertoire. The repertoire of adult Budgerigars includes a number of acoustically and functionally distinct calls (e.g., contact and alarm calls) and a long, complex warble song. Shortly after becoming independent, young Budgerigars begin imitating the calls of other Budgerigars. Their imitative ability is plastic at this stage: normally reared Budgerigar fledglings that were individually housed with Zebra Finches (*Taeniopygia guttata*) imitated Zebra Finch calls (Powell 1993). Warble song also appears around this same time, but the timing of its onset varies considerably among individuals. Both males and females warble, but males warble at a much higher rate than females (Farabaugh et al. 1992) and begin warbling slightly earlier. Around the time of independence, females may begin to give courtship-feeding solicitation calls. Thus, Budgerigars have a relatively complete adult repertoire shortly after fledging, a time when they are joining their first flock and finding a mate.

Adult Life

Flocking. Flocking is a common social structure of parrot species worldwide (see Forshaw 1989) and is the key feature of the adult parrot's social milieu. Flock size varies with habitat and food availability both within and among species. Parrot species that feed on depletable foods that are rare and clumped, such as the fruits of tropical fruiting trees, live in small, diurnal feeding flocks (Chapman et al. 1989). In contrast, species such as the Galah and the Budgerigar, which feed on widely dispersed and abundant foods such as seeds of grasses and shrubs, tend to travel in large flocks (Rowley and Chapman 1991). A few parrot species are solitary, such as the granivorous terrestrial Ground Parrot (*Pezoporus wallicus*), which is found only in the restricted heathland habitat of southeastern Australia (McFarland 1991a–c).

The advantages of flocking are numerous. Cannon (1984) suggested that flocking in Australian parrots is a response to the availability and predictability of resources, and that increased sociality enhances an individual's ability to locate and exploit resources in a heterogeneous environment. Joining a flock may be particularly advantageous for immature and less experienced birds, because they can benefit from older flock members' knowledge of the location and availability of important resources. Younger birds may need to learn to recognize and handle certain foods. For example, hand-reared Yellow-tailed Black-Cockatoos (*Calyptorhynchus funereus*) showed no interest in logs infested with moth larvae, but birds collected as adults excavated them immediately (McInnes and Carne 1978).

For Budgerigars, membership in a flock may be essential for individuals (especially young birds) to find their way during nomadic travels over vast distances. Older experienced birds may play a role in guiding the flock. There is an underlying continental pattern to Budgerigars' seasonal movements, and some evidence shows that birds return to sites where they previously bred successfully (Wyndham 1983). Budgerigar flocks also are very responsive to environmental change, and in any particular year, area, or habitat with atypical vegetative growth, their movements and breeding seasons may differ drastically from the usual pattern (Wyndham 1983).

Flock members also gain protection from predators. Westcott and Cockburn (1988) found that individual vigilance decreased and corporate vigilance (e.g., the probability that at least one bird is scanning for predators) increased with flock size in Galahs and Red-rumped Parrots (*Psephotus haematonotus*). This pattern is common in group-living birds and mammals (Elgar 1989). Predator avoidance appears to be an important function of flocking in Budgerigars. When watching Budgerigar flocks flying in the wild, one is struck by their tight formation, synchrony, quickness, and extreme wariness (Farabaugh pers. obs.). If one approaches a flock of Galahs feeding on the ground, the nearest birds lift and fly to the far end of the group; if one approaches a Budgerigar flock in a similar manner, however, the entire flock explodes into the air with a burst of contact and alarm calls (Farabaugh pers. obs.). Detection of predators may be particularly important in Australia, with its high diversity and abundance of aerial predators.

Flock social organization. Parrot flocks are usually not random associations of individuals. Studies of tagged birds have shown that the membership of flocks remains fairly stable within and between years in Galahs (Rowley 1980, 1983a, 1990), Pink Cockatoos (Rowley and Chapman 1986, 1991), and Slender-billed Black-Cockatoos (D. A. Saunders 1982). At times, especially during drought years, several distinct flocks may combine into larger flocks of several hundred (Pink Cockatoos) or several thousand birds (Western Corellas, Galahs, and some cockatoos in the genus *Calyptorhynchus;* Rowley and Chapman 1991). When conditions improve and the breeding season approaches, smaller flocks reappear. Flock membership is relatively stable in Pink Cockatoos, but some birds change from one neighboring flock to another (Rowley and Chapman 1991). The rate of change of membership probably varies across species. Rowley (1980, 1983a, 1990) described three different kinds of Galah social flocks that represent different life stages, and each type differs in its range of movement and stability of membership. Juvenile flocks, made up of the young of the year from an area, disperse widely; older nonbreeders form local flocks that wander nomadically over approximately 1000 km^2; and breeding pairs from a particular patch of woodland form a stable resident flock that remains within a 10-km radius of their nest trees, to which they return daily. The breeders join the locally nomadic flock when it is in their area. Galahs change flocks several times during their lives, and Galah flocks differ in the stability of their membership and the duration of their association. Whether Budgerigar flocks have a stable membership is unknown. In areas where Budgerigar densities are low, however, small, cohesive flocks of 20–30 birds occur (Farabaugh pers. obs.). It is possible that the vast flocks found in areas of high productivity are temporary amalgamations of such small flocks. Budgerigars may maintain stable flocks for a time, but the duration of these associations and the rate at which birds join and leave are unknown.

The essential social unit within parrot flocks is the mated pair. Most parrots exhibit obligate monogamy; the male helps feed the young and may also feed the female during incubation and brooding (Wyndham 1981, D. A. Saunders 1982, Skeate 1984, Snyder et al. 1987, Forshaw 1989, Beissinger and Waltman 1991, McFarland 1991b). In some species, mates share in nesting activities, including incubation (e.g., Galahs: Rowley 1990; Pink Cockatoos: Rowley and Chapman 1991). Many pairs remain together year-round, and experienced pairs have greater reproductive success than newly paired birds (Rowley 1983b). Divorce (i.e., replacement of a mate when the former mate is still living) is rare in Slender-billed Black-Cockatoos (1%: D. A. Saunders 1982) and Pink Cockatoos (one case in 75 pair-years: Rowley and Chapman 1991), and moderate in Galahs (8%: Rowley 1983a, b, 1990) and Western Corellas (15%: G. T. Smith 1991). More than 31% of Venezuelan Green-rumped Parrotlets (*Forpus passerinus*) change mates from one breeding season to the next, and at least 25% of the replacements involve divorce (Waltman and Beissinger 1992). Neither the rate of mate replacement nor the rate of divorce is known for Budgerigars. Their behavior in captivity (Masure and Allee 1934, Brockway 1964a, b, Trillmich 1976b) and observations of unbanded breeding Budgerigars in the wild (Wyndham 1980b) indicate that most have monogamous pair-bonds that are maintained from one breeding season to the next. Male domestic Budgerigars attempt to court females other than their mates, but they are significantly less likely to do so if the mate is in sight (Baltz and Clark 1994). A prolonged pair-bond is advantageous for an opportunistic breeder like the Budgerigar because experienced pairs can initiate breeding quickly when ecological conditions become favorable.

Vocal Behavior of Budgerigars

The Adult Vocal Repertoire

Learned vocalizations are thought to play a significant role in the formation and maintenance of social bonds in parrots, both within mated pairs and among the members of a flock. Budgerigars have an elaborate learned vocal repertoire with which they coordinate social and reproductive behavior (Brockway 1964a, b, 1969, Wyndham 1980b, Dooling 1986, Dooling et al. 1987c, Farabaugh et al. 1992, 1994). The repertoire consists of several distinct calls (contact, alarm, nest defense, courtship-feeding solicitation, agonistic harmonic chips, and nestling food-begging calls) and a long, complex song composed of soft, melodious warbling with loud chirps and squawks interjected. Warble song is quite different from the song of most songbirds: it is highly variable in structure both within and between individuals, it has no set length, and it lacks a stereotyped order of syllables. Warble song synchronizes reproductive behavior between mates (Brockway 1965, 1967, 1969). Hearing or performing warble song promotes full gonadal activity in both sexes.

Of the various calls in the Budgerigar vocal repertoire, the most interesting in terms of vocal communication and vocal learning is also the most frequently used: the contact call. In many avian species that flock for all or part of the year, flockmates develop a shared flock-specific call, usually a contact or distance call. Budgerigars give contact calls repeatedly when they are in flight, are separated from the flock, reunite with their mates after separation, or prepare for evening roost (Wyndham 1980b). Contact calls are strongly frequency-modulated narrow-band sounds that range between 2 and 4 kHz and have a duration of 100–300 milliseconds. The patterns of frequency modulation are extremely varied, but an individual can repeat a particular pattern with great precision from one rendition to the next. At any given time, a Budgerigar has a repertoire of one to several different patterns, or contact call types, but usually one or two types account for 95–100% of all the contact calls produced (Farabaugh et al. 1994). Budgerigars and other parrots use the contact call to maintain contact with the mate and flock (Hardy 1963, Brockway 1964a, b, Zann 1965, Brereton and Pidgeon 1966, Power 1966a, b, Cameron 1968, Rowley 1980, Wyndham 1980b, Pidgeon 1981, D. A. Saunders 1983). Galahs and Slender-billed Black-Cockatoos recognize their mates and offspring by their contact calls (Rowley 1980, D. A. Saunders 1983). Learned contact calls may help Budgerigars to locate, recognize, and maintain contact with mates and flockmates within the swirling cacophony of larger feeding assemblages. To investigate these possible functions of learned vocalizations in Budgerigars we examined the social process of contact call learning in adults and studied the Budgerigars' auditory and visual perceptual abilities by means of operant testing.

Social Interactions and Adult Contact Call Learning

If contact calls act to form and maintain social bonds within the flock, adult Budgerigars should be able to learn new contact calls, flocks should develop a shared flock contact call type, and call learning should be guided by social interactions. We investigated adult contact call learning in six adult male Budgerigars from different flocks (Farabaugh et al. 1994). To establish that each bird had a distinctive repertoire of contact call types (Fig. 6.2, left), we recorded their contact calls for several days. The birds were then placed into a large cage divided into two sections by a thin cloth partition. Three birds were placed on one side of the partition and three on the other, so that birds could hear but not see those on the other side. Over the next eight weeks, we recorded contact calls every three to five days from each bird. Within a week we found evidence of vocal learning: one bird

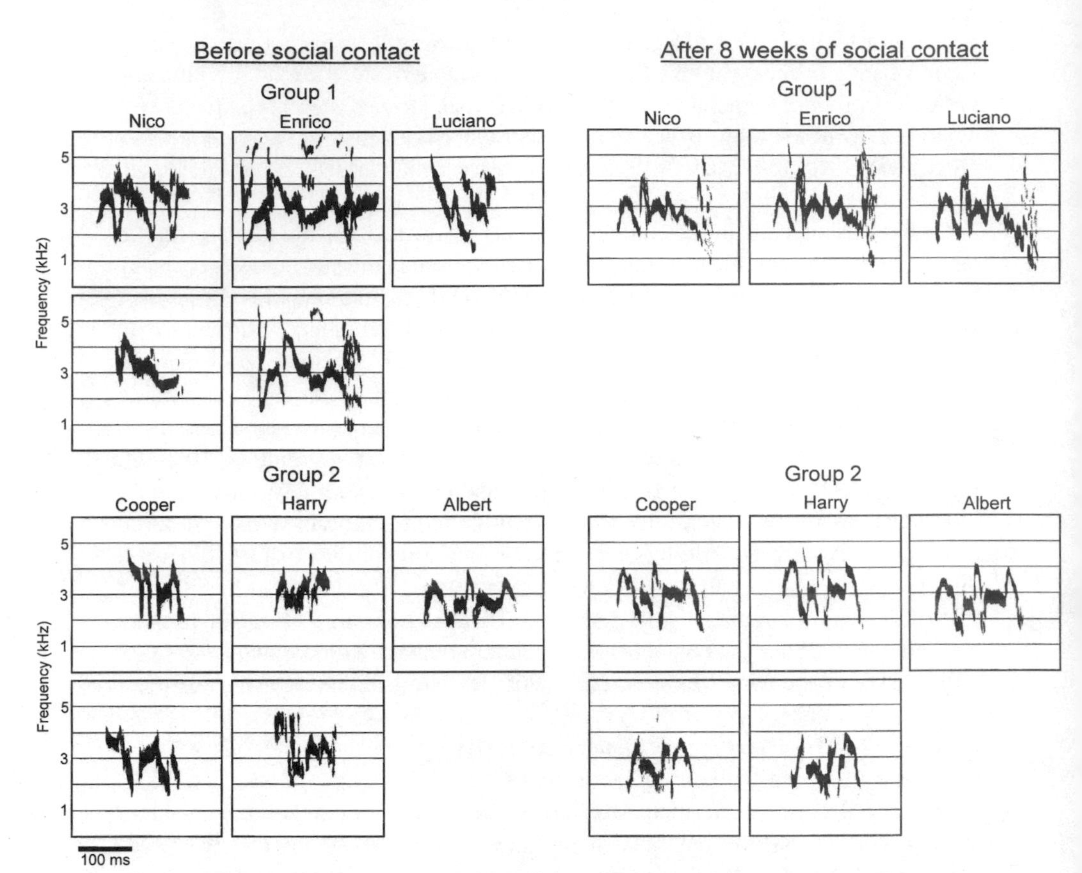

Fig. 6.2. Social interactions play a critical role in the development of flock-specific contact calls. Left panel: sonograms of the dominant contact calls of six male Budgerigars before social contact. Two birds (Luciano and Albert) had a single dominant contact call type that accounted for 97–99% of their recorded contact calls. Four birds (Nico, Enrico, Cooper, and Harry) each had two dominant contact call types that together accounted for 91–96% of their recorded contact calls. No contact call types were shared by all six birds. Right panel: The dominant contact calls of the same six birds after being housed for eight weeks in two groups of three birds, the groups in adjoining cages. Birds could see, hear, and physically interact with their groupmates but could only hear the birds in the other group. Each group of three birds now shared a dominant contact call type. Two of the birds in Group 2 (Cooper and Harry) had two dominant contact call types, and they shared both. Dominant contact-call types were different in the two groups. After Farabaugh et al. (1994).

sharing a contact call with another bird in the same group. Sharing began as imitation of another bird's call type but quickly developed into new call types composed of sections of call types from different birds. At the end of eight weeks, three birds on one side of the partition were sharing the same dominant contact call type, and three birds on the other side of the partition were sharing different ones (Fig. 6.2, right). There was only slight evidence of sharing across the partition (the dominant call of one group was recorded rarely from two birds in the other group). Thus, in the absence of social (but not aural) contact, vocal learning of contact calls was reduced greatly.

These results show that social interactions play a large (perhaps critical) role in learning contact calls. Socially directed contact call learning in Budgerigars corresponds closely with Pepperberg's (1985) social modeling theory of avian vocal learning. Young Budgerigars forming their first flock as they leave the crèche will soon develop a learned shared contact call. This shared call can act as a badge of flock membership and aids in synchronizing and coordinating the movements and activities of flock members (Treisman 1978). Some songbird species that form stable flocks for part of the year also show evidence that their species-specific contact calls are learned and that they can use these calls for group recognition (Mundinger 1970, 1979, M. S. Ficken et al. 1978, Mammen and Nowicki 1981, Nowicki 1983, 1989, Hailman et al. 1985). Social interactions also influence song learning in songbirds (see the reviews by Petrinovich 1988b and Baptista, this volume) and are particularly important in species that continue to learn new songs as adults. During its life, a parrot species such as the Galah moves through several different social groups—from family group to juvenile crèche, to juvenile flock, to local nomadic flock, to resident breeding flock (Rowley 1990). Field evidence shows that Galahs and other parrots share a common call type with their flockmates (D. A. Saunders 1983, Rowley and Chapman 1986). A Budgerigar may pass through even more social groups during its life and may have an even greater need for vocal plasticity in adulthood.

Budgerigar Perceptual Studies

Auditory Perceptual Foundations of Vocal Learning

Most of what we know about the acoustic basis of individual, group, and species recognition in birds has come from field playback studies (see the review by Falls 1992). Unfortunately, few such studies have been performed on parrots (Rowley 1980, D. A. Saunders 1983) and there have been none on Budgerigars. All the information relating to vocal recognition by Budgerigars comes from numerous laboratory investigations on hearing and auditory perception. Like most birds, Budgerigars hear best in a relatively narrow spectral region of about 1–5 kHz (Dooling 1991), with absolute thresholds approaching 3–5 dB sound pressure level (SPL) at about 3 kHz (Dooling and Saunders 1975). Temporal resolving power in Budgerigars and other small birds approaches the limits of human sensitivity (Dooling 1982, 1991), but Budgerigars appear to have enhanced spectral resolving power between 2 and 4 kHz (Dooling and Saunders 1975, Okanoya and Dooling 1987, 1991). Most energy in Budgerigar vocalizations is in the region of 1–5 kHz, with energy in contact calls in the region of 2–4 kHz (Dooling 1986).

Budgerigars discriminate readily among a wide variety of Budgerigar calls and show natural perceptual categories for the most common acoustic and functional classes of vocalizations in their repertoire (Dooling et al. 1987c). Budgerigars reared in acoustic and social isolation and then tested with these same calls showed subtle differences from normally reared birds in their perception of vocal categories (Dooling et al. 1990). Budgerigars also discriminated among Budgerigar contact call types more efficiently (they responded more rapidly) than among different Island Canary (*Serinus canaria*) or Zebra Finch contact calls. In other conditioning experiments requiring classification of contact calls from memory, Budgerigars were shown to remember and classify upward of 30 distinct contact call types nearly perfectly, even when the calls were significantly degraded, and there was no evidence that the auditory memory limit had been reached (Park and Dooling 1985, 1986). This performance level was maintained even when contact calls were increased in frequency by 300 Hz with most energy still in the region of 2–4 kHz, but not if the calls were increased by 2 kHz with most of the energy outside 2–4 kHz. These experiments showed that Budgerigars have excellent resolving power in the spectral region of 2–4 kHz and can make fine discriminations among frequency-modulated contact calls, and that they have the auditory capacity to learn, remember, and detect contact calls. Their auditory abilities may enable Budgerigars to detect, recognize, and locate their mates and flock in the confusion and noise of a vast feeding amalgamation.

Budgerigars clearly have the auditory abilities necessary for vocal recognition. They call more readily in response to their mate's call than to the calls of other Budgerigars (Ali et al. 1993). The limits of the Budgerigar's ability to recognize individuals by their contact calls were investigated by testing whether birds could discriminate among different birds' renditions of the same shared contact call type (S. D. Brown et al. 1988). In this experiment Budgerigars were presented with a set of four renditions of a shared contact call type by each of three birds that lived together (Fig. 6.3, top). The 12 calls sounded very similar. The three cagemates whose calls these were and three other Budgerigars in an adjoining cage were given an operant conditioning task designed to find the birds' natural perceptual categories for these calls. Calls were presented two at a time; the bird pecked a key if the calls were different and withheld a response if the calls were the same. In this task, birds pecked rapidly when they easily perceived the calls as different, and more slowly if they perceived the calls as similar; thus, the latency of their response was a measure of how similar the calls seemed to them. Only the cagemates consistently pecked more rapidly when the two calls presented were from different birds than when the calls were from the same bird, thus showing that cagemates could discriminate among individuals (Fig. 6.3, bottom left). All 12 calls sounded similar to the three other Budgerigars that had heard the calls but did not share them (Fig. 6.3, bottom right). A fine-grained acoustic analysis of these calls showed that subtle frequency and temporal differences among cagemates' calls was correlated with the cagemates' perceptual responses (S. D. Brown et al. 1988). These results reinforce the notion that Budgerigars, and possibly many other flock-living parrots, can use the learned shared contact call for both group and individual recognition. This ability may be advantageous for a

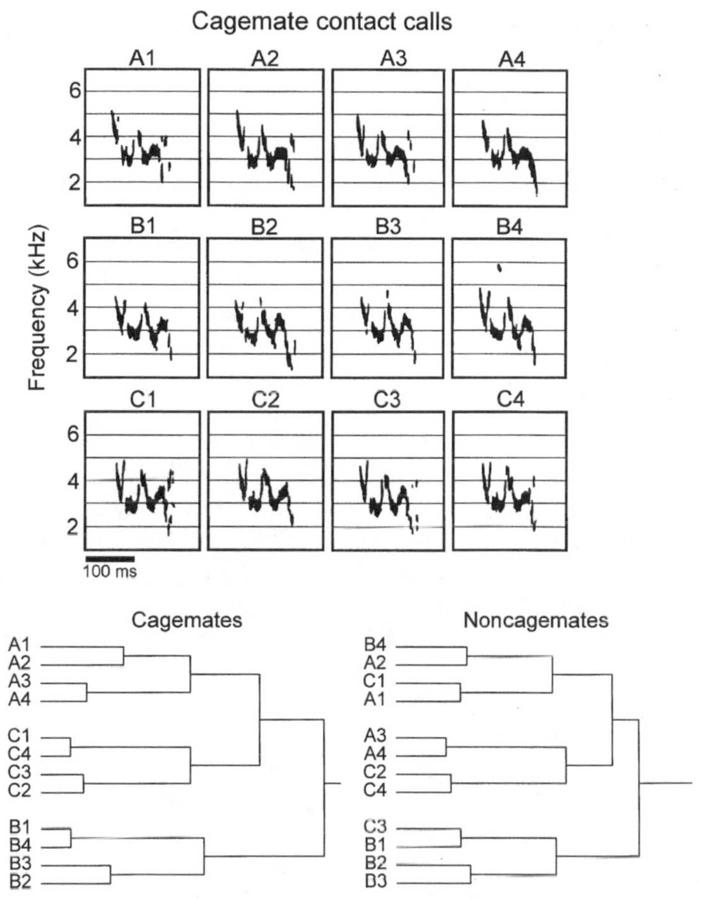

Fig. 6.3. Flock members can use shared contact calls for both individual and group recognition. Top: Sonograms of four renditions of a shared contact call type from each of three Budgerigars that lived in the same cage (birds A, B, and C). For example, A1–A4 are four contact calls recorded from bird A. These 12 calls were presented in an operant discrimination task to the three cagemates that produced the calls, and to three other Budgerigars (noncagemates) from an adjoining cage that did not have this call type in their repertoires. Bottom: Dendrograms from a hierarchical cluster analysis of the average response latencies of the three cagemates and three noncagemates discriminating among the 12 calls shown above. Shorter latencies reflect larger perceived differences in the stimuli and thus longer distances on the dendrogram. Cagemates responded consistently faster when the two test calls were from two different birds, and thus showed that they could discriminate among individuals by their contact calls, whereas noncagemates could not. After S. D. Brown et al. (1988).

flocking bird because the same call type can be used to find the flock and one's mate within the flock. Moreover, these data also indicate that this perceptual ability is learned. It is not simply a result of exposure to the calls, because noncagemates could hear the calls of the cagemates but failed to discriminate by individual identity. Rather, we concluded that nonacoustic factors involving production or use of these calls in an appropriate social context are what alters perception of the calls. Thus perceptual learning, like call learning, may require a social context.

Visual Basis of Individual Recognition in Budgerigars

Many parrots accompany their vocalizations with elaborate visual displays (Dilger 1960, Hardy 1963, Power 1967, D. A. Saunders 1974, Serpell 1981, Skeate 1984, Snyder et al. 1987, Waltman and Beissinger 1992). Under normal circumstances, individual recognition, social interaction, and vocal learning involve myriad nonacoustic cues. Rowley and Chapman's (1986) study of naturally cross-fostered Galahs suggested that visual cues may guide vocal learning throughout life. Although we know little of the interplay of vision and audition during vocal learning, some experimental results indicate that visual cues are important for recognition in Budgerigars. Trillmich (1976a) demonstrated that visual features of the face are important, and more recent operant experiments (S. D. Brown and Dooling 1992) showed that Budgerigars easily discriminated between Budgerigar and Zebra Finch faces and between the faces of male and female Budgerigars, but had considerable difficulty discriminating between the faces of male and female Zebra Finches (Fig. 6.4). Thus, Budgerigars are more sensitive to visual features of conspecific faces than they are to the faces of other species, such as Zebra Finches. In operant experiments using synthetic faces, S. D. Brown and Dooling (1993) showed that Budgerigars may use facial characteristics for recognition of age, sex, and individual identity. Interestingly, these cues became less salient when presented in inverted or scrambled faces, strongly suggesting that Budgerigars view the upright, unscrambled synthetic faces as Budgerigar faces rather than as a collection of unrelated visual cues. An isolate-reared Budgerigar tested with the same synthetic faces showed no advantage in discriminating among upright, unscrambled faces over inverted or scrambled faces. This preliminary result supports the idea that biologically important visual perceptual categories, like auditory perceptual categories, can be readily learned. Learned visual perception may be a factor in the misimprinting of cross-fostered Galahs. The young Galahs probably developed a visual perceptual category for the faces of the Pink Cockatoo foster parents rather than for the faces of their own species.

Vocal Communication in Budgerigars and Other Parrots

How well do Budgerigars represent vocal learning and communication in parrots? This question is difficult to answer because few comparably detailed data are available for other parrot species. We believe that Budgerigars exhibit some of the basic features of vocal communication in flock-living parrots: a repertoire consisting of an array of functionally distinct calls, a learned contact call that can be used

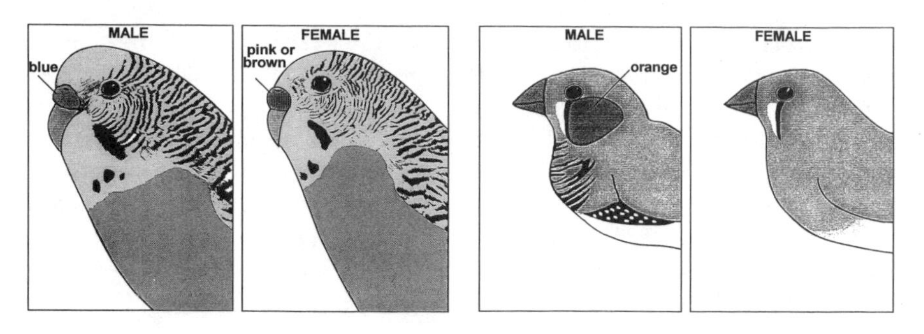

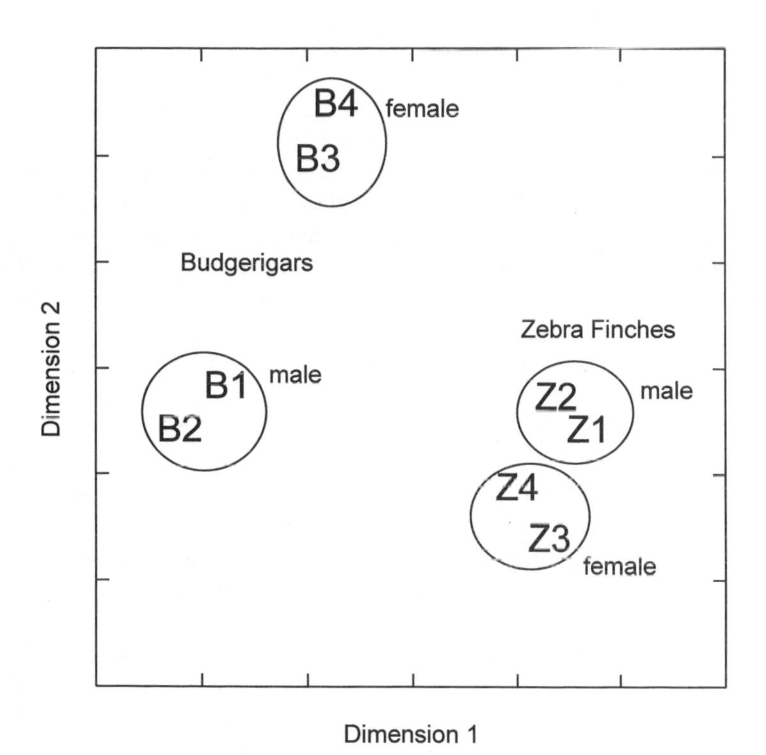

Fig. 6.4. Perception of Budgerigar faces by Budgerigars, as shown by multidimensional scaling. Top: Black-and-white schematic representations of Budgerigar and Zebra Finch faces used as stimuli in the visual perceptual tests. The actual stimuli were color photographic slides of two males and two females of each species. Bottom: two-dimensional spatial map of the combined response latencies of four Budgerigars tested on Budgerigar and Zebra Finch faces. In this mutidimensional solution, interstimulus distance accounts for an average of 77% of the variance in response latencies, with distances along the first and second dimensions accounting for 57% and 19%, respectively. Thus Budgerigars discriminated easily between Budgerigar and Zebra Finch faces (species well separated on the first dimension) and between male and female Budgerigar faces, but they discriminated with difficulty, if at all, between male and female Zebra Finch faces—stimuli that human subjects find easy to discriminate. After S. D. Brown and Dooling (1993).

for flock and individual recognition, and the ability to learn new vocalizations as adults. Other features of the Budgerigar repertoire, such as the size of the functional call repertoire and the presence of a complex song, may not be representative of parrots in general. Further, Budgerigars' vocal communication differs greatly from that of solitary parrots with very simple vocal repertoires, such as the Ground Parrot (McFarland 1991b), and from parrots that have very different mating systems, such as New Zealand's flightless nocturnal Kakapo (*Strigops hapbroptilus*), which uses leks (Merton et al. 1984). Male Kakapo advertise their presence on the lek by booming—a loud, rhythmic sound produced by extreme inflation of thoracic air sacs (Merton et al. 1984), possibly the most unusual parrot vocalization known.

Brereton (1971) considered the complexity of the vocal repertoire of Australian parrot species in relation to their social structure and concluded that species organized into small, stable flocks have more complex repertoires than extremely gregarious species such as the Budgerigar and the Galah. Brereton contended that the large repertoire of birds living in stable flocks reflects the greater amount of information needed to maintain such a complex social organization. Pidgeon (1981) found support for Brereton's idea by comparing repertoires from his own work with descriptions of vocal repertoires of other parrot species. Rowley (1990) pointed out that these analyses were limited by the subjective estimation of how many calls a species uses; one researcher may count a graded series as one call, and another may recognize the same series as two or more. It would be difficult to reanalyze the published studies of parrot vocal repertoires, because most include little or no sonographic analysis, no description of inter- or intraindividual variation, and no mention of which, if any, vocalizations were learned. Brereton's (1971) general conclusion may be true, but it has not yet been rigorously tested. Perhaps these questions would be best addressed by a combination of laboratory and field studies. Detailed recordings of an individual's complete repertoire are easiest to get in captivity, but essential data on social structures must come from the field.

What is so unusual about Budgerigar song? Warble song, or its equivalent, appears to be lacking in Australian parrots that have a stable annual breeding season (Zann 1965, Brereton and Pidgeon 1966, Cameron 1968, Rowley 1980, Pidgeon 1981, D. A. Saunders 1983, McFarland 1991b). Brockway (1965, 1967, 1969) made extensive investigations of the role of warble song in the synchronization of breeding between mates. Budgerigars can breed several times a year when conditions are good, so they may not use external cues such as daylength to regulate their breeding. Warble song may be an adaptation to stimulate gonadal development in nonseasonal breeders. More intensive studies are required to determine whether such complex vocalizations are actually lacking in other parrot species or are simply performed rarely. For example, female Budgerigars warble-sing much less frequently than males do, and their singing would probably go unnoticed by many observers in the field. No song has ever been reported for the seasonally breeding Galah (Pidgeon 1981, Rowley 1990). Rowley (1990) mentioned soft "murmurations" given by mated pairs or by parents to their nestlings, but he was unable to make sonograms. If these soft sounds represent Galah song, Budgerigars still differ in having a conspicuous song that males sing at a high rate. We must study other parrots that breed opportunistically to determine if high rates of song are common.

Vocal Communication in Parrots and Songbirds

Sonograms of parrot vocalizations may at first appear complex and unclassifiable to a researcher accustomed to working with songbird vocalizations. One is faced with a repertoire of calls that probably have distinct contexts and functions. Classification of the calls is a challenge because some functionally distinct calls may have an array of structural variants. Some parrot species may also have a complex song. Parrot song lacks the stereotypy of oscine song: it is composed of syllables that are variable in acoustic structure, it has no stereotyped syllable order, and it has no set song length. A complex oscine song such as that of the Common Starling (*Sturnus vulgaris*) is comparatively easy to analyze.

Perhaps the vocalizations of parrots and songbirds are so different because the neural control of vocal learning evolved independently in the two groups. The vocal control pathways of both parrots and songbirds include specialized nuclei in the forebrain that are lacking in avian species without vocal learning (e.g., suboscine passerines: Kroodsma and Konishi 1991; pigeons: Wild 1994a, 1994b; see reviews by Ball 1990, Brenowitz 1991a). Because of this similarity, it was once thought that the neural circuitry involved in learned vocalizations was homologous in parrots and songbirds (Paton et al. 1981, DeVoogd 1986). Recent neuroanatomical and neurochemical studies of these special vocal control nuclei in the Budgerigar, however, suggest that the similarity is superficial and that the vocal control nuclei of the two groups are not homologous (Ball 1994, Striedter 1994). Compared with songbirds, parrots' vocal control nuclei are in a different anatomical location (Paton et al. 1981), receive auditory input from a different auditory pathway (Brauth et al. 1987, 1994, Brauth and McHale 1988, Streidter 1994), and lack some of the steroid receptors that are abundant in the songbird vocal control nuclei. Gahr et al. (1993), for example, found no estrogen receptors in Budgerigar vocal control nuclei but did not look for androgen receptors (see Ball 1994). Other components of the vocal control pathway may be homologous in parrots and songbirds because these components have been found in all birds (including groups that lack vocal learning) that have been studied to date (Ball 1990, 1994, Brenowitz 1991a, Kroodsma and Konishi 1991, Wild 1994a, b). Connections of these homologous portions differ, however; innervation of the syrinx is lateralized in songbirds, but each side of the parrot syringeal muscle mass receives input from both right and left motor nuclei and right and left sides of the brain (Nottebohm 1976, Manogue and Nottebohm 1982, Brauth et al. 1994, Heaton et al. 1995).

It is not yet clear how these differences in the vocal control nuclei and pathways relate to differences in the learned vocalizations of parrots and songbirds, and this is an exciting area for future research. For example, the differences in the distribution of steroid receptors in the nuclei of the parrot vocal-learning pathways may reflect essential differences between parrots and songbirds with respect to hormonal control of learned vocalizations. In many songbird species, song is restricted to the male and song control nuclei are sexually dimorphic, a difference that is produced during ontogeny by steroid hormones (see the review in Arnold 1992). In songbird species in which both sexes sing in defense of a permanent territory (Farabaugh 1982), the size and steroid sensitivity of the vocal control nuclei differ little, if at all, between the sexes (Brenowitz et al. 1985, Brenowitz and Arnold 1985, Arnold et al. 1986). The vocal repertoires of male and female Budgerigars are very similar, and there is little sexual difference in the size of the brain nuclei associated with vocal learning, although detailed measurements have yet to be published. Singing rates differ, however: males sing much more often than females. This sexual difference appears to be the result of differences in circulating testosterone levels in adults rather than neural differences mediated by hormones during development. Less than two weeks after receiving testosterone implants, female Budgerigars increased their rate of warbling to malelike levels, showed changes in cere color to the male condition (blue), and exhibited malelike precopulatory behavior (Nespor et al. 1994). The development of neural pathways for vocal control and learning in parrots and songbirds appears to be controlled by different mechanisms.

So far we have stressed differences in vocal communication between parrots and songbirds, but intriguing functional similarities occur as well. Some songbirds that live in stable social flocks for at least part of the year learn contact calls as adults, develop flock-specific calls, and can discriminate their flockmates' calls from those of birds from other flocks (e.g., Black-capped Chickadees, *Parus atricapillus:* Mammen and Nowicki 1981, Nowicki 1983, 1989). In parrots and flock-living songbirds, learned contact calls function in the formation and maintenance of social bonds between flock members (Mundinger 1970, 1979, M. S. Ficken et al. 1978, Mammen and Nowicki 1981, Nowicki 1983, 1989, Hailman et al. 1985; for a review, see Farabaugh et al. 1994). Song in parrots is mainly affiliative and does not appear to have a territorial function. Songbirds with complex societies (e.g., group-living species) also have shared learned songs that function either partly (e.g., Australasian Magpies, *Gymnorhina tibicen:* E. D. Brown et al. 1988, E. D. Brown and Farabaugh 1992, Farabaugh et al. 1988) or mainly (e.g., American Crows, *Corvus brachyrhynchos:* E. D. Brown 1985) for group affiliation and group recognition. Not all parrot vocalizations are affiliative. Various species have learned vocal duets that are used in nest defense and defense of areas around the mate (Power 1966a, b, 1967, Todt 1975b, Mebes 1978, Arrowood 1988) and thus resemble the territorial duets of many songbirds (Farabaugh 1982).

Vocal learning in parrots continues into adulthood rather than being restricted to an early critical period, and learning and production are rarely separated in time (Dooling et al. 1987b, c, S. D. Brown et al. 1988, Farabaugh et al. 1994, Powell 1993). In this respect, parrots differ from the many songbirds that learn song early in life but don't produce it until the following spring (Konishi 1965, Marler 1970, 1987). Parrots resemble those songbird species that continue to learn as adults; however, parrots do not learn seasonally, and learning is rapid and can occur at any time of the year. Rapid vocal learning has been shown to occur in Black-capped Chickadees (Nowicki 1989) and several species of European siskins (Mundinger 1970), songbird species with learned calls. Vocal plasticity in adulthood is associated with changes in social environment in both parrots and songbirds.

Conclusions

This chapter integrates the data available on learning, perception, and development of vocal communication in Budgerigars and other parrots with what is known about their basic biology. Many important questions remain unanswered, and we encourage detailed studies of the vocal communication, natural history, behavior, and ecology of parrots. How representative are the Australian parrot species of parrots around the world? Do parrot species that differ greatly in size and stability of their social flocks differ in vocal complexity, too? What is the significance of stereotypy in contact calls and of variability in parrot song? Do the perceptual abilities of domestic Budgerigars occur in wild Budgerigars? If all Budgerigars have such perceptual abilities, do other parrot species have them, or are they a specialization found in species with very mobile flocks? How is the development of the neural learning pathways in parrots controlled? Is complex song lacking in parrot species that have stable breeding seasons?

At the time of this writing, Jack Bradbury and Timothy Wright (pers. comm.) are conducting a field study of vocal behavior in Neotropical parrots. After repeated observations and recordings of known individuals of four sympatric Costa Rican species (Yellow-naped Parrot, *Amazona auropalliata;* White-fronted Parrot, *Amazona albifrons;* Orange-fronted Parakeet, *Aratinga canicularis;* and Orange-chinned Parakeet, *Brotogeris jugularis*), they have found that the four species differ in flock size and stability, size and location of night roosts, food habits, and predator pressure. Yellow-naped Parrots have fixed communal roosts that have been used for decades; their contact calls are most similar between neighboring roosts and become increasingly different as the distance between roosts increases. Similar projects are sorely needed for other parrot species.

We also need to learn more about the well-studied Australian parrots. Field playback experiments of Budgerigar vocalizations could reveal important contextual variables that operant laboratory studies have missed (Falls 1992). On the other hand, laboratory studies allow excellent control, and the other Australian parrot species whose behavior and ecology have been so thoroughly studied in the wild (e.g., Galah, Slender-billed Black-Cockatoo, Pink Cockatoo, and Western Corella) would be very valuable for intensive laboratory research. Vocal behavior was not the main focus of these field studies and field techniques similar to those used by Bradbury and Wright would allow more detailed study of their vocal behavior.

More laboratory experiments using Budgerigars are needed to clarify some of the issues raised in this chapter. Vocal learning begins early in life for Budgerigars, and the extent to which social interactions in the nest influence the course of vocal learning in this species could be resolved by cross-fostering experiments. Another important issue is whether adult Budgerigars can recognize the voices of individual nestlings. If they can, at what nestling age is parental recognition of individuals possible? The results of studies of other domesticated birds indicate that years of selective breeding have affected their hearing sensitivities, and a comparison of the auditory sensitivities of domesticated and wild Budgerigars is therefore under way (Dooling unpubl. data).

Communication and vocal learning in Budgerigars and many other birds occur through a close combination of visual and auditory cues, yet little research has addressed both modalities. The technology now exists to examine cross-modal matching of voices and faces in the laboratory using operant conditioning experiments. Trillmich's (1976a) early experiments approached this problem, but the vast improvements in technology for manipulating visual and vocal stimuli can now allow more controlled and complex experimentation. Trillmich's studies were also conducted with domesticated Budgerigars, which exhibit vastly greater variation in plumage color and body shape than do wild Budgerigars; it would be interesting to examine the relative importance of visual cues in wild birds.

Parrots and songbirds represent two large taxonomic groups that have independently found neurobiological and behavioral solutions to the problem of evolving a modifiable acoustic communication system. These two groups thus provide a unique opportunity for investigating the selective factors that favored the evolution of vocal learning and its underlying neurological and psychological processes.

Acknowledgments

This work was supported by National Institutes of Health grant DC-00198 and MH-00982 to RJD and National Science Foundation grant IBN-9217175 to SMF. We thank Ian Rowley, Jack Bradbury, Martin Wild, and Beth Powell for kindly commenting on the manuscript. Greg Ball, Steve Brauth, James Heaton, and Martin Wild provided valuable discussions of comparative neurobiology in parrots and songbirds. We also thank the many graduate students and technicians who have worked on Budgerigar projects over the years. SMF is extremely grateful to Angus and Karen Emmott for their apparently limitless hospitality; without their help she would never have returned from the outback alive. SMF also thanks Jane

Hughes and Pete Mather; the School of Environmental Sciences at Griffith University; Wayne Longmore; Bruce and Mary Emmott; Tim and Jenny Palmer; Pete Kleinschmidt; the Australian and Queensland National Parks and Wildlife Services; the rangers at Lawn Hill National Park; and the residents of "Noonbah," "Tandara," and "Westerton" properties in central Queensland and "Gregory Downs Station" in northwest Queensland for their advice and discussion of wild Budgerigar habits and biology, and for their hospitality. Finally, we are extremely grateful that Ed Wyndham had the courage to study Budgerigar field ecology in the first place, and we wish him a steady recovery.

PART II VOCAL REPERTOIRES

Introduction

Vocal learning enables birds to develop extremely large and diverse vocal repertoires. Conversely, absence of learning, as in suboscines and many non-passerines, may limit repertoire size and complexity. Given that acoustic displays are used to "manage" social interactions crucial for a displaying bird's fitness, why do display repertoires vary so much from species to species?

The chapters in this section address some of the basic issues about display repertoires. Do birds classify display repertoires in the same way that humans do when we sort piles of sonograms? In our search for "types," what kinds of meaningful variation do we obscure? How are vocal displays used? Together, these chapters provide a foundation for launching varied studies about ecology and evolution of display repertoires.

In Chapter 7, Andy Horn and Bruce Falls use birdsong to revitalize a fundamental but neglected question in animal communication: How do animals decode their own displays? Armed with recent advances in psychology, they argue that the process is flexible, and that many song features—from note structure to whole singing performances—may be perceptual "gimmicks" that exploit this flexibility. Horn and Falls critically review playback techniques for testing their ideas and encourage use of these techniques to study the evolution of display repertoires and display codes more generally.

Jack Hailman and Penny Ficken explore the difficulties and rewards of analyzing total vocal repertoires and comparing them among closely related species. The authors show how cataloging vocal displays can obscure communicatively important types of variation. Vocal types often vary physically in frequency, duration, and amplitude, and also in combinatorial structure, patterns of spatial radiation, and other ways. The vocal types that we recognize may thus act like the uninformative carrier frequencies in radio broadcasting, with variations (modulations) of the carrier containing the important information. Based on their studies of vocalizations in Paridae, Hailman and Ficken propose that song in oscines may have evolved from various elements of the vocal repertoire, but from different elements in different species.

In a welcome departure from songbirds, Vincent Bretagnolle shows that questions about vocal displays are equally absorbing in a nonpasserine group, the petrels. All petrel species use only one or two major call types to accomplish myriad social functions. Unlike songbirds, petrels do not learn their vocalizations, yet, paradoxically, even neighboring populations may differ in vocal characteristics. In this highly original review, covering all the extant genera, Bretagnolle provides a host of answers and questions about petrel communication and opens opportunities for comparative analyses of many kinds.

7 Categorization and the Design of Signals: The Case of Song Repertoires

Andrew G. Horn and J. Bruce Falls

Communication works because different signals mean different things, and the communicators share the code (S. Green and Marler 1975, W. J. Smith 1977). Therefore, an important selective pressure on the design of signals is the ability of receivers to sort them into different categories (Marler 1982). Just as the rules that govern signal transmission through the environment have been used to explain the design features of signals (R. H. Wiley and Richards 1982, Klump, this volume), the psychological mechanisms of categorization might be used to explain the design of signal repertoires (Guilford and Dawkins 1991). Put simply, one can think of the structure of signals in a repertoire as containing psychological gimmicks that make categorization easier or harder for receivers, depending on the interests of the sender. By studying how animals categorize the displays they produce, we can find out what some of these gimmicks might be.

Even though categorization is a basic step in communication, surprisingly little research has been done on it. Relatively few papers document categorization or discuss its importance in animal communication, and few researchers have explicitly tested how animals categorize the different displays they receive (Marler 1982, Dooling et al. 1987c, 1992, Harnad 1987a, Owren 1990a, b).

The songs of birds may be second only to human speech in the number and diversity of their display categories. Birds are relatively easy to study and vary widely in their ecology, the degree to which their songs are divisible into song types, and the size of the resulting song repertoires. They provide an excellent model for studying the role of categorization in the evolution of display types. In this chapter we discuss the possible role of categorization in the design of display repertoires, using song repertoires as an example.

The Psychology of Categorization

Categorization, also called classification, is the sorting of the many stimuli that impinge on an animal into a few classes. Although little is known about how animals categorize signals, much is known about categorization by humans, mainly because of its importance in cognition and language (E. E. Smith and Medin 1981, Harnad 1987a). In this section we summarize the basic concepts discovered through this research in order to produce a list of psychological mechanisms that might be exploited in the design of animal signals, particularly birdsong.

Discrimination between Classes, Generalization Within

An ethologist who wanted to convince another that two display categories were structurally distinct would likely use an ANOVA-like test showing that the differences between classes were greater than the differences within them. Similarly, categorization by an animal involves treating differences within each class of stimuli as less important than the differences between the classes. The usual approach to studying the releasing properties of animal signals has been to focus less on the variance in signals than on the values that elicit maximal responses, that is, the mean or mode from the animal's point of view (Ehret 1987). In contrast, the study of categorization focuses both on these isolated peaks and also on generalizations within different stimuli and discrimination between them (Herrnstein 1982).

If receivers always responded similarly to similar signals and differently to different signals, then we could stop here in our search for perceptual gimmicks that signalers exploit to facilitate categorization. Signalers could ensure categorization simply by producing structurally distinct displays. This reasoning may be sound in most cases, but research into categorization has revealed two important caveats.

First, animals respond categorically to certain stimuli that vary in a continuous fashion (Fig. 7.1A; Owren 1990a). Even though the rainbow consists of a continuous array of light wavelengths, for example, humans see it as relatively discrete bands of colors. The functional significance of this "categorical perception" (and its very existence) has been hotly disputed (see Massaro 1989 and commentaries). For our purposes it is enough to note that where we see continuous variation in a signal, animals might see discrete categories, either because of perceptual constraints (as in our perception of colors) or because small differences in structure can have large consequences (as in our perception of the difference between driving just over and just under the speed limit). Signalers might therefore produce displays that appear very different to receivers by producing structurally similar signals on either side of a category boundary.

The second caveat is that stimuli that are structurally very different are often treated as the same (illustrated in Fig. 7.1B with four distinct stimuli that are lumped into two display types). To take a famous example: where humans see a pastoral landscape of trees, fields, birds, and sky, a tick senses only temperature gradients and butyric acid, the better to find its mammalian hosts (von Uexküll 1934). The tick's perspective probably stems from perceptual limitations of its sense organs. In vertebrates, though, differences within classes of stimuli usually

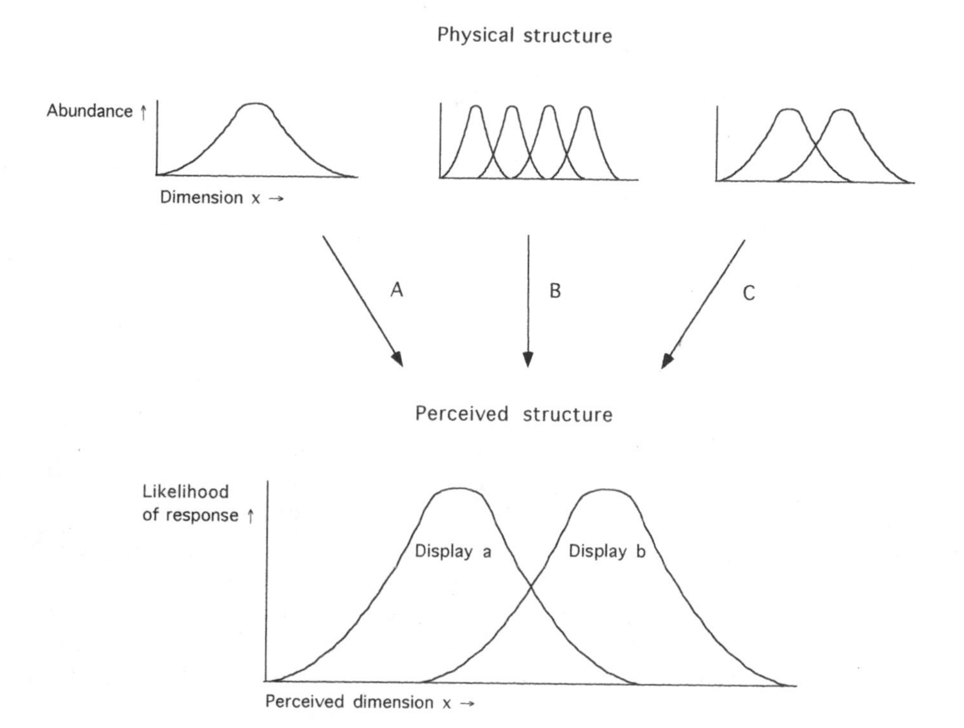

Fig. 7.1. Summary of the basic concepts in categorization. Three possible distributions of displays along a given dimension (e.g., frequency or duration) in the top panel are perceived as two distinct displays in the bottom panel in one of three ways: (A) a continuous distribution is split (categorical perception), (B) structurally distinct displays are lumped, or (C) structurally distinct displays are perceived as such.

are quite perceivable and, in different circumstances, might even form the basis for subcategories. An example of the formation of subcategories is the identification of different individuals that give a particular display type (see, e.g., Snowdon 1987, Stoddard, this volume). Thus, structurally distinct signals may well be treated as the same or different, depending on the circumstances.

To summarize, categorization treats variation within classes as less important than variation between them. Exactly how animals divide their perceptual world depends both on their perceptual limitations and on the functional significance of the stimuli, it cannot be divined simply by measuring signal characteristics. From a signaler's point of view, increasing the perceived difference between display types and decreasing the differences within them—conforming to a receiver's perception of categories (Fig. 7.1C)—will increase a signal's efficacy. In general, this requires signalers to produce structurally distinct signals. If a given stimulus dimension (e.g., wavelength in the case of colors) is perceived categorically, however, signals that straddle the category boundary will be seen as distinct even if they are structurally similar (Fig. 7.1A).

Dimensions, Form, and Features in Categorization

In this section we look more closely at aspects of signals that receivers might use to sort signals into different categories. Signals can differ in just one dimension, but most vary in several. How do receivers process this information? They might attend to one dimension, to several dimensions independently, or to a combination. Each of these aspects has different implications for signal design.

Categorization in one dimension. The simplest way for animals to categorize their displays is according to some simple criterion that divides them in one dimension. For example, a duration of 14 milliseconds could be the criterion dividing two syllable types, or, alternatively, 5 milliseconds and 20 milliseconds could be the "ideal" values for these syllable types, with responses to them becoming less likely as syllable durations move away from those values. Indeed, many song and syllable types are separable based on just one dimension (e.g., McGregor and Krebs 1982, D. A. Nelson and Marler 1989).

At first glance, the best dimensions to use to produce distinct displays would seem to be those in which receivers can perceive the subtlest differences—those that contain the greatest number of just noticeable differences (JNDs). Yet, subtleties birds might perceive in the lab are probably blurred by environmental noise in the field, particularly in certain parameters like amplitude (R. H. Wiley and Richards 1982). Also, the ability of animals to give a distinctive response to each category is more limited than their ability to tell them apart (e.g., Dooling 1982; also see below). If more refinement is required, the number of display types that can be usefully produced in any given dimension is reduced.

Categorization in multiple dimensions. The restriction on the number of displays might be partly overcome by producing displays that vary in several dimensions. Even in multivariate space, however, the more signals that are produced, the more similar (and therefore harder to distinguish) they will be. In general, the smaller the range of variation among stimuli, the harder they are to identify (Lockhead 1992). Moynihan (1970) first suggested that such constraints might put an upper limit on the size of species' display repertoires. The huge size of some song repertoires might be feasible because only the striking differences between successive song components, or their similarity to the listener's own songs (e.g., in song matching, discussed below), are important in communication. Birds that need to identify each song type (e.g., if each song type has a special meaning) might be restricted to smaller repertoires (see also Catchpole 1982) or might compensate for large repertoires by having especially differentiated song types.

Interactions between dimensions. Different dimensions often are not orthogonal but may interact; the easiest way to classify songs, for example, is to look at the "shape" of their sonograms and sort them into piles. Just as we recognize different tunes by the arrangement of the pitches of their notes relative to one another, birds may categorize different song types according to the relative frequencies of their notes. Such structure, based on relative rather than absolute cues, considerably increases the number of signals that can be distinguished, eases their categorization (because the signal itself contains the cues for recognition: Lockhead and Pomerantz 1991), and may be how we categorize most objects in everyday life (Rosch et al. 1976). Yet, despite its intuitive appeal, the importance of the concept of relative cues has only recently been appreciated in studies of categorization by humans (Lockhead and Pomerantz 1991) and the perception of birdsong (Weisman and Ratcliffe 1992).

So far, most of the work on the relative perception of song has dealt with interactions between two dimensions, which can be described using regression techniques. More complex interactions might be handled using principles borrowed from Gestalt psychologists (see Deutsch 1982 for application to music perception, Bregman 1990 for auditory perception, and Thorpe and Hall-Craggs 1976 for birdsong). For example, researchers might find that experiments that test how humans analyze auditory "scenes" (reviewed in Bregman 1990) can be readily adapted to test how birds separate songs from environmental noise.

Discrimination and Identification

Now that we have dealt with the nature of the stimuli that must be categorized, we turn our attention to the nature of the tasks the receiver must perform. Categorization tasks differ according to their time scale and to the sorts of standards that a subject must use to define class membership. Different sorts of signals are suited to different tasks. We begin with the fundamental (but not absolute) distinction between discrimination and identification.

Discrimination is simply telling whether two items are different; identification is telling what something is—that is, labeling it. Human subjects can label stimuli by naming them; animals can label stimuli by giving a particular response, such as striking a certain key or singing a certain song. Identification is obviously an example of categorization, and certain discriminations are, too, as long as discrimination between classes is stronger than discrimination within classes.

Discrimination and identification are not qualitatively distinct. In discrimination experiments, stimuli usually are presented simultaneously or in fairly rapid succession; in identification experiments, stimuli are usually presented in isolation. In general, subjects can make finer distinctions in discrimination tasks because the stimuli are presented closer together in time. Both sorts of tasks involve comparisons between stimuli, however. In discrimination, one stimulus is compared with another; in identification, a stimulus is compared with one stored in memory. Discrimination and identification are therefore better understood as relative and absolute discrimination (Harnad 1987b).

Animal signals may present receivers with either task, depending on the situation. For example, many species repeat one display type with slight variations before switching to another type (e.g., Horn and Falls 1991). Here, the task presented to receivers involves discrimination, which requires only that changes between display types be more noticeable than changes within types. In contrast, cases in which a single song carries a distinctive message involve identification and require conformity of the signal to a shared code (see, e.g., Kroodsma et al. 1989).

The distinction between discrimination and identification is seen more appropriately as a range of categorization tasks that differ according to their time scale. This range is dramatically illustrated by singing in Black-capped Chickadees (*Parus atricapillus;* Horn et al. 1992), whose songs always consist of a stereotyped two-note whistle. Each male can sing any frequency within the species-specific frequency range, but, within the course of a day, he sings the song only at particular well-spaced frequencies. Each frequency is usually repeated before a shift in frequency, making discrimination between successive songs straightforward. Identifying song variants over longer periods is more problematic, however. If one considered a male's entire output, one would say that each male sings one song type that varies continuously in frequency. If one identified song variants produced in only one day, however, one would conclude that each male has a repertoire of song types that differ only in frequency. Preliminary results suggest that individual frequencies produced during one day are used in song matching, whereas a lower frequency in general signals that a singer is more likely to escalate a territorial interaction (B. G. Hill and Lein 1987, Horn et al. 1992).

The standards with which animals compare stimuli in identification tasks is one of the most intriguing areas in cognitive psychology, mainly because these standards are not directly observable. Several theories have been proposed to explain what standards subjects might use (E. E. Smith and Medin 1981). Classical theory states that subjects search for necessary and sufficient features whose inclusion in a stimulus makes that stimulus part of a given category. For example, to categorize a stimulus as a square, one determines whether it has the feature "four equal sides." The exemplar theory states that subjects base their categories on stimuli that they have previously experienced. Similarity to these remembered objects determines which category new objects belong to. The prototype theory suggests that categories are grounded in generalized, "ideal" objects (for example, the category mean) that share properties with those that have been experienced, but are neither identical to them nor necessarily share particular properties with newly categorized objects. Similarity to these prototypes rather than to real objects determines which category objects fall into.

Any of the three models (classical, exemplar, or prototype) could apply to birds' recognition of song types. Birds might use distinctive features, such as frequency upsweeps or the presence of particular syllables, to assign songs to types. Alternatively, they might base their standard of a song type on specific examples, such as their own song, as studies of song matching suggest may be the case (see below). Finally, they might base their identification of song types on prototypes, which might incorporate the mean values of all the songs they have heard.

The Pervasive Importance of Attention

By now, the diversity of modes of categorization, features used, and tasks involved should be apparent. In this final section on the psychology of categorization, we discuss the role of attention in organizing the enormous variety of categorization mechanisms. By manipulating the attention receivers give to different features and to contrasts among signals, signalers gain considerable control over how their signals are processed. We consider three examples of the role of attention in categorization: in features, in standards, and in hierarchical levels used in categorizing.

First, to categorize stimuli, receivers might shift their attention among various features. For example, babies that are habituated to (i.e., stop looking at) objects of one color but varying shape are more likely to dishabituate (i.e., start looking again) in response to a novel color than to a novel shape. Thus, they attend more to novelty in the invariant dimension than in the variable dimension (L. B. Smith and Heise 1992). Similar effects seen in studies of nonhuman subjects (Mackintosh 1965) could be exploited by signalers, as in cases in which bird songs are repeated with more variation in some features than in others (Podos et al. 1992).

Second, receivers might also be encouraged to use different standards to compare songs depending on how singers deliver their song types. For example, if a particular song type is played to a male Western Meadowlark (*Sturnella neglecta*) that is not singing at the time, he will usually answer with that particular song type (provided the song type is in his repertoire). The common feature of the many functions that have been proposed for this song "matching" is that it invites the receiver to compare the response with its own most recent song (Guilford and Dawkins 1991). In other situations, such as in response to a silent intruder, meadowlarks that are singing steadily and repeating each song type many times abruptly change song types (Horn and Falls 1988a). This abrupt change in signal form against a background of repetition may direct the receiver's attention to the difference between the successive songs rather than to any contrast with some standard stored in long-term memory.

Finally, the attention of receivers might be directed to different levels of variation in the hierarchical structure of song, depending on how the song is delivered. Humans hear rapidly delivered sequences of short notes as rhythmic patterns and find individual elements hard to identify. Sequences of longer notes delivered more slowly, however, are heard as a collection of readily identifiable elements, and the overall pattern is hard to perceive (Warren 1993). Perhaps the briefer and more rapidly delivered the syllables in a song, the more likely the song as a whole is to be perceived as a unit. Syllables are more likely to be the important units in song perception for species in which syllables are long relative to the total song duration.

Playback as a Tool for Studying Categorization

We have reviewed several ways in which animals might categorize their displays, and we have examined how these mechanisms might be reflected in the structure and design of birdsong. Creative analyses of display structure might confirm some of these design features and reveal others. This type of research would certainly be valuable, and we encourage more studies of this kind.

A more fundamental question, however, is how animals themselves perceive their displays, particularly how, or indeed whether, they categorize them. If we do not know this, all our analyses of display structure are mere speculation. Indeed, many supposed design features of displays that appear to be exquisitely shaped to the "psychological landscape" (Guilford and Dawkins 1991) of receivers might have arisen entirely by chance.

Song types, for example, might arise without any categorization on the part of receivers. Some features of songs are inherently discontinuous, such as the jump from two- to three-note phrases in songs of Great Tits (*Parus major*). Discontinuous variation in songs might also be an epiphenomenon of song learning: simulations have shown that even slightly biased copying can produce repertoires of distinct songs (e.g., Kroodsma 1982b). Even the neat hierarchical structure of the most complex songs, so musical to our ears (Hartshorne 1973, R. Dawkins and Krebs 1978), is characteristic of many nonsignal behaviors—fly grooming, for example (R. Dawkins 1976b).

Playback experiments are perhaps the most important way to test whether the apparent design in animal signals really does affect the behavior of receivers. The structure of displays can be changed systematically and then played back to animals in the laboratory or field, to see when animals split or lump their displays and what features they use to do so.

In any experiment, the conclusions drawn depend largely on how the experiment was done. This caution is particularly true of experiments on categorization, because the way stimuli are categorized probably depends on the situation. We illustrate this point by describing three techniques that have been applied to categorization of song by birds: operant experiments, song-matching experiments, and playback of song sequences. We wish to provide both citations for readers interested specifically in song categorization and evaluations of each technique for readers who want to apply analogous methods to studies of other systems. Our main goal, however, is to show that any study of how animals categorize displays should maintain a broad perspective and use a variety of techniques.

Operant Experiments

Method. In operant experiments on song perception, a subject is trained to peck a key after hearing a certain song (by being rewarded with food) and not to peck that key after hearing other songs (often by having the lights turned off). The former stimulus is called the GO stimulus, the latter is called the NOGO stimulus. After training, subjects are tested on a range of exemplars to see whether they generalize the GO response within but not between classes of stimuli, that is, whether they categorize. The GO stimulus could also be a change between two classes of song, although this stimulus has not yet been tried.

Applications. An attempt to show categorization of song types was partly successful for one Great Tit (Shy et al. 1986) and has since been demonstrated more rigorously in Song Sparrows (Horning et al. 1993, Beecher et al. 1994a). The features Great Tits use for categorization were explored in detail by Weary (1989, 1990, 1991), who examined feature weighting and categorical perception of synthesized songs. Horning et al. (1993) studied the relative importance of the first and second halves of songs in discrimination by Song Sparrows, and Stoddard et al. (1992b) and Beecher et al. (1994a) explored the number of pairs of song types that Song Sparrows can discriminate.

Evaluation. Because the experimenter has tight control over training and testing conditions, operant experiments are excellent for examining the finer details of song categorization. The "dark side" of tightly controlling testing and training is that the results might be specific to the laboratory situation and have little relevance to natural interactions. The reinforcers used during training (usually food rewards and lights out) and the context in which birds are tested (an empty room with perfect listening conditions) may be very different from the natural conditions. Because birds are hard to train, operant experiments are also usually based on only a few birds trained on a few songs (but see Stoddard et al. 1992b, Beecher et al. 1994a).

A frequent defense of operant experiments is that they allow the researcher to separate whether birds are able to perceive a difference from whether they choose to treat the difference as significant (e.g., Horn 1992, Weary 1992). We think this argument has been overstated, however. It is true that operant experiments isolate the perception of song from its natural context, but only by replacing it with an unnatural context. Also, the notion of a "pure" perception that is thereby laid bare is questionable. Even the most elementary perceptual processes are now known to involve evaluation of relevance and direction of attention (see, e.g., L. B. Smith and Heise 1992). Outside of direct studies of the capabilities of sense organs (e.g., Dooling 1982), we question whether attempts to isolate perceptual and "higher-level" categorization will succeed (see also commentary by Ehret in Massaro 1989).

Nevertheless, the intensive control over training and testing conditions in operant experiments allows the researcher to test how birds categorize in a broad range of situations—when trained on a few or many exemplars, for example, or under different signal-to-noise ratios and reward regimes. The full potential of these methods is only beginning to be tapped (see Stoddard et al. 1992b, Beecher et al. 1994a).

Song Matching

Method. In many species, territorial males answer playback of a song of a particular type by singing the song type that was played to them (e.g., Falls et al. 1982, Falls 1985). In a sense, this response indicates categorization because the bird is treating some songs in the population as equivalent to its own song (within-class generalization), and others as different (between-class discrimination). Birds are forced to give categorical responses because the songs in their own repertoires are discrete by definition, but it is possible to test whether birds are "splitters" or "lumpers" by playing back various songs and seeing which are matched. Specifically, a song that is similar to one in the subject's repertoire but not of the same type should elicit this next-best song only if the subject lumps song types. The proportion of matching can also be used as a quantitative measure of how similarity among songs is perceived.

Applications. "Splitting" versus "lumping" experiments have been tried with Great Tits, Song Sparrows, and Western Meadowlarks (P. D. McArthur 1986, Falls et al. 1988, Falls, D. M. Weary, and Horn unpubl. data). Some of the structural features birds use to classify songs during matching (e.g., the ordering and timing of song elements) have been isolated for various species (Wolffgramm and Todt 1982, Falls et al. 1988, Weary et al. 1990a, Stoddard et al. 1992a). Experiments have shown that the proportion of matching declines as the similarity of a stimulus song to the subject's own song declines (e.g., Falls et al. 1982). This pattern suggests that birds use their own song as a standard when classifying the songs they hear, although some results suggest that neighbors' songs are also used (Falls 1985, Stoddard et al. 1992a).

Evaluation. Matching experiments are perhaps the most promising paradigm for studying how birds perceive their song types. They are easy and quick to perform, and they give unambiguous data in a fairly natural situation (Horn 1992). Their simple design affords some hope that comparable experiments can be conducted on different species. For example, the songs of subjects could be changed in standard ways (e.g., by one standard deviation, or by equal steps toward the next most similar song type in its repertoire) in several species to see whether the precision of matching correlates with the degree to which song types are differentiated.

At the same time, matching experiments need larger sample sizes and more quantitative comparisons between playback stimuli and response songs than have been used to date. If the species rarely matches or has a small repertoire (and therefore has a high expected proportion of matching by chance), large sample sizes are required just to demonstrate matching. Large samples are always needed if the proportion of matching is to be used as a quantitative measure of perceived similarity. Finally, detailed analyses of song structure are essential if these response gradients are to be compared with structural variation in songs. Without such comparisons, categorization cannot be rigorously demonstrated (see Falls et al. 1982, and Weary et al. 1990a for some initial attempts).

Responses to Song Sequences

Method. These experiments are of two types. In one, overall responses to two or more different sequences of songs are compared, using standard measures of response strength such as song rate and distance to the speaker. In the other design, one song is repeated until the subject habituates, and then the playback switches to a new song. If the subject considers the new song to be different from the first, then it should start responding again. This technique is the habituation-dishabituation paradigm so widely used in laboratory studies of perception (e.g., Dooling et al. 1987c).

Applications. Many experiments have demonstrated stronger responses to sequences of song types than to one repeated song, but few have directly pitted variation between song types against variation within song types (Kroodsma 1990). The two exceptions, which used Great Tits and Song Sparrows, suggest that the two levels of variation are equally effective in elevating responses: the subjects did not categorize songs by type (Stoddard et al. 1988, Falls, D. M. Weary, and Horn unpubl. data). In contrast, habituation-dishabituation experiments have revealed categorization of song types in Song Sparrows (Searcy et al. 1995). A few studies have tested higher levels of song variation, specifically whether sequences of very different song types stimulated stronger responses than sequences of similar song types (Horn and Falls 1988b, Falls et al. 1990, Falls, D. M. Weary, and Horn unpubl. data).

Evaluation. A wide range of stimuli can be tested in playback of song sequences, because experimenters can assemble those song collections of interest. Variation within and between both individuals and song types can be presented in intuitively appealing designs. The habituation-dishabituation paradigm in particular is widely applicable and hence is popular in studies of perception. But playback of a sequence of songs, especially if habituation is involved, often requires a prolonged, intense response at close quarters. In the absence of a real bird at the end of the speaker cable, such a situation is highly unnatural. Biologically valid results might best be obtained by focusing on long-distance responses (e.g., by playing from outside the territory; Stoddard et al. 1988) or on discrete responses that subjects make to a change in song stimuli (e.g., switches in song type by the subject; Falls et al. 1990).

Synopsis

We have reviewed only three of the many possible experimental paradigms for testing how animals categorize their displays (Horn 1992). Each has been applied with some success to the perception of song types, and each has its advantages and pitfalls. More important, each provides animals with very different tasks. In matching experiments, for example, the bird is asked to identify an isolated stimulus. The researcher is in effect asking, What song is this? (an identification task). In presentation of song sequences the bird is given more than one stimulus and is asked to compare them: Are these songs the same or different? (a discrimination task). We should not necessarily expect the same answers from each paradigm, because a wide range of categorization tasks might confront birds in nature.

If different playback techniques yield different answers, how do we test for relationships between signal design and categorization? The answer is probably obvious by now: any question about categorization must be approached with a wide variety of techniques that tackle questions in different but complementary ways. A clear picture of an animal's communication system begins to emerge only when the results of several experiments can be interpreted in the context of intensive observations of natural interactions. This point is the subject of the following section.

From Categorization of Displays to the Organization of Display Repertoires

Animals categorize their displays according to how the displays are used in natural interactions, and in turn different aspects of signal design may come into play in different contexts. Consider how a Western Meadowlark perceives its own songs (or how we think it does; Fig. 7.2). Each individual sings a set of distinct song types, some of which are shared with other birds. Within these shared song types, subtler variants produced by different individuals are recognized (Falls and d'Agincourt 1981, Falls 1985). Song-matching tests show that song types that can confuse humans are perceived as distinct by the birds (Falls et al. 1988); comparisons of responses to sequences of song types show that the birds recognize these similarities (Horn and Falls 1988b). Above the level of variation between song types, birds are also sensitive to differences between different combinations of song types: territorial males respond more strongly if sequences contain more switches between song types or a greater number of song types (Horn 1987, Horn and Falls 1988b). Each level of song variation comes into play in different situations (Horn and Falls 1988a, 1991).

Thus, the question of how displays within a repertoire become distinct, and whether animals perceive them as being distinct, quickly shifts into the broader issue of how animals organize the many levels and degrees of variation in their signal repertoires. The study of categorization by humans has followed a similar

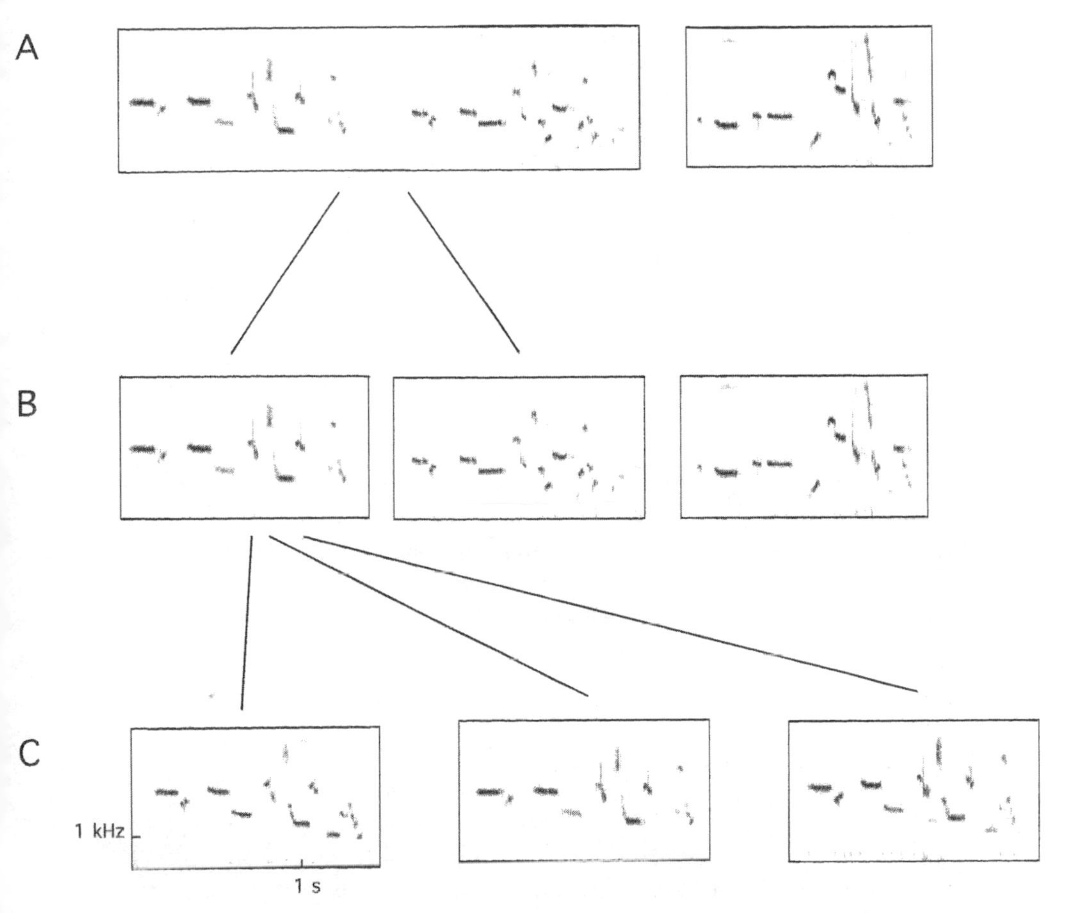

Fig. 7.2. Sonograms (analyzing filter bandwidth, 353 Hz) of songs of Western Meadowlarks (*Sturnella neglecta*), illustrating levels of song variation recognized by territorial males: (A) groups of similar song types, (B) song types, and (C) renditions of the same song type, as sung by different individuals.

path. Categorization is now viewed less as a hard-wired perceptual mechanism that sorts stimuli like letters at the post office than as a dynamic process in which different features and different sorts of comparisons are made depending on the function of the task being performed (Rosch et al. 1976, Lassaline et al. 1992). This insight affords several lessons for the biologist interested in the evolution of display repertoires. First, to emphasize yet again, no single playback experiment can determine how a particular species categorizes its displays. A range of experimental techniques must be used, directed by a thorough knowledge of how animals use their displays in natural interactions.

Second, display types should be seen not as hard-and-fast categories but as convenient fictions that give a provisional starting point for studies of communication. Pigeonholing displays into static types can hinder studies of communication. It keeps researchers from looking at the fringes of display categories, which is where dynamic selection for stereotypy and distinctiveness presumably occurs, and from seeing the evolutionary continuity between animals with strongly and weakly differentiated displays (e.g., between Great Tits, with "classically" discrete song types, and Black-capped Chickadees, with one widely varying song "type").

Third, focusing on the organization of display repertoires as a whole, rather than on individual display categories, challenges the general perception of codes in animal communication. We began this chapter by stating that communication is based on a shared code. If animals really are categorizing their signals in the ways we have suggested, however, then every display is defined relative to another. The classic model of animal communication—a code in which different displays can be matched to a message, like words to definitions in a dictionary—is only a special case of such comparisons. Instead of a single code linking universally shared displays to universally shared meanings, we in fact see several different codes depending on the situation. Again using birdsong as an example, in some species song types have different meanings, other species use song types mainly in matched countersinging, others use switches between song types to carry messages, and others use some combination of these codes (Horn 1992). Birdsong therefore offers a unique opportunity for a comparative study of communication codes.

Conclusions

Although ethologists have been classifying displays for a long time, only recently have researchers begun to examine how animals perceive the displays of their own species. At first glance, such studies seem to be methodological exercises. For example, playback experiments that test the perception of song types are usually introduced to support work that has already been done on the function of song repertoires. Correlations between the number of song types a male sings and his reproductive success seem more meaningful if another bird tallies that male's song types in the same way as a researcher tallies them.

We argue, however, that categorization studies also make an important contribution to our understanding of how signals evolve. Detailed descriptions of the physical relationships among signals in a repertoire can reveal design features that might help receivers organize otherwise confusing input. The efficacy of these features can be tested using playback experiments that examine responses to variation between display types rather than responses to the "average" display, as has been done in the past. In this way we can address the broader issue of how and why displays within an individual's or a species' repertoire become different.

Acknowledgments

We thank Danny Weary for invaluable help at every stage of writing this chapter, and the editors for patiently and enthusiastically providing thorough critiques. Alejandro Lynch, Pete McGregor, and Cindy Staicer also gave very useful comments on several drafts. Mike Beecher, Steve Nowicki, and Danny Weary generously provided drafts of unpublished work. AGH especially thanks Marty Leonard, Laura Horn, and Claire Horn for tolerance in the face of intolerability. JBF's research is supported by grants from the Natural Science and Engineering Research Council of Canada.

8 Comparative Analysis of Vocal Repertoires, with Reference to Chickadees

Jack P. Hailman and Millicent Sigler Ficken

It is probably true that individuals of all animal species communicate with one another in some fashion. The modes of communication range broadly from chemical signals in microscopic animals to exceedingly complex optical and acoustical signals in large, motile animals such as vertebrates. One of the most highly evolved and sophisticated communication systems known is avian vocal signaling. Although most researchers focus on song in oscine birds, we take a broader view, because vocalizations other than generally broadcast advertisement displays are important in the lives of the birds and are instructive with regard to principles of communication. By analogy, if one were to study verbal communication in *Homo sapiens* solely through recording and analyzing professorial lectures, a limited understanding of human language would be the result.

An important goal of zoosemiotics (the study of animal communication) is to understand what information is transferred between individuals and how it is encoded in signals. A seemingly straightforward approach to this immense task is to begin by inventorying species' repertoires, and indeed, such counts have been compiled and tabulated (e.g., Moynihan 1970). By comparing the richness of repertoires among species one could then formulate hypotheses as to how communication systems have evolved and how they are adapted to different species' ecological settings. Too little attention has been paid to the criteria used for listing repertoire elements as separate entities, however, and comparing mere counts of the number of signals used by different species could be highly misleading. Furthermore, the simple cataloging of signals could lead to a simplistic view of communication—namely, that each signal listed has some discrete meaning. Such a model of communication implies that once a complete list of signals and their meanings has been compiled, communication of that species is understood.

Studies of vocal repertoires in tits (Paridae) show some of the difficulties of compiling a straightforward list of vocal signals and of making comparisons among species. In any case, communication is not best understood by listing signal types with their meanings because not all information transferred is encoded in differences among signal types. By comparing vocal repertoires among closely related birds, we show that reasonable inferences about homologies are possible, and this comparison in turn allows glimpses of how vocal signaling is adapted to ecological differences among species.

Our examples come mainly from the subgenus Poecile (*Parus,* Paridae). The Poecile species are usually called chickadees in North America; elsewhere, all species of *Parus,* regardless of subgenus, are known as tits. Of the 10 subgenera of *Parus,* Poecile has the most species (Thielcke 1968), so it is especially appropriate for comparative studies. We selected Poecile to illustrate our points in this chapter because an extensive literature is available, including our own separate and joint studies, but most of the questions we ask and the approaches we devise to answer them are of a general nature and widely applicable.

How Many Types of Vocalizations Are in a Repertoire?

Determining the composition of a vocal repertoire is the first step in comparative studies aimed at examining similarities and differences among species. Although the question posed by the heading of this section initially seems simple and straightforward, we believe that it is badly formulated and probably inherently unanswerable. The question "how many?" suggests that vocalizations occur as simple qualitative types that can be easily counted. For example, the repertoire study of the Black-capped Chickadee (*Parus atricapillus*) by M. S. Ficken et al. (1978) lists by name 13 types of vocalizations (Fig. 8.1). As the authors made clear, however, these named units represent diverse entities. In order to appreciate their variety, we consider several of the named categories next.

Twitter (crackle). This "call," which sounds like a crackling fire and consists of a series of regularly spaced, broad-frequency pulses of extremely brief duration (Fig. 8.1D), is given between mates primarily in and around the nest cavity. Different recordings of the same individual vary appreciably only in the duration of the pulse train, and as the twitter (crackle) resembles no other sounds in the repertoire, it may prove to be a simple, qualitative type of vocalization. In this regard, however, it is atypical.

Single A. This vocalization is a small, chevron-shaped note on a sonogram (Fig. 8.1E), but it is far more variable acoustically than the twitter (crackle). Single A's closely resemble the A note of chick-a-dee calls (below), and it is not certain whether they should be considered a separate vocal category (e.g., Hailman et al. [1985] excluded them from analyses of chick-a-dee calls). Single A's vary widely in peak frequency and also in other ways: the ascending arm of the chevron may be emphasized with little or no descending arm after the peak, or exactly the opposite may occur, with virtually no ascending arm and an emphasized descending arm (compare the various notes in Fig. 8.1E). Furthermore, the frequency modulation of the arms (their slopes in sonograms) varies, so that calls also vary in total duration. The single A is therefore not a simple, discrete type of vocalization,

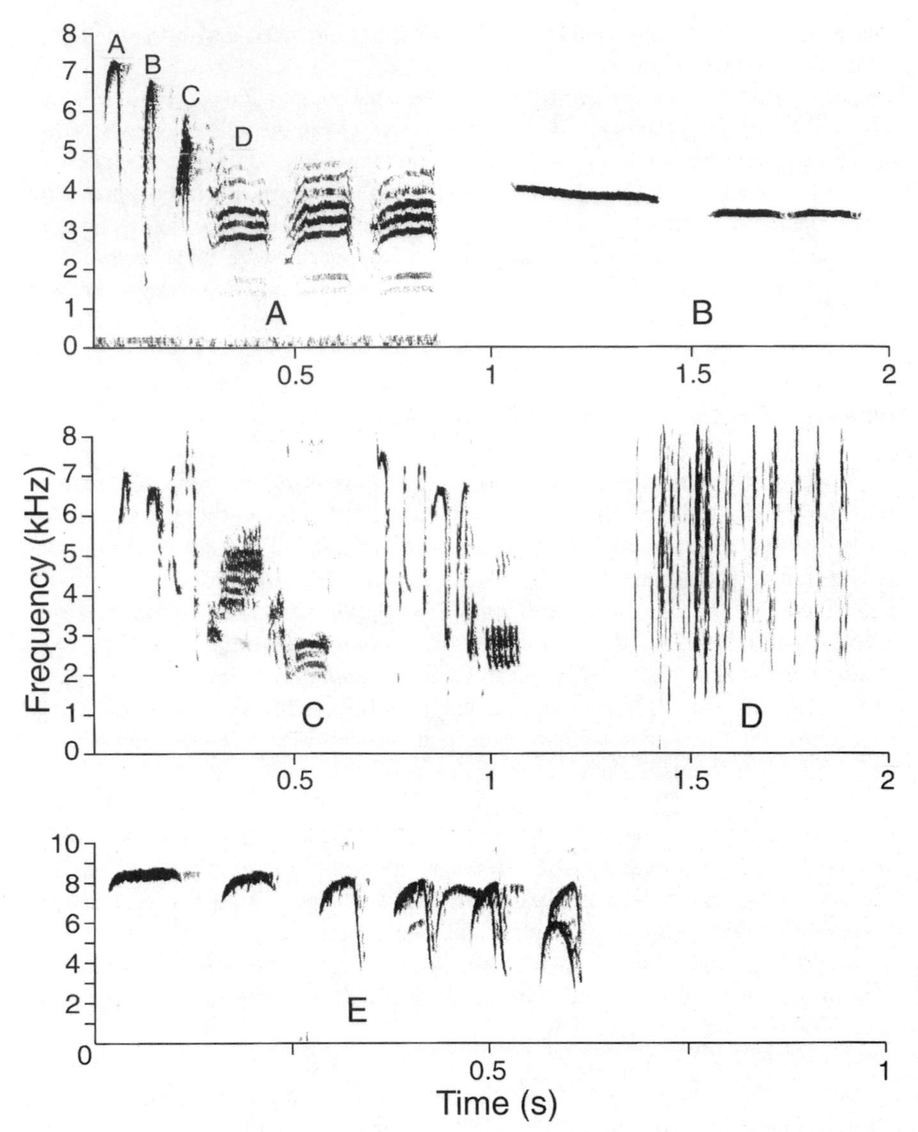

Fig. 8.1. Some elements of the Black-capped Chickadee's vocal repertoire: (A) chick-a-dee call of notes ABCDDD, (B) fee-bee ("song"), (C) two gargles showing the sharing of some note types between calls, (D) twitter (crackle); and (E) high (alarm) zees grading into A notes, with a second bird beginning to give A notes simultaneously with the fourth note of the first bird. All sonograms were produced on a Kay Digital Sonagraph 7800, analyzing filter bandwidth, 150 Hz.

but rather one that varies quantitatively in many ways, probably in the context of its use as well as in its acoustic structure. Whether or not all the forms of variation are continuous is a question answerable only with exhaustive quantitative study.

Begging dee and broken dee. Nestlings, and later fledglings, give calls characterized by harmonic-like frequency layers on a sonogram. Because parents often feed a young bird giving these calls, the vocalizations are known as begging dees. The female of a pair begins giving a similar call once the nest cavity has been excavated, and this vocalization has been called the broken dee. Are these two calls different, or should the latter be treated simply as the adult version of the former? That question seems to be rhetorical, for if two calls are age-dependent and never occur in a given individual at the same time in its life, no objective criterion exists for deciding whether the calls represent one or two types.

Fee-bee. The fee-bee vocalization has often been termed the "song" of the Black-capped Chickadee; although it is given primarily by males (M. S. Ficken et al. 1978), it does not have all the characteristics of song usually attributed to other kinds of songbirds, such as being the most acoustically complex vocalization in the repertoire (see, e.g., M. S. Ficken 1981, Hailman 1989). The fee-bee is simple in structure, consisting of two prolonged tones, the second at a lower acoustic frequency than the first (Fig. 8.1B). Thus there seems to be no song repertoire as there is in so many songbirds. Yet, careful work by Ratcliffe and Weisman (1985) proved that the fee and bee notes vary in acoustical frequency, and singing chickadees switch frequencies in a manner similar to how other songbirds switch song types within their song repertoires (Horn et al. 1992). So, is the fee-bee one type of vocalization or many? Furthermore, both sexes commonly give faint fee-bees (M. S. Ficken et al. 1978), which are separately named but acoustically similar to "ordinary" fee-bees except in loudness and context of occurrence. Perhaps amplitude differences affect message content.

Gargle. M. S. Ficken et al. (1978, p. 37) pointed out that "this is the most complex call of the chickadee, each call consisting of a series of two to nine short notes." Gargles are composed of diverse note types (Fig. 8.1C) drawn from a pool shared by members of a local population, each pool having about a dozen types of notes (e.g., M. S. Ficken and Weise 1984, M. S. Ficken et al. 1987). The notes used in a particular gargle are thus a sample of those available, and many different kinds of gargles may be recorded from a single individual. The gargle is therefore not a simple type of vocalization, but rather a category of many combinatorially different types. Although the types seem somewhat similar to the song types in song repertoires of many oscine species (M. S. Ficken et al. 1987), a major difference is that gargles are primarily close-range signals (M. S. Ficken 1981).

Chick-a-dee. The call for which the bird was onomatopoetically named is another combinatorial system, but markedly different from the gargle. Here, only four different types of notes (A, B, C, and D) are combined (Fig. 8.1A), with repetition, to form hundreds of qualitatively different types of calls (Hailman et al. 1985). If each different combination were counted as a separate type in the vocal repertoire, the ostensible size of the repertoire would be more than 10 times larger than if all chick-a-dee calls were counted together as just one type.

Mixed types. Chickadees can also mix types from the above list. For example, S. M. Smith (1991) mentioned a male that sang fee-D instead of fee-bee, the D being the D note of chick-a-dee calls, and one of us had a male on his study tract for two years that also sang fee-D instead of fee-bee (Hailman, unpubl. data). Dixon and Stefanski (1970, fig. 2) showed an utterance apparently composed of introductory notes from a chick-a-dee call followed by one fee note from the fee-bee. Clemmons (in press) also showed examples of "mixed" vocalizations from Black-capped Chickadees. These mixed vocal types may be abnormal utterances that have no meaning to social companions. In the absence of contrary evidence, however, we must also entertain the possibility that at least some of these vocalizations are not abnormal but simply rare. Perhaps to a chickadee the meaningful categories of utterance are parts of what we classify as types, such as the parts fee, bee, A, B, C, and D. In most of what the birds vocalize about, fee will go with bee and the lettered notes will go with one another, but perhaps sometimes the "correct" combination would be fee-D or some other mix.

The foregoing examples demonstrate that it is misleading to summarize a vocal repertoire by some number of "types" of vocalization. Furthermore, the types (or groups of types) that researchers define from spectrographic Gestalts and quantitative measurements might in fact bear little relation to what the birds themselves consider the same and different.

How Is Vocal Variety Characterized?

There is a perfectly straightforward, if theoretical, reason to expect that vocalizations will vary in all sorts of ways that prevent characterizing repertoires as some total number of entities, each encoding a different message. This reason concerns the researcher's conception of communication. If communication is viewed as a straightforward substitution code, then vocal type 1 might be assigned a "message" (in the terminology of W. J. Smith 1977) such as "I am aggressive," type 2 may be interpreted as "I see a hawk," and so on. Acoustic variation under such a model of communication is often treated as epiphenomenal—as meaningless jitter that is simply a manifestation of the inherent variation that characterizes all biological systems at every level.

At the other extreme is a quite different way of viewing vocal communication, easily explained by analogy with radio transmission. Radio communication uses a carrier signal on which is superimposed meaningful variations. The carrier frequency may identify the source of the transmission, but it provides little other information; it is the modulation of that signal that carries the information of interest. So could it be with avian vocal signals; the type of vocalization may be merely (or primarily) the carrier signal, and the variations of that type might encode much of the useful information.

We suspect that the truth of avian vocal communication lies somewhere between these extremes: apparent types of vocalizations do carry information, but the details, and in some cases perhaps the most important information, lie in the variations on a type. If this view has merit, the important task of a repertoire study becomes the characterization of variation. Without this characterization it will never be possible to decode the information carried by the signal. Some principles of variation can be illustrated with the vocal repertoire of the Black-capped Chickadee.

Types. Unlike radio transmission, in which the governmentally assigned broadcast frequency is largely arbitrary, vocal types in an avian repertoire differ in apparently nonarbitrary ways. For example, the high zees (alarm zees) used to indicate predator alarm and the D note of chick-a-dee calls used in mobbing fit Marler's (1955) dichotomy of acoustic structure for alarm and mobbing calls. It seems likely that the very type of signal uttered, regardless of its specific acoustic structure, is information laden. But to believe that types carry all of the important information is a mistake (see below).

Quantitative variants. As we pointed out above, even simple utterances represented by one continuous trace on a sonogram can vary in numerous ways. A study of predator-elicited calls in the Florida Jay (*Aphelocoma c. coerulescens*) shows the power of quantitative analysis in understanding this kind of variation (Elowson and Hailman 1991). The jay's calls differ in several ways, principally in duration, emphasized frequency bands, and change in acoustic frequency through the call. Four variables can be measured accurately from sonograms: (1) the midpoint frequency of highest amplitude at the middle of the call, (2) the lowest frequency at the start of the call, (3) the frequency and time of the highest minimum frequency in the call, and (4) the lowest frequency at the end of the call. From these measured variables a number of derived variables can be calculated, such as the slope of the minimum frequencies over time and the total duration of the call. Elowson and Hailman (1991) performed principal-components analyses to identify variables that would separate calls and found two kinds of separations: those of no overlap, thus defining qualitatively different call types, and continua with distinct modes of variation that defined different subtypes. In all, 12 different types and subtypes were identified, and the relationships among types and subtypes were explored by hierarchical cluster analysis.

One instructive result of these analyses demonstrates the importance of considering two or more measurable variables in concert. In some cases, univariate analysis of a derived variable clearly dichotomized certain types of calls, as in the slope of the minimum frequencies of the call over time (Fig. 8.2A). In other cases,

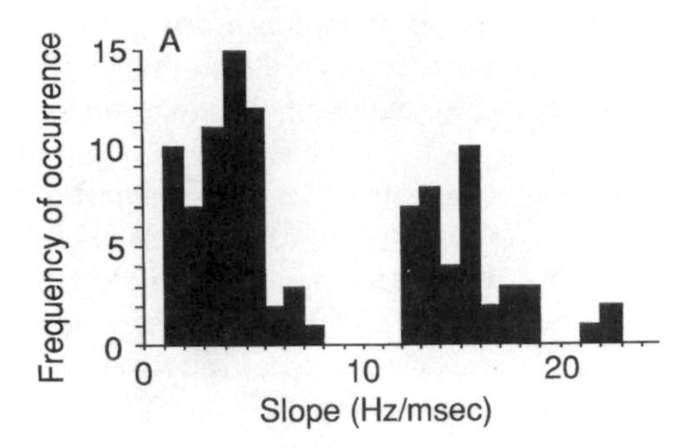

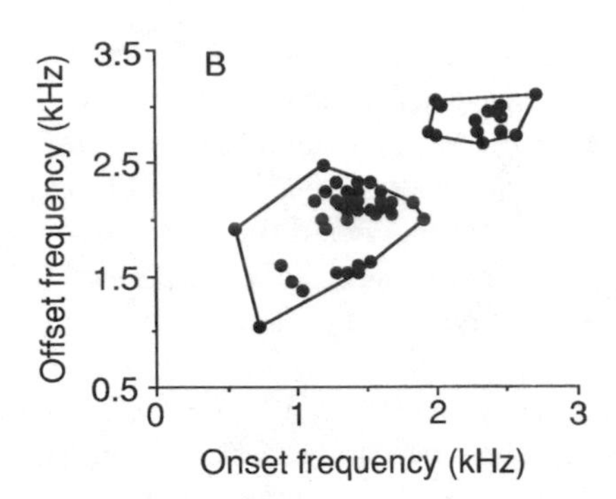

Fig.8.2. Two quantitative approaches to defining types from complex variation, illustrated by calls of the Florida Jay. (A) From measured acoustical variables—time and lowest frequency at the onset and offset of the calls—a derived variable (slope) was calculated that separated calls into two distinct types ("shallow" and "steep"). (B) More detailed analysis of shallow calls showed that both the onset and offset frequencies were continuously variable, but a bivariate plot separated them into nonoverlapping types ("low" and "high" shallow calls). We know of no comparable quantitative studies of tit vocalizations. After Elowson and Hailman (1991).

univariate analyses showed continuous variation, as in the lowest frequency at the onset of the call and the lowest offset frequency. If these two variables are plotted on Cartesian coordinates, however, they separate calls into two qualitatively distinct groups (Fig. 8.2B).

No comparable example appears to have been explored quantitatively in tits, although there are numerous cases for which such analyses might be appropriate. For example, M. S. Ficken (1990a) noted that the high (alarm) zee call of the Black-capped Chickadee may just be a higher-frequency version of the multiple A-note series (Fig. 8.1E).

Radiation pattern in space. A little-studied type of variation in vocalizations is the spatial pattern of their radiation. Witkin (1977) recorded chick-a-dee calls of Black-capped Chickadees facing toward and away from a microphone and found that the radiation patterns differed among note types. The B and D notes (see Fig. 8.1A) were as loud behind the bird as they were in front, but the A and C notes were noticeably softer behind the birds (by 4 and 2 dB sound pressure level [SPL], respectively, as measured 1 meter from the calling bird).

Combinatorial variants. As mentioned, the Black-capped Chickadee has two call systems that are combinatorial in nature, although they differ somewhat in their organization. Gargles are composed of a number of rapidly uttered notes (Fig. 8.1C) that are drawn from a "pool" of about two dozen note types shared by members of a local population. Pools of available note types differ geographically over distances as short as several kilometers, and hence might be termed micro-dialects, in contrast with dialects of song reported over much larger distances in several songbird species (e.g., M. S. Ficken and Weise 1984, M. S. Ficken et al. 1985, 1987). The syntactical rules governing gargle compositions appear to be simple: (1) a given note type typically appears just once in a given gargle, (2) successive notes tend to decrease in acoustic frequency through the vocalization, and (3) if a longer-duration, acoustically complex note is included (as in Fig. 8.1C), it is always placed in the terminal part of the call (M. S. Ficken and Popp 1992). A somewhat similar example is illustrated in Fig. 8.3 (below) with regard to buzzlike characteristics in different calls of another species.

Chick-a-dee calls are also combinatorial, but they are organized along different lines and have a stricter syntax than gargles. Calls vary considerably in total duration, being composed of 1 to at least 24 individual notes (Hailman et al. 1985). Markov chain analysis showed that note composition is "alphabetical" in the sense that the note types labeled A, B, C, and D in Fig. 8.1A almost always occur in that order (exceptions constitute less than 1% of calls). A given note type may be absent, given once, or repeated a variable number of times within calls, but the recurrences are always sequential. Further study using a combination of survivorship analysis and transitional frequencies showed that call duration is constrained. The longer the preceding A string, for example, the shorter the terminal D string and the higher the probability that the call will simply end after the A string (Hailman et al. 1987). Similar results apply to the lengths of B and C strings, but, curiously, regardless of the total number of A, B, and C notes combined, calls usually end with at least one D note, as if a terminal D note is semantically critical.

Another important characteristic of chick-a-dee calls is that they appear to be infinitely "open" to new types. A Zipf-Mandelbrot analysis plots log probability of occurrence of different call types as a function of log occurrence rank of those types. This analysis applied to chick-a-dee calls determined a curve that converged on a line of nonzero slope (Hailman et al. 1985). This result means that as the sample size of calls recorded is increased, one continues to find new (and rarer) call types, without any apparent bound. Among acoustic communication systems, only human language is also known to be open to new utterances. Only a few species with complex vocalizations, however, have been subjected to such analysis, so chickadees may not be unique.

Are Repertoire Elements Evolutionarily Correlated Characters?

In considering the evolution of vocal repertoires, researchers must be ever mindful of the possibility of correlated characters. Selection may act on vocal characteristics as a unit, and hence may impose constraints on how acoustic variation encodes information.

In parids, numerous examples of similarities occur among supposedly separate vocal types. As mentioned above, the two most complex calls of the Black-capped Chickadee are the gargle and chick-a-dee systems. Both are combinatorial although organized in somewhat different ways. As M. S. Ficken and Popp (1992) pointed out, however, in both types of calls the component notes show descending acoustic frequency within a call, and also tend to have note types with very brief durations and wide frequency range at the beginning and trills or banded notes at the end. Furthermore, three of the four major note types of the chick-a-dee call (A, B, and D) are similar in acoustic structure to three note types of gargles (R, F, and J, respectively). In gargles, these notes tend to follow the same rules as their counterparts in chick-a-dee calls (R at the beginning, F in the middle, and J at the end). Compare, for example, the terminal D notes shown in Fig. 8.1A with the terminal note in the first gargle of Fig. 8.1C. Another example of similar vocalizations in the Black-capped Chickadee involves the variable sees (sexual sees), most of which end with several note types found in gargles but in no other known elements of the repertoire.

In sum, it is unlikely that every type of variation in a vocal repertoire is directly related to its special communicative function. Rather, characteristics are likely to be correlated among vocal elements that reflect general principles of organization, thus providing a framework within which more specific communicative adaptations can be sought.

What Information Does Variation Encode?

Many researchers probably consider the meaning of a signal (the "message" of W. J. Smith 1977) to be the bottom line of communication analysis, so they attempt to reach that line quickly. Most experiments do not consider variation within vocal classes, and in some cases playbacks and other experiments are performed before variation has been adequately characterized (West and King, this volume). If one is satisfied with results from such studies, then the conception of animal communication is constrained to the viewpoint that vocal signals are mainly of a rather vague and general nature, and that birds generally do not transmit precise information. We believe it is more productive to assume that precise information is being encoded and that the ultimate task of the researcher is to discover what that information is. One is not likely to discover interesting natural phenomena if one begins with the preconception that they do not exist (even though that may be the case). We therefore turn to some examples of what appears to be known about the information encoded in vocalizations of the Black-capped Chickadee's repertoire.

The contexts of types. It seems unlikely that the information carried by any given type of signal in the chickadee repertoire has been fully characterized. This

Table 8.1. Contexts of use of vocal types by adult Black-capped Chickadees

Vocal type	Used by	Commonest contexts of use
Fee-bee	Mainly males	Territorial proclamation, serenading mate, other?
Faint fee-bee	Both sexes	Mates in vicinity of nest, esp. when no visual contact
Gargle	Mainly males	Mainly short-distance agonistic interactions
Subsong	Young birds only (?)	Alone (?) in summer after independence
Chick-a-dee	Both sexes	Wide variety of contexts involving mild alarm
Broken dee	Mated females	From completion of nest into early incubation
Variable see (sexual see)	Mated females	Near mate, including before copulation
Hiss	Females (only?)	When cornered, as in nest cavity ("snake mimicry")
Snarl	Mainly males (?)	In fights
Twitter (crackle)	Mated females (only?)	To mate near or in nest cavity
Tseet	Both sexes	Perched and in flight in a variety of situations
High zee (alarm zee)	Mainly males	When detecting predator or other alarming object
Scream	Both sexes (?)	When captured, as in banding
Squawk	Both sexes	In nest with young

Source: Modified from M. S. Ficken et al. (1978).

assertion follows from two facts: first, assignments of meanings ("messages") are often vague and general; and, second, in some cases more precise characterization has been proposed for variants of the types. Thus, the interpretations of signal types per se amount largely to statements of the behavioral contexts in which the signals are given, as summarized in Table 8.1. The table omits vocalizations of nestlings and fledglings (for which, see Clemmons and Howitz 1990), adds types not included in M. S. Ficken et al. (1978), and revises from the latter paper some of the contexts characterizing the vocalizations listed. Attempts to assign more precise meanings to the types usually founder because of readily observed exceptions.

Frequency shifts in fee-bees. The fee-bee is often cited as "the song" of the Black-capped Chickadee without explicit recognition that it is variable and that the variations could be information laden. Early observers reported hearing males shift the frequency of their fee-bee song during a bout of singing (e.g., A. A. Saunders 1935, Odum 1942), and such a shift was documented by Ratcliffe and Weisman (1985) in each of four captive males. Later recordings of predawn singing in wild birds showed that each male sang a few (two to four) frequencies and might sing up to 200 times before shifting frequency (Horn et al. 1992). The frequencies at which a male sang, however, were not fixed like types in a conventional song repertoire; rather, they changed over time. It appeared as if males were matching the frequencies of neighbors, and a playback experiment showed that males matched with high accuracy the frequencies of songs played to them.

Acoustic frequency and duration of high (alarm) zees. The variation found in both the acoustic frequency and the duration of high (alarm) zees appears to encode information. As indicated in Table 8.1, these calls are given when birds detect predators (as well as in other alarming contexts). Like many other elements in the repertoire, zees vary in acoustic frequency. M. S. Ficken and Witkin (1977) recorded zees given to three species of predators: Sharp-shinned Hawks (*Accipiter striatus*) flying within 40 meters of a chickadee flock, a Northern Saw-whet Owl (*Aegolius acadicus*) tethered near a feeder, and a mink (*Mustela vison*) on the ground that approached a feeder with chickadees nearby. Zees were also heard in response to a Northern Shrike (*Lanius excubitor*) near a feeder, but a perched Northern Goshawk (*Accipiter gentilis*) elicited only chick-a-dee calls and Red-tailed Hawks (*Buteo jamaicensis*) soaring in the distance were ignored. High zees are thus given when a truly threatening predator is detected but not to one that is not hunting (the goshawk) or to a raptor in the distance that does not ordinarily prey on small songbirds (the Red-tailed Hawk).

The main point here is that the high (alarm) zees differed in their highest acoustic frequency according to the predator (means and standard deviations): 8.87 ± 0.20 kHz in response to Sharp-shinned Hawks, 8.55 ± 0.19 kHz to the Saw-whet Owl, and 7.76 ± 0.17 kHz to the mink. Durations also varied, being 0.10 ± 0.03, 0.10 ± 0.02, and 0.07 ± 0.01 seconds, respectively, for the three predators. The mink, which was on the ground below the chickadees and thus less of a threat than the two avian predators, elicited shorter zees of lower acoustic frequency, suggesting that quantitative variation in zees encodes the degree of danger. Although these were field observations and therefore the sample size was small, they suggest a relationship between acoustic features of the call and the information encoded. Therefore, whereas the zee type per se does indicate that the caller has detected a potentially dangerous situation, the call could also act as a carrier vehicle for more precise information concerning the degree of danger.

Quantitative variation in chick-a-dee calls. The quantitative acoustic variation in chick-a-dee calls may encode information in addition to that encoded by the combinations of notes. Members of wild flocks have calls that are more similar to one another than to calls of chickadees from other flocks, and birds from different flocks captured and kept together changed their acoustical characteristics to be more like one another (Mammen and Nowicki 1981). Furthermore, playbacks of chick-a-dee calls at a feeder elicited different responses: playbacks of calls from the same flock caused no changes above baseline in ongoing behavior, whereas playback of calls from a different flock caused elevated calling and decreased foraging (Nowicki 1983). Chick-a-dee calls also vary sufficiently among individuals that the differences could be used for individual recognition (Mammen and Nowicki 1981).

The foregoing examples show that qualitative types of vocalizations per se are only part of the story: variants on the types can also carry useful information. Even in cases for which no analyses exist, the nature of variants suggests they carry information that could be used by companions of the caller. For example, a listener should be able to tell whether a caller is facing toward or away by comparing the amplitudes of notes within a chick-a-dee call, because the A and C notes radiate more strongly in front of a caller, and B and D notes radiate uniformly in space.

Do Related Species Have Similar Repertoires?

Despite diverse investigators, methods, and so on, studies of members of the subgenus Poecile show a remarkable similarity in the overall organization of their vocal repertoires (Table 8.2). At least in the sense that named types of vocalizations can be matched by general acoustic structure, the repertoires are clearly equivalent among these closely related species.

The subgenus Poecile provides the best material for comparative analysis of tit vocalizations because several species have been studied by different researchers working independently (see Table 8.2, and below). The first repertoire-like study of any tit that contained (a few) sonograms was by Gompertz (1961) on the Great Tit (*Parus major*), but no other species in the subgenus Parus appears to have been studied extensively. Similarly, the repertoire of the Blue Tit (*P. caeruleus*) was studied by Bijnens and Dhondt (1984), but the other member of its subgenus, Cyanistes, has not been studied. The repertoires of Great and Blue Tits obviously differ from those of the Poecile species (Table 8.2), as well as from one another, but detailed comparisons among subgenera would be premature at this time (see Hailman 1989 for a comparison of principal vocal types). Some vocalizations (e.g., song) of members of other subgenera have been studied, but not complete vocal repertoires (again, refer to Hailman 1989).

Although several researchers have independently studied the repertoires of Poecile species, some were aware of the work of the others, a fact that might have contributed to the similarities they reported. Nevertheless, this influence cannot have been present in every case, as the sequence of studies shows. Odum's (1942) classical paper on the Black-capped Chickadee was prespectrographic, rendering his verbal descriptions difficult to match with later objective evidence. The first sonograph-based repertoire studies in the subgenus Poecile were S. T. Smith's (1972) on the Carolina Chickadee (*Parus carolinensis*) and Haftorn's (1973) nearly simultaneous monograph on the Siberian Tit (*P. cinctus*). McLaren (1976) cited Smith's study but clearly conceived her repertoire of the Boreal Chickadee (*P. hudsonicus*) wholly independently. The first paper explicitly comparing one species' repertoire with other researchers' results on another was by M. S. Ficken et al. (1978) on the Black-capped Chickadee; those authors cited S. T. Smith (1972) and where possible adopted her terminology based on homologies judged

Table 8.2. Proposed vocal equivalents in the adult repertoires of six tit species (*Parus*, subgenus Poecile)

Black-capped Chickadee (M. S. Ficken et al. 1978)	Carolina Chickadee (S. T. Smith 1972)	Willow Tit (Haftorn 1993)	Mexican Chickadee (M. S. Ficken 1990 a,b)	Siberian Tit (Haftorn 1973)	Boreal Chickadee (McLaren 1976)
Fee-bee*	Song*	Song*	Peeta-peeta*, combinatorial song*	Trrryy*	Rapid musical call*
Faint fee-bee	Faint song*	Faint song		Soft trill#	
Gargle*	T-slink*, click-rasp*, rasp-slink*, tee-rasp*	Gargle*	Gargle*, ?trill#	Ptri-pyy, ptri-poi*, tjillup	Trilled call*, musical call*
Subsong*		Babble or subsong*			Warbled trilled & musical calls
Chick-a-dee (A*, B*, C*, D*)	Chick-a-dee (chip*, tee*, chick*, dee*)	Si-tää calls* and Pjä calls* (see text)	Chick-a-dee (A*, C*, D*; B* rare)	(Ti*, pst* tææ*, pev*, piv*, others?)	Chickadee (chit*, seep*, chicka*, dee*)
Broken dee*	Broken dee*	Pipae*		Begging#	Begging call*
Variable see* (sexual see)	Variable see*	Si-si-si	Variable see*	Sisisi[a]#	Solicitation[b]
Hiss*/snarl*		Hissing	Hiss*	Hiss#	Hiss*
Twitter* (crackle)		Twittering	Twitter*	Twitter#	Stutter
Tseet*	Tseet*	Sie*, sit*, and sit/ staccato* (see text)	Tseet*	Dytt#	Seep*
High zee* (alarm zee)	High see*	Zi-zi-zi alarm*	High zee*	Sisisi[a]#	Sharp seep
Scream#	Scream*/squeal*	=?Spitt*, distress			Squeal*
Squawk#				Squawk#	Squawk#

Note: The three species on the left may be especially closely related, as are the two on the right; the relationship of the Mexican Chickadee to the others is less certain. Vocal equivalents among species are based on acoustic similarities except for fee-bee equivalents (see text). Vocalization names in initial caps are those used in the reference cited; others are names assigned here. An asterisk (*) indicates sonograms in the reference cited; a pound sign (#) indicates that the authors of this chapter have recorded vocalizations that are not illustrated in the reference cited.

[a]Transcribed identically but noted in text that these are two different calls.

[b]Call differs between the sexes.

by acoustic structure and context of use. More recently, M. S. Ficken (1990a, b) described the vocal repertoire of the Mexican Chickadee (*P. sclateri*) and Haftorn (1990) reported on the Willow Tit (*P. montanus*). Thönen (1962) actually provided earlier repertoire information for the Willow Tit but included sonograms of only song. His paper and those of Ludescher (1973) and Haftorn (1990) were all subsumed into a fine paper by Haftorn (1993) that does present extensive sonograms and also traces the development of vocalizations in the Willow Tit. Finally, the original study of the Siberian Tit has been augmented by considerable new material (Hailman et al. 1994, Hailman and Haftorn 1995). In sum, extensive repertoire information exists for a half dozen Poecile species (Table 8.2).

Even within the subgenus Poecile, distinctions of phylogenetic relatedness are evident. The Carolina Chickadee is a geographic replacement for the Black-capped Chickadee; the two species hybridize in some areas of parapatric contact, and they are exceedingly difficult to separate by appearance. They are therefore assumed to be very closely related, despite the allozyme analysis by Gill et al. (1989), which is at odds with other taxonomic treatments of Poecile. Similarly, the Willow Tit of Eurasia resembles the Black-capped Chickadee so strongly that the two were formerly considered conspecific representatives on different continents. The AOU Check-list (1983) considers the three to constitute a probable superspecies. In parallel, the three brown-backed members of the genus—Siberian Tit, Boreal Chickadee, and Chestnut-backed Chickadee (*Parus rufescens*)—have long been considered to constitute a probable superspecies (see, e.g., AOU 1983). The Mexican Chickadee's relatives have always been uncertain, the species often being allied with the Black-capped Chickadee group (see, e.g., AOU 1983), but recent molecular evidence from mitochrondrial DNA places this form with the brown-backed tits (Gill et al. 1993). Table 8.2 therefore lists the Mexican Chickadee between the other two species groups.

We determined equivalent repertoire elements through similarities in acoustic structure (except for the proposed fee-bee equivalents discussed later). Insofar as we are aware, Table 8.2 represents the first explicit comparison of "total" vocal repertoires among closely related species of any birds, perhaps of any animals. It is clear to us that the striking similarity in repertoire organization among species is real, not an artifact of workers unduly influencing one another or of our forcing the data to fit into an ideal set of categories. This conclusion is strengthened by the fact that repertoires of *Parus* species in other subgenera cannot be fitted to the scheme of Table 8.2. For example, evident homologies to elements in the repertoires of the Great Tit (Gompertz 1961) or Blue Tit (Bijnens and Dhondt 1984) are elusive, although some general similarities can be found (Hailman 1989). Furthermore, our preliminary analysis of the Bridled Titmouse (*P. wollweberi*) of the subgenus Baeolophus suggests that, among other differences, this species lacks gargles or any equivalent. In sum, it appears to be the close relatedness among Poecile species that renders their vocal repertoires so readily comparable.

What Difficulties Arise in Equating Vocalizations?

Despite the successful comparison among Poecile species shown in Table 8.2, a major difficulty is encountered in proposing equivalences among vocal types of different species based on reports in the literature. Even if the authors showed many sonograms of variants within a recognized vocal type, rarely is the journal space available to characterize variation sufficiently well that readers can draw their own conclusions about how to designate types. Therefore a reader often cannot judge whether two species truly differ in some way, or, alternatively, whether two authors confronted with the same kinds of variation chose to divide and unite vocalizations differently when recognizing types. Future studies must move toward more quantitative analyses, as exemplified by the types shown in Fig. 8.2, if this ambiguity in typing is to be overcome.

An obvious example of the problem of recognizing types is the Carolina Chickadee's equivalents of the Black-capped Chickadee's gargle. S. T. Smith (1972) recognized vocal types that she called t-slink, click-rasp, slink-rasp, and tee-rasp, thus implying five kinds of notes (T, slink, click, rasp, and tee) that are combined in various ways. It seems clear from both her text and sonograms, however, that considerable variation exists and that many other note types could be recognized. Smith, of course, did not have the benefit of the detailed analyses of the gargle system in the Black-capped Chickadee published subsequently (M. S. Ficken and Weise 1984, M. S. Ficken et al. 1985, 1987, M. S. Ficken and Popp 1992). Although in hindsight it seems likely that the Carolina Chickadee has a gargle system essentially like that of the Black-capped Chickadee, the possibility remains that Smith's intuition correctly recognized four main types of gargles in the former species.

Chick-a-dee calls show that the problem with the gargles is typical rather than being an isolated example. S. T. Smith (1972) recognized five note types that are combined with one another in various ways—namely, high see, high tee, chick, dee, and loud tee—and stated (p. 46) that "High Tee, Chick, and Dee, in series, comprise the well-known 'Chick-a-dee' call." On this basis should we say that the species' chick-a-dee calls are composed of three note types or five? Furthermore, the published sonographic evidence raises other questions. For example, the high tee notes of Smith's fig. 2.15A (1972, p. 76) appear to resemble more closely the loud tee notes in the same sonogram than they do the ostensible high tee notes of her fig. 2.15C. One can with some confidence equate the Carolina Chickadee's dee with the Black-capped Chickadee's D note, and the former's chick with the latter's C note. Nevertheless, ready equivalences end there, and it is unclear how introductory chevron-shaped notes are to be sorted in the Carolina Chickadee.

Furthermore, not every named vocalization necessarily has an obvious equivalence among species. For example, S. T. Smith (1972) named a chip and an abrupt tee from the Carolina Chickadee, and Haftorn (1993) the plui of the Willow Tit. These calls were omitted from Table 8.2 because of their uncertain equivalences to vocal types in other species. Haftorn (1993) believed the plui to be a locally unique call occurring in Willow Tits only in his study area.

These kinds of problems, which are by no means restricted to comparisons between Black-capped and Carolina Chickadees, render judgments of equivalency among repertoires of even extremely closely related species fraught with dangers. Future studies will involve the analysis of large sample sizes of recordings from color-banded individuals in a wide variety of contexts. Even so, spectrographic analysis will not resolve all the problems, and arbitrary decisions will have to be used in describing repertoires, particularly for graded and highly variable vocalizations. The very fact that a comparison such as Table 8.2 is even possible not only reflects what must be true similarities, but also stands as a tribute to the thoroughness and objectivity of many field investigators.

How Do Species' Repertoires Differ?

Vocal types may differ among species in at least three ways. The first way is in types themselves. Furthermore, even equivalent—and presumably homologous—vocal types may differ in either acoustic structure or context of use. Both kinds of differences promise to reveal important aspects of how communication systems evolve, but even in the well-studied Poecile tits the comparative evidence is adequate merely to raise some tantalizing speculations.

Type differences. We emphasized above that without quantitative analysis, it is extremely difficult to decide whether reported differences in types are real species differences. Haftorn's (1993) study of the Willow Tit provides two cases in point. First, he distinguished si-tää contact calls from pjä agonistic calls, both of which are within the range of variation of the Black-capped Chickadee's chick-a-dee calls. Similarly, he distinguished between sit foraging calls and sie flight calls, both of which fall within the range of the Black-capped Chickadee's single A notes (Fig. 8.1E). We have no reason to question the likelihood that calls in two different contexts differ acoustically, although the exact way in which they differ remains to be clarified. Nonetheless, to resolve whether these are qualitatively different types of calls or information-laden variations of one type will require detailed quantitative study. In sum, it is possible that (1) the Willow Tit divides its repertoire into finer categories than the Black-capped Chickadee; (2) studies of the chickadee have failed to distinguish existing types matching those of the tit; or (3) in both species, si-tää/pjä (chick-a-dee) and sit/sie (A note) calls vary continuously, and one investigator emphasized subtypes as separate while another emphasized the continuity of the types.

Acoustic differences. The first three species listed in Table 8.2 (Black-capped Chickadee, Carolina Chickadee, and Willow Tit) live primarily in hardwood

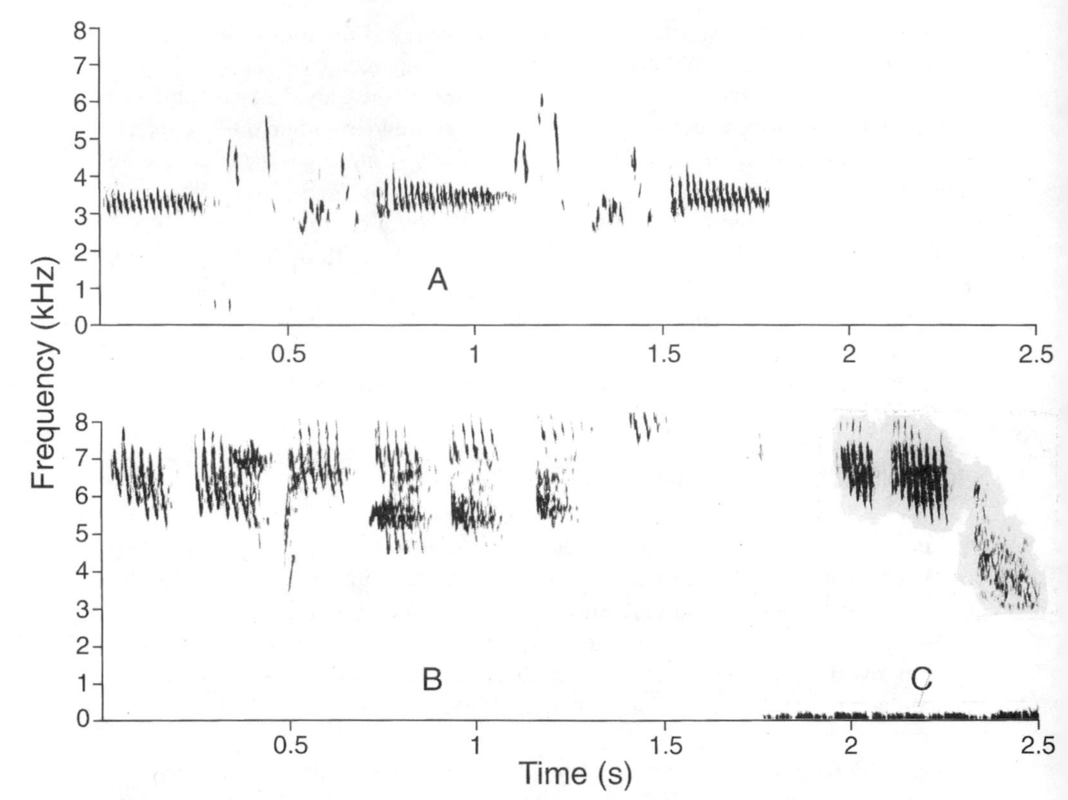

Fig. 8.3. Buzzlike quality in vocalizations of the Mexican Chickadee. (A) Repeated vocalization given in predawn singing, showing trills combined with some other elements. (B) High (alarm) zees given by a pair of birds in response to an *Accipiter* hawk flying high overhead. (C) Chick-a-dee call of notes AAD.

forests, and the other three (Mexican Chickadee, Siberian Tit, and Boreal Chickadee) inhabit primarily conifers. The second group has a noticeably more buzzlike quality to its vocalizations, apparent both to the unaided ear and in sonograms (Fig. 8.3). Here the issue of correlated characteristics again intervenes. The A notes of the chick-a-dee calls of the Mexican Chickadee (Fig. 8.3C) clearly differ from those of the Black-capped Chickadee (Fig. 8.1A), in that the former have a buzzing quality produced by rapid frequency modulation. The high (alarm) zees (Fig. 8.3B) and, perhaps to a lesser extent, the variable (sexual) sees also have this buzz structure, unlike the equivalent vocalizations of the Black-capped Chickadee. Furthermore, the male Mexican Chickadee also utters a repeated buzzing trill (Fig. 8.3A) during predawn singing; both the form of the call and its context are similar in the Siberian Tit (Haftorn 1973, Hailman et al. 1994).

One tempted to muse over the relation between acoustic structure and the habitat in which vocal signals are transmitted is faced with an obvious problem. The first three species are more closely related to one another than to any species in the second group, and the other three species similarly appear to be more closely related to one another than to any species in the first group. The sorting of similarities into those reflecting adaptations to habitat versus those stemming from common ancestry is a familiar problem to comparative zoologists. It is usually best resolved by seeing whether relatively unrelated species living in the same circumstances share traits by evolutionary convergence (Hailman 1988). Work on additional species of tits is clearly needed, and it might be instructive to do a large-scale comparison of coniferous versus deciduous forest species for a variety of taxa.

Usage differences. Characterizing similarities and differences with regard to the way homologous vocalizations are used among species is an even more difficult problem. It is especially so because the precise way vocalizations are used is often not understood for any species, much less for a whole comparative array of congeners. Comparisons among species, however, reveal both similarities and differences.

As was pointed out above, the high (alarm) zees of Black-capped Chickadees have a higher acoustic frequency when the danger is greater. Exactly the same relation occurs in the Mexican Chickadee (M. S. Ficken 1989). Calls given to hawks flying high in the distance (Fig. 8.3B) and to a Steller's Jay (*Cyanocitta stelleri*) that flew suddenly over the chickadee flock, ranged in highest frequency from 8.5 to 8.7 kHz (four incidents) whereas those given when detecting the dangerous Sharp-shinned Hawk within 30 meters were 9.0–9.2 kHz. Such calls are difficult to record in the field because of the rarity of a predator encounter, but even these small samples show a statistically reliable difference. Haftorn (1993, p. 275) remarked that the equivalent zi's of the Willow Tit vary in "call duration (number of notes), pitch or amplitude," and he emphasized that number of zi's in particular varies according to the degree of the caller's fear.

Contrast that similarity in the use of zees with the use of hiss calls among species. The hiss has long been known from a large number of tit species (e.g., Pickens 1928, C. H. Sibley 1955, Thielcke 1968), and in all reports is given only by birds in the nest (either females or nestlings) when the nest is disturbed. Observers have interpreted this call as mimicry of snakes, functioning to deter would-be nest predators. The Mexican Chickadee, however, uses the hiss in a wide variety of circumstances, especially when close to other birds in fall flocks and along with gargles during territorial encounters in the spring (M. S. Ficken 1990a). The only tendency toward common elements of use in the incidents described seemed to be hesitation to move because of some fear-provoking stimulus.

Another possible difference in usage among species relates to a little-studied type of call. The scream of the Black-capped Chickadee was recorded from a bird in the hand during banding and has not so far been recorded in any other circumstance (Hailman, unpubl. data). Haftorn (1993) described a nearly identical vocalization, termed the spitt, which the Willow Tit commonly gives in "alertness" and "alarm." Haftorn also mentioned (p. 267) a "distress" call, which might be contextually closer to the Black-capped Chickadee's scream (but he did not describe this call). The Willow Tit's spitt is commonly mixed with other types recognized by Haftorn (1993), a finding that supports the phenomenon of mixed types described earlier in this chapter for the Black-capped Chickadee.

What Is "Song" and How Did It Evolve?

The subject of song in Poecile tits has long been a source of contention. The first line in Table 8.2 contains the most songlike vocalizations of the species. The correspondences among species represented there were not based on acoustic similarity as in the other tabulated lines. Rather, after the other lines were determined for a given species, one prominent vocalization remained in every case, and it was placed on the same line with the Black-capped Chickadee's fee-bee. In some cases, but not all, the original author designated this vocalization the species' song. The "song" of the Carolina Chickadee is acoustically similar to the Black-capped Chickadee's fee-bee, but the other entries are structurally diverse.

"Song" in Poecile species. Authors have had difficulty fitting chickadee vocalizations into commonly accepted concepts of passerine song. Dixon and Stefanski (1970) showed that the fee-bee of the Black-capped Chickadee is used more to attract rivals than to repel them in territorial interactions (gargles being used in actual encounters). M. S. Ficken (1981) even wondered if song were being evolutionarily eliminated in this species. Haftorn (1973) reviewed the known functions of song and concluded that one of the gargle types of the Siberian Tit most closely fit the pattern of "song," but Holm (1982) argued that repeated trilling (Haftorn's trrryy in Table 8.2) was the song of this species. McLaren (1976) noted that the Boreal Chickadee uses no vocalization to advertise territory, so in this sense the species lacks song. Among uses of the described vocalizations, McLaren's rapid musical call is the most songlike.

This uncertain situation can be turned to analytical advantage by abandoning the preconception that vocal repertories can be dichotomized into song and calls. Hailman (1989) argued that a unitary view of song discourages questions about specific uses of vocalizations. When the concept of unitary song is abandoned, one can see much evidence in Poecile tits for three vocalization types—gargling, chick-a-dee calls, and the vocalization listed in the first line of Table 8.2—which together serve the functions apparently served by a single type of song vocalization in many other oscine birds (Hailman 1989).

Comparisons of the reputed song vocalizations of Poecile species show that they are acoustically diverse and therefore without obvious homologies. Three species—the Black-capped and Carolina Chickadees of Table 8.2 plus the closely related Mountain Chickadee (*P. gambeli*)—all have "whistled" pure-tone vocalizations that are clearly homologous. In sharp contrast, the Siberian Tit's trill (trryy) consists of rapidly modulated sounds of considerable frequency spread, actually resembling the terminal note of one type of gargle in this species (Hailman et al. 1994). The Boreal Chickadee uses its gargle homologue, the trilled call, in close agonistic encounters, but in more advertisement-like contexts gives the rapid musical trill, which consists essentially of the first part of the trilled call (McLaren 1976). Even more confusing is the Mexican Chickadee, which uses two types of "songs" (M. S. Ficken 1990b), one of which is combinatorial in ways reminiscent of chick-a-dee calls. In addition, Mexican Chickadees trill in their predawn serenades (previously undescribed; see Fig. 8.3A). This vocalization resembles the repeated two-note trilling of the Siberian Tit (Haftorn 1973, Hailman et al. 1994) but often includes a jumble of other notes as well, and so appears to be a form of gargle. Whatever else one may conclude from this acoustic diversity, it seems clear that song is not homologous among species of Poecile tits.

Predawn serenades. A wholly different aspect of singing also arises from repertoire studies. When the predawn serenades of a male Siberian Tit, given just outside the cavity in which the female was roosting, were recorded, it was found that the bird did not give just one kind of vocalization like unitary song (Hailman et al. 1994). Instead, he sang trills, gargles, chick-a-dee calls, and a variety of different call notes, all of which are also used in other circumstances. Although this phenomenon has gone largely unnoticed, it also occurs to a lesser extent in Black-capped Chickadees, whose predawn serenades include various vocal elements in addition to fee-bees (Hailman unpubl. data, Ficken unpubl. data, K. Otter pers. comm.). Similar serenading has been recorded in Willow Tits (Hailman et al. 1994). In fact, this phenomenon might be quite general; we have recorded similar occurrences in Mexican Chickadees and Bridled Titmice. It may go unnoticed by some researchers because they are concentrating on the "song" rather than on the total vocal repertoire.

Predawn singing at the nest appears to be primarily a precopulatory serenade of the female. Serenades at the nest cavity apparently were discovered in the Great Tit by R. Mace (1987a), who did not study the vocalizations used but did document the context. Serenading occurred before the female had emerged from the night's roost, and the pair copulated when she emerged. Mace noted that serenading increased as egg laying drew near, then fell off with the beginning of laying. The female Great Tit lays an egg before emerging from the night's roost; because the passage of the egg enlarges and clears out the oviduct, she may be especially fertile just after laying. Virtually the same events surrounding predawn nest serenades apply to the Siberian Tit (Hailman et al. 1994). Serenading begins at least several days before laying and falls off rapidly after the first few eggs are laid. The female of this species emerges from the night's roost, forages, then returns to the cavity to lay an egg. The male serenades her both before dawn (before she has emerged from the night's roost) and also while she is laying; in both circumstances, the pair flies off together quickly after the female's emergence, and copulation takes place.

Origin of song. The question of how song evolved in any oscine species is largely unexplored, presumably because song is regarded as so highly evolved that no primitive conditions exist to provide clues as to its origin. In this respect, avian song resembles human language, whose origins are equally elusive. If the Poecile tits represent a primitive condition, or even a secondarily derived simplification, they suggest a new hypothesis concerning the evolution of song (Hailman et al. 1994). We envision the ancestral male singing to his mate by uttering many of the various elements in his vocal repertoire, these carrying adequate markers for clear recognition of species and sex, and probably of the individual singer himself. Selection could consolidate some of these vocal elements into a specific vocal type promoted by reproductive hormones and thus used ultimately for both mate attraction and territorial defense. Predawn serenades of male tits, incorporating as they do vocal elements other than the reputed song of the species, may reflect the incomplete differentiation of a specific song vocalization. It is further possible that the singing directed to the mate and that used for territorial defense differ in composition, and only the vocalizations directed at the mate retain "nonsong" elements.

This new hypothesis concerning the evolution of song might help to clarify some otherwise troublesome aspects of parid vocalizations. For example, perhaps the vocalization type designated song bears no evident homology among some species within the same subgenus because song arose independently, and probably incompletely, from different vocal elements in different species. So, in some cases tit song resembles gargles, in other cases chick-a-dee calls, and perhaps in still others both or neither. Researchers may have experienced difficulty in identifying tit song both because it is not necessarily homologous among closely related Poecile tits and because it has not become as evolutionarily differentiated as it has in many other kinds of songbirds.

Finally, these studies of parids lead one to question the continued utility of the concept of song as a unitary phenomenon. Clearly, no single vocalization in Poecile tits serves all the functions usually attributed to song. For example, intrapair signals and territorial advertisement seem to use two or more different vocalizations. Unitary song in migratory passerines is apparently used by males both to defend a nesting territory from other males and to attract potential mates to it. In tits, which are not migratory, pair formation takes place within flocks before the birds become territorial in spring, and pair-bonds usually persist through the lives of the mates. Pair formation is therefore a phenomenon primarily of birds in their first year of life; it appears to begin as early as a few months after dispersal from the natal territory and may be a gradual process extending through the fall and winter (e.g., S. M. Smith 1991). The life history of tits is thus distinct from life histories of typical migratory north-temperate passerines, and the organization of vocal repertoires may reflect this difference. Males of migratory passerines must do things rapidly in order to breed successfully: establish a territory, attract a mate, and inseminate her. It may therefore be parsimonious to have a unitary song vocalization that serves many uses simultaneously. Tits, by contrast, have the leisure to form pairs over many months and to establish territories carefully in locales where they already know the habitat and the likely neighbors. One might therefore expect male tits to employ different vocalizations for pair formation, territorial advertisement, close boundary encounters, and precopulatory motivation of the female. It is this diversity of vocalizations that provides a window into the evolution of unitary song, and similar organizations of vocal repertoires should be looked for in other nonmigratory passerines such as jays (Corvidae).

What Else Can Comparative Studies Offer?

Comparisons among species can offer one more insight: the August Krogh principle. The Nobel Prize–winning physiologist articulated the principle that, for a given problem in biology, there may be just the right species or system with which to solve it. What *Drosophila* is to genetics or the giant squid axon is to neurophysiology, the Mexican Chickadee has been to decoding chick-a-dee calls. This species has proven instructive for two simple reasons: although the Mexican Chickadee uses four note types in chick-a-dee calls, one (the B note) is so rare as to be virtually absent, and the calls themselves have few component notes (Fig. 8.3C is typical in this regard). The call system is therefore markedly simpler than that of the Black-capped Chickadee.

Without recounting recently published evidence (M. S. Ficken et al. 1994), it is sufficient to say that note types can be correlated consistently with specific circumstances. In brief, A notes indicate a tendency for the caller to move some distance in space, C notes are given to a disturbing stimulus and show a tendency for the caller to alter direction of movement, and D notes connote a perched caller. These identifications do not explain chick-a-dee calling completely; we believe that repetitions of the same kind of note within a call may signal the "intensity" of behavior, and this hypothesis is potentially testable. We cannot say whether combinations of note types in a call encode something more than the simple concatenation of their separate meanings. But we do have a start on decoding this complicated, combinatorial vocal system, which is less tractable to analysis in the Black-capped Chickadee, in which it was first characterized.

Looking more widely in the genus *Parus,* many species other than Poecile tits appear to have combinatorial chick-a-dee-like calling systems (Hailman 1989). Recently analyzed calls of the Black-lored Tit (*Parus xanthogenys*) have revealed a new phenomenon: permuted note sequences (Hailman 1994). Chick-a-dee calls of Poecile tits appear always to string note types together in a fixed order, such as AABCDDD; exceptions are very rare (less than 1% of calls in the Black-capped Chickadee). In the Black-lored Tit, calls such as ACCCA or ACAABB occur regularly. Note types are not freely permuted, however; they are under constraints best characterized by the fact that B, C, and D notes rarely occur together in a call without at least one intervening A note. The A notes occur at the beginning or end of calls, or between strings of two other types of notes, or even within strings of one other type. The rare Black-lored Tit of the Indian subcontinent is in the same subgenus with the more accessible Great Tit of Eurasia, so comparative study of members of this group might give greater insight into the evolution of combinatorial calling systems.

Conclusions

The vocal repertoires of tits (*Parus*) in general, especially of chickadees and other tits of the subgenus Poecile, are perhaps the best-analyzed complex repertoires of any group of animals. As such they serve as a model for investigation of other kinds of species. Nevertheless, we view the initial analysis of repertoires as merely the foundation for future studies that will address interesting evolutionary and ecological questions about communication. The search for general principles must involve birds other than tits, and perhaps other kinds of vocal animals such as squirrels and primates among the mammals.

Consider, for example, the vocal variety of predawn nest serenades in tits. Spector (1992, p. 227) noted that the male Yellow Warbler (*Dendroica petechia)* "sings a sustained bout before sunrise that is rarely and only briefly equaled during the day for the number of song types used, the immediate variety (singing without immediate repetition of song types) with which they are sung, *the chiplike notes used between songs,* and the rapidity of delivery" (emphasis added). Dawn singing is more elaborate than singing at other times in other species as well (Spector 1992).

For the reasons discussed in this chapter, we reemphasize that an avian vocal repertoire is poorly characterized as a mere list of vocal types, like words in a dictionary. In fact, even when designating vocal types proves to be straightforward, simple typing discourages the analysis of both the variation within types and the often intimate relations among types. Ample evidence shows that at least some variations within a vocal type are not merely annoying jitter but important information codes. Characterizing such variation may therefore be the single most important key to understanding how vocal signals transmit their information, and what that information is.

Despite numerous problems besetting direct comparisons of repertoires among even very closely related congeners, equivalences can be drawn. The comparative method thus opens the door to tracing evolutionary divergence and specific ecological adaptations in avian vocal communication. Comparative studies must, however, heed the dual problems of phylogenetic dependence and correlated characters in attempting to identify adaptations.

We believe that the study of repertoires has already proven its worth in focusing attention more sharply on principles of how vocal signaling "really" works, and furthermore has proved instructive in providing new viewpoints concerning problems of specific vocal types (e.g., the origin of song). Comparative analysis not only promises a keener understanding of the evolution of signals and their ecologically adapted variations, but also uncovers species whose signals may be more amenable to experimental analysis than the species in which a phenomenon (e.g., combinatorial call structure) was first characterized. Despite the progress and promise, though, we deem it prudent to end with a cautionary note. Repertoire study—perhaps more than any other type of investigation of avian vocalizations—is immensely complicated because everything is connected with everything else. This fact leaves one with the distinct impression that one can never know enough about the whole to be truly certain of any one part.

Acknowledgments

Many collaborators, colleagues, students, and hosts have facilitated our studies of parid repertoires—too many to be named here; however, we especially thank Svein Haftorn and Elizabeth D. Hailman for extensive help, endless discussions, and faithful companionship in the field.

Note on Terminology

Agreement concerning uniform names for three vocalizations of Poecile tits (e.g., Tables 8.1 and 8.2) was reached among four current investigators on the subject (M. S. Ficken, S. Haftorn, J. Hailman, and M. McLaren). The new names were first used by Hailman et al. (1994) and Hailman and Haftorn (1995), and are provided as synonyms in this chapter. What is known variously in the literature as twitter and stutter was renamed crackle because it sounds like a crackling fire but does not sound like twittering or stuttering. High zees and variable sees were renamed alarm zees and sexual sees, respectively. We understand the undesirability of using functionally related terms such as *alarm* and *sexual,* but these are merely descriptors; authors may use the unadorned terms *zee* and *see* as alternatives. *High* is not a good distinguishing descriptor because sees are often higher in frequency than zees in all species studied. Similarly, *variable* is not a good distinguishing descriptor because zees as well as sees are variable in all the species studied. It should also be noted that onomatopoetic names are not unambiguous because their pronunciation depends on the phonemes of their language of origin. For example, the name *dytt* (Table 8.2) uses a Norwegian phoneme [y] that does not occur in English, so the word cannot be pronounced (properly) in English. It is therefore useful to have translatable names, such as gargle, or translatable descriptors, such as alarm, when designating avian vocalizations.

9 Acoustic Communication in a Group of Nonpasserine Birds, the Petrels

Vincent Bretagnolle

Animal communication has long been a subject of interest to ethologists (e.g., Tinbergen 1952, 1959), and although the concepts and theories have changed, the study of avian bioacoustics has played a central role in developing the field of animal communication (W. J. Smith 1977, R. Dawkins and Krebs 1978, Zahavi 1979a; reviewed in Krebs 1991). Birds communicate primarily through acoustic means: they use vocalizations and nonvocal sounds in territory establishment and defense, mate attraction, pair-bond maintenance, and parent-offspring relationships (Catchpole 1982, Kroodsma and Miller 1982, Searcy and Andersson 1986, Kroodsma and Byers 1991).

Unfortunately, although bird bioacoustics hold a central place in communication studies, the overwhelming majority of functional studies on bird vocalizations have been on passerines, especially oscines. This pattern is particularly true for the studies concerned with sexual selection, species-specific recognition, and geographic variation (Catchpole 1980, Krebs and Kroodsma 1980, Becker 1982, Mundinger 1982, Kroodsma et al. 1984). Though they comprise nearly half of all bird species, songbirds are not necessarily typical of the class Aves. First, though cultural transmission of vocal characteristics is strongly developed in oscine passerines, it is unknown in the suboscines (approx. 1000 species), and outside the Passeriformes has been reported only in parrots, hummingbirds, and perhaps some other groups (Kroodsma 1982a, Kroodsma and Baylis 1982, Baptista, this volume). Second, passerine vocalizations are frequently interpreted in terms of sexual selection theory because characteristics of male calls affect mate choice (Searcy and Andersson 1986, Lambrechts and Dhondt 1987, Alatalo et al. 1990). Oscines, however, usually have feeding territories, and females may be just as likely to base their choice on the quality of the male's territory as on his vocalizations (Radesäter et al. 1987, Arvidsson and Neergaard 1991). Third, in many oscine species, visual signals also play an important role in advertising behavior, either in territory contests or in female attraction (Searcy 1986), although the relative importance of visual and acoustic cues has seldom been determined (but see Yasukawa 1981b, 1990, Metz and Weatherhead 1991). Fourth, many oscines use the same songs both for territorial defense and for mate attraction and are thus susceptible to both intra- and intersexual selection pressures (Krebs and Kroodsma 1980; but see Morse 1970, M. S. Ficken and Ficken 1973, R. B. Payne 1979, Radesäter et al. 1987, Radesäter and Jakobsson 1988).

Petrels, which constitute three of the four families of the Procellariiformes, offer an interesting comparison with the passerines, and with research on communication in general. They are a monophyletic taxon with many species (C. G. Sibley et al. 1988, Warham 1990), thus allowing comparative studies, and they feed exclusively on pelagic marine resources and therefore do not hold feeding territories. Though they actively defend their burrows from intruders of their own and other species, males do not guard territories. Female choice, if it occurs, should thus be based primarily on the characteristics of the males or, possibly, of the nest sites. Petrels show delayed sexual maturity (petrels usually breed for the first time at four to six years old: Warham 1990); during the prebreeding period, pair formation always takes several years. Delayed pairing offers the opportunity to study the temporal progression of the pair formation process as well as the possible effects of age and experience on mate choice (e.g., Bretagnolle 1989a). Finally, acoustics are the sole channel used by burrowing petrels for communicating between mates and rivals, a situation that is highly unusual in birds and results from their strictly nocturnal and fossorial habits (Bretagnolle 1990a, MacNeil et al. 1993).

In this chapter I provide the first comprehensive review of what is known about petrel vocalizations, covering all extant genera. I detail both the information content (message) of these vocalizations and their major functions, point out that research techniques must be adapted to this type of bird, and explore how petrel bioacoustics may provide interesting insights into areas of bird communication research.

Life Histories and Ethology of Petrels

Systematics. The order Procellariiformes comprises four families: Diomedeidae (albatrosses), Procellariidae (fulmars, gadfly petrels, prions, and shearwaters), Hydrobatidae (storm-petrels), and Pelecanoididae (diving-petrels). Warham (1990) took the term *petrel* to mean any procellariiform, but I follow the more usual restricted sense and exclude albatrosses. The term *burrowing petrels* refers to all petrels except the fulmar group. Petrels represent a significant proportion (approx. 30%) of the world's seabirds, comprising between 90 and 101 species (Jouanin and Mougin 1979, Howard and Moore 1980, C. G. Sibley and Monroe 1990, Warham 1990). The systematics and taxonomy of petrels are still unsettled (Warham 1990); the following information and names are from Warham (1990), except when specified.

Life histories. Petrels exhibit extremely diverse ways of life (Jouventin and Mougin 1981, Croxall 1984), ranging from the purely coastal species such as the diving petrels (*Pelecanoides*) to the strictly pelagic biennial breeders such as the White-headed Petrel (*Pterodroma lessonii*). All species are strictly monogamous, with very high partner fidelity between years (e.g., 93% in Cory's Shearwater, *Calonectris diomedea:* Mougin et al. 1987), and are colonial, though coloniality and fidelity vary according to species. Petrels also show delayed sexual maturity, varying from 2 years (*Pelecanoides*) to 12 (*Macronectes*) years; pair formation takes place during the several years before the birds start to breed.

Which channel is used for communicating? Nearly all species of petrels are either strictly or mostly nocturnal on their breeding grounds (Bretagnolle 1990a, Warham 1990, MacNeil et al. 1993). Petrels breed within deep burrows, which further limits their opportunities to see each other, and they strongly avoid moonlit nights (Watanuki 1986, Bretagnolle 1990a). Optical signals are thus totally absent from the communicative behavior of petrels at their colonies, but not at sea, where they are active during daytime (Bretagnolle 1993). The only exceptions are the six strictly diurnal species in the fulmar group, which use both optical and acoustic signals, as do the albatrosses (Luders 1977, Bretagnolle 1988, 1989a). Petrels have well-developed olfaction (Bang 1966), which they use to locate food (review in Verheyden and Jouventin 1994). Whether petrels also use their olfaction for communication is still under debate, but it seems unlikely (Hutchison and Wenzel 1980, Bretagnolle 1990b, unpubl. data; but see Grubb 1974). Colony location and homing were previously attributed to olfactory navigation (Grubb 1974), but more recent evidence is contradictory: according to James (1986), petrels use only visual cues to locate their burrows. Tactile communication has never been investigated but is apparently restricted to mutual preening (Bretagnolle unpubl. data). Therefore, sound is by far the dominant channel for communication in the petrels.

The Nature and Diversity of Petrel Vocal Repertoires

Petrels rely on sound for communication, so their acoustic repertoires would be expected to include different calls for different contexts of communication. In this section, I review the various types of vocalizations and discuss how petrels' nearly total reliance on acoustic signals may have affected their vocal repertoires.

Sound recordings and analysis. I tape-recorded petrel calls during more than 20 field trips between 1984 and 1994. My methods of analysis are described elsewhere (Bretagnolle 1989b, Bretagnolle and Lequette 1990, Richard 1991, Genevois and Bretagnolle 1994). During the years 1984–1994, I studied 45 species of petrels in the field, covering all 23 genera except for *Thalassoica* and *Halocyptena* (sound recordings of these genera and some other species were made available to me by other workers).

Description of vocal repertoires. The meaning of repertoires, how they should be studied, and how they should be presented are subjects of some debate (see

Schleidt et al. 1984, E. H. Miller 1988, Hailman and Ficken, this volume). For convenience, I distinguish "major" and "minor" calls: major calls are those used for pairing (including sexual and agonistic contexts), and minor calls are all others. This distinction roughly corresponds to the distinction between song and calls that is often made for passerines (see also Hailman 1989). To minimize bias introduced by the well-known large genera, I summarize our current knowledge of vocal repertoires by genus (Table 9.1). The complete vocal repertoire of Bulwer's Petrel (*Bulweria bulwerii*) is provided as an example (Fig. 9.1; see also Figs. 9.2 and 9.3 for additional examples of calls, and Bretagnolle 1988, 1989b for complete descriptions of vocal repertoires).

Without exception, all petrel genera have one or more minor calls and one or two major calls. These calls include the following six categories.

1. *Food-begging calls* apparently exist in the chicks of all species investigated so far, though little attention has been given to them (but see Brooke 1986, Bretagnolle 1988, 1989b, Bretagnolle and Thibault in press). Typically, food-begging calls are given during feeding events. They have also a submissive connotation, as suggested by the fact that they are often given when the chick is disturbed (e.g., by a human observer), and because similar calls are uttered by the adults of several species in apparently submissive contexts (Bretagnolle 1989b; *Halobaena:* Bretagnolle unpubl. data; *Pachyptila:* A. Tennyson in Marchant and Higgins 1990). Food-begging calls of the chicks are structurally very similar across species. Interestingly, chicks of all species of the fulmar group (no data are available on *Thalassoica* chick calls) have an additional call used exclusively for the purpose of food begging (Bretagnolle 1988, 1990c), a character they share with the albatrosses (Bretagnolle unpubl. data).

2. *Copulation calls* are apparently restricted to the shearwaters and some species of the fulmar group (Bretagnolle and Lequette 1990; Audubon's Shearwater [*Puffinus lherminieri*], *Macronectes, Fulmarus:* Bretagnolle unpubl. data). These calls are given during copulation but not before, so they cannot be considered precopulatory behaviors. This pattern differs from that found in most other seabird orders, which do have precopulatory displays (Tinbergen 1959, van Tets 1965, J. B. Nelson 1978, Jouventin 1982).

3. *Agonistic calls* are minor calls used in agonistic interactions. They are common in petrels, but are not found in all genera or apparently even in all species within the genera in which they do occur. They are present in several species of the genera *Oceanodroma* (Taoka et al. 1988, 1989b), *Pterodroma* (Grant et al. 1983, Bretagnolle 1995), *Procellaria* (Brooke 1986, Warham 1988a), and *Bulweria* (Fig. 9.1).

4. *Distress calls,* such as the calls birds make when handled, occur primarily in storm petrels (all species so far investigated) and some (possibly all) gadfly petrels. No alarm calls (i.e., calls given toward predators to alert conspecific birds) have so far been discovered in petrels, though an alarm visual display is known for *Macronectes* (Bretagnolle 1988). Warham's (1988a) statement that *Procellaria* has alarm calls is probably incorrect; these calls are more likely to be agonistic calls (cf. Brooke 1986, Bretagnolle unpubl. data).

Table 9.1. The vocal repertoires of petrels

Genus	No. of species	No. of species studied[a]	Food-begging call	Copulation call	Agonistic call	Distress call	Contact call	No. of minor calls	No. of major calls	Visual displays	Sources[b]
					FULMARS						
Macronectes	2	2	2	1	1	0	0	4	2	Yes	1,2
Fulmarus	2	2	2	1	1	0	1	5	1	Yes	3,4
Thalassoica	1	1	?	?	?	?	1	>1?	1	Yes	4
Daption	1	1	2	0	1	0	1	4	1	Yes	4
Pagodroma	1	1	2	1	1	0	1	5	1	Yes	4,5
					GADFLY PETRELS						
Lugensa	1	1	1?	?	0?	?	0?	>1	1	No	4
Pterodroma	24[c]	7	1	?	1 (2?)	1	1	>4	2 (3?)	No	4,6–12
Pseudobulweria[c]	4	1	?	?	1?	0	0	>1?	1 (2?)	No	4
					PRIONS						
Halobaena	1	1	1	0	0	0	1	2	1	No	13–15
Pachyptila	6	5	1	0	0	0	1	2	1	No	4,13,16

					SHEARWATERS						
Bulweria	2	1	1	0	1	1	0	3	2	No	4,17
Procellaria	4	4	1	?	1	0	0	>2	2	No	4,18,19
Calonectris	2	1	1	1	0	0	0	2	1	No	20
Puffinus	16[c]	9	1	1	0	0	0	2	1	No	4,21–24
					STORM-PETRELS						
Oceanites	2	1	1	0	0	1	0	2	2	No	25
Garrodia	1	1	1	?	0	0	0	>1?	2	No	4
Pelagodroma	1	1	1	?	0	0?	0	>1?	2?	No	4
Fregetta	2	1	?	?	0	1	0	>1?	2	No	4
Nesofregetta	1	1	1	?	?	1	?	>2?	2?	No	4
Hydrobates	1	1	1	0	0	1	0	2	2	No	4,26,27
Halocyptena	1	1	?	?	?	?	?	?	2	No	4
Oceanodroma	10[c]	3	1	0	1	0	0	2	2	No	4,28–31
					DIVING-PETRELS						
Pelecanoides	4	2	1	0	0	0	0	1	1	No	4

Note: For each genus, the number of call types is given for each category mentioned in the text. Note the differences between genera in the number of major calls.

[a]Including published and unpublished studies.

[b]*Sources:* 1, Bretagnolle 1988; 2, Bretagnolle 1989a; 3, Luders 1977; 4, Bretagnolle unpubl. data; 5, Guillotin and Jouventin 1980; 6, Grant et al. 1983; 7, Warham 1988b; 8, Warham 1979; 9, Bretagnolle 1995; 10, Bretagnolle and Attié 1991; 11, Tomkins and Milne 1991; 12, Warham et al. 1977; 13, Bretagnolle 1990c; 14, Genevois and Bretagnolle 1994, 1995; 15, Bretagnolle et al. unpubl. ms.; 16, Bretagnolle et al. 1990; 17, James and Robertson 1985a; 18, Brooke 1986; 19, Warham 1988b; 20, Bretagnolle and Lequette 1990; 21, James and Robertson, 1985b; 22, Brooke 1990; 23, Storey 1984; 24, James 1985; 25, Bretagnolle 1989b; 26, James 1983; 27, James 1984; 28, Taoka et al. 1988; 29, Taoka et al. 1989a; 30, Taoka et al. 1989c; 31, James and Robertson 1985c.

[c]Discordant with Warham's (1990) checklist.

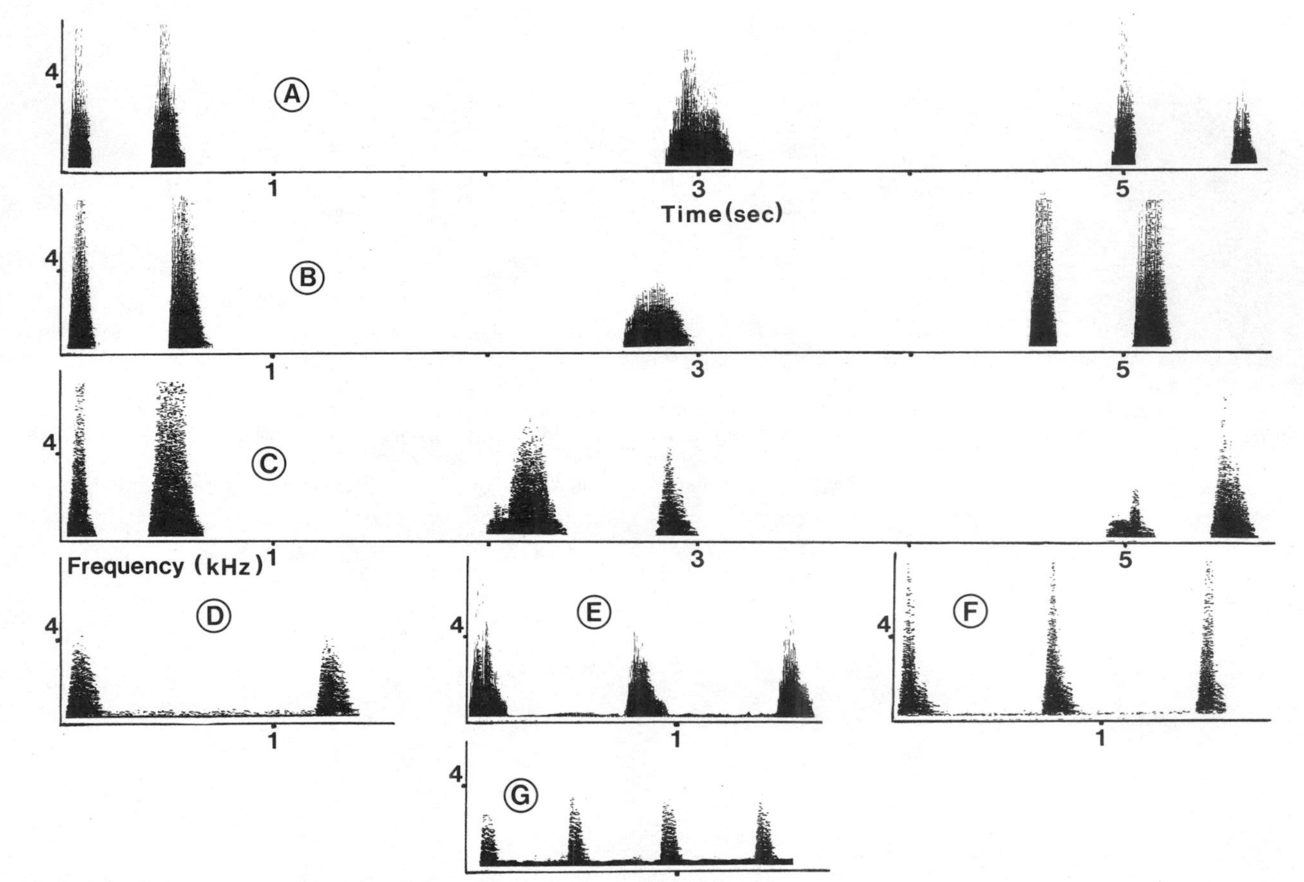

Fig. 9.1. The vocal repertoire of the Bulwer's Petrel (*Bulweria bulwerii*) contains two major calls (duet call and fast repeat call) and three minor calls (slow repeat call, an agonistic call; and food-begging call and distress call, not shown). Birds were recorded on Salvages Island, in the North Atlantic Ocean, with a Uher 4400 tape recorder and a Sennheiser MD 421 microphone. Sonograms were printed on a Kay Elemetrics 6061B Sonagraph, using narrow-band (45 Hz) and wide-band (300 Hz) filters. (A–C): Three examples of duet calls (A, B, two different males; C, female). (D–F): Three examples of slow repeat call (D, E, two males; F, female). (G) Fast repeat call (female).

5. *Contact calls* are short, repetitive calls given by birds in flight, presumably to avoid collisions. They are known in several genera, particularly *Pterodroma* (Grant et al. 1983, Bretagnolle and Attié 1991, Tomkins and Milne 1991).

6. *Major calls* are given primarily during courtship. Approximately half the genera have a single major call in their vocal repertoires; other genera have two. A typical example of the one-call type is Cory's Shearwater (Bretagnolle and Lequette 1990); Wilson's Storm-Petrel (*Oceanites oceanicus;* Bretagnolle 1989b) is typical of the two-call type. In the repertoires of one-call type species, the single call is always given by both sexes. In contrast, in two-call type species, one type is shared by both sexes, and the other is given either by the male only (e.g., *Oceanites;* Bretagnolle 1989b) or by both sexes (e.g., *Oceanodroma:* Taoka et al. 1988, 1989c; *Bulweria:* Fig. 9.1).

Acoustic repertoires of petrels compared with those of other birds. As expected, petrel vocal repertoires include calls that are used in all general contexts of communication: sexual, agonistic, and parent-offspring relations. As petrels have an "obligate" vocal communication system, it might be further predicted that their repertoires have achieved higher levels of diversity than those of other birds. Comparing petrel call diversity (e.g., number, variability in physical structure) with that of other birds is, unfortunately, impossible because complete descriptions of vocal repertoires that cover orders, or even families, are extremely scarce (but see W. J. Smith 1971, Jouventin 1982, Hailman 1989, Hailman and Ficken, this volume). Petrel repertoires typically include six to eight different call types, which certainly does not exceed the usual range of passerine repertoires. In fact, many passerine species have much larger repertoires (see, e.g., Marler 1956, Gompertz 1961, W. J. Smith 1977, Bijnens and Dhondt 1984), apparently because a full range of minor calls is lacking in petrels. Penguins have about as many calls in their repertoire as petrels do (three to six), but they also use optical signals (Jouventin 1982). Gulls, which also use a wide range of optical signals, have more acoustic signals than petrels (Tinbergen 1953, 1959). The only other seabirds that are strictly nocturnal on their colonies are several auklets from the North Pacific, and these also apparently have acoustic repertoires larger than those of petrels (e.g., nine calls in the Ancient Murrelet, *Synthliboramphus antiquus:* Jones et al. 1989). Also, diurnal petrels do not show impoverished acoustic repertoires compared with the burrowing petrels (Table 9.1, Bretagnolle unpubl. data). Thus, there has been no major diversification of calls in burrowing petrels, perhaps because of phylogenetic inertia. As I suggest below, however, subtle variations in major calls can serve to convey a great diversity of information.

Ontogeny and Physical Structure of Calls, and Possible Effects of the Environment

In the preceding section I pointed out that petrels have several call types in their repertoires. In this section, after establishing that calls are apparently not learned in petrels, I focus on the physical structure of the vocalizations and suggest several factors that may have shaped their diversity.

Ontogeny. The question of whether petrels learn their calls has not been carefully investigated. Cross-fostering experiments suggest that no vocal learning occurs. In several groups, chicks are able to develop and produce the adult call before fledging (e.g., *Pagodroma, Halobaena, Macronectes, Calonectris diomedea*). In a cross-fostering experiment within the fulmar group, five chicks of the Snow Petrel (*Pagodroma nivea*) reared by Cape Petrels (*Daption capense*) developed the call of their own species (Bretagnolle unpubl. data). James (1985a) claimed that calls of Manx Shearwaters (*Puffinus puffinus*) showed a pattern of change over six years that suggested the calls were culturally transmitted. The change was minor, however, and genetic drift is an alternative explanation. Slater (1991) also questioned James's results and suggested that the use of different tape recorders might explain his observations. Furthermore, penguins, the closest order to the Procellariiformes phylogenetically, do not learn their vocalizations (Jouventin 1982). Thus, no evidence suggests vocal learning by petrels.

Physical structure. Sonograms of calls have now been published for approximately 32 petrel species (see Marchant and Higgins 1990, and Table 9.1 for references). In most cases, however, sonograms are not available for the complete repertoire. The physical structure of petrel calls is extremely diverse, with a fundamental frequency and several harmonics (Fig. 9.2). Some species have no detectable harmonics (e.g., *Fregetta* and some *Pterodroma*), but this is rare; others have no clear or detectable fundamental frequency (e.g., *Oceanites, Pagodroma,* and some calls of *Procellaria, Hydrobates,* and *Oceanodroma*). In the latter cases, the spectral structure of the call can be very complex with rapid frequency and amplitude modulations, and with broad-band noise structures. The calls of several species are apparently made up mainly of noise (Fig. 9.2). Unlike passerine calls, petrel calls lack rapid amplitude modulation (except *Pseudobulweria:* Bretagnolle unpubl. data) and complex and rapid frequency modulation (except some calls of *Procellaria;* Brooke 1986, Warham 1988a). Last, in nearly all species, calls are temporally subdivided into distinct units or syllables (see examples in Figs. 9.2, 9.3). However, *Fregetta,* some *Pelecanoides,* and some *Puffinus* use only one syllable in their calls. Some species use syllables of two or three stereotyped durations (i.e., brief and long syllables); their position within the call may vary between individuals but not within, and this may be a primitive form of syntax (*Halobaena, Pachyptila*).

Two examples of environmental constraints. Petrels usually breed in huge colonies on small islands, where the sound levels created by the sea and the wind are high. Moreover, petrel colonies are extremely noisy because many birds are calling at the same time (see, e.g., R. H. Wiley 1976, Robisson 1991). Background noise alters signal detection and localization as well as sound propagation. As

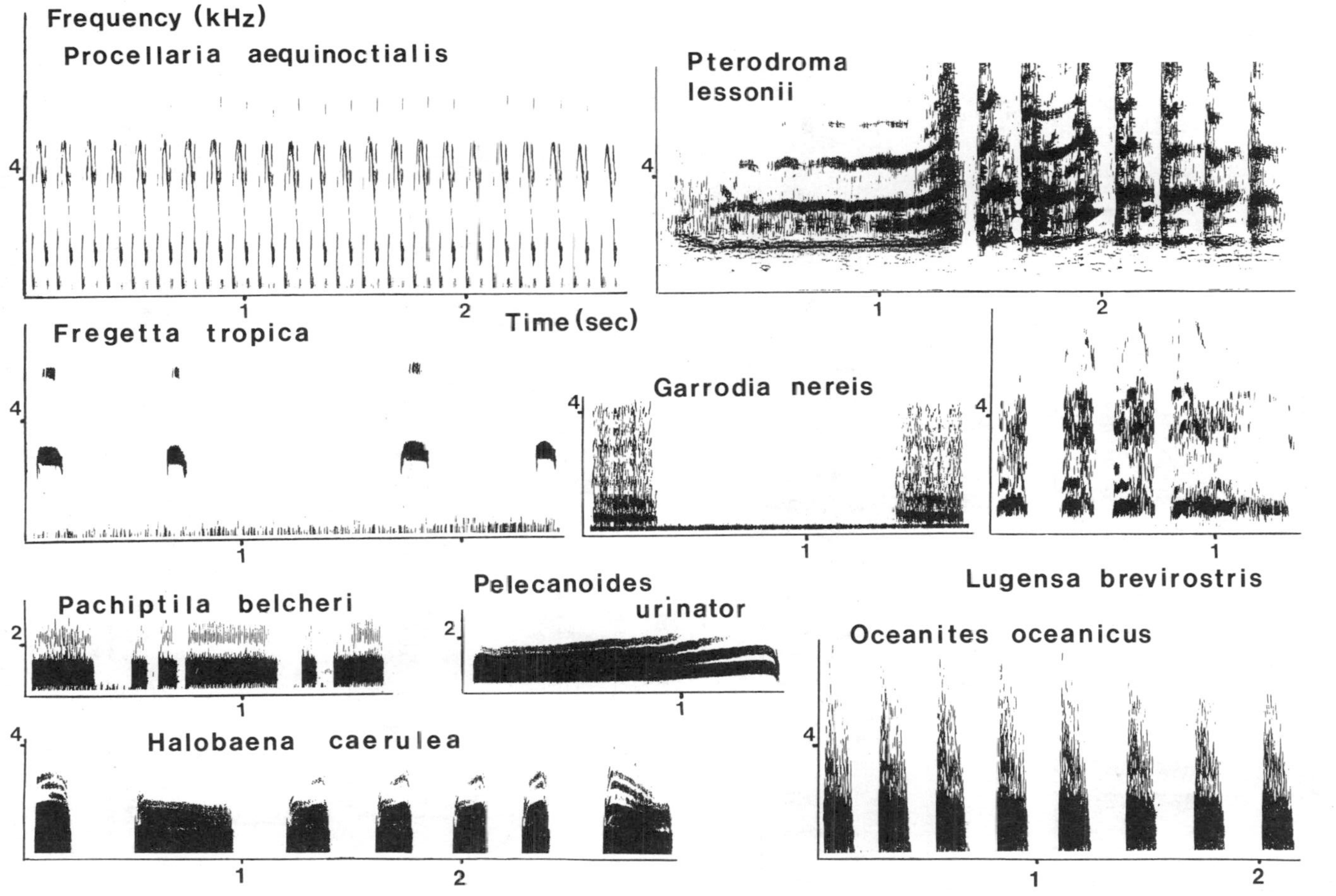

Fig. 9.2. Sonograms of major calls of some subantarctic species of petrels. Only male calls are represented, and where males have two calls, only the sexual call is shown. Note the distinctiveness of species. Note also that *Pachyptila belcheri* and *Oceanites oceanicus* show some characteristics of broad-band noise. Sonograms were produced on a Kay 6061B Sonagraph with a wide-band analyzing filter (300 Hz).

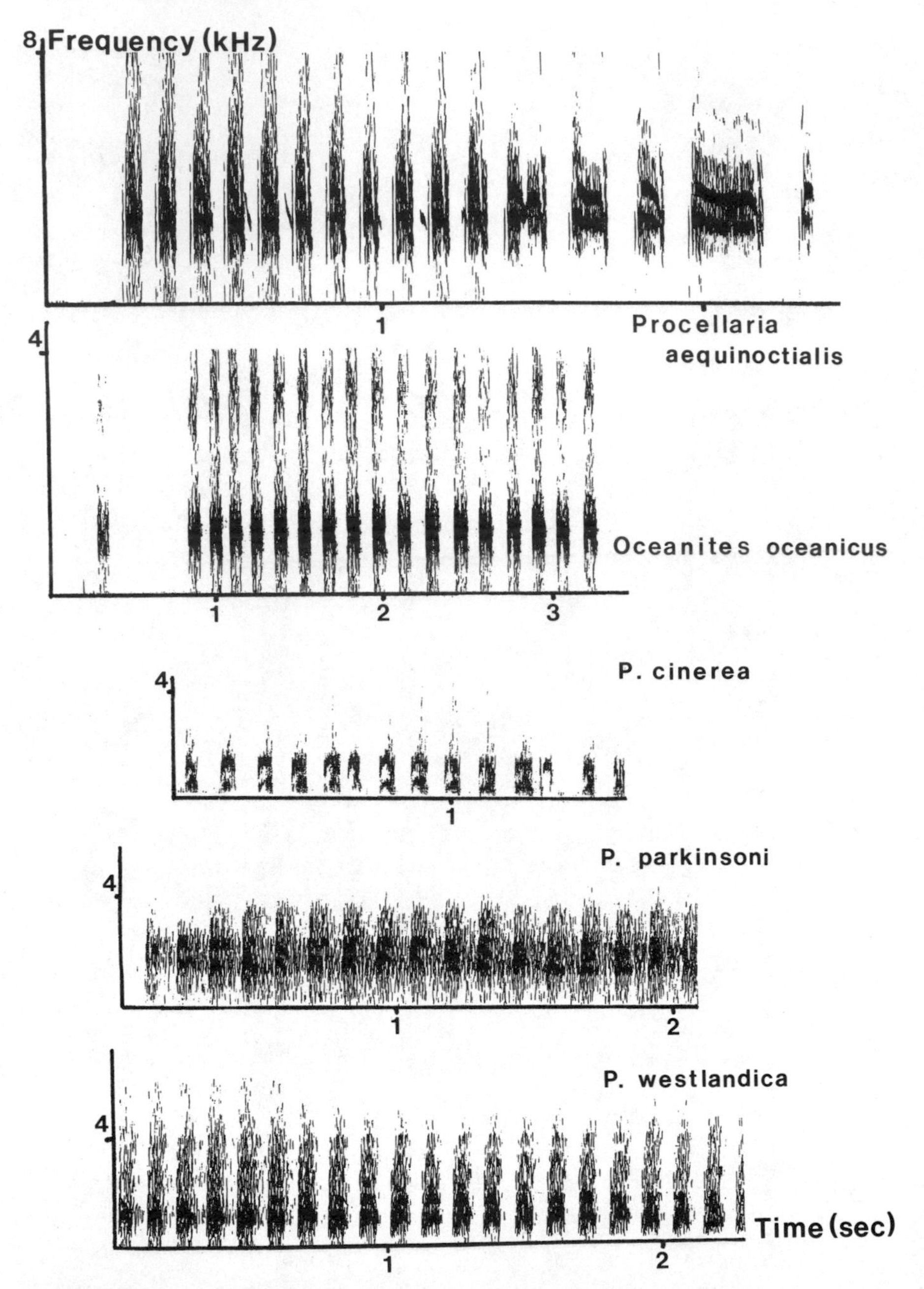

Fig. 9.3. Environmental constraint on petrel calls. Male advertising call of five different species from two genera (*Oceanites oceanicus* and the four *Procellaria* species: *cinerea, aequinoctialis, parkinsoni,* and *westlandica*). A wide frequency band and repetition of syllables may improve detectability and reduce degradation. Note the structural similarities between *Oceanites* and *Procellaria* in calls, irrespective of phylogeny (the two genera belong to different families), suggesting convergence. Note also that within *Procellaria,* the advertising calls of the male are similar, suggesting phylogenetic inertia.

individuals must be located among the other thousands of birds within the colony, one would expect that the physical structure of the signal would improve detectability and reduce attenuation (R. H. Wiley and Richards 1982, Robisson et al. 1993). Many petrel species have calls with wide spectra (the result of many harmonics or noise) as well as repeated syllables (Fig. 9.3), two acoustic parameters that improve detectability (R. H. Wiley and Richards 1982).

Unfortunately, having highly detectable signals is a risky strategy, because colonial breeding also attracts predators (Bretagnolle in press). In their breeding colonies, petrels suffer from avian predators such as gulls and Brown Skuas (*Catharacta lonnbergi;* Watanuki 1986, Mougeot et al. unpubl. ms.). The latter in particular hunt using acoustic cues provided by their prey (F. Mougeot and V. Bretagnolle unpubl. data). Predation risk is undoubtedly one reason why petrels are nocturnal on their breeding grounds and avoid moonlit nights (Watanuki 1986, Bretagnolle 1990a, MacNeil et al. 1993). To further reduce predation risk, several species perform part of the pairing in flight, hence the existence of aerial calling (see above). But species subject to intense predation might be expected to use signals whose physical structure would further reduce their detectability from predators. The existence of "scrambled" signals in petrels (i.e., the presence of a noise component) and the reduction of the number of syllables might be antipredator strategies (Figs. 9.2 and 9.3).

Pair Formation and Sexual Selection

Songs have a dual function in passerines: territorial defense and female choice (e.g., Searcy and Andersson 1986, Falls 1988). These functions are expected of petrel calls, as pair formation relies almost exclusively on calls. As I detail below, playback experiments have actually established both intersexual and intrasexual communication.

Colony choice and pair formation. It has been suggested that colonial species use acoustic and optical stimuli to synchronize their breeding activity (following the Darling hypothesis: Darling 1938; review in Gochfeld 1980) or to attract conspecific birds (Wittenberger and Hunt 1985). Two experimental studies provide support for the notion that an assessment of vocal activity is used by petrels in the process of joining a colony (Podolsky and Kress 1989, 1992). The most conclusive evidence comes from general observations; it is well known by birders (especially those who want to catch storm-petrels) that petrels are strongly attracted by conspecific vocalizations, and tape lures have been used extensively to attract petrels to mist nets (Furness and Baillie 1975, Ainley et al. 1976).

All burrowing petrels apparently follow the same pattern in pairing. First, young birds of both sexes visit the colonies, though they may not land, and call actively (see, e.g., James 1985b). Later, immature males try to establish or occupy a burrow; once they get one, they call within it, at its entrance, and possibly in flight to attract unpaired females (Storey 1984, James 1985b, Bretagnolle and Lequette 1990). In some species, such as the storm-petrels, a specialized call is involved—one of the major calls (Taoka et al. 1988, 1989b, Bretagnolle 1989b). This stage may occur entirely in flight in some gadfly petrels (*Pterodroma*). Then, depending on the species, females who are either in flight (Storey 1984, James 1985b, Bretagnolle 1990c, Bretagnolle and Lequette 1990) or on the ground (Bretagnolle 1989b, Brooke 1986) may engage in courtship duetting with males on the ground. In the last stage, females enter the burrow and duet with the occupant (see, e.g., Bretagnolle et al. unpubl. ms.). In several species, these stages may correspond to different cohorts (birds of successive ages; Serventy 1967, Brooke 1990, Bretagnolle unpubl. data for Blue Petrel). In any case, pair formation takes at least two seasons.

Playback experiments. Petrels are nocturnal on their breeding grounds, and they are colonial breeders. For these two reasons the design generally advocated for playback experiments is inappropriate, because it relies on territorial behavior and uses optical signals or movements to score responses (e.g., Emlen 1972, Catchpole 1989, Kroodsma 1989a, c, Searcy 1989). Taking advantage of the peculiar colonial and nocturnal behavior of petrels, researchers have used two different playback techniques to either (1) catch or attract the birds (James 1983, Taoka et al. 1989a, Bretagnolle and Robisson 1991, Podolsky and Kress 1992) or (2) elicit calling as a response (James 1984, Taoka et al. 1988, 1989b, Bretagnolle 1989b, Bretagnolle and Lequette 1990). In the first type of experiment, the relative attractiveness of the different played-back calls is scored by catching or counting the nonbreeders that respond to the loudspeaker by either flying over it or calling. In the second type of playback, birds are subjected to playback tests at their breeding burrows. Brooding birds are silent, and a response is scored when the bird calls in response to the playback.

Mate attraction and evidence that calls function in intersexual communication. Very few studies have demonstrated that petrels use calls for mate attraction, and none have documented mate choice via calls (for detailed experimental studies, see Storey 1984, James 1985b, Bretagnolle et al. unpubl. ms.). If petrel calls are to serve intersexual functions, the following predictions should be upheld: (1) there are sexual differences in the calls, (2) sound differences are perceived, (3) male calls are attractive to nonbreeding females, and (4) a selected male trait exists.

Much evidence now supports the first three predictions. Sex differences in petrel voices have been known for a long time (Brooke 1978). The first review discussed 11 species, of which 5 were stated as not showing sex differences (James and Robertson 1985a). Two of those species are in fact sexually dimorphic (Wedged-tailed Shearwater, *Puffinus pacificus:* pers. obs.; and Bulwer's Petrel; Fig. 9.1). In the case of the two *Procellaria* petrels, James and Robertson (1985a)

Table 9.2. Individual and sexual signatures in the petrels

Genus	Species	Individual signature (coefficient of variation)[a]	Sexual signature (parameter involved)
Fulmarus	*glacialoides*	2.85	
Daption	*capense*	3.37	
Pagodroma	*nivea*	1.65	Frequency, tempo
Pterodroma	*lessonii*	1.57	—
Pterodroma	*mollis*	4.03	Frequency
Halobaena	*caerulea*	4.95	Syntax
Pachyptila	*desolata*	3.92	Syntax
Pachyptila	*belcheri*	3.03	Syntax
Bulweria	*bulwerii*	6.77	Frequency, tempo
Calonectris	*diomedea*	3.82	Frequency
Puffinus	*puffinus*	1.63	Spectrum
Puffinus	*yelkouan*[b]	1.16	Spectrum
Oceanites	*oceanicus*	4.69	Frequency, tempo
Pelagodroma	*marina*	2.37	—
Fregetta	*tropica*	1.18	Frequency
Hydrobates	*pelagicus*	2.46	Spectrum
Oceanodroma	*leucorhoa*	6.71	Spectrum
Pelecanoides	*georgicus*	3.98	Spectrum, syntax
Pelecanoides	*urinatrix*	1.68	Syntax

Notes: Individual signatures are given as an index. Sexual signature indicates the general acoustic variable involved in sex differences (frequency, value of the fundamental frequency; tempo, temporal variables such as rhythm or syllable durations; syntax, number of syllables per call or a difference in the ordination of long and brief syllables; spectrum, a combination of frequency and amplitude variables).

[c]oefficients of variation (Jouventin 1982, Bretagnolle 1989b, Bretagnolle and Lequette 1990)—a large ratio indicates that calls are highly variable between individuals but highly stereotyped within individuals. From Bretagnolle and C. Rabouam (unpubl. data).

cite "Brooke (in prep)" as stating that the species lack sexual dimorphism. But Brooke himself did not write that, because no statistical analysis of sexual dimorphism was in his paper (Brooke 1986). Moreover, Warham suspected sex differences in *P. aequinoctialis* and confirmed them in *P. westlandica* (Warham 1988a). Additional examples not considered by James and Robertson (1985a) are given in Table 9.2. Because sex differences have been detected in most species investigated so far, it is extremely likely that all species do exhibit such differences, although it is more obvious to the human ear in the case of aerial calling species and in burrowing petrels than in diurnal fulmars (James and Robertson 1985a, unpubl. data). Interestingly, sex differences in voice are variably coded on temporal, frequency, or even syntactic parameters according to species (see Table 9.2).

Petrels do perceive sexual differences in voice (James 1985b, Bretagnolle

Two lines are missing from the footnotes to Table 9.2, on page 173. Footnotes a and b should read as follows:

[a]Values given are ratios between the population coefficient of variation and the mean of individual coefficients of variation (Jouventin 1982, Bretagnolle 1989b, Bretagnolle and Lequette 1990)—a large ratio indicates that calls are highly variable between individuals but highly stereotyped within individuals. From Bretagnolle and C. Rabouam (unpubl. data).

[b]Warham (1990) lists *P. yelkouan* as a subspecies of *P. puffinus*.

1989b, Bretagnolle and Lequette 1990, Brooke 1990), and male calls do attract nonbreeders (Furness and Baillie 1975, James 1983, Storey 1984, Fowler et al. 1986, Warham 1988a, Bretagnolle 1989a, Podolsky and Kress 1989, 1992). That male calls attract females especially was suggested for one species and proven for another (Bretagnolle 1989b, Bretagnolle et al. unpubl. ms.).

Finally, the possibility has been investigated that petrel calls convey information about individual quality. Male body weight (or condition) may be a good criterion for female choice in petrels because it is an indicator of fat reserves, which often increase with age and experience (Brooke 1990, Weimerskirch 1992), and because it may be correlated with lifetime reproductive success (e.g., Bryant 1988, Scott 1988). Genevois and Bretagnolle (1994) proved that information on male body weight was conveyed in the calls of the Blue Petrel, because there was a significant positive correlation between body condition and temporal parameters of the call. A similar result has been found for Cory's Shearwater (Bretagnolle and Thibault unpubl. data). It is not known whether the females take this information into account.

Burrow defense and evidence that calls function in intrasexual communication. The same type of call can elicit very different responses when played back to nonbreeders (attraction) and breeders (territorial reaction: Bretagnolle and Lequette 1990, Bretagnolle et al. unpubl. ms.). Moreover, in many petrels, birds of a given sex respond only to calls of birds of the same sex (and to their own mates), thus demonstrating the agonistic function of petrel calls (e.g., Taoka et al. 1988, 1989b, Bretagnolle and Lequette 1990, Brooke 1990). Interestingly, female-female competition is strong in petrels, probably as a result of long-term pair-bonding, and females' reaction to female calls is of the same magnitude as males' reaction to male calls (Bretagnolle and Lequette 1990, Bretagnolle et al. unpubl. ms.). Calls apparently contain information relevant to their agonistic content; such "motivational" messages have been found in Wilson's Storm-Petrel (Bretagnolle 1989b). Similarly, Cory's Shearwater, most *Puffinus,* the Blue Petrel, and the prions (*Pachyptila*) increase the modal frequency of their calls in territorial contacts (Bretagnolle and Lequette 1990, Bretagnolle and Genevois unpubl. data).

Sexual selection. Calls thus have a function in sexual advertisement and intersexual selection in petrels as well as in intrasexual competition, just as in passerines (see Krebs and Kroodsma 1980, Searcy 1986). But the fact that calls can function in sexual communication does not necessarily mean that sexual selection is occurring (Andersson 1994). Whether differential mating success results from calling differences must be clarified, though petrels, with their very long mating periods, might prove to be difficult subjects. Curiously, two phenomena found in passerines have not been documented in petrels: neighbor recognition (Brooke 1986; Bretagnolle unpubl. data for Blue Petrel) and extra-pair copulations (F. M. Hunter et al. 1992, Swatschek et al. 1994; but see Austin et al. 1993).

A Major Constraint on Petrel Calls: Coding and Decoding Messages within a Single Call

Half the genera of petrels have a single major call in their repertoire that is used both in sexual and agonistic contexts. Below, I discuss how these major calls also encode identifying messages (see W. J. Smith 1977) that identify species, populations, and individuals.

Evidence for individual recognition. Individual recognition is common in birds; it has been reported in at least 136 species (Rabouam and Bretagnolle unpubl. ms., Stoddard, this volume). Mate recognition is especially relevant for long-lived and monogamous seabirds because partners usually pair for life, and they breed in colonies (see Beecher 1989). Individual signatures have been widely documented in seabirds (Falls 1982, Jouventin 1982, Rabouam and Bretagnolle unpubl. ms.), though site tenacity can also facilitate mate fidelity (Morse and Kress 1984). Individual signatures occur in at least 19 petrel species (Table 9.2; Brooke 1978, Guillotin and Jouventin 1980, Bretagnolle 1989b, Taoka and Okumura 1989, Bretagnolle and Lequette 1990, Bretagnolle and Rabouam unpubl. data). In all studies to date, individual signatures have been demonstrated for temporal variables but not for frequency or amplitude variables. Investigations of genera that have complex spectra (e.g., *Procellaria*) might lead to different conclusions. Acoustic recognition of the mate has been experimentally established for five species (Brooke 1978, 1986, Taoka and Okumura 1989, Bretagnolle and Lequette 1990; Bretagnolle unpubl. data for Blue Petrel).

Evidence for species recognition and geographic variation. Species recognition by voice has been widely documented in birds (review in Becker 1982). In the total absence of optical cues call discrimination and reproductive isolation would be expected to occur between sympatric petrel species—petrel calls should be especially distinct in conditions of sympatry. Petrel calls of sympatric species do differ greatly (Warham 1988a, Bretagnolle et al. 1990, 1991; see also Fig. 9.2). In fact, they are so species specific that they can be used to determine taxonomic relationships (Bretagnolle et al. 1990, Bretagnolle 1995). But character displacement, with calls more distinct in sympatry than in allopatry, has not yet been documented.

Petrels breed on isolated and remote oceanic islands that provide natural geographic isolation; moreover, petrels are highly philopatric (Weimerskirch et al. 1985, Mougin et al. 1988, Ovenden et al. 1991, Austin et al. 1994). These two characteristics should promote genetic drift, and thus geographic variation. Geographic variation has been detected in petrel morphometrics (Power and Ainley 1986, Massa and Lo Valvo 1986, Bretagnolle et al. 1991, Bretagnolle 1995), coloration (Ainley 1980, Clancey et al. 1981), and genetics (Randi et al. 1989, Ovenden et al. 1991, Austin et al. 1994). Geographic variation is also common in petrel vocalizations and has been found in all families and most genera (James

1985a, Bretagnolle 1989b, Bretagnolle and Lequette 1990, Bretagnolle et al. 1991, Tomkins and Milne 1991, Bretagnolle 1995, Bretagnolle and Genevois unpubl. ms. a, b).

Coding and decoding different messages within a single vocalization. Individual recognition requires not only individual consistency but also interindividual variation in the call (Falls 1982, Rabouam and Bretagnolle unpubl. ms.). Species-specific recognition demands that the same call be recognizable to all conspecific individuals. A similar paradox applies for the coding of geographic variation, because population recognition requires within-population consistency (conflicting with interindividual variation) and between-population variation (thus conflicting with species-specific recognition). How, then, can the various messages be encoded together in the call without confusion?

The answer apparently lies in both the coding and the decoding of a message. Detailed analysis of Blue Petrel calls revealed that (1) individual stereotypy and quality are coded on some temporal parameters, especially those at the end of the call (Genevois and Bretagnolle 1994, Bretagnolle and Genevois unpubl. ms. b); (2) geographic variation is coded on both temporal parameters (except those at the beginning of the call) and frequency parameters (Bretagnolle and Genevois unpubl. ms. a); and (3) sexual dimorphism is coded in syntactic parameters (Bretagnolle 1990c). The various messages seem therefore to be coded in different parts of the call or in different parameters. This pattern is also found in Cory's Shearwater (Bretagnolle and Lequette 1990). Further, in this and other species, the meaning (e.g., agonistic vs. sexual) of the same call varies according to the status of the receiver bird (nonbreeder vs. breeder) and its location (flying vs. landed; Storey 1984, James 1985b, Brooke 1990, Bretagnolle et al. unpubl. ms.).

Conclusions

The ecology and evolution of vocal signals: phylogenetic inertia and selective pressures. Petrels are a good model for investigating the evolution of bird vocalizations because the diversity of their acoustic repertoires and the parallel diversity of their life histories suggest a causal link. Below I list five hypotheses that are relevant both to the evolution of petrel calls and to more general evolutionary questions about bird communication.

1. Petrels do not learn their calls. Thus, one may expect to find more divergence between closely related species when they are in sympatry than when in allopatry (see the review in E. H. Miller 1982). Character divergence is not yet documented, perhaps because in sympatry birds are more acute at distinguishing their own species from others (Bretagnolle and Robisson 1991, Bretagnolle unpubl. data).

2. Dialects occur in songbirds as a consequence of song learning, but microgeographic variation also occurs in petrels, even in populations only 2 kilometers

apart (Bretagnolle and Genevois unpubl. ms. a). The origin of geographic variation in petrel calls may be genetic, so petrels might be useful subjects for testing the population marker hypothesis (Nottebohm 1969b, M. C. Baker 1982).

3. Individual recognition probably exists in all petrel species but might be more pronounced in highly colonial species or species showing high mate fidelity.

4. Some petrel species have one major call, others two. Repertoire size (e.g., number of major calls) is not independent of phylogeny (see Table 9.1), but other selective pressures (e.g., reduction of ambiguous signals) might also be present.

5. Some petrel species use aerial calling extensively (e.g., some *Puffinus* and all *Pterodroma*), but others do not. James and Robertson (1985a) and Brooke (1986) discussed aerial calling and suggested relationships between aerial calling and sexual vocal differences. Above, I suggested that predation is an important selective pressure. Curiously, the presence or absence of aerial calling is not consistent within petrel lineages; for instance, species belonging to the same genus can differ greatly (e.g., *Procellaria:* Warham 1988a).

Some of these evolutionary hypotheses are difficult to test because no accepted phylogeny is available at the genus level for Procellariiformes (Bretagnolle 1993). Thus, comparative analyses taking into account the nonindependence of species points cannot be undertaken at present (see Harvey and Pagel 1991). Improved understanding of petrel systematics is a major goal for future research and will, I hope, permit us to distinguish between phylogenetic and environmental effects on the evolution of petrel calls.

Acknowledgments

I thank all the people who have worked with me: L. Ackerman, V. Barthelemy, E. Buffard, B. Fillon, F. Genevois, F. Mougeot, C. Rabouam, and especially F. Genevois, for our numerous discussions. P. Duncan, S. J. G. Hall, and J. Warham read and improved a first draft of the manuscript and made many corrections on the English. D. E. Kroodsma and E. H. Miller made extensive comments and constructive criticisms that improved the chapter greatly.

PART III VOCAL VARIATION IN TIME AND SPACE

Introduction

Vocal displays vary in many ways. During ontogeny, most vocalizations (especially learned ones) are highly variable, especially early in development (see Part I). Even minor variations in successive vocalizations can be communicatively significant, and the number of displays an individual uses can be in the thousands (see Part II). In this section, the focus is on another level of variation: how vocal displays vary through time and space.

In Chapter 10, which is about learned song traditions, Alejandro Lynch analyzes patterns of differentiation within an evolutionary framework. He identifies analogues of the evolutionary forces responsible for gene frequency changes (mutation, migration, drift, selection) and adapts methods of population genetics to estimate parameters of population structure, using data from natural populations as illustrations. Lynch's approach helps to reveal the factors responsible for the origin and maintenance of song diversity within and among songbird populations.

Bob Payne's chapter is on long-term cultural evolution of song in Indigo Buntings (*Passerina cyanea*). Payne recognizes two kinds of cultural evolution: changes within song lineages and replacements between song lineages. Changes within a lineage are the cultural counterparts of phylogeny and include both the gradual accumulation of differences through innovation or "error" and the splitting into sublineages. Song changes between lineages are the cultural counterparts of natural selection. Changes within a bunting population over only 5–10 years were great and involved a high turnover of the song pool through cultural extinction and replacement by immigrants.

Jochen Martens's scope is geologic in time and continental in space. In his chapter Martens describes regiodialects over large areas of the Palearctic zone. Dialects within regiolects share features that set them apart from other regiolects. Regiolects meet in narrow zones of secondary contact that coincide with distributional limits of subspecies or semispecies. Though they are maintained by cultural mechanisms, regiolects were established in Quaternary refugia, so they have remained intact for extremely long periods. Martens illustrates stages of regiolect development using examples from tits (*Parus*), warblers (*Phylloscopus*), buntings (*Emberiza*), and tree-creepers (*Certhia*).

Another contrast with the burgeoning studies on songbirds is Ted Miller's chapter on spatial variation in vocal and nonvocal displays of shorebirds. Acoustic displays are evolutionarily conservative and especially valuable as systematic characters in this group, even well above the species level. Miller discusses examples of display differentiation between closely related species of snipe (*Gallinago*), plovers (*Charadrius* and *Pluvialis*), and sandpipers (*Calidris*). He also discusses homologous features of shorebird displays and how rates of evolutionary change may be influenced by sexual selection.

The last two chapters in this section take a broader sweep at the globe. Drawing on his years of research in North America and the Neotropics, Gene Morton contrasts styles of vocal communication and their behavioral significance in the temperate and tropical zones. He discusses a variety of topics, including duetting, female singing, and factors that might reduce sex role divergence in the Tropics. Major latitudinal differences in sex roles, song functions, year-long territoriality, and breeding synchrony may also relate to latitudinal differences in extra-pair behavior, such as extra-pair fertilizations. Morton urges greater biogeographic breadth in research on bird sounds.

The final chapter in this section, by Don Kroodsma, Jacques Vielliard, and Gary Stiles, also focuses on the Neotropics. Avian diversity in the Neotropics is extraordinary, and to a large extent unquantified, but vocalizations can be used as a critical initial guide in documenting distributions and even new taxa. Furthermore, the Tropics in general and the Neotropics in particular offer extraordinary opportunities for studying bioacoustic phenomena related to duetting, mimicry, interspecific interactions, dawn song, strategies of development, sexual selection, and much more. If we are ever to understand acoustic communication in birds, we must work outside the north temperate zone and on groups other than songbirds. This chapter is an invitation to explore the diverse communities of the Neotropics.

10 The Population Memetics of Birdsong

Alejandro Lynch

Cultural evolution can be defined as the change in the frequency of cultural traits through differential transmission from one generation to the next. Such nongenetic transmission of traits or behaviors across generations can be a powerful source of variation, so it is important to understand the processes involved. It has been argued that the evolution of cultural traits is driven by processes analogous to those involved in biological (or genetic) evolution (Alexander 1980, Mundinger 1980, Cavalli-Sforza and Feldman 1981, Boyd and Richerson 1985). Many songbirds acquire their specific song patterns by social learning, and birdsong is therefore often cited as one of the best examples of culture in nonhuman species (Bonner 1980, Mundinger 1980, Slater 1986).

Although studies of birdsong diversity usually invoke one or more evolutionary forces to explain the variation within or among populations, in general no attempt is made to explain cultural evolution quantitatively in terms of the interaction of all possible evolutionary factors. In order to describe birdsong diversity in an explicit evolutionary framework, we must first identify cultural analogues of the well-known evolutionary forces responsible for changes in gene frequency (mutation, migration, drift, and selection). Once these forces are identified, we can take advantage of the well-developed mathematical theory of population genetics to make inferences about the relative importance of these forces in the origin and maintenance of song diversity within and among populations.

In this chapter I discuss the main concepts of cultural evolution by drawing analogies with biological evolution. I describe models adopted from population genetics that are useful in estimating important parameters of population structure (e.g., mutation and migration rates, and memetic diversity within and among populations), illustrating the use of these models with data from a number of passerine species.

Cultural Evolution of Birdsong

I begin by discussing important aspects of the transmission of bird songs as they relate to cultural evolution. I then identify analogues of the well-known population genetic processes involved in gene frequency changes (mutation, drift, migration, and selection) that can be applied to birdsong evolution. Last, I indicate useful population genetics models that can be borrowed to estimate important parameters of population structure (e.g., mutation and migration rates, diversity within populations, and differentiation among populations). Throughout the section I emphasize some important differences between genetic and cultural evolution.

Transmission of Cultural Traits

The unit of cultural transmission. Before we can analyze cultural change, we need to determine what is being transmitted. Richard Dawkins (1976a) coined the term *meme* for the unit of cultural transmission, defining it as "an entity that is capable of being transmitted from one brain to another" (p. 196). We can apply this concept to birdsong and define a song meme as a song pattern that is transmitted from one bird to another during the learning process (Lynch et al. 1989).

In some cases whole songs are transmitted (see, e.g., Slater et al. 1980, Beecher et al. 1994b); in others, birds copy individual syllables or blocks of linked syllables from different songs and recombine them with other syllables or blocks to form a new song (Marler and Peters 1982b, Jenkins and Baker 1984). A meme can therefore be an individual syllable, a group of linked syllables, or a whole song. In any event, these memes are replicators (*sensu* R. Dawkins 1976a), and it is with their evolutionary fate that I am concerned: What is the rate of origin of new memes? How much movement of memes occurs among populations? Are some memes more successful than others at producing copies of themselves? I am concerned with the fate of the individuals that carry these memes (the carriers or vehicles) inasmuch as it affects the survival of the replicators.

The unit of change: continuous or discontinuous? Birdsong evolution can be studied in two conceptually different ways. First, one can classify songs or song elements, by visual inspection, into discrete categories (e.g., Thompson 1970, Slater and Ince 1979). Usually this task is not difficult because songs and song elements can typically be split into discrete types with high interobserver reliability. Even if the variation is continuous, perceptual thresholds might make this continuous variation appear to be discontinuous (D. A. Nelson and Marler 1989, Horn and Falls, this volume). The second approach is to take a number of linear measurements from sonograms and perform some form of statistical analysis (e.g., R. B. Payne 1978a, M. L. Hunter and Krebs 1979, Martindale 1980). The main problem with this approach is that it is too crude to detect homologies in song elements due to common descent. In this chapter I therefore take the former approach, although a theory of cultural evolution does not depend on the method of description of song patterns. Of course, using continuous variation to characterize song diversity requires the development of a different theoretical framework, one more akin to the theory of quantitative genetics.

Mode of transmission. The mode of transmission across generations is a major difference between cultural and biological traits. Genetic transmission is usually strictly vertical (from parent to offspring), whereas cultural transmission can be vertical, horizontal (between members of the same generation), or oblique (from members of a given generation to members of succeeding generations who are not direct descendants; Cavalli-Sforza and Feldman 1981). Most songbirds learn their songs either before natal dispersal or from their neighbors during their first breeding season (Kroodsma 1982a). It is therefore likely that all three modes of transmission occur, although oblique transmission is probably the most common.

Mutation

Mutational input is the ultimate source of variation, so the rate of mutation is an important factor determining the amount of diversity within and among populations. Song transmission is not always accurate: sometimes birds produce imperfect copies of their models. Poor copies can arise from copy errors resulting from limited exposure to the model, they may represent variations on a theme (improvisation), or they may be inventions of totally new types. All these processes can be subsumed under the rubric "cultural mutations" (Jenkins 1978).

Lynch et al. (1989) distinguished two types of cultural mutations. *Point mutations* are changes in the morphology of the basic elements that compose a song (referred to as elemental improvisation by Marler and Peters [1982b]). *Recombinations* are the rearrangements of these basic elements within and among songs (combinatorial improvisations of Marler and Peters [1982b]).

In general, cultural mutation is a common phenomenon; it occurs in every generation during the transmission process (see, e.g., Jenkins 1978, Slater and Ince 1979). In contrast, biological, or gene, mutation is a rare event. The high rate of cultural mutation makes song a good subject for evolutionary studies because (1) populations will attain equilibrium between mutation and drift relatively quickly after a disruption of their steady state (e.g., after a population bottleneck), and (2) long-term studies can follow the fate of variants as they arise and spread in a population through time.

Drift

When song patterns are transmitted from generation to generation through social learning, some may be lost by accident or sampling error, particularly in small populations. The frequency of a song meme is therefore subject to sampling variation between successive generations. In general, drift reduces the diversity within populations and increases differentiation among populations. An important type of sampling drift occurs during severe reductions in population size, such as during the colonization of new areas by a small number of individuals. During such a founder event, birds will likely carry a small set of the total meme pool existing in the original population (see, e.g., A. J. Baker and Jenkins 1987).

Effective population size. The strength of drift depends on population size. Most population genetics models make a number of unrealistic assumptions in connection with the size of the population and therefore with the amount of drift that is expected to occur. For example, it is often assumed that population size is constant from generation to generation, that generations are nonoverlapping, and that all individuals have an equal probability of producing offspring. To offset such unrealistic assumptions, Wright (1931) developed the concept of effective population size (N_e), which can be defined as the size of an idealized population that satisfies the above assumptions and has the same diversity as the actual population. The effective population size is usually smaller than the census population size.

In the ideal population there are no fluctuations in population size. If the size of a real population is not constant from generation to generation, then its effective population size will be determined by the harmonic mean of the size in each generation:

$$N_e = \frac{t}{\sum_{i=1}^{t} \frac{1}{N_i}} \qquad (10.1)$$

where t is the number of generations of population fluctuation and N_i is the population size at generation i (M. Kimura 1983). Note that the harmonic mean is dominated by the smallest numbers, and thus extreme fluctuations in population sizes will markedly reduce the effective size.

Ideally, each meme has an equal probability of being copied, so the variance in copying success has approximately a Poisson distribution (Crow and Kimura 1970). The actual number of copies derived from individual memes will vary, but all memes have the same expected number of descendants. In reality, not all memes have the same probability of being copied. For example, individuals differ in their life spans, so the song memes of some individuals will have a higher chance of being transmitted simply because those birds live (and presumably sing) longer. A high proportion of memes is thus derived from a small number of models, reducing the effective population size. Also, if generations are overlapping and the age distribution is stable, the effective population size per generation can be obtained by

$$N_e = \frac{NL}{EV_k} \qquad (10.2)$$

where N is the census population size, L is the generation length (average age of memes when they are copied), E is the life expectancy (or mean age at death), and V_k is the variance in copying success (W. G. Hill 1979).

Migration

Populations that are separated tend to diverge. If contact occurs between those populations, however, memes can spread from one to the other and the populations will remain more or less similar.

The effectiveness of migration as a homogenizing factor depends in part on when song is learned. If birds learn at least some of their songs before natal dispersal, then they will carry these songs with them when they move. If birds learn their songs after they disperse, however, or if they learn new songs after they move to a new area and do not use any songs learned before dispersing, then song elements will not diffuse even if there is extensive contact between populations. Given the usually high rate of cultural mutations, migration rates, unless they are exceptionally high, cannot prevent the accumulation of new mutants in different populations before they are dispersed throughout the entire region (Spieth 1974, Lynch et al. 1989).

Selection

Some traits reinforce their own persistence and spread; others do not and eventually disappear. Cultural selection can be defined as the probability that a given meme will be accepted by a young bird. The selective value is determined by how acceptable a meme is, and it refers to the survival of the memes, not their vehicles. On the other hand, the possession of one meme or another might affect the Darwinian fitness of an individual bird (the carrier or vehicle).

The fitness of a meme is determined by its propensity to survive and spread from host to host. The relationship between the cultural fitness of a given meme and the Darwinian fitness of its carrier can be complex. For example, the possession of a particular meme might increase the Darwinian fitness of the carrier, which in turn might increase the probability of the carrier spreading its memes. Alternatively, if the probability of a song being copied depends on the frequency with which it is uttered, and if successful males reduce their song output after mating but unsuccessful males continue singing, young nestlings and fledglings could be preferentially exposed to the songs of these unpaired males and might tend to copy their songs.

A number of studies have suggested that song memes are functionally equivalent and that the relative frequencies of memes reflect chance events that made some memes more frequent than others (e.g., Slater et al. 1980, R. B. Payne et al. 1981, McGregor and Krebs 1982, Horn and Falls 1988c, Catchpole and Rowell 1993, Lynch and Baker 1993). The results of these studies imply that there is no selection (or very weak selection) of some memes over others, and that we can therefore consider memes as effectively neutral, in the same sense as neutral alleles (M. Kimura 1983). In this case, the diversity of memes within populations can be explained by an interaction of mutation and migration, both of which introduce new variants, and drift, which tends to eliminate them. It should be stressed that the neutral theory does not ignore selective constraints, and thus is consistent with the presence of purifying selection, which would eliminate most of the deleterious mutants as they arose and would not contribute much to diversity in the population (Lynch and Baker 1993).

Two potentially important selective forces in the evolution of song memes are the social and acoustic environments. The social environment can affect the fate of memes in a frequency-dependent manner. For example, common variants might be copied disproportionately often relative to their frequency in the population ("majority-type advantage" or "conformist bias"; see, e.g., Beecher et al. 1994b). This process would lead to a few predominant types, thus decreasing diversity. Alternatively, if rare variants were copied with higher probability, there would be a more even distribution of types and a concomitant increase in diversity ("rare-type advantage"; Boyd and Richerson 1985).

The acoustic properties of the environment might influence the type and number of song elements that exist within a population. For example, types that do not transmit well or are potentially confused may be at a disadvantage (Hansen 1979). Here, the acoustic environment will influence the "realized" mutation rate by affecting the strength of purifying selection (Lynch and Baker 1993).

Differences between populations could also be related to differences in the acoustic environment. For example, if different environments have different acoustic properties (see, e.g., R. H. Wiley and Richards 1982), directional selection can promote the acoustic differentiation of populations (e.g., Nottebohm 1975, M. L. Hunter and Krebs 1979). Also, it has been argued that populations inhabiting areas with depauperate avifaunas are under less pressure to produce distinctive songs and are therefore more variable (Marler 1960). Note, however, that this process involves different amounts of purifying selection in different populations, and it is therefore consistent with a neutral model (Lynch and Baker 1993).

Estimating Parameters of Population Structure

Two complementary approaches can be used to study song diversity. We can describe patterns of variation within and among populations and then make inferences about the processes that led to these patterns by estimating population parameters such as meme diversity, meme flow, and selection coefficients. This is the approach I use for most of the analyses reported here. Alternatively, given the high rate of change of song elements, it is possible to study this dynamic process directly through time by following the fate of cultural traits as they arise and spread in a population (see, e.g., Jenkins 1978, R. B. Payne et al. 1981, Payne, this volume).

Diversity within populations. One way to study meme diversity is to consider song memes as alleles at a single locus. Given the wide variety of song elements in most species, the two most appropriate models are the infinite-alleles model (M.

Kimura and Crow 1964) and the K-alleles model (M. Kimura 1968). The infinite-alleles model assumes a sufficiently large number of possible distinct mutants so that each new mutant is of a type not already represented in the population. The K-alleles model allows for recurrent mutation and assumes a fixed number of potential types, K. The rate of mutation from one type to any of the others is $\mu/(K-1)$, where μ is the total mutation rate for a given type. This model is appropriate when similar syllable types can develop independently; for example, when birds are limited in the diversity of sounds they can produce (see, e.g., Tubaro 1991).

If S different memes occur in a population of size N, and if the frequency of the kth meme is p_k, then we can define meme identity, I, as

$$I = \sum_{k=1}^{S} p_k^2 \qquad (10.3)$$

This parameter is called homozygosity or gene identity in population genetics (e.g., Nei 1973) and represents the probability that two randomly chosen memes are identical. We can obtain an unbiased estimate of I by

$$\hat{I} = \frac{n \sum_{k=1}^{s} x_k^2 - 1}{n - 1} \qquad (10.4)$$

where n is the sample size, s is the number of memes in the sample, and x_k is the frequency of the kth meme in the sample (Nei and Roychoudhury 1974, Gregorius 1987). A useful measure of diversity derived from I is

$$s_e = \frac{1}{I} \qquad (10.5)$$

which is called the effective number of alleles in population genetics (M. Kimura and Crow 1964); by analogy, s_e can be called the effective number of memes (Lynch and Baker 1993). It corresponds to the number of different memes that would exist in an ideal population with the same s_e as the real population, and in which all memes had equal frequency $1/S$. This measure is a better index of diversity than the actual number of observed memes in a sample, s_a. The actual number depends greatly on sample size, whereas the effective number is largely independent of sample size. The difference in dependence with sample size arises because larger samples will detect a greater proportion of rare memes, which increase the number of observed memes but contribute very little to s_e. An obvious estimator of s_e is $\hat{s}_e = 1/\hat{I}$.

This measure of diversity (s_e) under the infinite-alleles model attains an equilibrium value when the input of new types by immigration and mutation is balanced by the elimination of types by sampling drift:

$$s_e \approx 2N_e + 1 \qquad (10.6)$$

where N_e is the effective population size and v represents the combined effects of mutation (μ) and migration (m) rates. Under a K-alleles model, the relationship is

$$s_e \approx \frac{2N_e v \dfrac{K}{K-1} + 1}{2N_e v \dfrac{1}{K-1} + 1} \tag{10.7}$$

where K is the number of different potential memes (M. Kimura 1968, Crow and Kimura 1970).

Using either equation 10.6 or 10.7 as appropriate, one can estimate $N_e v$, the number of new memes entering the population each generation via mutation and migration. It should be noted, however, that if the memes fit an infinite-alleles neutral model, Ewens (1972) showed that a better way to estimate $N_e v$ is to use the total number of memes and the number of distinct memes in the sample (see equation 10.8 below).

The use of this method to obtain estimates of mutation and migration is limited in several ways. First, we cannot distinguish between new variants that arise from mutations and those that enter the population by immigration. Also, to be able to estimate the rate of input of new variants (v) we need to know the effective population size. Finally, this method of estimating $N_e v$ is biased because it assumes that all elements that enter the population differ from preexisting ones (barring recurrent mutation). This assumption is true for mutation, but it might not be so for migration, especially when mutation rates are low and migration rates are high, because types will be shared between populations and some of the immigrant memes will therefore go undetected.

Test of neutrality. A theory of neutral mutations predicts that variation in a population depends on a balance between the origin of new forms through mutation and immigration, and the extinction of existing forms by random drift. In such a case, the vast majority of existing forms are functionally equivalent. Ewens (1972) showed that the total number of distinct alleles (memes, in our case), s, in a sample and the sample size, n, are sufficient to give an expected distribution of allele (or meme) frequencies in a neutral infinite-alleles model. The relationship between the expected number of memes in a sample of size n and the population parameter $\theta = 2N_e v$ is

$$E(s) = \sum_{i=0}^{n-1} \frac{\theta}{\theta + i} \tag{10.8}$$

where θ is changed iteratively until $E(s)$ equals s. A number of tests can be used to determine whether the observed sample corresponds with the expected values of the model. Watterson (1978) showed that a good test is to compare the observed and expected values of I. The observed value can be obtained using equation 10.4; using equation 10.8 we can obtain θ and from it the expected value of I using the relation $E(I) = 1 / (\theta + 1)$. The distribution of I can be obtained by simulation, and from it confidence limits can be calculated (e.g., Whittam et al. 1983, Lynch and Baker 1993).

Differentiation among populations. Most songbird species display marked geographic variation in their songs. The analysis of such population structure can help explain the relative importance of different evolutionary factors in the origin and maintenance of memetic diversity. We can measure the degree of population subdivision by comparing the probability of obtaining two memes that are the same (or different) in one population relative to the same probability for two memes in different populations. Two such measures were introduced by Latter (1973). The first, ϕ, related to the coefficient of kinship and to G_{ST}, is defined as

$$\phi = \frac{I_W - I_B}{1 - I_B} \tag{10.9}$$

where I_W is the expected probability of identity of memes within a population, and I_B is the expected probability of identity among memes drawn from different populations. This measure, in an island model of population structure (i.e., migration can occur between any two populations) and under a neutral model, depends on the rates of migration (m) and mutation (μ) and the effective population size (N_e) according to the following equation:

$$\phi \approx \frac{1}{2N_e(m + \mu) + 1} \tag{10.10}$$

The degree of differentiation, as measured by ϕ, is inversely related to m, μ, and N_e. An increase in m will reduce differentiation because there is more meme interchange between populations, and an increase in N_e will reduce the amount of differentiation caused by drift. An increase in μ also will decrease differentiation, which is somewhat counterintuitive and can make this measure inadequate as an index of population differentiation in certain circumstances (see Lynch and Baker 1994 for details).

The second measure of differentiation, γ, called the measure of mutational divergence by Latter (1973), is defined as

$$\gamma = 1 - \frac{I_B}{I_W} \tag{10.11}$$

This measure, in an island model of migration and under a neutral model, is a function of mutational input, which tends to differentiate populations, and the opposing effects of migration, and is independent of population size:

$$\gamma \approx \frac{1}{\dfrac{m}{(d-1)\mu} + 1} \tag{10.12}$$

where d is the number of populations in the region. The probabilities of identity can be estimated using the meme frequencies in each population, corrected for sample size, using the following equations:

$$\hat{I}_W = \frac{1}{d} \sum_{i=1}^{d} \frac{(n_i \sum_{k=1}^{s} x_{ik}^2 - 1)}{n_i - 1} \tag{10.13a}$$

$$\hat{I}_B = \frac{1}{d(d-1)} \sum_{i \neq j}^{d} \sum_{k=1}^{s} x_{ik} x_{jk} \tag{10.13b}$$

$$\hat{I}_T = \frac{\hat{I}_W + \hat{I}_B(d-1)}{d} \tag{10.13c}$$

where n_i is the sample size from population i, s is the total number of memes in the region, and x_{ik} and x_{jk} are the frequencies of the kth meme in samples i and j, respectively (Lynch 1991, Lynch and Baker 1994). Although these results are based on an island model, which represents an extreme of long-range dispersal, a two-dimensional steppingstone model of population structure (in which only adjacent populations exchange migrants) results in similar patterns of differentiation (M. Kimura and Maruyama 1971).

Estimates of meme flow. Slatkin (1985) developed a method to estimate levels of gene flow using the frequency of alleles (memes, in this case) restricted to a single population. If levels of meme flow are high, then only memes that occur at low frequencies will be restricted to just one population ("private" memes). Memes at higher frequencies will have an increased likelihood of spreading to other populations. Conversely, if levels of flow are low, even memes with higher frequencies can be unique to a population. Slatkin's method uses the formula

$$\ln \bar{p}(1) = a \ln (N_e m) + b \tag{10.14}$$

where $\bar{p}(1)$ is the average frequency of alleles unique to a single population, and $a = -0.505$ and $b = -2.44$ are empirically derived constants. The estimates obtained depend on sample size, and Slatkin therefore proposed an approximate correction method. Under his approach, the estimate obtained using equation 10.14 is adjusted by the ratio of 50 (the sample size in Slatkin's simulations) to the average sample size of the populations being investigated. The value of N_em is an estimate of the average number of immigrant memes per generation. The relationship of this value to the number of migrating birds is usually complex and depends on factors such as the timing of song learning and number of song memes per bird.

Applications

Data from wild populations of songbirds can be used to illustrate the types of analyses that can be performed and the inferences that can be drawn based on the theory described above. I first use data from the literature on the frequency of song memes in one or more populations to obtain estimates of various important parameters of population structure and then describe a transfer experiment that allows a more direct estimation of some of these parameters.

Testing the Neutrality of Song Memes

I obtained data from several sources for nine species: Song Sparrow (*Melospiza melodia:* Harris and Lemon 1972), Winter Wren (*Troglodytes troglodytes:* Catchpole and Rowell 1993), Chaffinch (*Fringilla coelebs*) from Britain (Slater et al. 1980) and the Atlantic islands (Lynch and Baker 1993), Great Tit (*Parus major:* McGregor and Krebs 1982), Indigo Bunting (*Passerina cyanea:* Thompson 1970), Lazuli Bunting (*Passerina amoena:* Thompson 1976), Dark-eyed Junco (*Junco hyemalis:* L. Williams and MacRoberts 1977, 1978), Wood Thrush (*Catharus mustelinus:* C. L. Whitney 1992), and Eurasian Tree Sparrow (*Passer montanus:* A. L. Lang unpubl. data). Data on the frequency of song types were available for the Great Tit, Winter Wren, Chaffinch, Dark-eyed Junco, and Wood Thrush; data in the form of frequency of syllables were available for the Lazuli Bunting, Eurasian Tree Sparrow, and Song Sparrow; and both types of data were available for the Indigo Bunting.

Assuming an infinite-alleles model, I found good agreement between the observed and expected values of meme identity, I (see Fig. 10.1), suggesting that within-population diversity is essentially neutral. This result implies that most variation is maintained by immigration and mutational input, on the one hand, and random extinction of memes, on the other, whereby chance plays the major role determining the fate of memes. This agreement is remarkable considering the roughness of the data and the probable violations of some of the assumptions of the infinite-alleles model. For example, some species likely have a limited repertoire of song elements from which to choose (e.g., Wood Thrush, Indigo Bunting, and Lazuli Bunting), and a K-alleles model would therefore be more appropriate. Also, the model assumes that types are transmitted independently of each other. Songs are more likely to be transmitted independently than elements within a

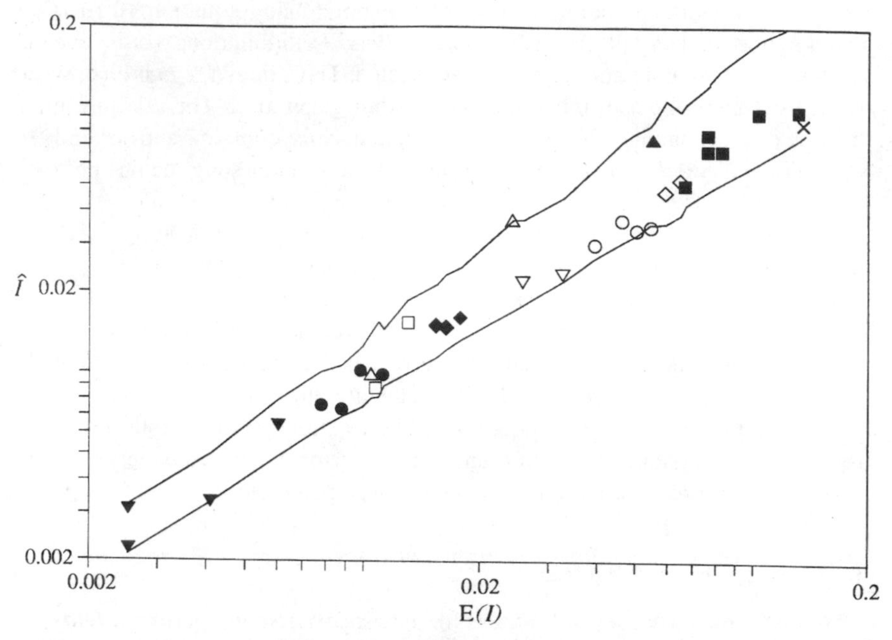

Fig. 10.1. Log-log plot of observed ($\hat{I}$) and expected E(I) meme identity, indicating good agreement between the observed distribution of meme frequencies and the expected distribution under a neutral model. ■, Great Tit; □, Dark-eyed Junco; ●, Eurasian Tree Sparrow; ○, Lazuli Bunting; ◆, Song Sparrow; ◇, Wood Thrush; ▲, Chaffinch (Great Britain); ▼, Chaffinch (Atlantic islands); △, Indigo Bunting (songs); ▽, Indigo Bunting (syllables); ×, Winter Wren. Solid lines represent 95% confidence limits.

song, which tend to be linked, and this linkage would tend to decrease the observed values of I (Lynch and Baker 1993).

Within-Population Diversity

Comparison of diversity between regions. Lynch and Baker (1993) compared levels of variation between continental and Atlantic island Chaffinch populations. At the level of syllables (memes of length 1), no difference was found in diversity (as measured by $\hat{s}_e$) between populations from Iberia ($\hat{s}_{e1}$ = 23.4) and the Canary Islands ($\hat{s}_{e1}$ = 22.7). For memes of length 5, however, populations in the Canary Islands had much higher levels of diversity ($\hat{s}_{e5}$ = 318.2 vs. $\hat{s}_{e5}$ = 43.8). These results suggest that the rate of point mutation is the same in both regions, but the rate of recombination is higher in the Atlantic islands. Lynch and Baker (1993) suggested that this increased recombination is the result of looser syntax in the songs on the islands, perhaps because of the depauperate avifauna, and a concomitant relaxation in selective pressure for distinctive songs.

K-alleles model. Some species might have a limited repertoire of songs or song elements from which to choose, and a *K*-alleles model would thus be more appropriate than the infinite-alleles model. Whitney's data on Wood Thrushes illustrate how equation 10.7 can be used to estimate the number of new memes entering the population per generation in such a case (C. L. Whitney 1992). Wood Thrushes seem to have a limited number of song types, and most songs from the entire species' range can be categorized based on a key developed from a single population (C. L. Whitney and Miller 1987). The key can classify 25 distinct song types, so I use 30 as a rough estimate of K, the total number of song types. The average effective number of memes is $\hat{s}_e = 19.9$, so assuming $K = 30$ and using equation 10.7, we obtain an estimate of the number of new songs entering the population per generation, $N_e v$, of 27.6. This value likely represents mostly input by mutation rather than immigration, because given the limited number of types available, most immigrant types will already exist in the population.

Rate of input of new variants. As a final example I will estimate the rate of input of new variants by mutation and immigration of songs (v) in a population of Great Tits studied by McGregor and Krebs (1982). The estimate of $N_e v$ averaged over the six years of their study, obtained using equation 10.8, is 5.25, and the harmonic mean of the number of songs in the population is 75.83 (equation 10.1). If we assume a constant population size in which each song has an equal probability of being copied, the distribution of copying success (the number of songs derived from a particular one) is approximated by a Poisson distribution with mean and variance of 1. Using values of variance in copying success $V_k = 1$, life expectancy $E = 2.5$, and generation length $L = 1.5$ years (Perrins 1979), I calculated the effective population size using equation 10.2 as $N_e = 45.52$. Therefore, the rate of input of new songs, v, is 0.12. This value is very similar to the value $v =$ 0.13 obtained for a British population of Chaffinches using data from Slater et al. (1980). Using a different approach, Slater et al. (1980) obtained an estimate of $v =$ 0.15. Considering that the rates of migration and mutation of songs seem to be of the same order of magnitude (see below), these results indicate that the mutation rate of songs in these species is very high, approximately 0.01–0.10. This high rate of cultural mutation can potentially create significant population subdivision in spite of moderate to high migration rates.

Among-Population Differentiation

To examine among-population differentiation, I obtained data for four species for which several samples were available: Eurasian Tree Sparrow (A. L. Lang unpubl. data), Wood Thrush (C. L. Whitney and Miller 1987), Lazuli Bunting (Thompson 1976), and Chaffinch from New Zealand (Lynch et al. 1989). I estimated the amount of differentiation using γ, and the amount of meme flow ($N_e m$) using the method of rare alleles (equation 10.14; Slatkin 1985). Using equation 10.10 we can estimate the number of new songs entering the population per generation ($N_e v$), and given that the rate of input of new variants is the result of the combined effect of mutation and migration ($v = \mu + m$), we can obtain the number of new mutants per generation, $N_e \mu$.

Table 10.1. Patterns of memetic differentiation in four songbird species

Species	No. of populations	No. of immigrants per generation ($N_e m$)	No. of mutants per generation ($N_e \mu$)	Mutational divergence (γ)
Wood Thrush	6	31.9	27.8	0.12
Lazuli Bunting	5	37.1	26.4	0.43
Eurasian Tree Sparrow	8	33.2	25.8	0.77
Chaffinch	18	9.6	25.5	0.64

Table 10.1 shows the number of populations sampled for each species and the estimates of the number of new immigrants per generation ($N_e m$), the number of new mutants per generation ($N_e \mu$), and the mutational divergence among localities (γ). In general, the rates of mutation and migration are of the same order of magnitude, indicating that roughly equal numbers of memes enter a population by mutation and meme flow. This means that at least in these species, we cannot ignore the effect of mutation in subdivided populations, as is commonly done in population genetics (e.g., Crow and Kimura 1970). One should be cautious when comparing levels of differentiation across species, however, for two reasons: the geographic scale of these studies differs, and song elements are classified subjectively. Both factors will affect the observed diversity and therefore the mutation rate estimates.

Transfer Experiments

So far I have used indirect evidence based on the patterns of variation within and among populations to suggest that song elements are neutral, that is, that different variants are functionally equivalent (see also Lynch and Baker 1986, 1993, Lynch et al. 1989). Here I present results from a pilot transfer experiment that provide direct evidence for the neutrality of Chaffinch songs in terms of the fitness effects on their carriers.

Thirty male Chaffinches were captured in Karioi, North Island of New Zealand, and transferred to Kohwai bush, near Kaikoura, South Island (a distance of 380 kilometers), in September 1987. Their songs and calls were recorded either before capture or while in captivity. The following spring three of these birds were sighted near where they had been released. All were still singing their North Island songs in unmodified form, but two of them had modified their contact call to resemble the local variant (Fig. 10.2). These two birds had established territories, and at least one of them had mated. The third bird, who was still using his North Island contact call, had not established a territory.

These results suggest that the distinctive North Island songs are functionally equivalent to the South Island ones in the South Island social milieu (that is, it does not matter what one sings as long as one sings). Contact calls, on the other

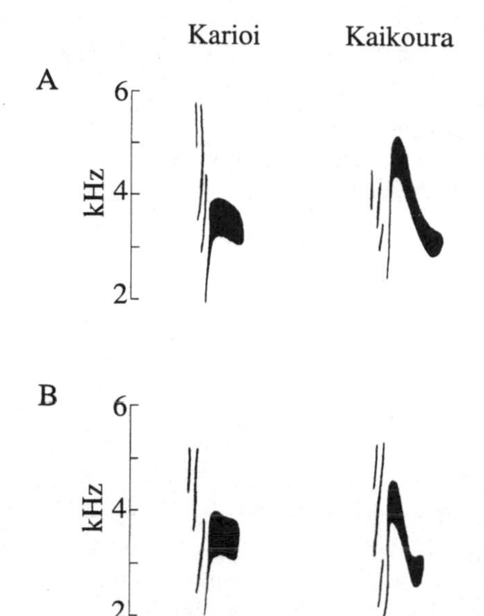

Fig. 10.2. Ink tracings of sonograms (300-Hz analyzing-filter bandwidth) of representative Chaffinch contact calls from two populations in New Zealand, showing typical calls for the two populations (A), and how one bird (male CH-M) changed his call when he was moved from Karioi (B, left) to Kaikoura (B, right).

hand, are probably important in mate acquisition and social interactions with conspecific birds in general, and pressure to conform to the local call dialect is apparently strong. Although I do not have any information on the reproductive success of these males, the evidence presented here suggests that the particular song variants sung by a male Chaffinch do not affect his ability to obtain a territory and a mate (see also R. B. Payne et al. 1988, R. B. Payne and Westneat 1988). Whether these immigrant song elements are copied by local young birds with probability equal to their frequency in the population, or whether they are discriminated against and therefore driven to extinction, needs to be investigated.

An extension of this experimental approach that seems promising would be to introduce birds with distinct song elements into a population and to follow the diffusion of the new cultural variants through time (see, e.g., Dobzhansky and Wright 1947, A. E. M. Baker 1981). By combining this experimental work with analytical theory and computer simulations relating to the spread of innovations (e.g., Hägerstrand 1967), we could test important aspects pertaining to the theory of cultural evolution of birdsong, such as neutrality of song variants and their rate of diffusion, patterns of song transmission across generations, and linkage of elements in a song.

Conclusions

I have attempted to incorporate the processes of cultural evolution into the study of birdsong. I believe that this approach is promising, although more work is needed before it will be possible to make general statements about the relative importance of the different evolutionary factors in shaping the diversity of bird songs. In what follows, I suggest some tasks for the future.

The use of the memetic approach advanced here depends on being able to classify songs or song elements into discrete categories. A major drawback of visual classifications of songs or song elements is their subjectivity, which makes comparisons among studies very difficult. Different investigators classify song elements using different criteria, which influences the estimated mutation rate. A more objective method of classification of song memes needs to be developed; perhaps digitization of sound patterns (see, e.g., E. H. Miller 1979a) combined with methods of pattern recognition could help in this area.

I have borrowed extensively from population genetics theory. Although I believe the models presented here are a good first approximation, birdsong evolution departs in some important respects from genetic evolution, and these models need to be refined. For example, I assumed that memes are independent. At least two potential sources of linkage occur between memes, however: song elements are linked within songs, and songs might be linked within a repertoire. It might be more appropriate to consider the storage of memes not at just one locus but at a variable number of loci, within each of which syntactical rules affect both the number of elements that can exist and the linkages among them. Also, because some population genetics models assume very low mutation rates, it is difficult to apply them with confidence to the study of birdsong evolution. We need more specific mathematical models or computer simulations that take the peculiarities of the cultural transmission of birdsong into account.

It is crucial to conduct long-term studies of banded birds to corroborate the estimates of population structure obtained by indirect methods (Payne, this volume). In particular, transfer experiments like the one described above can be effective to this end because they provide "marker" memes whose fate can be followed through time and space.

Because song elements have high rates of cultural mutation, birdsong diversity departs in some important respects from genetic diversity as revealed by traditionally used markers such as protein-encoding genes. The recent discovery of hypervariable regions of the genome such as the control region of the mitochondrial DNA molecule (e.g., Vigilant et al. 1989), however, raises the intriguing possibility of detecting parallels between the patterns of diversity of these genetic markers and song elements. Comparative studies between these two types of characters could be very illuminating.

I end with a speculative scenario for the origin and maintenance of diversity in bird songs. The levels of memetic diversity within populations in a number of songbird species seem to be consistent with a random drift–migration–mutation model. I tentatively postulate that each species has a potential mutation rate that is adjusted in different populations according to the acoustic properties of the environment. In this scenario, then, the main importance of selection is to weed out unsuitable types that appear by mutation or immigration. In general, populations will tend to diverge randomly because normal levels of meme flow are not high enough to prevent the accumulation of new variants in different populations that result from the high rates of meme mutation. If the acoustic environment changes, however, or if individuals of a given population colonize a new habitat with different acoustic properties, then directional selection can act on the neutral or nearly neutral variants already in the population. In population genetics this phenomenon is termed the Dykhuizen-Hartl effect (Dykhuizen and Hartl 1980, M. Kimura 1983). On the other hand, vocalizations that are important in social situations and that require conformation to a local standard might be subject to majority-type advantage (Lynch unpubl. data). This type of frequency-dependent selection will be important in determining which variant predominates in a given area, although different variants might become dominant in different areas.

Acknowledgments

Mike Dennison, Andy Horn, and Anna Toline provided many useful comments on a previous version of this chapter. I thank Tony Lang for giving me access to his unpublished data, and Geoff Plunkett for assistance during the fieldwork in New Zealand. Don Kroodsma's and Ted Miller's thorough editing greatly improved the intelligibility of this chapter. This is Contribution no. 5 of the Population Memetics Group.

11 Song Traditions in Indigo Buntings: Origin, Improvisation, Dispersal, and Extinction in Cultural Evolution

Robert B. Payne

Bird songs can change either within a behavior lineage when improvisations accumulate in a song tradition, or among behavior lineages when new songs appear and others disappear. Two kinds of cultural evolution occur in song: one tracks the changes within a tradition, the other parallels the demographic processes within a population. Changes in song in bird populations may involve both kinds of processes. The current diversity of songs within a species may involve changes in time within a cultural lineage, branching from a common lineage, dispersal of song lineages between populations, and even a mix of cultural lineages in the songs of an individual bird. It is possible to build a general theory of cultural evolution that includes both approaches.

Within a behavior lineage, similarities of songs over time may be due either to continuity of a particular theme or to independent innovations or improvisations on a basic theme. An accumulation of improvisations ("errors" or cultural "mutations") can change a song lineage and can also split a lineage into two or more descendant lines. These kinds of changes within lineages are phylogenetic changes (both serial change within a lineage as a song continues through time, and cladistic divergence of a lineage into a set of derivative branches; Ruvolo 1987).

Between behavior lineages, songs within a population may change over time as old song themes disappear and new ones are introduced by immigrants (Lemon 1975, Slater and Ince 1979, Ince et al. 1980). Such dynamic changes, in which one kind of song replaces another in a population, are the cultural counterpart of genetic changes due to the differential success of distinct lineages, as in genetic drift and natural selection. The mechanisms of change in a dynamic turnover of song lineages involve song copying, with social interactions between individuals, as well as the immigration and extinction of lineages within a population.

Song traditions are also of interest to the concept of cultural coevolution (Cavalli-Sforza and Feldman 1981, Durham 1991), the idea that cultural and genetic traits enhance the transmission of each other. Cultural coevolution assumes that culture and genes are closely associated in a developmental sense. When a young bird learns its song from its father, or when genes affect whether a bird learns a certain song, the genes and songs are passed on together, in a process of coevolution of genes and song traditions.

Although birdsong traditions illustrate the general processes of cultural transmission and evolution (Slater 1986, 1989), few field studies have shown song transmission between individual birds, and traditions have seldom been observed continuously for more than a generation or two (R. B. Payne 1985, R. B. Payne et al. 1987, R. B. Payne and Payne 1993b). Many studies of song variation in space have been interpreted as change over time. When songs actually were recorded over time, the singing birds were unmarked and the songs were not sampled from year to year.

This chapter explores some questions about the continuity and change of song traditions in birds, incorporating observations from a long-term study of Indigo Buntings (*Passerina cyanea*). The answers to these questions will help us create a touchstone for the study of cultural evolution. Does a song change in cultural transmission along a lineage, and does it change in a regular, clocklike manner? What is the importance of cultural extinction and of the introduction of new songs by dispersal? How do new songs originate? How important is improvisation or modification of song? Is change within the song pool of a population described by a phylogenetic model, a dynamic model, or both? How can we assess the association of genetic and cultural change? What is the probability of song continuity and change in individual birds and in populations, and how do these parameters compare with the demographic parameters for the population? Finally, what is known about song continuity and change in other songbirds, and what observations are needed to determine the processes of song transmission and cultural evolution of song?

Songs in a Population of Indigo Buntings

Methods and Study Areas

My colleagues and I observed Indigo Buntings from 1977 to 1994 in southern Michigan (R. B. Payne 1981a, 1982, 1983a, b, 1989, 1991, 1992, R. B. Payne et al. 1981, 1987, 1988, R. B. Payne and Westneat 1988, R. B. Payne and Payne 1989, 1990, 1993a, b, 1996, in press). One study area (600–1200 ha) was at the E. S. George Reserve, where songs had been recorded in earlier years (Thompson 1970). The other was near Niles (200–400 ha).

Buntings arrive in the study areas in May and early June, and they nest until late August. As the previously unbanded males were netted, they were aged by plumage and color-banded for individual recognition. Males breed in their first (yearling) year, and many are successful; some move from one territory to another during the breeding season. Females also breed as yearlings; they build the nest, incubate, and feed the nestlings. Pairs often raise two and sometimes three broods in a season. Their singing sites and nests were plotted on a map marked with a 100-meter grid (R. B. Payne 1989, 1992).

To document their behavior, we recorded 80–180 singing resident males in each area several times each season. Yearlings were recorded each week because we wanted to document how their songs changed over time. The older adults were recorded less often because we found they rarely changed their songs. Yearlings generally did not sing the song of their father (R. B. Payne et al. 1987), and their first song of the season often was unlike the song of any male that was recorded on the area in their natal year (R. B. Payne and Payne, in press). Fewer than 20% of the buntings that bred in each study area were born there (R. B. Payne and Payne 1989, R. B. Payne 1991). By the end of their first season, about 80% of the yearling males had changed their behavior to match the song theme of a local neighbor. The others retained their first song and did not match a local song theme (R. B. Payne et al. 1987, 1988, R. B. Payne and Payne 1993b). Because some males had more than one song in a season, our sample size of cases, in which each song theme was counted separately for each male, was 100–280 in a year.

A bunting song theme is a series of pairs of "song figures," or "elements." Males recorded hundreds of times typically used only one song theme each, and their renditions of the song differed only in the deletion of terminal elements. A few males had two songs, but these were usually variations on the same theme. Song elements from songs of all males were scored from sonograms with the element catalog of Thompson (1970). Sequences of elements were then compared using the computer program MATCH, which used the number of elements in the same sequence as a criterion of a match of song (R. B. Payne et al. 1981). Songs of two birds were called a "match" if they had at least three different elements in common and these were in the same sequence. In practice, if they matched three elements in series they usually matched five or six, which is a typical song length of a bunting (Thompson 1970, Emlen 1972, Shiovitz 1975). The criterion of three elements allowed classification of short songs, and it also gave a standard by which we could match two songs with an improvisation in the substitution, addition, or deletion of a song element. MATCH was also run with an option of one change in four elements, so songs could match with at least three of four in a sequence. The probability is low that songs without a recent common origin will match by chance because wild buntings have a species repertoire of about 100 song elements that are geographically widespread (Shiovitz and Thompson 1970, Emlen 1971a). Songs rarely matched between our two study areas (R. B. Payne et al. 1981), and matches usually were between males on adjacent territories, as Thompson also found (1970).

Breeding success was determined as the number of young buntings that fledged from nests on a male's territory (R. B. Payne et al. 1988). We analyzed breeding success from 1980 through 1987; all records for 1986 and 1987 were from males born by 1984 or of known kinship lineage. Survival was determined by seasonal resightings of birds and identification by their color bands (R. B. Payne and Payne 1990).

Continuity and Change within a Behavior Lineage

Continuity in song themes occurs when the same theme is found from year to year. At the George Reserve, we documented continuity for 54 different song themes, each representing a behavioral lineage that was matched by two or more buntings. The males that transmitted a song theme were almost always adults, and males that copied a song theme were usually yearlings, though males of all ages were sometimes transmitters and copiers (R. B. Payne and Payne 1993b).

In practice, we documented continuity when a song was a match across years in the same local neighborhood, and the birds with the song were color-banded and their songs were recorded from year to year (Figs. 11.1, 11.2). Some songs were transmitted with little change, but others changed in the form of a song element or in a substitution, deletion, or rearrangement of elements. Changes usually involved a yearling male improvising on the song of a neighbor rather than an older male altering its song from year to year, though a few adult males did this (Fig. 11.2). From 1 to 12 males had a song theme in a given year (Figs. 11.1, 11.2); the largest number of males that matched a song in a year was 22, and the average was 3 or 4 (R. B. Payne et al. 1988).

Diagrams like those in Figs. 11.1B and 11.2B were made for all song themes that were copied and that continued into the next year. For songs that we first recorded from 1978 to 1984 and followed to their last appearance (from 1979 to 1994), there were 341 song-years, and 287 examples of a song reappearing in successive years but none of a song skipping a year and then reappearing. In only four cases (<2%) did a song reappear when we did not see a banded male with that song return from the previous year. In one case he may have been there earlier in the season, in another the song was reintroduced into the core area by another male (the theme persisted just outside the main study area), and in the other two cases an unbanded adult was probably the unbanded male that we had recorded with the song in the previous year.

The minimum number of birds that were involved in the continuity of a song for 2 years or more varied from one (a male was copied, but he returned with the song himself in the next year) to a series of seven birds (a song that persisted for 12 years). For example, the song depicted in Fig. 11.1 was kept alive by one bird from 1984 to 1994, and the song in Fig. 11.2 was continued by at least four birds from 1978 to 1994 (one from 1978 to 1981, one from 1981 to 1984, one from 1984 to 1987, and one from 1987 through 1994). In 18 of the 54 song themes, one bird returned for 2–6 years and maintained the song; when he no longer returned, neither did the copiers, so the song disappeared. The number of transmission events needed to account for the continuity of a song ranged from none when an adult returned with the song to six (mean, 1.38 ± 1.36 SD), over an average song survival of 6.30 years (± 3.68 SD; range: 2–16 years).

Copied songs sometimes changed as one bird improvised on the song of its model (Figs. 11.1, 11.2). Most song lineages showed minor changes when the songs were traced across three or more cultural generations. When these changes

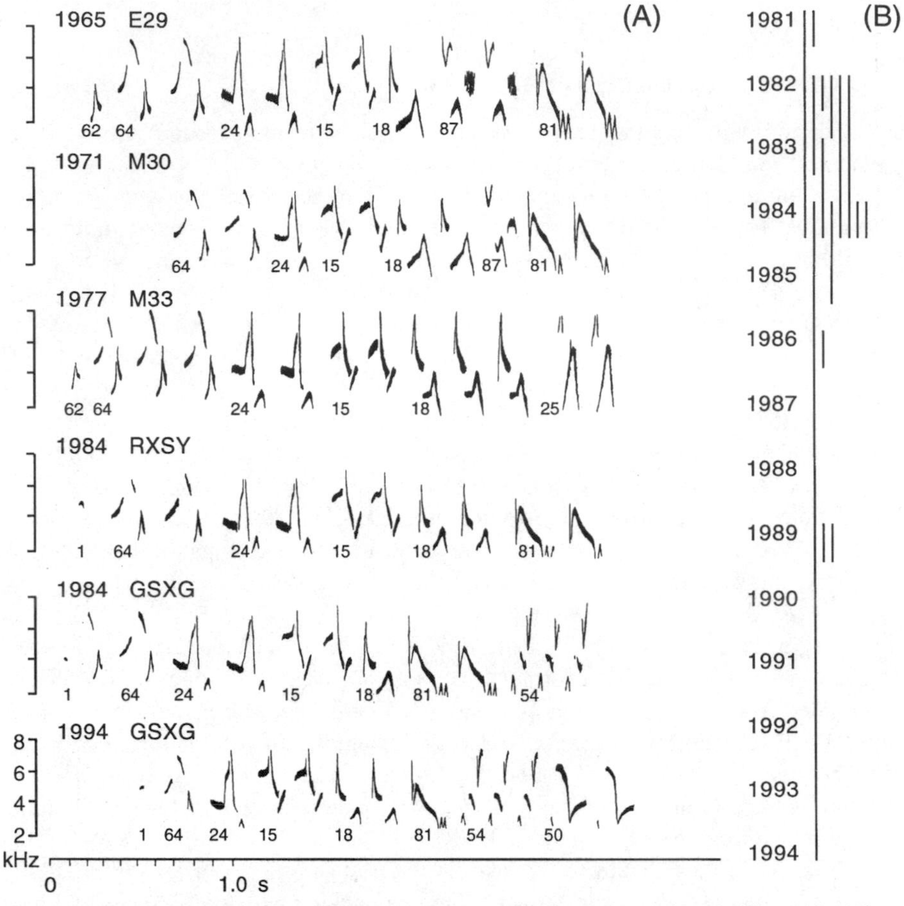

Fig. 11.1. (A) Continuity of a song tradition of Indigo Buntings at the George Reserve (song b of Figs. 11.4, 11.5). The song was first recorded in 1965–1969 and 1971, within 200 meters of grid T25 and near M30 (see Figs. 11.4, 11.5); see Thompson 1970, 1972. It was recorded in 1977 near M30 from a bird captured as a tutor for an experiment (R. B. Payne 1981a). In 1982 the song was near X32 (800 meters from T25, 1000 meters from M30), where it continued with male GSXG for 11 years; in most years he was the only male recorded singing the song. The number of repetitions of a song element was consistent for many individuals, but GSXG varied repetitions within a song bout and between seasons. (B) Continuity within a song lineage. Each vertical line shows the years when a particular male was observed and his songs were recorded.

Fig. 11.2. (A) Continuity and a series of changes within a song tradition (song A) at the George Reserve near grid T25 (see Figs. 11.4, 11.5). The song was recorded at the same site by W. L. Thompson in 1965 and 1969–1972. It was recorded from banded males each year from 1978–1994. Note the accumulation of changes in the song through time. Note also the two songs in 1984 of male RWGX (the last part of his song was styled on that of another neighbor with a sequence of elements that partly overlapped the A sequence), and his return (1985–1987) with a mixed song that recombined elements of his 1984 songs; male GBXY copied the mixed song. Note also the divergence of the song into two lineages with the song of male YXRO, who copied neighbor GXGB when YXRO was an adult immigrant in 1985. YXRO had song elements perhaps retained from the previous year, when males with elements 41, 87, and 25 sang in other areas near the reserve; or perhaps he acquired his song in 1985 from male GOXB, who had the sequence 41-25-26-87 of song elements in grid R27. Across years the continuity of song A in a sequence of three elements (allowing a change of one in a series of four elements) is seen from one bird to the next, but across nearly 30 years it gradually evolved, so that the 1965 songs do not match the 1994 songs. (B) Continuity within a song lineage. The vertical lines show the years when each male was observed and his songs were recorded. The number of males that continued the song is shown by the number present in one year that returned in the next.

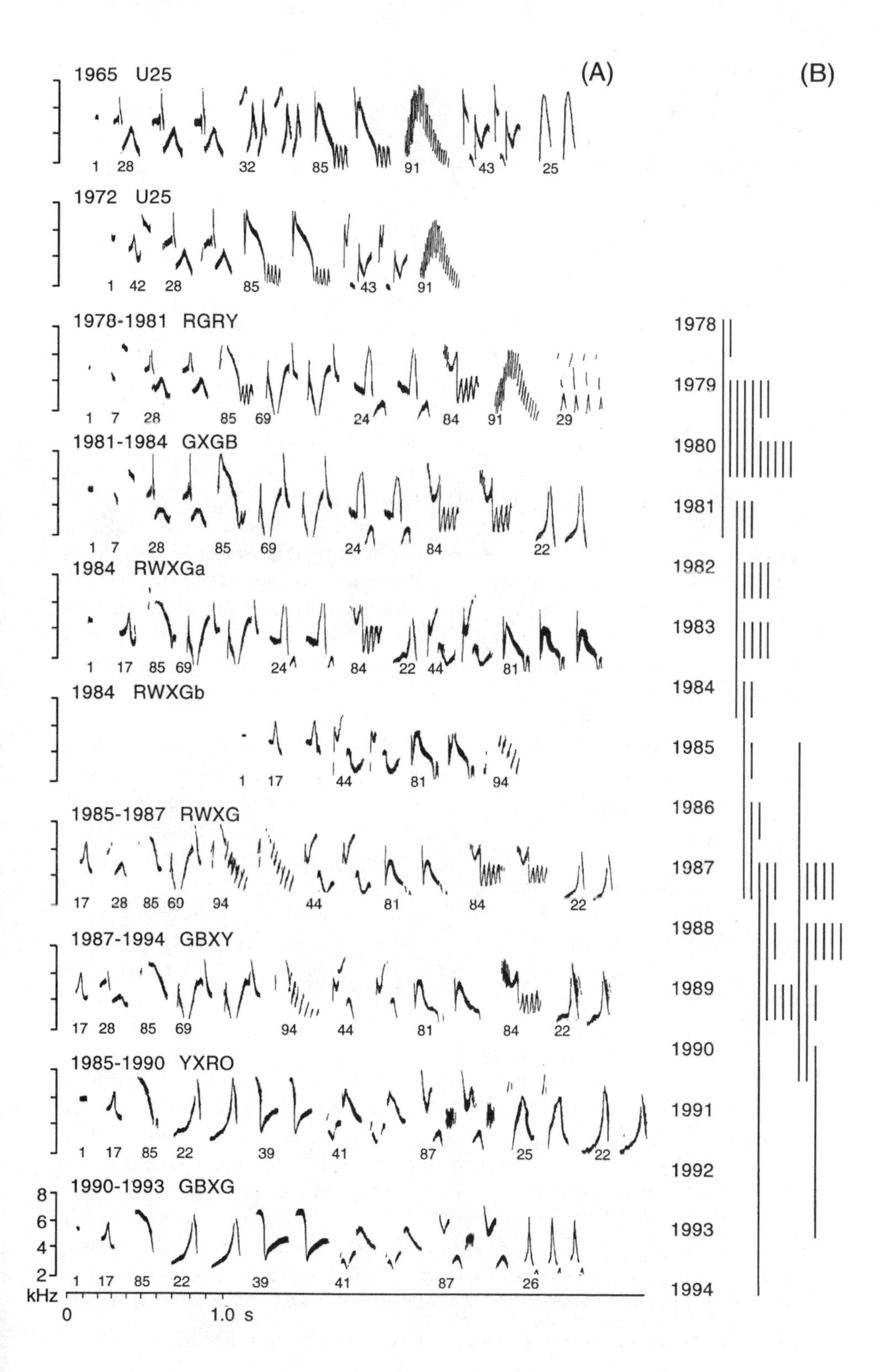

(A)
(B)
1965 U25
1 28 32 85 91 43 25
1972 U25
1 42 28 85 43 91
1978-1981 RGRY
1 7 28 85 69 24 84 91 29
1981-1984 GXGB
1 7 28 85 69 24 84 22
1984 RWXGa
1 17 85 69 24 84 22 44 81
1984 RWXGb
1 17 44 81 94
1985-1987 RWXG
17 28 85 69 94 44 81 84 22
1987-1994 GBXY
17 28 85 69 94 44 81 84 22
1985-1990 YXRO
1 17 85 22 39 41 87 25 22
1990-1993 GBXG
1 17 85 22 39 41 87 26
8
6
4
2
kHz
0
1.0 s
1978
1979
1980
1981
1982
1983
1984
1985
1986
1987
1988
1989
1990
1991
1992
1993
1994

accumulated and more than one song element varied, some of these continuing song theme traditions were not recognized as the same by the MATCH criteria when the data were selectively sampled at intervals of several years.

Changes among Songs: Immigration and Extinction

Changes in the population's "song pool" resulted also from the immigration of new songs and the extinction of others. At the George Reserve, about 60% of the songs were new in each year from 1979 to 1984 (Fig. 11.3). Most new songs did not reappear the next year, so most of the 1970s songs were gone by the early 1980s. We documented a similar continuity of some songs and turnover of others at Niles (Table 11.1), so the pattern appears to be general for bunting populations.

The cultural diversity of the new songs usually disappeared by the next year for one of four reasons: (1) A male with a new song failed to return the next year. Most songs were lost in this way, and most losses involved yearlings, as survival from the first to the second year is lower than in later years (R. B. Payne and Payne 1990). Overall, nearly half of the extinctions of songs that were new in the years 1979–1985 disappeared when the birds that sang them did not return the next year. (2) A yearling male switched from one song to another within a season, usually when he changed to match the song of a neighbor (R. B. Payne 1982, R. B. Payne et al. 1988, R. B. Payne and Payne 1993b). (3) A yearling retained a song through his first season, then returned with a different song, usually one of a neighbor from his first season (R. B. Payne et al. 1988). (4) Songs were lost when males returned one year or more and sang the same song but were never copied. Most new songs in the population were lost within a year or a few years because fewer than 20% of new songs were copied by another male in the study area.

Other new songs persisted, however, either because the founder male returned with the song the next year or because the song was copied by another male that returned. When the songs were copied, lineages were established, so some new songs from early in our study persisted into the later years. The song pool in 1985, for example, included many songs that first appeared from 1979 to 1984, and the 1988 songs could be traced back to songs that had been introduced in six of the nine years from 1979 through 1987 (Fig. 11.3). From 1977 through 1994, two song themes persisted and were recorded in every year, and two other song themes that first appeared by 1984 were still present in 1994.

The new song themes introduced by yearling males were apparently learned outside the study area rather than being improvised by the yearlings (Table 11.2). A few new songs were traced to sites within 1 kilometer of the core area, and the yearling males had apparently learned the songs at those outlying areas before moving onto our core study area. Other songs likely came from more remote neighborhoods. Another reason for suspecting that the new songs were not innovated on the spot is that captive buntings reared in the absence of other singing males failed to develop recognizable bunting songs. Although such deprived birds sang, their songs lacked the elements and paired sequences of elements of the wild

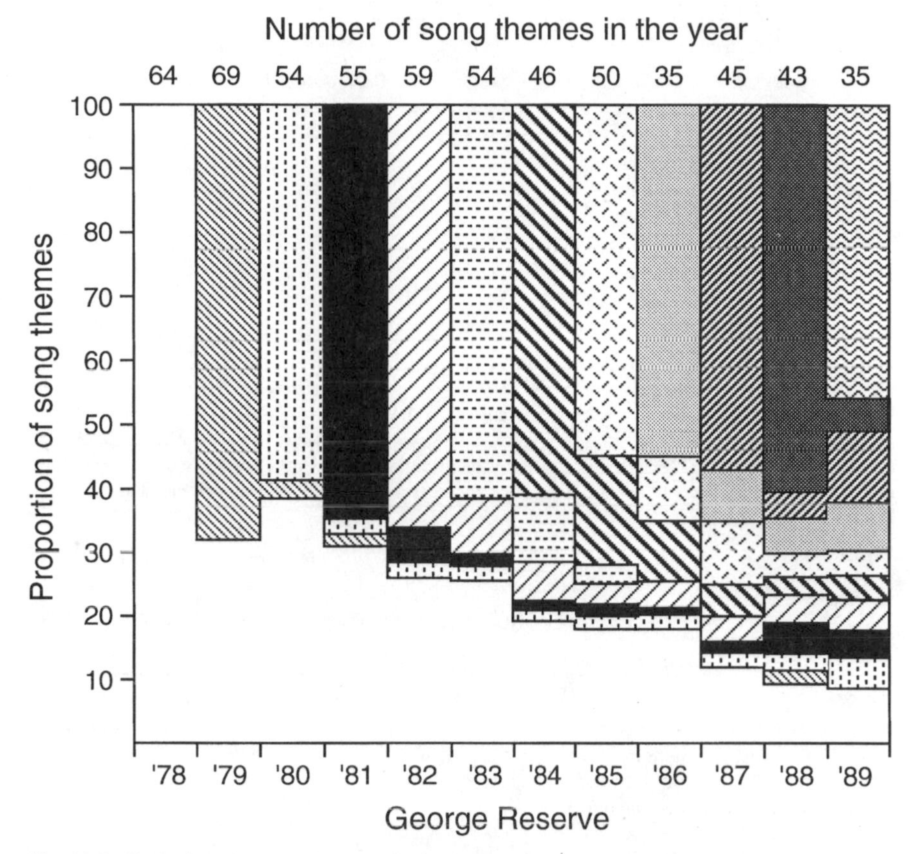

Fig. 11.3. Continuity and turnover of songs at the George Reserve from 1978 to 1989. The white bars represent the songs that were present in 1978 and that persisted through 1989. The other bars show the songs that were new in each year from 1979 through 1989 and whether they persisted through later years. A few songs survived for 10 years or longer, most new songs were lost by the next year, and a few songs became established as local traditions.

buntings (Rice and Thompson 1968, R. B. Payne 1981a). In playback tests, wild buntings failed to respond to these isolate songs (Shiovitz 1975). The first songs of wild yearlings were unlike the songs of these naive birds and consisted for the most part of normal bunting song elements.

Most new songs that were copied by another male were introduced not by yearlings but by immigrant adults (Table 11.2). Of the 28 songs not present in 1979 that were copied and matched by another male in 1980 or 1985, 61% were first recorded either from a new adult ($N = 16$) or from a yearling male that was copied when he returned as an adult ($N = 1$). Also, the few yearling males that matched a song with other yearlings (and not with a local adult) usually had territories more than 400 meters apart or near the margin of the study area, and I suspect the source of their songs was a male (most likely an adult) near but not in the study area ($N = 6$; Figs. 11.4, 11.5). We were able to trace 5 of the 28 new

Table 11.1. Proportion (percentage) of new songs that appeared in two populations of Indigo Buntings in southern Michigan, and the cultural success of the new songs

	Population	
	George Reserve	Niles
Song not copied		
Male did not return next year	49	53
Male changed song to match a local song	24	21
Male returned with a different song	3	5
Male returned with same song, never copied	3	5
Song copied by another male		
Male was copied in his first year	16	13
Male returned and was copied in a later year	3	4
Total different songs[a]	237	77

[a]This includes song themes that were first recorded in the study area from 1979 through 1984, but excludes song themes recorded in earlier years. Some new song themes were recorded in more than one year; these were counted only for the first year. The new song themes for the George Reserve are indicated in shaded areas of Fig. 11.3, where the numbers indicate both old song themes and new song themes for each year.

Table 11.2. Origin of new song themes of Indigo Buntings on the George Reserve in southern Michigan, 1980 and 1985

Categories	No. of songs	Song themes[a]
Song shared by two or more local male buntings in 1980, 1985, or both years		
Immigrant adult, song not recorded elsewhere	11	e, i, j, n, q, r, u, v, a′, e′, h′
Immigrant adult, song recorded nearby	3	c*, d, t
Immigrant adult, not copied in earlier year	2	a, p
Yearling, song recorded nearby	2	f′, m)
Yearling and uncaptured, song not found nearby	2	c′, g
Yearlings on area and outside area	1	d′
Yearlings, marginal[b], song shared by several males	4	g′, w, f, h
Yearlings, remote[c], learned from common source?	2	k, b′; c′, f′, m
Yearling, copied in a later adult year	1	b
Song not shared in 1979, 1980, or 1985		
Adult, first year in area	3 in 1980, 3 in 1985	
Adult, returning from previous year	1 in 1980, 7 in 1985	
Yearling and uncaptured (1980)	22 (13 by males resident for >10 days)	
Yearling and uncaptured (1985)	27 (11 by males resident for >10 days)	

Note: The table excludes males that changed one song and matched another local song within the season.

[a]Letters refer to songs on the maps in Figs. 11.4 and 11.5.

[b]Marginal songs are those recorded only on the margin of the study area.

[c]Remote males with same song were on territories ≥400 meters apart.

*Song was present 1 kilometer distant in 1964; also was 500 meters distant in 1977, and was rediscovered in 1981 outside the 1979 study area.

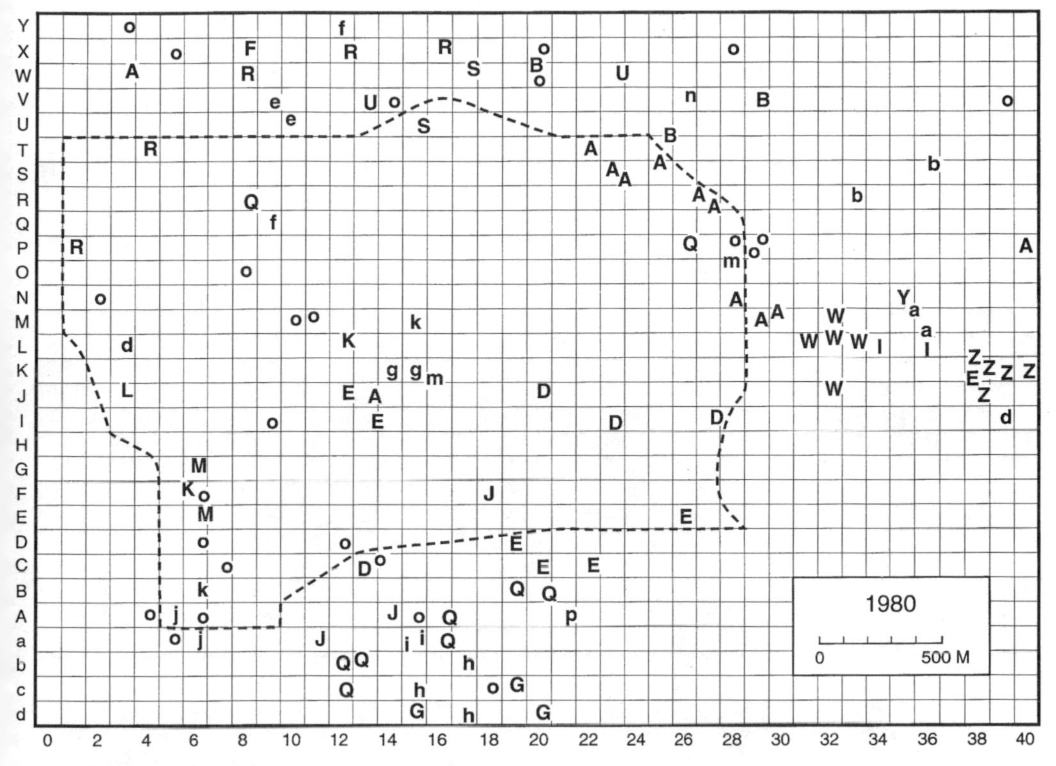

Fig. 11.4. Distribution of song themes at the George Reserve in 1980. Each letter refers to the song of a male on his territory. When a male had more than one song, the song he sang for most of the season is shown. Songs in uppercase letters were also there in 1979 (map in R. B. Payne et al. 1988). Songs denoted by lowercase letters (a–p) were matched by two or more males in 1980 or 1985 and were first recorded in 1980. Songs indicated by "o" were unique in the area; a few of these were matched in other years. A few "o" songs in 1979 (see R. B. Payne et al. 1988) were copied in 1980 or later years and are indicated by the letters of the matched songs.

shared song themes to a source within 1 kilometer of the core area, but most of the themes probably originated at greater distances. Most of the males that introduced new themes that were later copied were first captured as adults, presumably in their first adult year after they dispersed from another breeding area (R. B. Payne et al. 1988). Two males that we did not catch and band introduced two song themes too.

We documented how marked males dispersed locally the year after they learned their song, and in the process how they dispersed their song themes. Several cases illustrate this process. (1) Song A: male GOXO copied song A at location R27 as a yearling in 1979 (R. B. Payne 1982). He returned to R27 on 20 May 1980, and the next day he moved 2 kilometers away to W4 (Fig. 11.4), where he remained

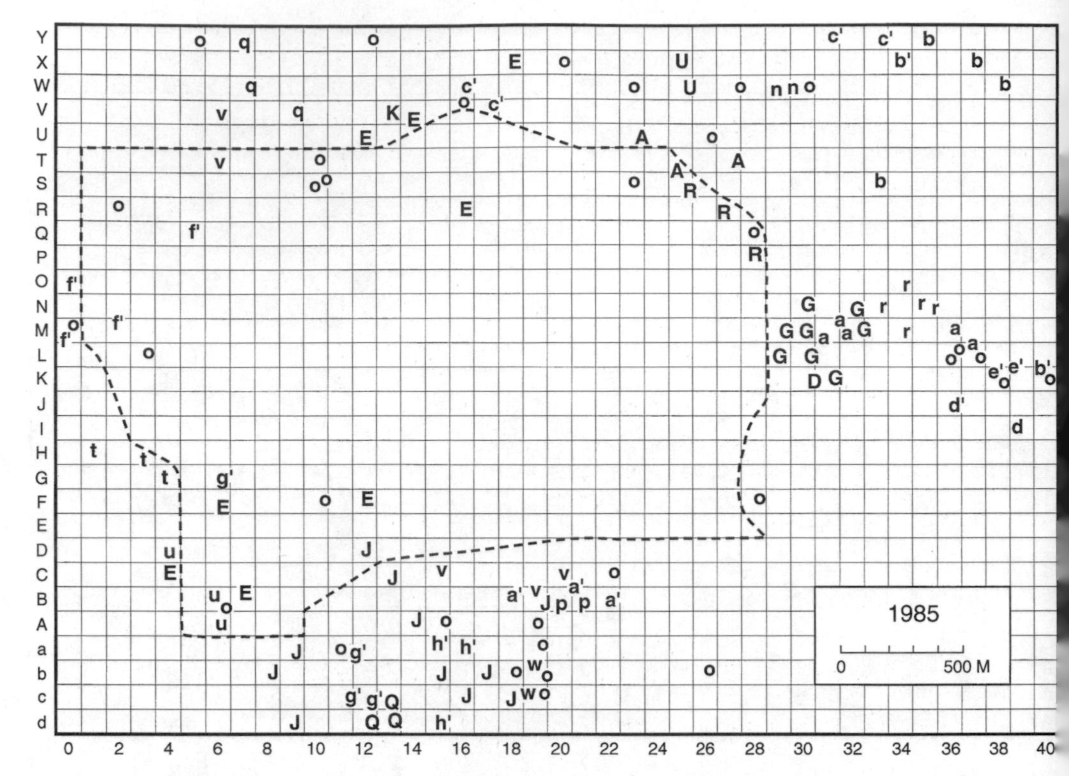

Fig. 11.5. Distribution of song themes at the George Reserve in 1985. Legend as in Fig. 11.4. Songs denoted by lowercase letters were matched by two or more males in 1980 or 1985 and were recorded first in 1980 (a–p) through 1985 (q–w); songs marked with a prime (′) were first recorded there after 1980. Songs indicated by "o" were unique in the area; a few of these were matched in other years. A few "o" songs in 1979 (in R. B. Payne et al. 1988) were copied in 1985 or other years and are indicated by the letters of the matched songs.

for 70 days. (2) Song G was heard only south of the reserve from 1979 through 1982, where male WXOW copied it in d18 as a yearling in 1981. In 1982 he settled in K31, 2 kilometers from his earlier site, and there he was copied by other males in 1982–1985 (Figs. 11.4, 11.5). South of the reserve, song G disappeared in 1983, but east of the reserve it persisted through 1994. (3) Song D was copied by yearling male BRSX in J25 in 1982. He returned from 1983 through 1987. In 1985 he moved to K30, two territories away from where he had acquired the song. (4) The last example illustrates the origin and movement of a song theme of mixed source. Song E was a mixed song (EK = E in Fig. 11.5) that male BGYX acquired as a yearling. He copied the first half of songs E and K from bilingual male (i.e., he had two songs themes rather than one) GXGR in U13 in 1982, and he returned to C4 in 1984. BGYX's version of song E had a three-element sequence of song K, which occurred in the southwest corner of the reserve in 1978–1979 (R. B. Payne et al. 1981). The mixed song originated in 1980 when GXGR learned the songs of both an E and a K neighbor in grid K13. GXGR had two songs in his adult years; he settled in U13 in 1981, where he sang through 1985 (Fig. 11.5). Male BGYX in his first adult year took the mixed song to C4. In each of these four examples, a male had acquired the song as a yearling and then, in a subsequent breeding season, dispersed both himself and his copied song from 2 to more than 10 territories.

In the last half of our study period, we documented a decline in both study areas in both the number of new songs and the number of all songs. From 1985 to 1989 the proportion of new songs was only 40–50% at the George Reserve, for example, where it had been 60% in earlier years (Fig. 11.3). This decline in new songs is explained by the decline in the number of buntings. In the later years, fewer yearling males were seen, and the older males (and their songs) that did not return were not replaced by new males. Although this decline coincided with a decline of Indigo Buntings in the Breeding Bird Survey counts in eastern North America, on a regional scale this decline was only 6.9% for the period 1978–1988, or -0.69% per year (J. R. Sauer and Droge 1992). The decline in buntings on the study areas during these years was 60%, far greater than the regional change, which suggests that local changes in habitat affected the number of buntings in the study areas.

In summary, the songs in a population of buntings undergo a gradual turnover. New songs are gained every year when yearling and new adult males immigrate, and songs are lost when yearling males switch their songs to match the song of a neighbor, or when males that sing a particular theme fail to return. Some themes are successfully introduced into the population by new adults, and the themes are established when they are copied by yearling males. The new songs then persist when the adults or the yearling males that copied them return from year to year. Songs become extinct as often as expected from the number of birds that match a song, given the survival of males and the probability that a male will switch songs within a season or between years (R. B. Payne et al. 1988).

Cultural Survival

Our archives of song theme traditions make it possible to estimate the time and generation scales of cultural change. I estimated cultural survival of bunting songs by comparing the proportion of songs recorded in an earlier period with those that survived to a later one. I used a model of exponential decay, in which the number of song themes that survive is described by the expression

$$S = B\,e^{-kt}$$

In this model, S is the number of songs at the end of a sample period, B is the initial number of songs in the population, e is the base of natural logarithms, k is the annual rate of change, and t is the number of years between samples. The half-life, $t_{.50}$, of songs (the period over which 50% of the songs survive) is estimated from the following expression:

$$t_{.50} = \frac{\ln 2}{k}$$

A few songs that were recorded at the reserve in the 1960s (Thompson 1970) were recorded again in the late 1970s. From the proportion of old songs that persisted over a 15-year period I estimated a half-life of 3.8 years for the cultural survival of songs (R. B. Payne et al. 1981).

Although the song themes are transmitted across generations and therefore are traditions (Shils 1981), the assumptions of the exponential decay model affect the validity of this estimate of cultural survival. First, the model assumes that all songs are equally likely to survive. In reality, however, the survival of songs is influenced by both the age of the singer and the number of males that sing the song. Survival of an adult's song, for example, is more likely than survival of a yearling's first-season song (R. B. Payne et al. 1988). Also, a male is more likely to keep a song from year to year if other males match it; if he is the only male with the song, then he is more likely to change. In addition, the probability that a song will return the next year is greater for songs sung by many males. The probability that no male will return with a particular song is the last term of a negative binomial expansion, where the exponent is the number of males with the song in a given year, and the survival term is the mean for the population (R. B. Payne et al. 1988, R. B. Payne and Payne 1990, R. B. Payne 1992). Finally, an estimate of cultural survival based on songs with at least three of four song elements continuing in the same sequence is low because a song can be transmitted with its elements rearranged from their sequence in the song of the bird that is copied (Fig. 11.2), and the accumulation of these changes over several cultural transmissions can be substantial. One song of 1965 thought to be gone by 1979 (R. B. Payne et al. 1981) in fact had persisted; but the structure and sequence of its elements had changed so much that its continuity was overlooked.

Because the birds were color-banded and their songs were recorded every year, we could trace the survival of each song lineage and compare that survival with the survival of the birds. At the reserve, 24 different song themes were matched by two or more birds in 1979, and at least one of the two birds was an adult in each case. Nine of the 24 songs survived to 1985 (Figs. 11.4, 11.5), and $t_{.50} = 4.23$ years. In 1979, an additional 35 themes were given as the last song of the season (by a yearling or an adult) but did not match those of other adults. Combining these unmatched themes with the matched themes to estimate the cultural half-life gives $t_{.50} = 2.21$ years. For comparison, the annual survival, P_s, of male buntings at the reserve is 0.52 ± 0.02 SE years, and the expectation of further life, S_o, is 2.08 years (R. B. Payne and Payne 1990, R. B. Payne 1992), so $t_{.50} = 1.33$ years. The cultural survival of a matched song is three times greater than the survival of the males in the population. In contrast, the survival of the unmatched songs of the yearling males is lower because these males switch their song within a season, stay in the population only a short time, or return with a different song the next year (R. B. Payne et al. 1988).

Continuity and Change of Songs in a Population

The continuity and change of song traditions can be observed directly along with the survival and dispersal of marked birds. The effects of song changes by individual buntings and population changes in the song pool (Table 11.3) can be summarized as follows. Whether a male keeps his song varies with his age; most first songs of yearlings are dropped when the yearling settles on a territory and copies a local resident neighbor. Whether a bird is copied also depends on his age: songs of most yearlings are not copied, for example, but songs of immigrant adult males are often copied, either in their first adult year in the study area or later. Whether a male is copied also depends on his date of arrival, because early birds are more likely to transmit their songs. Whether a male disperses varies with his breeding success; males that are unsuccessful one year often settle on a new territory the next year. The risk of cultural extinction is high because song neighborhoods are small, but for two reasons a song is more likely to persist if it is sung by an adult: yearling males are more likely to switch songs between years, and older males are more likely to survive and return to the same territory. Also, a song is more likely to persist if it is sung by several males rather than by only one, because survival of a song depends on the survival and return of at least one male. Local extinctions of songs are balanced by immigration. The songs also change through improvisation, as when one male copies another and modifies the song, and songs sometimes change when a male combines elements of songs of more than one model male. Although the song traditions continue from year to year, often with an accumulation of minor changes, most differences in the songs over a series of years originate from the introduction of new songs by immigrants.

Cultural Evolution and Birdsong

Two Models of Cultural Evolution: Changes within a Lineage and Changes among Lineages

"Cultural evolution" implies a change in learned behaviors in populations over time. Behavior traits can change both when minor differences accumulate within a lineage (by improvisation and by the divergence of lineages) and when lineages vary in their cultural success. The changes among lineages are either stochastic (founder effects and sampling errors in small song neighborhoods) or causally associated with the songs (a cultural counterpart of natural selection). These changes involve the loss of some behaviors from a population and their replacement by new ones. Although minor changes accumulate within song lineages in the buntings, more changes at the population level result from demographic and social dynamic processes, including dispersal of birds with new songs and differences in the cultural success of the alternative songs.

Table 11.3. Components of cultural continuity (song transmission, ST) and change (cultural evolution, CE) in Indigo Buntings

Variable	Factors affecting variable	Effects on ST, CE	References
		POPULATION LEVEL OF ANALYSIS	
Survival	Age	Adult survival is higher than yearling survival	a
Age	—	Survival, change of song, dispersal	a, b
Density	Habitat; social atraction	High neighborhood density, greater likelihood of being copied (ST)	c
Neighborhood age distribution	—	ST varies with number of yearling males to copy song	c
Dispersal	Age; previous breeding success	Immigration (CE), movement of song theme from site to site (ST, CE)	b, f
		INDIVIDUAL MALE LEVEL OF ANALYSIS	
Previous breeding success	Age	Little effect (ST)	d
Breeding success when copied	Age (not pair duration)	Little evidence of prior assessment of fitness (ST)	c
Arrival date in spring	Age	Early arrival increases ST	c
Duration on territory	Age	None	c, e
Breeding plumage	Birth date	Bluer yearlings are more likely than browner males to transmit their song	c, e
Song switching	Age; *N* neighbors that match a local song theme	Loss of immigrant songs, continuity of resident songs	d, f
Body size	Age; size of parents (heritability)	None	c, g
		BEHAVIORAL LEVEL OF ANALYSIS	
Improvisation of song detail	Age of male	Accumulation of change (CE)	f
	Number of model males a yearling copies	Possibly more improvisations when male styles its songs after multiple song models (ST, CE)	f
Innovation of song theme	Age; experience	Naive birds develop their own abnormal songs, but no field evidence of transmission of isolate songs (ST)	h, i

References: a, R. B. Payne and Payne 1990; b, R. B. Payne and Payne 1993a, c; R. B. Payne and Payne 1993b; d, R. B. Payne et al. 1988; e, R. B. Payne 1989, 1992; f, this chapter (Figs. 11.1, 11.2); g, R. B. Payne and Payne 1989; h, Rice and Thompson 1968; i, R. B. Payne 1981a.

A within-lineage model of song transmission and cultural evolution requires four assumptions: (1) song is learned, (2) similarity over time is due to a local continuity of a lineage, (3) change is due to accumulation of improvisations within a song tradition, and (4) variations in song at one time indicate sources of change analogous to variation in chromosomal structure or nucleotide sequence as sources of material for genetic evolution (Lemon 1966, 1975, R. B. Payne 1973a, 1985, Baptista 1975, 1977, Slater and Ince 1979, Ince et al. 1980, M. C. Baker and Thompson 1985, M. C. Baker 1987). These assumptions are not always valid, and other explanations of song continuity are possible. Apparent continuity could result from innovation or improvisation on a common theme, and apparent local continuity with some change could result from local extinction of some songs and the establishment of similar songs by immigration from other local source populations. The variety of behavior and population biology in songbirds suggests that we should work out the details for each species, perhaps with a test of song development in each species, even though all songbirds appear to learn their songs (Kroodsma 1982a). Also, the initial models of cultural change do not distinguish between changes within a tradition and replacements with alternative behaviors (Slater et al. 1980, Cavalli-Sforza and Feldman 1981, J. M. Williams and Slater 1990). Modeling cultural change in a realistic manner will require observations to determine whether individual birds disperse and learn their songs away from their natal area, and whether songs in a population change by improvisations within a tradition or by replacements through extinction and dispersal.

Geographic and Historical Data in Studies of Cultural Evolution

Nonhistorical approaches to historical problems rely on data taken at a single time rather than on a known genealogy. The assumptions of such approaches have been reviewed and criticized in cultural anthropology and historical linguistics. For a parallel situation in evolutionary interpretation of data in molecular genetics, see Wilson et al. (1977), Gillespie (1991), and Avise (1994).

Studies of cultural evolution have often considered changes within a lineage (both a gradual accumulation of changes of behavior, and a divergence of lineages) rather than the changes associated with the dispersal of lineages into a population or the differential success and extinction of lineages (Watson and Galton 1875, Lotka 1931, Yasuda et al. 1974, Renfrew 1987, Yoffee 1990, Robb 1993). Differentiation among lineages resembles the natural selection of alternative genetic traits. This resemblance has biased the choice of a historical model, because cultural anthropologists and historical linguists avoid "biologizing" their fields with formal parallels to natural selection. Early models of cultural evolution were based on the same assumptions (Cavalli-Sforza and Feldman 1981, Ruvolo 1987). Nevertheless, we also recognize that a language present at any time in history is the result of both a past accumulation of changes within a lineage and a combination of elements from different lineages, as is the case with English and many other languages (Swadesh 1951, Weinrich et al. 1968, Sankoff and Sankoff 1973, Ruvolo 1987, Bateman et al. 1990, Hoch 1991, T. McArthur 1992). The assumptions of the nonhistorical models as they apply to geographic data are summarized below.

1. The rate of change has been constant through time. Small modifications within a behavior tradition may occur regularly by a "cultural clock," an analog to the "molecular clock" model of the accumulation of neutral mutations in molecular genetics (see, e.g., Wilson et al. 1977). The model of a clock, with a constant accumulation of mutations across generations, is often assumed (e.g., C. G. Sibley and Ahlquist 1990), but the assumption is not always valid because rates of genetic change sometimes differ among genetic lineages (Wu and Li 1985, Gillespie 1991, Martin and Palumbi 1993). Although there is no "universal molecular clock," small genetic changes within a clade such as the songbirds do accumulate in an approximate "local molecular clock" (Bledsoe 1987a, b). In human history, a clock has been proposed for "cultural mutations" in the change and replacement of words within a language tradition. Such a clock has been assumed in analyses of limited historical records to infer times of divergence within language families (Swadesh 1952, Crowley 1992, Lehmann 1992). A more compelling case for gradual cultural evolution can be made from historical or archaeological records in humans (Weinrich et al. 1968, Cavalli-Sforza et al. 1988, Cavalli-Sforza 1991, Renfrew 1992, J. H. Moore 1994a). The rates of change in birdsong are not well known, but we do know that species differ considerably. Local song dialects may develop in Chaffinches (*Fringilla coelebs*) in less than 300 years (Sick 1939, Armstrong 1963), for example, while song dialects in Indigo Buntings can diverge from a common ancestral song theme in fewer than 10 years (Fig. 11.2).

In birdsong as in other behavior, marked individuals provide the most robust data for documenting the rates and patterns of cultural evolution. The annual rate of song change in buntings differed from the mean survival of a generation of buntings because the song themes were often continued by certain long-lived males, and survival of the song themes followed these individuals and not a population statistic (Figs. 11.1, 11.2). In some instances, improvisations accumulate within a tradition and diverge, as in Indigo Buntings and perhaps in other species that have not yet been observed with long-term studies of individuals (Mundinger 1980, 1982, Lynch and Baker 1993).

2. Culture diverges through time. Cultures may have diverged over time, and their ontogenetic pattern can be described in terms of a branching tree. This assumes historical relationships of diverging cultural traits (Cavalli-Sforza et al. 1988, Bateman et al. 1990, Renfrew 1994) much like species divergences in the course of biological evolution. On the other hand, alternative processes can lead to resemblances among lineages. These processes include convergence and traits acquired from other sources outside a lineage, such as the "borrowed" traits of human language (Ruvolo 1987, Bateman et al. 1990).

In the song of species such as the Indigo Bunting, the history of behavior traditions within a population is better described not as a branching tree but as a reticulate web or a radial star, in which dispersal and immigration explain much of the variation. Variation in the pattern of song transmission occurs even at the level of the individual bird. When a male's songs are modified, each by a different copier, and the song variants of the copiers are copied in turn, or when a male copies elements, phrases, and sequences from several males (Fig. 11.2A; McGregor and Krebs 1982, Thielcke and Krome 1989, Beecher et al. 1994b, Lemon et al. 1994), then a lineage has diverged into descendant lineages.

3. Individuals and their cultural traits move on a small scale, and this local dispersal is responsible for the geographic distributions of traits. A diffusion model assumes that individuals and their traits move short distances from one generation to the next, and that the accumulation of distances over time leads to the observed spatial pattern of traits (Cavalli-Sforza and Feldman 1981, Aoki 1987, Piazza et al. 1987, Cavalli-Sforza et al. 1988, 1994, Gillespie 1989, 1991, Sokal et al. 1991). This model is appropriate to describe the history of many cultural traits, but not the traits that enter a population through long-distance dispersal.

The idea that songs can be used as markers of local dispersal between neighboring populations has been suggested for many songbirds (Thielcke 1962, 1965b, 1971, 1974, 1983, 1984, 1987, Thielcke and Linsenmair 1963, Tretzel 1965, 1967, Conrads 1966, 1976, 1977, 1984, Conrads and Conrads 1971, Baptista 1975, 1990a, Mundinger 1975, 1982, Baptista and King 1980, Kreutzer 1981, Baptista and Johnson 1982, Jenkins and Baker 1984, M. C. Baker and Thompson 1985, Lemon et al. 1985, A. J. Baker and Jenkins 1987, M. C. Baker 1987, Shackell et al. 1988, Glaubrecht 1989, 1991, Lynch et al. 1989, Slater 1989). On the other hand, the geographic patterns of bird songs can result from long-distance dispersal. Observations of individuals' dispersal from birth site to breeding site ("natal dispersal") or from one breeding site to another ("breeding dispersal") show that the distances are highly variable: a few birds move much farther than the median distance (Barrowclough 1980, Shields 1982, Bauer 1987, R. B. Payne 1990a, R. B. Payne and Payne 1993a). The dispersal distances have been determined for a few populations whose songs differ geographically (Blanchard 1941, M. C. Baker and Mewaldt 1978, R. B. Payne 1991). Insofar as birds actually disperse considerable distances between the site where they are born or where they learn a song and the site where they breed (Kroodsma 1974, 1978a, R. B. Payne 1981a, b, 1985, 1991, McGregor and Krebs 1982, 1989, Adret-Hausberger 1986, R. B. Payne et al. 1987, Baptista and Morton 1988, McGregor et al. 1988, Petrinovich 1988a), it is not possible to estimate the rate of cultural change of songs in time from their microgeographic variation in space.

Song Traditions and Cultural Evolution in Other Songbirds

Both continuity and change occur over time in the song behavior of several kinds of songbirds. The general processes of song transmission and cultural evolution are not in doubt, but the details remain to be determined for many species. In a population of Indigo Buntings, for example, dispersal histories determined the cultural diversity of songs. Nevertheless, long-term studies of other species have shown that song continuity is more common than replacements among lineages or changes within song lineages, especially when the number of birds in a song neighborhood is large. Also, in many species the microgeographic distribution of songs has persisted longer than expected on the basis of the average survival of the birds (Kreutzer 1972, R. B. Payne 1978a, b, 1981c, McGregor 1980, Bjerke and Bjerke 1981, Mundinger 1982, Adret-Hausberger 1983, 1986, McGregor and Thompson 1988, C. L. Whitney 1992).

Local songs have been documented over many years in resident White-crowned Sparrows (*Zonotrichia leucophrys nuttalli*) in coastal California. Blanchard (1941) made song sketches in the 1930s that are still recognizable as the local song themes of the Berkeley hills population. Recordings made there over about 30 years (1950–1980) reveal great stability in song structure (Marler and Tamura 1962, Marler 1970, Baptista 1975, Trainer 1983). The local song dialects and their microgeographic distributions have also remained stable in some other populations of this songbird species (M. C. Baker 1975, 1982, Baptista 1975, Baptista and King 1980, Baptista and Morton 1982, 1988, Petrinovich 1988a).

Continuity of song traditions is also known for creepers (*Certhia*). Thielcke recorded Eurasian Tree-Creepers (*C. familiaris*) over many years in Germany. He found that local populations differed in their song structure; the songs remained stable for more than 20 years, and their geographic distribution persisted over the same period (Thielcke 1961b, 1962, 1965b, 1984, 1992). Short-toed Tree-Creepers (*C. brachydactyla*) also retained their local song themes over many years (Thielcke 1961b, 1987).

Local song differences in songbirds were first recognized in Chaffinches (Promptoff 1930, Mayr 1942, Armstrong 1963). In a long-term test of the cultural stability of Chaffinch songs, Baron von Pernau trained Chaffinches with the songs of Tree Pipits (*Anthus trivialis*) and then released them between the years 1704 and 1720 near Rosenau, Germany. Some 250 years later, no trace of pipit song could be found in the Chaffinches at Rosenau (Thielcke 1988a, b). Like the results of shorter-term studies of Chaffinch song (Ince et al. 1980), such a difference suggests cultural evolution of song, but it cannot be determined whether the changes occurred within or among cultural lineages.

In some songbirds, cultural evolution proceeds rapidly within a song lineage or tradition both when young birds first learn their song, and between years and within a season. Rapid changes in song by a bird and matched changes between neighbors have been recorded in Village Indigobirds (*Vidua chalybeata*) and Yellow-rumped Caciques (*Cacicus cela;* R. B. Payne 1985, Trainer 1989). In the indigobirds, each song in the repertoire of a male changes through improvisation on the theme from year to year and even within a season. These changes accumulate for at least four years, a period longer than the life span of most males in the population (R. B. Payne and Payne 1977). The song improvisations of the male with the highest breeding success are copied by his neighbors (R. B. Payne 1985).

The other mode of cultural evolution in this species involves dispersal and multiple song lineages. When males disperse, they stop singing their old song themes and adopt the local themes. As a result, the songs of immigrants become extinct in the new area, whereas the shared songs of the established neighborhoods last for several years (R. B. Payne 1985).

The variation within and among song themes in a population sometimes has been interpreted as a local cultural accumulation of change even when no change was observed through time (Lemon 1975, Slater and Ince 1979, McGregor and Krebs 1982, Slater et al. 1984, A. J. Baker and Jenkins 1987, Lynch and Baker 1993). In Village Indigobirds, variation within a song theme was first described in these terms. Songs of neighboring males differ in the forms of an element or in the addition, deletion, or sequence of elements. These variations suggested the sources of cultural change within a song tradition (R. B. Payne 1973a). The hypothesis was tested and confirmed with observations of color-marked males recorded from year to year. As predicted, they modified their songs to match the song variants of the most successful breeding male (R. B. Payne 1985). Because every species has a unique history of behavior, we cannot generalize from this species to all songbirds, and we should follow up the observation of song variants at one time with continuing observations over the lifetimes of individual birds for each species.

Coevolution of Song Traditions and Genes

In the coevolution of cultural and biological traits, learned behaviors and genetic traits are closely associated in a developmental sense, and changes in behavior affect gene frequencies or vice versa (Cavalli-Sforza and Feldman 1981, Aoki 1986, 1989, Durham 1991). Song traditions and genes can be passed along together when a bird learns its father's song. If song traditions coevolve in this manner, two predictions can be made: first, birds with the alternative behaviors should differ genetically; second, birds with these behaviors should differ in their reproductive success.

Neither prediction is strongly supported by the available evidence. First, little evidence exists that songs and genes are transmitted together. Molecular genetic data are not generally available for birds with the alternative song themes, but in the classical sense of "genetic," no common genetic transmission occurs along family lineages. No genetic or size differences were found among Indigo Buntings with the different song themes (R. B. Payne and Westneat 1988). When we recorded both father and son in the field, we found that the young males dispersed first and then learned their song themes after they settled on a new territory (R. B. Payne et al. 1987, R. B. Payne and Payne 1993b). Also, in Bewick's Wrens (*Thryomanes bewickii:* Kroodsma 1974), Marsh Wrens (*Cistothorus palustris:* Verner 1976), Saddlebacks (*Philesturnus carunculatus:* Jenkins 1978), White-crowned Sparrows (Baptista and Morton 1988, Petrinovich 1988a), and Corn Buntings (*Miliaria calandra:* McGregor et al. 1988), song themes do not pass from father to son but are copied later, after the son disperses and settles next to a neighbor. As the songs are not transmitted from father to son, there is no common developmental link between the cultural and genetic lineages.

Second, there is little evidence that acquiring a certain song theme affects a male's breeding success. A male Indigo Bunting might gain in establishing and holding a territory when he has the song of a neighbor if he affects the behavior of the male he copied, or if he deceives other males into recognizing him as the older male (males discriminate between individuals' songs on playback: Emlen 1971b), but neither of these possible advantages was shown for the males that matched the song of a neighbor (R. B. Payne 1983a). Also, mean breeding success was the same when males were compared among song themes, both within a season and over their lives, and breeding success did not differ between males that matched a common song and males that did not (R. B. Payne et al. 1988, R. B. Payne and Westneat 1988). As suggested for Chaffinches (Slater et al. 1980), the bunting song themes appear to be "selectively neutral," at least apart from their local social context.

Similarly, there is no compelling evidence that songs and genes have coevolved in most other species. In Great Tits (*Parus major*), annual survival and lifetime reproductive success vary with the number of song themes in a male's repertoire (McGregor et al. 1981). Neither song themes nor repertoire size are similar in father and son, and whether a male matches a song with a neighbor does not affect his reproductive success (McGregor and Krebs 1982, 1984a, 1989). In contrast, the young male Medium Ground-Finch (*Geospiza fortis*) usually acquires his song from his father (Millington and Price 1985, Gibbs 1990). Males with one of the four song themes had a reproductive advantage over males with one of the other song themes. Over a few years, the rare song increased: here "cultural evolution" was a change in the frequencies of song themes in a population (Gibbs 1990). It remains to be seen whether the song themes change permanently, either through extinction or through divergence into separate lineages.

The observations available for songbirds in general do not support the association of song development and genetic differences, an association necessary for gene-culture coevolution. We need more field observations of marked birds, their dispersal, their reproductive success, and their songs over their lifetimes before we can compare the causes, rates, and consequences of cultural evolution in song among species of songbirds.

Conclusions

Song themes of Indigo Buntings are transmitted when yearling males copy the song themes of older adult males. The songs are behavior traditions that are passed from one cultural generation to the next. Certain traditions are known to have continued for 30 years in one population. Changes within a song tradition involve an accumulation of improvisations through time. The direct observation of continuity of a tradition with gradual change through time shows that song variants that are observed at any given time are the sources of cultural change (R. B. Payne 1973a, Lemon 1975, Slater and Ince 1979, Slater et al. 1980). Although changes occur within a lineage of songs, and this is a mechanism of cultural evolution in the buntings, the degree of change varies among song lineages, and there is little support for cultural drift with a constant rate of change according to a cultural clock.

The songs of Indigo Buntings also change at the population level when immigrant males introduce new songs into the area and its local song pool, when resident territorial males switch from one song to another, and when no males with a particular song return the next year. The cultural transmission of songs does not follow the paternal family lineages; rather, it occurs after birds disperse to their breeding site. The immigration of new birds and the failure of males to return to the area the next year were responsible for most of the changes we observed in the song themes. Within seven years, most songs disappeared and were replaced by the new songs of immigrant adult males, and dispersal resulted in a nearly complete turnover of the song pool in the population.

The probability of song continuity and change can be described and compared among species in a model of cultural evolution, and the cultural parameters can be compared with the demographic parameters of a population. Cultural change and evolution among song lineages in Indigo Buntings do not imply a coevolution of behavior and genetics, because the songs are not transmitted along kinship lineages, the genes and the songs do not covary, and reproductive success does not vary among males with different song themes. Songs appear, change, and disappear as one song lineage replaces another. Whether Indigo Buntings are representative in the transmission and cultural evolution of song will be known only when more species have been observed. Songbirds exhibit diverse songs and social behaviors, and we are still searching for the appropriate hypotheses and functions of songs to explain the variation among extant songbird species. The challenge, I believe, is to develop theories of cultural continuity and change for songbird species that differ in their phylogenetic relationships, mating systems, population density, and dispersal, and to test the theories with long-term observations of the individual birds.

The diversity of songs present within a population may be markers of the past. Whether the songs mark a divergence of cultural lineages from an ancestral song within a population or a pool of many immigrations into the population can be determined only through observations of individual birds. Comparison of birdsong with human languages points out a similarity: cultural evolution is known directly only when touchstone historical records are available, and our attempts to reconstruct the past beyond these records are problematic (Weinrich et al. 1968, Bateman et al. 1990, Hoch 1991, Lehmann 1992, Nichols 1992, Renfrew 1992, J. H. Moore 1994a, b). The rich historical accounts of the English language tell us that the diversity of elements originates from many sources, and linguists point out that many human speech communities have borrowed their sounds, grammars, and words across ethnic lineages (Burling 1992, T. McArthur 1992). Much as the elements of a human language have spread from one cultural lineage into another cultural lineage and have given rise to the cultural diversity within a speech community, so does the diversity of songs within a songbird population, such as that of the Indigo Bunting, trace its origin to many cultural lineages that result from birds immigrating into the population. This view emphasizes the need for historical records of dispersal. Also, we are increasingly aware of the risk of extinction of cultural traits in small populations of humans (Robb 1991, Hale et al. 1992), much as we saw in the local extinction of song themes in the Indigo Buntings. Indirect estimates of cultural traditions can be made in humans from our language traces and in birds from their songs, but our tests of these estimates will depend on direct evidence through the actual histories of the cultural lineages.

Acknowledgments

Laura Payne and 30 other biologists in our research group have observed the buntings and recorded their songs. Laura Payne coded the song elements, and Kent L. Fiala wrote the program MATCH to trace the continuity of songs in space and time. W. L. Thompson provided sonograms and the coded songs of Indigo Buntings at the George Reserve from his studies in the 1960s and early 1970s. The Museum of Zoology, University of Michigan, made available the E. S. George Reserve. Amtrak and many landowners allowed us access to their lands at Niles. For comments on the manuscript I thank D. E. Kroodsma, C. S. Parr, and J. L. Woods. Our research has been supported by the National Science Foundation.

12 Vocalizations and Speciation of Palearctic Birds

Jochen Martens

The vocalizations of birds are a means by which individuals of the same species communicate; all the members of a species can recognize conspecific birds by their voices. Vocalization differences between species are among the most powerful isolating mechanisms known. The demarcation of territories and the formation of mating pairs are practically always linked to sound production; hence the vocalizations are usually constant throughout large populations, even if the populations are distributed over extensive geographic regions. Presumably, therefore, all individuals of a species are at least potentially capable of selecting one another as mates. If the characteristics of their vocalizations change, however, the consequences are profound. All variant characteristics that lie beyond the limits for communication within the species likely make it impossible for their bearers to reproduce, so those individuals are lost to the population as gene carriers.

Nevertheless, vocalizations, especially the songs of songbirds, can vary considerably in the dialects they form. Dialects are limited to groups of individuals and can be regarded as temporal or spatial variants of the auditory communication system (Wickler 1986). Among passerines, dialects are handed down to the next generation by learning (Lynch, this volume, Payne, this volume). Territorial songs can exist in several dialect variations, each sung by part of a population. They thus form a geographic mosaic in which individual dialects can overlap one another. Dialects with such a small-scale distribution are usually so structured as to preserve the distinguishing characteristics of the species.

More important in the present context are song variants encompassing extensive subpopulations of a species and all individuals within this large range. These variants are often conspicuously different from one another, even to the human ear, and their characteristics frequently do not match the reaction-eliciting template of other large populations of the same species, or match it only in part. These variants over large areas—called institutions (Mundinger 1980) or regiolects (Tembrock 1984)—are often referred to as macrogeographic variation. When the carriers of different institutions come into contact with one another during the mating season, the consequences are substantial. In the extreme case, which is not uncommon, individuals from the two populations cannot mate with one another because the crucial auditory chain reactions between male and female cannot proceed. In such cases, isolating mechanisms have arisen, forming barriers that separate the two populations genetically from each other. This observation relates directly to the species concept most widely accepted in zoology: individuals are considered members of a species only if they are able, at least potentially, to mate with other members of the species under natural conditions (Mayr 1993). If this prerequisite is not met, the two populations are regarded as separate species, however large or small their morphological or genetic differences may be.

It is a long way, however, from the small-scale dialect that does not interfere with communication to the ethological-acoustic barrier of the regiolects. Songbirds in particular have passed through this route, often with the eventual production of new species. No less frequently, though, evolution has stopped this progress along the way, and the ethological differences present no serious obstacle to mating between members of neighboring populations.

The central question with which I am concerned in this chapter is how regiolects arise. We now know that the song tradition among songbirds is an extremely stabilizing element within the parental population (Thielcke 1973a). Although learning processes routinely allow small variants, the basic parameters of the territorial song stay remarkably uniform in space, across large geographic regions and also through time (although we can infer that only indirectly). To grasp the principles of speciation from an acoustic-ethological viewpoint, then, we must understand how the stabilizing song tradition, which maintains the cohesion of all individuals of a species as a biological unit, can be breached. Evidently some mechanisms loosen the strict tradition and perhaps occasionally override it altogether. What are they? To answer this question, we must study how vocalizations are altered in space and time. Furthermore, we should determine what happens in the contact zones between different regiolects with respect to the sounds themselves and to population dynamics. Can it be that regiolects in themselves are isolating mechanisms? In this chapter I document the individual stages from dialect to regiolect and ultimately to independent species, using examples related to Palearctic birds.

Song Constancy over Large Regions

Species without regiolects. The Willow Warbler (*Phylloscopus trochilus*) ranges from central Europe almost to the Pacific, a west-east distance of at least 7000 kilometers. Its song is highly variable from one individual to the next. Songs vary from population to population, too, so that birds from distant populations share few song notes. A descending verse melody, an important characteristic of Willow Warbler song to the human ear, is the same in Europe and Siberia (Martens 1993). Playback experiments show, however, that no regiolects occur in the large transpalearctic area of the Willow Warbler. Hence, the whole species is acoustically unitary.

Another impressive example is the territorial song of the Coal Tit (*Parus ater*), which is uniform from North Africa to Japan (Thielcke 1969, Martens 1975, 1993). This vocal uniformity is all the more striking in view of the marked geographic variation in the Coal Tit's plumage; in fact, its representative in the western Himalayas is so different in appearance that it is considered a separate species, the Black-crested Tit (*P. melanolophus*). Nevertheless, that tit's song is like the songs of all the Coal Tits. After a hybrid zone with the Coal Tit was discovered in the central Himalayas (Diesselhorst and Martens 1972, Martens 1975), the two forms came to be recognized as conspecific. Thus, exceedingly constant territorial song classifies all Coal Tit populations as a single species, which includes the Black-crested Tit. The song has changed very little over either long periods or across vast geographical spaces. When Coal Tits are raised with no opportunity to learn their song, however, the typical song is immediately lost, and the deprived individuals are unable to communicate with free-living conspecific birds (Thielcke 1973b).

Broadly distributed regiolects. In other species, regional differences in the song can be considerable, but each variant is constant over large parts of the range. The Common Chiffchaff (*Phylloscopus collybita*) is a good example. Some regiolects extend from southwestern to eastern Europe and from the Urals to eastern Siberia, while others are restricted to the Iberian Peninsula and to the Canary Islands. Still more impressive are the large regiolect regions of the Great Tit (*Parus major*): one extends from north Africa to eastern Siberia, and others encompass large areas of southern East Asia. The Willow Tit (*Parus montanus*) has both large and small regiolect areas. A relatively large region of uniform song occurs from eastern Europe to the Pacific, for example, but considerably smaller regions also occur, such as the ones limited to central and northwestern Europe, to the Alps and adjacent mountains, and to several parts of China (Thönen 1962, Martens and Nazarenko 1993, Martens et al. 1995).

Hybrid Zones Where Regiolects Come into Contact

What are the effects of differing regiolects? The carrier populations can adjoin one another parapatrically, either in a very narrow contact zone that can be nearly punctate or linearly elongated, or in overlap zones of varying extent. Let us look at the species mentioned above again, from this viewpoint.

The four subspecies groups of the Great Tit show highly diverse coloration and wing-to-tail proportions as well as strongly differing acoustic characters (Fig. 12.1), as demonstrated by, among other things, the playback of recorded songs. Male Great Tits in Germany belong to the *major* group and give strong territorial responses to song from western Siberia, 5000 kilometers away, where representatives of the *major* subspecies group also live. Vocalizations from the other subspecies groups, however, differ considerably: whistles with only slightly changing

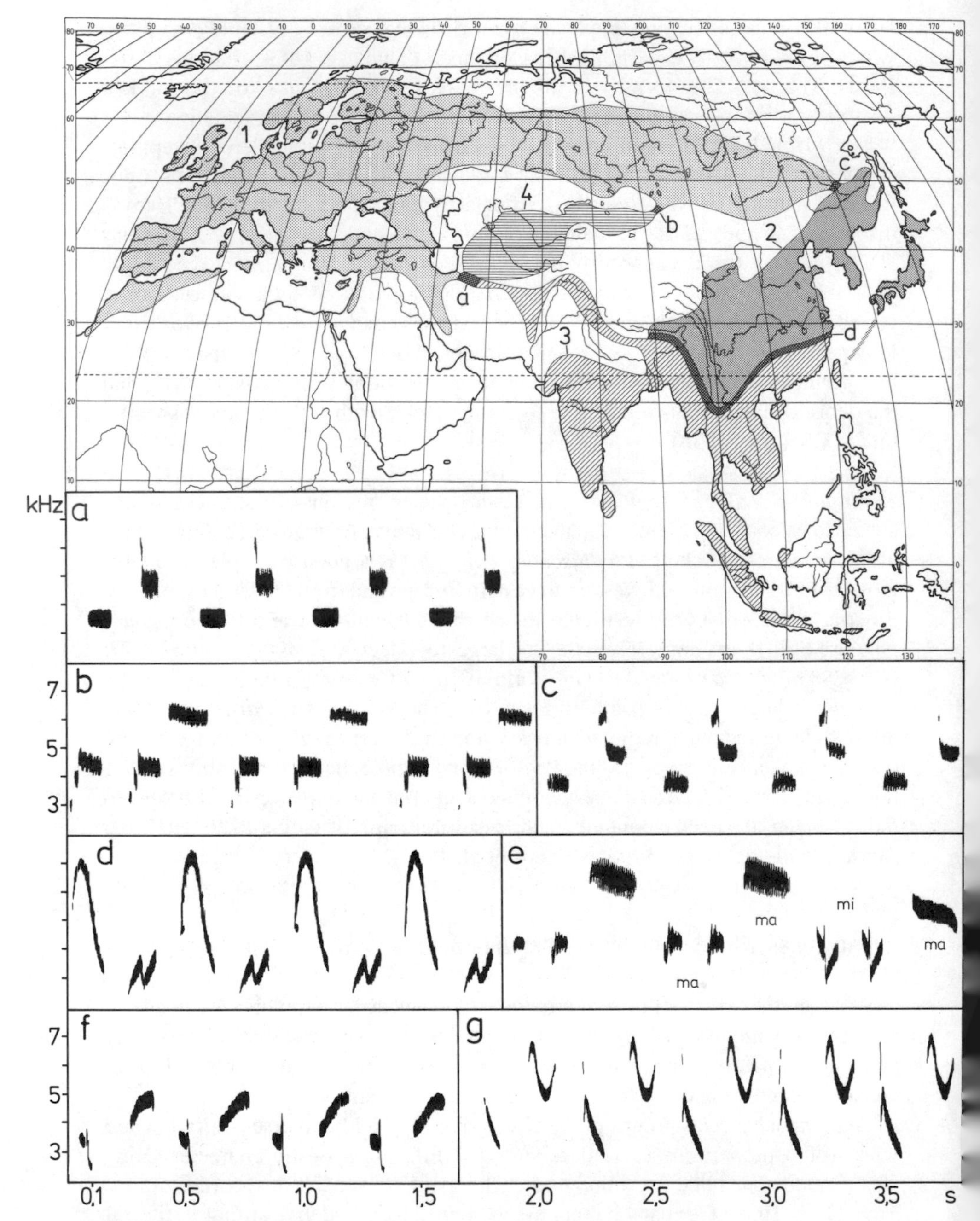

Fig. 12.1. Regiolect distribution and territorial song of the subspecies groups in the *Parus major* complex, showing areas of secondary contact, song characteristics of subspecies groups, and mixed song (e). The map shows the areas of the (1) *major,* (2) *minor,* (3) *cinereus,* and (4) *bokharensis* subspecies groups. Zones of secondary contact and overlap, mostly including areas of hybridization: *major, cinereus,* and *bokharensis* groups, northeastern Iran and

frequency in *major* and *bokharensis* groups (Fig. 12.1a–c, f) are replaced in the *cinereus* and *minor* groups by rapidly falling and rising note forms (Fig. 12.1d–g). In playback experiments the latter were not accepted by central European *major* tits in the wild. It would be interesting to know the relative influences of coloration and vocalization at points where the representatives of the various subspecies groups come into contact with one another (Fig. 12.1, map). In Iran, where the *major, bokharensis,* and *cinereus* groups are contiguous, mixed populations of locally varying structure seem to have arisen ("*intermedius*"). In southwestern Mongolia hybridization occurs between *major* and *bokharensis* (Eck and Piechocki 1977); unfortunately, nothing is known about the acoustic behavior of birds in this contact zone. In any case, *major* and *bokharensis* Great Tits have partly similar, sometimes nearly identical, songs (confirmed in playback experiments), so that hybridization at the conjunction of their areas is particularly likely.

A quite different situation occurs in eastern Siberia, where tits of the *major* and *minor* subspecies groups meet at the middle Amur (Fig. 12.1, map area c). The two local subspecies also differ ecologically there. *Major* ranges widely, exclusively in association with human habitations and open agricultural land; because of this ecological preference it has reached eastern Siberia only recently. In contrast, *minor* lives in the original semiopen, hilly woodlands. Many birds in this strictly delimited, ecologically defined contact zone, especially in the villages, are hybrids and live only in the immediately adjacent agrarian areas (Formozov et al. 1993, A. A. Nazarenko and B. Petri pers. comm.). Clearly, the enormous vocal differences between *major* and *minor,* as a result of which the songs are no longer mutually understandable in the allopatric region, are not an effective isolating factor. The plumage patterns of the two forms are nearly the same in the distribution of light and dark patterns, though not in coloration, and in the relatively sparse populations near the boundary this similarity may have been decisive for the formation of mixed pairs. Different hybrids, and even the same individual, can sing pure *major* verses, pure *minor* verses, or verses in which *major* and *minor* notes are combined (Fig. 12.1e). Unchanged *major* and *minor* note groups are always preserved in these verses.

In European textbooks the Great Tits are routinely cited as an example of circular distribution and speciation at a distance. This assessment is not correct. The four subspecies groups that form the ring area and the "bridge" across it (Fig. 12.1) meet secondarily in contact zones. No long-lasting, continuous distribution

---

Turkmeniya; *major* and *bokharensis* groups, southwestern Mongolia; *major* and *minor* groups, eastern Siberia; *minor* and *cinereus* groups, southern China. The recent secondary contact area of the *major* group in Kazakhstan and Kirghizia, mostly within *bokharensis* area, is not given here. Map drawn from various sources. Sonograms of territorial song. (a–g) The *major* group: a, southern Germany; b, western Siberia, Novosibirsk; c, eastern Siberia near the eastern distributional limit (see map); d, *minor* group, Ussuriland; e, hybrid area in eastern Siberia, Oblutshje, strophe built up of *major* (ma) and *minor* (mi) notes; f, *bokharensis* group, Kirghizia; g, *cinereus* group, Nepal. Recordings by J. Martens except a, c, and e (B. Petri). Sonograms in all figures were prepared on a Kay Elemetrics DSP 5500 Sonagraph; original prints are used throughout.

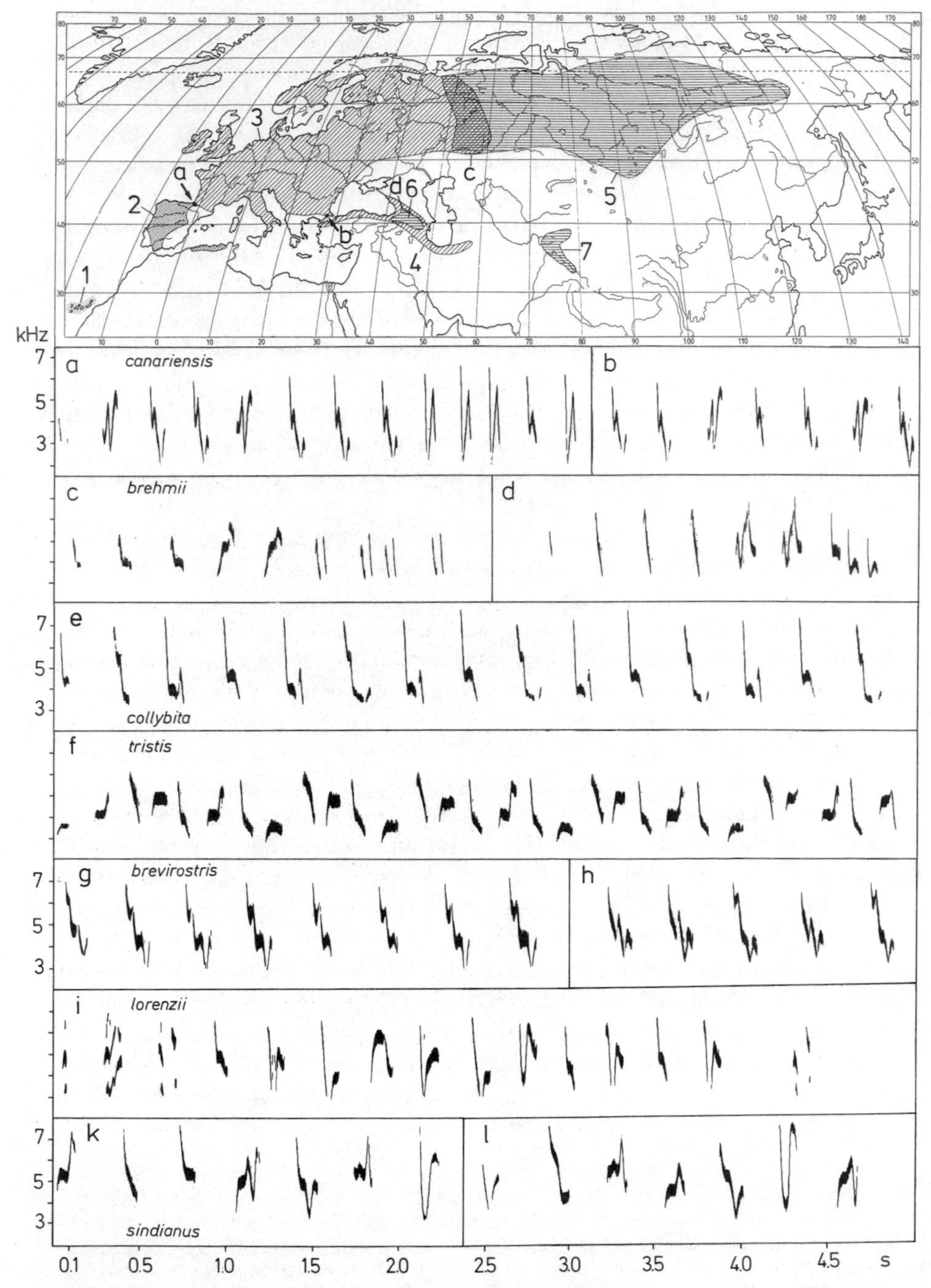

Fig. 12.2. Regiolect distribution, zones of secondary contact, and territorial songs of the *Phylloscopus* [*collybita*] superspecies, showing that regiolect distribution is congruent with that of species and subspecies. The map shows the areas of (1) *Ph. canariensis* (Canary Islands); (2) *Ph. brehmii* (Iberia and northern Africa); (3, 4) *Ph. collybita* (3,

of this group of forms exists (Eck 1980). But the *major* and *minor* Great Tits were certainly well on the way to becoming separate species (as are two other forms, *cinereus* and *bokharensis*).

Recently, a dramatic change occurred in the pattern of distribution of Great Tits in central Asia, again resulting from human activities. Around 1960, in Alma-Ata, Kazakhstan, and in Bishkek, Kirghizia, at the northern fringe of the Tien Shan Mountains, hundreds of *major* Great Tits from Siberia were released on several occasions. They spread rapidly and now intermingle with the resident *bokharensis* Great Tits, which are very different in coloration but only slightly different in their vocalizations. In some localities the two seem to mix (Formozov et al. 1993), but at other places they have so far lived side by side without mixing, as I found in 1993 in the Tcharyn Valley, a Kazakh river oasis. There the immigrants from the group of yellowish green *major* Great Tits have now become the more common form. Also, the Issyk Kul basin has been invaded; *major* tits are now common and no competing *Parus* species occur there except the smaller *P. cyanus* and *P. montanus*.

Parapatry of Regiolects and Local Hybrid Barriers

The Common Chiffchaff. Now we return to the Common Chiffchaff, whose distribution is subdivided into many regiolect regions (Fig. 12.2). It is easy to discern common structures in the sonograms of most of the regiolect forms. Notes of sharply falling frequency are widespread, and many songs are composed exclusively of such notes (Fig. 12.2a, b, e, g, h). Often the descending part is inflected so that the element is described as a "descent-and-knee" note—a note type that strongly descends in frequency, then runs horizontally for a few milliseconds before descending again (Fig. 12.2e). In some cases this type is accompanied by a second note type with a rising frequency, a regiolect character found only in the eastern Palearctic (Fig. 12.2f, i, k, l). These regiolects also include other note types, making the songs more varied.

Some of the response-triggering parameters in the Common Chiffchaff group are known. In nominate *collybita,* the initial frequency must be 6–8 kHz and the descending part can change only in a narrow range of frequency-to-time ratios. The proper strict "slope" must be preserved (Becker et al. 1980b). The ascending notes characteristic of the eastern Palearctic are not "understood" by these chiffchaffs in playback experiments.

---

northern area *c. collybita, c. abietinus,* central to eastern Europe; 4, southern area *c. brevirostris, c. caucasicus, c. menzbieri*) in west to east order, disjunct from Turkey to Turkmeniya; (5) *Ph. tristis* (eastern Europe to eastern Siberia); (6, 7) *Ph. sindianus* (6, *s. lorenzii* Caucasus; 7, *s. sindianus* Karakoram). Zones of secondary contact and overlap: (a) *Ph. brehmii* and *c. collybita,* western Pyrenees; (b) *c. collybita* and *c. brevirostris,* northeastern Greece; (c) *c. abietinus* and *tristis;* (d) *c. caucasicus* and *s. lorenzii,* Caucasus. Map drawn from various sources. Sonograms of territorial song: (a, b) *Ph. c. canariensis,* Tenerife; (c, d) *Ph. brehmii,* Spain; (e) *Ph. c. collybita,* Germany; (f) *Ph. tristis,* western Siberia; (g, h) *Ph. c. brevirostris,* northwestern Turkey; (i) *Ph. s. lorenzii,* Caucasus; (k, l) *Ph. s. sindianus,* northwestern India. Recordings by J. Martens except c and d (G. Thielcke) and g and h (P. S. Hansen).

So far, we know much about three regions in which different chiffchaff regiolects come into contact. The best known is the minute area "a" in southwestern France where the subspecies *Phylloscopus collybita brehmii* and *Ph. c. collybita* meet (Fig. 12.2). The songs of *brehmii* (Spain, Fig. 12.2c,d) and *collybita* (France, Fig. 12.2e) are very dissimilar in syntax and note form; no greater differences can be found in the whole chiffchaff complex. *Phylloscopus c. collybita* uses exclusively irregular sequences of descent-and-knee notes, whereas the songs of *brehmii* have a fixed syntax comprising entirely different note types (Thielcke and Linsenmair 1963). Nevertheless, in playback tests *brehmii* responded to *collybita* song (Thielcke et al. 1978), apparently because *brehmii* also has a song with descent-and-knee notes, though it is performed only when the bird is extremely aroused, as in a territorial dispute. In the *brehmii-collybita* contact region only about a third of the pairs are mixed, considerably fewer than would be expected by chance (Salomon 1987, 1989). In addition to what appears to be acoustically mediated pair formation (no well-defined morphological characters distinguish *brehmii* and *collybita*), a genetic effect has come into play: by way of the maternal line, as revealed by cytochrome-b tests, female hybrids are sterile (Helbig et al. 1993). Because of the partly interrupted gene flow (male hybrids may be fertile) and female hybrid sterility, each form has thus attained species rank, according to a central criterion of the biospecies concept.

A broad zone of contact and even overlap has been found for chiffchaffs in eastern Europe, where the subspecies *Ph. collybita abietinus* and *Ph. collybita tristis* meet (Fig. 12.2, map area c). The territorial songs of these two subspecies (Fig. 12.2e, f) sound very different to the human ear (Martens and Meincke 1989). Hybrids are known, as well as mixed singers (Marova and Leonovitch 1993), but the extent of hybridization has not yet been determined. Remarkably, both song forms exist side by side in many places; here the singers have probably reached the species level.

A particularly noteworthy restriction of gene flow has been discovered in the third regiolect contact zone, in the Caucasus, between the (sub)species *Ph. collybita caucasicus* and *Ph. collybita lorenzii*[1] (Fig. 12.2, map area d). The two are separated vertically, *caucasicus* living in the lower-altitude belt and *lorenzii* in the upper belt. At the few places where their areas are contiguous, hybridization has not so far been detected, though the territorial songs are relatively similar (Martens 1982). In these two vicariants, again, the territorial songs differ in degree of complexity, *caucasicus* having only descent-and-knee notes and *lorenzii*'s song including ascending notes. It is quite clear that here, too, closely related regiolect carriers have formed a species boundary.

The chiffchaff complex, with its many subtle morphological variations and acoustic regiolects, illustrates vividly that only a detailed knowledge of the behavior of the regiolect carriers in the contact zones can reveal isolating mechanisms that are not obvious through morphological analysis alone. Acoustic

[1]Editors' note: *Phylloscopus lorenzii* is recognized as a species by C. G. Sibley and Monroe (1993).

characters considered merely in themselves, however, and not embedded in a matrix of morphological, ecological, and genetic characters, cannot reliably document species boundaries. Our studies of chiffchaff regiolects, together with the recent focus on additional sets of characters, including the genetic ones revealed by cytochrome-b analysis, now justify subdivision of the chiffchaff complex into four, perhaps even five, independent biological species (Helbig et al. 1996, Martens and Eck 1995).

Area overlap, species status, and the development of characters. From the studies described above, I conclude that populations become "species" only when the (acoustic) isolating mechanisms are so well developed that they no longer permit pair formation between the carriers of regiolects. Only then, given that morphological and ecological differences also exist, can one expect complete prevention of gene flow and an extension of the overlap region. But how do the acoustic characters in these cases become so pronounced? Does the song of close relatives retain homologous structures even at an advanced stage in the species formation process, or does the progressive development rapidly eliminate the previously shared features?

The Blue Tit (*Parus caeruleus*) and Azure Tit (*P. cyanus*) can help answer this question (Fig. 12.3). Both tits belong to the superspecies *P.* [*caeruleus*] and are descended from a common ancestor; in parts of eastern Europe they occur together. As would be expected, the territorial songs of central European Blue Tits and west Siberian Azure Tits are distinct from one another (Fig. 12.3 d, d′, e). The two species nevertheless share certain note types, to which the Blue Tits responded in playback experiments (Martens and Schottler 1991). In the song of the Azure Tit these note types have apparently retained much of the original frequency modulation, but the whole frequency range has been lowered by 2–3 kHz. This frequency difference may be a purely physical effect, with the Azure Tit being unable to produce frequencies as high as those of the smaller Blue Tit (see Wallschläger 1980).

The superspecies comprising Blue and Azure Tits has still more to teach us. Marked morphological and acoustic subdivisions occur within each species. The Blue Tits of North Africa and the Canary Islands are more intensely colored, and their vocalizations depart widely from those of the European form. The songs are composed of note groups, several of which are produced in sequence to form a verse (Fig. 12.3a–c). This syntax is also found in almost all other *Parus* species (Thielcke 1968) and hence may be a plesiomorphic character. Nevertheless, the song syntax has been completely changed in the remaining Blue Tits as well as in the Azure Tits of Europe and Asia (Fig. 12.3d–k). One can speculate that the African and Canary Island Blue Tits are a particularly old group within the superspecies. Furthermore, there is an astonishing resemblance between the song forms of all European-Asian Blue and Azure Tits with respect to syntax and the sequence and frequency modulation of the notes (Fig. 12.3d, d′, e–k). These surprising similarities apply especially to two geographically distant populations

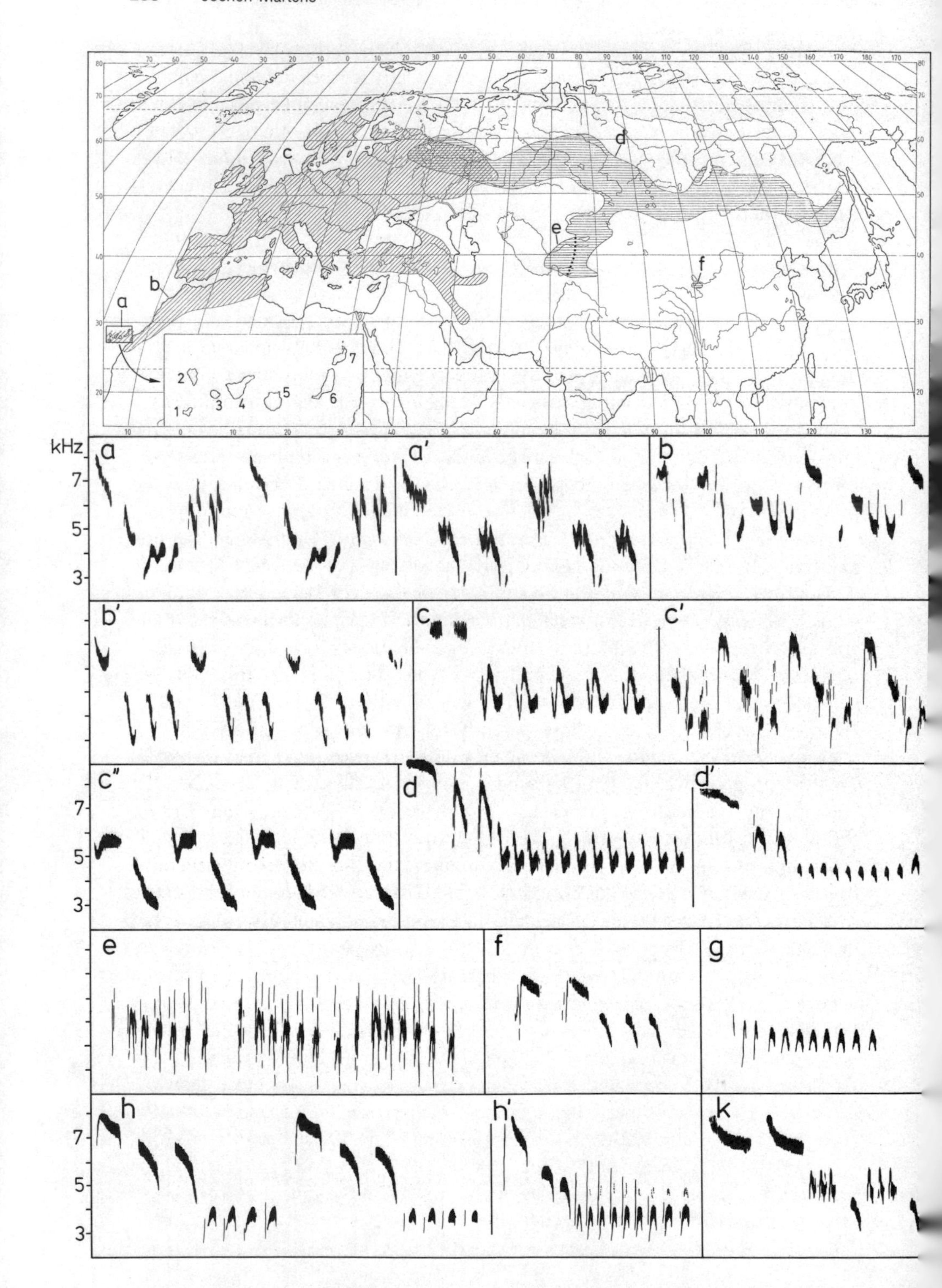
a
b
c
d
e
f
1
2
3
4
5
6
7
kHz
7
5
3
a
a'
b
b'
c
c'
c''
d
d'
e
f
g
h
h'
k

of the superspecies: the Blue Tits of Europe and the Near East and a small Azure Tit (*P. cyanus flavipectus*) of central Asia are vocally almost indistinguishable (Fig. 12.3 d, d′ and h, h′). A striking feature of *flavipectus,* like all *caeruleus* forms, is the yellow underside, which is absent in (almost) all other Azure Tits. Can it be that morphology and acoustics indicate a common history of European Blue Tits and the relict central Asian *flavipectus* Azure Tit, manifest in characters only slightly modified in allopatry? Many facts support this view.

Two buntings, the Yellowhammer (*Emberiza citrinella*) and the Pine Bunting (*E. leucocephalos*), which live together in western Siberia, provide additional information but also show how multilayered the problem is (Fig. 12.4). The two use extremely similar songs, even including the final buzz of the verse (Fig. 12.4a–e). But the songs have subtle differences, not immediately apparent in the sonogram but easily audible. The buzz in *leucocephalos* is more distinctly subdivided into clicklike single pulses, whereas in *citrinella* it amounts to a nearly continuous, rough whistle with many overtones (Fig. 12.4a–e). These differences (Löhrl 1967, Wallschläger 1983) may serve as isolating mechanisms. Males of both species observed singing together in a tree in western Siberia seemed to ignore one another, suggesting that each could tell its own song from the other's (Mauersberger 1971). Details of the isolating mechanisms are unknown, however; furthermore, local hybridization between the two species has been reported (Johansen 1954). The great similarities in their vocalizations, which nevertheless seem to include distinguishing characters, contrast sharply with the substantial differences in color and pattern of the two buntings. The morphological differences should not be underestimated as possible isolating mechanisms (Panov 1989, p. 217).

Constancy of Acoustic Characters across Species Boundaries

Even when species boundaries between two songbird populations are evident at the point of secondary contact, the vocalizations of the two groups often remain very similar. The chiffchaffs, Blue and Azure Tits, Yellowhammer, and Pine Bunting all provide examples of this. Homologous ethological structures derived

Fig. 12.3. Regiolect distribution of territorial song in the *Parus* [*caeruleus*] superspecies, showing that song syntax in the North Africa and Canary Islands *ultramarinus* group (a–c″) differs from that of all the other superspecies taxa, and showing the congruent verse forms (d,d′ and h,h′) in the *caeruleus* and *flavipectus* groups. The map shows the areas of *P. caeruleus* (a–c): a, Canary Island part with the islands (1–7) of El Hierro, La Palma, Gomera, Tenerife, Gran Canaria, Fuerteventura, and Lanzarote; b, north African part; c, European and west Asian part; and the area of *P. cyanus* (d–f): d, north Asian part (*cyanus* group); *flavipectus* group: e, *flavipectus* s. str. (central Asia); f, *berezowskii* (northern China). Map drawn from various sources. Sonograms of territorial song: (a–d) *P. caeruleus* with *ultramarinus* subspecies group (a–c″) and *caeruleus* subspecies group (d,d′); (e–k) *P. cyanus* with *cyanus* subspecies group (e–g, k) and *flavipectus* subspecies group (h,h′). (a–c) *ultramarinus* group, Canary Islands: a,a′, Tenerife; b,b′, Gran Canaria; c,c″, Lanzarote; d,d′, *caeruleus* group, southern Germany; (e–g, k) *cyanus* group: e, western Siberia, Novosibirsk; f, Issyk Kul, Kirghizia; g, southeastern Kazakhstan, Tcharyn; k, Ussuriland; h,h′, *flavipectus* group, Kirghizia, Chatkalskij Alatau. Recordings by J. Martens.

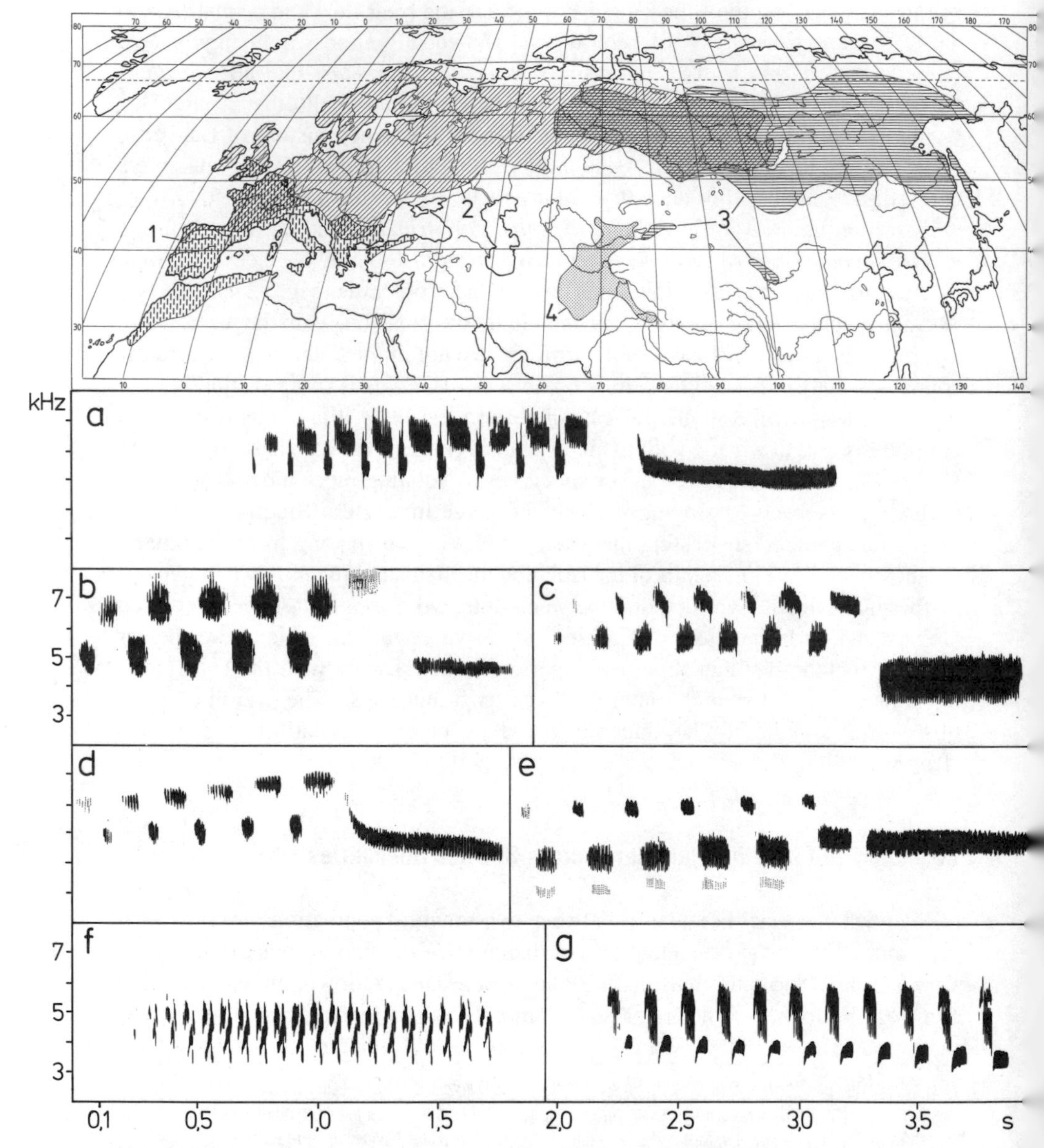

Fig. 12.4. Distribution and territorial song in the subgenus *Emberiza* (*Emberiza*), showing partial distributional overlap and peripheral isolates (area 3). All verse types (except final buzz) have common syntax and similar notes. (1) *E. cirlus.* (2) *E. citrinella.* (3) *E. leucocephalos* with peripheral isolates of nominate *leucocephalos* (western) and subspecies *fronto* (southeastern). (4) *E. stewarti.* Map drawn from various sources. Sonograms of territorial song. (a, b) *E. citrinella:* a, southern Siberia, Tuva, Sayan; b, southwestern Germany. (c–e) *E. leucocephalos:* c, southern Siberia, Tuva, Ereni; d, Kirghizia, peripheral isolate southeast of Issyk Kul Lake, see no. 3 on map; e, *E. l. fronto,* peripheral isolate in northern China, see no. 3 on map. (f) *E. cirlus,* southwestern France. (g) *E. stewarti,* Kirghizia, Chatkalskij Alatau. Recordings by J. Martens except a, c (B. Veprintsev, V. Leonovitch, from disk Melodiya C90 24177 003), and e (A. Gebauer).

from a common ancestor thus remain demonstrable in many cases. The effective isolating mechanisms are mostly based on small changes in the frequency modulation of individual notes. The institutions of the chiffchaff songs (Fig. 12.2) constitute a particularly nice experimental illustration.

But how do vocalizations continue to change when the species formation process has been completed, when no more hybrids are produced and the ecological requirements of the two recently evolved species have diverged further? Here we return to the two buntings *Emberiza citrinella* and *E. leucocephalos.* Among their closest relatives are two other species, *E. cirlus* in the western Mediterranean (Fig. 12.4, map 1) and *E. stewarti* in central Asia (Fig. 12.4, map 4). These four species form the subgenus *Emberiza.* All four have in fact retained distinct acoustic similarities. They all use strongly frequency-modulated notes and note groups combined to form verses (Fig. 12.4a–g), a verse form not otherwise found in the many species of the genus *Emberiza.* Even if the final buzz was omitted in playback experiments, the verses of *E. cirlus* and *E. citrinella* were easily distinguished by the birds themselves, even in sympatry (Kreutzer and Güttinger 1991). This example shows that the basic structures of acoustic parameters can persist over long phases of evolution. At the same time, however, subtle changes are established that eventually come to play a crucial role as isolating mechanisms, regardless of how conspicuous the morphological differences that have meanwhile developed.

Reasons for Acoustic Changes

Climatic Changes in the Quaternary

Limitations on acoustic characters. We saw above that subspecies can distinctly modify their species' song, making it a more or less effective isolating mechanism. On the other hand, we encountered examples of songs that remained very similar, even across species boundaries (chiffchaff, Blue and Azure Tits, buntings of subgenus *Emberiza*) and consequently over long periods. Both phenomena demand an explanation. Our basic point of departure is that the songs of passerines are learned, though the contribution of learning may vary because a genetic basis often also exists. In any case, learning shapes the fine details of the local regiolect (Becker 1990).

I know of no mechanism that permits the vocalizations of a continuously and extensively distributed species to undergo acoustic modifications beyond the degree of local dialects (Thielcke 1970a). Tradition operates so precisely, as is evident in the example of the Coal Tit, that the conspicuous morphological variants (which we assign to subspecies) can be more numerous than the learned regiolects. Regiolect carriers are sharply delimited geographically. Where two regiolects meet, they are in secondary contact and have not arisen in the zone of parapatry. Instead, I attribute their origins to the changing climatic conditions in the Quaternary. The vegetation belts in Eurasia were greatly altered by climatic influences during the various phases of the Pleistocene, and the forested regions had to retreat into southern refuges, whose locations in Eurasia are well known (Frenzel 1967, 1968, Haffer 1982). As these small, often widely separated forests were produced like isolated islands, the numbers of individuals in the bird populations inhabiting them were greatly diminished. From a genetic viewpoint the consequences are well known: restriction of the gene pool with a bottleneck effect, rapid recombination of genes, and the spread of positive mutants.

Many decades ago Stresemann (1919) found a relationship between Quaternary climatic fluctuations and the evolution of the two European tree-creepers, *Certhia brachydactyla* and *C. familiaris.* Can we now transfer this scenario to acoustic characters? Well-founded interpretations are available only for the learned components. If the tradition of learning is maintained in the home population, changes in song forms should be rare, as suggested by the examples of the allopatric subareas of Blue and Azure Tits, as well as the Coal and Black-crested Tits. However, we must also take other situations into account. Perhaps Pleistocene forest refuges were not optimal habitats. The populations must have been sparsely distributed and the individual pairs far from one another. In these circumstances it would have been hard for the young birds to become familiar with the entire repertoire of the refuge's population. Special local developments, which could result when the tradition was not adequately passed on, might later (after the climate improved so that the population enlarged and became homogeneously distributed) have become established throughout the refuge. In consequence, the populations gave rise to a uniform regiolect, and the initial stage of speciation was reached.

Examples from the present epoch. Recent examples of this mechanism of "flawed and incomplete" tradition support this thesis. The Short-toed Tree-Creeper (*Certhia brachydactyla*) occupies a closed area in central Europe north of the Pyrenees, where it is a common species. Here its song is extraordinarily uniform, with insignificant dialects (Thielcke 1961b). In Spain, south of the Pyrenees, where in historic times the woodlands have also been progressively reduced and locally thinned out, the species has been split into small, separate populations. The local song forms of the isolated populations vary distinctly. In North Africa the disparities are particularly marked; the song there does not conform to the central European norm, and in playback experiments the African birds did not recognize the European form (Thielcke 1973a).

Let us reconstruct an especially striking example of regiolect distribution with this hypothesis. The Willow Tit (*Parus montanus*) is clearly subdivided according to regiolects in Eurasia (Thönen 1962, Martens and Nazarenko 1993, Martens et al. 1995). Three of them are as follows. From Scandinavia almost to the Pacific the Willow Tits sing Siberian song. Each male produces two note types: whistles at a single frequency, a sequence of which forms a verse, and notes with descending frequency, ending in a very short whistle component. In Europe apart from

Scandinavia, these two note types have separated and are represented in geographically distinct populations. The horizontal whistles (alpine song) occur nearly exclusively in the Alps and Carpathians, and only the modulated note (lowland song) occurs in the rest of Europe. The explanation of this remarkably sharp geographical division is immediately apparent. In a southern Alpine refuge and in another on the Balkan Peninsula, singers of Siberian song in different, extremely sparse populations each lost one of the elements in the repertoire, perhaps by chance, resulting in lowland and alpine songs. Siberian song itself also persisted during this period, perhaps in a much larger relict area with a high density of individuals, and changed little. This hypothesis has since been corroborated by breeding experiments. When the young of lowland singers were raised with no experience of the song, they developed only alpine song (Heckershoff 1979, B. Schröder pers. comm.). The alpine song is evidently based on a stronger genetic component than is the lowland song. Loss of the lowland note as a result of interrupted tradition can thus indeed convert the tits to alpine singers.

These regiolects have proved to be sturdy barriers, for neither alpine nor lowland song is understood as conspecific in the allopatric regiolect region. The same applies even in some contact zones. In Switzerland these song forms can be represented on the two slopes of a valley and yet be mutually exclusive, so that no mixed singers are detectable (Thönen 1962). In the foothills of the Bavarian Alps, however, a broad hybrid zone has formed. There the alpine and lowland songs have combined to reproduce the Siberian song—in this case, "secondarily" Siberian (Martens and Nazarenko 1993).

Acoustic Changes on Islands

So far we have considered bird populations on large land masses. We have seen that the processes of acoustic restructuring, like the morphological ones, operate slowly. For rapid acoustic changes to take place, the vocal tradition must be interrupted.

But what happens on oceanic islands? Again, the Blue Tits provide a good example. Blue Tits live on all seven main islands of the Canary archipelago (Fig. 12.3), most of which are of volcanic origin. On the islands near the African coast—Lanzarote and Fuerteventura—the populations are tiny, at most 100 individuals each. Blue Tits came to the islands across the ocean from North Africa, as indicated by their coloration and, at least on a few islands, their songs. Their repertoire closely resembles that of the Blue Tits of Morocco (Becker et al. 1980a). La Palma, the westernmost island, was evidently colonized by unimprinted birds, for their territorial song today consists mainly of nonlearned calls (Schottler 1993); hence their repertoire is very limited. The acoustic differences between the island populations today are considerable, so much so that in playback experiments they may show little or no mutual understanding (Schottler 1993). By assuming that song elements are passed on accurately by tradition over long periods, Schottler was actually able to chart the immigration routes of the Blue Tits through the archipelago on the basis of acoustic characters.

When islands are first colonized by small populations, parts of the species' song repertoire are lost. Later the repertoire may expand again secondarily, usually by errors in learning or simply by invention of new elements. Examples of this process can be found on the Canary Islands, and the situation is easily simulated in experiments. A small population in the laboratory will pick up notes foreign to its natural song when it hears them played by a tape recorder; later, unimprinted young birds learn them by listening to the "reimprinted" elders. Becker (1978) demonstrated this process over two generations with caged Marsh Tits (*Parus palustris*). Young Eurasian Bullfinches (*Pyrrhula pyrrhula*) also readily incorporate human whistles or canary verses into their song and pass the new song on to their young, even to grandsons (Nicolai 1959). In view of these experiments, even though as yet they cover only four generations, it is easy to imagine how new regiolects can arise by loss of tradition.

Hypotheses

Let us now combine the above observations to form hypotheses. Thielcke (1973a, 1983) postulated the "withdrawal of learning" hypothesis as follows. Young birds emigrate from their home population before their acoustic imprinting has been completed; in a newly colonized region they develop a new acoustic repertoire, which after only a few generations differs distinctly from the song of the home population. In the process, crucial steps toward the production of two independent species have been taken, and the changes act as a "pacemaker of evolution" (Thielcke 1970a). According to the data so far available, rapid modification of the song is likely to occur only during the colonization of islands. There, in geographic seclusion, what appeared within a few generations can remain undisturbed or even develop (slowly) further. In continental regions the conditions are less favorable for rapid change in vocalizations. Peripheral populations on continents seem not to be sufficiently isolated, so unimprinted birds that colonize peripheral areas cannot maintain their vocal distinctiveness. As we have seen, the regiolects usually, but certainly not always, closely match the distribution of subspecies. Thus, comparatively long times are necessary for the development of both vocal and morphological character complexes. In the Pine Bunting, for example, two peripheral isolates, one belonging to nominate *leucocephalos,* one to subspecies *fronto* (Fig. 12.4, map 3), do not show regiolects. Long persistence of acoustic characters is indicated again.

I therefore propose to modify the original hypothesis formulated by Thielcke into a general "acoustic character shift" hypothesis with two subdivisions. 1. "Withdrawal of learning" refers to loss of a repertoire during rapid, permanent initial colonization of islands or perhaps of isolated continental areas. The colonists are a small group, and they bring with them only small parts of the learned home repertoire or none at all. 2. "Character shift in small populations" would then refer to conditions of extremely sparse distribution in an area with inadequate tradition of parts of the acoustic repertoire. This situation applied to populations in isolated areas of continental regions mainly during the Quaternary, but never to isolated individual birds or small groups of new immigrants.

The first process is indeed swift, though it remains an open question how long it takes to actually delimit distinct species. The second is slow and can be regarded as parallel to morphological modification. These changes in vocal tradition are still rapid, however, in comparison with speciation processes in nonpasserines.

Discussion

The territorial songs of songbirds show a great discrepancy between extreme constancy in space and time and variability such that a large- or small-scale mosaic of different regiolects is formed. Both constancy and change have great biological significance. When vocalizations remain unchanged over long periods—such as in the geologically young phase of the Quaternary—a songbird species hardly has a chance to divide into several species. The Coal Tit (*Parus ater*) illustrates this case. Its populations are morphologically variable from North Africa to Japan, and one population was even given species status, but the geographic uniformity of the song and the production of hybrids point to a single species.

The evolution of new species always implies interrupted gene flow. Among ethological isolating mechanisms, vocalizations are the most important. Morphological differences are unnecessary for speciation, as revealed by the two chiffchaffs *Phylloscopus collybita* and *Ph. brehmii.*

If this reasoning is correct, we should find a range of effectiveness in isolating mechanisms in the wild, from those that only insignificantly impede gene flow to those that erect such solid barriers that they insurmountably separate two gene pools. And indeed we have found such a wide range among the songbird examples. Much evidence suggests that songbirds have undergone this development in the very recent geological past.

The tradition of song characters in successive generations is extraordinarily precise and allows only a few variants. In most cases, the variants are all within the limits of the acoustic norm for the species as a whole. Modification over time requires special conditions, which probably prevailed often during the Pleistocene. Only when the entire area of a species is split into parts geographically distant from one another do acoustic traditions change. For the Coal Tit even that has not been enough, however. Although its range was fragmented into Pleistocene refuges, its acoustic tradition has probably never been disturbed. It is necessary for the population density to be very low in a refuge, so that the individual pairs are largely isolated, in order for the tradition to be loosened and regiolects to form. Only after postglacial expansion of the areas is the actual biological weight of these changes revealed. Often they turn out to have been so profound that they now delimit separate units, or species. In other cases the evolved differences were insufficient to produce such separation.

This hypothesis is nicely corroborated by the fact that regiolect populations, whatever their taxonomic rank, are also morphologically differentiated. It follows that as characters passed on by learning processes were being modified, so were those anchored in the genome. This pattern is a strong indication that these events have temporally consistent dimensions, and it gives evidence of the age of the regiolects.

Rapid, nearly sudden change in acoustic characters can be expected, according to the concepts we have now developed, only when tradition is broken off abruptly so that the young birds invading previously uncolonized regions have been imprinted incompletely or not at all. These events happen mainly on islands. The hypothesis I have developed here covers the case in which conspicuous regiolects develop on islands within a few generations, a process confirmed by the Blue Tits of the Canary Islands.

Conclusions

The hypotheses formulated in this chapter are not all founded on and confirmed by a consistent body of facts. Let us now turn to some considerations for future work.

Origin of the regiolects. If it is true that regiolects arose mainly in glacial refuges, regiolect contact zones should be concentrated in particular regions of Eurasia; there, well-developed subspecies or even young species ought to meet one another. More effort must be directed toward locating such zones of secondary contact. Striking subspecies boundaries can serve as guidelines. Promising regions are in southwestern Europe, around the boundary between Europe and Asia, in the Caucasus, in central Asia, and in the Himalayas. The relationships of Palearctic populations to those in the Nearctic also deserve more attention. The *Parus montanus–P. atricapillus* complex and the Eurasian and American treecreepers (*Certhia*) are two good examples.

Aging. Our reasoning so far leads to the conclusion that many regiolects, as they first came into being, corresponded to the geographical subspecies they represented. Aging can be further established by genetic investigation, especially by DNA sequencing. In fact, vocal, morphological, and genetic analyses of *Phylloscopus* species have given consistent results (Helbig et al. 1993, 1996). These interdisciplinary approaches open broad new prospects for future work.

Species-rich genera. Previous studies have been restricted to a few genera, usually those easily analyzed because their vocalizations are simple. Additional genera comprising many species and diverse forms should also (increasingly) be included, even species with complicated songs such as those of the genus *Turdus*. Genera particularly worth investigating are *Oenanthe, Emberiza, Carduelis, Passer, Certhia, Parus, Sitta, Carpodacus, Corvus, Regulus,* and *Sylvia*.

Calls. All the previous considerations were related to songs. Calls, however, are no less important in the acoustic life of birds. Many calls develop without learning, but others seem to be learned, at least in their local details. In the five island populations of the Canary Island chiffchaffs, the contact calls on each island are different, though the songs are the same (Henning et al. 1994). Are these calls learned or not? What is their role in the process of speciation? These questions are completely open.

Nonsongbirds. These birds are usually neglected in questions of acoustic geographical variability because their vocalizations are regarded as invariable (R. B. Payne 1982, Martens 1993). How far back in nonsongbirds is it possible to trace the morphological and acoustic separations that resulted in new species? Here we should emphasize the study of those taxa that are in the process of marked radiation. The genera *Larus, Falco,* and *Dendrocopos* are good candidates. Furthermore, it is necessary to extend the area considered to the Nearctic, and in the case of *Falco* nearly worldwide. Can it be that here, as in the songbirds, morphological and acoustic differentiation keep pace with one another? As yet hardly anything is known about the time frame for vocal changes in nonsongbirds.

Physical constraints. The theses we have constructed entirely ignore ecological factors, but R. H. Wiley and Richards (1978) showed that vocalizations are very likely affected by physical factors in the animal's biotope and adapt accordingly. This process is also relevant to speciation. Two *Phylloscopus* species of the eastern Palearctic—*tenellipes* in eastern Siberia and *borealoides* in Japan and Sakhalin—are practically indistinguishable morphologically. They are thought to be closely related, but their songs cannot be derived from one another. *Phylloscopus borealoides* uses whistles in a high-frequency range (Martens 1989) that are adapted to noisy environments, like those near plunging mountain streams (Martens and Geduldig 1990). *Phylloscopus magnirostris* of the Himalayas sings the same kind of song in the same habitat, though in this case we know of no closely related sibling species. Perhaps such special adaptations can lead to especially rapid acoustic modification in separate refuges; more data are needed to test this hypothesis.

Genetic restriction. We also need more information about the strength of the genetic predisposition for particular vocalizations. Considerable differences are found even within the genus *Parus,* as demonstrated by the Coal Tit and the Willow Tit. Depending on the strength of the genetic restriction, tradition can be either delicate or stable. Experiments in this category should put much more emphasis on raising birds, especially from the egg, in isolation. Such results will help explain the rates at which songs change and also lead to a better understanding of the vocalizations of nonsongbirds.

Acknowledgments

With S. Eck I discussed systematic problems. A. Gebauer, P. S. Hansen, B. Petri, and G. Thielcke provided important tape recordings. The following provided financial support for field trips: German Research Society; German Academic Exchange Service; the former Soviet Academy of Sciences and their subsequent organizations in Russia, Kirghizia, and Kazakhstan; and Feldbausch Foundation, Fachbereich Biologie at Mainz University. Sincere thanks are due to all friends, colleagues, and organizations.

13 Acoustic Differentiation and Speciation in Shorebirds

Edward H. Miller

Shorebirds offer rich opportunities for ecological and evolutionary research because of their high ecological and social diversity, worldwide distribution, and taxonomic richness (Hayman et al. 1986). The group has therefore received much attention from systematists, who have resolved many relationships and revealed many patterns using an array of anatomical, biochemical, and genetic characters (Bock 1958, Jehl 1968, Strauch 1978, C. G. Sibley and Ahlquist 1990, Chu 1994, 1995). To date, behavioral characters have been used little in systematic studies of shorebirds, though behavior (especially communication) has proven to be invaluable in systematic studies of birds and other taxa (R. B. Payne 1986, von Helversen and von Helversen 1994, Bretagnolle, this volume, Martens, this volume). Shorebirds have a great variety of striking optical and acoustic displays, which should prove valuable in exploring species relationships and detailing patterns of evolutionary differentiation (E. H. Miller 1984, 1992, Ward 1992). The purpose of this chapter is to illustrate the usefulness of vocal and nonvocal acoustic displays in resolving species limits and relationships among closely related shorebird species.

Shorebirds include many closely related species that are difficult to distinguish by external features but can be distinguished easily by vocalizations, even by their nonbreeding call notes (Hayman et al. 1986, Paulson 1993). Intriguingly, and paradoxically, shorebirds also seem to exhibit great evolutionary conservatism in the form and use of acoustic displays. Evolutionary conservatism is suggested by several observations. First, only minor geographic variation in shorebird vocalizations has been noted over large distances and between disjunct populations (E. H. Miller and Baker 1980, E. H. Miller 1983a, 1986, E. H. Miller et al. 1983). Second, acoustic displays of closely related species sometimes are extremely similar, even between species that diverged from one another long ago (e.g., Green and Solitary Sandpipers, *Tringa ochropus* and *T. solitaria:* Oring 1968; Long-billed and Short-billed Dowitchers, *Limnodromus scolopaceus* and *L. griseus:* E. H. Miller et al. 1984). Conservatism is suggested also by the unifor-

Dedicated to the memory of William W. H. (Bill) Gunn.

mity present in acoustic displays across entire higher taxa, such as the oystercatchers (Haematopodidae; A. J. Baker 1974, E. H. Miller and Baker 1980, Cramp 1983, A. J. Baker and Hockey 1984). Evolutionary conservatism seems paradoxical in light of the presumed need for species specificity, which makes shorebirds an especially interesting group for investigating speciation and the origin of isolating mechanisms.

In this chapter, I discuss breeding displays in some related species of shorebirds. I emphasize long-distance nuptial displays for two reasons. First, such displays likely evolve quickly through sexual selection (Mayr 1963, West-Eberhard 1983, Eberhard 1985, Butlin and Ritchie 1994). Indeed, because long-distance nuptial displays simultaneously serve in species recognition and mating competition, sexual selection may actually accelerate speciation (R. B. Payne 1986, Coyne et al. 1988, Andersson 1994, Coyne 1994, von Helversen and von Helversen 1994). Second, long-range displays are not complicated by optical display components and fine acoustic features that commonly evolve for short-range communication (E. H. Miller et al. 1988, Bain 1992). Therefore, long-range displays should be prone to geographic variation and should be sensitive markers of species limits. It is important to note that long-range nuptial displays are not the exclusive domain of males; for example, females of the polyandrous Eurasian Dotterel (*Eudromias morinellus*) "in winnow-glide display flights . . . give rhythmic sequences of *peeps* which act as songs and carry long distances"—a display form strikingly similar to one used by males of most monogamous plover species (Nethersole-Thompson and Nethersole-Thompson 1986, p. 309; Fig. 13.2A, below; see Owens et al. 1994). Such nuptial displays should tell us much about species' distinctiveness and phylogenetic relationships, regardless of the displaying sex.

Background Accounts: Systematics and Acoustics

Gallinago *snipe.* There are about 15 species of snipe. The superspecies *Gallinago gallinago* includes the forms *gallinago* (Eurasia) and *delicata* (North America). These forms (usually treated as subspecies of *G. gallinago*) differ in plumage, number of tail feathers, and other features (Tuck 1972, Cramp 1983). Another subspecies, *G. g. faeroeensis,* is recognized for Iceland and for the Faeroe, Orkney, and Shetland Islands (Cramp 1983, Hayman et al. 1986). Sometimes other taxa are referred to as subspecies of *G. gallinago,* including South American (*G. paraguaiae*) and African (*G. nigripennis*) Snipe (Tuck 1972, Devort et al. 1990).

A loud nonvocal display (drumming) is widespread in snipe and was also found in the group's ancestral species (Miskelly 1987, 1990). Drumming occurs during the repeated dives in males' spectacular aerial displays during the breeding season (Fig. 13.1; Tuck 1972, Glutz et al. 1977, Reddig 1978, 1981, Cramp 1983, Byrkjedal 1990, Devort et al. 1990). The sound is generated by vibration of the outer

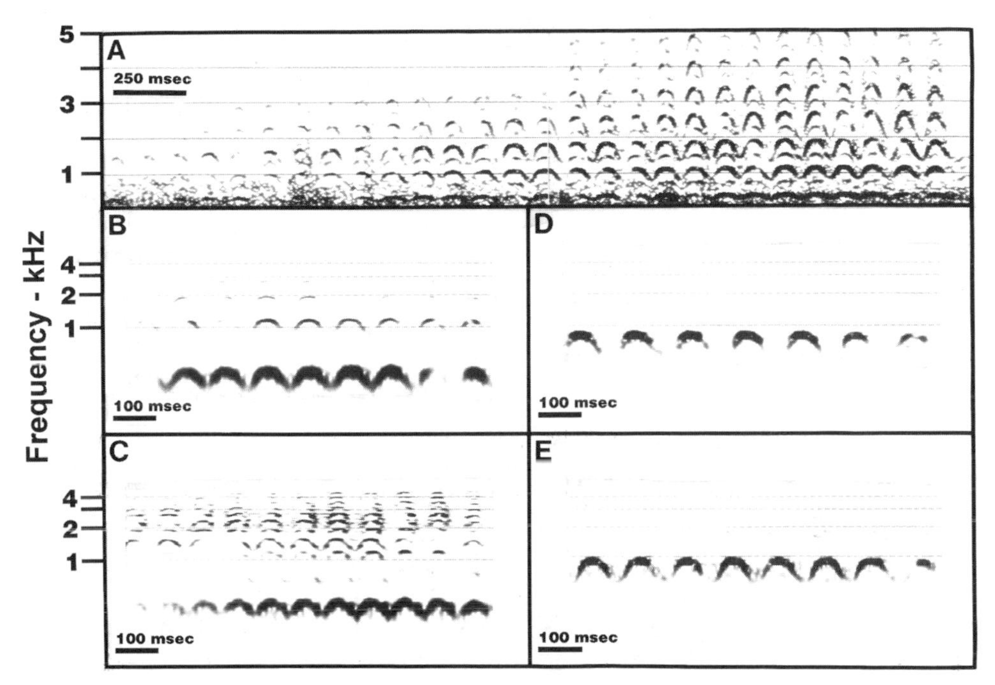

Fig. 13.1. The nonvocal drumming display of Common Snipe differs strongly between the disjunct forms *G. g. gallinago* and *G. g. delicata.* (A) Entire drum from *G. g. gallinago* (Iceland). (B) Last second of A, on a different time scale. (C) Last second of drum from *G. g. gallinago* (Russia). (D, E) Last second of drums from *G. g. delicata* in British Columbia (D) and Manitoba (E). Analyzing filter bandwidth, approximately 50 Hz. Analyses were done over the frequency range 100–5000 Hz (note that a logarithmic frequency scale was used for parts B–E). *Sources:* All recordings except C are by E. H. Miller; A, B, near Raufarhöfn, Iceland; C, Veprintsev 1982, recorded by B. N. Veprintsev and V. Leonovitch, Semzha, Kanin Peninsula, White Sea, Russia; D, near Creston, British Columbia; E, near Churchill, Manitoba.

rectrices in the airstream modified by the set wings. Breeding males also utter several kinds of loud vocalizations in the air and on the ground (Tuck 1972, Grudzien 1976, Glutz et al. 1977, Bergmann and Helb 1982, Cramp 1983, Nakamura and Shigemori 1990). Drumming in *gallinago* and *delicata* is a long, eerie sound. It begins softly and increases in loudness and frequency as a dive progresses, reaching a crescendo just before a dive ends (Fig. 13.1A). Each drum is several seconds long and is pulsed and rich in harmonics.

Charadrius *plovers.* Plovers are a large and diverse family (Charadriidae) with many interesting systematic puzzles. One such puzzle concerns Common Ringed (*Charadrius hiaticula*) and Semipalmated (*Ch. semipalmatus*) Plovers, sibling species whose relationship to one another is unresolved. They are treated as separate species in most accounts, although they do hybridize (N. G. Smith 1969). The Common Ringed Plover has a Holarctic distribution that includes Canadian high-arctic populations. The Semipalmated Plover is confined to North America; it has a generally subarctic distribution (locally sympatric in places with Common

Ringed Plover; C. G. Sibley and Monroe 1990), although it breeds south as far as Oregon (Paulson 1993).

Breeding males of many *Charadrius* species engage in conspicuous display flights that include striking optical signals (slow exaggerated wingbeats, lateral rocking, etc.) and loud vocalizations. Common Ringed and Semipalmated Plovers utter several kinds of vocalizations in their display flights, including a loud, rhythmically repeated call (RRC). A more complex vocalization (song) is sometimes given during display flights and invariably given at the end of a display.

Calidris *sandpipers.* Calidridine sandpipers (24 species) are a monophyletic group within the Scolopacidae. They have diverse mating systems, and their breeding biology has been studied extensively (Pitelka et al. 1974). Several species complexes exist, including one (referred to as "AMP" hereafter) that comprises Dunlin (*Calidris alpina*), Purple Sandpiper (*C. maritima*), and Rock Sandpiper (*C. ptilocnemis*). It is widely assumed that *maritima* is most closely related to *ptilocnemis* (e.g., Cramp 1983, C. G. Sibley and Monroe 1990). The three species show extensive geographic variation in size and plumage (Todd 1953, S. F. MacLean and Holmes 1971, Browning 1977, Greenwood 1979, 1986, Cramp 1983, Wenink et al. 1993, 1994, 1996, Wenink and Baker 1996).

Many calidridine species have aerial displays that are convergent with those of *Charadrius* plovers (Cramp 1983, E. H. Miller 1984, 1992), and they have two call types that are given in similar circumstances. For descriptive purposes the two types are referred to as RRCs and songs, though they almost certainly are not homologous between the Charadriidae and Scolopacidae.

Pluvialis *plovers.* Four species of *Pluvialis* are recognized: *apricaria, dominica, fulva,* and *squatarola.* The Grey Plover (*P. squatarola*) is sometimes placed in its own genus (*Squatarola*). The American Golden-Plover (*P. dominica*) and Pacific Golden-Plover (*P. fulva*) are closely related and often are considered conspecific (C. G. Sibley and Monroe 1990). They differ greatly in habitat, distribution, migration, and vocalization, however, and should be considered separate species (Connors et al. 1993; sonograms in Greenewalt 1968, Tikhonov and Fokin 1981, E. H. Miller 1984, G. L. MacLean 1985, Connors et al. 1993, Marchant and Higgins 1993, and Johnson and Connors 1996). The phylogeny of *Pluvialis* is only partly known. The Grey Plover is the least derived *Pluvialis* species, but it is not known whether *P. apricaria* diverged before or after the sibling species *P. dominica* and *P. fulva* (A. J. Baker pers. comm.).

Male *Pluvialis* engage in dramatic aerial displays during the breeding period (Drury 1961, E. G. F. Sauer 1962, Cramp 1983, E. H. Miller 1984, 1992, Byrkjedal 1996) that serve both to attract mates and to proclaim territory occupancy. As in *Charadrius,* the displays include conspicuous optical signals (slow exaggerated wingbeats, lateral rocking, etc.) and loud vocalizations. As in *Charadrius* and the calidridines mentioned, there are two major classes of vocalizations: RRCs and song. These probably are homologous to the same classes recognized for *Charadrius,* but not scolopacids.

Homologous Features

Background. Phylogenetic analysis requires the identification of homologous features. Methods for selecting homologous features and deciding whether they are ancestral (plesiomorphous) or derived (apomorphous) for a group have been treated extensively (Eldredge and Cracraft 1980, E. O. Wiley 1981, D. R. Brooks and McLennan 1991). The concept of homology is best developed for anatomical characters because most phylogenetic analyses have been based on them (Hall 1994). In light of early ethological interest in the phylogeny of display behavior, however, it seems surprising that ethological characters have been used so little in phylogenetic analyses (R. B. Payne 1986, D. R. Brooks and McLennan 1991, Wenzel 1992, Gittleman and Decker 1994).

Adolf Remane proposed three criteria for recognizing homologous characters: "(1) similarity of position in an organ system, (2) special quality (e.g., commonalities in fine structure or development), and (3) continuity through intermediate forms" (D. R. Brooks and McLennan, 1991, p. 7). These criteria are a useful starting point for considering homologous features of shorebird sounds. Examples for the first two follow; the intermediate forms (item 3) are straightforward and are not discussed here.

Remane's criterion of position. This criterion has a natural extension in the sequential structuring of many sounds. The sequence of song parts in *Pluvialis* plovers provides an example. The brief part ("y" in Fig. 13.4B, C, below) in the song of American and Pacific Golden-Plovers differs structurally but can be judged as homologous based on its position between the two parts (x, z) that precede and follow it. Other examples are in the RRCs of *Charadrius* and the song of dowitchers, Least Sandpiper (*Calidris minutilla*), and Purple Sandpiper (described below).

Remane's criterion of special quality. Slight continuous changes in the acoustic domains of time, frequency, and amplitude presumably can evolve easily, hence continuous quantitative measures of them (e.g., modulation rate, vocalization duration) are likeliest to be phylogenetically informative around the species level. Categorical variables (e.g., harmonic structure, pure tone) hold more promise for phylogenetic analysis at higher levels. The closely related American and Pacific Golden-Plovers differ greatly in RRC duration, for example, yet the calls are organized identically as sequences of homologous parts (Fig. 13.3C–F, below). Because the parts can be distinguished qualitatively, Remane's criterion of special quality can be applied. Examples include (1) the harmonic richness (pulsing) in the first part of RRCs in Semipalmated Plover, Common Ringed Plover, American Golden-Plover, and Pacific Golden-Plover (marked in Figs. 13.2C, F, 13.3E, F); (2) the rhythmic and quasi-rhythmic frequency modulation (FM) in the long element of song in American and Pacific Golden-Plovers (marked in Fig. 13.4); and (3) the crescendo near the end of snipe drums (Fig. 13.1A).

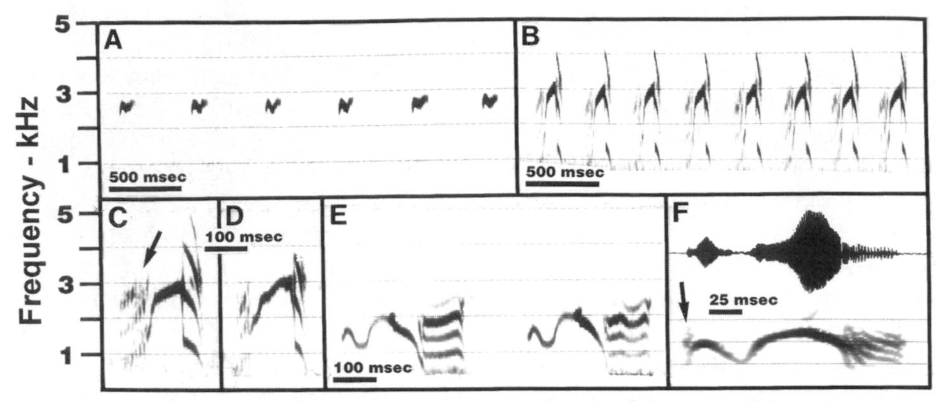

Fig. 13.2. Aerial display calls (rhythmically repeated calls, or RRCs) occur widely in plovers, regardless of their mating system, and show both strong similarities and distinctive differences between related species. (A) Part of long sequence of calls by female Eurasian Dotterel in nuptial display flight. (B) The same as A, but for male Semipalmated Plover (British Columbia). (C) Single call from B. (D) Single call from long sequence by male Semipalmated Plover (Alaska). (E) Two successive calls from long sequence by male Common Ringed Plover (Spitzbergen). (F) Amplitude envelope (upper) and corresponding sonogram (lower) of a single call from a long sequence by a male Common Ringed Plover (Russia). The introductory harmonically rich (pulsed) parts are marked by arrows in C and F (compare Fig. 13.3E, F). Analyzing filter bandwidth, 100–115 Hz. Analyses were done over the frequency ranges 100–5000 Hz (A–E) and 100–8000 Hz (F). *Sources:* A, Veprintsev (1982), Maria Pronchishcheva Bay, Taimyr Peninsula, Russia; B, C, recorded by E. H. Miller, "Haines Triangle," British Columbia, between Haines, Alaska, and Haines Junction, Yukon; D, recorded by P. G. Connors, Penny River, Alaska; E, recorded by E. H. Miller, Longyearbyen, Svalbard, Norway; F, Veprintsev (1982), recorded by B. N. Veprintsev and V. Leonovitch, Krest Bay and Eul'kal', Chukotka, Russia (note that this calling sequence on Veprintsev's record appears to be at the twice the correct speed; the analyses shown in part F are based on this assumption).

Because of their complexity and consequent resistance to rapid evolutionary change, *relationships* among acoustic features should prove to be particularly informative in phylogenetic analysis at levels above the species. Relationships can be characterized in quantitative measures of covariation or in more general terms such as AM-FM coupling. Relationships can also be manifest as emergent properties (e.g., sequential organization). A simple form of sequential organization is directional change over time, as illustrated by "sequential grading" in amplitude, frequency, and duration of elements in snipe drums. More complex sequential organization is the ordering of homologous parts within RRCs (e.g., of *Charadrius* and *Pluvialis;* Figs. 13.2, 13.3), and of song parts in dowitchers and Least Sandpiper (Fig. 13.5). Complex relational properties occur at higher hierarchical levels, too; for example, "embedded grading," with graded sequences nested within other graded sequences. An example is the song of Dunlin and Rock Sandpiper, which shows grading both within each song unit and over an entire song. Another example is the long pulsed elements in the song of dowitchers, Least Sandpiper, and Purple Sandpiper, which decline in frequency successively both within and across song units (Fig. 13.5; see p. 253).

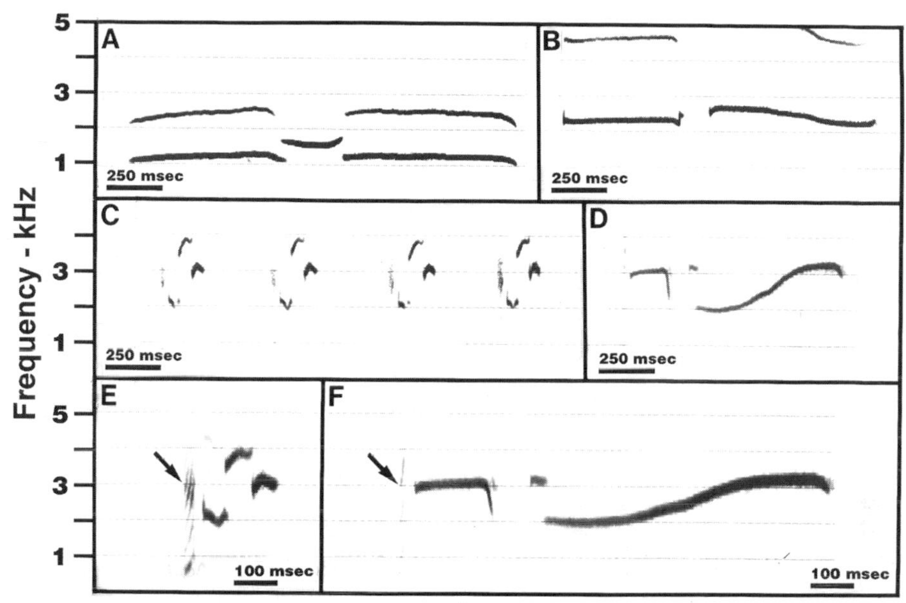

Fig. 13.3. Differences and similarities between RRCs of *Pluvialis* plovers; note the large species differences in fine frequency and temporal characteristics. (A) Single RRC by Grey Plover (Russia). (B) Single RRC by Eurasian Golden-Plover (Russia). (C) Four RRCs by American Golden-Plover (Alaska). (D) Single RRC by Pacific Golden-Plover (Alaska). (E) Single RRC by Lesser Golden-Plover (last RRC in C, on different time scale). (F) Single RRC by Pacific Golden-Plover (RRC in D, on different time scale). Analyzing filter bandwidth, 57 Hz. Analyses were done over the frequency range 100–5000 Hz. *Sources:* A, recorded by I. Byrkjedal, near Sabettayakha River, Yamal, Russia; B, recorded by B. Veprintsev and V. Leonovitch, Semzha, Kanin Peninsula, Russia; C, recorded by E. H. Miller, Canning River delta, Alaska; D, recorded by P. G. Connors, Seward Peninsula, Alaska.

Microevolutionary Processes and Patterns

Transmission distance. Long-distance displays must be physically adapted to withstand various forms of degradation and attenuation to reach and be perceived accurately by receivers; hence they should be simpler than short-distance displays, all else being equal (Bain 1992). This principle is well illustrated by acoustic displays of shorebirds that inhabit open, windy environments. These displays show two common adaptations: narrow bandwidth and rhythmic repetition (Schleidt 1973a, E. H. Miller 1984). Many shorebird displays do not conform to this pattern, however, and are extremely complex; examples are shown in Figs. 13.2–13.6. Problems with transmitting complex sounds over long distances can be reduced through frequent repetition, especially rhythmic repetition, and especially if the display form varies little across repetitions. Exceptions to this generalization occur too, however. An example is the Least Seedsnipe (*Thinocorus rumicivorus*), males of which rise infrequently in flight display, then utter a single complex song as they glide to the ground (G. L. Maclean 1969, E. H. Miller 1996). It can be concluded that adaptations for long-range transmission do not

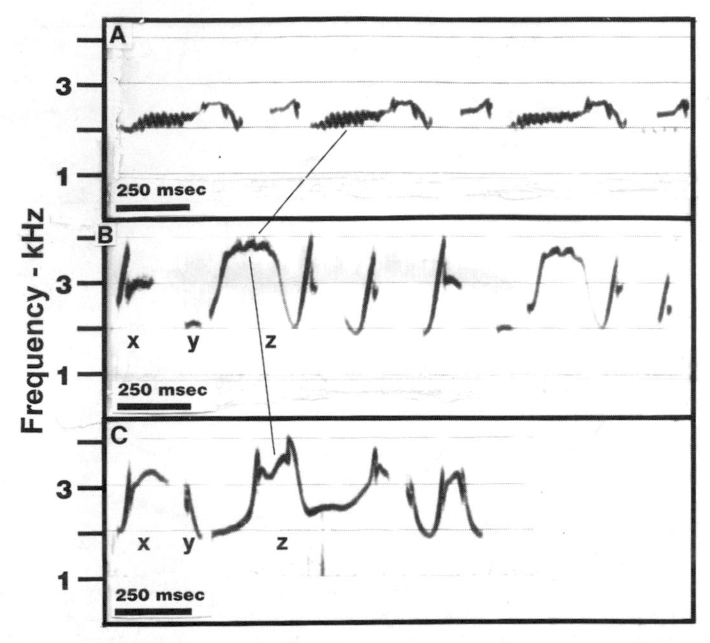

Fig. 13.4. Differences and structural correspondences in song of different species of *Pluvialis* plovers. (A) Eurasian Golden-Plover (Russia). (B) American Golden-Plover (Alaska). (C) Pacific Golden-Plover (Alaska). Corresponding elements in B and C are marked as x, y, and z; the thin lines between panels connect corresponding frequency-modulated regions in song of the three species. Analyzing filter bandwidth, 58 Hz. Analyses were done over the frequency range 100–5000 Hz. *Sources:* A, recorded by B. Veprintsev and V. Leonovitch, Semzha, Kanin Peninsula, Russia; B, recorded by P. G. Connors, Seward Peninsula, Alaska; C, recorded by E. H. Miller, Nunivak Island, Alaska.

account for physical complexity or most specific physical attributes of shorebird nuptial sounds. The principal evolutionary constraint set by long transmission distance presumably is on acoustic variability, not complexity. This generalization appears to apply even at the level of syntax (e.g., in RRC and Song organization), as long- and short-range vocalizations do not differ appreciably in syntactic complexity (such a comparison is confounded by the multidimensionality of short-range communication, however; Bain 1992).

Geographic variation. Many shorebird species are widely distributed, have disjunct populations, or are permanent residents—attributes that seem ideal for promoting geographic differentiation in displays. Again, shorebirds do not conform to expectation. Parental alarm and piping vocalizations of the nonmigratory American Oystercatcher (*Haematopus palliatus*) are similar in disjunct populations in Argentina and Massachusetts (E. H. Miller and Baker 1980). Nuptial flight song of male Short-billed Dowitchers is essentially identical in disjunct breeding populations in northwestern North America and Labrador (E. H. Miller et al. 1983). RRCs and song of Dunlin, Semipalmated Sandpiper (*Calidris pusilla*), and American Golden-Plover are extremely similar in Alaska and Manitoba

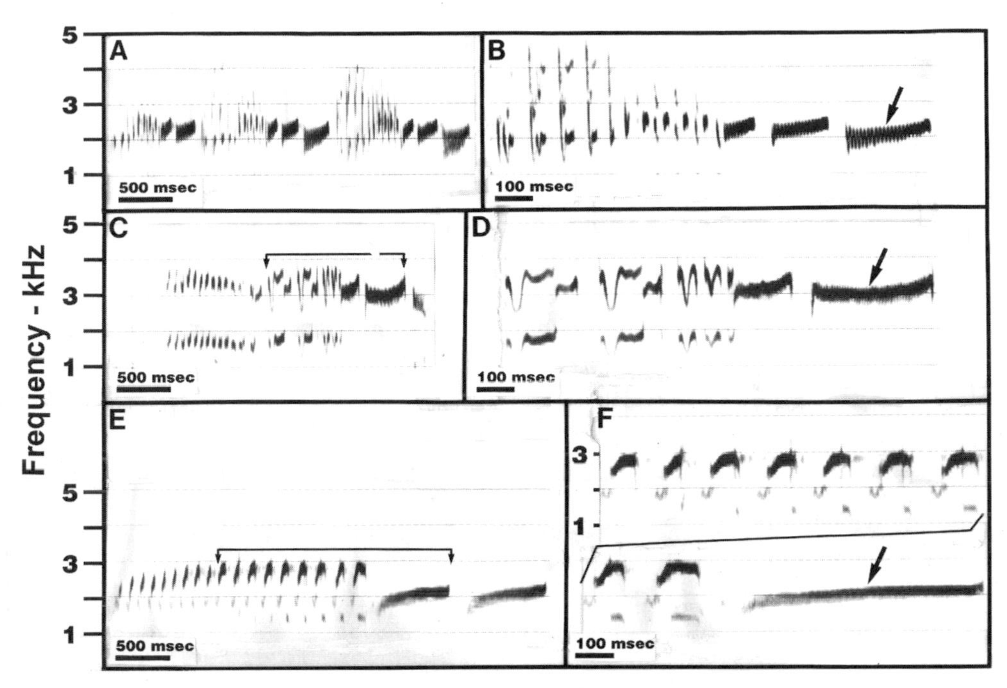

Fig. 13.5. Evolutionarily ancient song organization in Scolopacidae, showing structural similarities judged to be plesiomorphous (see text). (A) Short-billed Dowitcher (British Columbia), last three song units in five-unit sequence. (B) Central of part A, shown on a different time scale. (C) Least Sandpiper (Manitoba), single song unit with introductory trill (a variable feature; the part between arrows is shown in D). (D) Marked part of C, shown on a different time scale. (E) Purple Sandpiper (Iceland), single song (The part between arrows is shown in F). (F) Marked part of E, shown on a different time scale. The terminal long pulsed elements are judged to be homologous across species; one example for each species is marked in B, D, and F (compare Fig. 13.6). Analyzing filter bandwidth, 115 Hz. Analyses were done over the frequency range 100–5000 Hz. *Sources:* All recordings are by E. H. Miller; A, B, "Haines Triangle," British Columbia, between Haines, Alaska, and Haines Junction, Yukon; C, D, near Churchill, Manitoba; E, F, near Raufarhöfn, Iceland.

populations (E. H. Miller 1983a, Connors et al. 1993), and those of the Least Sandpiper show only weak frequency differences between Alaska and Nova Scotia, paralleling differences in body size (E. H. Miller 1986). Drumming of the snipe *G. g. gallinago* and *G. g. delicata* likewise is uniform throughout the vast breeding range of each form (see next section). By itself, the lack of geographic variation in shorebird nuptial displays does not weaken the prediction that sexually selected traits will vary geographically. That prediction is a comparative one and needs to be tested by comparing sexually selected displays (e.g., RRCs and song) with display types that are not sexually selected (e.g., parental alarm) within a species. This matter is discussed further below.

The low geographic variation in shorebird acoustic displays suggests that little vocal "learning" takes place. Thus, acoustic displays should reflect historical patterns and phylogeny fairly clearly, and should be useful in resolving systematic problems such as those involving sibling species.

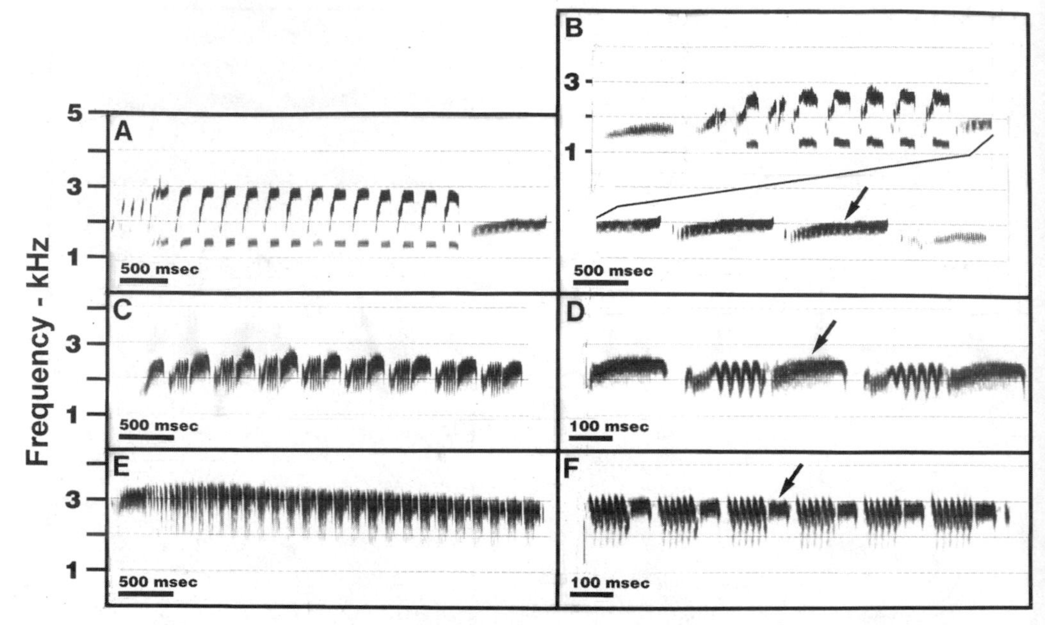

Fig. 13.6. Song structure within a species complex of calidridine sandpipers. (A) Purple Sandpiper (Iceland), single song with introductory trill (a variable feature). (B) Purple Sandpiper (Spitzbergen), single song embedded in long series of RRCs (single RRCs are shown before and after the song). (C) Rock Sandpiper (Alaska). (D) Last second of C, shown on different time scale. (E) Dunlin (Alaska). (F) Last second of E, shown on a different time scale. The terminal long pulsed elements in B (one is marked) are judged to be homologous to those marked in D and F (compare Fig. 13.5). Analyzing filter bandwidth, 115 Hz. Analyses were done over the frequency range 100–5000 Hz. *Sources:* All recordings are by E. H. Miller; A, near Raufarhöfn, Iceland; B, Longyearbyen, Svalbard, Norway; C, D, St. Paul Island, Pribilof Islands, Alaska; E, F, Nunivak Island, Alaska.

Sibling species. Nuptial displays are sensitive indicators of population differentiation and speciation at the level of sibling species, as the following examples make clear. The snipe forms *G. g. gallinago* and *G. g. delicata* usually are treated as subspecies of the species *G. gallinago,* as mentioned above. Thönen (1968) noted a striking difference in snipe drumming between north-coastal Alaska (*delicata*) and western Europe (*gallinago*), being lower in both frequency and modulation rate in the latter.[1] In *gallinago,* the fundamental frequency is approximately 350–400 Hz around the drum's crescendo, and there is a rich frequency spectrum with strong emphasis on odd harmonics (Fig. 13.1B, C). The fundamental frequency in *delicata* is about twice as high, and energy simply falls off progressively with frequency (Fig. 13.1D, E). In *gallinago,* the pulse (modulation) rate is much higher and pulses are correspondingly briefer.

The differences between *gallinago* and *delicata* drumming are strong, involve

[1]Other sonograms are in Grudzien (1976), Thielcke (1976), Glutz et al. (1977), Jellis (1977), Reddig (1978, 1981), Bergmann and Helb (1982), Cramp (1983), and Nethersole-Thompson and Nethersole-Thompson (1986).

several prominent acoustic characteristics, and are consistent throughout the vast breeding ranges of the two forms. Therefore it seems reasonable to view *gallinago* and *delicata* as separate species, pending information on nuptial vocalizations, playback experiments, and breeding in areas of local sympatry (western Aleutian Islands? D. D. Gibson pers. comm.).

The closely related Common Ringed and Semipalmated Plovers occasionally are treated as conspecific, yet their nuptial vocalizations differ greatly. RRCs of the Common Ringed Plover are approximately 230–240 milliseconds long and are repeated about every 120–130 milliseconds (Fig. 13.2). RRCs begin with a brief pulsed portion (not visible in many sonograms: compare Fig. 13.2E, F), followed by a frequency-modulated nonharmonic section, and end with a harmonically rich section (Fig. 13.2E, F). The pulsing that corresponds to the harmonic richness of the final call part is evident in the waveform (Fig. 13.2F, upper; see Watkins 1967).[2]

RRC elements in the Semipalmated Plover correspond exactly to those just described, though they differ in structure. RRCs of this species are briefer and uttered more quickly (Fig. 13.2B). Each RRC is introduced by a rapidly pulsed section followed by two tonal parts (Fig. 13.2C, D). The first of these (the call's central part) rises in frequency from approximately 1500 to 3400 Hz (high values from Queen Charlotte Islands, British Columbia), then declines to the starting frequency of the last part of the call. The frequency contour of this middle part varies geographically but is always characterized by having no harmonics. In contrast, the terminal part is always harmonically rich.

The RRCs of Common Ringed and Semipalmated Plovers differ in call duration, call-part duration, interval between calls, frequency, and frequency contour of the central part. These multiple differences, taken together with the minor geographic variation in RRCs over the extensive breeding range of each form, suggest that the forms should be treated as separate species. This view is strengthened by the observation that even flight calls by nonbreeding birds on migration or wintering grounds "are . . . sufficiently distinct . . . to allow confident identification on call alone" (Hayman et al. 1986, p. 283, Paulson 1993). As for snipe, song structure and areas of local sympatry need to be investigated; song seems to differ greatly between the two forms (E. H. Miller pers. obs.).

Phylogenetic Analysis

It is useful to begin phylogenetic analyses at low levels, where relationships are likeliest to be expressed. Such a conservative approach seems called for in the present case, as differences are apparent even between sibling species.

[2]For other sonograms of this call type, see Glutz et al. (1975, p. 103, figs. b, c), Bergmann and Helb (1982, p. 130, fig. b), and Cramp (1983, p. 138, figs. IV, V, and VII). The call type may be variable in this species, as some published sonograms do not show the terminal harmonically rich portion (e.g. compare the first two calls in fig. IV with others there, and with those in figs. V and VII in Cramp 1983).

Gallinago *snipe.* Drumming seems to be a sensitive indicator of differentiation between closely related taxa, but it also offers promise for investigating phylogenetic relationships and evolutionary patterns more generally. Evolutionary divergence above the level of sibling species is surprisingly great. For example, *G. paraguaiae* exhibits differentiation of elements within each drum, with alternating pulse couplets and singlets (E. H. Miller pers. obs.). Drumming in several other snipe species also has become elaborated beyond a simple series of pulses. In the Pintail (*G. stenura*) and Swinhoe's (*G. megala*) Snipe, each drum ends distinctively with changed intervals between pulses and a long terminal element (Labutin et al. 1982, Veprintsev 1982, Cramp 1983, Byrkjedal 1990). Other striking species differences in drumming occur, though published analyses do not permit detailed comparisons (Terborgh and Weske 1972, G. L. Maclean 1985, Nakamura and Shigemori 1990). Species differences in drumming probably reflect (in part) species differences in number, shape, and size of rectrices, and in how the rectrices are used in drumming—for example, in the temporal pattern and extent of tail spreading (Byrkjedal 1990).

Charadrius *plovers.* Close relatives of the Common Ringed and Semipalmated Plovers are the Little Ringed Plover (*Ch. dubius*), Killdeer (*Ch. vociferus*), Long-billed Plover (*Ch. placidus*), Wilson's Plover (*Ch. wilsonia*), and Piping Plover (*Ch. melodus;* Bock 1958, Taylor 1978). Published sonograms for the Little Ringed Plover reveal acoustic features astonishingly similar to those described above, and these are likely homologous: calls have a tripartite structure, and the parts have the attributes described above and occur in the same sequence (Glutz et al. 1975, p. 153, fig. 13.E, Bergmann and Helb 1982, p. 131, fig a, Cramp 1983, p. 126, fig. 13.IX). Homologous calls in the Killdeer may correspond only to the central call part recognized here, but more detailed sonograms are needed to confirm this possibility (Bursian 1971, R. E. Phillips 1972). No sonograms of this call type have been published for Wilson's or Piping Plovers, and no sonograms of Long-billed Plover vocalizations have been published at all (E. H. Miller 1992).

The direction of evolutionary change in acoustic characters can be deduced because *Ch. semipalmatus* was derived from one stock of the widespread *Ch. hiaticula* (Taylor 1978). Acoustic changes that occurred during the differentiation of *Ch. semipalmatus* can be summarized as follows: calls became briefer, were uttered more slowly, and became higher in frequency; the pulsed introductory portion of calls became longer and more pronounced; central and terminal call parts became briefer; and the central call part became simpler. Aside from these general patterns and relationships, little can be said because of the dearth of published material. As with *Gallinago,* however, acoustic characters seem to have great potential for resolving species relationships at various levels, and also for tracing the nature of evolutionary change.

Calidris *sandpipers.* Rhythmically repeated calls differ strongly among the AMP complex species in quantitative features, but those features are uninformative regarding the species' relationships to one another (E. H. Miller pers. obs.). Other vocal classes must be used to explore relationships.

Ancestral and derived features of AMP song can be deduced by considering song structure in two other scolopacids, the Short-billed Dowitcher and the Least Sandpiper.[3] Dowitcher song consists of several rapidly repeated units (Fig. 13.5A). Each song unit begins with a complex trill and concludes with several long frequency-modulated elements. The long terminal elements increase successively in duration, and they decrease in carrier frequency successively both within and across song units (Fig. 13.5B). Least Sandpiper song is extremely similar to this: each of the repeated song units starts with a trill and ends with several frequency-modulated (pulsed) elements. As in dowitchers, the terminal song unit elements increase successively in duration, and they decline successively in carrier frequency both within (Fig. 13.5D) and across song units. Other similarities are suggested in fine features of trill elements. Finally, consider the song of the Purple Sandpiper, which also begins with a trill and ends with several long frequency-modulated (pulsed) elements (Figs. 13.5E, F, 13.6A, B). The song (unit) is rarely repeated in succession, however, as it is in the other two species.[4] Furthermore, the introductory trill and terminal pulsed elements often are uttered in isolation from one another (Cramp 1983, p. 352, figs. V, VI), lending confusion as to what constitutes song (Cramp 1983). Trill elements in Purple Sandpiper song start at a low frequency and shift to higher frequency, as in the songs of the Short-billed Dowitcher and Least Sandpiper (Fig. 13.5B, D, F). If these structural and organizational (syntactical) similarities of song are interpreted as ancient homologies, then Purple Sandpiper song is ancestral within the AMP complex.

Song structure within AMP has two forms: (1) the form found in the song of the Purple Sandpiper (Figs. 13.5E, F, 13.6A, B), and (2) the form found in Dunlin and Rock Sandpiper song. The song of the latter two species is a long series of repeated, complex units incorporating slow and fast FM (Fig. 13.6C–F). The units are temporally separated from one another and show progressive changes in duration, frequency, and other characteristics over a song. The units of Rock Sandpiper song are about 350–550 milliseconds long and have several distinct parts: an initial pulse, brief rapid FM followed by some long slow pulses, and a long terminal part with rapid FM. The terminal part is remarkably similar to the long FM elements that terminate song in the Short-billed Dowitcher and the Least and Purple Sandpipers described above (Fig. 13.6D). Dunlin song is very similar, but its elements are briefer, more rapid, and higher in frequency. Dunlin song units have only two parts, one corresponding to the slowly pulsed part of Rock Sand-

[3]For references to published sonograms of Least Sandpiper vocalizations, see E. H. Miller (1984, 1992, 1995). For dowitcher vocalizations, see Greenewalt (1968) and E. H. Miller et al. (1983, 1984).
[4]In my detailed description of Least Sandpiper song, I overlooked its routinely compound nature and referred to each song unit (as recognized here) as song (E. H. Miller 1983b). This feature of song organization became apparent only through comparison with dowitchers (E. H. Miller et al. 1983, 1984).

piper song elements, the other corresponding to the terminal part with rapid FM (Fig. 13.6F).

Several interpretations can now be made:

1. Purple Sandpiper song is homologous to the repeated unit in Short-billed Dowitcher and Least Sandpiper song.
2. The repeated unit in Rock Sandpiper and Dunlin song is homologous to the entire song of Purple Sandpiper (hence to the repeated song units of Short-billed Dowitcher and Least Sandpiper).
3. The long terminal part with rapid FM in Rock Sandpiper and Dunlin song units corresponds to the long terminal element with rapid FM in song of the other species.
4. Purple Sandpiper song has a uniquely derived condition of loose association of parts, with song trills and song FM elements often occurring by themselves.
5. Song of Rock Sandpiper and Dunlin is derived relative to that of Purple Sandpiper.

A third class of vocalization resolves the relationship between Rock Sandpiper and Dunlin. The cricket vocalization is a distinctive call given on the ground, mainly by males (E. H. Miller pers. obs.). It is uttered rhythmically in bouts or long sequences and may be audible for tens of meters. Sometimes calling males stand on a hummock and give this call for long periods, usually while the mate (or prospective mate) is nearby. The cricket call is used also in various kinds of short-range communication, as during nest-scraping displays. The cricket call has a similar two-part structure in Purple and Rock Sandpipers but is distinctive in having only one part (and in other ways) in Dunlin (E. H. Miller pers. obs.). The cricket call's structure therefore suggests that the Dunlin is the most highly derived species within the AMP complex.

The preceding analysis established that the Purple Sandpiper is the least derived species within AMP, and the Dunlin is the most derived. Furthermore, Rock Sandpiper vocalizations are distinctively different from those of the Purple Sandpiper. In light of these observations, it is untenable to consider Rock and Purple Sandpipers conspecific, and it may be unreasonable even to view them as parts of a superspecies (Cramp 1983, p. 355, C. G. Sibley and Monroe 1990, p. 241).

Pluvialis *plovers.* The RRCs and song of *Pluvialis* show a range of differentiation that illustrates both extreme evolutionary conservatism and rapid evolutionary change. *Charadrius* can serve as an outgroup for deducing ancestral and derived features of *Pluvialis. Charadrius* RRCs exhibit variable harmonic richness, a multipartite structure, sudden frequency shifts, and pulsing. The last three attributes all appear in *Pluvialis* RRCs, so they can be considered ancestral to the genus and must be very old (Fig. 13.3; Connors et al. 1993).

The RRCs of the Grey Plover are uttered very slowly (approx. 5–6 per minute), and those of the Eurasian Golden-Plover (*P. apricaria*) fairly rapidly (approx. 30 per minute; Fig. 13.3). The absence of an introductory pulsed element is a derived condition (Fig. 13.3A, B). The RRC of the Grey Plover has three tonal elements, which presumably is the ancestral state; that of the Eurasian Golden-Plover has only two, which therefore is derived. In the American Golden-Plover, RRCs are brief (approx. 200 milliseconds) and uttered very rapidly (about 2 per *second*)—about 20 times as fast as in the Grey Plover (Fig. 13.3C)! They begin with a brief pulsed element—the "twisted rope" note of Connors et al. (1993)—that is followed by several brief tonal parts at different frequencies (Fig. 13.3C, E). The RRCs of the Pacific Golden-Plover are much longer (about 1 second) and are uttered much more slowly (some 15–25 per minute). However, they are identical in structure: a brief introductory pulsed element followed by three tonal elements at different frequencies (Fig. 13.3D, F). (Note that the introductory element in both species, and the middle tonal element in *P. fulva* RRCs, are not apparent in most published sonograms.)

Its song structure underscores the uniqueness of the Grey Plover. The song is extremely long and includes an introductory trill followed by a series of long tonal notes with some sudden frequency shifts (Byrkjedal 1996, E. H. Miller pers. obs.). No structural correspondences with the song of the other species are apparent. In contrast, homologous song features seem obvious in the other species, particularly between the American and Pacific Golden-Plovers (Fig. 13.3).

In summary, *Pluvialis* RRCs have differentiated greatly in temporal properties, and this is a prevalent trend (E. H. Miller 1986). Homologous parts of RRCs are readily identifiable, however, and part sequences have been conserved. Fewer homologies in song can be identified; some are attributes (e.g., quasi-rhythmic FM) and others lie in the sequential structure.

Evolutionary rates. The preceding examples all suggest strong evolutionary conservatism in some attributes of acoustic structure. It is of interest to estimate the rate at which some of those attributes evolved. To do this, divergence times and acoustic attributes must be known. Dowitchers can serve as an example.

Long-billed and Short-billed Dowitchers diverged from one another approximately 4 million years (MY) ago (Avise and Zink 1988). On various measures, song differences between them average 22% (E. H. Miller et al. 1984, table 1). Assume, then, that the two species have diverged equally from their common ancestor (i.e., each has diverged 11% over 4 MY). Rate of evolutionary change in song can be judged relative to a factor of e per million years (= 1 darwin; Futuyma 1986). Thus, dowitcher song has changed at a rate of (0.11/4/2.718), or about 10 *milli*darwins—roughly 1% per million years! This figure is comparable with low average rates of morphological evolution based on the fossil record (e.g., 40 millidarwins for molar height in Tertiary horses), and contrasts with high rates routinely found for colonizing species and laboratory populations (Futuyma 1986). Rates of evolutionary change must have been much slower for other properties of shorebird displays, such as the frequency modulation, harmonic richness, and sequential ordering in the examples provided above.

West-Eberhard's (1983) scenario for speciation and the evolution of sexually selected characters leads to the prediction that displays with different functions will evolve at different rates. Insufficient evidence exists to judge whether differences in evolutionary rates like those just noted can be explained in that way. Some supportive evidence comes from parental alarm calls versus song and RRCs in calidridine sandpipers. Most calidridine species have two forms of parental alarm—a trill and a single note (E. H. Miller 1984, 1985b, 1992). Alarm call forms are immediately recognizable in species as distantly related as the Surfbird (*Aphriza virgata*), Baird's Sandpiper (*Calidris bairdii*), and the Least Sandpiper, though they differ in simple quantitative features (E. H. Miller 1985b, E. H. Miller et al. 1987, 1988). In contrast, the songs of those species—though they occur in extremely similar contexts—are so different in structure that homologous features cannot even be identified. A similar contrast between similar alarm calls and dissimilar nuptial vocalizations occurs between the American and Pacific Golden-Plovers (Connors et al. 1993, Johnson and Connors 1996).

Conclusions

Sexual selection has acted on shorebirds to produce nuptial sounds of great complexity, even for sounds used principally over long distances. Many of these sounds are similar in related species, but others are so different that homologous characters cannot be detected. Display types with different functions (e.g., parental alarm) show greater similarities, and hence evolved more slowly (E. H. Miller 1984). This range of variation underscores the importance of sexual selection in effecting species-distinctive mating displays, and also points to a wealth of material for systematics research on shorebirds. Several lines of investigation may be particularly fruitful.

Like passerines, many shorebirds show strong interspecific differences in their displays; in fact, this characteristic led workers to suggest that the usefulness of vocalizations in passerine systematics is limited to species-level questions (W. E. Lanyon 1969, R. B. Payne 1986). In contrast, numerous features of nuptial acoustic displays in shorebirds are evolutionarily conservative and should be useful in revealing relationships well above the species level. Examples include entire display classes, such as drumming in *Gallinago* and semisnipe (*Coenocorypha*); high-level organizational properties, such as embedded grading in dowitcher and calidridine song; sequential organization, as in plover RRCs and calidridine song; individual elements, as illustrated by corresponding parts of plover RRCs and calidridine song; and specific attributes such as pulsing in the terminal elements of calidridine song and tonality in parts of plover RRCs. If nonnuptial acoustic displays are examined together with nuptial ones, it seems likely that some shorebird relationships can be resolved even well above the genus level. In particular, descriptions of vocalizations from outside the breeding period, and from chicks and females, should be sought (Saether 1994).

Vocal characters frequently are vital for resolving the systematic status of cryptic or sibling bird species (W. E. Lanyon 1969, R. B. Payne 1986, N. K.

Johnson 1994), and the same applies to nuptial acoustic displays of shorebirds, despite the evolutionary conservatism just outlined. At higher levels of phylogenetic analysis, it should prove valuable to couple behavioral and genetic analyses. Such an approach will permit evolutionary trends to be traced and evolutionary rates to be estimated—rates of change both in different display classes and in different levels of acoustic structure. The resulting information should reveal a great deal about the action of sexual selection in speciation and about behavioral evolution generally (e.g., regarding evolutionary constraints; McKitrick 1993, von Helversen and von Helversen 1994).

Information and research needs for shorebird communication are similar to needs in other areas of field biology. Better and more extensive information is needed about display characteristics, when displays occur seasonally and during the day, the ecological settings of displays, and the meanings and functions of displays. Such information is needed simply for developing techniques to detect and enumerate shorebirds. Some groups of shorebirds are still essentially unknown yet likely are of conservation concern; examples are woodcock and snipe (E. H. Miller 1992, 1995). It is crucial to determine which members of these groups are distinct species, and knowledge of acoustic display structure is essential to accomplish this (R. B. Payne 1986, N. K. Johnson 1994). The evidence presented above on snipe underscores how little we know about species distinctiveness even in widely distributed and seemingly well-known taxa. Thus, research on the systematics of taxonomically poorly known groups for which such information is of practical conservation concern should have the highest priority.

Acknowledgments

My interest in shorebird acoustics was kindled by the generosity and support of Ian McLaren, John Fentress, and Bruce Moore (Dalhousie University) many years ago. I remain indebted to them. For help with recordings, analyses, ideas, and literature used in this chapter, I thank Alan Baker, Jeffery Boswall, Ingvar Byrkjedal, Pete Connors, Dan Gibson, Bill Gunn, Wally Johnson, Ian Jones, Brina Kessel, Don Kroodsma, Pete Myers, Elin Pierce, John Richardson, Brian Stushnoff, Ron Summers, Boris Veprintsev, and Michael Wilson.

14 A Comparison of Vocal Behavior among Tropical and Temperate Passerine Birds

Eugene S. Morton

Avian diversity is far richer in the Tropics than in the temperate zones, yet most research on the evolution and function of birdsong focuses on temperate zone species. This geographic bias reflects the distribution of the biologists who study birdsong, but the general validity of such a restricted focus is open to question. Are temperate zone biologists describing general phenomena or the peculiarities of temperate zone life histories? Do the Tropics simply have more species than the temperate zone, or do the selection forces on some tropical birds differ qualitatively from those acting on temperate birds? Based on my knowledge of Neotropical and Nearctic birds, I believe that the Tropics are different, both in the number of species present and the details of their life histories. Consider, for example, that most Neotropical passerines are suboscines and therefore do not learn songs (Kroodsma 1982a). Fifty percent of the 383 breeding passerine species in Panama are suboscines, and the percentage is even higher in South America. Entire large families—including the tyrannid flycatchers, antbirds, ovenbirds, woodcreepers, cotingas, and manakins—constitute these tropical groups. In contrast, temperate passerines are primarily songbirds, which learn their songs. Mating systems also differ in the Tropics. Most tropical species appear to have socially monogamous breeding systems like those of most temperate passerines, whether they learn songs or not. But the Tropics also feature breeding systems not common in temperate areas, such as helpers, communal breeding, and leks.

In this chapter I attempt to explain the prevalence of a unique tropical singing style, duetting, and the widespread occurrence of female singing, in duets with males and alone. Duetting is common in tropical birds, but it is not tied to any particular tropical social system. Duetting species come from monogamous and cooperative breeding social systems and from a wide taxonomic range, including many species that do not learn songs (Farabaugh 1982). Why, then, is duetting so common in the Tropics? This question is a convenient starting point from which to discuss major differences in life history evolution that may have influenced the evolution of birdsong. My explanation of female singing and duetting focuses on new information and new concepts of social systems affected by latitude. An appreciation of these latitudinal effects should provide biologists with new insight about the diversity of birdsong. There are many untraveled research avenues open in the Tropics and few biologists to follow them at present.

Duetting, Pair-bonds, and Territoriality

Pair-bonds and territoriality differ fundamentally between the temperate and tropical zones. Temperate zone breeding seasons coincide with intense territorial behavior. The time devoted to pair formation and reproduction is a short pulse followed by a long nonbreeding period during which birds stop or reduce territorial behavior, or migrate out of breeding ranges altogether. In sharp contrast, tropical territorial behavior and breeding seasons are long-term efforts (J. R. Baker 1938). Territories are often defended year-round, and no more intensively during the breeding than the nonbreeding season. Song is not highly correlated with pairing or reproduction, but rather is used in territorial defense throughout the year. Furthermore, singing and territorial defense are not entirely, or even largely, male behaviors.

Most of the world's passerine birds fit this tropical pattern. In Panama, at least 60% are permanently pair-bonded on year-round territories (E. S. Morton 1980). Species there include both suboscines (e.g., antbirds, woodcreepers, and flycatchers) and oscines (e.g., wrens, tanagers, and finches). Turnover of individuals is low, and even with the arrival of newcomers the boundaries of existing territories do not change (Greenberg and Gradwohl 1986). The neighborhoods of many tropical birds are thus highly stable through time.

With year-round territoriality and permanent pair-bonds, the roles of the sexes converge in tropical birds. Singing by females is but one manifestation of this sex role convergence. The songs of males and females differ in the vast majority of tropical species and duetting indicates territorial defense against same-sexed individuals, not a form of cooperation between pair members (Farabaugh 1982). In a few species, however, females and males sing identical songs (e.g., Zapata Sparrow, *Torreornis inexpectata:* E. S. Morton and Gonzalez 1985; Blue-black Grosbeak, *Cyanocompsa cyanoides:* Morton and Derrickson unpubl. data; Cuban Red-winged Blackbird, *Agelaius phoeniceus:* Whittingham et al. 1992, Morton unpubl. data). Some form of pair cooperation might be expected in those species because the individuals appear to forgo same-sex competition for a truly cooperative defense of territory. Other manifestations of sex role convergence are found in parental care: in the Tropics males build nests, incubate eggs, brood young, and feed offspring as much as their mates do (Skutch 1969).

The Dusky Antbird (*Cercomacra tyrannina*) is typical of many permanently territorial and pair-bonded, insectivorous tropical species. In Panama, these antbirds breed chiefly in the May–October part of the rainy season. Pair members may duet or sing alone, with sex-specific songs (Fig. 14.1). Either pair member will abandon its mate if an opening occurs on a nearby territory of apparently higher quality (E. S. Morton and Derrickson in press). Within five minutes of

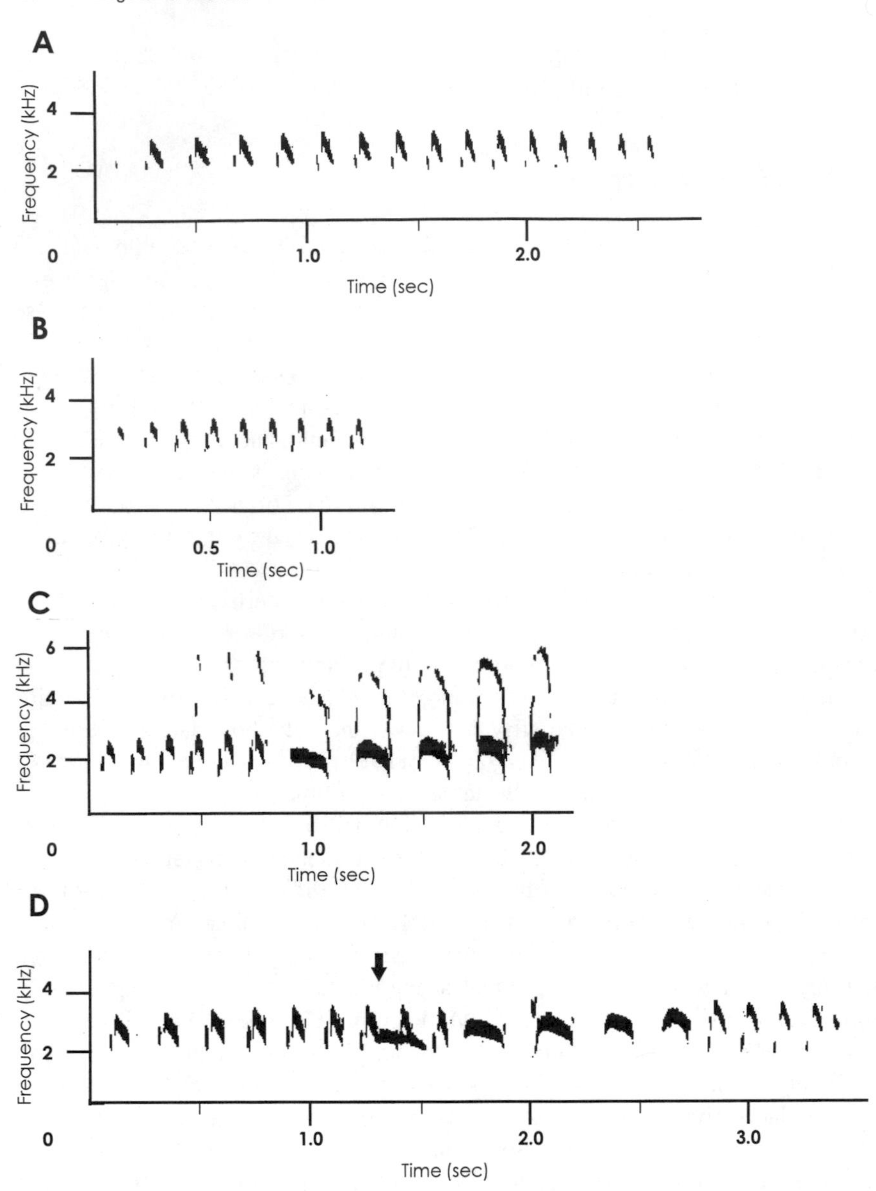

Fig. 14.1. Sonograms of songs of male and female Dusky Antbirds (*Cercomacra tyrannina*), illustrating the difference in songs used by paired and unpaired birds. (A) Male courtship song is a long, rapid twitter, variable in length. (B) Female courtship song type I consists of a trill and is similar to a male courtship song. This form is given about 10% as often as female courtship song II. (C) Female courtship song II is the more common of the two female courtship songs. It begins with a malelike trill and ends in a typical female rising series of notes. (D) Typical duet, with the male beginning, interrupted by the longer and rising notes of the female. Just after the second female note, the male gives a call note that sounds like *tidit.* Sung by either sex during a song by the mate, *tidit* may serve, like a duet song, to keep the mate from switching to courtship song.

its mate's disappearance, the unpaired male or female begins to sing a different song (Fig. 14.1). My colleagues and I induced the song by experimentally removing mates, but human observers might spend many months before hearing a natural occurrence of this very different song, which has not been previously described. This "courtship song" often attracts a new mate, either a floater or one from an established territory (Morton and Derrickson unpubl. data); however, birds of both sexes also remained unpaired for the two-month duration of our study.

Thus the need to defend a permanent territory as a single individual, regardless of sex, selects for egalitarian sex roles, including female singing, in many tropical species. New pair-bonds can form within minutes, and the new pair duets and forages together in a way indistinguishable from long-established pairs. In spite of this potential for rapid change, however, many pairs stay together for life. With 90% adult annual survival (Morton and Derrickson unpubl. data), neighborhoods are highly stable.

This common tropical territorial system contrasts sharply with the short-term breeding season territories and pair-bonds found in most temperate species. The short breeding season and short, temporary pair-bonds and territorial defense confine singing behavior to the breeding season. The seasons of reproduction, territoriality, and, consequently, singing coincide. This pattern is atypical of song use in lower latitudes. Furthermore, sex roles in the temperate zone are unusually divergent: males sing and set up and defend territories, and females construct the nest and incubate. Temperate climatic constraints on birds may have affected sex role evolution, sexual selection, territorial behavior, and food habits by causing reproduction and territoriality to coincide. Researchers' undue focus on temperate oscines contributes to a temperate zone bias in interpretations of song function and evolution because the function of birdsong in the temperate zone is probably related to these climatic constraints.

The concentration of singing and breeding into a short season makes it especially difficult to determine the function of song in temperate zone birds. One form the temperate zone bias takes is the premise that birdsong generally functions to attract mates. In an erudite discussion of song function, Cronin (1991, p. 220) placed birdsong in the debate between Darwin and Wallace: Darwin thought elaborate birdsong's purpose was to "charm the female," whereas Wallace held that it was a part of competition between males. We can recognize this old debate still. If tropical species were included in the argument, however, singing behavior would be described, generally, as directed to same-sexed individuals and as functioning to defend territory against same-sexed birds. The temperate bias led to the belief that antiphonal duetting in tropical birds was a pair-bonding and pair-bond maintenance mechanism rather than a manifestation of territorial aggression toward same-sexed birds (Farabaugh 1982).

From a tropical perspective, a question arises about whether birdsong in temperate zone birds actually does attract mates: has this premise been adequately tested? The degree to which tropical permanently paired species use courtship songs like the one described for the Dusky Antbird is unknown. Most, I suspect, use nonsong vocalizations in the same way the permanently territorial and pair-bonded Carolina Wren (*Thryothorus ludovicianus*) does: widowed males use a close contact *tsuck* call note to court females, and sing for intrasexual territorial defense (E. S. Morton 1982). Is it possible that mate attraction and singing are correlated and that a causative relationship exists between song and territorial defense?

I suspect most modern biologists side with Wallace in believing that birdsong woos females not because of its beauty but because it signals a male's fitness. How do females assess a male via his song? Earlier, I suggested that a female's assessment of a male is based on the time he spends singing, not on his song per se (E. S. Morton 1986). If the *amount* of singing is correlated with food availability on territories, females might use the time a male spends singing to evaluate his, and his territory's, quality. Of course, the male's song might also signify the presence of a rare resource such as a secondary cavity for nesting (Eriksson and Wallin 1986). Females, in other words, use song in species recognition and in assessing males' quality (or control of a resource vital to them).

Recent research provides experimental support for mate choice based on the amount of time spent singing (Gottlander 1987, Radesäter et al. 1987, Reid 1987, Alatalo et al. 1990, Arvidsson and Neergaard 1991). Reid, for example, found that food supplementation resulted both in more singing by males (showing that food limited singing time) and in earlier pairing. The postulated difference in how sexes assess song might influence male singing behavior in the American Redstart (*Setophaga ruticilla*): if females assess the amount of singing, this might explain why males with repertoires repeat a single song (repeat mode) before pairing, and afterward use more song types (serial mode; Lemon et al. 1987). Females should be better at ascribing song output to a particular male if he sings one song type, especially when, as is the case in some paruline warblers, repeat mode songs are individually distinctive. In species with dense breeding populations coupled with high breeding synchrony, selection might favor single songs for female assessment of male song output. In contrast with repeat mode songs, serial mode songs are shared by males in a population as local dialects. In breeding habitats with highly uniform food productivity, individual song *complexity* may permit females to favor older males (Catchpole 1980). When singing is confined to the reproductive period, the conflation of potential song functions with reproduction produces a complex mix of possibilities indeed.

The preceding paragraph describes temperate zone singing. But the mate-attracting functions of song do not seem to be generally applicable in the Tropics. Where singing occurrs in both sexes, as is typical in the Tropics, its mate-attraction functions are not highlighted so extensively. In contrast, the emphasis is on territorial maintenance. We therefore need more research on permanently paired and territorial species to determine the extent to which song functions in mate attraction. It seems likely that when mate attraction is usually a one-time event in a bird's life, the forces affecting song evolution are balanced toward sources of selection other than mate attraction, particularly if nonsong call notes are used in pair-bonding.

Why is singing largely the male's role in temperate zone birds, even in recent tropical derivatives? For example, the Carolina Wren, distributed from Guatemala to Maine, retains its year-round territoriality and pair-bonding in northern latitudes. But only the males sing, even though the numerous tropical *Thryothorus* species all have singing females. In the next section, I suggest a reason for the fundamental difference in birdsong use and function between tropical and temperate passerines.

Breeding Synchrony and the Evolution of Extra-pair Fertilization Behavior

I believe that birdsong is indirectly related to extra-pair fertilization (EPF) behavior. The differences between temperate and tropical birdsong are related to latitudinal differences in the likelihood that EPF behavior will evolve.

Extra-pair fertilizations are increasingly recognized as important and widespread components of the mating systems of socially monogamous and polygynous birds. The tendency for such fertilizations to evolve is related to breeding synchrony and the density of fertile females in a population. The authors of both major reviews of EPF behavior (Westneat et al. 1990, Birkhead and Møller 1992) believe that the evolution of EPF behavior is favored when females are asynchronously fertile. Males are fertile throughout the breeding season, but females are fertile only around the time of egg laying. If few females are fertile simultaneously, the operational sex ratio in favor of males should result in fertile females being subject to many EPF attempts (Birkhead and Biggins 1987). If this pattern *were* true, then EPF behavior should be rampant in the Tropics, where long breeding seasons and high nest predation rates produce asynchronous female fertility relative to the temperate zone.

On the contrary, however, EPF behavior is tied to synchrony, not asynchrony, in female fertility (Stutchbury and Morton 1995). When DNA fingerprinting results were coupled with an analysis of relative testis size (Harvey and Harcourt 1984, Møller 1991a) to assess the importance of extra-pair fertilizations, a dramatic difference in breeding synchrony was found between temperate zone migratory songbirds and tropical songbirds that defend territories year-round. The average proportion of females in a population that are fertile on the same day of the breeding season can be compared with a synchrony index (Kempenaers 1993). This index for temperate zone species with high EPF rates ranged from 17 to 58%. The synchrony index for a population of Dusky Antbirds in Panama was 8%, and none of the 12 broods analyzed with DNA fingerprinting had extra-pair young (Fleischer et al. in press). Breeding tropical songbirds had significantly smaller testes than did temperate zone songbirds known to have high EPF rates, but testis size did not differ significantly between tropical birds and temperate zone songbirds with low EPF rates. DNA studies combined with relative testis size information support the view that synchronous female fertility promotes the evolution of EPF behavior as a breeding strategy.

Female fertility is more synchronous in temperate than tropical areas, suggesting that extra-pair mating strategies may differ significantly on a latitudinal basis. Specifically, the short temperate breeding season coupled with synchronous arrival and egg laying should favor both the evolution of EPF behavior and behavioral correlates, such as semicolonial settlement patterns, that favor EPF success in males *and* females. It is equally important, from our perspective on song evolution, that females can exert much control over extra-pair mating behavior by selecting or rejecting males, including their own social mates (S. M. Smith 1988, R. H. Wagner 1991, Kempenaers et al. 1992, Lifjeld and Robertson 1992).

Birdsong in temperate passerines possesses two features that, while not entirely unique, differ strongly from the situation in nontemperate regions: song is strongly tied to breeding territoriality, as described above, and the breeding system itself is adapted to favor extra-pair reproduction. Recent DNA fingerprinting studies on Hooded (*Wilsonia citrina*), Wilson's (*W. pusilla*), and Kentucky (*Oporornis formosus*) Warblers found EPF rates of around 50% (Stutchbury et al. 1994, Bereson et al. in press, M. V. McDonald, J. Rhymer, R. C. Fleischer unpubl. data), matching those found earlier in Indigo Buntings (*Passerina cyanea*), Purple Martins (*Progne subis*), and other species (E. S. Morton et al. 1990, Westneat et al. 1990, Birkhead and Møller 1992). Furthermore, the EPF rates were highest for first nesting attempts, when female fertility is most synchronized (Stutchbury et al. 1994). This pattern suggests that temperate zone birdsong is designed to function in an EPF breeding system, wherein a male successful in EPFs garners more than 50% of his fitness from such fertilizations. Unsuccessful males lose fitness to successful males. If we categorized temperate birds within the range of breeding systems found in tropical birds, most would be termed lek breeders (see R. H. Wagner 1993)! The design features of temperate zone birdsong must promote extra-pair mating and help guard paternity. Because most tropical passerines are asynchronous breeders, the relatively synchronous, EPF-adapted temperate species are thus not representative of birds in general.

Determining the features of birdsong adapted specifically to a high EPF system should provide a rich avenue for future research. To illustrate, I return to an earlier example from paruline warblers with multisong repertoires. Males sing a single song in a repeat mode so that unpaired females can assess the individual male's territorial quality and health. I now amend the earlier discussion: by singing one individualistic song, a male is identified individually each time he sings, enhancing females' quests to determine the amount of time he is able to spend singing. The emphasis is on *females,* not the male's one social mate. A further amendment involves switching to serial mode, a repertoire of songs shared with neighbors in which a male threatens neighboring males through distance ranging (E. S. Morton 1982), keeping them from his fertile mate as much as possible (Lemon et al. 1987, Godard 1991, see also Spector 1992).

Unlike the temperate zone with its climatic constraints, the Tropics offer species with both high and low breeding synchronies for direct comparison. The Clay-colored Thrush (*Turdus grayi*), for example, breeds synchronously in Panama (synchrony index of 39%) at the onset of the dry season (E. S. Morton 1971). All the females within a population lay their first clutches within 7–14 days, much like the average degree of synchrony in temperate zone species (Stutchbury and Morton 1995), but populations separated by less than 10 kilometers may begin breeding a month or more apart (E. S. Morton 1971, 1983, Dyrcz 1983). This species has a large relative testis size, suggesting a high rate of EPFs (Stutchbury and Morton 1995). Males start singing simultaneously in local populations, beginning in a dawn chorus at 5:00 A.M. and continuing throughout the day for the duration of the three- to four-month breeding season. Each male has an individually distinctive song, which suggests that the "monogamous" males are also trying to attract other females for copulation. Singing begins with an abruptness and daylong vigor reminiscent of a temperate zone spring. Also, males sing from elevated perches, a characteristic of many temperate zone birds (R. H. Wiley and Richards 1982). Sex roles in this thrush are markedly different, with males singing and females building nests and incubating. Males defend against other males that attempt to sing on their territories, but nonsinging males and all females are tolerated. Feeding areas are not defended. Adults wander widely to feed on fruit or to hunt for invertebrates for their nestlings. The large testis size and synchronous female fertility in Clay-colored Thrushes predict an abundance of extra-pair fertilizations and support the idea that breeding synchrony, not simply latitude, is responsible for the evolution of EPF behavior. Indeed, preliminary DNA fingerprinting results show three of eight thrush broods (38%) had extra-pair young (B. J. Stutchbury, Morton, and W. H. Piper unpubl. data). Future research should add more taxonomic diversity to the comparative database on singing and EPF behavior. At this point, the strong similarity to temperate bird singing and the sex role divergence in the Clay-colored Thrush suggest similar underlying selection pressures.

Song function and evolution in tropical birds with low breeding synchrony are likely to emphasize territorial maintenance, which favors sex role convergence, whereas song in temperate or tropical species with concentrated female fertility emphasizes female assessment and guarding paternity, which favors sex role divergence. What might this idea predict with relation to song structure, function, and variability?

Factors Associated with Ranging, Comparing Tropical and Temperate Birds

Ranging theory offers several hypotheses to be tested. Any general theory of birdsong evolution must be directed to the level at which selection operates on birdsong—the level of individual conspecific birds interacting through song and other vocalizations. For example, stating that type A songs of a paruline warbler attract mates (i.e., function intersexually) is not an adequate explanation for their function because the explanation is not at the level of individuals interacting. The explanation I suggested above for the existence of repeat (or type A) songs allows for experiments to test the competitive ability of males, relative to one another, to attract females.

Likewise, it is important to know precisely what birds are perceiving when they listen to conspecific song. It is not adequate to say simply that a song is detectable. Ranging (E. S. Morton 1982, 1986) is based on the realization that frequency-dependent sound propagation, or the distance to which a song is *detectable,* is not as important as *how close a singer seems to be* to the perceivers of its songs. Singer closeness is judged based on how degraded the song has become during its travel rather than on how loud it is. Degradation is caused by reflections, reverberations, and differential attenuation of a song's component frequencies (Richards and Wiley 1980, Richards 1981). A loud song, detectable over a long distance because of high source amplitude, may still be ignored by listening birds if it has become degraded.

The only example known of a correlation between frequency-dependent sound propagation and birdsong was found in tropical forest near the ground. Sound frequencies at 1500–2500 Hz propagate particularly well near the ground, and the birds there have song frequencies near 2200 Hz (E. S. Morton 1975). Furthermore, tropical forest birds sing from where they forage even if they forage on the ground. But the song frequency–singing height–attenuation correlation in this case was due to the great acoustic homogeneity (temperature, humidity, and wind gradients are almost nonexistent under the canopy; Allee 1926) found near the ground in tropical forests. In this special case, factors affecting frequency-dependent sound propagation and sound degradation converged and the birds conformed to simple acoustic attenuation rules (E. S. Morton 1986).

Temperate zone forests are acoustically more heterogeneous. Consequently, temperate zone forest birds, and birds living in windy and sunny open habitats everywhere, sing from higher perches or while in flight. But their goal is not to get better frequency-dependent sound propagation to increase detectability; rather, it is to avoid acoustic degradation—to sound closer to perceivers.

Birds do not perceive general aspects of degradation to make judgments about a singer's distance (E. S. Morton 1986). They assess distance from conspecific birds by comparing the degree of degradation of incoming songs with their own remembered undegraded version. Therefore, if both singer and assessor have the sound in their repertoires, ranging can occur.

Selection favors the ability of singers to manage the behavior of assessors (Hennessy et al. 1981, Owings and Hennessy 1984), not to provide them with abstract information. Assessors, of course, "want" to be able to range and thus ignore songs that are not a threat to them. The attempt by singers to manage assessor behavior creates what I call the listener-singer dichotomy with reference to birdsong (E. S. Morton 1986). Thus the interests of a singer and those of his listeners often differ. A male would surely "hope" that his singing keeps other males from fertilizing his mate (threat, sound close by matched countersinging), but his mate is making decisions about whether to accept extra-pair copulations (assessing amount of singing). A Carolina Wren in midwinter would not "want" to interrupt its foraging to defend its territory against a singing male that appears to be, but is not, close (listener does not have song in repertoire; E. S. Morton 1982, Shy and Morton 1986, Strain and Mumme 1988).

The results of singer-listener conflicts weigh heavily in differences between tropical and temperate birds in singing behavior because neighborhood stability and EPF behavior differ latitudinally and are important influences on the ways singers attempt to manage listeners' behavior. For example, in the Dusky Antbird, the only nonoscine passerine so far tested for its ability to range, both sexes are able to range (E. S. Morton and Derrickson in press). Dusky Antbirds use ranging to provide "honest" distance cues to listeners because they use song to defend their territories against intrusions by same-sex individuals. Because of low breeding synchrony, the cost to males intruding into other territories is likely to be higher than any payoff in fitness (Stutchbury and Morton 1995). The lack of extra-pair fertilizations in Dusky Antbirds suggests that females do not pursue extra-pair copulations, for unknown reasons. Therefore, EPF behavior is not likely to evolve; its absence in turn allows other sources of selection to favor sex role convergence.

Sex role convergence in communication may also be manifest in plumage. It is no coincidence, for example, that female Cuban Red-winged Blackbirds not only sing "male" songs and have a monogamous pairing system but are also black, like males (Whittingham et al. 1992). If female fertility is asynchronous in Cuban Red-winged Blackbirds, I would predict that EPFs are rare.

Stable territory neighborhoods promote the use of a management strategy that provides assessors with honest distance assessment through ranging. In such neighborhoods, all of a bird's life history, not just one season, is tied to its territory ownership. If a bird defends a territory, challengers can assume that it has high resource-holding potential (G. A. Parker 1974). A song shared and rangeable by all members of the population would therefore be favored by selection because such a song says, "I am here, 10 meters away and getting closer with each rendition" and is threatening when coupled with high resource-holding potential. Whether it be a California coastline or a tropical forest, stable neighborhoods are correlated with dialects in song-learning species, and with shared songs attained through genetic control in nonoscines. Because stable neighborhoods are common in the Tropics, most oscines there exhibit dialects.

Tropical birds offer an almost untapped source of research questions that differ qualitatively from those offered by the temperate zones. White-breasted Wood-Wrens (*Henicorhina leucosticta*) provide yet another example. In Panama, females share a dialect repertoire of 4–5 song types, whereas males have large repertoires (>30 song types), very few of which are shared by adjacent males (Morton unpubl. data). Could male turnover be much higher than female turnover, producing a stable neighborhood for females but an unstable one for males?

Conclusions

Song use in tropical and temperate avian breeding systems differs fundamentally in four respects: (1) In many tropical species, singing occurs during both breeding and nonbreeding periods, characteristically all year long. (2) In temperate latitudes female fertility is generally synchronous because of climatic constraints, resulting in selection favoring extra-pair fertilization; female fertility may be either synchronous or asynchronous in tropical birds because of the potentially long breeding period. (3) Asynchronous female fertility in the majority of tropical passerines has reduced selection for EPF behavior. When coupled with year-round territoriality, in which either males or females may defend territories alone before acquiring a mate, low EPF behavior has led to convergence in sex roles, including long-distance territorial defense and sometimes mate attraction via song in females. (4) Stable territorial neighborhoods in the Neotropics favor the use of ranging to threaten conspecific birds, resulting in shared songs and "honest" distance cues provided by singers; shared songs may or may not be learned.

These latitudinal differences need more emphasis in future research on birdsong, particularly on the effects of demographics on song evolution (stable vs. unstable neighborhoods). In the Tropics, female singing is but one element of sex role convergence, but it is a major latitudinal distinction. The suggestion that low EPF behavior underlies sex role convergence in tropical birds in turn provides a new conceptual view to contrast birdsong evolution in the temperate zone. I believe that the high incidence of EPF behavior in temperate zones has influenced the evolution of birdsong there, and that fundamentally different forces operate in temperate and tropical zones.

Acknowledgments

I thank D. E. Kroodsma, D. Spector, and B. J. Stutchbury for excellent suggestions for improving the manuscript, and K. C. Derrickson, R. Fleischer, A. Sangmeister, and C. Tarr for research collaboration on the Dusky Antbird. The Smithsonian Tropical Research Institute provided logistical help in Panama. Much of the research discussed in this chapter was sponsored by a Scholarly Studies Grant from the Smithsonian Institution.

15 Study of Bird Sounds in the Neotropics: Urgency and Opportunity

Donald E. Kroodsma, Jacques M. E. Vielliard, and F. Gary Stiles

Research on bird vocalizations most often focuses on songbirds of the north-temperate zone. At least 15 of the chapters in this volume, for example, are almost exclusively about songbirds, and only 3 focus on nonpasserines (Chapters 6, 9, and 13). The remaining chapters address more general issues that encompass all taxa. The overall geographic bias is also illustrated by a perusal of the addresses of the contributors to this volume; it would seem that study of avian bioacoustics is reserved primarily for North Americans and Europeans, and what they write about the Tropics is derived from brief forays outside the Holarctic Zone.

But the urgency and opportunity for studying vocal signals in birds is unparalleled in the Tropics, especially in the Neotropics, where the diversity of species is truly extraordinary (e.g., Haffer 1990) and much less known than in the Old World Tropics. The urgency lies in the severe and mounting threats to many natural habitats and their birds, such that many species might be lost before we can even document their existence, let alone study them adequately (Bierregaard and Lovejoy 1989, Collar et al. 1992, Prum 1994). Bird sounds provide the most efficient means by which the complex avifauna of the Neotropics can be surveyed (see T. A. Parker 1991, Riede 1993), and those sounds, when recorded and archived appropriately (Kroodsma et al., this volume), also provide excellent systematic characters for understanding species limits and relationships (see, e.g., Braun and Parker 1985). For the vast majority of species, essentially nothing is known about their ontogeny, vocal repertoires, geographic variation, how vocalizations are used in communication, or even the barest essentials of their life histories. Almost every evolutionary experiment imaginable seems to have been played out among Neotropical species, and countless opportunities exist for studying every aspect of acoustic communication.

In this chapter, we encourage amateur and professional students of bird vocalizations to face the challenges and opportunities of the Neotropics (see also Buckley et al. 1985). Those who accept the invitation can contribute to our understanding of bird diversity and may well widen our horizons regarding strategies of avian communication. Answers to many questions based on studies of a few temperate zone species may well need revision, and numerous new questions are undoubtedly waiting to be asked (Morton, this volume).

The Urgency

Neotropical Biodiversity

Avian diversity in the Neotropics remains to a large extent unknown and probably vastly underestimated (see Prum 1994). New species continue to be described, especially from South America, including owls (Fitzpatrick and O'Neill 1986, Koenig and Straneck 1989, Vielliard 1989), nightjars (Lencioni-Neto 1994), parakeets (Ridgely and Robbins 1988), antbirds (Fitzpatrick and Willard 1990, S. M. Lanyon et al. 1990), antpittas (Schulenberg and Williams 1982, Graves et al. 1983, Stiles 1992), cotingas (Robbins et al. 1994), tapaculos (B. M. Whitney 1994b), canasteros (Vielliard 1990b), wrens (T. A. Parker and O'Neill 1985), and tanagers (Schulenberg and Binford 1985), to mention only a few.

Perhaps even more important than the new species is the growing realization that many allopatric populations previously considered subspecies are entitled to species status (whether one applies the biological or phylogenetic species concept). Current classifications actually conceal much of the avian diversity of the Neotropics. Through most of the twentieth century, "every isolated 'species' was . . . scrutinized for the possibility that it was simply a 'geographic representative' of some other species, in which case it was reduced to the rank of subspecies" (Mayr 1982, p. 594). Through the efforts of Hellmayr (1929), J. T. Zimmer (1926), J. L. Peters (1931–1986), and Meyer de Schauensee (1966), all authors of important surveys of Neotropical bird classification, this philosophy has played a dominant role in shaping our understanding of Neotropical avian diversity. Unfortunately, subspecies are often viewed by temperate zone–based ornithologists as an unimportant category of diversity (e.g., see discussions in A. R. Phillips 1981, Parkes 1982, E. H. Miller and Scudder 1994), and information for contributors to ornithological journals actually discourages subspecific identification (Anon. 1992). As a result, the broadly defined polytypic species of, say, Hellmayr or Meyer de Schauensee have tended to divert attention from the active speciation processes within them.

Careful scrutiny of such wide-ranging polytypic species has repeatedly revealed that they comprise two or more independent evolutionary units that have definitely or probably obtained species rank. For example, Meyer de Schauensee (1970) recognized only one species of Slaty Antshrike (*Thamnophilus punctatus*), and Ridgely and Tudor (1994) recognized two; in a recent study of this complex, however, Isler et al. (in press) found that species status was warranted, using conservative criteria, for six to eight allopatric populations. The Blue-tailed Emerald (*Chlorostilbon mellisugus*) of J. T. Zimmer (1926) was recently shown to comprise at least eight allospecies (Howell 1993, Stiles 1996). Willis (1992a) reported that three, not two, sibling species of antthrushes (*Chamaeza*) occur in eastern Brazil; and not one but two species of *Microcerculus* wrens are found in Costa Rica (Stiles 1983, 1984). The White-fronted Manakin (*Pipra serena*) comprises at least two distinct phylogenetic or biological species (Prum 1994), as does the poorwill formerly named *Nyctiphrynus ocellatus* (Robbins and Ridgely 1992). The Streaked Saltator (*Saltator albicollis*) is two species, not one (Seutin et al. 1993). "There are dozens of other legitimate examples of differentiated allopatric taxa that are currently recognized as subspecies in the Neotropical suboscines alone (e.g., *Chloropipo holochlora litae, . . . Laniisoma elegans buckleyi, Machaeropterus regulus striolatus, Pyroderus scutatus occidentalis, Rupicola peruviana sanguinolenta; sensu* Snow 1979)" (Prum 1994, pp. 699–700). Furthermore, molecular differentiation among many isolated subspecies, and even among unrecognized populations, is often much greater in Neotropical species than among species of Nearctic passerines (Capparella 1988, 1991, Hackett and Rosenberg 1990, Hackett 1993).

Conversely, of course, a number of currently recognized morphological species may be only geographical races of a single biological species. The Golden-crowned Warbler (*Basileuterus hypoleucus*) and the Yellow-bellied Warbler (*B. culicivorus*) are probably conspecific, for example, because they interbreed freely at contact zones, seem to have the same song forms, and respond fully to the other's songs and calls (Silva 1991). Matching our taxonomy with the true biodiversity in the Neotropics remains a major challenge.

In the absence of data on behavior, vocalizations, ecology, and biochemical characters, the taxonomy of Neotropical birds continues to be based, of necessity, mainly on plumage characters. Many recent studies have revealed that speciation may not be accompanied by much differentiation in plumage, however, and many good species were previously recognized only as subspecies by this criterion. Very often, the first clue to the true status of these species was provided by a difference in vocalizations, which in turn prompted the discovery of morphological, behavioral, or biochemical differences; examples include *Empidonax* flycatchers (R. C. Stein 1963), *Formicarius* antthrushes (Howell 1994), and *Microcerculus* wrens (Stiles 1983, 1984). The experimental analysis of vocal differences played a major role in permitting W. E. Lanyon (1967, 1978) to unscramble the taxonomy of the very difficult tyrannid genus *Myiarchus*. Because recordings of bird sounds can be obtained from many members of a population more easily and efficiently than can, say, museum specimens or blood samples for DNA work, the study of vocalizations is a quicker and more effective way to pinpoint possible problems of species-level taxonomy (see also R. B. Payne 1986).

Systematists who study songbirds, which learn their songs, sometimes despair at the high levels of geographic variation, often expressed as local dialects. Yet, at least in the Holarctic region, vast geographic areas often contain a homogeneity of song that betrays past evolutionary histories (Martens, this volume). In North America, songs of Marsh Wrens (*Cistothorus palustris;* Kroodsma 1989d) and

Winter Wrens (*Troglodytes troglodytes;* Kroodsma and Momose 1991) provide two examples. In those species, as in the *Microcerculus* wrens discussed by Stiles (1984), certain aspects of the learned song and its manner of delivery are so distinctive that the differences identify different phylogenetic species. In north-temperate zones, these "regiolects" (Martens, this volume) seem to be a consequence of the postglacial expansion of populations. Glacial effects were indirect in the Tropics, and the frequency of such regiolects there remains to be determined.

Nonlearned vocalizations are especially useful as an aid in identifying evolutionary units in the suboscines and most nonpasserines (excluding, at the least, most hummingbirds and parrots). Among the suboscines, for example, all the evidence suggests that vocal learning does not occur (Kroodsma, this volume), but several caveats are important here (West and King, this volume). Song development has been studied in only three suboscines, all tyrannids of the north-temperate zone. Certain tropical flycatchers (e.g., *Mecocerculus leucophrys*) have more complex vocabularies that warrant further study in this regard. Furthermore, no one has raised suboscines from the egg to demonstrate unequivocally that young suboscines never need to hear adults. Nevertheless, several factors combine to convince one that vocalizations are reliable cues of evolutionary units among suboscines: the coevolution of vocalizations and genes among *Empidonax* flycatchers (N. K. Johnson and Marten 1988); the coevolution of plumage, morphology, and vocalizations among woodcreepers (Marantz 1992); the lack of vocal learning after leaving the nest (Kroodsma, this volume); and the distinctly different brains of suboscines and oscines, which are presumably related to the differences in vocal development (Nottebohm 1980b, Kroodsma and Konishi 1991). It is the study of bird sounds, easily and quickly recorded from so many individuals, that holds the greatest promise for a rapid assay of the species-level taxonomy of flycatchers, antbirds, woodcreepers, tapaculos, cotingas, and other suboscines (see W. E. Lanyon 1978, Braun and Parker 1985, Vielliard 1990a, B. M. Whitney 1992, 1994a, Willis 1992b, B. M. Whitney and Pacheco 1994), as well as other groups in which vocalizations appear not to be learned (e.g., owls, doves, woodpeckers: Marshall 1978, Copi and Vielliard 1990).

Vocalizations can be the key not only to identifying taxonomic problems but also to censusing Neotropical bird communities. Nowhere else in the world is such a remarkable diversity of birds found (Ridgely and Tudor 1989, 1994). In Amazonia, for example, 500 or more bird species occur at some localities (Haffer 1990). Avifaunal complexity increases as barriers such as rivers (Haffer 1992) and mountain ranges (Janzen 1967) restrict dispersal and increase endemism. In complex tropical forests, most birds are much easier to hear than to see, and attempts to census birds in such habitats will be woefully inadequate if the observer is unable to recognize vocalizations. Only through the use of vocalizations will truly comparative studies of the diversity and richness of bird communities be possible. Those of us who listen to and tape-record birdsong, amateurs and professional biologists alike, have extraordinary opportunities to contribute to the knowledge of the evolutionary history of this most complex avifauna.

Accompanying these exciting opportunities is an urgent necessity. Understanding species limits for birds has enormous consequences for conservation biology (see, e.g., Eldredge 1992, Zink 1993) because bird-oriented conservation efforts can only be as good as the estimates of avian diversity. Given our incomplete knowledge of Neotropical diversity, it is likely that species are disappearing before their existence has been documented. Furthermore, given the polytypic nature of so many currently recognized species, it is likely that our failure to appreciate the true nature of differentiation among populations will lead to the loss of many evolutionary units. Two of the Slaty Antshrike groups occur, for example, in valleys of northern Peru where no biological refuges exist (Isler et al. in press). Many Andean forms are faced with imminent extinction with hardly an obituary to chronicle their demise, perhaps because, at least in part, the forms are not currently recognized as species. As habitats are fragmented (see, e.g., Bierregaard and Lovejoy 1989) and then disappear, the birds (and other flora and fauna, of course) are increasingly threatened, as was so well documented by Collar et al. (1992). It was recognition of the urgency of these matters that led Conservation International to establish its Rapid Assessment Team (RAP Team), which uses bird sounds as the key to chart avian diversity (T. A. Parker 1991). Thus, although a better understanding of bird systematics is a worthy goal in itself, that understanding is especially critical for helping to establish conservation priorities.

Archives of Neotropical sounds. Several archives already have sizable collections of recordings from the Neotropics (see Appendix). The Library of Natural Sounds at Cornell University has a substantial collection, based largely on the recording efforts of Paul Schwartz and Ted Parker. Significant holdings are also found in the Arquivo Sonoro Neotropical (Campinas, Brazil), the British Library of Wildlife Sounds (London), and the Bioacoustics Laboratory and Archives (University of Florida). These substantial collections, supplemented by many private collections, could serve as the basis for a multitude of scientific studies.

Coverage of nearly all species is inadequate, however. Many species have yet to be recorded, and many others are represented by only a single recording. Even for well-recorded species (e.g., M. Isler and P. Isler have 130 recordings of *Thamnophilus doliatus*), huge geographic holes remain in the sample. Well-organized efforts to increase samples from little-known species and from poorly sampled regions will help tremendously in the quest to understand avian diversity in the Neotropics.

Coverage is inadequate in at least two other ways as well. First, "songs" are overrepresented. The nonlearned songs of suboscines may be especially useful for systematic analysis, but nonlearned calls—not learned songs—may be more useful for analyzing oscines and hummingbirds, for example. It is important in these efforts that homologous vocalizations be identified (Vielliard 1992) and that the ontogeny of those vocalizations be understood. Some vocalizations of tropical birds are given only rarely, too, such as the vocalization given by a Dusky Antbird (*Cercomacra tyrannina*) that has lost its mate (Morton, this volume). Second, disappointingly little information exists for the behavioral context in which the recordings were obtained. If Neotropical species are to be identified *and* understood, the quality of the information accompanying the recordings must improve.

Resources for research in the Neotropics. A number of resources are available for those who would work in the Neotropics. The *Directory of Neotropical Ornithology* (Rosenberg and Wiedenfeld 1993) provides a wealth of information on government agencies, universities, and other organizations; personal contacts in the different countries and in North America are also provided. Lists of field stations where research can be based are available for several countries (Castner 1990).

Field guides are increasingly available for countries throughout the Neotropics (e.g., Costa Rica, Panama, Venezuela, Colombia, Brazil). Ridgely and Tudor's (1989, 1994) guides to the oscines and suboscines of South America provide broad coverage and can be supplemented by books with more restricted coverage (e.g., Isler and Isler 1987). More practical and specific guides are also available for several countries (e.g., Forrester 1993 for Brazil, C. Green 1991 for Ecuador), as are a variety of audio guides (Kettle 1991). Hardy's pioneering efforts with his Voices of the New World series are invaluable learning tools that include tapes of tinamous, cuckoos, pigeons and doves, nightjars, owls, trogons, woodcreepers, mimids, wrens, vireos, thrushes, wood-warblers, jays, and others (ARA Records, P.O. Box 12347, Gainesville, Fla. 32604). The Cornell Library of Natural Sounds (LNS) has produced tapes for eastern Ecuador, the Peruvian rainforest, and a Costa Rican cloudforest, and will soon have tapes for parrots, antbirds, and other Neotropical groups. (A fine selection of Brazilian bird sounds is also widely available as the background to the movie *Out of Africa.*)

What to do? A variety of efforts can be made to improve our understanding of Neotropical bird sounds (see also Strahl 1992, Foster 1993, and Jenkinson 1993, who urge biologists to work in the Neotropics). A huge role exists for amateur recordists, for example, who are playing an increasing role in recording birds and donating their tapes to major archives. Techniques of sound recording can be readily learned at workshops, such as the week-long course provided annually by the Library of Natural Sounds, and several archives have recording equipment available for loan. Working in association with a major archive or with other professionals, amateurs can be directed to species or locations of greatest need.

Professional biologists who do not study bird sounds can also contribute. These biologists often visit remote regions, and they have great opportunities to enhance collections from poorly sampled regions. Again, training and equipment can often be arranged from major archives.

Whenever possible, attempts should also be made to collaborate with local biologists (Foster 1993). Local scientists and naturalists are often extremely knowledgeable about local birds and their vocalizations; collaboration with local experts can therefore both greatly enrich a research project and improve the probability of its success. Such collaboration can also stimulate local scientists and naturalists to record and publish their own findings. Also, access to equipment and other technology is often limiting in Latin America, and collaborative efforts can help provide resources needed to do the research.

In short, we encourage both professionals and amateurs, and both visitors to and residents of the Neotropics, to focus on bird sounds and thereby help to document some of richest bird communities on this planet. The bird communities, in turn, serve as rapid indicators of other flora and fauna, which are often less easy to survey. As Conservation International's RAP Team recognizes, it is the sounds of birds that can be so important in surveying and saving appropriate regions of the Neotropics (T. A. Parker 1991).

Extraordinary Opportunities

In the Neotropics, the opportunities to make significant and novel contributions to the study of bird communication are staggering. To whet the curiosity of biologists, especially those early in their careers, we here provide a brief selection of potentially interesting problems worthy of further study.

Patterns of speciation in this diverse avifauna deserve further study. To what extent have mountains (Janzen 1967), rivers (Capparella 1988, 1991), and isolated refugia (Haffer 1992) contributed to the development of the avifauna? How has the complex topography of the Andes and the recency of its final uplifting affected speciation? The forming and breaking of bridges with Central America and other changes associated with sea level fluctuations have contributed to the immense array of taxonomic problems at the species and genera levels. Hints are provided largely by distributional limits of different taxa and by congruence in geographical patterns of speciation among different groups (Haffer 1985, Cracraft and Prum 1988, Nores 1992, Brumfield and Caparella in press). What role do vocalizations play in speciation, and how well do vocalizations now reflect these historical processes?

The role of interspecific interactions in signal evolution can be studied nowhere better than in species-rich tropical communities (e.g., Terborgh et al. 1990). What constraints on signals within such rich assemblages are derived from the necessity for species identification? Does membership in mixed-species flocks cause convergence or divergence in vocalizations, and in which vocalizations? How does the overall sound environment affect signal evolution and temporal patterns of singing? Also, ranges of congeneric species often abut one another, as if these species in some way limit one another's distribution (Terborgh 1971). Three species of antthrush (*Formicarius*), for example, replace each other altitudinally on the Caribbean slope of Costa Rica. Their songs are different, though certain calls seem indistinguishable; each species also seems to respond to the songs of the others, and they seem to be interspecifically territorial where they co-occur. How do such zones of contact influence evolution of vocal signals? Do vocalizations converge at zones of contact, for example, as they sometimes do among temperate zone songbirds (e.g., Indigo and Lazuli Buntings, *Passerina cyanea* and *P. amoena:* Emlen et al. 1975)? Or do signals diverge, thus offering good examples of character displacement? Or do the birds simply learn to recognize each other's vocalizations and adjust behaviorally? The patterns and roles of vocal signals in these interactions within and between communities are unknown.

Duetting is far more frequent in tropical birds than in temperate zone birds (e.g., Farabaugh 1982, Morton, this volume). The quantity and quality of females' contributions to the duets and their precise timing can be remarkable. Increased singing by females is probably related to year-round residents' increased investment in a mate and territory, but why is there such a diversity of intra-pair communication strategies among groups? Why are the vocal exchanges of some species rather ordinary, and why are some wrens (e.g., *Thryothorus, Campylorhynchus*) such accomplished duetters, at least by temperate zone standards? The full song chorus of the high Andean Rufous Wren (*Cinnycerthia unirufa*) is truly exceptional: up to 10 or more individuals in a cooperative group each contribute different song phrases to the joint effort (much like members of a symphony orchestra tuning instruments before a performance). Questions abound: Why is there a high diversity of songs among individuals? How many different songs does each individual contribute to the chorus? Do male and female contributions differ? How is it decided which song is used and when? What is the ontogeny of such duetting behavior? What are the intergroup functions of such a display, or the consequences of the diversity of songs? How does the behavior differ in the related Sepia-brown Wren (*C. peruana*)? Among some other species, the ontogeny of song and the extent of geographic variation seem to differ for males and females (Levin 1988). Why? Our *whys* should fuel a host of descriptive studies of how vocalizations are used by different species in different social environments in the Tropics.

Some tropical species engage in extensive *mimicry.* The thrush *Turdus lawrenceii* imitates a large number of other species (Vielliard 1982, Hardy and Parker 1985). The female Thick-billed Euphonia (*Euphonia laniirostris*) uses the mobbing calls of other species when her nest is threatened, apparently attracting individuals of the model species to help her ward off predators at the nest. The male *Euphonia violacea* also includes the alarm calls of other species in his song (Snow 1974, E. S. Morton 1976, Remsen 1976). Other examples of mimicry probably remain to be documented. Will closer scrutiny of these species improve our understanding of the evolution of vocal mimicry (Baylis 1982)?

Song schedules of tropical birds are also puzzling. Some species sing all day long, but many sing only at dawn. Why are vocal displays in so many species restricted to a brief period at dawn? And why do so many flycatchers and woodcreepers have special songs used at dawn but not later in the day? Is the temporal sequence with which birds enter the dawn chorus more or less patterned than it is in temperate zones birds? (See also Staicer et al., this volume.)

General strategies of vocal development and, more specifically, how vocal learning evolved remain a puzzle (Nottebohm 1972, Kroodsma, this volume), but the diversity of tropical species could supply some answers. Among the thousand or so suboscine species, perhaps some differ in the extent to which they rely on the sounds of others for normal development. Hummingbirds and parrots may provide many new insights, too. In hummingbirds, vocal learning was first inferred from the microgeographic distribution of songs in the Little Hermit (*Phaethornis longuemareus;* Snow 1968, R. H. Wiley 1971). Similar data now exist for *Ph. eurynome, Ph. pretrei, Colibri serrirostris, Augastes lumachellus* (Vielliard 1983), *Ph. superciliosus* (Stiles and Wolf 1979), and for *C. coruscans* and *C. thalassinus* (S. L. L. Gaunt et al. 1994), and this learning ability was demonstrated experimentally with *Calypte anna* (Baptista and Schuchmann 1990). Patterns of cultural evolution in the songs of the lekking *Ph. superciliosus* (L. L. Wolf and F. G. Stiles unpubl. data) appear similar to those of the Indigo Bunting (Payne, this volume). Learning from neighbors is not universal among hummingbirds, however, because individuals of some species have complex, individually distinctive songs (e.g., many *Colibri* and *Augastes* species; see Vielliard 1983 for examples). Some parrots learn vocalizations in artificial environments (Pepperberg 1990), but one cannot assume that all parrots learn their vocalizations. The suboscines, hummingbirds, and parrots include a great variety of species among which to search for correlations between styles of vocal development and other life history parameters, an approach that will surely help inform us not only about strategies of vocal development but also about the evolution of vocal learning itself. Styles of vocal development need to be studied in other groups, too; the elaborate and prolonged songs of some jacamars (*Galbula* and *Brachygalba,* e.g.) are worth exploring for dialects, which would implicate vocal learning.

Sexual selection and the evolution of vocal displays are topics for which Neotropical groups offer countless opportunities. The cotingas are an intriguing group, for example (Snow 1982, Stiles and Skutch 1989, Trail and Donahue 1991): their life histories are diverse, and their plumage and visual displays are varied. Some species lek and have remarkable vocalizations (e.g., Capuchinbird, *Perissocephalus tricolor*), some do not lek and have less striking voices (e.g., White-browed Purpletuft, *Lodopleura isabellae*), and some do not lek but still have remarkable voices (e.g., berryeaters, *Carpornis*). What are the relationships among vocal displays, visual displays, and mating systems in these cotingas?

Hummingbirds, too, more than 300 species, offer many opportunities for study. The males typically advertise and court females, either singly or in leks. Hummingbird songs are often persistent and piercing, and males sing in a variety of social situations. In the genus *Phaethornis,* for example, *Ph. pretrei* sings in isolation; *Ph. eurynome* sings alone but often in aural contact with neighbors (as in an "exploded" lek); *Ph. longuemareus, Ph. superciliosus, Ph. guy, Ph. anthophilus,* and *Ph. griseogularis* sing on well-spaced perches and seem to maintain only aural contact with their neighbors (Snow 1968, R. H. Wiley 1971, Stiles and Wolf 1979); and *Ph. ruber* and *Ph. idaliae* occasionally sing alone but usually sing in "closed" leks in visual contact with at least one neighbor. The songs of *Eupetomena macroura* are especially fascinating because the males use a territorial song in isolation and a completely different song form in leks (Silva and Vielliard 1988). The same pattern occurs in male *Chalybura urochrysia,* which use a short, explosive song when holding flower-centered territories during the breeding season but a longer, softer, rather lilting song on feeding territories outside the breeding season. The organization and evolution of vocal and visual display repertoires among hummingbirds are largely unknown; good data exist only for *Calypte anna* (e.g., Stiles 1982) and a few species of hermits. Most studies, too, have focused on species that lek in the understory or in open habitats where close approach is possible; many hummingbird species that inhabit the forest canopy or middle levels (e.g., *Elvira, Lampornis, Eupherusa*), however, have very complex songs that have rarely, if ever, been recorded, much less studied in detail.

The range of complexity in hummingbird song is certainly intriguing. Many hummingbirds don't sing at all in the usual sense (Smith, this volume), such as North American members of *Selasphorus* and the two species of *Archilochus;* in Costa Rica, *Selasphorus flammula* sings, but *S. scintilla* does not. A wide range of complexity occurs within certain genera; for example, in Costa Rica, the songs of *Amazilia amabilis* of the Atlantic slope consist of repetitions of a single note or variants thereof, whereas those of *A. decora* of the Pacific slope (sometimes considered a subspecies of *A. amabilis*) are highly complex and variable; the songs of several other species (e.g., *A. tzacatl, A. rutila,* and *A. fimbriata*) are intermediate in complexity. Two species stand out as remarkable singers with long, elaborate, and loud songs: *Campylopterus excellens* of Oaxaca, Mexico, and *Sephanoides fernandensis* of the Juan Fernandez Islands (Stiles pers. obs.). Another strikingly complex song is that of *Threnetes ruckeri* on the Pacific slope of Costa Rica; the long, elaborate songs there are a striking contrast to the simpler, syncopated phrases of the same species on the Caribbean slope. The range of variation—among individuals, among populations of the same species, and among species—is poorly understood among hummingbirds, and careful study will yield insights into patterns of vocal development (e.g., extent of learning or improvisation), geographic variation, sexual selection, and a variety of other biological phenomena.

Mating systems involving intense sexual selection occur not only among cotingas and hummingbirds but also among other groups, such as manakins (Trainer and McDonald 1993, McDonald and Potts 1994) and flycatchers (Westcott and Smith 1994). How vocal behavior and its development are related to sexual selection in these social systems is largely unknown.

The role of vocalizations in complex animal societies can be studied nowhere better than in sedentary species in the Neotropics. Species with highly social behavior abound in the Tropics, and their vocal communication systems are especially intriguing. The Guira Cuckoo (*Guira guira*), for example, lives in social groups, has a large vocal repertoire, and combines these sounds to signal different probabilities of flight (Fandiño-Mariño 1989). Depending on the nature of the predator and the threat, different calls are used to elicit appropriate reactions (flying, freezing, dropping to ground). All the signals appear to develop from a single physical pattern seen early in life, and the derivations seem to follow the structural-functional rules proposed by E. S. Morton (1977). In the complexity of their use, these signals seem to rival the vocal communication systems found among some primates (Seyfarth et al. 1980).

Patterns of geographic variation among Neotropical birds also invite further study. In wrens of the genus *Microcerculus* (Stiles 1983, 1984), for example, both song and morphology change regionally in Costa Rica. But from Panama to northern South America, plumage changes but song seems not to. And in Peru, song changes dramatically but plumage does not. The voices of many flycatchers suggest regional differentiation, too, though data on morphology are lacking (e.g., songs of *Megarynchus pitangua* and *Myiozetetes similis* of Costa Rica are different from those of Colombia). The songs are so different among some populations of songbirds that it is difficult to believe the birds belong to the same species. The White-breasted Wood-Wren (*Henicorhina leucosticta*), for example, typically has a short phrase of four to seven clear notes, with successive phrases separated by several seconds or more, and males repeat the same phrase many times before switching to another (Stiles and Skutch 1989); on the Pacific slopes in the Choco, however, males have an incredibly loud, ringing crescendo, much longer and with a totally different cadence. The Orange-billed Sparrow (*Arremon aurantiirostris*) in Central America sings a song that consists of a medley of high, thin, slurred whistles (Stiles and Skutch 1989), but in the Magdalena Valley of Colombia the song consists of a sharp, insectlike double buzz. Widespread species such as the House Wren (*Troglodytes aedon*), which occurs throughout the Americas, would be worthy of close study. *Zonotrichia capensis* also occurs in a variety of habitats, from sea level to above 3500 meters, and from Mexico to Tierra del Fuego (Nottebohm and Selander 1972, Tubaro et al. 1993). Understanding the geographic patterns of vocalizations will not only help identify species limits, but will also contribute to a fundamental knowledge of how and why vocalizations change over both space and time (see Chapters 10–14, this volume).

Abundant opportunities exist in the Neotropics for comparative work, both in relation to well-studied North American species and within species-rich tropical groups. After all the vocal work on the Red-winged Blackbird (*Agelaius phoeniceus*) that has been done in North America, for example, perhaps the next step in understanding that species should be to study the several related *Agelaius* of South America (Whittingham et al. 1992). And the Neotropics offer a large number of *Turdus* species with a wide variety of vocal behaviors for comparisons with the well-studied *T. migratorius* of North America. Or consider the remarkable number of species in certain passerine groups within the Neotropics: antbirds, manakins, spinetails, tapaculos, flycatchers, woodcreepers, finches, seedeaters, swallows, tanagers, thrushes, warblers, wrens, and more. More than 50 South American genera have 10 or more species, and some genera have two dozen or more (e.g., *Synallaxis, Tangara, Sporophila;* C. G. Sibley and Monroe 1990). Furthermore, in many groups, more species are likely to be described in the future

(Isler et al. in press). Each species is in itself an evolutionary experiment simply waiting to be described and understood. The range of vocal behaviors among the tanagers and their finchlike relatives, for example, presents abundant opportunities; like hummingbirds, some tanagers have complex songs, but some gaudy species of *Tangara* have never been reported to "sing" at all. In these diverse evolutionary experiments among different lineages, how has the evolution of vocal signals proceeded? How do signals develop? How many signals exist? What are the relationships between repertoire organization and other life history variables? Have similar evolutionary experiments among different lineages produced similar results?

Unique groups, such as tinamous, provide a host of unanswered questions. Tinamous are an old group with many species believed to be related to the ratites (C. G. Sibley and Ahlquist 1990, Caspers et al. 1994). Sexual roles are reversed in this group; males incubate eggs and rear the chicks, and the (sometimes polyandrous) females compete for territories and mates. How is the diversity of the sounds produced by tinamous (Hardy et al. 1993) related to their social organization? How well does "song" structure (e.g., patterns of modulation) match patterns that would be expected in different habitats? Few tape recordings have been made of birds of known sex, and the vocal roles of males and females are essentially unknown. In some species, such as the widely distributed Grey (*Tinamus tao*) and Cinereous (*Crypturellus cinereus*) Tinamous, the songs are surprisingly uniform over considerable geographic expanses. In other species, however, such as the Great Tinamou (*T. major*), songs differ in pattern but not in tonal quality between the Caribbean lowlands (e.g., La Selva) and the Pacific slope (e.g., the Osa Peninsula) of Costa Rica. How and why vocalizations change over geographic space are unknown in this group (see also Hardy et al. 1993), but the roles of habitat structure and vocal development are key issues to be resolved.

Parrots, too, are a diverse but little-known group. More than 100 species of macaws, parrots, and parakeets occur in South America. As Farabaugh and Dooling (this volume) point out, relatively little has been published about the vocal behavior of these fascinating species in nature. The high intelligence of some is indisputable (see, e.g., Pepperberg 1990), but the role of learned vocal signals in natural social situations is largely unknown. Many species in the Psittacidae are endangered, and the opportunities to study some of them in nature are quickly disappearing.

Conclusions

We believe that the enormously diverse Neotropical avifauna represents the greatest set of challenges and opportunities for students of bird vocalizations for now and well into the twenty-first century. The possibilities for contributing to the knowledge of avian communication strategies are virtually endless, and the brief selection of interesting groups and questions we offer here barely scratches the surface. The taxonomy of Neotropical birds is still incompletely known, and the vocal repertoires of the vast majority of species are at best sketchily known. We are confident that a more thorough and systematic documentation of their vocalizations will not only produce a veritable flood of new questions of theoretical interest but will also provide a valuable tool both for rapid assessment of avian diversity and for its conservation.

Acknowledgments

This chapter contains many uncited personal observations by JMEV and FGS from their years of living in the Neotropics; references are provided where they are available. We thank those who offered invaluable advice during the preparation of this chapter: Greg Budney, Mario Cohn-Haft, Mort and Phyllis Isler, Curtis Marantz, Ted Miller, Gene Morton, Van Remsen, Mark Robbins, Tom Schulenberg, and Bret Whitney. We thank the chancellor of Universidade Estadual de Campinas, Brazil, for funding a collaboration between JMEV and DEK. For other funding, we thank the Brazilian Academy of Science and the National Council of Research (JMEV) and the National Science Foundation (IBN-9111666 to DEK). JMEV thanks collaborators Cecilia Copi, Maria Luisa da Silva, and Wesley R. Silva.

PART **IV** CONTROL AND RECOGNITION OF VOCALIZATIONS

Introduction

The focus in this section is on how vocal displays are produced and detected. The neuroendrocrine system that controls vocalization in birds has emerged as an excellent model for studying the interaction between brain and behavior, and this topic alone could fill an entire volume. The neural displays from the brain are translated into vocalizations by the syrinx. Syringes and morphological structures vary from species to species, but they also vary among individuals of a given population, and clues about the physiological and morphological constraints on signal structure may be available in birds' vocalizations. How these signals are detected by other individuals is the final link in this sequence. The task of detecting simple or complex displays in a noisy environment is challenging, so adaptations for detection and perception by receivers seem likely.

The first chapter in this section, by Eliot Brenowitz and Don Kroodsma, explores the neuroendocrine regulation of singing behavior. Topics include the neural basis of song learning, the development of sex differences in song regions of the brain, brain correlates of behavioral plasticity, and evolution of the neuroendocrine song control system. The authors review previous and current research and also outline an agenda for important future studies. To provide a fuller understanding of the evolution and diversity of neuroendocrinology and vocalization, Brenowitz and Kroodsma urge future researchers to rely less on traditional study species (the Island Canary, *Serinus canarius,* and the Zebra Finch, *Taenopygia guttata*) and instead use a variety of other taxa.

In his chapter, Marcel Lambrechts identifies some widespread yet enigmatic features of bird vocalizations. He argues that gradual drift in song structure and alterations in percentage performance time reflect constraints on the motor system rather than differences in motivation, as has been widely assumed. Constraints may thus play an important role in the organization of vocalizations. If constraints are important, then several components of song variation, such as song repertoires and the complexity of song types, may be adaptive mechanisms that increase the precision of singing. Lambrechts suggests research projects that will reveal how physiology and morphology might constrain characteristics of singing.

Georg Klump discusses how vocal displays are detected and recognized by birds in their naturally noisy environments. Birds must not only detect that a display has occurred, they must also determine specifically where it was emitted and by whom. Such tasks are not simple, and the consequences of inaccurate perception may be costly. Klump shows how we can advance this field of study by moving from artificially simple laboratory environments to complex and variable natural environments. Special factors to consider include the spectro-temporal pattern of masking background noise in the natural environment, the directionality of the bird's auditory system, and the previous experiences and memories of receivers.

Song is an important form of sexual communication in songbirds for both mate attraction and intermale competition, but we know little about whether male and female birds perceive song differently. Laurene Ratcliffe and Ken Otter investigate some proximate and ultimate reasons for suspecting that song recognition and perception differ between the sexes. They consider the benefits of accurate perception and discuss neurobiological differences between the sexes. Using current evidence for sex differences in song perception and their own research on Black-capped Chickadees (*Parus atricapillus*), Ratcliffe and Otter consider whether sex differences could shape song evolution. The authors discuss some of the difficulties in interpreting current approaches to studying song perception and suggest profitable avenues for future research.

The final chapter in this section considers a special class of display detection, neighbor recognition. Philip Stoddard critically reviews experimental approaches for testing neighbor recognition and reveals many shortcomings. Traditional field playback experiments do not permit discrimination between two biologically important situations: (1) the subject failed to recognize differences in displays, and (2) the subject recognized differences but did not respond differently to them. Also, reliable interpretation of experiments depends on consistent differences in the territorial threat level of different classes of singers. Where these differences vary unpredictably, playback experiments lose their power to explain patterns of recognition. Stoddard also shows that song sharing influences birds' ability to recognize neighbors, whereas repertoire size does not. Neighbor recognition appears to be universal, but Stoddard identifies some natural situations that need additional study.

16 The Neuroethology of Birdsong

Eliot A. Brenowitz and Donald E. Kroodsma

Animal behavior traditionally has been studied from either a mechanistic or a functional approach (e.g., Curio 1994). Over the past two decades, however, biologists have increasingly recognized that a complete understanding of behavior requires that these two perspectives be integrated. Physiological mechanisms constrain the evolution of behavior, and these same mechanisms evolve to regulate behaviors that influence survival and reproduction. This integrative approach to behavior is characteristic of the discipline of neuroethology, which has undergone much recent expansion.

Animal communication has been an especially productive topic of research for neuroethologists for several reasons. (1) Communication signals used in reproduction evolve in response to sexual selection and typically play a clear role in intrasexual competition and mate attraction. (2) These signals tend to be species specific in form, which provides opportunities for comparative studies of underlying mechanisms. (3) Communication signals are often highly stereotyped in structure. This stereotypy increases the likelihood of drawing clear connections between aspects of an animal's physiology and the regulation of discrete components of the behavior. (4) The evolution of signals in any sensory modality is subject to the laws of physics as they affect the production, transmission, and reception of these stimuli. We can therefore identify constraints that have influenced the evolution of signals. (5) Both the production and the detection of signals are controlled by discrete, specialized neural circuits. (6) Communication is commonly linked to reproduction, and both the behavior and the underlying neural pathways are therefore often sensitive to the influence of the sex hormones secreted by the gonads. These factors together make communication systems particularly amenable to study from an integrative approach, and it is thus not surprising that many of the most productive model systems in neuroethology focus on animal communication.

Birdsong has emerged as one of the leading model systems for neuroethologists. It shares the general benefits of animal communication described above, but song is also of interest because it is a learned behavior in songbirds, parrots, and hummingbirds. Great taxonomic diversity occurs in various aspects of vocal learning, including the time of life when it takes place, the extent to which song is imitated or improvised, the number of vocalizations that are learned (i.e., repertoire size), and whether it occurs in one or both sexes. Such diversity presents unequaled opportunities for comparative studies of the relationship between the structure and function of brain regions and vocal behavior. In this chapter we review the current status of research on the neuroethology of birdsong and raise issues that should be addressed by future studies.

The Song Control System in the Oscine Brain

The development and production of song in oscine passerines are regulated by a discrete network of interconnected nuclei in the brain (Fig. 16.1). This song control system is organized into two main pathways that play different roles. One circuit, the main descending motor pathway, regulates the production of song. This pathway consists of neuronal projections from the high vocal center (HVC) to the robust nucleus of the archistriatum (RA) in the forebrain, and from the RA to the tracheosyringeal part of the hypoglossal motor nucleus in the brainstem (nXIIts). Motor neurons in the nXIIts send their axons to the muscles of the sound-producing organ, the syrinx. When these motor neurons are stimulated, the syringeal muscles contract and change the tension on the tympaniform membranes, which act as the sound source; contraction of the syringeal muscles also changes the shape and length of the vocal tract, which can influence the frequency composition of sounds produced by the membranes. If nuclei in the motor pathway are inactivated, a bird may adopt appropriate posture and beak movements but produce no song. Activity of some neurons in this pathway tends to be synchronized with singing behavior (Nottebohm et al. 1976, McCasland 1987, Nowicki and Marler 1988, Suthers 1990, Vicario 1991a, Wild 1993, Vu et al. 1994).

The second main pathway, that of the anterior forebrain (Doupe 1993), is believed to be essential for song learning and recognition (see next section). This pathway consists of projections from the HVC to area X, then to nucleus DLM in the thalamus, from the DLM to the lateral portion of the magnocellular nucleus of the anterior neostriatum (LMAN), and finally to the RA (Nottebohm et al. 1976, Bottjer et al. 1989). In addition, LMAN neurons that project to the RA also project to area X, thus providing the potential for feedback within this pathway (Perera et al. 1995, Vates and Nottebohm 1995).

The Anterior Forebrain Pathway and Song Learning

The timing of song learning differs greatly among, and sometimes within, oscine species. At one extreme are species in which sensory learning is restricted to the first year of life, and songs do not change thereafter; such species are referred to as age-limited learners (Marler and Peters 1987). The other extreme consists of "open-ended" species in which song structure can change considerably

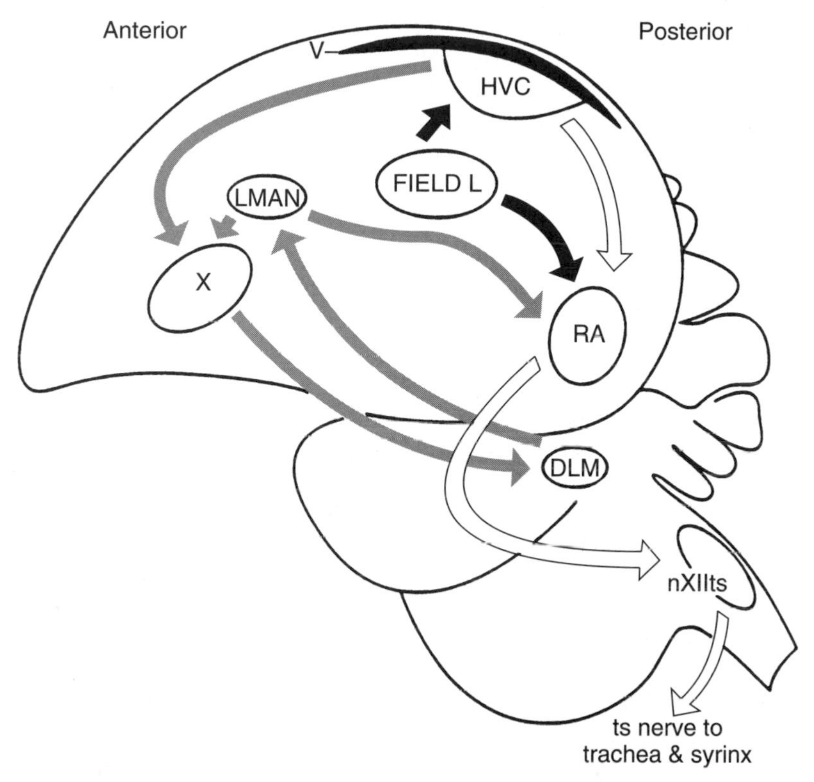

Fig. 16.1. A schematic sagittal drawing of the songbird brain showing anterograde projections of major nuclei in the song system. The white arrows indicate the main descending motor pathway for the control of song production, including the HVC, RA, and nXIIts. The gray arrows indicate the anterior forebrain pathway from HVC to RA, via X, DLM, and LMAN. LMAN also projects to area X. This pathway conveys auditory information within the song system and is important for song learning. Field L is a forebrain auditory region that projects to the HVC and RA (black arrows). Abbreviations: V, ventricle; HVC, high vocal center; RA, robust nucleus of the archistriatum; X, area X; LMAN, lateral portion of the magnocellular nucleus of the anterior neostriatum; DLM, medial portion of the dorsolateral nucleus of the thalamus; nXIIts, the tracheosyringeal part of the hypoglossal nucleus. Reproduced with permission from Nottebohm et al. (1990).

beyond the first year of life. Numerous other species show temporal patterns of song learning that lie between these extremes. The ecological correlates of such variation in the timing of song learning are poorly known and merit further study (e.g., Kroodsma and Pickert 1980, Kroodsma, this volume). These species differences are especially important to remember, however, when one is searching for relationships between the development of neural circuits and song learning.

Several kinds of evidence suggest that the anterior forebrain pathway plays a dominant role in the process of song learning (see Doupe 1993 for a more detailed review). Essentially all studies of song learning thus far have used the Zebra Finch (*Taenopygia guttata*), an age-limited learner, and, to a lesser extent, the Island Canary (*Serinus canaria*), an open-ended learner.

The first suggestion of the role of this pathway in song learning came from studies of the effects of lesions. Inactivation of the LMAN, DLM, or area X in adults does not disrupt previously crystallized song, whereas the same lesions in juvenile male Zebra Finches prevent the development of normal song (Bottjer et al. 1984, Sohrabji et al. 1989, Scharff and Nottebohm 1991, Halsema and Bottjer 1992). If the LMAN is lesioned in juvenile males, they produce songs with an aberrant but stable structure. In contrast, juvenile males with lesions in area X persist in producing songs that are plastic in structure, as though they are unable to crystallize. Data from adult canaries, which can develop new songs as adults, appear similar to data from juvenile Zebra Finches. Lesioning the LMAN of adult male canaries in mid-September, when song is seasonally plastic in structure, leads to a progressive decline in syllable diversity (Nottebohm et al. 1990).

The two major neural pathways develop at different times, and the timing suggests that these pathways play roles in different phases of song learning. The development of the anterior forebrain pathway coincides with the initial sensory phase of song learning. Canaries already have most of the connections in this pathway before hatching: HVC neurons that project to area X, DLM neurons that project to the LMAN, and LMAN neurons that project to the RA (Alvarez-Buylla et al. 1988, Alvarez-Buylla 1992). In male Zebra Finches, functional connections are already present between all of the nuclei in the anterior forebrain pathway by 15 days after hatching, when sensory song learning is occurring (F. Johnson and Bottjer 1992, E. Nordeen et al. 1992).

The main descending motor pathway develops after the anterior forebrain pathway and plays an important role in the sensorimotor phase of song learning. When a young male Zebra Finch first begins to sing, 25–35 days after hatching, connections between the HVC and the RA in this pathway are just being initiated (Konishi and Akutagawa 1985). The HVC neurons that project directly to the RA continue to be born after hatching and into adulthood in both Zebra Finches and canaries (K. Nordeen and Nordeen 1988, Alvarez-Buylla 1992). The addition of new RA-projecting neurons to the HVC occurs at a higher rate in adult canaries than in adult Zebra Finches, perhaps reflecting the difference between open-ended and age-limited song learning.

These correlations in timing, of course, do not necessarily indicate that the observed changes in brain structures either cause or result from song learning (Doupe 1993). As an example, Song Sparrows (*Melospiza melodia*) in western Washington State continue to sing well-structured songs in the winter, after their song nuclei have undergone a seasonal regression in size (G. T. Smith et al. 1995). This observation shows that birds can produce crystallized song even if their song nuclei are incompletely developed.

The anterior forebrain pathway may be necessary for song learning because it conveys auditory information within the song system. To acquire a sensory model of song, a bird must hear the songs of adult conspecific birds. Auditory feedback from self-generated song is essential for sensorimotor learning and for maintenance of crystallized song (Konishi 1965, Marler 1991c, K. Nordeen and Nordeen 1992). One auditory region in the forebrain, field L, projects to the HVC, providing a direct link between the auditory and song systems (Kelley and Nottebohm

1979, Fortune and Margoliash 1992). Neurons in both song control pathways, including those in area X, the DLM, and the LMAN, as well as in the HVC and RA, respond to auditory stimuli and discharge best or exclusively in adult males to the individual bird's own song (Margoliash 1983, 1986, Doupe and Konishi 1991, Margoliash and Fortune 1992, Vicario and Yohay 1993). Furthermore, within the anterior forebrain pathway, the selectivity of neuronal responses for a bird's own song increases as one progresses from area X to the DLM to the LMAN (Doupe and Konishi 1991).

The nucleus RA may represent the locus for the observed interaction between the sensory model of song acquired early in life and the bird's subsequent production of song. Individual neurons in the RA receive input from both the LMAN and the HVC (Herrmann and Arnold 1991). This convergence of input on the RA suggests that during subsong and plastic song, a bird's vocal output is mediated by the HVC-to-RA projection. The practice song may be "compared" at the level of RA neurons with information on the song "template" acquired during sensory learning and encoded in the anterior forebrain pathway.

A logical extension of this scenario is that after sensory learning, a model of the tutor song is stored in one or more brain regions that have access to the song system. The search for the site of the putative song template in the brain is in its preliminary stages. Current data suggest that the song template, if it exists, is not stored within the forebrain song nuclei as an exact copy of the tutor song(s). After sensory learning, but either before or just after the onset of sensorimotor learning, neurons in the HVC of White-crowned Sparrows (*Zonotrichia leucophrys*) and in area X and the LMAN of Zebra Finches do not respond preferentially to the tutor song over other conspecific songs (Doupe 1993, Volman 1993). Furthermore, in juvenile White-crowned Sparrows, HVC neurons begin to respond selectively only during plastic song, and then they respond more strongly to the bird's own developing song than to the tutor song. It thus seems that the selectivity of neurons in the HVC is determined by a bird's own song production rather than by the song model to which it was exposed during sensory learning. An alternative possibility is that the tutor song is stored as a template in other, as yet undetermined, areas of the brain. Perhaps the tutor song is stored in a distributed form across a population of neurons in presinging juveniles, such that no single cell would demonstrate response selectivity to the entire tutor song. Rather, single neurons may be tuned to specific song features rather than to the entire song.

Much progress has been made in understanding the role of nuclei in the anterior forebrain pathway in song learning by Zebra Finches and canaries. They are only 2 of the roughly 4000 species of songbirds, however, and by themselves cannot provide a complete picture of the neural basis of song learning. Zebra Finches and canaries differ from each other in many ways—in song repertoire size, in timing of song learning, in body size, in phylogeny, and in many life history features. We must be aware that an attribute of song behavior that we attempt to relate to aspects of neural development may not have a cause-and-effect basis. For example, a comparison of these two species suggested that seasonal changes in the size of the song control nuclei were related to whether songs were learned early or throughout life (see, e.g., Nottebohm 1981). Only by studying another age-limited song-learning species was it possible to refute that hypothesis (Brenowitz et al. 1991); seasonal plasticity of the song nuclei need not be correlated with the development of new songs. Further, certain features of song learning in the Zebra Finch and canary may make them less than ideal subjects. In the Zebra Finch, for example, the sensory and sensorimotor phases of song learning overlap to a considerable extent.

Further progress in this field will require researchers to take advantage of the multitude of evolutionary experiments in song control provided by the diverse oscine lineages. To expand on work done with the Zebra Finch we might, for example, study closely related estrildid finches in which song learning continues later in life than it does in the Zebra Finch. Similarly, we might investigate fringillids, which are closely related to canaries but learn songs only during a limited period early in life. Finding an age-limited learner in which the sensory and sensorimotor phases do not overlap extensively would help resolve some of the limitations inherent in using the Zebra Finch.

On a broader level, investigators should face the challenge of explaining species differences in the timing and mode of song learning at the level of function in the anterior forebrain pathway. Can we, for example, identify interspecific differences in the development and function of these brain nuclei in species that learn song by imitation of a model and those that learn by improvisation? Does the developmental timing of this pathway differ between age-limited learning species in which song learning is truly restricted to early critical periods and species in which a subsequent period of action-based modification of song occurs in young adults (Marler 1990, D. A. Nelson 1992b, Beecher et al. 1994b)? Investigation of carefully chosen species will provide new insights and help to resolve questions raised but not answered by studies of male Zebra Finches and canaries.

Future studies should also address the possible neuroendocrine basis of *perceptual* song learning by female birds. Even in species in which females do not actually sing, they may still learn to recognize song produced by conspecific males. Female song perception is important in the context of species recognition, and perhaps also in mate selection (Kroodsma 1988a, Slater 1989, Searcy 1992b). Some evidence suggests that nonsinging female oscines must learn to recognize conspecific song (M. C. Baker et al. 1981a, Balaban 1988a), though this issue has not been systematically addressed.

Although the extent to which females must learn is unknown, they apparently use their song control nuculei to recognize song. Females that do not sing generally have the same song control nuclei as males, albeit of smaller size (Nottebohm and Arnold 1976). Lesioning part of the HVC makes adult female canaries less selective and induces them to perform copulation-solicitation displays to heterospecific as well as conspecific songs (Brenowitz 1991a). We do not know, however, whether females use the same brain regions in learning to recognize song that males use in learning to produce song. The degree to which neurons in female song nuclei respond selectively to conspecific song or respond differentially among the songs of different conspecific males (e.g., mate vs. other males) is also unknown. This topic lends itself to a comparative approach. For example, are there behavioral or neuroendocrine differences between females of species in which males are age-limited song learners as opposed to open-ended song learners? The entire subject of female song learning merits further study.

Sex Differences in Song Behavior and the Song Control System

The extent to which females sing varies considerably among species. At one extreme are species such as the Zebra Finch and Carolina Wren (*Thryothorus ludovicianus*), in which only males normally sing. At the other extreme are species such as the Bay Wren (*T. nigricapillus*) and Buff-breasted Wren (*T. leucotis*), in which males and females contribute equally to antiphonal song duets (Farabaugh 1982, Levin 1988). Between these extremes are species in which female song is present but is typically less complex in structure and occurs less commonly than male song. Female canaries occasionally sing spontaneously, but only rarely and then with simple songs (Pesch and Güttinger 1985). Females of the East African White-browed Robin-Chat (*Cossypha heuglini*) duet with males but include only about one-eighth as many different types of song syllables in their repertoires as males do (Todt et al. 1981, Hultsch 1983). Another duetting species in which females have less complex song repertoires is the Rufous-and-white Wren (*T. rufalbus*). Females of this species sing only about half as many different types of songs as males do (Farabaugh 1982). The ecological and social correlates of this extensive variation in female song behavior are largely unknown (Farabaugh 1982).

The variation between species in song behavior is accompanied by concomitant variation in the neural song control system (Fig. 16.2). In species in which only males sing, pronounced sexual dimorphisms occur in the structure of the song nuclei. In Zebra Finch females, for example, a well-defined area X is lacking, the remaining forebrain nuclei are much smaller, and HVC neurons do not form synaptic connections with RA neurons (Nottebohm and Arnold 1976, Konishi and Akutagawa 1985). In species in which females are able to sing, brains of males and females have the same network of song nuclei. The degree to which the sexes of any species differ in the size of their song nuclei corresponds closely with the extent to which they differ in the complexity of song behavior (Fig. 16.2). Thus, the volumes of song nuclei in the duetting Bay and Buff-breasted Wrens do not differ between the sexes, and intermediate degrees of sex difference are found in the song nuclei of canaries, White-browed Robin-Chats, and Rufous-and-white Wrens (Nottebohm and Arnold 1976, Brenowitz et al. 1985, Arnold et al. 1986, Brenowitz and Arnold 1986).

Such extensive interspecific variation in sex differences of song behavior and song nuclei presents interesting opportunities for comparative studies of song learning and the development of the song control system. As will be discussed

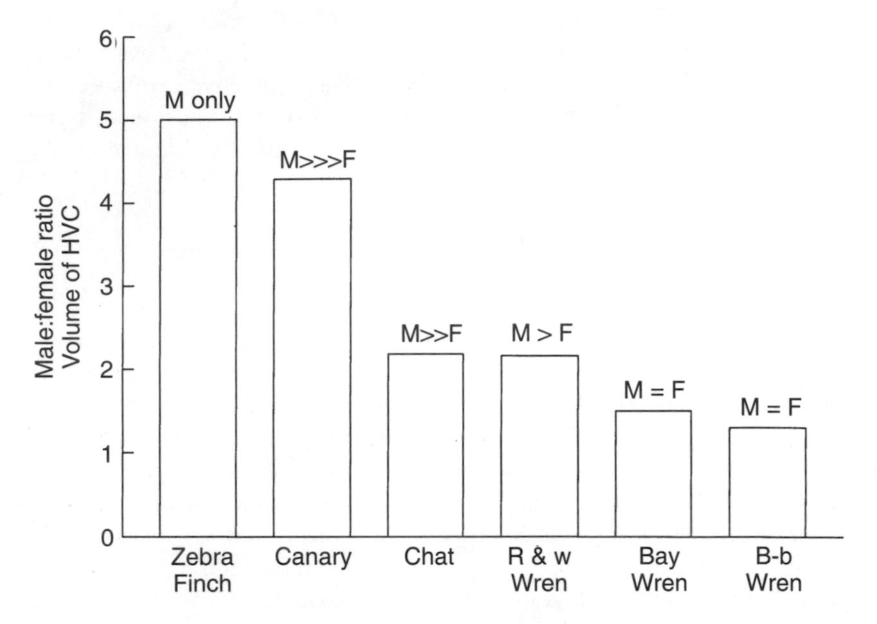

Fig. 16.2. Male:female ratios for the volume of the high vocal center (HVC) in the Zebra Finch, canary, White-browed Robin-Chat, Rufous-and-white Wren, Bay Wren, and Buff-breasted Wren. The letters on top of each histogram indicate the relative occurrence or complexity of song behavior in males (M) and females (F) of each species. The graded pattern of sex differences in the size of the HVC, as well as other song nuclei, corresponds to a graded pattern of sex differences in song behavior across these species.

below, gonadal steroid hormones have an important influence on development of the song control networks in the male brain. Do these hormones play a role in masculinization of the song nuclei in singing females, and if so, what prevents masculinization of other regions of the brain that regulate typical female reproductive behavior? Does song learning in females follow the same developmental stages and time course as learning in males? These and related questions merit attention in future studies.

Steroid Hormones and the Oscine Song Control System

The developmental organization of the song system is strongly influenced by the secretion of steroid hormones from the gonads and by the metabolism of these hormones in the brain. In Zebra Finches, testosterone secreted by the gonad in young males is converted to estradiol in the brain (Schlinger and Arnold 1992a, b), and this estradiol appears to masculinize the song control system (Gurney and Konishi 1980, K. Nordeen et al. 1986, H. Simpson and Vicario 1991, Schlinger and Arnold 1992a, b). At critical periods of development, estradiol prevents the death of neurons in the nucleus RA and stimulates the addition of new neurons in the HVC and area X (Konishi and Akutagawa 1987, E. Nordeen and Nordeen 1989).

The gonadal steroids also appear to affect the song-learning process. Nottebohm (1969a) castrated a male Chaffinch (*Fringilla coelebs*) at six months of age, before it developed crystallized song. When he treated this bird with testosterone at two years of age, it learned to copy a conspecific tutor song presented at that time. Song learning in normal Chaffinches does not occur past the first year of life. Male Swamp Sparrows (*Melospiza georgiana*) and Song Sparrows castrated by four weeks posthatching acquired a sensory model of conspecific song and went through the subsong and early plastic song stages of sensorimotor learning. The castrated birds did not, however, achieve crystallized song (Marler et al. 1988). Similarly, Zebra Finches castrated as juveniles developed abnormal songs that were variable in structure (Bottjer and Hewer 1992). These results suggest that testosterone is not necessary for sensory learning or early motor learning but is critical for the development of stable motor patterns of song. Testosterone may stimulate song crystallization by inducing changes in the distribution of synaptic inputs from the LMAN and HVC to neurons in the RA (Herrmann and Arnold 1991, Mooney 1992, Doupe 1993).

Steroid hormones are also important in the activation of song behavior in adult birds. Castration reduces or completely eliminates song production in males. Implanting castrated males with testosterone reinstates song (Arnold 1975, Harding et al. 1983, Pröve 1983). Seasonal changes in song production are correlated with changes in circulating levels of testosterone. In many species, adult females can be stimulated to sing when implanted with testosterone (e.g., Nottebohm 1980a).

The sex steroids appear to exert their organizational and activational effects by acting directly on neurons in song regions of the brain. Cells in all the song nuclei other than area X contain receptors that bind testosterone or its metabolites (Arnold et al. 1976, Brenowitz 1991b, Gahr et al. 1993). The hormone-receptor conjugate binds, in turn, to the chromatin DNA in the nucleus and activates transcription of selected genes. The consequent changes in protein synthesis are the mechanism by which steroids influence the structure and function of the song system.

The role, if any, that steroid hormones play in the development of female song systems is largely unknown. Is fine endocrine control responsible, for example, for the brain differences between singing and nonsinging females of different species? Or do steroid hormones influence perceptual song learning by females? Both singing and nonsinging females of various species have receptors for steroid hormones in their song nuclei, just as males do, and sex differences in the number of steroid-sensitive cells correlate well with sex differences in the complexity of song (see Brenowitz 1991b for a review). These observations suggest that gonadal steroids influence the development and/or activation of the song system in females, but this topic clearly needs further study.

The degree of endocrine control of the song system and other behaviors may vary considerably among species. In some tropical oscines, for example, aggressive territorial behavior is relatively dissociated from the secretion of steroid hormones by the gonads (Levin and Wingfield 1992). Perhaps the activation of the song system in these birds is also relatively independent of steroid secretion. Some preliminary insights on this issue come from studies of the Bay Wren, a Neotropical song-duetting species. Both male and female Bay Wrens use song to defend territories, and both have circulating levels of testosterone much lower than those typical of temperate zone species. Furthermore, male and female wrens continue to sing and respond aggressively to simulated territorial intrusions even after removal of their gonads and treatment with antihormonal agents (Levin and Wingfield 1992). Despite their extremely low levels of circulating testosterone, both adult male and female Bay Wrens have concentrations of steroid hormone receptors in their song nuclei that are as high as those in canaries and Zebra Finches, in which testosterone is known to activate song behavior (Arnold et al. 1976, Brenowitz and Arnold 1985, 1989, 1992). The presence of steroid hormone receptors in Bay Wren song nuclei may indicate that extremely low testosterone levels are adequate to activate song behavior in this species, that steroids play a role in development of the song system or in song learning in juveniles, or both. Song Sparrows in Washington State commonly sing during the autumn and winter, when the gonads are highly regressed and circulating testosterone levels are extremely low (Wingfield and Hahn 1994). Future studies should investigate the mechanisms, endocrine or otherwise, that regulate the organization and activation of the song system in species in which song behavior may be dissociated from circulating gonadal steroid levels.

Comparative Studies of the Neuroendocrine Control of Vocal Behavior

Vocal learning also occurs in parrots and hummingbirds (Nottebohm 1972, Baylis 1982, Kroodsma 1982b, Kroodsma and Baylis 1982, Dooling et al. 1987b, Baptista and Schuchmann 1990). Do parrots and hummingbirds have vocal control systems in their forebrains similar to those of songbirds?

The brain of the one parrot species that has been studied (Budgerigar, *Melopsittacus undulatus:* Paton et al. 1981, Striedter 1994) offers some intriguing contrasts with oscine brains. In both songbirds and Budgerigars the main descending motor pathway consists of connections between nuclei in the same two anatomical divisions of the forebrain (the neostriatum and the archistriatum), which in turn project to the nXIIts in the hindbrain. But there are many differences in the two systems. The absolute positions of the forebrain vocal nuclei, for example, differ between the two groups. Oscine and Budgerigar song circuits also differ in their cell structure, connections, and chemical properties. Furthermore, auditory input to the vocal control system arises from field L in songbirds but from other regions of the forebrain in Budgerigars. Finally, forebrain vocal control nuclei in the songbird brain can be labeled selectively by antibodies raised against gonadal steroid hormone receptors and markers for several neurotransmitter chemicals, but the same procedures fail to selectively label forebrain nuclei in the Budgerigar (Ball 1990, Gahr et al. 1993, Striedter 1994). Functionally, the oscine circuit clearly regulates vocal learning and production, but experiments to determine whether a similar system occurs in the Budgerigar remain to be done.

Budgerigars also have three brain nuclei that are comparable to area X, the DLM, and the LMAN found in the anterior forebrain pathway of songbirds (Paton et al. 1981, Striedter 1994). The Budgerigar nuclei and the three songbird nuclei are found in the same regions of the brain, and they are similarly interconnected. Furthermore, the Budgerigar nucleus comparable to the LMAN projects to the region comparable to the RA, as in the songbird brain. The Budgerigar and oscine circuits differ in several ways, however. Unlike songbirds, the region comparable to area X is not connected to the region comparable to the HVC in the Budgerigar. Also, the Budgerigar region comparable to the LMAN apparently lacks receptors for androgenic steroid hormones and other neurochemicals that distinguish this nucleus in songbirds.

Preliminary analysis indicates that Anna's Hummingbird (*Calypte anna*), which also learns to sing, has forebrain regions similar in appearance to the RA and LMAN of oscines (M. Gahr unpubl. data). These regions are not well differentiated in the brains of Ruby-throated Hummingbirds (*Archilochus colubris*), which lack well-developed song. As with the Budgerigars, it has yet to be demonstrated in hummingbirds that these brain regions play a role in vocal learning and song production.

Evolution of the Vocal Control System

The preceding comparative analysis raises the question of whether the vocal control systems of songbirds, parrots, and hummingbirds are homologous or homoplasous. The involvement of the same anatomical divisions of the brain and the similar patterns of interconnections between these areas in all three groups might be interpreted as evidence that the systems are homologous (e.g., DeVoogd 1986). The differences between the systems discussed above, however, especially in their chemical properties, have been viewed as evidence that the vocal control systems of parrots and songbirds evolved independently (Striedter 1994). Support for this latter view comes from the fact that well-defined forebrain vocal nuclei have not so far been observed in any avian groups other than the songbirds, parrots, and hummingbirds (Nottebohm 1980b, Ball 1990, Brenowitz 1991b). The groups that have been studied include the galliforms, columbiforms, and suboscine passerines (Karten and Hodos 1967, Bonke et al. 1979, Kroodsma and Konishi 1991). If the vocal control systems of Budgerigars, hummingbirds, and songbirds did in fact evolve independently, then it is all the more striking that they involve many of the same brain regions. Such convergence might point to fundamental constraints on the manner in which avian brains can integrate auditory feedback with the control of the vocal motor apparatus, a necessary component of vocal learning. These constraints could be developmental and/or phylogenetic in origin. In all three groups, the evolution of vocal systems may have involved independent elaboration of circuits already present in a rudimentary form in ancestral birds (Brenowitz 1991b). A better understanding of how brain systems evolved, and of the selective forces for vocal learning itself, will require a broad-scale phylogenetic analysis (Striedter 1994).

Perhaps a more tractable question is, When in the phyletic lineage leading to modern songbirds did the hormone-sensitive song control system in their forebrains first arise? If this system appeared with the origin of the passeriform lineage, then one might expect to find at least rudimentary vocal nuclei in the brains of suboscine passerines. Although a comprehensive search has not yet been conducted, preliminary studies of representatives from three suboscine families have failed to detect any forebrain song nuclei. Standard tissue-staining techniques did not reveal discrete clusters of cells comparable to the HVC, RA, LMAN, or area X in the brains of four species of tyrannid flycatchers (Nottebohm 1980a, Kroodsma and Konishi 1991, T. DeVoogd unpubl. data), the furnariids *Asthenes hudsoni* and *Synallaxis frontalis* (Nottebohm 1980a), or the thamnophilid Slaty Antshrike (*Thamnophilus punctatus:* Brenowitz unpubl. data). Nor did preliminary analyses of three species of tyrannid flycatchers and the Slaty Antshrike indicate the presence of steroid hormone receptors in regions of the brain where the HVC, RA, and LMAN are found in oscines (Gahr et al. 1993, Brenowitz unpubl. data). A more detailed analysis of the brains of these species is required, but the available data suggest that these birds lack a vocal control pathway similar to that of the oscine passerines.

Pending more comprehensive analysis of suboscine passerine brains, one can tentatively conclude on the basis of existing data that the hormone-sensitive song system arose only with the origin of the oscine lineage. This network of forebrain song nuclei has been observed in at least 57 oscine species in 10 families (Certhiidae, Corvidae, Fringillidae, Hirundinidae, Laniidae, Muscicapidae, Paridae, Passeridae, Sturnidae, and Sylviidae), five superfamilies, and the two major songbird divisions Corvida and Passerida (classification of C. G. Sibley et al. 1988; Nottebohm 1980a, 1987, Brenowitz et al. 1985, 1991, Arnold et al. 1986, Brenowitz and Arnold 1986, Ball 1990, DeVoogd et al. 1993, Brenowitz unpubl. data). The only oscine superfamily not yet analyzed is the meliphagoids, which include fairywrens, honeyeaters, pardalotes, thornbills, and scrub-wrens. Steroid hormone receptors have been searched for and detected in the forebrain song nuclei of 19 oscine species (Arnold et al. 1976, Zigmond et al. 1980, Brenowitz and Arnold 1985, 1989, 1992, Gahr et al. 1993, Brenowitz, A. P. Arnold, P. Loesche unpubl. data). In these species, androgen receptors are found in the HVC, RA, and LMAN in the forebrain; in the midbrain song nucleus ICo (intercollicular nucleus); and in the nXIIts. Estrogen receptors occur in the HVC and ICo of these species.

Comparative analysis of the song control system in different songbird taxa indicates that this system is very uniform in morphology and chemical properties across taxa. Extreme diversity is found within and between taxa in various aspects of song behavior, however, including the production of song by the sexes, repertoire size, and the developmental timing of song learning. Three attributes of the song system may enable the production of extensive behavioral diversity by this highly conserved network of brain nuclei. First, the network appears to function solely in regulating song-related behavior. The devotion of the song system to song behavior provides more flexibility for evolutionary modification of such factors as neuron number and developmental timing of the brain circuits than might be true if this network also functioned in contexts unrelated to song. Second, gonadal steroid hormones have a profound influence on the development and activation of these circuits. Patterns of hormone secretion and metabolism show great diversity across avian taxa in such aspects as developmental timing, seasonality, and sex (see, e.g., Wingfield and Moore 1987). This diversity implies that hormone secretion and metabolism are evolutionarily plastic traits. Relatively small changes in hormone release and metabolism, in turn, can have dramatic effects on song control networks and song behavior. Third, song is a learned behavior and is thus subject to rapid modification via cultural evolution. Together these three attributes of the oscine song control system may provide the plasticity that has facilitated the diverse expression of song behavior across taxa.

This comparative analysis suggests that the hormone-sensitive forebrain song system is a trait found among all oscine lineages. That observation is consistent with the hypothesis that the song control network first arose very early in the evolution of the oscine lineage. Indeed, it is intriguing to speculate that the initial development of this steroid-sensitive neural system was a definitive event in the evolutionary origin of the songbirds. A systematic phylogenetic analysis of how song is controlled among passerine birds could reveal how the songbird system evolved.

Brain Space for Learned Song

One of the most distinctive features of birdsong is the great range of complexity among species. In species such as the White-throated Sparrow (*Zonotrichia albicollis*), a male produces only a single, simple type of song (Borror and Gunn 1965). In contrast, a male Brown Thrasher (*Toxostoma rufum*) has a repertoire of several thousand different types of songs (Kroodsma and Parker 1977). Between these two extremes of repertoire size lie the majority of bird species. Among species with song repertoires, some species enhance song complexity by delivering different song types in rapid succession, and others repeat each song type several times before switching to another (see Kroodsma 1982b).

Song repertoires are thought to evolve in response to sexual selection and to play a role in territorial defense or advertisement for a mate (e.g., Catchpole 1986,

Searcy and Andersson 1986, Kroodsma and Byers 1991). Among some species, for example, repertoire size and population density appear to be positively correlated, suggesting that competition for territories or mates in dense populations promotes larger signal repertoires (Kroodsma 1983a). When territorial males are replaced by loudspeakers broadcasting recorded song, other males trespass at a lower rate, which suggests that large song repertoires may enhance the effectiveness of song in deterring trespassing (Krebs et al. 1978, Yasukawa 1981a). With playbacks directed at specific territorial birds, males habituate more rapidly to presentation of a single song type than to a number of songs (Krebs 1976, Yasukawa 1981a, Searcy et al. 1994).

Females are also sensitive to the variety of male songs, and larger song repertoires may be more attractive to females than smaller repertoires. Male Sedge Warblers (*Acrocephalus schoenobaenus*) with large repertoires mate earlier in the breeding season than do males with small repertoires (Catchpole 1980). Similarly, Great Reed-Warbler (*A. arundinaceus*) males with large repertoires have more mates than males with smaller repertoires (Catchpole 1986). Isolated female canaries build nests and lay eggs more rapidly in response to playback of increased song variability (Kroodsma 1976a). Females of several species also perform more copulation-solicitation displays in the laboratory in response to tape recordings with increased song variability (reviewed by Searcy 1992a).

Selection for large song repertoires has led to specializations in the song control nuclei. Nottebohm (1981) found that the volumes of the HVC and RA in adult male canaries were positively correlated with song repertoire size. Similar correlations between the size of song nuclei and repertoire size were also reported for female canaries induced to sing by testosterone implants (Nottebohm 1980c), and for sibling species of Marsh Wren (*Cistothorus palustris;* Canady et al. 1984, Kroodsma and Canady 1985, Brenowitz et al. 1994). Differences in the volumes of song nuclei between birds with different-sized song repertoires appear to reflect differences in both the number and size of cells in these nuclei (Brenowitz et al. 1994). In a broad comparative analysis of 41 oscine species in nine families, DeVoogd et al. (1993) found a significant correlation between song repertoire size and the volume of the HVC relative to overall brain volume. The results of that study are consistent with the idea that the relationship between brain space and song learning arose early in oscine phylogeny and has been maintained throughout subsequent divergences of oscine lineages.

Further evidence of a consistent functional relationship between brain space and repertoire size is provided by congeneric species that differ in repertoire size (Brenowitz and Arnold 1986). Male Rufous-and-white Wrens and Bay Wrens have similar song repertoire sizes (Farabaugh 1982, Levin 1988) and also show no significant size differences in their song nuclei. Repertoires of female Rufous-and-white Wrens, however, are less than half as large as those of female Bay Wrens, and all of the song nuclei of female Rufous-and-white Wrens are significantly smaller than those of female Bay Wrens. Thus, the "rules" by which the song control circuits encode song repertoires are clearly conserved in speciation events.

Overall, the available data suggest a deterministic relationship between song behavior and brain space: learning more songs is associated with having more and bigger neurons. None of the studies cited above determined the causal direction of this relationship, although three hypotheses can be proposed: (1) the number of songs learned by a male determines the size of his song nuclei, (2) the size of song nuclei determines the number of songs that a male learns, or (3) other factors determine both the size of song nuclei and the number of songs learned (see below; Nottebohm 1981, Brenowitz and Arnold 1986).

In an initial attempt to discriminate among these hypotheses, we used eastern Marsh Wrens (Brenowitz et al. 1995). In nature, males sing about 50 song types apiece (Kroodsma and Canady 1985), but in the laboratory repertoire size can be controlled by tutoring with selected repertoire sizes. The hand-reared males in our experiment were assigned to one of two groups; one group was tutored with a small repertoire (5 song types) and the other with a large repertoire (45 song types). The birds' early tutoring experience successfully determined their adult song behavior. The following spring the first group had only 5 or 6 song types, but the second group had 35 to 46 types. Despite the pronounced differences in adult song-repertoire size between these two groups, however, we found no significant differences between them in the mean volumes of the song nuclei HVC and RA (Fig. 16.3). Thus, males that produced small song repertoires were just as likely to have large song nuclei as males that had large repertoires.

The results of that study refute hypothesis 1, which says that early learning experience is the main determinant of the size of the song nuclei. The correlation between song repertoire size and the volumes of the HVC and RA found among Marsh Wrens in nature (Canady et al. 1984, Kroodsma and Canady 1985, Brenowitz et al. 1994) implies that the size of the song nuclei sets an upper limit to the number of songs that a bird can learn (see, e.g., Nottebohm 1981). Then what factors determine the size of the song nuclei? One possible answer stems from Schwabl's (1993) observation that female canaries supply different amounts of testosterone to the eggs within a clutch. The level of aggressive behavior shown by the chicks after hatching is correlated with the amount of maternal testosterone in each egg. Given the role of testosterone and its metabolites in development of the song nuclei, as discussed above, perhaps the different levels of maternal testosterone given to the eggs also cause differential development of the song nuclei before and after hatching. The degree to which development of the song nuclei is stimulated by maternal testosterone could then determine the number of song syllables or types that the young bird is able to learn. This hypothesis could be tested by measuring or manipulating the levels of testosterone in developing eggs and then testing for differential development of song nuclei and song-learning ability.

As described above, male birds may gain an advantage in mate attraction or competition with other males by producing large song repertoires. According to our Marsh Wren study, the maximum number of song types in a male's repertoire, in turn, could be determined by the size of the song nuclei. Taken together, these observations raise the speculation that, to the extent that potential mates and

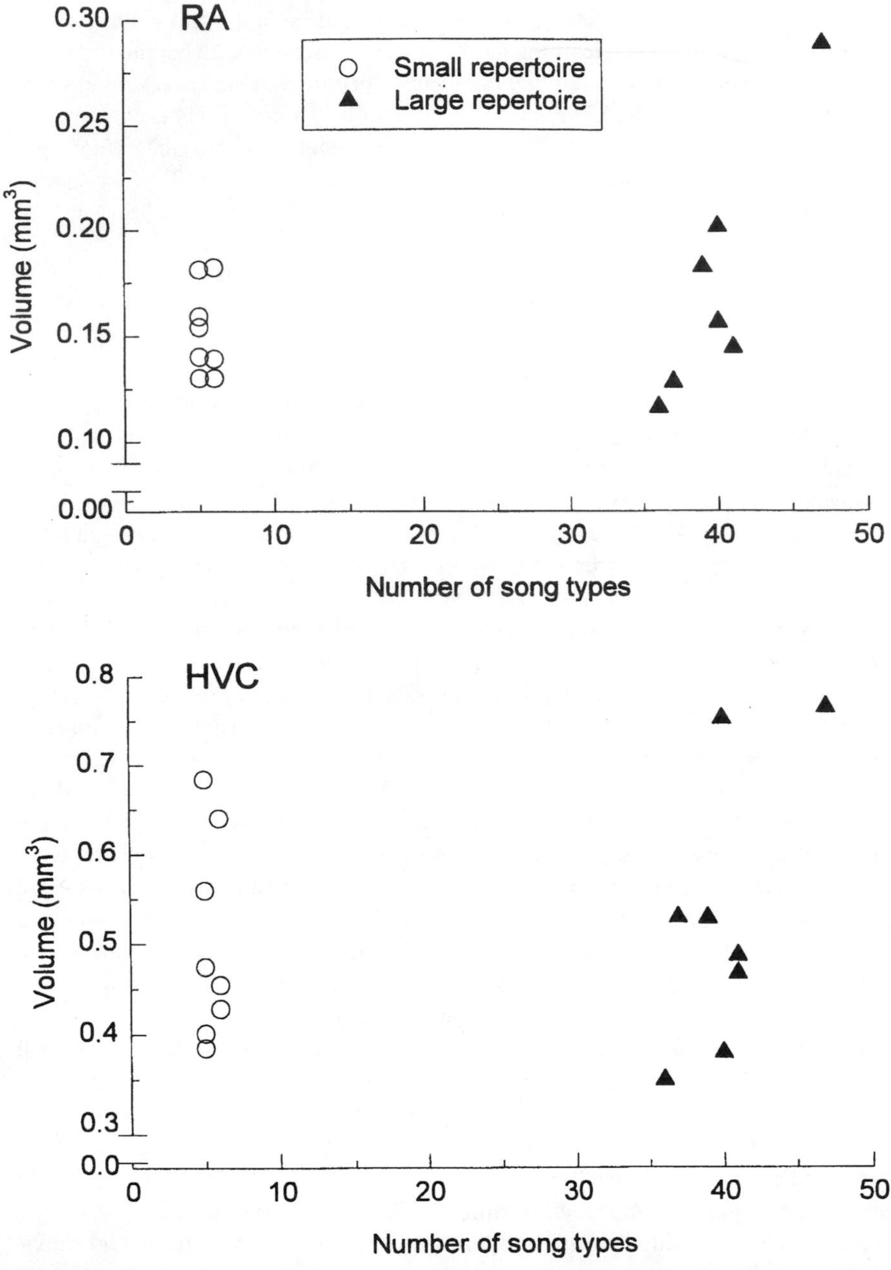

Fig. 16.3. The relationship between size of song nuclei and song repertoire size in male Marsh Wrens tutored early in life with either 5 (○) or 45 (▲) song types. The volumes of the RA (top panel) and the HVC (bottom panel) on one side of the brain are indicated for eight males in each song-tutoring group as a function of the number of song types produced by each male in his adult repertoire. Note the lack of overlap in song repertoire sizes of the males in the two different tutoring groups, and the extensive overlap in the size of the song nuclei in the brains of these males.

competitors impose directional selection on birds to develop large song repertoires, they also select for males that can devote more brain space to learning and producing song. One would expect developmental, functional, and energetic constraints on the extent to which sexual selection can drive this process in the brain, given that song is only one of the many functions regulated by the brain. These constraints may in turn set an absolute limit on the maximum song repertoire size that a species can evolve without a concomitant increase in overall brain size. One (untested) prediction is that species with extremely large song repertoires, such as the mimids, have larger brains than closely related species similar in body size but with smaller repertoires.

Plasticity in the Song System

One of the distinguishing features of oscine song behavior is its flexibility. The most striking expression of this plasticity is that songs are learned, and individuals are able to learn a variety of conspecific dialects and even the songs of other species. Songbirds may be unique among nonhuman animals, with the possible exception of cetaceans, in their obligate learning of species-specific communication signals (see Brenowitz 1994). This flexibility in vocal development confers the advantage of adapting communication behavior to important social and ecological conditions the birds encounter (see Chapters 1–6, this volume).

Behavioral plasticity in song is shown in a number of ways. The development of new song patterns continues into adulthood in several species. Even species that do not learn to *produce* new songs as adults, however, show adult plasticity in other aspects of song behavior. Territorial birds must learn to recognize the songs of their neighbors each year, for example (Falls 1982), and during aggressive interactions at territorial borders, birds of many species rapidly switch song types so as to match their competitors (Beecher, this volume). Birds may also learn to recognize the specific songs produced by their mates (Stoddard, this volume). Seasonal changes in the production of song occur in Arctic, temperate, and subtropical species that have distinct breeding seasons, with changes ranging from a decrease in song rate to a complete cessation of song activity after breeding (see, e.g., Brenowitz et al. 1991). Species that completely stop singing outside the breeding season may also go through a period of plastic song when they first start singing again the next year.

Plasticity also occurs in the song control system of the oscine brain. As discussed above, the structure of the song control network in juvenile birds changes dramatically during song learning. In adult birds presumptive neurons continue to be born in the walls of the lateral ventricle and to migrate throughout the forebrain. Some of these newly generated neurons are incorporated into the HVC and appear to replace existing neurons (see, e.g., Nottebohm 1987, Alvarez-Buylla 1992, Kirn and Nottebohm 1993). In Zebra Finches, the rate at which these new neurons are added to the HVC decreases at about the same time the sensory phase of song learning ends (K. Nordeen and Nordeen 1988). In male canaries, the rate at which new neurons are incorporated into the HVC varies seasonally and is greater in the fall, when song repertoires are modified, than in the spring, when song is stable (Alvarez-Buylla et al. 1990). The addition of new neurons to regions of the forebrain outside the HVC does *not* vary seasonally in canaries (Alvarez-Buylla 1994). These data from Zebra Finches and canaries are consistent with the idea that the addition of new neurons to the HVC is functionally related to the development of song, though this relationship has yet to be directly demonstrated.

Pronounced seasonal plasticity can be found in other attributes of the song nuclei as well. The volumes of the HVC, RA, area X, and nXII have been reported to be up to 70% larger in the brains of breeding than nonbreeding birds in the canary (Nottebohm 1981), Red-winged Blackbird (*Agelaius phoeniceus:* Kirn et al. 1989), Orange Bishop (*Euplectes franciscanus:* Arai et al. 1989), Rufous-sided Towhee (*Pipilo erythrophthalmus:* Brenowitz et al. 1991), House Sparrow (*Passer domesticus:* Rucker and Cassone 1991), Gambel's White-crowned Sparrow (G. T. Smith et al. 1995b), Song Sparrow (G. T. Smith et al. 1995a), and Common Starling (*Sturnus vulgaris:* Bernard and Ball 1993). Corresponding seasonal differences in the size and number of neurons also occur in these brain regions (DeVoogd et al. 1985, Brenowitz et al. 1991, K. Hill and DeVoogd 1991). Seasonal plasticity of the song nuclei occurs both in age-limited and open-ended song-learning species, and in species with single-song-type and multiple-song-type repertoires (Brenowitz et al. 1991).

Lack of seasonal plasticity in the song nuclei of a seasonally breeding bird has been reported only in laboratory-reared White-crowned Sparrows (*Zonotrichia leucophrys nuttalli:* M. C. Baker et al. 1984). Seasonal plasticity is seen, however, in the song nuclei of the Gambel's race of White-crowned Sparrows (G. T. Smith et al. 1995b). To verify the observation of M. C. Baker et al. (1984), it is thus important that song behavior, circulating hormone levels, and song nuclei be analyzed in wild Nuttall's White-crowned Sparrows.

The functional significance of seasonal plasticity of song nuclei in the brain is not clear, although several hypotheses can be proposed. (1) It is related to seasonal changes in the quantity or quality of song production. Birds sing at higher rates and with greater stereotypy during than after the breeding season, and changes in the brain may simply reflect these behavioral changes. (2) It provides a mechanism for modifying *perceptual* memories of song (Alvarez-Buylla et al. 1990, Cynx and Nottebohm 1992a). This process could be relevant in the contexts of learning to recognize the songs of territorial neighbors and mates. (3) It enables birds to develop new song patterns as adults; that is, it provides the structural plasticity for open-ended song learning (Nottebohm 1981). (4) We must also consider the possibility that seasonal changes in brain structure are not functionally related to song behavior in a direct manner. Seasonal plasticity of the song nuclei might occur as a passive consequence of seasonal changes in the circulating concentrations of gonadal steroid hormones, to which the song control networks are very sensitive.

Data from one or more of the eight species in which seasonal plasticity of song nuclei has been studied thus far support each of these four hypotheses. Perhaps no single explanation exists for the functional role of seasonal plasticity in the song control system. As with so many other questions raised in this review, our understanding of the phenomenon of seasonal plasticity in the oscine brain would benefit from a carefully designed comparative approach.

Hypothesis 1 could be tested by comparing pairs of subspecies or congeneric species that differ in the seasonality of their song production. The distributions of both the Yellow Warbler (*Dendroica petechia*) and the House Wren (*Troglodytes aedon*), for example, extend from North America to the Neotropics, and one might expect their song production to be much more seasonal in high temperate latitudes. It would be interesting to compare seasonal plasticity of song nuclei in birds at different latitudes.

Hypothesis 2 posits seasonal changes in perceptual memories of song in birds. This hypothesis could be tested using operant conditioning techniques to measure the ability of birds to learn to discriminate song types (see, e.g., Cynx and Nottebohm 1992a). In temperate latitudes most birds do not use song to defend territories beyond the breeding season. In the Tropics, birds may remain on territories and continue to sing throughout the year (Farabaugh 1982). If seasonal plasticity of the song nuclei does provide a mechanism for acquiring perceptual memories of the songs of territorial neighbors, then one might expect to observe less pronounced seasonal changes in the ability to learn song discriminations among tropical birds than in closely related temperate zone species.

Nottebohm (1981) proposed that seasonal plasticity of the song nuclei is a mechanism by which adult birds develop new song patterns (hypothesis 3). The occurrence of equally pronounced seasonal changes in the morphology of song nuclei in age-limited song-learning species suggests that neural plasticity is not sufficient to produce behavioral song plasticity (Brenowitz et al. 1991). The synaptic plasticity associated with seasonal changes in the song nuclei may, however, be necessary for behavioral plasticity. One could predict, then, that seasonal plasticity should be present in the song control systems of all open-ended song learners.

Hypothesis 4, which states that seasonal plasticity of song nuclei is a passive consequence of changes in circulating steroid hormone concentrations, may be the most difficult to test. Ideally, one wants to investigate species that show seasonal changes in one or more aspects of their song behavior but in which song is relatively dissociated from circulating steroid concentrations even during the breeding season. The White-browed Sparrow-Weaver (*Plocepasser mahali*) of Africa may be such a species (Wingfield and Lewis 1993).

Conclusions

Throughout this review we have identified unresolved issues in the study of avian vocal control systems. We believe that the following 10 questions, in particular, should be addressed by future studies:

1. What are the ecological correlates of individual and species differences in the timing of vocal learning, and what are the consequences for the neuroendocrine control of vocal behavior?
2. Where and how are sensory models of song (i.e., song templates) stored in the brain before the onset of song production?
3. Do the development and function of the song control circuits differ between species that learn song by imitation as opposed to by improvisation?
4. What mechanisms regulate the organization and activation of vocal control systems in species in which vocal behavior is dissociated from circulating gonadal steroid levels?
5. What physiological factors determine the size of song nuclei and thus the maximum song repertoire size?
6. To what extent must nonsinging female birds learn to recognize conspecific male song, and what is the neural basis for such perceptual learning?
7. What factors regulate the development of the song control circuits in species in which females sing?
8. Are the vocal control systems found in the brains of songbirds, parrots, and hummingbirds homologous or homoplasous?
9. What roles do neurogenesis and neuronal replacement in the adult brain play in song behavior?
10. What are the mechanisms and functional consequences of seasonal plasticity in the song control system?

In addressing these questions, we must begin to make good use of the full range of diversity in vocal behavior found among the avian taxa. There are about 4000 species of songbirds alone, and each one can be viewed as a natural experiment for testing hypotheses about relationships between neuroendocrine mechanisms and song behavior. The extensive diversity of song behavior that occurs among the many songbird taxa provides extraordinarily rich material for researchers interested in the mechanisms of song. The full potential of the comparative approach, however, has yet to be realized in studies of the song control system. Concentration on selected model species such as the Zebra Finch and canary has been productive and was a reasonable first approach to study of the song control system, but we believe the time has come for investigators to adopt a more explicitly comparative approach toward this exciting system.

Acknowledgments

DEK is supported by National Science Foundation grant BNS-9111666; EAB is supported by National Science Foundation grant IBN 9120540, National Institutes of Health grant MH 53032, and an Alfred P. Sloan Research Fellowship.

17 Organization of Birdsong and Constraints on Performance

Marcel M. Lambrechts

Song, like any animal trait, is influenced by many ultimate and proximate factors. Ultimately, birdsong likely evolves through sexual selection to increase success in male-male competition or in procuring mates. At the proximate level, different external and internal factors determine the phenotypic expression of song (e.g., ecological, social, physiological, morphological, and genetic factors). Song is the result of interactions among different selective pressures and constraints working on its detection, memorization, development, production, and transmission (see, e.g., E. S. Morton 1975, R. H. Wiley and Richards 1978, Kroodsma and Miller 1982, Ryan and Brenowitz 1985, Halliday 1987, Partridge and Endler 1987, Read and Weary 1992, Oyama 1993).

Most researchers who study songbirds focus on the causes and consequences of song learning, roles of the brain and auditory system in song development, and the functional significance of song type structure and song repertoire size (see, e.g., reviews in Kroodsma and Miller 1982, Marler and Terrace 1984, Slater 1989, McGregor 1991, Nottebohm 1993). Although motor constraints probably influence vocal performance in different animal taxa (e.g., Ryan and Brenowitz 1985, Halliday 1987, Zemlin 1988), little research has been done on the effects of motor constraints (e.g., energetic, physiological, or morphological) in birds. A full understanding of variation in song will be achieved only after many proximate factors are understood, so we must take an interdisciplinary approach to examine the effects of motor constraints on development, performance, and organization of birdsong. In this chapter I discuss recently discovered features of song performance that may reflect motor constraints, focusing on song performance in Great Tits (*Parus major*).

Song Organization in the Great Tit

Song organization and complexity are highly variable among songbirds, ranging from species with brief, repetitive songs of one type to species with long, versatile songs and large song repertoires (Hartshorne 1973). Great Tits have a

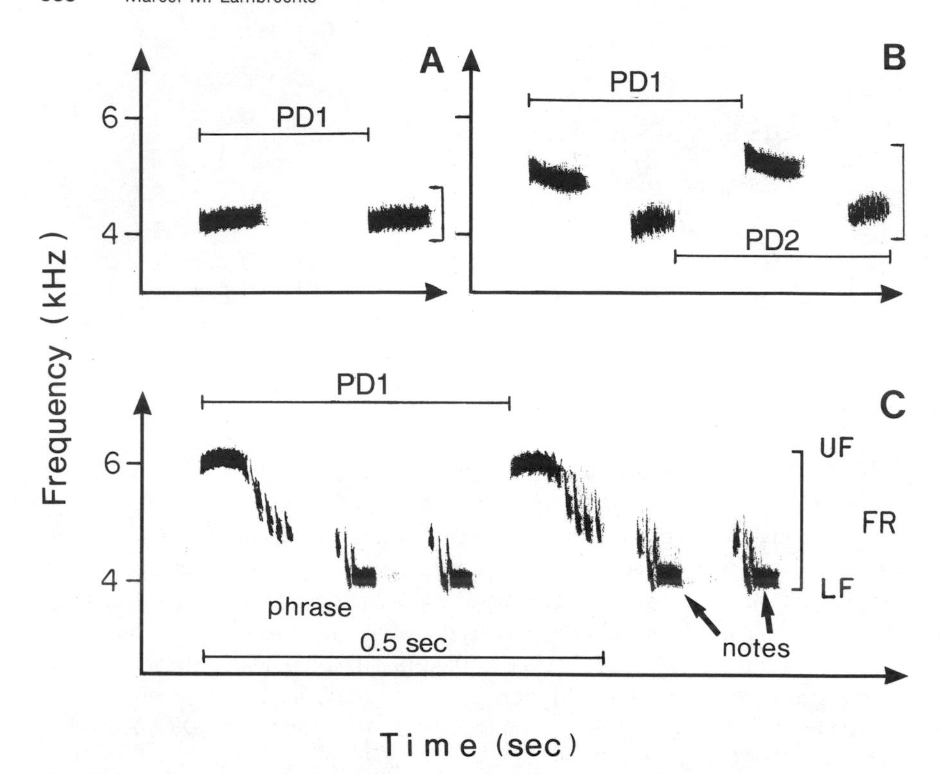

Fig. 17.1. Three song types produced by the same Great Tit. Phrase structure is used to classify song types. A phrase is a unit of one (A), two (B), or more than two notes (C) that is rapidly repeated in a stereotyped fashion within a song. Here songs are cut off after two phrases, although up to 20 phrases may be sung before a song is stopped. Phrase measures: PD, phrase duration plus interphrase interval; FR, frequency range; UF, upper frequency; LF, lower frequency. PD can start from different notes in a phrase (PD1 vs. PD2, see sonogram B), which is useful to obtain a measure of PD of the terminal phrase in a song (see Fig. 17.2). Great Tit phrases and songs usually share the same frequency attributes. Sonograms are copies of Kay Sonagraph model 6061 B sonograms with a filter bandwidth of 300 Hz.

simple song (Fig. 17.1), a fact that has facilitated studies of variation in song performance within and across individuals. In a typical song, or "strophe" (Bergmann and Helb 1982, Lambrechts and Dhondt 1987), one to five notes are grouped into a phrase, and 1 to 20 phrases are repeated rapidly. Most songs last one to five seconds and are followed by another song a few seconds later. Great Tits often sing songs of one type in a bout for several minutes, then switch to another kind of song (another song type). Individual male Great Tits possess one to eight song types, each with a distinctive phrase structure. Some individuals share song types because individual birds copy (learn) song types of other Great Tits (McGregor and Krebs 1982, Lambrechts and Dhondt 1988; Figs. 17.1, 17.2).

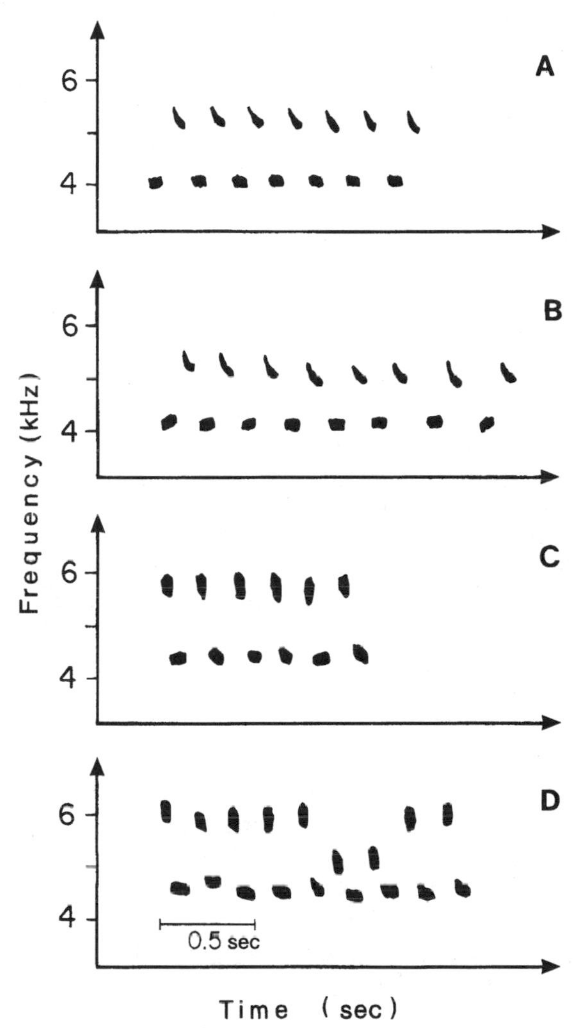

Fig. 17.2. Four songs without or with abrupt changes in time or frequency attributes. (A) Song with seven phrases having similar phrase durations (no drift). (B) Song with eight phrases with a significant prolongation of the phrase duration (drift) and narrowing of frequency toward the end of the song. (C) Song with six phrases having few changes in frequency range. (D) Song with nine phrases with varying frequency ranges (SFP). Note that abrupt changes in frequency are found especially in the high-frequency notes of a phrase (see D). Sonograms are copies from a Unigon spectrum analyzer model 4500, Uniscan.

Constraints on Time Components of Song

Song length and drift within songs. The song output of individual birds can differ considerably from one moment to the next. This variation in song duration or rate is often attributed to variation in the tendency or motivation to sing (e.g., Weary et al. 1988). For instance, song output may increase after a male's mate is removed from the territory (see, e.g., Otter and Ratcliffe 1993) or when territorial males are presented with playback of conspecific song (e.g., Weary et al. 1988).

Great Tits show consistent individual differences both in the number of phrases per song and in song duration (Lambrechts and Dhondt 1987, Weary et al. 1990b, McGregor and Horn 1991). In both the Belgian and the English populations, some males average more than 10 phrases per song, but others sing only about 4, even using the same song type at the same time of year. Males may increase song duration somewhat toward the end of the breeding season, but individual differences in the number of phrases per song and in song duration usually persist all season long (February–May). Males also differ consistently in the number of phrases per song from year to year, although some males increase song duration after their yearling season (Lambrechts and Dhondt 1986). Song duration remains relatively constant even when males are presented with a playback stimulus (Lambrechts and Dhondt 1987, Lambrechts 1992). Significant individual differences in song duration have been reported in many other songbird species, including Blue Tits (*Parus caeruleus:* Bijnens 1988), Coal Tits (*Parus ater:* McGregor 1991), Wood Nuthatches (*Sitta europaea:* McGregor 1991), and Eurasian Chiffchaffs (*Phylloscopus collybita:* McGregor 1988a).

Within individual Great Tits, the total duration of sound per song (the total duration of all notes constituting a song) is similar across different song types (Lambrechts unpubl. data). This pattern has consequences for organization within songs: long phrases or phrases with many notes are repeated fewer times than are brief phrases or phrases with few notes (Lambrechts and Dhondt 1987). Furthermore, within songs, long notes are separated by long internote intervals (Lambrechts unpubl. data). These relationships among time components in songs are probably widespread both within and between species (e.g., Kreutzer 1985, Lambrechts unpubl. data).

Individuals that produce many phrases per song also produce long songs, and vice versa. Because of drift (Lemon 1975), however, the number of phrases per song is not an exact measure of song duration. *Drift* describes the tendency for cadence to become slower toward the end of a song (Lambrechts and Dhondt 1987; Figs. 17.2, 17.3). For example, for two-note songs, phrases average about 13 milliseconds longer near the end of a song than at the beginning because the internote intervals become longer (Lambrechts and Dhondt 1987). McGregor (1991) reported an increase of about 12 milliseconds in internote intervals between the beginning and end of a song in English Great Tits. Drift is found in many songs of a single song bout. As a consequence of drift, phrase duration increases toward the end of a song but becomes briefer again (the cadence increases again) a few seconds later when a new song begins.

Drift in phrase duration varies among individuals, even for songs of one type and with the same number of phrases (Fig. 17.3). For example, individuals that normally sing long songs show less drift than do individuals that normally sing short songs when both sing one song type with the same number of phrases.

Motivation does not seem to explain variation in drift or song duration. If motivation influenced these factors, one would expect the long songs of highly motivated singers to have less drift than the short songs, which are typically sung with less motivation. I determined, however, that long songs show greater drift

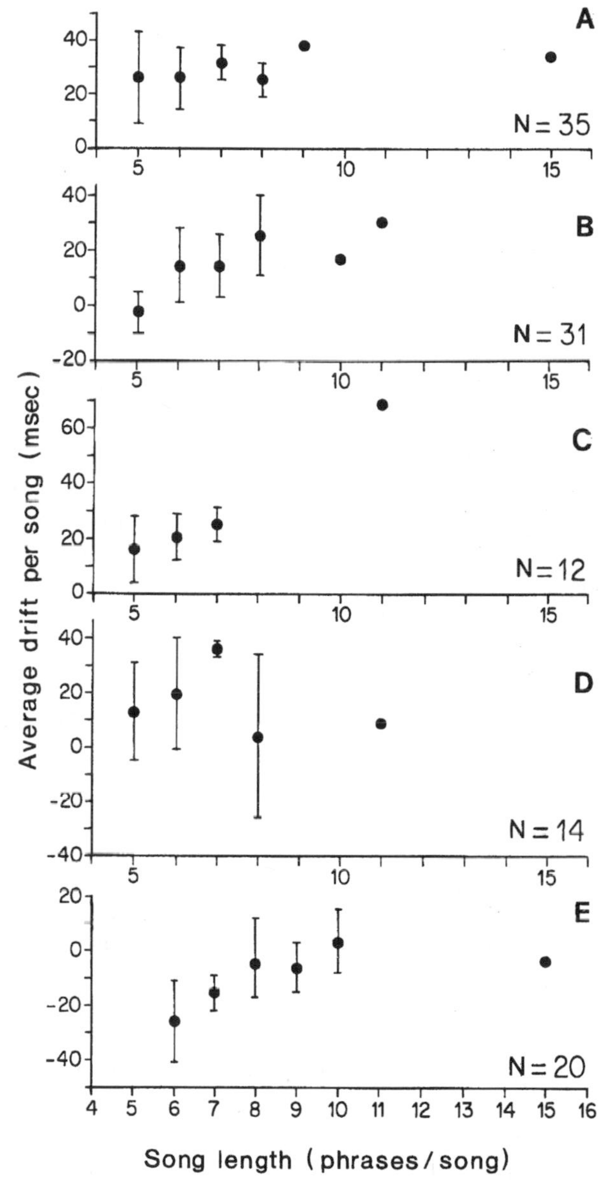

Fig. 17.3. Average drift (± SD) per song tends to increase with song length (number of phrases/song), as shown for five song bouts (A–E) of the same song type from five different birds. Drift per song is calculated by the difference between (1) the sum of the PD durations of a song's two terminal phrases, and (2) the sum of the PD durations of two initial phrases of a song (PD is shown in Fig. 17.1). In bout c, for example, PDs of terminal phrases are on average 9 milliseconds longer (18 milliseconds divided by 2) than PDs of initial phrases in songs with 5 phrases, and PDs of terminal phrases are 35 milliseconds longer than PDs of initial phrases in a song with 11 phrases. The first phrase in a song was not considered. For all other phrases in a song, measure PD2 was used (i.e., duration of interphrase interval plus duration of the following phrase; see Fig. 17.1B). Terminal phrases are sometimes shorter than initial phrases, which is expressed as a negative drift on the Y axis. Note that Great Tits show remarkable individual variation in this measure of drift. Drift per song could also be expressed as a regression of phrase duration plotted against phrase number (see Lambrechts and Dhondt 1987).

than short songs. When I compared the drift in songs of various durations within a single bout, I found that four of five individuals showed a trend toward greater drift in long songs (Fig. 17.3). One individual tended to decrease phrase duration during a song, but the decrease was more pronounced in short than in long songs (Fig. 17.3E). Therefore, factors other than motivation, such as performance constraints, may limit how many phrases occur per song; for instance songs may be stopped after a certain level of drift is reached (Lambrechts and Dhondt 1987).

The percentage of time spent singing within song bouts. The percentage of time a bird spends singing (PTS) in a bout can be calculated by dividing song duration by the sum of the song duration plus the following intersong interval (Lambrechts and Dhondt 1988, Weary et al. 1991). Short-term alterations in PTS can be quantified with this measurement as well.

Great Tits show considerable individual differences in how much time they spend singing in song bouts, and these differences persist when the birds respond to playback (Lambrechts 1988, Lambrechts and Dhondt 1988). Individual differences in song rate (e.g., number of songs per minute) have also been reported in other songbird species (e.g., Greig-Smith 1982, Radesäter et al. 1987).

Both song duration and intersong interval vary within Great Tit song bouts (Lambrechts and Dhondt 1988). In some individuals, song duration may vary an order of magnitude from a few seconds up to 20 seconds or more, and the PTS can range from 40 to 80%, on average. Furthermore, Great Tits often show a significant decrease in song duration, a significant increase in the intersong interval, or both toward the end of a song bout.

If motivation to sing a song type is the main source of variation in the song features described above, the PTS should decrease relatively weakly in bouts with a relatively high initial PTS. In fact, the opposite situation prevails when bouts of the same individual are compared: a high initial PTS is followed by a more marked decrease in song duration or increase in intersong interval than when the initial PTS is low (Lambrechts and Dhondt 1988, Weary et al. 1991). These results support the notion that song features are not greatly influenced by motivation, but rather that Great Tits are somehow limited in their song performance.

Does Energy or Physiology Constrain Time Components of Song?

Energetic or other physiological factors limit vocal performance in many animal species (Halliday 1987). In anurans, for example, calling is supported exclusively by aerobic metabolism, so oxygen consumption is a reliable measure of energy expenditure during vocal performance. Calling can result in more than a 20-fold increase in oxygen consumption above resting rates, and thus can be energetically quite expensive (see, e.g., Bucher et al. 1982, Ryan 1985, Prestwich et al. 1989, K. D. Wells and Taigen 1989).

In gray treefrogs (*Hyla versicolor*), tradeoffs occur between call duration and call rate (K. D. Wells and Taigen 1986). Oxygen consumption is similar in individuals that produce long calls at a slow rate and those that make brief calls at a fast rate. Individuals with long calls, however, have significantly briefer calling periods (in hours). K. D. Wells and Taigen (1986) therefore hypothesized that continuous activity of the muscles of the motor system responsible for calling results in a higher rate of glycogen use than muscle contractions that are interrupted more frequently. Thus, intramuscular glycogen reserves might limit vocal performance.

It is often suggested that song production in birds, too, may be energetically costly and limited by energetic constraints. First, like anurans and insects (Halliday 1987), birds produce loud vocalizations despite their relatively small body size (see, e.g., Brackenbury 1979, A. S. Gaunt 1987). Second, song rate increases as food availability increases (e.g., Searcy 1979, Cuthill and Macdonald 1990). Third, at low ambient temperatures birds must expend more energy to maintain body temperature. Song output therefore decreases during or after a cold period, and positive effects of food provisioning on song output are strong at low ambient temperatures (e.g., Garson and Hunter 1979, Gottlander 1987, Reid 1987). Fourth, parasite loads can weaken an animal, and therefore may reduce song output. Barn Swallows (*Hirundo rustica*) exposed to high loads of ectoparasites decreased their song rates (Møller 1991c). Finally, birds may have less energy available for singing than amphibians because birds have higher metabolic rates (e.g., Bennett and Harvey 1987, K. D. Wells and Taigen 1989). Bird species with high resting metabolic rates (RMRs) seem to have low song rates, independent of body size and corrected for taxonomic status (Read and Weary 1992). All these results suggest that singing in birds may be directly or indirectly limited by energy availability, but alternative explanations of the observed relationships cannot be excluded. Birds may, for instance, decrease their song rates during cold periods not because singing is energetically expensive but because high song rates are incompatible with efficient food-searching behavior.

Direct physiological measures of energy expenditure by birds during vocal performance are rare, probably because high vocalization rates are difficult to obtain in metabolic chambers (Gaunt 1987). In Carolina Wrens (*Thryothorus ludovicianus*), singing behavior resulted in a fivefold increase in oxygen consumption above resting rates (Eberhardt 1994), though the oxygen consumption measurements did not control for other (perhaps more expensive) activities performed during the song bouts (Horn et al. 1995, S. L. L. Gaunt pers. comm.). In domesticated Red Junglefowl (*Gallus gallus*), dominant males produce an average of 0.3–0.5 crows per minute, rates much lower than the song or note rates observed in many songbirds. In roosters fed ad libitum, crowing is energetically cheap because oxygen consumption during crowing is lower than oxygen consumption during other less energetically expensive activities, such as feeding. Unlike frogs and insects, roosters seem to consume little energy during crowing compared with other activities (Horn et al. 1995). To simulate field constraints (e.g., the dawn chorus), however, oxygen consumption measurements must be made in songbirds with unusually high song performances that have been exposed to severe food constraints.

Sound production in birds requires an airstream that circulates through the lung–air sac system and passes over syringeal membranes and through the upper respiratory tract. The contraction of respiratory, syringeal, postural, and other muscles, and the morphological characteristics of the vocal system are responsible for the temporal organization and fine acoustic structure of vocalizations (Greenewalt 1968, A. S. Gaunt 1987, Brackenbury 1989, Hartley 1990, Vicario 1991b). Perhaps physical or physiological constraints on the muscles or neurons responsible for sound production determine some parts of the temporal structure of birdsong.

Limitations may exist on the rate and intensity at which muscles can be contracted during note production and subsequently recover to their initial state. For example, muscle contraction may be more intense or may last longer in long notes than in short notes (e.g., pulses or trills). Muscle recovery may therefore be slower after the production of a long note, resulting in a longer internote interval. Thus, physical constraints on muscles may explain why longer notes are separated by longer internote intervals, and vice versa.

Constraints on temporal components of vocalizations in songbirds may reflect respiratory or physiological limitations of the motor system (see Hailman et al. 1987, Kreutzer 1990, Dhondt and Lambrechts 1991), and, as in frogs (K. D. Wells and Taigen 1989), neuromuscular fatigue may be responsible for a rapid decrease in song output throughout song bouts (Bekoff 1988b, Lambrechts and Dhondt 1988; but see Weary et al. 1988, 1991). If fatigue is a factor, the performance characteristics described above, such as drift in Great Tit songs, will likely be found in other rapid repetitive behavior driven by bursts of muscle contraction that vary considerably in length or rate; for instance, drift may occur in nonvocal drumming in animals (McGregor 1991), in the D strings of chickadee calls (Hailman et al. 1987), and in the repetitive tail movements of European newts (*Triturus:* A. J. Green 1991).

Whether singing is physiologically exhausting remains an unresolved problem (e.g., Lambrechts 1988, Weary et al. 1991). One could argue that ventilatory or syringeal muscles are adapted for endurance, but this endurance may depend on the level at which individuals perform and the level of performance to which they have been trained. In humans, for instance, the muscles of the legs are probably adapted for endurance, which explains why humans can run or walk long distances. But strong individual differences exist in the ability to run (e.g., trained vs. untrained); well-trained sprinters, for example, have great difficulty finishing a 10-kilometer run or a marathon. Great Tits normally sing about four months per year; their songs usually last up to 5 seconds and are separated by longer intervals (on average 20–40%; Weary et al. 1988). During the fertile period of the female, however, the PTS can be extremely high at dawn (on average 40–80%; see above), and some males produce—for a brief period—songs more than 20 seconds long and with intersong intervals of less than half a second. In addition, Great Tits sing mostly in a very stereotyped fashion, which means that the same muscles and neurons are continuously used during vocal performance. Because extreme song performances in Great Tits are rare, and training opportunities are therefore rare, the existence of neuromuscular exhaustion during extreme song performances cannot be rejected in this species.

Constraints on Composition and Frequency Components of Song Types

The frequency of notes and their rates of repetition in songs provide evidence of certain constraints in how songs are constructed. Song notes are not distributed randomly within phrases of Great Tit song types (Lambrechts unpubl. ms.). Great Tit song types can be classified into three groups: one-note song types (one note per phrase), two-note types, and more-than-two-note types. An example of each is shown in Fig. 17.1 (see also sonograms in McGregor and Krebs 1982). One-note song types have a frequency of about 4 kHz, and a single note is repeated relatively quickly throughout a song, so phrases are brief. In two-note song types, the lower frequency per type averages 3.8 kHz, and the upper frequency per type averages 5.1 kHz; more-than-two-note types average 3.8–6.0 kHz (see LF and UF in Fig. 17.1). Two-note and more-than-two-note song types thus have a similar lower frequency, but the upper frequency is significantly lower in two-note song types. Two-note types therefore have a narrower frequency range (see also M. L. Hunter and Krebs 1979). Note rates differ at different frequencies within song types, and also differ among song types (e.g., Fig. 17.1). Notes are at about 4 kHz when uttered at high rates (2.6–3.4 notes/second), as in one-note song types, some two-note types, and the low-frequency notes of more-than-two-note types; this is the species' commonest frequency. Low note rates (about 2.4 notes/second), which occur in high-frequency notes of more-than-two-note song types, are dominant at other frequencies. A relationship between frequency and percentage performance time also exists when calls and songs are compared. In Great Tits, common calls (e.g., scolding) or songs often have low frequencies. Rarer calls (e.g., alarm or copulation calls) have substantially higher frequencies (Gompertz 1961, Eyckerman 1979, Lambrechts unpubl. data).

Consecutive performances of a repetitive behavior are rarely identical, possibly because of constraints on motor control or inherent variation due to the many (e.g., physiological) processes involved in behavior (see, e.g., Schleidt 1974, Podos et al. 1992). In Great Tits, drift is highly variable within and among individuals. Furthermore, consecutive notes within songs differ in acoustical features such as form, intensity, and frequency (Podos et al. 1992; Fig. 17.2). Each song is therefore structurally unique, and each defined song type can be considered as a probabilistic group consisting of an "infinite" number of song variations (Podos et al. 1992).

In yearling Great Tits with crystallized song types and in older males, the frequency range of phrases may show abrupt modifications within songs, a phenomenon that I term song frequency plasticity (SFP). This plasticity can be quantified easily: in 7% of about 2600 songs of 34 individuals, the frequency range of consecutive phrases within songs differed by at least 300 Hz, and sometimes by more than 800 Hz (Lambrechts unpubl. ms., Fig. 17.2D). More interesting, Great Tit SFP seems to be related to song type structure (Lambrechts unpubl. ms.). It was especially pronounced when high-frequency notes clearly deviated from the species' commonest frequency (about 4 kHz) and were produced at relatively high rates (Fig. 17.2D). Song-frequency plasticity was less common in song types with notes closer to 4 kHz, regardless of note rate, and in song types with high-frequency notes produced at low rates. The percentage of songs with considerable SFP changed when individuals switched song types. Song types that deviated considerably from the average frequency or rate of note repetition often showed sharp changes in SFP. These data all suggest that some song types are difficult to repeat in a stereotyped fashion.

Songs with high SFP are rare in adult Great Tit song repertoires. Perhaps birds acquire song types that can be produced in a stereotyped fashion during the song-learning process, for instance through "selective attrition" of song types (Marler and Peters 1982c) and action-based learning (Marler 1990). Young birds are known to produce more songs than they eventually crystallize, and perhaps the songs that are culled from the repertoire are the more difficult ones, those that involve extreme frequencies, rates of note repetition, or song durations that cannot be produced consistently (cf. Lambrechts and Dhondt 1990). If this is true, then young birds should produce songs with a much wider frequency range or a longer duration than is present in crystallized song (Marler and Peters 1982b). Furthermore, to be crystallized, some song types may require more rehearsal than others. If this is so, phenomena such as SFP and drift should be less common toward the end of the singing season, maybe because of a fine adjustment of song structure.

Does Physiology or Morphology Constrain Frequency Components of Song?

Two individuals cannot possess identical vocal performance systems, so each individual has unique voice characteristics at some level (Tosi 1979). The size of the vocal apparatus varies in different animal species, and bigger animals tend to produce lower-frequency sounds. At least in this rough way, sound characteristics are influenced by morphological characteristics of the motor system. Such an influence occurs both within and across species (e.g., Ryan and Brenowitz 1985, Zemlin 1988).

The human larynx operates most efficiently at the emphasized frequency of normal speakers, which differs among individuals and is related to body size. In adult human singers, a particular frequency may be described as the "natural level" or the "optimum pitch level," which is situated at about the lower quarter of the total singing range. For each human singer a "voice register" also exists, consisting of a series of sounds over a frequency range that are performed with the same mechanical principle and have the same "quality." The upper or lower limits of the voice register are determined by the appearance of an abrupt modification in quality—breaks or voice transitions, for example. At very high frequencies the

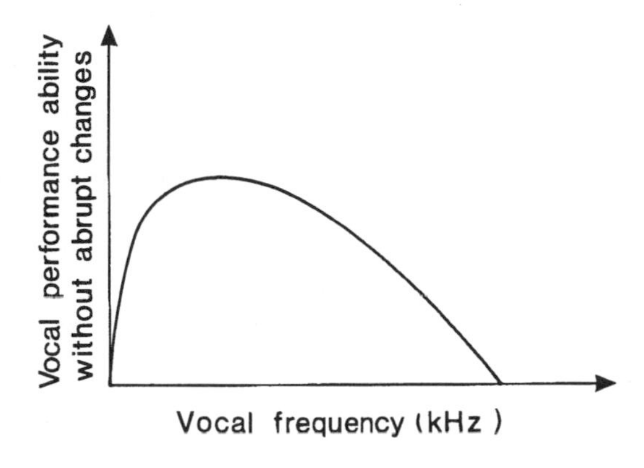

Fig. 17.4. Model of a relationship between singing ability and vocal frequency in adult birds. Performances above the curve result in abrupt vocal changes. The position of the curve on the X axis will vary within and among species. The form of the curve will depend on the acoustic features of interest (e.g., form, volume) and the species (see text).

voice becomes breathy, a phenomenon that has been attributed to a failure of the vocal folds (review in Zemlin 1988).

Similar frequency-related phenomena may exist in Great Tits. The commonest frequency of Great Tit song (4 kHz) is situated at the lower part of the frequency range (3–7 kHz). Notes produced at relatively high rates show abrupt modifications in frequency characteristics, especially at frequencies far from 4 kHz. Perhaps a voice register like the one in humans also exists in Great Tits.

I hypothesize that performances without abrupt changes in voice characteristics will occur over a relatively narrow frequency range (Fig. 17.4). Extreme performances of stereotyped, repetitive sounds (e.g., with loud intensity, high note rates, or long vocalizations at high rates) would be likeliest in this range, and less likely outside it. Furthermore, performance abilities should be best in about one-fourth of the total frequency range of the individual. At that frequency, syringeal membranes are probably relaxed and muscle contractions are minimized, and that frequency would be close to or at the frequency most often produced by birds. Abrupt changes in voice characteristics at high vocal frequencies could occur because muscle contraction is more intense at extreme vocal frequencies, or because more muscles are involved. Also, different types of vocalizations (e.g., calls vs. songs) may require different mechanisms of sound generation (see the reviews in A. S. Gaunt 1987, Brackenbury 1989), and performance abilities may be higher with one mechanism than another.

To determine optimal pitch levels for vocal performance and voice registers, one could attempt to teach the same or different hand-reared birds the same (artificial) tutor song type, holding note rate constant but raising and lowering the overall frequency. Tutor and learned song copies should differ outside the voice register of the hand-reared birds, and copies should be rather accurate if the songs are within the voice register. Therefore, song copies of hand-reared Great Tits should have high frequency plasticity at frequencies far from 4 kHz, even though the tutor songs do not show SFP at extreme frequencies. Also, observed voice registers should be narrower with high rates of note repetition (see above).

Motor Constraints and the Evolution of Birdsong

Energy allocation and song performance. If singing is time-consuming, then song performance and organization should be limited by the energy used in more critical, energetically expensive activities (e.g., Halliday 1987, Partridge and Endler 1987). The energy used in basic physiological processes—to maintain body temperature, to survive, and to reproduce—certainly differs among individuals both within and among species, and the adaptive allocation of energy likely has influenced the evolution of birdsong. Thus, bird species with high resting metabolic rates have lower song rates and continuity than species with lower resting metabolic rates (see above).

Constraints of the vocal organ and the antiexhaustion hypothesis for song switching. Control of the airstream through respiratory muscle contraction may govern the temporal organization of vocalizations, whereas variation in the position of the syrinx and its membranes may account for fine structure (e.g., modulations). The songbird syrinx has four or more pairs of intrinsic muscles working independently, permitting fine and rapid adjustment of the syrinx and of the structure of consecutive notes (A. S. Gaunt 1983, 1987, Brackenbury 1989).

Some species sing with "immediate" variety, switching repeatedly between song types (ABCDE . . . ; e.g., Sedge Warbler, *Acrocephalus schoenobaenus*). These species often possess long songs and complex song repertoires of more than 100 types. Other species sing with "eventual" variety, repeating a single type in a bout before a new type is introduced (AA BB CC; e.g,. Great Tits). The latter species often have short songs and possess small repertoires of up to 10 song types. Furthermore, some species have single-song repertoires (AA . . . ; e.g., Alder Flycatcher, *Empidonax alnorum;* see Hartshorne 1973).

A significant decrease in percentage of performance time throughout song bouts could reflect neuromuscular exhaustion in singers that produce long songs in a stereotyped fashion and at high rates. Because different muscles and neurons are involved in producing different sounds (A. S. Gaunt and Gaunt 1985b, Suthers and Hector 1985, Vicario 1991b), switching song types could be an adaptive mechanism to reduce neuromuscular exhaustion (Lambrechts and Dhondt 1988). Thus, when performance time is high, a significant decrease in the percentage of time spent singing is followed by a recovery after the bird switches to a new song type (Lambrechts 1988, Lambrechts and Dhondt 1988).

Song repertoires themselves might be an adaptation to reduce neuromuscular fatigue (Lambrechts and Dhondt 1988). Alternatively, song repertoires may have evolved for other reasons, with song switching as an antiexhaustion mechanism evolving later. Also, significant decreases in percentage performance time might be evidence that fatigue is occurring. If birds do not sing at their maximal ability, however, the decrease in percentage of performance time could be a mechanism to reduce neuromuscular exhaustion so that a singing performance can continue for a longer time.

Neuromuscular exhaustion can be expected in species singing with eventual variety that make extensive use of the same neuromuscular system to produce the songs. Because neuromuscular exhaustion should occur especially during prolonged and extensive muscle exercise (e.g., K. D. Wells and Taigen 1986), fatigue should occur especially in bout singers that vary considerably in both song duration and intersong interval, such as Great Tits, Tufted Titmice (*Parus bicolor*), and Cirl Buntings (*Emberiza cirlus*). For instance, Great Tit songs are usually 1–5 seconds in duration, but some males can produce songs of more than 20 seconds duration (Lambrechts and Dhondt 1988). In bouts that begin with long songs and short intersong intervals, the percentage of time spent singing should decrease rapidly, and abrupt modifications in voice characteristics (e.g., SFP) should be observed more frequently (Lambrechts and Dhondt 1988, 1990).

Species with single-song repertoires may have evolved special song performance adaptations to allow stereotyped singing over a longer period because they do not have the option to switch to a new song type. Therefore, stereotyped singers with single-song repertoires should have briefer songs with longer intersong intervals than stereotyped singers with repertoires.

High switching rates probably evolved in species that switch types repeatedly for reasons other than reducing neuromuscular exhaustion, because the switching rate is high and the number of repetitions per bout is low. Complex repertoires and high rates of switching may allow longer songs to be produced over longer periods, however, because stereotyped muscle activity and rate at the level of each muscle are reduced (Lambrechts and Dhondt 1988). In addition, special adaptations in note form, motor system morphology, sound-generating mechanisms, or respiration patterns (e.g., minibreaths vs. pulsatile expirations; A. S. Gaunt 1987, Hartley 1990) may have evolved to allow the performance of extremely long songs. In Common Grasshopper Warblers (*Locustella naevia*), for example, stereotyped songs are much longer (up to 95 seconds: Schild 1986) than in Great Tits, but the notes (pulses) are much briefer (Brackenbury 1978, Schild 1986). Perhaps the warbler pulses are physiologically less "exhausting" than the Great Tit notes. Note form may therefore influence performance ability and song duration.

Constraints on the composition of vocalizations: the antiplasticity hypothesis. We assume that for efficient communication to take place, the within-repertoire variation is large and birds ensure the distinctiveness of the signals in their repertoires (Kroodsma 1982b). Distinctiveness among signals can be achieved within repertoires if different vocalization types (notes, songs, or calls) differ in a consistent way from one another in structure (e.g., frequency components), and if the probability of abrupt vocal plasticity within a type (e.g., SFP) is minimized.

I hypothesize that when individuals deviate from individual-specific (or species-specific) frequencies to increase within-repertoire variation, they must change certain singing characteristics to sing with a minimum of plasticity. Males can thus decrease either the percentage of time spent singing, the rate of note repetition, or the intensity of sound at an extreme frequency. At 6 kHz, for example, male Great Tits produce on average about 2.4 notes per second, because a much higher note rate (e.g., 4 notes/second) would result in a higher proportion of notes with abrupt alterations in frequency characteristics (SFP). Males can also reduce plasticity, such as drift or SFP, at frequency extremes if they produce briefer songs at a lower rate. These adjustments in singing behaviors at extreme frequencies may be adaptive mechanisms to facilitate stereotyped singing over a broad frequency range.

A slow note rate at a given extreme frequency may be an adaptation to reduce song plasticity at that frequency (see above). Hence, at extreme frequencies, one-note song types cannot produce a consistent song pattern because a slow rate of note repetition would break up stereotyped songs into series of isolated notes. Songs do not break up, however, if one or more notes with a common frequency are sung in between notes with an extreme frequency. I therefore suggest that complex song types with combinations of notes at different frequencies have evolved both to preserve song structure by shortening internote intervals between consecutive notes and to minimize song plasticity at these frequencies by increasing internote intervals at extreme frequencies (long glissando-shaped notes may have evolved for similar reasons). This adaptive hypothesis of song type complexity may explain why Great Tits evolved more-than-two-note song types, and why they did not evolve one-note types at extreme frequencies.

Conclusions

Whereas neurobiological investigations have begun to link learning, hearing, and brain morphology with vocal organization, relatively little is known about the physiological and morphological mechanisms that generate sound (e.g., A. S. Gaunt 1987). Only recently has it been suggested that limitations of the sound-generating apparatus may be partly responsible for the way birds perform and organize their vocalizations. I suggest that the causes and consequences of motor constraints need to be studied using six interdisciplinary approaches.

1. Comparative studies within and across species are needed to establish the relationships among time attributes (e.g., song duration, drift, percentage of performance time, switching rate of note types) and frequency attributes (e.g., frequency range, glissando-shaped notes) of vocalizations, and thus to test predictions about nonrandom delivery of notes within and across vocalizations (see above). The following questions are just a small sample of those that need answers: Are long songs separated by short intersong intervals at the start of a song bout, as predicted by the motivation hypothesis? Are long songs separated by long intersong intervals at the end of a bout as predicted by the fatigue hypothesis? Are abrupt vocal changes such as drift and SFP more likely at the end of a song bout and less likely in short songs? Is the rate of note repetition slower or the sound intensity lower at extreme frequencies than at common frequencies? Is the fre-

quency range of the total song repertoire broader in birds singing with immediate variety than in birds singing with eventual variety? Are long notes separated by long internote intervals, and does note form influence this relationship (e.g., long glissando-shaped notes may have relatively shorter internote intervals)?

2. Voice registers and optimal pitch levels for vocal performance need to be described. This will require detailed descriptive studies of variations in note structure over a broad frequency range, and learning experiments with hand-reared birds copying well-defined song types without abrupt vocal changes (see above).

3. Learning experiments with hand-reared birds should determine if, and what kind of, song types are difficult to repeat in a stereotyped fashion. Studies of song ontogeny may establish whether song types that are difficult to perform are excluded from the adult repertoire, and whether rehearsal or fine adjustments in song structure decrease abrupt vocal changes in these song types. Individual birds therefore must be followed throughout the singing season.

4. Performance features within song bouts should be examined in relation to food availability, ambient temperature, parasite loads, and physical fitness of the singer. A bird's health or nutritional status could be measured and experimentally manipulated (cf. Gustafsson 1993). Birds in poor condition may show more drift or sing briefer songs than healthy birds. The dawn chorus is an interesting period to study because food constraints may limit performance time (Lambrechts 1988, Cuthill and Macdonald 1990).

5. Physiological studies of the muscles responsible for sound generation and modulation are required to establish causal relationships among song performance, neuromuscular fatigue, muscle glycogen reserves, oxygen consumption, motor control, sound-generating mechanisms, and other factors. Because muscular fatigue is probably related to muscle fiber structure and muscle development, histological research might determine the types of muscle fibers used in vocal performance (e.g., Suthers and Hector 1985), and experimental work might establish positive relationships between muscle development and performance components.

6. Playback experiments can be used to investigate the consequences of motor constraints for intraspecific communication. Drift, song duration, performance time, song rate, and abrupt vocal plasticity may reflect "male quality," and therefore may be used in male-male competition or in female attraction (see Lambrechts 1992). The antiplasticity hypothesis predicts that increased abrupt vocal plasticity within vocalization types reduces the ability to discriminate among vocalization types or among individuals (e.g., neighbor-stranger discrimination). This prediction needs to be verified.

Researchers studying birdsong focus on the causes and consequences of between-song-type variation (e.g., song dialects, song sharing), but within-song-type variation (e.g., drift, SFP) is rarely studied, and its causes are not understood. I therefore finally suggest that a multidisciplinary study of within-song-type variation may help to determine the role of motor constraints in the development, production, and organization of birdsong, and in intraspecific acoustic communication.

Acknowledgments

I thank D. E. Kroodsma and E. H. Miller for their advice and valuable comments on the manuscript, and their invitation to write this chapter, and A. A. Dhondt and R. F. Verheyen for the use of spectrum analyzers at Antwerp University.

18 Bird Communication in the Noisy World

Georg M. Klump

Birds, with their elaborate vocal repertoire and their prominent use of acoustic signals in communication, have been excellent subjects for studying how signals are transmitted and perceived in noisy environments and how the structure of vocalizations can be optimized to achieve this goal. Some signals, such as the high-pitched aerial predator calls of many small birds, are designed to be transmitted only short distances or to a limited audience (Marler 1955, Klump et al. 1986). Other signals, however, such as songs used in territorial defense or mate attraction, presumably evolved under selection pressures to maximize transmission distance (but see Lemon et al. 1981). The maximum transmission distance limits the possibility of communication to a particular volume. This volume, in which a signal can be detected and recognized by the receiver, is the "active space" of the signaler (Marten and Marler 1977, Brenowitz 1982c).

The factors that determine the active space have been an important issue in research on acoustic communication in birds (E. S. Morton 1975, Dooling 1982, R. H. Wiley and Richards 1982). In their review, R. H. Wiley and Richards (1982) comprehensively covered all aspects of the physical determinants of acoustic signal transmission that lead to the frequency-dependent attenuation of the signal with increasing distance from the emitter. When the signal's amplitude is reduced to a level equal to the sensory threshold of the receiver, the maximum transmission distance of the signal has been reached. The absolute signal amplitude, however, is not the only important component that determines the maximum distance from the sender at which the signal can be detected. The level and the spectral and temporal characteristics of background noise are other major determinants of the active space for a bird's vocal signal. Because noise is ubiquitous in birds' natural environments, its masking effect on acoustic signals frequently limits their communication over long distances (see, e.g., Brenowitz 1982c, Ryan and Brenowitz 1985, Dabelsteen et al. 1993).

Our knowledge of how birds perceive signals in noise has grown notably in many respects since the early 1980s. Dooling (1982), in his review on auditory perception in birds, could refer to studies of hearing in noise for only a single avian species, the Budgerigar (*Melopsittacus undulatus*). Today, data on thresholds for signals masked by broad-band noise are available for more than 10 bird species (reviewed below). These data permit us to estimate the signal-to-noise ratio necessary for the detection of signals in the natural environment. Together with results from experiments in which signals were masked by narrow-band noise, this information allows us to infer the bandwidth of the perceptual channels in the birds' auditory system—that is, to deduce the filter characteristics of the bird's auditory system.

Perception studies have progressed beyond just measuring the frequency selectivity of birds' auditory systems to the study of other factors that determine the audibility threshold for signals in a masking background noise. Three mechanisms that could considerably increase perception distances for acoustic signals in the natural environment are discussed below: (1) integrating information across different frequency channels in the auditory system, (2) using the temporal pattern in the background noise, and (3) exploiting the directionality of the peripheral auditory system to increase signal-to-noise ratios in perception.

The theoretical concepts applied to the study of detection of signals in noise can be extended to a discussion of the recognition or classification of signals. In a recognition task, a bird has to discriminate one type of signal against a background of other types of signals. A typical example would be the discrimination between conspecific and heterospecific songs. In this chapter I show that the responses of birds in the context of song recognition can be explained within the framework provided by signal detection theory, a concept that has been useful in analyzing the detection of signals in noise.

The Mechanism of the Detection of Signals in Noise

Most researchers studying signal detection in noise assume that the recipient compares the perceptual input during two time periods, one with only a noise background and one when a signal has been added to the noise. Because the receiver has no a priori knowledge when a signal has been sent, it must continuously scan the perceptual input for signals of interest. The key components of such a scanning mechanism are an encoder for the signal amplitude, followed by a device that computes a running average of the perceived signal energy (e.g., a leaky integrator). The output of this device feeds into a decision element that reports the presumed occurrence of a signal. In this energy detector model, the decision element presumably will detect a particular increment in the energy of the input and interpret this increment as a signal event. Often, this increment threshold of the detector is assumed to be 3 dB, the level increase at which the signal and the background noise have the same energy at the detection threshold (i.e., the signal-to-noise ratio is 1, which is equivalent to a signal-to-noise ratio of 0 dB). Some results of behavioral studies in the laboratory suggest, however, that even lower signal-to-noise ratios (e.g., a signal-to-noise ratio of –4 dB) may be sufficient for birds to detect signals (see Dooling and Searcy 1981, Klump and Okanoya 1991).

Because the acoustic signals of interest to a bird are usually limited to a certain frequency range, a receiver can considerably increase the signal-to-noise ratio in its auditory system. By analyzing the perceptual input separately in different frequency bands, a receiver can improve signal detection. The bird's peripheral auditory system provides for such a frequency analysis (for a review, see Manley 1990) and can be viewed as a set of band-pass filters that cover the total range of hearing. The energy detector model can also be applied to the perceptual input in the individual analysis channels provided by these band-pass filters. How wide these filters are, what signal-to-noise ratio is critical for the detection of signals in individual frequency channels, and how the information from different frequency bands is then combined are the major questions in the laboratory studies on signal detection in noise.

The Active Space of Red-winged Blackbird Song: A Case Study

Few field studies have experimentally studied how birdsong is detected amid the background noise in the natural environment. Brenowitz (1982c), who carried out such a study on Red-winged Blackbirds (*Agelaius phoeniceus*), emphasized the difficulty of deciding in a field setting whether a bird did not respond because it failed to detect the signal or because it was not motivated to repond even though it detected the signal (see also Stoddard, this volume). Brenowitz addressed this problem by using song signals of an amplitude sufficient to elicit a response when played to a territory holder without masking by background noise (Brenowitz 1982b). He then added broad-band noise to the playback. Songs were presented at the same intensity as before, but now at various signal-to-noise ratios, ranging from 0 to 24 dB, measured in the 4-kHz octave band in which most of the spectral energy in the song is found. The birds did not respond when the signal-to-noise ratio was 0 dB, but at a signal-to-noise ratio of 3 dB the strength of the response was similar to that measured in playback of an unmasked song. The sharp transition between no response and full response suggests that detection and recognition of the song, rather than the bird's disposition to respond, determined the outcome of the experiment.

How do the results of this field study relate to laboratory data on signal detection in noise? Hienz and Sachs (1987) determined the critical signal-to-noise ratio that Red-winged Blackbirds need to detect pure tones in continuous broad-band noise (for details on how such studies are done, see below). From their data it is possible to calculate the signal-to-noise ratio needed for the detection of a 4-kHz pure tone in wide-band noise as it would be measured in the 4-kHz octave band, thus making the laboratory results comparable with those obtained by Brenowitz in the field (1982c). The signal-to-noise ratio at the detection threshold predicted by the laboratory study was –5 dB, lower than the 3-dB threshold reported by Brenowitz (the calculations are based on a critical masking ratio of 29.5 dB that was determined for 4-kHz tones with a continuous noise masker of a spectrum level of 24 dB sound pressure level [SPL]; Hienz and Sachs 1987).

This discrepancy may be explained in two ways. First, the bandwidth of the song Brenowitz used as a stimulus in the field differed from the bandwidth of the pure tone used in the laboratory. The sound energy of the song probably was spread over more than one frequency channel of the Red-winged Blackbird's auditory system, whereas all sound energy of the pure-tone signal was concentrated in a single frequency channel. As a result, the effective signal-to-noise ratio within a single frequency channel would have been higher for the pure tone than for the song. Furthermore, the 4-kHz analysis channel in the Red-winged Blackbird's auditory system probably has a much narrower bandwidth than the bandwidth of the 4-kHz octave band used by Brenowitz (1982c) in defining the signal-to-noise ratio. This difference would also result in a higher signal-to-noise ratio measurement in the field study using the song as a stimulus than in the laboratory study using a tone. Second, if a blackbird in the field is to respond, it must not only detect the signal but also must recognize it as the song of a Red-winged Blackbird. Perhaps a higher signal-to-noise ratio is necessary for recognition than for simple detection. Laboratory studies of song recognition in background noise, which would allow a direct comparison of the signal-to-noise ratios necessary for recognition versus detection, have yet to be done.

Common Procedures Used in Laboratory Studies on Signal Detection in Noise

The most detailed knowledge about the mechanisms involved in the detection of signals in noise comes from standardized experiments performed in acoustically controlled laboratory environments. Rather than relying on the bird's natural propensity to respond (as it would, e.g., in territorial interactions), researchers in the laboratory use operant paradigms to control the subject's motivation to respond and to manipulate the criterion on which the decision process is based. The theoretical framework for these psychophysical experiments is provided by the signal detection theory (e.g., see D. M. Green and Swets 1966, McNicol 1972). In this theory, a discrimination measure, d′, is derived that represents the perceptual difference between two stimulus conditions (Fig. 18.1). The discrimination measure thus provides a "currency" with which to compare results from different experiments and modeling efforts (for the potential application of the signal-detection theory and d′ in field studies, see R. H. Wiley and Richards 1982).

Two procedures have been used to determine d′. The first is the GO/NOGO procedure, in which the subject must perform an operant GO task when it hears a signal in the background noise but should not give the GO response (i.e., should

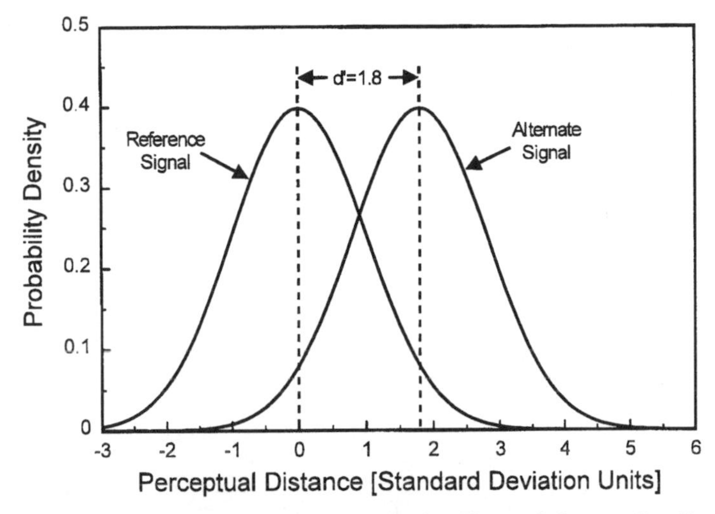

Fig. 18.1. The discrimination measure frequently used to describe differences in the perception of two stimulus conditions, d′ is derived in signal detection theory by a model shown here. Two probability-density functions, approximated by normal distributions, summarize the statistical distribution of states (e.g., levels of excitation) in the perceptual system. One normal distribution represents the perceptual states elicited by the reference signal; the other distribution represents states elicited by an alternate signal that differs from the reference and is discriminated. In the model, the distribution of states elicited by the reference signal (e.g., resulting from the perception of a background noise alone) is plotted with mean = 0 and SD = 1. The probability-density function of the states in the perceptual system for the alternate signal condition (e.g., elicited by a signal embedded in the noise) is then plotted on the scale calculated for the reference distribution. The difference between the means of the two distributions (indicated by the dashed lines) is d′. It can be inferred from the observation of the subject's behavioral responses in the reference condition and in the alternate signal condition (D. M. Green and Swets 1966).

show a NOGO response) when it hears the background noise alone. Thus, when a subject correctly reports the occurrence of a signal in the background noise, it scores a "hit." When a subject incorrectly reports a signal when the background noise alone is presented, it scores a "false alarm." The discrimination measure d′ is then determined by the percentage of hits and false alarms. For example, a hit rate of 52% and a false alarm rate of 4% would indicate a d′ of 1.8 (Table 18.1). A larger hit rate with the same false alarm rate would indicate a d′ larger than 1.8. Threshold values d′ = 1.0 or d′ = 1.8 are commonly used when two stimulus conditions in psychoacoustical studies are defined as being perceived differently. A higher value of d′ at threshold means a more conservative threshold estimate. Another commonly used definition of a threshold in a GO/NOGO procedure is a hit rate of 50% and a relatively low false alarm rate (usually well below 20%). This approach ensures that at a perceptual threshold d′ is mostly above 1.0 (Table 18.1). If the rate of false alarms and the percentage of hits at the perception threshold are given, the value of d′ can be calculated for comparison with results from other studies.

A second procedure that has been successfully applied in psychophysical studies is an operant paradigm in which the animal is presented with a choice between

Table 18.1. Behavioral response measures that define discrimination or detection thresholds for GO/NOGO and two-alternative-forced-choice (2AFC) procedures

False alarms (%)	d′ = 1	d′ = 1.8
1	9.2	29.9
2	14.6	40.0
3	18.9	46.8
4	22.6	52.0
5	26.0	56.2
6	29.0	59.7
7	31.7	62.7
8	34.3	65.4
9	36.7	67.7
10	38.9	69.8
11	41.0	71.7
12	43.1	73.4
13	45.0	75.0
14	46.8	76.4
15	48.6	77.7
20	56.3	83.1
% correct in 2AFC procedure	76.0	89.9

Notes: Frequently used threshold criteria are values of 1.0 or 1.8 for the discrimination measure d′ that is derived from signal-detection theory (see Fig. 18.1). In a GO/NOGO procedure, d′ can be calculated from the percentage of false alarms and hits. This table assesses the performance of a bird in a behavioral experiment. If a subject shows a false alarm rate of 5% in a GO/NOGO procedure, for example, it should correctly report the GO stimulus in more than 26% of the trials for above-threshold performance (criterion d′ = 1). In a 2AFC procedure, d′ can be calculated solely on the basis of the percentage of correct responses; the critical values for above threshold preformance are given in the table (for a more detailed table of d′ in relation to hit and false alarm rates, see Swets 1964).

two alternatives (e.g., go left/go right). In this two-alternative-forced-choice (2AFC) procedure, the probability of a correct response with no discrimination between the stimulus conditions is 50%. The discrimination measure can be calculated on the basis of the percentage of correct responses. Typically, either 71 or 75% correct responses are used as a threshold criterion (71 and 75% correct responses correspond to a d′ of 0.78 and 0.95, respectively).

Both the GO/NOGO procedure and the 2AFC procedure can be applied using two different methods of stimulus presentation: the method of constant stimuli and the method of limits. In the former, the stimulus parameter (e.g., the signal-to-noise ratio in an experiment studying detection of a sound in a noise background) is varied randomly between trials within a certain range. In the method of limits, the experimenter tries to track a threshold by systematically increasing or decreasing the signal-to-noise ratio from one trial to the next in a staircase procedure. When carefully applied, both methods yield reliable detection thresholds. Readers interested in a more detailed discussion of the psychophysical procedures, which is beyond the scope of this chapter, are referred to the contributions in *Methods in Comparative Bioacoustics,* edited by G. M. Klump et al. (e.g., Dooling and Okanoya 1995, Okanoya 1995).

The Bandwidth of Bird Auditory Filters: Critical Masking Ratios and Critical Bandwidths

From a physiologist's point of view, the birds' auditory system is composed of a set of band-pass filters that separate the frequencies in the sound spectrum. Listening through these filters, the bird selectively detects the signals in its noisy environment. The auditory band-pass filters are thought to represent the tuning characteristics of the sensory cells (the hair cells) on the papilla basilaris in the bird's inner ear (Manley 1990). The filters let a part of the background noise spectrum pass through as is determined by their filter function, but the band-passed noise masks the signals for which the filter is optimally tuned. Fletcher (1940), who pioneered the studies of frequency selectivity in the human auditory system, suggested that the bandwidth of the band-pass filters in the auditory system can be calculated if one assumes that the filter function has a rectangular shape and that, at the detection threshold, the energy of the signal passing through the filter is equal to the energy of the masker. To make computations easy, researchers commonly use a pure tone as the signal and a Gaussian white noise as the masker, in which the energy in the spectrum is distributed evenly. This approach provides a constant density level of the noise, which is called the spectrum level if the sound intensity is defined with reference to 1-Hz-wide bands of the noise. The ratio of the intensity of the signal to the spectrum level of the masker at the detection threshold, called the critical masking ratio, is then taken as a measure of the frequency selectivity of the filter. For the determination of critical masking ratios, the total bandwidth of the masking noise ideally should be large enough to cover the hearing range of the species under study. However, it is also sufficient if the bandwidth is larger than the real bandwidth of the auditory filter under study (not just of its theoretical equivalent with a rectangular band-pass shape that lets the same amount of energy pass).

Critical masking ratios provide a measure of the signal-to-noise ratio necessary for the detection of a sound in a background noise and these have now been determined for more than 10 bird species (Fig. 18.2). These ratios can be determined relatively easily and are the most frequently used measure of the frequency selectivity of a bird's auditory system. In general, the critical masking ratio (CR) is frequency dependent. At 1 kHz, the median CR for all birds studied was 23.2 dB; at 2 and 4 kHz the birds' median CRs were 24.0 and 28.8 dB, respectively. In most bird species the frequency dependence of the CR follows the pattern observed in mammals, increasing monotonically with increasing frequency (Fig. 18.2); the increase in the CR averages about 2–3 dB per octave. Some bird species, however, show a nonmonotonic relationship between frequency and CR. A number of studies have confirmed the unusual relationship between CR and frequency in the Budgerigar (Dooling and Saunders 1975, J. C. Saunders et al. 1979, Okanoya and Dooling 1987), in which the CR monotonically decreases with increasing frequency up to about 3 kHz, and then rises steeply above this frequency. A morphological study of the Budgerigar's inner ear, however, re-

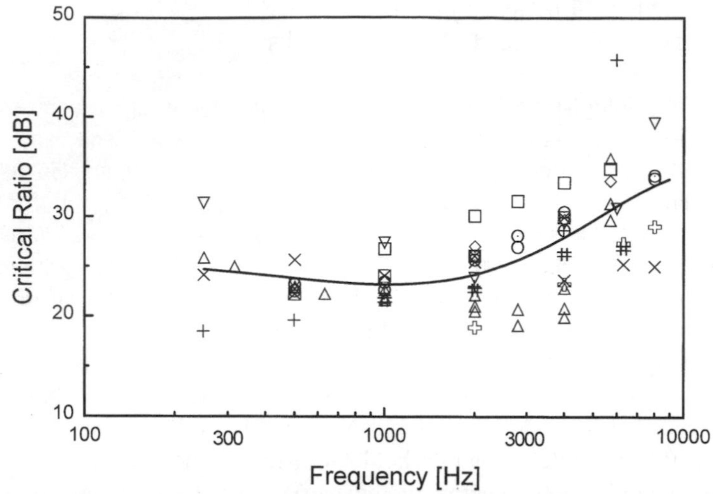

Fig. 18.2. Critical ratios for 10 bird species in relation to the frequency of the test signal, and the least-squares fit of a seventh-order polynomial to the data (solid line = "average bird"). The critical ratio is a measure of a bird's ability to detect a pure-tone signal in a wide-band background noise. It is the signal-to-noise ratio at the masked threshold if the sound intensity of the noise is given as the spectrum level. Smaller values indicate better discrimination ability. In combination with knowledge of the levels of background noise, the critical ratio can be used to predict masked thresholds in the natural environment (see Figure 18.3). Critical ratio data determined in continous background noise are shown for the Budgerigar (△ *Melopsittacus undulatus:* Dooling and Saunders 1975, J. C. Saunders et al. 1979, Okanoya and Dooling 1987), the Cockatiel (⊠ *Nymphicus hollandicus:* Okanoya and Dooling 1987), the Rock Pigeon (+ *Columba livia:* Hienz and Sachs 1987), the Zebra Finch (◇ *Taeniopygia guttata:* Okanoya and Dooling 1987), the Island Canary (□ *Serinus canaria:* Okanoya and Dooling 1987), the Great Tit (× *Parus major:* Klump and Curio 1983, U. Langemann, B. Gauger, and Klump unpubl. data), the Song Sparrow (○ *Melospiza melodia:* Okanoya and Dooling 1988), the Swamp Sparrow (⊙ *Melospiza georgiana:* Okanoya and Dooling 1988), the Red-winged Blackbird (▽ *Agelaius phoeniceus:* Hienz and Sachs 1987), the Barn Owl (✙ *Tyto alba:* M. L. Dyson and Klump unpubl. data), and the Common Starling (# *Sturnus vulgaris:* Langemann et al. 1995). There is considerable interspecific variation.

vealed no specializations correlating with this unusual CR function (Manley et al. 1993). In the Red-winged Blackbird, the Brown-headed Cowbird (*Molothrus ater:* Hienz and Sachs 1987), and the Great Tit (*Parus major:* Klump and Curio 1983, U. Langemann, B. Gauger, and Klump unpubl. data), the CR does not vary much with frequency.

Because we assume equal energy in the signal and the background noise at the detection threshold, the corresponding critical-ratio bandwidths are only an estimate of the real bandwidths of the auditory system's filters. In humans, this assumption holds only for frequencies below 500 Hz. At higher frequencies the signal energy at the detection threshold is less than expected from the equal energy assumption. The difference may be up to 6 dB at frequencies of 6 kHz and above (see, e.g., Zwicker et al. 1957); when averaged over the human's whole hearing range, it is about 4 dB. Estimated filter bandwidths are thus often too small when determined from critical masking ratios. We should measure the auditory filter bandwidths in birds more accurately, especially if the goal is to predict perceptual limits on the basis of a functional model of a bird's auditory system.

A more direct estimation of the auditory filter bandwidth (also called the critical bandwidth or critical band) compares the signal-to-noise ratios at threshold for signals in masking noise of different bandwidths (for a review, see Scharf 1970). If the spectrum level of the noise is kept constant while the bandwidth is reduced, and if the noise bandwidth is wider than the bandwidth of the auditory filter, we will observe a constant signal-to-noise ratio at the detection threshold for the signal. When the bandwidth of the masking noise is reduced below the bandwidth of the auditory filter while the spectrum level of the masker is kept constant, the signal-to-noise ratio at the detection threshold will decrease because less energy of the masker falls inside the filter's pass-band. This method has been used to determine the width of the auditory filters in the Budgerigar (J. C. Saunders et al. 1978, 1979) and the Common Starling (*Sturnus vulgaris:* Langemann et al. 1995). In the Budgerigar, the critical ratios underestimate the bandwidth of the auditory filters by a factor of about 2.5, so the error made in estimating the filter bandwidth using critical ratios is similar to the error made in humans (see above). In the Common Starling, however, the bandwidths of auditory filters that are estimated from critical ratios are in good accord with the bandwidths determined directly by band narrowing.

A theoretical model of auditory masking (Buus in press) also predicts a decrease in the threshold with increasing signal duration, a phenomenon called temporal summation. Indeed, a considerable improvement in the detectability of signals in noise with increasing signal duration has been found in the Budgerigar (Dooling and Searcy 1985). For signals with a duration of less than 200 milliseconds, the Budgerigar's threshold for the detection of a 2.86-kHz tone in noise was reduced by about 3 dB per doubling of the signal duration. Temporal summation in birds seems to be widespread (e.g., Dooling 1979, Dooling and Searcy 1985, Klump and Maier 1990) and will probably prove to have a great influence on the detection of vocal signals in the natural environment. It could explain the adaptive value of Eurasian Blackbirds' (*Turdus merula*) use of elements of a mean duration of 200 milliseconds in the long-ranging motif part of the song (Dabelsteen 1984b), and why the aerial predator calls of many small songbirds, addressed to their flockmates, have a duration of about 500 milliseconds (Klump and Shalter 1984).

Because critical-masking ratios in birds are relatively independent of the level of the masking noise (see, e.g., Okanoya and Dooling 1987, 1988), this measure can be used for a rough estimate of masked thresholds of birds in the natural environment if the spectrum and level of the background noise are known. The average spectrum of wind-generated noise measured in a deciduous temperate oak forest at moderate wind speeds, for example, has most of the energy at low frequencies, and the spectrum level of the noise steadily declines with increasing frequency (Fig. 18.3). If we add the critical masking ratios (e.g., the "average" bird CR function) to the spectrum level of the masking noise, we arrive at the predicted masked thresholds (upper curve in Fig. 18.3). In our example, the

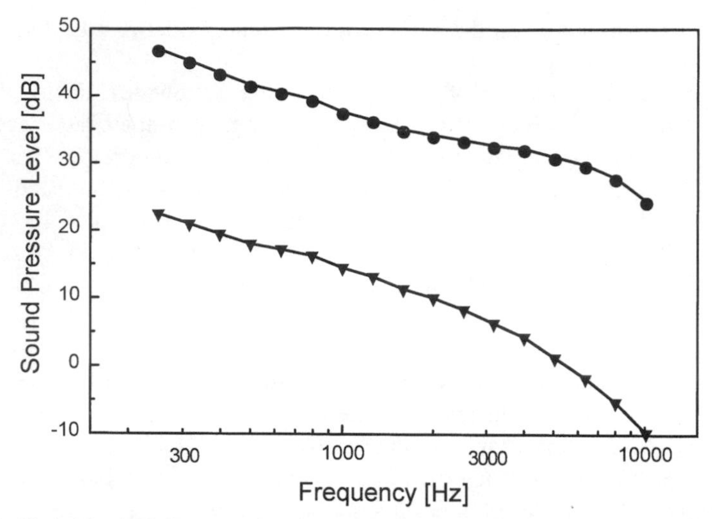

Fig. 18.3. Masked thresholds (upper curve) can be estimated if the spectrum of the background noise in the natural environment and the bird's critical ratio are known. The lower curve shows mean data of the spectrum level of background noise (i.e., the sound intensity, in 1-Hz bands of noise) computed from hourly measurements of the sound pressure level in third-octave bands recorded in a deciduous wood near Witten, Germany, throughout a summer day with moderate windspeeds (1–5 Beaufort, Klump unpubl. data). The masked thresholds shown in the upper curve result from adding the bird's critical ratio (here the "average bird" function shown in Fig. 18.2 was used) to the spectrum level of the wind-generated masking noise.

masked thresholds are between 30 and 40 dB for the range of sound frequencies found in bird vocal signals. Birds with a CR function that is relatively constant over a wide range of frequencies would experience masked thresholds that parallel the spectrum level curve of the noise (e.g., a Great Tit would benefit from using very high frequency vocal signals; Klump and Curio 1983). The low-pass characteristic of the noise in the natural environment may explain why so often the dominant frequencies in the vocalizations of bird species lie above the frequencies to which their hearing system is most sensitive (e.g., Konishi 1970, Dooling et al. 1971, Dooling et al. 1978). It is the masked thresholds, not the absolute thresholds, that determine the hearing system's capabilities in the noisy world.

The steady calling of insects (especially in the Tropics) represents another source of noise that has a profound influence on birds' optimal frequencies for communication. Because noise generated by insects consists predominantly of high-frequency components, it will set an upper limit on the suitable frequencies for communication. Thus, the noise generated by wind plus insects will leave the birds with a spectral sound window in which to try to fit their signals (Brenowitz 1982c, Ryan and Brenowitz 1985).

Although the psychophysically determined critical masking ratios and the measurements of the level of background noise provide us with the most complete data set presently available for estimating masked thresholds in the natural environment, we ignore a number of factors that affect masked thresholds if we look only at these two measures. The stimulus conditions that have been used in the laboratory experiments are rather artificial and may deviate from conditions in the bird's real acoustic world. The effects described in the next two sections can have considerable influence on estimates of masked thresholds, and clearly more research is needed before we arrive at an accurate account of masking in the bird's natural environment.

Spectro-temporal Release from Masking

One factor we have ignored so far is the statistical variation in the amplitude of sound signals in the natural environment. The wide-band noise so commonly used in laboratory experiments has a relatively constant envelope, but signals transmitted over some distance in the natural environment may show considerable fluctuations in amplitude (Richards and Wiley 1980, R. H. Wiley and Richards 1982). These amplitude fluctuations, mainly the result of air turbulence caused by heat or wind, have been viewed as a constraint on the transmission of certain characteristics of vocalizations, such as rapid amplitude modulations (R. H. Wiley and Richards 1982). The amplitude fluctuations are, however, imposed not only on the signals of interest but also on the noise signals that form the background in which the bird has to detect a call or song. Humans can improve their detection of signals by exploiting amplitude fluctuations in the background noise (reviews by B. J. C. Moore 1990, 1992). A considerable release from masking is seen in humans when a masking wide-band Gaussian noise is replaced with an amplitude-modulated noise of the same bandwidth and long-term energy (see, e.g., J. W. Hall et al. 1984, Schooneveldt and Moore 1987). J. W. Hall et al. (1984) first described this release from masking in a signal detection task by coherent (i.e., simultaneous and in-phase) amplitude fluctuations in different frequency bands of a masker, and called the effect "comodulation masking release." The authors suggested that the receiver might improve the detection of signals by using the correlated amplitude fluctuations of the masker in different frequency channels of the auditory system (represented by the critical bands).

In general, the models that try to explain comodulation masking release form two classes. The first class emphasizes the comparison of the incoming information in different frequency channels of the auditory system—that is, in different critical bands. It is suggested that the detection of a signal added to one of the channels may be enhanced by a cross-spectral comparison of the time course of the signal amplitude in different analysis channels. If a masking signal shows correlated amplitude fluctuations in different frequency channels, the addition of a signal to one of the channels will reduce the correlation between the amplitude fluctuations in the different channels, which then is detected by the auditory system (the model proposed by Buus [1985] involves cancellation of the correlated envelope fluctuations in different frequency channels). Alternatively, with correlated amplitude fluctuations in different frequency channels, a low amplitude of a masker in one channel could predict a good time to listen for a signal embedded in the masker in another frequency channel (the "dip-listening hypothesis": Buus 1985).

The second class of models emphasizes cues available within a single frequency channel. Schooneveldt and B. J. C. Moore (1989) suggested, for example, that the auditory system detects the signal by observing a change in the modulation pattern of the envelope of the acoustic signal embedded in the masker. A signal in a comodulated masker may be detected by the observed steadiness in amplitude for the duration of the signal. Other authors suggested that the temporal masking pattern is a major factor for comodulation masking release (e.g., Gralla 1993). If the energy in different frequency bands of a masking noise is coherently amplitude modulated, then the envelope of the masker will show large amplitude fluctuations. At times of large amplitudes of the masker, the amount of masking will be large, and at times of small amplitudes the amount of masking will be small (given that the forward masking effect of large masker amplitudes declines sufficiently rapidly with time). The auditory system then exploits this variance in the amount of masking and detects more signals than it would in a more steady masker, in which fewer and shorter episodes with low masker amplitudes occur. Results from experiments on humans suggest that both within- and between-channel cues are important in determining the amount of masking release.

Birds experience a considerable release from masking under the same conditions in which comodulation masking release is found in humans (Klump and Langemann 1995). The amount of masking release in relation both to the bandwidth of the masker and to the rate of envelope fluctuations in the Common Starling, for example, is similar to that found in humans (Fig. 18.4; Schooneveldt and Moore 1987). For rates of amplitude fluctuations comparable with those imposed on signals during transmission in the birds' habitat (envelope frequencies of 25 Hz or below; Richards and Wiley 1980), an average release from masking of 15 dB was found. Thus, if birds use this ability to exploit the temporal fluctuations of a masker in nature, they will be able to detect signals in background noise at much greater distances.

The results of speech intelligibility tests in humans (Fastl 1993) provide further evidence that receivers can improve recognition by exploiting the temporal structure of a masking noise. For example, if speech tokens are presented in noise with a speechlike spectrum and with a spectrum of the envelope frequencies typically found in speech, the signal-to-noise ratio necessary for the recognition of speech elements is reduced by 10 dB. In the control for this experiment, the masker was a continuous noise with a speechlike spectrum lacking the envelope fluctuations found in speech, and the signal-to-noise ratio resulting in 50% correctly identified speech tokens was about −10 dB (i.e., the overall level of the signal was 10 dB below the level of the masker). When the masker was competing speech, the signal-to-noise ratio necessary for the recognition was reduced even more, to −18 dB. Speech intelligibility tests in humans also show that repetition of a signal can

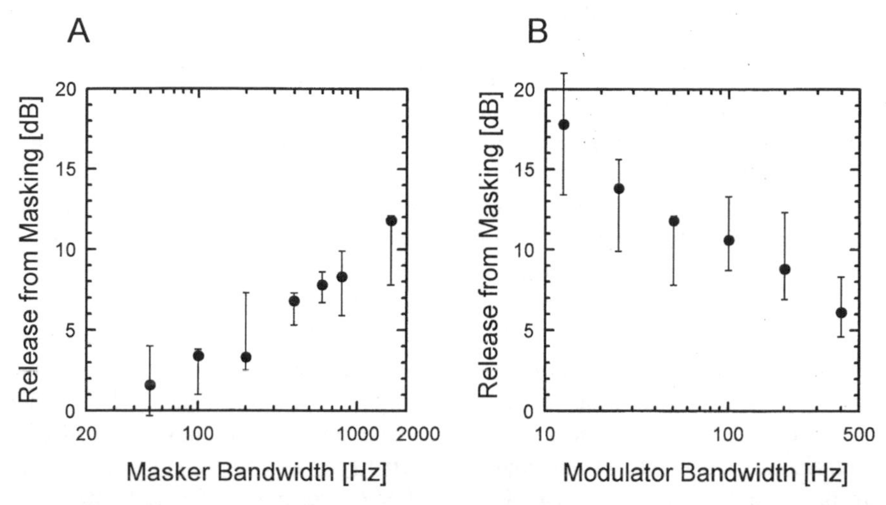

Fig. 18.4. Signals embedded in noise can be detected more easily (i.e., a masking release is found) when a temporal structure is imposed on the noise by modulating its envelope (e.g., with low-pass noise). Such envelope fluctuations may be caused by air turbulence, for example, and their dominant frequency range in the natural environment is below 50 Hz (see Richards and Wiley 1980). (A) The release from masking increases with the bandwidth of the masking noise. Illustrated are median values and ranges of data from five Common Starlings (the modulator was a 50-Hz low-passed noise). (B) The release from masking decreases with an increasing upper-frequency limit of the low-pass modulator, here characterized by the modulator bandwidth. Median values and ranges of data are from the same birds; the masker bandwidth was kept constant at 1600 Hz. Thus, low-pass modulators with a low cutoff frequency result in a larger release from masking than modulators with a high cutoff frequency (data from Klump and Langemann 1995).

improve recognition. When speech tokens were presented three times rather than only once in a speechlike masker, the signal-to-noise ratio necessary for a constant recognition performance was reduced by an additional 5–8 dB (Fastl 1993). Similar principles may apply to the perception of song elements by birds in their natural habitat.

The songs numerous individuals of different species sing during the dawn chorus create a noise that is temporally structured and prevalent (see, e.g., E. S. Morton 1975, Henwood and Fabrick 1979, Kacelnik and Krebs 1982). To avoid the intense acoustic competition and the consequent masking, many species restrict their calling to certain periods in the dawn chorus (e.g., Aschoff and Wever 1962, Cody and Brown 1969), and they may shift their song emission to reduce acoustic interference (M. S. Ficken et al. 1974). As the studies cited in Fastl (1993) indicate, an individual that repeats a signal in a time-structured masking noise can extend the range of its communication considerably. With more information on how birds perceive vocalizations in a natural, time-varying masking noise, we will be able to better understand the adaptive value of specific signaling patterns and more accurately predict communication distances.

Spatial Release from Masking

Signals of interest to birds and masking noises often originate from spatially separated sources. In this case, animals can make use of their directional hearing to improve their sensitivity for the detection of the signals in noise. Spatial release from masking in free-field sound presentation has been shown to occur in a wide range of vertebrates, including green treefrogs (*Hyla cinerea:* Schwartz and Gerhardt 1989), Budgerigars (Larsen et al. 1994), ferrets (*Mustela putorius:* Hine et al. 1994), and humans (Saberi et al. 1991).

Two factors may be involved in the observed release from masking. The first factor, which already comes into effect when a signal received by one ear alone is analyzed separately, is the relationship between the angle of sound incidence and the mechanical transduction of the free-field sound to the excitation of the inner ear. Although bird ears, unlike mammalian ears, are generally not equipped with pinnae, they show a directional pattern of sensitivity to sound that results mainly from the sound shadow provided by the head (see, e.g., Schwartzkopff 1952, Lewald 1990, Klump and Larsen 1992; for a review of studies that suggest an additional contribution of the interaural canal in birds to the directional pattern of sensitivity, see Klump and Larsen 1992). The Barn Owl (*Tyto alba*), a specialist that can rely exclusively on its sense of hearing for prey capture, is an exception (Konishi 1973) in having a facial ruff of feathers that functions like a pair of mammalian pinnae in amplifying sound from some directions, thus creating a directional pattern of sensitivity (Coles and Guppy 1988, Moiseff 1989).

A release from masking will occur if the signal is from a direction to which the ear is more sensitive and the noise is from a direction to which the ear is less sensitive. The reduction in masking attributable to this physical directional sensitivity of the ear is frequency dependent. In the Rock Pigeon (*Columba livia*), for example, the maximum difference in the attenuation of incident sounds originating from different directions lies between 1 and 5 dB at frequencies between 250 and 500 Hz and is about 20 dB for frequencies between 4 and 6 kHz (Lewald 1990). These data were obtained by measuring cochlear microphonics, electrical inner-ear potentials generated by the sensory cells that represent the acoustical signal, much like an electrical microphone signal. In the Eurasian Bullfinch (*Pyrrhula pyrrhula*), this maximum difference in the cochlear microphonics is about 10 dB at 3200 Hz (Schwartzkopff 1952). In the Common Starling, the directional pattern of attenuation of incident sound, as shown by measurements with a probe microphone at the entrance of the auditory meatus, reveals maximum differences that range from 2 dB at 1 kHz to about 15 dB at 8 kHz (Klump and Larsen 1992; directly measuring the directionality of the starling's tympanum vibrations with a laser vibrometer gave similar results).

The second factor that can be involved in spatial release from masking is a result of the neural analysis of binaural signals in the auditory pathway. If the signal and the masker differ in their interaural phase relationship (e.g., the signal originates from a source in front of the listener and the noise is from a source on the side), release from masking can be considerable, up to about 15 dB. This binaural masking-level difference (often abbreviated as BMLD or MLD; B. J. C. Moore 1989) is thought to be the result of the neural computation mechanisms involved in sound localization (see, e.g., Colburn 1977). Humans show a significant BMLD for broad-band signals with sharp amplitude transients (e.g., click trains) and for low-frequency sinusoids (e.g., 500 Hz tones). The BMLD in humans decreases with increasing frequency of the tone signal and is only 2–3 dB for frequencies above 1500 Hz (for a review, see B. J. C. Moore 1989). This decrease may occur because the coding of the phase information in the auditory nerve deteriorates with increasing frequency. A masking level difference resulting from an interaural neural comparison of phase is probably less important in songbirds; in the species studied so far, phase coding in the auditory nerve is mostly restricted to frequencies below 3–4 kHz (e.g., in the Common Starling; see Gleich and Narins 1988). In the Barn Owl, however, which shows phase coding up to 9 kHz (W. E. Sullivan and Konishi 1984, Köppl 1994), release from masking by means of binaural phase comparison should be possible even at high frequencies. Indeed, Takahashi and Keller (1992) showed that dynamic changes in the interaural phase of a signal embedded in a background noise, which simulate the motion of a sound source, enhance the accuracy of spatial coding in the Barn Owl's auditory midbrain even at frequencies close to 8 kHz. Their study demonstrates how the analysis of interaural phase relationships in signals can result in improved spatial coding, which may lead to a better signal-to-noise ratio in the detection and recognition of sounds.

In summary, spatial release from masking can result in a considerable improvement in the signal-to-noise ratio for the receiver. The amount of improvement depends on the spectral composition of the signal and the angular separation between the signal and the noise source as heard at the position of the receiver. Despite the potential importance of spatial release from masking for the detection of signals in the "real world," only a single investigation explicitly studied this problem in birds. Behavioral measurements in the Budgerigar showed that the spatial release from masking can be up to about 12 dB at the best frequency in its audiogram (2.86 kHz; see Larsen et al. 1994). Behavioral data from other bird species are lacking, but the physical measurements of the intensity differences between signals arriving at the ear from different directions reported above give at least an indication of the available cues.

The Recognition of Signals

The detection of a signal by the receiver is only the first step in perception, although it has been a main focus of research on how birds perceive sounds. Detection of the signal alone, however, in many cases is not sufficient to elicit the receiver's response. Field researchers who study how birds respond to song playbacks always include the second step, recognition, in the design of their experiments. Experiments that apparently are aimed at defining the maximum detection distance of a certain song element include the step of recognition, too; for example, the active space in song transmission often is determined by observing territorial responses of receivers (e.g., Brenowitz 1982c).

The recognition of signals is also influenced by noise. In this context, however, the significance of noise is not restricted to the effect that the ratio between the signal energy and the energy of a masking background noise has on perception, as, for example, in studies of critical masking ratios and release from masking (see above). In the more general terms of information theory (Shannon and Weaver 1949), noise can be viewed as any random variation in the parameters used by the receiver to classify and recognize signals. This variation may be introduced by the sender of the signal, when the signal is traveling from the sender to the receiver, or in the perceptual system of the receiver.

Signal detection theory and the associated discrimination measure, d′, can also be applied to general questions of species or individual recognition (R. H. Wiley and Richards 1982). A receiver trying to determine if a song is from a conspecific or heterospecific individual must take into account the statistical distributions describing the random variation in song parameters in the population of conspecific individuals and the song parameters of sympatric alien species. Dabelsteen (1985) developed the "room for variation" hypothesis in discussing the role of variation in multiple song parameters for species recognition in Eurasian Blackbirds. D. A. Nelson and Marler (1990) used the concept of "signal space" in discussing species recognition and the role of variation in the songs of conspecific and heterospecific individuals. To be reliably discriminated, conspecific songs must be sufficiently distinct from those of other species in the multivariate signal space.

The songs an individual learns may provide an important basis for song recognition. In that perceptual process, the individual bird's own song has been shown to be an especially potent stimulus in eliciting a neuronal response in some brain areas (see, e.g., Margoliash 1986, Volman 1993). Birds may use either the songs they learn to sing or songs they hear frequently (familiar songs; e.g., the songs of territory neighbors) as a reference when judging song degradation (see, e.g., E. S. Morton 1982, McGregor and Avery 1986). If these familiar songs also form the reference for species recognition, their statistical variation will determine the variance in the distribution of the perceived reference signals. This reference distribution corresponds to the probability density function for the presentation of noise alone in the signal detection example given in Fig. 18.1. The standard deviation of this reference distribution is the unit of the discrimination measure, d′.

Researchers studying song-based species recognition in the field generally find a significant change in the response of birds if song parameters are modified by approximately 2–3 standard deviations from the local population mean (e.g., Dabelsteen and Pedersen 1985, 1988a, 1992, 1993, D. A. Nelson 1988, 1989, Naugler and Ratcliffe 1992). In terms of signal detection theory, these results amount to a d′ of about 2–3 at the threshold for the detection of a deviation from the conspecific song, assuming that the reference is formed by familiar songs of conspecific birds from the local population. The difference in song parameters between the species-specific standard song and the variant used in the playback experiment sufficient to elicit a different response has been termed the just-meaningful difference (JMD: D. A. Nelson and Marler 1990). The JMD is usually larger than the just-noticeable difference (JND), which is generally measured in psychoacoustic experiments that determine difference limens. In the context of species recognition, for example, the JMD for variations in the frequency of the playback signal in Field Sparrows (*Spizella pusilla*) was about 15–20% of the maximum frequency (D. A. Nelson 1988, 1989). Dabelsteen and Pedersen (1985, 1988a, 1993) reported similar data in studies of blackbirds. In contrast, JNDs for frequencies in birds generally fall in the range of 0.5–2% (Fay 1988).

The observed frequency difference at the response threshold for JMDs and JNDs may differ by an order of magnitude, but these JMDs and JNDs may be rather similar with respect to the value of d′ at the threshold. Common Starlings, for example, seem to be about average in their ability to discriminate frequencies, and they discriminated pure tones that deviated from a reference signal by about 1.8% with a d′ of 2, and by 2.8% with an average d′ of 3 (range of reference frequencies, 1.0–6.35 kHz, median data: U. Langemann and Klump unpubl. data). Thus, the performance of birds expressed by the discrimination measure for signal detection does not not differ much between psychoacoustic studies in the laboratory and behavioral studies in the field. Just noticeable differences may differ from just meaningful differences simply because the distribution of reference signals that forms the basis for the decision process of the birds in the field is more variable (i.e., noisy) than in the laboratory.

Conclusions

Building on results from psychoacoustic studies in the acoustically simple environment of the laboratory, this chapter has demonstrated the effects of noise on the detectability of birds' vocal signals. Critical masking ratios, now known from more than 10 bird species, have been used to assess the influence of noise on the detectability of signals in the natural environment. This assessment can be only a first-order approximation of perception in the real world, however, because factors leading to a significant release from masking have to be taken into account. Two factors discussed above, for example, are the spectro-temporal pattern of the masking background noise in the natural environment and the directionality of the auditory system, which may lead to a considerable spatial release from masking.

Signal detection theory, which provides the basis for studies of masking, can also be applied to studies of the recognition or classification of signals. In studies of masking, the perceptual input provided by the background noise alone forms the reference against which signals presented in the noise are discriminated. In recognition studies, however, the reference is typically provided by the distribution of previously perceived signals of a specific class (e.g., the songs of conspecific birds or of certain individuals in the field, or the reference signals played as a background in psychoacoustic studies in the laboratory). I have shown here that, in the terms of signal detection theory, the discrimination performance of birds is remarkably constant across different contexts, while with respect to the parameters used (e.g., frequency) the discrimination thresholds can vary considerably. Thus, the apparent discrepancy between JNDs and JMDs disappears if we use a common perceptual measure. A much wider application of signal detection theory in behavioral research would permit comparison of results from studies ranging from psychophysics to the evaluation of complex behavioral responses (see also R. H. Wiley and Richards 1982).

We now have a good understanding of some of the factors that influence the perception of signals in a noisy environment, and we can compute a rough estimate for the detectability of signals, but we do not know the impact of these factors on bird communication in the natural habitat. One reason is that measurements of the temporal and spectral properties of the noise background in the natural environment are still scarce, and they often integrate only over a short time span (e.g., a day) or are made under specific meteorological conditions (e.g., for technical reasons they concentrate on periods with little wind). "Prime time" in bird communication, the dawn chorus, is especially worthy of study because the spectro-temporal pattern of masking then may have a major effect on the perception of song (Staicer et al., this volume).

The process of song perception is a challenge to the bird, whose task in the natural environment is not simply to determine whether any additional signal was broadcast by some sender (i.e., the mere detection of a sound), but rather which signal was emitted, by whom, and often where (i.e., the signal has to be detected and recognized, and often it is advantageous for the receiver to locate the sender). How the recognition and classification of signals are affected by background noise has been studied rarely in the field and not at all under controlled laboratory conditions. Nor do we know how the experience and memory processes of the receiver, which constitute the internal reference in his nervous system, will affect the classification of signals. The psychoacoustic tools needed to study these questions under controlled laboratory conditions are now available. Modern computer methods will allow us to bring a virtual acoustic environment to the laboratory, and by simulating natural environments we will come closer to understanding how birds perceive sounds in their natural world.

Acknowledgments

Søren Buus suggested applying signal detection theory to the question of the relation between JNDs and JMDs. I thank T. Dabelsteen, R. J. Dooling, U. Langemann, and especially D. E. Kroodsma for their helpful comments on a previous version of this chapter. This research was supported by a grant from the Deutsche Forschungsgemeinschaft within the SFB 204 "Gehör."

19 Sex Differences in Song Recognition

Laurene Ratcliffe and Ken Otter

Sexual communication involves the exchange of information between prospective mates, and often, though indirectly, between individuals competing to attract mates. Studies of sexual signaling have contributed greatly to our understanding of major evolutionary processes such as sexual selection and speciation (Ryan and Rand 1993). Ironically, songbirds feature prominently in such theory, even though remarkably little is known about how song receivers (usually females) respond to variation in sexual signaling (usually by males; Becker 1982, Catchpole 1982, Balaban 1988a, Searcy 1990). Unlike insects and anurans, female songbirds are frustratingly covert in the decision-making process that leads to the expression of their courtship preferences.

Recent technical advances, especially the development of the copulation-solicitation assay, allow researchers to assess the features of male song and singing behavior to which females attend (King et al. 1980, King and West 1987b, Searcy 1992a). Such work is crucial for evaluating hypotheses about the origin and significance of female choice on the evolution of male song (Searcy 1992b, Searcy and Yasukawa, this volume). But in many ways we have barely begun to appreciate the complexity of female songbird behavior (Baptista and Gaunt 1994), and much remains to be learned about females as song receivers. For example, how important is song (relative to other cues) in mate choice? If song is influential, what features of song structure or pattern of delivery make it detectable, discriminable, or memorable (Guilford and Dawkins 1991)? In many species males sing to both male and female audiences (McGregor 1991). But are the demands of male and female receivers similar? Historically, as in other areas of animal behavior (Zuk 1993), researchers have either mostly ignored female audiences or assumed that female and male interests were congruent. Recently this view has been challenged by the suggestion that male and female songbirds may have different listening interests, for both ecological and evolutionary reasons. If this is so, it is reasonable to hypothesize that mechanisms of song recognition might differ in fundamental ways between males and females (Dabelsteen and Pedersen 1988a, 1993, Searcy and Brenowitz 1988, Searcy 1990).

In this chapter we explore the sex difference hypothesis and consider its implications for the study of the selective forces shaping song and singing behavior. In discussing the question of sex differences, we are motivated by increasing evidence from behavioral and genetic fingerprinting studies that show that females of some socially monogamous species actively pursue multiple matings (e.g., S. M. Smith 1988, Kempenaers et al. 1992, Lifjeld and Robertson 1992, Otter et al. 1994a). If female songbirds use male song to assess their mates relative to potential extra-pair males, then selection could favor an enhanced ability to detect and remember subtle individual differences in vocal performance (Hamilton 1990, J. M. Hutchinson et al. 1993, M. S. Sullivan 1994). Thus, a second goal of this chapter is to encourage researchers to expand their thinking about the range of auditory tasks faced by female songbirds.

After a review of current hypotheses and supporting evidence regarding sex differences in song recognition, we discuss some of the thornier issues and pitfalls endemic to such work, using data from our ongoing studies of Black-capped Chickadees (*Parus atricapillus*) to illustrate our points. Finally, we suggest a number of areas in which female song-recognition behavior could be profitably investigated. We assume readers are already familiar with the extensive literature showing that male song serves in intersexual communication and has been shaped by female choice (see Searcy and Yasukawa, this volume).

Sex Differences in Song Recognition

Why Might Sex Differences Exist?

Like any *why* question in biology, this question can be answered at more than one level. We begin with a review of functional (ultimate) arguments, followed by a discussion of proximate mechanisms.

The ultimate cause. Sex differences in song recognition might arise for a number of reasons. For example, song might develop in only one sex and function exclusively in intrasexual communication, or sensory systems might be shaped by other sex-specific selection pressures, such as predation. Sex differences could also arise if the intra- and intersexual aspects of song function were divided between different song forms, as appears to be the case in some paruline warblers (Spector 1992).

Although each of these hypotheses has merit, virtually all the literature to date centers on just two ideas: the sound degradation hypothesis and the risk-of-investment hypothesis (reviewed by Searcy 1990, Dabelsteen and Pedersen 1993). Both hypotheses have been applied to species whose song is multipurpose (the same songs function in both territory defense and mating), and both hypotheses examine how differential selection intensity on male and female receivers might influence species recognition.

The risk-of-investment hypothesis was named by Dabelsteen and Pedersen (1993) to describe the argument presented by Searcy and Brenowitz (1988) and Searcy (1990). The hypothesis states that in species in which males use song in intrasexual aggressive signaling and females use song in courtship, the costs of errors in species recognition will be borne disproportionately by females. Females that fail to discriminate heterospecific from conspecific males risk hybrid mating, whereas males that respond aggressively to a heterospecific intruder incur relatively minor costs in wasted time and effort. This asymmetry should give rise to females that discriminate conspecific song better than males do. To recognize conspecific birds, for example, females might use more song features than males, or females might be less tolerant of deviation from standard, species-specific "norms" (Shiovitz and Lemon 1980). The pattern of female and male sensitivity to variation in song features might also differ. A key feature of the risk-of-investment hypothesis is its directionality (females are more discriminating than males). Moreover, Searcy and Brenowitz (1988) suggested that these predictions should be applicable to other animals and signaling systems besides birdsong—anywhere the sexes show asymmetric reproductive investment (Trivers 1972).

In contrast, the sound degradation hypothesis (Dabelsteen and Pedersen 1988a, 1993) is nondirectional and makes no assumptions about sex differences in the costs of receiver error. This hypothesis states that sex differences in song decoding may be expected if males and females of the same species, by virtue of having different behavior or habitat requirements, hear songs that are degraded differently in their passage through the environment (Dabelsteen et al. 1993). Each sex, it is argued, should be most sensitive to variation in those features of song that best resist the effects of attenuation and degradation for a given receiver location. In practice, males and females might rely on different features of song for species recognition or have different tolerances for signal variability. Because the sound degradation hypothesis requires that males and females inhabit different sound environments, it seems a less likely general explanation for sex differences in song recognition than does risk of investment. It should be kept in mind, however, that the two hypotheses are not necessarily mutually exclusive, and additional variables (e.g., the importance of nonsong cues in species recognition) may influence the evolution of sex differences in song decoding (Dabelsteen and Pedersen 1993).

Both hypotheses consider the demands of species recognition. But it seems logical to suppose that selection might also favor sex differences in the ability to discriminate individual differences. For example, the net benefits to females that choose mates (Ryan and Rand 1993) might be greater than the net benefits to males that simply discriminate neighbors from strangers. We suggest that Searcy and Brenowitz's (1988) risk-of-investment hypothesis can be extended to include the demands of individual quality recognition. In particular, in species in which females mate with both social partners and extra-pair males (Birkhead and Møller 1993), we might expect females to be better than males at discriminating song features correlated with genetic quality or resource-holding potential.

The proximate cause. The causal basis for cognitive variation between the sexes is still unknown (D. Kimura 1992). We do not yet understand what mechanisms in songbirds give rise to sex differences in responsiveness to acoustic stimuli. Perceptual differences could result from differential sensitivity of males and females to sound input, perhaps mediated through different neural structures in the ascending auditory tracts (H. E. Williams 1985). It has also been suggested that processing differences could result if male and female brains organize and retrieve information differently. Recent evidence from mammals and birds indicates that sexual dimorphism in the size of the hippocampal complex is correlated with sex differences in spatial ability and ecological factors influencing the use of space (e.g., Jacobs et al. 1990, D. F. Sherry et al. 1993). Neurophysiologists studying songbirds have predicted that sex differences in song recognition are to be expected for species in which the brain pathways for control of song differ between males and females (Nottebohm et al. 1990). This prediction assumes that dimorphism in the song control system leads to dimorphic song-processing ability. Put more simply, the sex with larger, better-connected song control nuclei (usually male) is expected to show better discrimination, at least of conspecific songs, than the sex with smaller nuclei (usually female). This view remains to be reconciled with the functional hypotheses described above that predict good or even better song discrimination by females than males.

It is beyond the scope of this chapter to review the extensive literature on neurobiological correlates of song production and recognition (see Brenowitz and Kroodsma, this volume), but one point does deserve mention. Although sexual dimorphism in song quality and the neurophysiology of song production are strongly correlated (see, e.g., Brenowitz and Arnold 1986, Baptista et al. 1993b), the evidence for sex differences in the neurophysiology of song processing is still lacking (see the reviews by DeVoogd 1991, and Cynx and Nottebohm 1992a). Two studies examined how the female song system responds to sound. H. E. Williams (1985), using artificial sound stimuli, found that parts of the motor pathways that respond to sound in male Zebra Finches (*Taeniopygia guttata*) do not respond in females. Based on this evidence she predicted that females would probably be inferior to males in song discrimination tests (see below for the results of such tests). Brenowitz (1991b) asked whether a song nucleus associated with song production in male Island Canaries (*Serinus canaria*) is important for conspecific song recognition by females. Intact females and females with unilateral lesions of the song nucleus higher vocal center (HVC) preferred conspecific song over heterospecific song, but females with a bilaterally lesioned HVC gave similar sexual responses to both types of stimuli. It is not clear whether lesioning eliminates the ability to discriminate between songs or influences the ability to respond to those differences. The study also raises the question of whether male and female brains necessarily organize auditory information the same way (C. L. Williams et al. [1990] presented evidence that male and female rats acquire and use spatial information differently). Clearly more work on female song neurophysiology, especially song processing, is required to identify the proximate neural mechanisms controlling discrimination in both sexes.

Sex differences in the morphology of neural song pathways are known to arise from the organizational effects of steroid hormones during early development (S. Goldman and Nottebohm 1983). Remarkably little is known about how environmental factors affect avian neurogenesis and subsequently shape behavioral response. Recently Cynx and Nottebohm (1992a, b) pointed out that sex differences in song recognition could also arise from seasonal variation in adult hormone levels, and H. E. Williams et al. (1993) described social factors that result in differential learning. These questions need further investigation.

Is There Evidence for Sex Differences in Song Recognition?

The literature on sex differences in song recognition began about 15 years ago. Many of the earlier studies (and some recent ones) are fraught with methodological problems that make it difficult to evaluate the soundness of their conclusions. These difficulties include comparing the responses of hand-reared females with the responses of wild males, possible effects of hormone implants on patterns of female sexual preference, the use of single exemplars of test songs, the failure to control adequately for other factors affecting test songs such as sound quality, the failure to test males and females with the same songs in reciprocal experimental designs, and the failure to consider alternative explanations for the results. A key difficulty has been the inability to test males and females in identical experimental contexts in which song perception is isolated from song function (see below for more discussion and suggestions for resolution of this point). In the following review, we flag studies whose conclusions we think should be accepted with caution, with the aim of encouraging better experiments.

Strictly defined, the term *song recognition* refers to the identification of a song as belonging to a class of appropriate conspecific stimuli. Operationally, however, the term is often used more loosely to describe a differential response connoting song preference (Balaban 1988a). Song recognition studies frequently measure responses to variation in song macrostructure (features thought to function in the identification of species, subspecies, and populations; e.g., temporal and frequency parameters, syllable structure, and syntax). Studies that examine responses to variation in song microstructure (features connoting individual identity and quality; e.g., phrase length, syllable complexity, and fine structure) are much less common, and few studies examine both (see the review by Clayton 1990b). The literature on sex differences in song recognition is still meager, and most researchers choose to examine responses to macrostructural variation. In the review that follows, we include only the results of studies that tested male and female responses to the same (or similar) test stimuli. It is important to keep in mind that most of the inferences about sex differences have been based on indirect comparisons of males and females in different testing contexts (discussed below).

Song duration. Some evidence suggests that, compared with males, females respond more weakly to partial songs than complete songs. Removal of the introductory notes from a male Red-winged Blackbird's (*Agelaius phoeniceus*) song diminishes the female's sexual response (Searcy and Brenowitz 1988, Searcy 1990) but does not diminish the aggressive responses of territorial males (Beletsky et al. 1980, Brenowitz 1982b). Females discriminate a partial song from a complete song even when both have been re-recorded with natural habitat sounds as background, so the sex difference is probably not an artifact of testing captive females in sound chambers at distances less than 1 meter from the loudspeaker. Female Red-winged Blackbirds might use more song features than males in species song recognition (Searcy 1990); as Dabelsteen and Pedersen (1993) pointed out, however, the possibility that females are responding to variation in the amount of song per unit time on test tapes also needs to be investigated.

White-throated Sparrow (*Zonotrichia albicollis*) males also appear to be less sensitive to song duration than females. Falls (1963) observed that territorial males did not seem to respond differently to songs of three or six notes than to a song of a single note and concluded that song duration and note number are not critical features for eliciting male aggressive response. This conclusion must be interpreted with caution, however. The songs used in these particular trials were not controlled for overall duration, and all elicited less response than normal song, perhaps because they also lacked the species-specific pitch change characteristics of normal song. In comparison with the data for males, it is clear that the sexual response of female White-throated Sparrows is influenced by the number of notes in a song, independent of song rate. Wasserman and Cigliano (1991) found stronger responses to five-note songs played at 4 songs per minute than to two-note songs played at the same rate. In recent tests we found that five-note songs (at 4/minute) were more stimulating than two-note songs played at 10 songs per minute (J. Roosdahl and Ratcliffe unpubl. data). Moreover, females did not discriminate between two-note songs played at 4 songs per minute and 10 songs per minute. Collectively, these data confirm that female White-throated Sparrows are less stimulated by partial (two-note) songs than by longer (five-note) songs.

Temporal patterning of syllables. The evidence that males and females respond differently to variation in temporal patterning of syllables is mixed. Territorial male Swamp Sparrows (*Melospiza georgiana*) did not respond differently to a synthetic copy of natural Swamp Sparrow song (swamp synthetic) and a song containing different Swamp Sparrow syllables arranged in a Song Sparrow (*M. melodia*) temporal pattern (complex; S. S. Peters et al. 1980, Searcy et al. 1981a). In contrast, captive females responded sexually more to swamp synthetic than to complex song (Searcy et al. 1981b), leading the authors to conclude that females respond preferentially to conspecific song patterning, but males do not. This conclusion should be treated with caution, however. Only a single exemplar of each song was used in the studies; moreover the two stimuli varied in syllable complexity as well as timing, and only three females responded in these tests.

Searcy et al. (1982) found no sex difference in response to Swamp Sparrow songs that varied in the temporal rate of syllable presentation, as neither males nor females discriminated among test songs.

The most detailed experiments on sex differences in song recognition have employed Eurasian Blackbirds (*Turdus merula*). Dabelsteen and Pedersen (1993) showed that for the introductory "motif," males and females responded similarly to songs with altered syllable duration. In contrast, manipulation of pause duration between syllables significantly influenced male aggression but had little effect on female sexual response.

Frequency range and ratio. Some evidence suggests that males and females are sensitive to different parts of the frequency range of conspecific song. Female Eurasian Blackbirds responded strongly to songs in which the frequency had been increased up to and beyond the species' natural limit, whereas males discriminated against such songs (Dabelsteen and Pedersen 1985, 1988a, 1993). At the lower end of the frequency spectrum, no sex differences were apparent. Females, like males, gave significantly weaker responses to song motifs with the frequency level lowered by 0.5 kHz–1.0 kHz.

Female White-throated Sparrows may also be more sensitive than males to frequency variation in the upper range of conspecific song, but this suggestion is based on a single experiment with limited numbers of test songs (S. B. Meek, Ratcliffe, and R. Weisman unpubl. data). Previous work on males suggested that they discriminate against songs outside the species' frequency range (Falls 1963), and that they use absolute frequency cues within the species' range for neighbor-stranger discrimination (R. J. Brooks and Falls 1975a). When we tested strangers' songs, we found that variation in the absolute frequency of song within the species' normal range had no discernible effect on the strength of the aggressive response (Hurly et al. 1990). Males do discriminate against songs with altered frequency ratios (the pitch interval between adjacent notes), independent of absolute frequency (Hurly et al. 1992). When females were presented with the same stimuli used for males, they were more sexually responsive to songs of higher rather than lower absolute frequency, and they showed better discrimination of frequency ratio in high-frequency songs than low-frequency songs (S. B. Meek, Ratcliffe, and R. Weisman unpubl. data).

Syntax. We are not aware of any evidence of sex differences in ability to discriminate among songs based on syntax. In fact, male and female Red-winged Blackbirds (Beletsky et al. 1980, Searcy 1990) and Eurasian Blackbirds (Dabelsteen and Pedersen 1993) all failed to respond differently to conspecific songs with altered or normal syllable order. Thus, syllable syntax does not appear to be a salient feature for song recognition in Red-winged Blackbirds or Eurasian Blackbirds.

Male and female Swamp Sparrows, on the other hand, are sensitive to song syntax, but no compelling evidence indicates a sex difference. Balaban (1988a) found that both male and female Swamp Sparrows in a New York population discriminated between natural local song and song from a northern Minnesota population. These two dialects differ primarily in the order of note categories within syllables. When local New York songs were rearranged into Minnesota syntax, and Minnesota songs arranged into New York syntax, New York males responded more aggressively to the local songs with the foreign syntax, whereas females responded sexually more strongly to the foreign songs with the local syntax. Although these data suggest that songs with local syntax are particularly evocative for females, Balaban (1988b) correctly cautioned against concluding that males and females attend to different parts of songs. For example, it is clear that the New York females also attended to note origin, because their syntax preferences were stronger when test songs were formed with local notes than with foreign notes.

Syllable structure. Some data suggest that males and females respond differently to variation in conspecific syllable structure. Red-winged Blackbird males responded aggressively to a Mockingbird (*Mimus polyglottos*) imitation of conspecific song (Brenowitz 1982a, b), but females discriminated such songs from normal controls (Searcy and Brenowitz 1988, Searcy 1990). In this case, females were probably using the fine structure of syllables to make the discrimination. With respect to dialects, Brenowitz (1983) showed that New York Red-winged Blackbird males could distinguish California songs from local ones only if the pulse rates of the foreign songs fell outside the range of the New York population's rates. Searcy (1990), however, working with Pennsylvania females, found that they discriminated even those foreign songs with pulse rates within the local range. It is tempting to conclude that Red-winged Blackbird females demonstrate enhanced discrimination of syllable structure. We concur with Searcy (1990), however, that given the differences between the two studies (population, song stimuli), it is premature to conclude that the sex differences are reliable. More reciprocal tests of male and female dialect discrimination are needed.

Female, but not male, Eurasian Blackbirds disregarded systematic experimental manipulation of amplitude modulation in motif syllables (Dabelsteen and Pedersen 1993). This sex difference is consistent with the finding that amplitude functions of blackbird song motifs incur considerably more blurring during transmission to females near the ground than to males in the canopy (Dabelsteen et al. 1993).

A growing number of operant conditioning studies with Zebra Finches suggest that males discriminate conspecific song syllables better than females do. Zebra Finch song is used only in an intersexual context, so this difference, if real, is intriguing. Tests of timbre perception of a single syllable showed that both males and females solved the perceptual task in the same way; males learned significantly faster than females, however (reviewed by Nottebohm et al. 1990). Cynx and Nottebohm (1992a) asked whether males and females could learn to discriminate between songs of two familiar males (differing in syllable structure); females required significantly more trials than males to reach criterion in this task, too. Interestingly, in a third task, involving discrimination between two unfamiliar songs, no significant sex difference in the speed of acquisition of the task was apparent.

Interpreting the Data: Some Caveats

Overall, these results suggest that male and female songbirds respond differently to some variants of conspecific song, at least in the very different contexts of aggression and courtship. Support for the sound degradation hypothesis is based mainly on data from a single species, the Eurasian Blackbird, in which males sing from elevated song posts and female receivers perch in dense vegetation near the ground. In this case, features of song like pause duration and amplitude patterning of syllables (features that degrade considerably in transmission to females) seem to have less influence on female response than on male response. We need many more data to evaluate whether differential sound transmission pressures constitute a major selective force for the evolution of sex differences in song recognition. We should also consider that in many species, male-male communication is long range, whereas males sing to females at close range. Surprisingly few data exist on sex differences in receiver location or signal reception. The emerging technology for simultaneous monitoring of signalers and receivers should help us collect this information (Smith, this volume, McGregor and Dabelsteen, this volume).

How good is the support for the risk-of-investment hypothesis? The data from the Red-winged Blackbird and White-throated Sparrow studies suggest that males are less responsive than females to variation in some stereotyped features of conspecific song, although neither study adequately controlled for the possible effects of sex differences in receiver location. The data from Swamp Sparrows appear equivocal. As for Zebra Finches, the performance differences between females and males on some tasks may reflect sex differences in the relevance of the song discrimination task rather than inherent differences in song-processing ability (Cynx and Nottebohm 1992a). Overall, the risk-of-investment hypothesis remains a useful framework for generating research on song discrimination by females, but we agree with Searcy (1990) that more rigorous experiments are required.

The many problems that surround the interpretation of sex differences in these experiments were thoroughly discussed by Balaban (1988a), Searcy (1990), and Dabelsteen and Pedersen (1993), among others. A chief difficulty is methodological: the most common ways of testing song recognition (e.g., playback to territorial males and sexual responses of females) are not direct measures of song perception; nor are the two contexts directly comparable. Moreover, failure to distinguish between stimuli may be motivationally mediated and does not necessarily mean failure to attend to differences (Weary 1992). Operant experiments can provide a useful avenue of alternative inquiry because they allow researchers to test song perception independent of song function, and habituation does not occur. We strongly encourage researchers to include and report data from both sexes when carrying out such studies.

Operant experiments have some limitations, however. One important consideration is the structure of test stimuli. To be really useful, test stimuli should be ecologically relevant (e.g., in mimicking the natural levels of background noise associated with song production) and presented in as natural a context as possible (e.g., embed the feature of interest, such as a syllable with altered timbre, within a song containing normal syllables; Cynx and Nottebohm 1992a). Investigators should also be aware that some aspects of song perception may remain constant across motivational contexts and others are unique to particular motivational contexts. For example, female Brown-headed Cowbirds (*Molothrus ater*) were sensitive to both the beginning and ending of conspecific song in a copulation-solicitation assay, but they seemed sensitive only to the beginning of a song in a food-rewarded experiment (Johnsrude et al. 1994). In other words, the strategies birds use to solve operant tasks must be taken into consideration. Finally, even in operant experiments, it may be difficult to design a test that is equally valid for males and females. In this regard, Nottebohm et al. (1990) cautioned that sex differences in apparent discrimination could result from different abilities to produce the required operant response. We are not aware of any data regarding this possibility.

Relative Pitch Recognition in Male and Female Black-capped Chickadees

For some years we and our colleagues have been investigating the phenomenon of relative pitch recognition in songbirds (Ratcliffe and Weisman 1992). Unlike absolute pitch perception, which involves the ability to identify the frequency of a single note without reference to an external standard, relative pitch perception refers to the ability to perceive frequency relationships between adjacent notes. A study of male Black-capped Chickadees (*Parus atricapillus*), a nonmigratory species common in North America (S. M. Smith 1991), provided one of the first examples of relative pitch recognition in animals (Weisman and Ratcliffe 1989). But as we show below, consideration of female chickadee behavior has been crucial to our understanding of the evolution of pitch and pitch stereotypy in the song of this species.

Background. Chickadees have a complex call repertoire shared by both sexes (M. S. Ficken et al. 1978, Hailman and Ficken, this volume). Advertising song, a tonal two-note *fee bee* (or sometimes *fee bee yee;* Fig. 19.1A), is sung mostly by males, during the dawn chorus and in daytime territorial contests (Shackleton and Ratcliffe 1994). Singing during the dawn chorus is also directed partly at females (Otter and Ratcliffe 1993). We found that two characteristic kinds of frequency

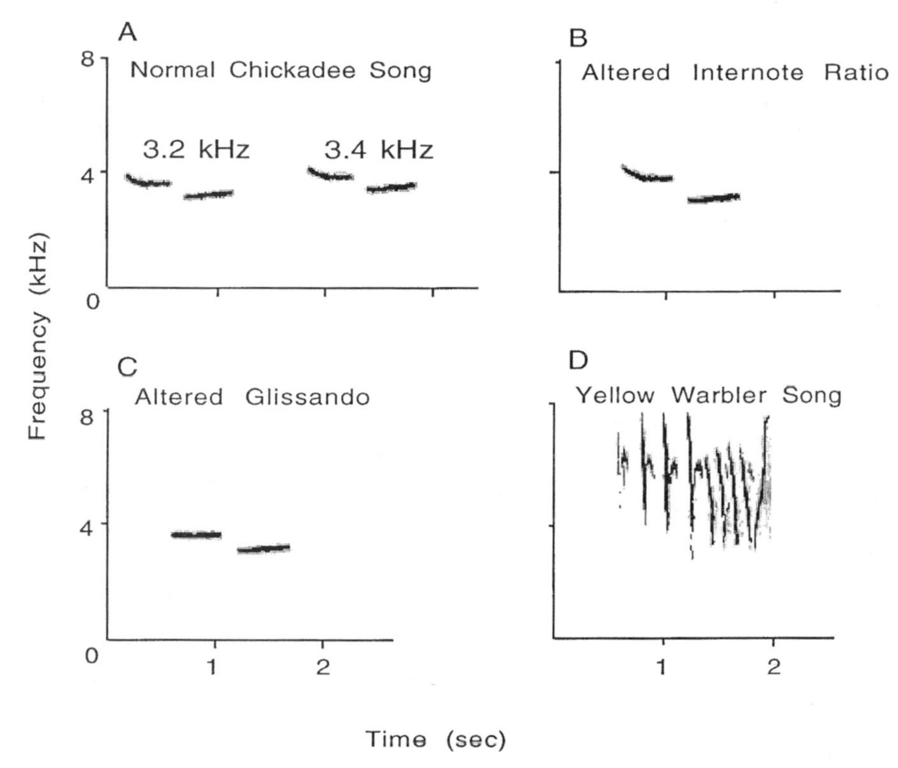

Fig. 19.1. Female Black-capped Chickadees responded sexually more strongly to normal conspecific songs at two different absolute frequencies (A) than to heterospecific song (D). Values in A refer to frequencies of the *bee* note. Chickadee songs with an increased frequency interval (B; altered internote ratio) and altered glissando (C) also elicited weaker sexual responses when they were played back before normal song. Sonograms were produced on a Kay Elemetrics DSP Sonagraph (analyzing filter bandwidth 59 Hz), then scanned into CANVAS and printed with a Macintosh printer.

change in *fee bees* are highly stereotyped: the minor downward sweep from the start to end of the *fee* note (which we call glissando), and the larger frequency drop from the end of *fee* to the start of *bee* (the frequency interval). Both vary less than 2% within and among birds, over wide geographic areas, even when individuals shift the absolute frequency of the entire song up or down within the species-common frequency range (Ratcliffe and Weisman 1985, Weisman et al. 1990). For both the glissando and the frequency interval, the change in frequency over time is better predicted by a ratio function (i.e., *fee* start/*fee* end and *fee* end/*bee* start, respectively) than a difference function (i.e., *fee* start – *fee* end and *fee* end – *bee* start, respectively).

Such stereotypy suggested that chickadees might use relative pitch as a feature in conspecific song recognition, but our results give a mixed picture. In playback experiments, territorial males responded less aggressively to songs with "flattened" glissandos (in which the frequency ratio was reduced by >3SD below the species' mean, to a value of 1.0; Shackleton et al. 1992). Our results with manipulated frequency intervals were different, however. Even though males in the laboratory can detect aberrant intervals (Weisman and Ratcliffe 1989), males in the field showed little evidence of such discrimination (Ratcliffe and Weisman 1992). This negative result was repeated in a second field season (Ratcliffe unpubl. data) in which we used 20 new subjects and the same experimental procedure and song-synthesis procedure as Shackleton et al. (1992). Like Searcy and Brenowitz (1988), we wondered why males would produce such stereotyped songs, yet tolerate versions in which a major structural feature (the frequency interval) deviated considerably from the species' standard. Moreover, male chickadees reared in social isolation produce songs with aberrant frequency structure, suggesting that stereotypy is learned in some way (Shackleton and Ratcliffe 1993). To attempt to resolve this puzzle, we focused on the role of female chickadees as song receivers. We used the copulation-solicitation assay to determine whether females distinguish normal songs from those with altered glissandos and altered frequency intervals.

Methods. We trapped adult female chickadees in early winter near Kingston, Ontario, Canada, and brought them into breeding condition by lengthening the photoperiod and implanting them with 17-β-estradiol (Otter 1993). All females were maintained on breeding light conditions (16L:8D) until they were tested (five to six weeks for birds captured in January and February; one week for birds captured in April). Birds were housed in individual sound-attenuation chambers and tested beginning the third day following implantation.

Four playback song stimuli (two versions of each, for a total of eight songs) were synthesized using SoundEdit Pro software (Macromedia, Inc., for Macintosh). The two versions of each song were synthesized independently, using the same program, to provide some subtle differences in amplitude variation. Length of notes, internote spacing, relative amplitude within notes (including the amplitude drop in the middle of *bee*), and frequency parameters of the synthetic songs were based on species averages (Weisman et al. 1990). The first two stimuli were conspecific song with a normal, species-typical glissando and frequency interval (normal songs; Fig. 19.1A). These two normal songs had frequencies (at start of *bee*) of 3400 and 3200 Hz, respectively, corresponding to high and mid-frequency songs of wild birds (Otter and Ratcliffe 1993). Females' responses to normal songs were compared with responses elicited by songs with increased frequency intervals (altered ratio, Fig. 19.1B) and with the glissando removed (altered glissando, Fig. 19.1C). These alterations represent changes in the frequency ratios >3 SD, outside the species' range. Songs of two Yellow Warblers (*Dendroica petechia*), recorded locally, served as the heterospecific control (Fig. 19.1D).

Each female was tested daily, between 8:00 A.M. and 12:00 P.M., with a normal (unaltered) conspecific song and either an altered or a heterospecific song, with two hours between each test (Table 19.1). The same stimulus pair was presented

Table 19.1. An example of the experimental design, showing the order of stimulus presentation to Black-capped Chickadees over six days of testing (Fig. 19.1 shows sonograms of the stimulus songs)

	Stimulus pair 1		Stimulus pair 2		Stimulus pair 3	
	Day 1	Day 2	Day 3	Day 4	Day 5	Day 6
Bird 1	A1 vs. D2	D1 vs. A2	A2 vs. B2	B1 vs. A1	C1 vs. A2	A1 vs. C2
Bird 2	A2 vs. D2	D1 vs. A1	C1 vs. A2	A1 vs. C2	A2 vs. B1	B2 vs. A1
Bird 3	A2 vs. D1	D2 vs. A1	B1 vs. A1	A2 vs. B2	A1 vs. C2	C1 vs. A2
Bird 4	A1 vs. D1	D2 vs. A2	A1 vs. C2	C1 vs. A2	B2 vs. A1	A2 vs. B1

Notes: A, normal chickadee song; B, altered internote ratio; C, altered glissando; D, Yellow Warbler song. The order of the stimuli on a given day represents the order in which the stimuli were presented to the bird. There are two versions (1 or 2) or each stimulus type.

the following day, with the order of songs reversed between days. All birds were tested with the following stimulus pairs: (1) normal conspecific song versus Yellow Warbler, (2) normal versus altered ratio song, and (3) normal versus altered glissando song. It should be noted that for the second stimulus pair (normal vs. altered ratio), we controlled for any effects on response that might be the result of the absolute frequency, rather than the relative frequency, of the *fee* and *bee* notes in altered song. That is, the altered ratio song had a *fee* note corresponding to that in the normal 3400 Hz song and a *bee* note corresponding to that in the normal 3200 Hz song; we paired one of the two normal songs (3400 Hz and 3200 Hz) with the altered ratio song on each day.

At the start of each day, we presented each female with one of three randomly chosen, freeze-dried model chickadees for 5–10 minutes. This served to visually stimulate the females and make them more likely to respond to song playback. Females were exposed to the models only once per day, and the models were removed before playback. Each playback consisted of two minutes of silence (PRE), two minutes of playback (at 10 songs/minute, TRIAL), followed by another two minutes of silence (POST). Thus, each female was exposed to two playback trials per day (at 20 songs per trial, for a total of 40 songs) for six days. The design of the experiment was such that each female was tested with each version of each song once, except for the normal conspecific songs. Females were tested with both versions of normal song three times each (Table 19.1).

Trials were videotaped through a window in the door of the test chamber and scored by an observer naive to the design and predictions of the experiment. Response measures included the total number of copulation displays and total display duration.

Results. With the exception of one display in one trial, females did not display during the PRE or POST period; displays occurred only in response to song presentation (TRIAL period). Five birds completed all matched pairs tests; a sixth completed only the normal conspecific versus heterospecific test. We found no

Table 19.2. The number and duration of displays given by female Black-capped Chickadees to normal chickadee songs versus (1) heterospecific control, (2) song with altered internote frequency ratio, and (3) song with altered glissando

Stimulus pair	Presentation order			
	Normal	Altered	Altered	Normal
1. Chickadee (conspecific) vs. Yellow Warbler (heterospecific)				
Number of displays	6.8 ± 3.4	1.2 ± 0.8[a]	0	5.2 ± 2.8[a]
Duration of displays (sec)	15.6 ± 8.6	2.2 ± 1.5[a]	0	11.7 ± 8.5[a]
2. Normal ratio vs. altered ratio				
Number of displays	5.2 ± 3.2	4.2 ± 1.5	3.4 ± 1.0	4.6 ± 2.2
Duration of displays (sec)	24.0 ± 21.0	10.2 ± 6.5[b]	3.7 ± 2.2	6.6 ± 3.9[a]
3. Normal glissando vs. altered glissando				
Number of displays	3.8 ± 2.3	4.2 ± 1.7[b]	1.8 ± 1.1	4.2 ± 1.7[a]
Duration of displays (sec)	7.2 ± 5.7	6.2 ± 4.0	3.5 ± 2.4	7.8 ± 5.1

Notes: Values shown are means ± SE based on a single value from each of six birds (1) or five birds (2, and 3). Data are separated by order of stimulus presentation. Within days, birds were presented with a stimulus pair; the second day the same stimulus pair was presented in the reverse order.
[a]Comparison between stimulus pairs within day is significant at $P < 0.05$ (Wilcoxon signed rank, 2-tailed tests).
[b]Comparison of response to altered stimulus between days is significant at $P < 0.05$ (Wilcoxon signed rank).

effect of song version for any stimulus type, in either number of displays or display duration (Wilcoxon signed ranks, all $P > 0.1$, 2-tailed tests).

Females showed a clear sexual preference for normal conspecific song over heterospecific song. Table 19.2 gives the mean responses of six females (based on a single value from each bird) to both stimulus types, according to whether normal conspecific song was presented before or after heterospecific song. In both cases, all six females gave more copulation displays and displayed longer in response to normal chickadee song than to Yellow Warbler song (Wilcoxon signed ranks, all $T = 0$, all z values < -2.03, all $P < 0.05$, 2-tailed tests). We found no effect of stimulus order on responses (Wilcoxon signed ranks, all $P > 0.1$).

Females also discriminated between normal song and songs with altered ratio (Table 19.2), although the results in this case were complicated by order effects. Females displayed longer to altered ratio songs played as the second stimulus of the morning than as the first ($T = 0$, $z = -2.00$, $P < 0.05$). Altered ratio songs elicited significantly shorter displays than normal song only when they were presented as the first stimulus ($T = 0$, $z = -2.00$, $P < 0.05$). Females also gave fewer displays to altered ratio song, but the difference was not significant ($P > 0.1$).

Similarly, females gave more displays to altered glissando song when it was the second trial of the morning than when it was the first ($T = 0$, $z = -2.03$, $P < 0.05$; Table 19.2). When the normal song stimulus preceded altered glissando, females showed no differential response ($T = 2$, $z = -0.54$, $P > 0.1$); when the altered glissando was played first, it elicited significantly fewer displays than normal song ($T = 0$, $z = -2.03$, $P < 0.05$). Females also averaged shorter displays to altered glissando songs, but the difference was not significant ($T = 3$, $z = -1.20$, $P > 0.1$).

Discussion. Despite the small sample sizes, our data support the hypothesis that female Black-capped Chickadees discriminate among conspecific songs with altered within- and between-note frequency ratios. When we presented song with altered internote frequency ratios, or with altered glissando, as the first of the paired stimuli of the morning, females displayed for less time, or gave fewer displays, respectively, in comparison with their responses to normal chickadee song. This study, along with one on White-throated Sparrows (S. B. Meek, Ratcliffe, and R. Weisman unpubl. data), provides some of the first evidence for relative pitch recognition in female songbirds. The results suggest that female chickadees prefer songs with species-stereotypic relative pitch characteristics. Because we tested songs at only one extreme of variation in relative pitch, our conclusion must be tempered until a broader range of stimuli can be tested.

The important result is that females tested in a sexual context discriminated against a song of a type (altered ratio) that was tolerated by males tested in a territorial context. We think the difference in male and female behavior is real, not a result of experimental artifact in the fieldwork, because males from the same population that discriminated against altered glissandos (Shackleton et al. 1992) failed to discriminate songs with ratios altered to the same degree as those used in these female tests. As with previous studies of sex differences, we cannot conclude that females are more discriminating than males, because the two testing contexts are not comparable. We think it unlikely that the results are attributable to a sex difference in perception per se. Captive male chickadees can detect songs with altered ratios (Weisman and Ratcliffe 1989, Weary and Weisman 1991); unfortunately, females have not yet been tested in operant trials. One possible explanation is that female chickadees may place more weight on ratio features of song than males do because this feature is useful in mate assessment. A male's ability to produce species-typical frequency intervals may be shaped more by the pressures of intersexual selection through female choice than by pressures of signaling in male-male interactions. Studies identifying how variation in other features of male song influence female chickadee sexual response are currently in progress.

Conclusions

The study of sex differences in song recognition is still a relatively new field, emerging primarily from the explosive growth in avian neuroethology. In common with researchers studying sexual dimorphism in the brains of other vertebrates, we lack a cogent theoretical framework for understanding why sex differences in recognition might evolve. Behavioral ecologists can contribute significantly to this area by carrying out field studies that identify sources of selection on singing behavior and song reception. From our review of the literature, it is clear that conclusions about sex differences in songbird recognition rest on a meager empirical base. Many of the studies reporting sex differences have methodological flaws, or the results are open to alternative interpretations. One of the chief obstacles in this work has been the lack of methodology for testing song recognition independent of song function.

Despite these limitations, the literature does suggest that the males and females of some species may rely on different features of song, or at least weight the same features differently, in communicating with conspecific birds. Most work has been on songbirds of northern latitudes, using species in which females rarely or never sing and the same song serves in both territorial defense and mate attraction. Females appear to be as adept at discriminating among different song variants as males, and in some cases even more so, which supports the risk-of-investment hypothesis. Thus, failure to produce song need not be correlated with diminished perceptual ability.

A multitude of questions about the origins, development, and fitness consequences of sex differences in song recognition remain unanswered. Many of these present daunting technical challenges, but several avenues can be usefully pursued with established procedures. At the perceptual level, sexual differences in the effects of early learning on male and female song perception and later female sexual preference need to be compared (H. E. Williams et al. 1993). Nottebohm et al. (1990) pointed out that it is unclear how songbirds process learned sounds such as conspecific song. Moreover, they stressed that the processes of perception leading to a preference for conspecific song need not be the same in males and females. Thus, it would be useful to find out what factors guide female decisions about appropriate song models early in development. Obviously this approach need not be limited to questions about conspecific identity, but could also incorporate questions relating to individual preference.

Neither do we know much about sex differences in ecological sources of selection on song reception. Male and female Eurasian Blackbirds inhabit distinct microhabitats (Dabelsteen and Pedersen 1988a, 1993), but whether such environmental partitioning occurs in other species is unknown. Distance to signaler could be important even when males and females occupy the same habitat. For example, males often court females at close range with songs that also act as long-distance territorial signals. In this case, females might be less tolerant than males of song features that degrade quickly with distance. On the other hand, even if song functions in close-range courtship, females living in dense habitats may use distant singing as an accurate signal of the density of conspecific males (Eriksson and Wallin 1986), and song output as a signal of individual male availability (R. Mace 1986). The characteristics of an individual male's songs, such as absolute pitch, might even provide information to females about the habitat, and thus the territory quality (Wass 1988). If song does function in long-distance communication between males and females, it would not be surprising if females tolerate degraded songs as well as males. Clearly we need to know more about how the songs of various species degrade over distance through different habitats, as well as details of the location and sex of intended receivers.

Finally, we need to expand the theory and develop ways of testing the risk-of-investment hypothesis that are not confounded by sex differences in the experimental context. One way might be to look at species in which females sing "male" song for territorial defense and also respond to the same songs sexually (e.g., White-throated Sparrows: A. M. Houtman pers. comm.; White-crowned Sparrows, *Zonotrichia leucophrys:* Baptista et al. 1993b; many tropical duetting species: Brenowitz and Arnold 1985). If females tested in both contexts were more discriminating than males, it would argue for sex differences in higher-order processing unrelated to the immediate testing situation.

Acknowledgments

We thank the Queen's University Biological Station for excellent logistical support of our ongoing studies of chickadees, and Ron Weisman for generous provision of space and equipment for testing captive birds. J. Roosdahl, C. Naugler, and S. Leech expertly assisted with data collection and analysis. We also thank Don Kroodsma, Raleigh Robertson, and Barrie Frost for comments on various portions of the manuscript. This study was supported by a Natural Sciences and Engineering Research Council of Canada operating grant to LR and a Queen's Graduate Award to KO.

20 Vocal Recognition of Neighbors by Territorial Passerines

Philip Kraft Stoddard

The study of individual vocal recognition in birds has been prompted by two convergent sets of interests. Ethologists first tested vocal recognition as part of a general program to explore levels of song organization and perception (Marler 1960). Successful demonstration of species' vocal recognition then prompted studies of song-dialect discrimination, neighbor-stranger song discrimination, and discrimination between songs of individual neighbors. Probably the most captivating idea that developed from these studies was the proposal that large repertoires interfere with birds' ability to recognize their neighbors by their songs (Kroodsma 1976b, Krebs and Kroodsma 1980, Falls 1982). This intriguing hypothesis led Falls (1982) to ask whether the process could work in reverse—whether the advantage of remaining recognizable to one's neighbors acts to constrain the evolution of ever-increasing repertoires (the repertoire constraint hypothesis).

A second line of inquiry at about the same time followed the rise of sociobiology theory. Observers of animal social behavior were quick to recognize individual recognition as a prerequisite to complex social relations ranging from parental care to cooperation among unrelated individuals. Parent-offspring recognition, for instance, is now considered essential to prevent misdirection of parental care among synchronous social breeders with mobile offspring (Colgan 1983, Beecher 1991). Among unrelated individuals, cooperation is believed to be a stable strategy only when repeated interactions occur among familiar, recognized individuals (Trivers 1971). So basic is individual recognition to the daily functioning of complex social groups that most researchers investigating primate societies and avian cooperative breeding systems consider it patently absurd to even question the existence of individual recognition, either vocal or visual. The same argument applied to territorial songbirds and parents of many bird species is not so compelling. Indeed, territorial songbirds apparently make frequent mistakes in visual recognition (Nice 1943), and carefully documented examples of misdirected parental care have been found when multiple broods of offspring come together unexpectedly (Beecher et al. 1981).

Studies of individual recognition in oscine birds have followed two main avenues of inquiry: neighbor recognition in territorial oscines with different-sized repertoires, and kin recognition among communally breeding birds with varying potential for misdirected parental care. The processes of recognition appear to involve vocal learning in both cases, but the determinants of individuality in signal production are distinctly different. Individual differences among the contact calls used in parent-offspring recognition appear to have a strong genetic basis (Medvin et al. 1992), whereas the songs used for territorial defense and mate attraction are learned from other individuals. Parent-offspring recognition and the direction of parental care receive a superb treatment in a recent volume on kin recognition (Beecher 1991), so I shall focus on individual recognition among territorial songbirds. The two main points I make in this chapter are that (1) our current field methodologies for studying song perception are so confounded with social effects that studies published to date do not allow meaningful comparisons among species, and (2) individual vocal recognition of neighbors is the general rule among territorial oscines, irrespective of song repertoires.

Field studies of neighbor recognition encounter a special problem: the behavioral assays used to date have neither the power to describe the *relative* degree of recognition ability nor the power to say that recognition does not exist. Insufficient sample sizes at the taxonomic level (Garland and Adolph 1994) and confounding social effects prevent the comparative method from giving us additional leverage on questions of relative recognition ability (Weary et al. 1992 is an exception to the sample size problem). The essence of the overall problem is as follows: If a bird fails to display different behavior in response to two different songs, it may be because (1) the bird cannot tell the difference between the songs (fails to recognize), (2) playback of those two songs simulates social situations to which it responds the same way, or (3) the researcher fails to evoke or detect the bird's response. Comparative studies of neighbor recognition among species with different repertoire sizes have been founded on the assumption that social conditions, such as territoriality, are similar among related species, and thus differences in playback response must be due to differences in recognition ability. We also assume that differences in social conditions are the evolutionary forces underlying differences in vocal complexity among closely related species (Kroodsma 1977c, Searcy and Andersson 1986). Our logic has a critical flaw. How can differences in social conditions be responsible for differences among repertoire sizes, yet conveniently disappear when we want to compare studies of recognition?

Methods and Assumptions of Neighbor Recognition Studies

Recognition is the awareness of identity and the perceptual discrimination between identities. Awareness, however, is singularly difficult to observe or measure because it is a mental process (Griffin 1981). Instead, we can observe and measure the behavioral outcome when an animal recognizes differences in identity. If an animal recognizes differences between stimuli (e.g., songs of different singers) it may react differently to each—an observable act of discrimination. But it is important to remember that an animal can also recognize a difference between stimuli and show no outward sign.

Field Assays of Neighbor Recognition

Neighbor recognition studies use playback experiments designed to promote a simulated interaction between a territorial bird and its (usually his) neighbors. These experiments hinge on the "dear enemy" effect, the tendency of territorial birds to display reduced aggression to familiar or trusted individuals, generally neighbors. The rationale is that a territory holder has little cause for alarm if a neighbor sings from his own territory, where he normally sings. On the other hand, if an unfamiliar bird—a stranger—sings from the same location, presumably that bird is an intruder seeking a territory insertion or takeover. The resident therefore should respond weakly to a neighbor singing from a familiar location but strongly to a stranger singing from the same location. We can simulate these two song conditions by playing recordings of neighbor and stranger songs from the neighbor's location. We then expect that if the territory holder can tell the neighbor songs from stranger songs (the perceptual process of recognition), he will respond more strongly to the songs of the more threatening bird (the behavioral process of discrimination). A significant difference in response to neighbor and stranger song classes is evidence that the listener can discriminate between songs of neighbors and strangers.

Discrimination of individual neighbors by song is a finer discrimination than discrimination of neighbors from strangers. Discrimination of neighbors from strangers requires only discrimination among songs of two classes, familiar and unfamiliar, whereas discrimination of individual neighbors requires classification of familiar songs into as many sets as there are neighbors or neighboring territories. Falls and Brooks (1975) developed a method for demonstrating recognition of individual neighbors by song. They also used the neighbor-stranger discrimination assay to contrast aggressive responses of territorial male White-throated Sparrows (scientific names are in Table 20.1) to playback of neighbor and stranger songs. In this test, however, both song classes were played from two locations: the territory boundary shared with the neighbor (N boundary), and the opposite boundary where neither song would be expected. Presumably the neighbor would sing at the opposite boundary only if he were shifting his territory and thus were untrustworthy. White-throated Sparrows showed characteristic neighbor-stranger discrimination (hereafter N-S discrimination) at the regular shared boundary, N, but not at the opposite boundary. By demonstrating that the relative response to neighbor songs varies according to their playback territory, Falls and Brooks inferred White-throated Sparrows to be capable of discriminating among the songs of individual neighbors. Their method has proven robust and has been used to demonstrate individual recognition of neighbors in Great Tits, European

Table 20.1. Published field studies of neighbor recognition in the Passeriformes organized by family (according to C. G. Sibley and Ahlquist 1990)

Family	Species	Common name	NSD	NND	Repertoire size	References
Vireonidae	*Vireo olivaceus*	Red-eyed Vireo	—	0	50	13
Muscicapidae	*Catharus fuscescens*	Veery	+	—	3.5	32
	Erithacus rubecula	European Robin	+	+	175	4,5,(17)
Certhiidae	*Campylorhynchus nuchalis*	Stripe-backed Wren	+	+	5	34
	Thryothorus ludovicianus	Carolina Wren	+	—	25	27
Paridae	*Parus major*	Great Tit	+	+	3.5	11,18,19,21
	P. bicolor	Tufted Titmouse	+	0	10	25
Fringillidae	*Fringilla coelebs*	Chaffinch	+	—	2.5	22
	Passerina cyanea	Indigo Bunting	+	—	1	2,7
	Emberiza citrinella	Yellowhammer	+	—	2	15
	Pipilo erythrophthalmus	Rufous-sided Towhee	+	—	6	23
	Spizella pusilla	Field Sparrow	+	—	2	14
	Zonotrichia albicollis	White-throated Sparrow	+	+	1	6,9
	Z. leucophrys	White-crowned Sparrow	+	—	1	1
	Melospiza georgiana	Swamp Sparrow	+	—	3.5	26
	M. melodia	Song Sparrow	+	+	9	20,(16,26),28–30
	Dendroica petechia	Yellow Warbler	+	—	13	32
	Wilsonia citrina	Hooded Warbler	+	+	5.5	12
	Setophaga ruticilla	American Redstart	+	—	3	32
	Seiurus aurocapillus	Ovenbird	+	—	1	33
	Geothlypis trichas	Common Yellowthroat	+	—	1	35
	Icteria virens	Yellow-breasted Chat	+	—	"large"	24
	Sturnella magna	Eastern Meadowlark	0	—	55	10
	S. neglecta	Western Meadowlark	+	—	8	8,10
	Agelaius phoeniceus	Red-winged Blackbird	+	—	6	36
	A. phoeniceus	Red-winged Blackbird, female	0	—	[1]	3

Notes: NSD, neighbor-stranger discrimination, NND, neighbor-neighbor discrimination. Repertoire sizes are approximate averages. Studies with findings contrary to the tabled entry are cited in parentheses. Square brackets denote that the vocalization is a "call," not a learned song.

References: (1) M. C. Baker et al. 1981b, (2) Belcher and Thompson 1969, (3) Beletsky 1983, (4) Brémond 1968, (5) Brindley 1991, (6) R. J. Brooks and Falls 1975b, (7) Emlen 1971b, (8) Falls 1985, (9) Falls and Brooks 1975, (10) Falls and d'Agincourt 1981, (11) Falls et al. 1982, (12) Godard 1991, (13) Godard 1993, (14) P. Goldman 1973, (15) Hansen 1984, (16) Harris and Lemon 1976, (17) Hoelzel 1986, (18) Järvi et al. 1977, (19) Krebs 1971, (20) Kroodsma 1976b, (21) McGregor and Avery 1986, (22) Pickstock and Krebs 1980, (23) Richards 1979, (24) Ritchison 1988, (25) Schroeder and Wiley 1983a, (26) Searcy et al. 1981c, (27) Shy and Morton 1986, (28) Stoddard et al. 1990, (29) Stoddard et al. 1991, (30) Stoddard et al. 1992b, (31) Weary et al. 1987, (32) Weary et al. 1992, (33) Weeden and Falls 1959, (34) R. H. Wiley and Wiley 1977, (35) Wunderle 1978, (36) Yasukawa et al. 1982.

Robins, Song Sparrows, and even an electric fish (*Gymnotus carapo;* McGregor and Avery 1986, Brindley 1991, Stoddard et al. 1991, McGregor and Westby 1992).

Researchers have had mixed success demonstrating recognition of individual neighbors using a conceptually simpler playback design in which a neighbor song is played at the regular and opposite boundaries. If response intensity varies according to the boundary at which the neighbor's song is played, then the experimenter infers that the responder associates that song with a particular neighbor, which is evidence of individual neighbor recognition. This method was used successfully to demonstrate discrimination among individual neighbors (hereafter N-N discrimination) in Stripe-backed Wrens (R. H. Wiley and Wiley 1977) and Hooded Warblers (Godard 1991), but was unsuccessful with Red-eyed Vireos (Godard 1993) and Song Sparrows (Stoddard and M. D. Beecher unpubl. data). R. H. Wiley and Wiley (1977) suggested that the original method of Falls and Brooks (1975)—comparing location-specific differential response to both neighbor and stranger songs—was somewhat more powerful than their own more direct method of comparing response to neighbor songs played at the regular and opposite boundaries. I concur with this view, having tried both methods with Song Sparrows and obtained significant response differences only with the original Falls and Brooks method (Stoddard et al. 1991, Stoddard and M. D. Beecher unpubl. data).

Assumptions in the Interpretation of Playback Results

A researcher can fail to find vocal discrimination in four separate ways, only one of which is actually the absence of vocal recognition (Fig. 20.1). Without extensive follow-up study, it is impossible to tell whether failure to detect neighbor discrimination (N-S or N-N) reflects failed recognition or some other factor. Recognition studies, many of them published, support most of the paths shown in Fig. 20.1.

Past interpretation of neighbor recognition playback experiments has rested on three implicit assumptions. First, any significant behavioral discrimination implies perceptual discrimination—recognition of differences. Second, if the bird *can* discriminate perceptually between songs, it *will* do so behaviorally. Third, different behaviors reflect different intensities of response. Let us consider each of these assumptions in light of the fundamental distinction between recognition and discrimination described above.

Assumption 1. Any significant behavioral discrimination implies recognition of differences. A bird cannot discriminate behaviorally unless it recognizes a difference between stimuli. This principle appears in Fig. 20.1 as the single path (5) between recognition and discrimination. If a researcher can demonstrate behavioral discrimination between song classes, he can assert that the subject birds can recognize the difference between those song classes—provided, of course, the song classes were adequately defined and represented by the stimuli (Kroodsma 1986, 1990).

Assumption 2. If the bird can discriminate perceptually between songs (recognize the difference), it will do so behaviorally. The absence of behavioral

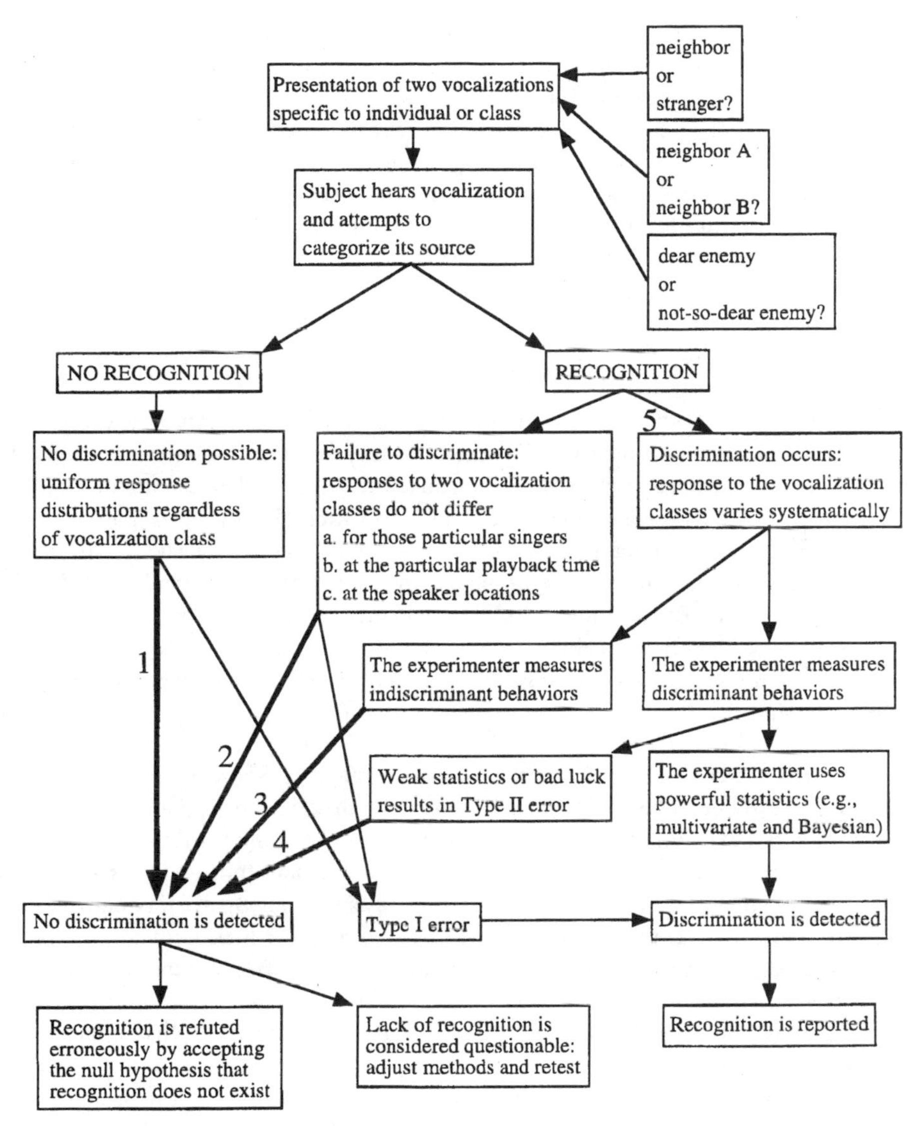

Fig. 20.1. Playback experiment designed to test for neighbor recognition. Note the three pathways (boldface 2–4) by which an investigator can fail to obtain evidence of neighbor recognition when it exists. Only boldface pathway 1 represents the absence of recognition.

discrimination therefore implies the inability to perceive a difference between songs. This assumption is valid if failed recognition is the only reason why a bird might not display differential response to songs of different individuals (Fig. 20.1, path 1). Unfortunately, discrimination might not be observed when recognition exists, for at least three other reasons (Fig. 20.1, paths 2–4). Failure to discriminate is considered again later.

Assumption 3. Different behaviors reflect different intensities of response. Past studies of N-S discrimination used the *number* of significant response measures as an indicator of recognition ability, such that a greater number of significant measures was taken as an indicator of stronger recognition than was a lesser number (Kroodsma 1976b, Falls and d'Agincourt 1981, Searcy et al. 1981c). As I shall discuss later, if we want to use multiple-response measures, we must somehow account for the different weighting of each behavior and the fact that none of these responses will be statistically independent of the others.

Why Absence of Song Discrimination Does Not Imply Failure to Recognize Song Class Differences

Unfortunately for the experimenter, birds can recognize differences yet fail to show behavioral discrimination for a variety of reasons (Fig. 20.1, path 2). Birds may fail to discriminate between two vocalization classes when the appropriate responses to those vocalization classes do not differ (1) for those particular singers, (2) at the particular playback time, or (3) at the chosen speaker locations. When results of a "neighbor recognition" test are negative, extensive knowledge of the individual birds is required to know whether one is truly observing failed recognition or merely a breakdown in neighbor relations. Following is a brief discussion of each condition (Fig. 20.1, path 2, a, b, c) that leads to failed discrimination.

Failure to discriminate between two particular singers. Some individuals may never trust their neighbors and thus always react with equal strength toward songs of neighbors and strangers or toward songs of different neighbors (Temeles 1994). Unmated male Song Sparrows, for instance, are a constant threat to their mated neighbors (Arcese 1989b). Playback experiments using Song Sparrows frequently fail to show a weaker response by mated males to songs of their unmated neighbors, regardless of playback location (Stoddard and M. D. Beecher unpubl. data).

Failure to discriminate at a particular time. Birds that normally find strangers more threatening than stable, mated neighbors sometimes have reason to distrust those neighbors as well. In the spring of 1988, cat predation on a high-quality territory at my Seattle field site created a "black hole" into which five male territorial Song Sparrows vanished over a two-month period. As each bird disappeared, a cascade of neighbors shifted over one territory to occupy the successively better vacant territories. Territories were in constant flux, and fights among neighbors were common. That spring the residents around the black hole responded strongly to both neighbor and stranger songs on the N boundary even though some of those same individuals showed strong N-S discrimination the year before and the year after (Stoddard et al. 1990, 1991). It seems unlikely that those birds temporarily forgot their neighbors' songs, but it is entirely plausible that they temporarily lost trust in their neighbors. Until boundaries stabilized, the appropriate behavioral response was the same for songs of neighbors and strangers.

Failure to discriminate at the playback speaker locations. Speaker placement has great influence over the illusion created in any playback experiment. A mismatch between song classes and playback locations, either intentional or inadvertent, may affect both the qualitative and quantitative responses of the subjects. The few neighbor recognition studies (dear enemy studies) that have experimented systematically with speaker placement have obtained stronger responses to all songs inside the territory than on the border, and greater differences in response to neighbor and stranger songs on the N border than inside the territory (Falls and Brooks 1975, Wunderle 1978, Ritchison 1988, Stoddard et al. 1991). Presumably, songs played inside the territory elicit the highest response because they simulate more closely the sort of intrusion that leads to loss of territory in those species. Songs played from the boundary may represent a challenge to be met or ignored, depending on the singer.

In some cases closely related species differ qualitatively in their response to neighbor and stranger songs at a given location. Response to neighbor and stranger songs at different locations has been compared in two closely related emberizines, the White-throated Sparrow and the Song Sparrow (Falls and Brooks 1975, Stoddard et al. 1991). Both species showed strong N-S discrimination on the N boundary and no N-S discrimination on the opposite boundary, indicating their ability to recognize individual neighbors (Fig. 20.2). Both species also showed an elevated response to both song classes at the territory center, but White-throated Sparrows maintained strong N-S discrimination whereas Song Sparrows showed no difference in response to neighbor and stranger songs, responding maximally to both. The different responses of these two species point not to differences in recognition but to differences in the threat posed by neighbors.

Behavioral measures of discrimination vary in their reliability (Fig. 20.1, path 3). For instance, response latency and song rate are popular measures of response intensity, but both pose significant problems in interpretation. Delayed latency to respond to a playback may indicate that a song stimulus had a low threat value, but it may also mean that the song did not transmit well through the environment, or that by chance the subject was far from the playback speaker when it first heard the playback. A low number of songs elicited by a playback may indicate that a bird was not threatened by the playback song, or it may indicate just the opposite—the bird was so agitated that instead of pausing to sing, it remained in constant flight, searching for its virtual challenger (Stoddard et al. 1988). Such possibilities underscore the importance of choosing unambiguous response measures.

Sometimes birds discriminate between song classes, but weak statistics or plain bad luck interferes with the experimenter's ability to detect song discrimination (Fig. 20.1, path 4). A P value equal to 0.10 indicates that in only 1 out of 10 similar

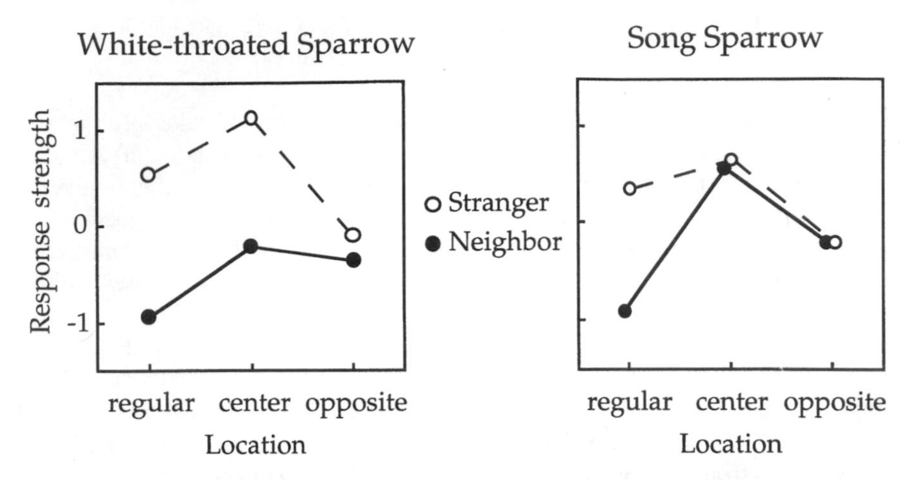

Fig. 20.2. Response of territorial male White-throated Sparrows and Song Sparrows to stranger and neighbor songs played from three locations (Falls and Brooks 1975, Stoddard et al. 1991). Speaker locations were the regular boundary shared with the neighbor whose song was used, the approximate center of the territory, and the opposite boundary shared with a different neighbor. The key difference between the species is that White-throated Sparrows maintained the neighbor-stranger response difference at the territory center, whereas Song Sparrows did not. The data shown are normalized composite scores for the number of flights and the closest approach to the playback speaker. This approximation of the first principal component of response intensity was made using the published means, so no error estimates are shown.

instances would the results be due to chance effects, yet such results are generally not publishable (Gerhardt 1992). I have often obtained spurious results when, unknown to me, a territory boundary shifted between the time when I mapped a territory and when I completed a playback series. Such "bad luck" can be reduced by remapping territory boundaries the day before a playback trial. Another source of bad luck appeared in the form of one territorial male Song Sparrow who consistently gave a response opposite that of every other field subject. To my continuous aggravation, there was no legitimate reason to exclude this bird from the sample; he held a large territory in the study tract for four years.

Considerations in the Use of Multiple-Response Measures in Neighbor Recognition Playback Experiments

The use of multiple-response measures to playback stimuli has its roots in classic ethology. If different behaviors have different release thresholds, then a stronger response is accompanied by the display of a greater number of aggressive behaviors. I see two problems with this approach as applied to studies of vocal recognition. First, the different responses are not given independently. Behaviors may be correlated, as is the case for latency of approach and number of flights, or mutually exclusive: low-intensity responses may cease when the bird switches to a higher-level response. For instance, number of songs is often used as an indicator of response intensity, but for many species silent searching often precedes an attack on an intruder. The number of songs during the playback period might thus reflect either a lower-level response than nearest approach distance to speaker or the number of flights during the playback period. If the study is done in a way that promotes a low-intensity response, such as a habituation-dishabituation paradigm (e.g., Searcy et al. 1994), then number of songs might be the best measure of response intensity.

The second problem with taking multiple measures in any playback study is that often the researcher does not know a priori which behavioral responses are appropriate. Equal weighting of inappropriate responses dilutes any real effect and weakens the ability to detect differences in behavior. The weighting issue has long been recognized as a problem in numerical taxonomy, in which a meaningful outcome depends on a scoring system that is both pertinent and accurate. I often use a pilot study to determine which measures best indicate a differential response to different stimuli. My main study then uses only those measures that showed promise in the pilot, avoiding problems of meaningless or conflicting measures. McGregor (1992) offered another solution to the problem: multivariate statistics. The multivariate technique of principal components analysis (PCA) produces independent measures of response intensity that weight the original measures by their contribution to the response variance. The multivariate approach thus eliminates both the problem of correlated measures and the use of multiple response measures.

Reliance on nonindependent multiple measures before multivariate analysis came into use may have strengthened the impression that repertoires interfere with individual recognition by song. In his 1976b paper on N-S discrimination in Song Sparrows, Kroodsma noted that differential responses to neighbor and stranger songs were significantly different for the number of flights and average distance to the speaker. Three other measures were not significant. Kroodsma observed that N-S discrimination seemed "less polarized" in Song Sparrows than in species in which individuals sing only a single song. Although he cautioned against comparing studies with different methodologies, he was sufficiently swayed by this contrast to title a paper "The Effect of Large Song Repertoires on Neighbor 'Recognition' in Male Song Sparrows." Searcy et al. (1981c) and Falls and d'Agincourt (1981) addressed this concern by comparing NSD in congeners tested with identical methods, but again took the number of significant differences in correlated response measures as indicative of relative recognition ability.

Interpretation of Negative Neighbor-Stranger Playback Results: The Confounded "Dear Enemy" Effect

Recognition is much easier to prove than to disprove using field experiments. When we detect evidence that birds discriminate between songs of neighbors, Fisherian statistical methods allow us to reject the null hypothesis that recognition does not exist. We cannot, however, formally accept a null hypothesis simply because we failed to reject it. Nonetheless, researchers using Fisherian statistics continue to embrace the null hypothesis when their results turn up no evidence of neighbor discrimination. Most recently, Godard (1993) interpreted negative data on neighbor song discrimination as evidence that Red-eyed Vireos cannot recognize neighbors by song, and then dismissed alternative explanations in the discussion without testing them explicitly (e.g., the assertion that territories were stable even though birds in the study were not individually banded). Bayesian methods of analysis are better for weighing the likelihood of different effects underlying response to playback (Gerhardt 1992), but no method frees researchers from the necessity of keeping all our assumptions explicit and valid.

No hard-and-fast rule declares that birds must trust neighbors on their territories any more than they trust unfamiliar individuals in the same locations. Indeed, territorial neighbors may be seeking to enlarge their territories or obtain mates by annexing the territories of our playback subjects. In such cases, the neighbor is not such a dear enemy after all, and the experiment most likely will not reveal vocal discrimination. The study of individual recognition using assays of territorial behavior hinges on an understanding of the underlying territorial dynamics. Failure to acknowledge this dependency has produced a strange divergence in interpretation of data. Ironically, studies designed by bioacousticians to investigate neighbor recognition have been reinterpreted by behavioral ecologists as studies of territorial behavior, which indeed they are (Ydenberg et al. 1988, Temeles 1994). Whereas studies that failed to show N-S discrimination were generally interpreted as failed recognition by birdsong researchers, these same studies are interpreted in the territoriality literature as indicative of untrustworthy neighbors. Either failed recognition or untrustworthy neighbors is a plausible explanation for failure to discriminate neighbors from strangers, and thus neither interpretation can be supported conclusively with a conventional playback study.

If male-male competition is a driving force behind the evolution of song repertoires (Searcy and Andersson 1986), then larger repertoires should be a direct consequence of greater competition among males, in particular among neighboring males. Greater competition among neighbors should, in turn, affect the degree to which a neighbor will be tolerated singing on the wrong side of a common territory boundary. Thus, even if birds with sizable song repertoires can recognize their neighbors by song, they should harbor an elevated mistrust of those neighbors as well. Researchers should therefore have a harder time eliciting reliable discrimination behavior in birds with extensive song repertoires, regardless of the birds' underlying recognition abilities.

I put forward one example that is consistent with the hypothesis that expanded song repertoires reflect heightened competition among neighbors rather than reduced neighbor recognition abilities. Remember that Song Sparrows (with song repertoires) showed N-S discrimination only at the correct border, whereas White-throated Sparrows (with single song types) showed N-S discrimination at both the correct neighbor border and the territory center (Fig. 20.2). This difference leads to the prediction that neighbors are a greater territorial threat to male Song Sparrows than to male White-throated Sparrows. A study of a nonmigratory Song

Sparrow population found that neighbors were indeed a significant territorial threat, accounting for one-fifth (21%, $N = 198$) of the aggressive territory takeovers and about half (53%, $N = 94$) of the nonaggressive replacements (Arcese 1989b). Data on aggressive takeover are not available for White-throated Sparrows, but nonaggressive takeovers were observed after artificial removal of 34 territorial males (D. J. Loncke and J. B. Falls unpubl. data); neighbors accounted for only 15% of the replacements. These data suggest that male Song Sparrows exert greater territorial pressure on their neighbors than do male White-throated Sparrows. Of course, to evaluate this hypothesis properly, we need comparable observations on aggressive takeover in related species that differ in their repertoire sizes (Garland and Adolph 1994).

How do we interpret comparative field studies of N-S discrimination if the responses may be indicative of differences in recognition abilities, territorial competition from neighbors, or some combination of the two? Comparative studies of differences in N-S discrimination between Eastern and Western Meadowlarks (Falls and d'Agincourt 1981) and between Swamp and Song Sparrows (Searcy et al. 1981c) both found stronger N-S discrimination in the species with smaller repertoires. Were these differences due to underlying differences in recognition ability or to differences in the threat from neighbors?

If conventional assays of neighbor recognition are so confounded with territorial effects, is there any reliable way to detect a weakness or failure of neighbor recognition? Two approaches are noteworthy: demonstration of differences in discrimination of different song classes, and psychoacoustic operant conditioning.

Pioneering the first of these approaches, McGregor and Avery (1986) demonstrated that Great Tits readily discriminated the unshared songs of neighbors but confused the shared songs. These authors demonstrated one prediction of the repertoire constraint hypothesis, that similar songs among repertoires should be confused. McGregor and Avery were able to demonstrate reduced recognition of shared songs because individual Great Tits could in fact discriminate unshared songs of the same neighbors.

The other approach for detecting weakness in vocal recognition ability uses operant conditioning and psychophysical analysis of auditory perception. Operant tests of song discrimination work because they are divorced from the confounds of territorial behavior (Beecher and Stoddard 1990). Operant tests in the lab can reveal sensory capabilities that would permit individual recognition of neighbors. They can also yield data that indicate which features of songs are recognized or confused (Stoddard et al. 1992b, Horning et al. 1993, Beecher et al. 1994a). Further, operant tests can demonstrate recognition capabilities in species that will not discriminate among vocalizations either in natural social interactions or in field playback experiments (e.g., Loesche et al. 1991). Psychoacoustic operant techniques let workers test perceptual hypotheses about vocal recognition, such as whether a bird has sufficient memory and discrimination abilities to recognize neighbors by song. Operant tests are best used to complement field studies of recognition (Beecher and Stoddard 1990). For instance, if we found that a species has the perceptual capability to recognize individuals by their vocalizations, then without field data to the contrary we might erroneously conclude that vocal discrimination occurs under natural conditions. Operant tests require a huge time investment and are not undertaken lightly. Nonetheless, they may be the only way to resolve the question of neighbor recognition in the few species in which field tests have not revealed song discrimination.

Do Song Repertoires Interfere with Recognition of Neighbors?

Kroodsma (1976b) was the first to suggest that song repertoires interfere with N-S discrimination, noting that Song Sparrows showed a weak differential response to songs of neighbors and strangers. Krebs and Kroodsma (1980) and Falls (1982) postulated a theoretical basis that explained why an increase in a singer's repertoire size should impair recognition by its neighbors (the repertoire constraint hypothesis): As repertoire size increases, (1) the perceptual memory of the listener is taxed; (2) each song type will be sung proportionately less, and the neighbors will therefore be given less opportunity to learn the song types, assuming song output remains constant; and (3) differences between song types diminish and songs of individuals become more alike, further confounding listeners. Comparative analysis of neighbor recognition reveals a striking pattern consistent with this hypothesis: oscines with single song types or small song repertoires show strong N-S discrimination, those with medium-sized song repertoires of approximately 10 songs per individual show moderate N-S discrimination, and those with large song repertoires do not show N-S discrimination (Falls 1982).

Following Falls's review (1982), the repertoire constraint hypothesis became a focus of research. Several species with intermediate repertoires, including Song Sparrows, were shown to be fully capable of discriminating among the songs of individual neighbors. Others were even shown to remember songs of individual neighbors from one year to the next (McGregor and Avery 1986, Godard 1991). In all but 3 of the 25 oscine species tested to date (Table 20.1), males can discriminate between songs of neighbors and unfamiliar individuals in the same population (i.e., N-S discrimination). Six of the eight species tested on the finer discrimination of individual neighbors (N-N discrimination) showed this capability as well.

What are we to make of the three outstanding cases of failed discrimination? Eastern Meadowlarks failed to show discrimination between songs of neighbors and strangers (Falls and d'Agincourt 1981). Tufted Titmice discriminated between the songs of neighbors and strangers, but neither they nor Red-eyed Vireos were found to discriminate between the songs of different neighbors (Schroeder and Wiley 1983a, Godard 1993). Do these species possess particular vocal attributes that interfere with recognition, as suggested by the authors of those studies? Is it significant or coincidental that Eastern Meadowlarks and Red-eyed Vireos sing large repertoires?

I argued above that conventional field studies do not provide reliable evidence that a bird does not recognize its neighbor, and thus we have no data capable of addressing the question of whether repertoires affect neighbor recognition. One might argue that the point is moot because, even when accepted at face value, the preponderance of positive results weigh against a general repertoire effect on neighbor recognition. Weary et al. (1992) compared song repertoire sizes with the reported degree of neighbor discrimination, controlling for phylogeny and accepting all published field studies at face value. They detected no apparent relationship of repertoire size to strength of N-S or N-N discrimination. Their argument was weakened, however, by the underrepresentation of species with large repertoires. Based on a critical evaluation of the literature, I suggest that song recognition among oscine neighbors is the rule for species with small and medium repertoires (1–25 song types/individual), but that exceptions may exist among species with larger repertoires. Until we employ a valid method to refute recognition, we cannot know for sure.

If repertoires ever interfere with individual recognition, their effect should be felt most in birds with extremely large repertoires—say, more than 100 song types. Recognition might persist nonetheless, coded in less variable features of the repertoire. Individuality among singers of large repertoires could be coded in features common to every song type, or it could be coded in characteristic sequences of songs. Common Nightingales (*Luscinia megarhynchos*), for example, cycle through their repertoires in a fixed order (Todt et al. 1979). More difficult to recognize would be birds that sing large repertoires with eventual variety, such as Eastern Meadowlarks (Falls and d'Agincourt 1981), because listeners would have to listen to hundreds of songs to discern a characteristic sequence. European Robins have large repertoires (100–250 song elements per individual), but unlike meadowlarks, the robins sing with immediate variety. Playback experiments using natural recorded song sequences as stimuli (i.e., many song types in each playback trial) showed that European Robins can discriminate between neighbors and strangers (Brémond 1968) and among individual neighbors as well (Brindley 1991). Unfortunately, no attempt was made to determine whether European Robins recognized the song sequence or the song elements themselves. Playbacks using synthetic repertoires would be required to determine the mechanism of vocal recognition in this species.

I can find nothing wrong with the theoretical arguments for why repertoires *should* interfere with song recognition (Krebs and Kroodsma 1980, Falls 1982). Investigation of the perceptual tasks involved, however, reveals an impressive capacity for rote memorization of songs that may be independent of repertoire size. In operant perceptual experiments, male Song Sparrows learned to discriminate 64 conspecific songs into two artificial repertoires, a task at least as difficult for a bird as discriminating among all its neighbors' songs (Stoddard et al. 1992b). A wild-caught male Golden-crowned Sparrow (*Zonotrichia atricapilla*) learned to discriminate among the same 64 Song Sparrow songs with equal facility (Stoddard 1989), even though a Golden-crowned Sparrow's repertoire consists of a single song with a simple structure. These discriminations were performed among songs chosen from many repertoires and arranged to minimize potential use of any possible similarities among the vocal tract characteristics of individual singers or the noise signatures of particular territories or recording rigs.

If birds have distinctive voice qualities that transcend song types, a singer might be recognizable regardless of its repertoire size. Nowicki (1987) demonstrated resonance in the oscine vocal tract that, if individually distinctive, could produce individual vocal signatures on accurately learned (i.e., copied) song types. H. E. Williams et al. (1989) identified syringeal suppression of resonance as the basis for individually distinctive vocalizations in Zebra Finches (*Taeniopygia guttata*). Suthers (1994) found that individual variation in the syrinx of Oilbirds (*Steatornis caripensis*), a nonpasserine species, produces individual-specific vocal signatures in the formant structure of their vocalizations. Similar distinctiveness has yet to be demonstrated in the vocal tract morphology of a passerine, although indirect evidence suggests that such variation exists. Alternatively, systematic copying errors may result in individual biases within the learned song repertoires (Weary et al. 1990b). Thus learning biases may produce the same end result as vocal tract signatures.

Experiments seeking evidence that putative voice traits are perceived by listeners have produced mixed results. Repeated efforts have failed to detect use or perception of possible voice signatures within the repertoires of Song Sparrows (Beecher et al. 1994a, Stoddard and M. D. Beecher unpubl. data). Weary and Krebs (1992) reported that Great Tits showed perceptual sensitivity to features common to all songs in an individual's repertoire, but their sample sizes were small and their experimental design did not exclude birds' use of extraneous environmental features (West and King, this volume). The results of this study are intriguing nonetheless, and replication would be welcomed using methodological controls such as those recommended by Lambrechts and Dhondt (1995).

Does Song Sharing Affect Recognition of Neighbors?

The combination of territorial site fidelity and song learning often results in neighborhoods of birds that share song types. Shared songs of neighbors should be especially difficult to distinguish because of their similarity, which can be extreme (e.g., R. B. Payne 1982, Falls et al. 1988). Many, if not most, species for which N-S and N-N discrimination have been demonstrated also show extensive song sharing among neighbors. Song sharing might even facilitate N-S discrimination because neighbors will tend to sound more alike than birds from farther away. Theory predicts, however, that song similarity should hinder recognition of individuals by their song (Krebs and Kroodsma 1980, Falls 1982), so neighbors that share songs should be especially difficult to discriminate from one another, at least by those songs common to their repertoires. We can break this problem into two questions: (1) Are shared neighbor songs sufficiently similar to reduce the accuracy of individual recognition? and (2) Even if shared songs can be discriminated, do they require more time for the discriminator to learn?

Schroeder and Wiley (1983a) suggested that song sharing might interfere with recognition among neighbors. They readily demonstrated N-S discrimination among Tufted Titmice but were unable to demonstrate recognition of individual neighbors. Tufted Titmice show close and extensive song sharing among neighbors that might interfere with N-N discrimination (Schroeder and Wiley 1983a). Song sharing by neighbors was later found to interfere with N-N discrimination in Great Tits (McGregor and Avery 1986). Before we can say with certainty that shared songs interfere with recognition of neighbors, we need studies that find positive discrimination when birds are tested on songs that are *not* shared, and weaker or failed discrimination when they are tested on shared songs from the same individuals. McGregor and Avery's (1986) study of neighbor recognition in Great Tits is so far the only field study in which the results were broken down this way. Using the method of contrasting neighbor-stranger playbacks at the regular and opposite boundaries (Falls and Brooks 1975), McGregor and Avery (1986) found evidence both for discrimination of neighbor and stranger songs and for reduced discrimination of shared song types. In an earlier song-matching study, Great Tits responded differently to playbacks of song types they shared with neighbors and strangers, indicating that they can discriminate both shared song types and songs of neighbors and strangers (Falls et al. 1982). Shared song types thus appear to interfere with neighbor recognition but do not present a permanent obstacle to song discrimination.

Territorial male Song Sparrows in the field readily discriminate songs of established neighbors, shared and unshared alike (Stoddard et al. 1991). Laboratory studies, however, suggest that shared songs may delay the perceptual song-learning process required for neighbor recognition. In operant song classification studies, Song Sparrows initially confused shared song types (Beecher et al. 1994a). If birds were rewarded for discriminating between shared types, however, they readily learned to tell them apart, taking about a third longer to learn the discriminations than they did to learn unique song types (Stoddard et al. 1992b). Likewise, the birds treated variations within song types as similar but readily learned to discriminate among variations when reinforcement was provided.

Acoustic mimicry hypotheses (J. L. Craig and Jenkins 1982, R. B. Payne 1982, 1983a) gain support from the findings that Great Tits and Song Sparrows are confused to some degree by shared song types. Even if shared songs do not constitute an absolute impediment to neighbor discrimination, their slowing of the perceptual song-learning process may serve as help or hindrance to a floater seeking a territory, depending on his prior investment in song learning. Floater male Song Sparrows maintain stable home ranges that overlap the territories of several adult males (Arcese 1990). Floaters learn to sing the shared songs of these males and then often obtain their first territory by inserting themselves between the territories of their tutors (Beecher et al. 1994b). Although the new bird is ultimately recognized, he may be mistaken initially by each neighbor for the other, and in this way may reduce aggression from his neighbors, especially if he establishes a territory while his neighbors are otherwise occupied caring for young. Shared neighbor songs can also work against floaters, making it harder for them to obtain a territory outside their limited home ranges. Furthermore, unless a floater has been resident for some time, he will find it difficult to distinguish one neighbor from another if they all sound alike. Inability to discriminate consistently between a vulnerable resident and a strong one could pose a serious problem to a floater seeking to establish a territory in an unfamiliar neighborhood, and shared songs might thus provide mimetic defense for territorial residents. The reluctance of some species with song repertoires to match neighbor song (Falls et al. 1982, Falls 1985, Stoddard et al. 1992a) could prevent active comparison by floaters, making it harder for them to distinguish between neighbors.

True Recognition or Habituation to Neighbor Song?

The reduced response to neighbors that we take as evidence of neighbor recognition might be caused by listener habituation to familiar songs coming from familiar locations (Falls 1982). The dear enemy effect of reduced aggression to playback of neighbor songs on the neighbor territory is consistent with this mechanism (Table 20.1). Habituation has been favored as the most parsimonious explanation for such reduced aggression, but a recent experiment suggests that responses toward neighbors may be governed by more cognitive processes as well. Male Song Sparrows do not reply to playback of shared neighbor song with a matching song type (type matching), but instead reply with another song that they share with that neighbor (repertoire matching; Beecher et al. in press, Beecher, this volume). Repertoire matching cannot be accounted for by habituation; it requires memorization of each song in a neighbor's repertoire plus association of individual songs with individual singers. Operant experiments show that such capabilities are within the perceptual and memory capacities of Song Sparrows (Stoddard et al. 1992b, Beecher et al. 1994a).

Conclusions

Falls's (1982) review of vocal recognition in birds focused attention on the question of oscine repertoire interference in neighbor recognition. From the mid-1980s to the mid-1990s, researchers demonstrated that oscines with various repertoire sizes can recognize their neighbors by their songs. From this body of research have emerged new questions to be addressed in the next decade. I list five below.

1. Is neighbor vocal recognition universal among territorial birds? In particular, the universality of neighbor vocal recognition among oscines remains challenged by eventual variety singers with large repertoires (e.g., Eastern Meadowlark).

Even though current field methods allow for only unidirectional inference, when executed with care field studies can confirm the existence of recognition. Conventional field techniques should still be used to probe for neighbor recognition in untested groups. Alternate techniques must be developed to refute the existence of neighbor recognition in species that fail to show the dear enemy response to field playbacks.

Relatively little is known of neighbor vocal recognition by species that do not learn their songs, although vocal recognition of relatives appears common in colonially breeding species, and N-S discrimination has been demonstrated for a few territorial nonpasserines (Falls 1982). We know nothing about neighbor song recognition among the suboscines. Those studied to date (some species of tyrant flycatchers) develop normal songs without social interaction or auditory feedback (Kroodsma 1989b, Kroodsma and Konishi 1991). Some tyrant flycatchers possess elaborate vocal repertoires (W. J. Smith and Smith in press a, b), however, and would be interesting subjects for studies of vocal recognition.

We know embarrassingly little about the ability of female passerines to recognize male or female songs (Beletsky 1983, Ratcliffe and Otter, this volume). Females that do not sing may still have a strong reproductive interest in identifying males by their songs, especially in species whose males sing large repertoires to court females. Females might discriminate perceptually among songs of different males without displaying malelike differences in aggression, so careful attention to female behavior will be necessary to investigate recognition processes in the neglected sex.

2. How does vocal mimicry affect neighbor recognition? Do young birds that sing the local song types pass themselves off as established neighbors? Two studies suggested that shared song types confuse avian listeners (McGregor and Avery 1986, Beecher et al. 1994a), but how long does such confusion last? Researchers probing this question should remember a complication in detecting the dear enemy effect: poor recognition and neighbor mistrust may produce similar results in playback experiments.

3. Do birds recognize singers by features other than the time-frequency structure of songs? Higher and lower levels of repertoire organization could also code the identity of the singer. For species that sing large repertoires with immediate variety, such as European Robins, Common Nightingales, and Marsh Wrens (*Cistothorus palustris;* Verner 1976), song order presents an additional cue for individual recognition. Field playbacks manipulating the order and origin (singer) of song elements would be useful for investigating the relative contribution of these features to vocal identity. Likewise, vocal signature traits transcending all songs in the repertoire could signal identity of a singer, countering the confusing effects of vocal mimicry. Such traits are physiologically possible (Nowicki 1987, Suthers 1994), but it is not yet possible to say whether they exist and are perceived (Weary et al. 1990b, Weary and Krebs 1992, Beecher et al. 1994a, Lambrechts and Dhondt 1995).

4. What are the neural pathways involved in recognizing neighbors' songs?

Studies of neuroanatomy and neural development have focused on the "production side" of song learning. Studies relating to song perception have addressed the sensory feedback pathway by which a bird refines its own singing. Neuroethologists have just begun to investigate how birds perceive the songs of other individuals (Brenowitz 1991b, Brenowitz et al. 1994) and have yet to identify the neural substrates believed to underlie neighbor discrimination (Margoliash 1987).

5. Are annual cycles of forebrain neurogenesis and arborization involved in song recognition? Adult neurogenesis and seasonal changes in dendritic arborization are not restricted to species that learn to sing new songs every season (Brenowitz et al. 1991). But are these neural changes critical for song recognition, a form of perceptual song learning? In many oscines, such as the emberizine sparrows, the sensitive window for learning to produce songs closes in the first year (Marler and Peters 1977, Beecher et al. 1994b), but adults retain the lifelong ability to learn to discriminate among the songs of new neighbors (Stoddard et al. 1992b). Cynx and Nottebohm (1992a) reported seasonal variation in the speed with which Zebra Finches acquired new song discriminations in operant testing. Either of these behavioral trends could be affected by seasonal changes in the architecture of the song system, and both deserve further consideration.

Acknowledgments

J. Bruce Falls contributed to this chapter with unpublished data and critical discussion. Peter Arcese generously shared his expertise while I was his guest on Mandarte Island, showing me complex social behavior in Song Sparrows that depended on individual recognition. I extend special thanks to Mike Beecher for many fine years of collaboration and support.

PART V THE BEHAVIOR OF COMMUNICATING

Introduction

The first four parts of this volume cover how vocalizations develop, how many occur, how they vary in space and time, and how they are controlled and perceived by the brain. In the end, however, we are interested in how vocalizations (and nonvocal acoustic displays) are used between individuals in natural circumstances, the topic of Part V.

John Smith describes the kinds of information that a singing bird makes available about its behavior. By simulating territorial intrusions with a flexible playback procedure, Smith searches for correlations between a bird's vocal behavior and its actions. Across a variety of species, he finds that vocalizations are often strongly correlated with behavior and therefore provide socially valuable information to listeners. The distribution of behavior among phylogenetic lineages challenges the common distinction between song and nonsong vocalizations, reveals that different groups may use markedly different display patterns for similar functions, and shows how variables such as social circumstances can affect the needs (costs and benefits) of displaying. The form of interactive playback used by Smith can clearly be elaborated to address a broad array of important questions.

Torben Dabelsteen and Peter McGregor offer a complementary chapter on the topic of interactive playback. Viewing communication as dynamic and interactive, these authors suggest some approaches to describing and studying social interactions. They argue that a strong test of display function requires playback that can realistically simulate a bird in natural circumstances. Such a simulation, impossible with traditional tape-loop playback, is now achievable through digital technology. The authors discuss interactive playback experiments and outline some challenges (such as how to interpret rapid variations in vocal responses and the use of individual response strategies by test birds). Digital interactive playback holds great promise for research on the behavior of communicating, and Dabelsteen and McGregor make some intriguing suggestions for future applications.

Most thinking about communication considers the roles of one displayer and one receiver, but Peter McGregor and Torben Dabelsteen in their second chapter discuss how these dyadic exchanges occur in a larger social network in which many individuals display and many receive. This chapter explores communication networks from the viewpoints of senders and receivers, and examines how participants in networks adapt their sending and receiving. Studying several simultaneously active senders and receivers presents certain challenges, and the authors describe technology that can be used to study such networks. Expanding beyond models of dyadic communication will present great technical and conceptual challenges but should contribute greatly to our understanding of animal communication.

Networks of senders and receivers are exemplified by the dawn chorus, a time of frenzied vocal behavior and the subject of the chapter by Cindy Staicer, David Spector, and Andy Horn, who place the phenomenon into a broader perspective. The authors describe features that characterize the dawn chorus across a wide range of bird species, review patterns in timing of the chorus, discuss how dawn and daytime singing often differ qualitatively, show that dawn and dusk choruses of tropical species are largely restricted to the breeding season, and suggest an association between territoriality and the dawn chorus. Staicer, Spector, and Horn recognize three classes of explanations for the dawn chorus (intrinsic, social, and environmental) and conclude that no single explanation is adequate. Rich with hypotheses and ideas, this chapter is an important contribution to understanding temporal patterns of communication and their adaptive bases.

The displays male birds use to influence female choice have been of special interest since Darwin's time, and they are the topic of the final chapter in this section. Bill Searcy and Ken Yasukawa first review the available evidence for female preferences and then address what they feel is the most important general question for research on sexual selection: How do mating preferences evolve? Hypotheses for the evolution of mating preferences are discussed and evaluated on the basis of available data on song preferences in birds. Searcy and Yasukawa conclude that song preferences have been favored because of direct benefits to females in the form of resources and paternal care, and indirect benefits in the form of good genes. By contrast, evidence for Fisherian coevolution of male song and female preferences is weak. Searcy and Yasukawa identify four research areas that are significant for furthering our understanding of sexual selection and birdsong.

21 Using Interactive Playback to Study How Songs and Singing Contribute to Communication about Behavior

W. John Smith

Songs and patterned singing performances are signaling specializations that help to increase the orderliness of animals' interactions. A considerable part of the information made available by most animal signaling renders a signaler's behavior more predictable than it would otherwise be (W. J. Smith 1977). Unfortunately, we know relatively little about the information that singing, among the most noticeable kinds of signaling, makes available about behavior. Most often, the behavior of singing individuals is quite simple. They are typically alone and not, except for singing, doing much that is obviously interactional. Singing helps them keep in touch with distant mates or neighbors with whom they must reaffirm relationships, negotiate, or arrange for closer encounters (W. J. Smith 1991). No individuals respond during most bouts of singing, however, and observations of singers thus provide only scattered clues to the aspects of the singer's behavior that are being made predictable.

Although playback of recorded songs has advanced our knowledge of the information provided by songs, most such experiments seek to understand only information about who and where the singers are (e.g., Stoddard, this volume). Most other playback experiments focus on the effectiveness of different songs or different numbers of songs in eliciting responses (reviewed in Kroodsma and Miller 1982). Designed to compare effects, the stimulus presentations are highly constrained. Further, when subjects respond to playback, they tend not to continue singing but to perform other signals. As a result, even with the results of such playback experiments we have only a limited and fragmentary knowledge of how the enormous diversity of songs and patterns of singing contribute to social order.

To make this diversity fully apparent requires agreement on terms. To say that birds and many other animals from insects to whales "sing" is a metaphor. No general agreement exists among zoologists on what the word implies. Some arguments favor continued multiple-meaning use (Spector 1994), but others favor a single concept rooted in attempts to understand communication, because whatever properties singing has, its adaptiveness derives from its capacity to influence recipients by making information available. Historical use confounds two fundamentally different kinds of signaling devices. In this chapter I refer to one of these, the sustained and quasi-rhythmic production of signal units, as "singing." The special contributions of singing to social events derive both from its continuity and from the ways basic signal units are incorporated into a singing bout. The basic units used most frequently in singing bouts are here called "songs" (W. J. Smith 1991).

What can a flexible playback technique based on actively interacting with birds help us to learn about the information that singing and songs provide about a singer's behavior? The question is not, How do birds respond to different playback stimuli? but As birds interact in various ways with the playback stimuli, does their selection of songs and patterns of singing correlate with (and thus inform about) their different kinds of behavior?

The initial focus of experiments that Anne Marie Smith and I did was to learn whether different song forms within a species' repertoire provide behaviorally distinctive information, and whether such information is sufficiently fundamental to be common to the songs of birds of diverse evolutionary lineages. We have found some widespread kinds of information. The apparently mosaic phylogenetic distribution of species whose songs provide these kinds of information is puzzling, however, and has led us to consider both potential alternative means of providing information about behavior while singing and the possibility that species with different social systems and ecological regimes may accomplish substantially different tasks with singing. We have also found many ways in which to elaborate the basic interactive playback procedure, and I describe several of them late in the chapter. I then review a number of questions arising from the work and compare the main kinds of information uncovered thus far with general classes of information that were previously apparent in studies of animal communication.

The Underlying Perspective

Misunderstandings abound in the ethological literature about communication and other social behavior. At the core of many current problems are differing assumptions about issues such as deception and the place of selfish behavior in organized social relationships and groups. Other problems arise from attempts to infer motivations or other internal states of signaling individuals.

Even the basic term *information* is sometimes misunderstood. As I use the term, information is what makes things and events predictable in any way. It is an abstract property, not a thing. Information is essential for all choices, and every action of an animal can be viewed as involving choices among alternative acts. Signaled information is important in social events because enhanced predictability influences how recipients of the information choose to act (W. J. Smith 1977, pp. 2, 193, 1986a).

No specific cognitive model is invoked in saying that information, by enhancing predictability, facilitates choices. Any action, including signaling and responding to a signal, is a choice, a selection from among possible actions. This extremely general sense of choosing is also applicable to electrons, molecules, cells, and computers. Thus, choices need not involve conscious awareness, intentionality, or the like. Indeed, much of humans' signaling and responding occurs without conscious awareness. For instance, the choice to smile or to respond to a smile during conversation is more often made without deliberation than with. Analyses of the information made available by singing or most other signaling need not deal with cognitive processes or any other internal states.

Further, even though the effects of an animal obtaining information can be enormous, specification of the kinds of information that are signaled is neutral with respect to causality and function. Functional interpretations of any kind (e.g., about how signaled information might contribute to attempts to "manage" or "manipulate" responding individuals) involve contextual variables and inferences not implicit in the description of the information itself.

The act of signaling does not, by itself, usually enable an individual to control an interaction. Each participant vies with the others for control. The result is often an exchange of signals that facilitates joint movement toward some resolution of the participants' different agendas (and thus is, in effect, a negotiation as described in Hinde 1985). An individual most likely cannot consistently manipulate other individuals such as mates or neighbors, who can be expected to detect its attempts and resist (W. J. Smith 1986a).

Communicating—the process of signaling and eliciting responses—is complex and flexible. Both signaling and responding are always conditional and probabilistic. Signaling does not entail full sharing of the signaler's information, but only that information that (on average) tends to elicit responses useful to the signaler (W. J. Smith 1986a). Responses are not automatic (signals do not simply "release" responses, an old notion implicit in oversimplified models of managing and manipulating). Instead, a responding individual is influenced by information from multiple sources contextual to a signal, some concurrent and some remembered (W. J. Smith 1977).

Further, that individuals share information when they signal does not imply that they necessarily share tangible resources. An individual that shares with others the information that it is likely to attack, for example, may dissuade them from trying to appropriate its food, space, or mate. The dissuaded individuals avoid being attacked, and both they and the signaler profit, in spite of differential access to resources.

Interactions are commonly orderly even when they do not lead to equal apportioning of resources. Social relationships and groups are stable much of the time even when there are vast inequalities among the participants. What is necessary is that each individual gains more from its social behavior than it would gain from such alternatives as bucking the status quo or being solitary. Order and stability require constraints on the extent to which individuals can withhold information, signal unreliably, or try to control or manipulate other individuals (W. J. Smith 1986a). Especially in interactions among familiar individuals, most signaled information must support reliable predictions that make it adaptive to respond.

Our initial task of analyzing the information made available—of analyzing what is made more predictable—by any form of signaling is purely descriptive. Different kinds of analyses are needed to investigate internal states or functions, analyses that are made easier if the signaled information has been interpreted first.

The Initial Comparative Project

Tyrannids and the origins of the playback procedure. We were led to develop flexible interactive procedures by the need to keep Eastern Wood-Pewees (Tyrannidae: *Contopus virens*) singing and actively searching for a simulated intruder. In order to produce extended interactions with continued attempts to confront, we found that the stimuli need to be both timed to the pace and nature of each subject's actions and moved locally to simulate jockeying for position. Our solution was to have the experimenter carry and operate mobile playback equipment and respond interactively to each subject. Although the presence of a human may limit how closely some subjects approach, the vast majority respond and search vigorously for the simulated intruder for several minutes.

In daytime singing during the breeding season, each Eastern Wood-Pewee typically uses two distinct song forms (Table 21.1). One is uttered about eight seconds apart in strings terminated by an utterance of the other (e.g., AAAAB). The number of A songs in a string is inversely proportional to the extent to which a singer can be seen to actively seek out another bird (a male or female, or even a bird of another species) by approaching it, its nest, or its territorial boundary. Diverse interactions follow the approach, including attacks, fights, sexual chases, copulations, and feeding a mate. The shortest strings of A songs come with close approach before more direct interaction or when close interactions falter. The longest strings come with solitary foraging. Intrusions we simulated by interactive playback elicited approach and searching with short strings of A songs from all subjects. Playback enabled us to elicit extended singing and to understand better how the two song forms are used during infrequently observed territorial confrontations (W. J. Smith 1988).

We obtained similar results with the Eastern Kingbird (*Tyrannus tyrannus*), a distantly related flycatcher that also bases its daytime singing on two song forms (W. J. Smith and Smith 1992). The kingbird's song forms do not resemble the pewee's, however, and the kingbird's B incorporates the A form (Table 21.1). Further, singing kingbirds produce no obviously nonrandom string lengths. Lone kingbirds sing primarily A songs with an occasional B. B songs become prominent when a singer is actively interacting with its mate, neighbors, or predators. All subjects that approached interactive playbacks of either A or B songs sang predominantly B songs, as predicted. This result cannot be explained by any form of matching, or avoidance of matching, of playback stimuli (see below).

Pewees and kingbirds uttering primarily song form A tend to do little other than singing that may lead to close interaction with another individual. Often they

Table 21.1. Structures of song forms used by selected tyrannids, oscines (songbirds), and nonpasserines

			Relation of B to A				
			Dissimilar unit(s)		B = A + dissimilar affix		
Species	A (common with minimal activity)	B (increases with more active social behavior)	Vocal	Nonvocal	Prefix	Suffix	Both
Tyrannids							
Contopus virens	Whistled *pee-ah-wee*	Whistled *wee-ooo*	x				
Tyrannus tyrannus	Harsh *zeer*	Chatter-*zeer*			x		
Myiarchus crinitus	*Weep*	*Wit, weihp, churr,* & other song forms	x				
Fluvicola pica	Nasal *reeur*	Buzz-squeak	x				
Myiozetetes cayanensis	Plaintive *feeeee*	Series of shrill *kee*	x				
Pitangus lictor	Nasal *drayz*	*Dray-zee-zee*				x	
Oscines							
Seiurus motacilla	Accelerating whistles	A + diverse twitters				x	
S. aurocapillus	Series of *tee-chur* components	Jumbled series, including some A					x
Dendroica nigrescens[a]	Buzzy series	More complex series	x				
Passerina cyanea	Series of high components, many paired	A + further components				x	
Sporophila bouvronides	Trill of 1–3 parts	Trill + a variable suffix (& sometimes modified recursions)				x	
Melospiza georgiana	Simple trill	Complex set of distinctive components, (i) prefixed to A or (ii) independent	x(ii)		x(i)		
Polioptila caerulea	Faint, brief, nasal *pwee*	Bursts of chirps, twitters, trills, squeaks, & buzzes	x				
P. plumbea	(i) Louder series of clear whistles, & (ii) faint, brief, nasal *pwee*	Similar to *P. caerulea*	x				
Turdus merula[b]	Warbles motif & brief twittering	Shorter motif & longer twitters				x	
Nonpasserines							
Tapera naevia	Whistled *sem-fim* components	Rising series of *wee*	x				
Melanerpes carolinus	*Kwirr* bursts	(i) Longer bursts of A, (ii) drumming		x(ii)			x(i)

[a]From Morrison and Hardy (1983).
[b]From Dabelsteen and Pederson (1990).

forage or do other maintenance behavior. Conversely, singers uttering a high proportion of the less common B song form take more initiative. They actively approach other birds or sites where encounters are relatively likely to develop. While still singing, most stop partway in the approach, leaving the next response to other individuals. Some individuals, however, take further initiative by, for example, attacking or mounting.

I must emphasize that the "interactional initiative" just mentioned is defined entirely in terms of observable behavior. The wording aids comparisons of a broad class of behavior across events and in different species. No attribution of intent or motivational states and no functional interpretations are implied by "actively seizing interactional initiative" or "deferring initiative to other individuals."

The Great Crested Flycatcher (*Myiarchus crinitus*) provides similar information about behavior, though it uses about a dozen different vocalizations and its singing is more complex than that of either the kingbird or the pewee (W. J. Smith and Smith in press a, b). Our observations and playback again showed that the most common song form (A) correlates with minimal interactional initiative. It is dropped (if already being sung) on close approach to mates, opponents, or playback. Several other vocalizations are also used with approach, each in a distinctive way. For example, one song form occurs in loose interactions and with initial searching for playback. It is replaced by another during continued searching and closer interacting. Yet another is sung by subjects while confronting opponents or persistently searching close to playback. This flycatcher thus uses several B songs to make finer distinctions during its more active interactional behavior.

All of the 17 species of tyrannids (6 are shown in Table 21.1) with which we have done extensive playback trials have songs that correlate with different extents of interactional initiative. Several of these species also provided other information, as yet analyzed only for the Great Crested Flycatcher. The song this flycatcher utters most during initial approaches to playback also occurs with especially unstable behavior in nonexperimental social events in which uncertainty is unusually high—situations in which a singer may suddenly change what it is doing. The singer's options include pausing and watching (sometimes silently), further approach, and withdrawing (which we elicited by approaching with playbacks). That a song form correlates with such moments is especially interesting because abrupt behavioral shifts are unusual during most singing.

Oscines. Like diverse suboscine tyrannids, songbirds from various lineages also have distinct song forms, some correlated with approach and some with other behavior, thus providing information about different extents of interactional initiative. At least some of the seven species we have studied also have ways of providing information about relatively unstable behavior.

The songs of four of these species share gross structural similarities and behavioral correlates with the A and B songs of the Eastern Kingbird (Table 21.1). The paruline Louisiana Waterthrush (*Seiurus motacilla*) often sings both its A and B songs incompletely and thereby scales the information these provide about behavior. Shortened A songs are most regularly sung by lone waterthrushes, whereas A songs uttered in approaches to playback are usually completed (W. J. Smith and Smith in press c). The B song form, especially its longer versions, predominates in approach flights and persistent close searching. Sets of brief B components and related vocalizations are uttered in moments of indecisive behavior. The variable A and B song forms of emberizine finches such as the north-temperate-breeding Indigo Bunting (*Passerina cyanea*) and the Lesson's Seedeater (*Sporophila bouvronides*) of highly seasonal tropical grasslands are also grossly comparable to the kingbird's, and Dabelsteen and Pedersen (1990) reported a related pattern for the Eurasian Blackbird (*Turdus merula*).

Another emberizine, the Swamp Sparrow (*Melospiza georgiana*), shares the affixing procedure that differentiates A from B song forms (Table 21.1), but each individual has two to four distinct A song forms. A bird does not select among them to differentiate levels of interactional initiative. The complex "affix" is unlike the simple trilled A songs and, as in the kingbird and thrush, often is uttered as an independent song. It may be repeated and even sung in flight by birds approaching playback, and in only some of these repetitions is it affixed to A segments (W. J. Smith and A. M. Smith unpubl. data). In other, infrequent events it is prefixed to trills uttered in flight displays (Nowicki et al. [1991] recognized it as a structurally "alternative category" of songs).

Even more complex singing is used by two sylviids with which we have completed fieldwork, the migrant Blue-gray Gnatcatcher (*Polioptila caerulea*) and the resident Tropical Gnatcatcher (*P. plumbea*). The migrant has a faint, nasal A song (resembling the begging of its fledglings) and a less common assortment of B songs (Table 21.1), variably packaged in repeated bursts, that predominate during close approach to playback. The resident *P. plumbea* also has nasal songs, but it more commonly selects among several louder series of clear components unlike songs of *P. caerulea*. Tropical Gnatcatchers approaching playback are slow to change from their clear songs (many do not change), but some add nasal songs and, when close, switch to brief chirping B songs resembling those of *P. caerulea*.

Nonpasserines. We also found song forms that provide information about different extents of interactional initiative and behavioral instability in non-passerines, species that are only distantly related to the passerine tyrannids and songbirds. We have completed trials with a cuckoo and a woodpecker, but birds of many lineages remain unstudied.

Individual Striped Cuckoos (*Tapera naevia;* W. J. Smith and Smith unpubl. ms.) sing, during the breeding season, an A song form (Table 21.1) abundantly when individuals are not interacting closely. B songs are uttered in closer counter-singing and by subjects that have closely approached playback. Before a cuckoo that has approached playback switches to B songs, it usually utters faint bursts of variable, apparently condensed A and B song forms. These brief vocalizations are also uttered by some subjects reverting from the B to the A song form and correlate with times of change and unstable behavior.

The Red-bellied Woodpecker (*Melanerpes carolinus*) is especially interesting. It usually sings different song forms on breeding and wintering territories, and has been quicker to respond when breeding. Its song is then a *kwirr* and is usually uttered in bursts: primarily short bursts by lone individuals, and longer bursts with approach to playback. During closest approach, many subjects switch wholly or partly to drumming, a percussive song form (W. J. Smith and A. M. Smith unpubl. data).

Songs and singing in "stay apart contests." A kind of singing we have begun to explore with interactive playback often occurs in vigorous vocal exchanges between separated individuals. For example, some tyrannids have distinctive song forms commonly called dawn songs (see also Staicer et al., this volume). These songs appear to provide information about countersinging contests rather than close confrontations. Although these songs occur primarily at dawn, they also occur in both crepuscular periods and occasionally on moonlit nights or before dark thunderstorms. Further, dawn songs sometimes continue through late morning, at least when newly arrived migrants are rapidly establishing territories. Singing with dawn songs is usually highly ordered sequentially, vigorous (especially if several individuals are mutually audible), and maximally rapid.

Not only do dawn-song singers not approach each other closely (if at all), no subject began dawn songs after approaching playback. Further, in more than a dozen trials in which individuals of five species were singing dawn songs before playback, subjects either continued these songs and did not approach, or approached and became silent (or greatly curtailed the songs).

A distinctive feature of circumstances in which dawn songs predominate may be an urgent and persistent lack of information about other individuals. Low light may prevent singers from seeing each other or taking much action. Most daylight cases of dawn songs occur when developing social relationships are still fluid. Not yet knowing how committed its opponents are, an individual may more economically begin assessing their challenges through probing and contesting vocally rather than through physical contact. Singing contests could develop if all opponents lacked sufficient information to commit to attack behavior. No participant need then expect to be abruptly attacked.

The information provided appears to be that each contestant is prepared to sustain countersinging rather than take the offensive. The duration, consistency, and rate of each bird's performance relative to other individuals may be key measures. Losers might slow or cease countersinging first, and winners might sometimes approach and intrude on quiet neighbors. These predictions could be tested by temporarily removing individuals, substituting interactive playback for their performances, varying the playback, and monitoring the neighbors. Support for the prediction would reveal this countersinging to be fundamentally similar to the formalized, noncontact contests of species such as fighting fish (*Betta splendens:* M. J. A. Simpson 1968, see interpretation by W. J. Smith 1986a, p. 76) and red deer (*Cervus elaphus:* Clutton-Brock and Albon 1979). Maynard Smith

(1974) named these contests "displays" but violated the established use of the term in ethology (see W. J. Smith 1977).

Dawn songs and singing present many puzzles. Not all tyrannids have differentiated dawn songs. Although most tyrannids sing rapidly and rhythmically before sunrise, many use the same song forms that are sung in the daytime. The Eastern Phoebe (*Sayornis phoebe*), for instance, rapidly alternates its main daytime song forms at dawn (W. J. Smith 1969, 1977). Observations suggest that the shifting proportions provide information about a singer's bias toward interactional or noninteractional behavior in a way similar to that of the Eastern Wood-Pewee (see above). Yet, our playback to phoebes usually elicited alternation of songs without approach. The phoebe and pewee thus use singing differently. Pewees, although they approach and challenge with their two main daytime songs, are unlike phoebes in having a third song form that accounts for up to half of their dawn countersinging (W. Craig 1943). Phoebes do not usually closely confront birds with their two main songs, but countersing with them at dawn and later.

A further complication is that special dawn song forms are not restricted to countersinging contests. For instance, male Eastern Kingbirds may utter dawn songs when consorting closely with their mates before nesting, and the females may utter some, too. Unlike the sunrise performances, numerous kinds of vocalizations are used in these male-female interactions, mostly arrhythmically. Apparently the rapid, highly rhythmic, regular sequencing of songs provides important information in countersinging contests.

Dawn singing is characteristic of a wide array of birds. It has been much studied in several species of paruline wood-warblers, although not yet with interactive playback. Paruline patterns differ in many ways from those of tyrannids (see Staicer et al., this volume).

Functional implications. Consider next some probable functions of making information available about levels and kinds of interactional initiative, or about behavioral stability and alternatives to approach. Most singing is loud. Its continuing, quasi-rhythmic structure makes the timing of songs predictable. These features suggest that singing may often enable social arrangements to be made, reaffirmed, or opened for negotiation (e.g., for a signaling exchange that may resolve some contested issue) at a distance (W. J. Smith 1991).

Why should an animal announce that it is taking only limited interactional initiative and thus effectively defer further initiative to others? This may be the most abundant information that singing provides about behavior, and it may enable individuals to affirm their presence and readiness to respond to distant, silent, or unseen neighbors or mates. The status quo need not be disturbed. Singers could unassertively forestall intrusions by posting claims of territorial ownership and "showing the flag," and also offer restrained readiness to attend to mates when needed. An individual can escalate by announcing that it is actively seizing the initiative without immediately making physical contact. It sings as a negotiating step, in effect daring opponents to challenge. As a step in courting, a singer that escalates requires its mate or potential mate to reveal its readiness to interact. Providing information about unstable behavior can also be a negotiating or courting ploy. It can force choice on the other individual while providing little information about the option (if any) a singer may favor. It can also be a probe to elicit information while sharing only a minimum.

Further Development of the Comparative Project

We initially sought species whose different song forms correlate with different behavior and asked whether this pattern has a wide phylogenetic distribution. We found many species that did not appear to use songs in this way. Questions arose about the diversity and evolution of song structures and singing, the relation of various kinds of signals to one another, the information made available, the ways songs contribute to interactional behavior, and ways we might elaborate interactive techniques.

The Diversity of Songs and Other Signaling Specializations

In this section I consider distinctions that facilitate analyses of songs and singing. The first two distinctions are between songs and related vocalizations and between songs that do and do not correlate differently with behavior. The vocalizations in all such cases are signaling units with basically comparable structures. The three subsequent distinctions, however, are among fundamentally different structural classes, each class best characterized as comprising a separate "repertoire" of signaling specializations (W. J. Smith 1986b, 1990).

Most species have from four to six repertoires, each important because its distinctive structural properties give it special capabilities for providing information. The "displays" repertoire comprises the basic signal units of each species and includes each song form and nonsong vocal forms. Distinctions in structure between song and nonsong can be large or small. Two other repertoires are sets of procedures for performing displays: one set for modifying the structures of the basic units, and one for combining displays into specialized sequences (e.g., procedures for combining songs into sustained singing performances). Additional repertoires involve basic signal units that cannot be performed by an individual acting alone (e.g., human handshakes and bird duets). Such mutually performed units are called "formalized interactions" (W. J. Smith 1977). As with the display units of single individuals, repertoires of rules are used to modify structures and sequence these joint performances.

Song and nonsong vocal forms. The distinction between song and nonsong vocalizations is open to interpretation. Disagreements exist about the assignment of many vocal displays (e.g., see W. J. Smith 1991, Spector 1994). Although I define songs as vocalizations that occur primarily in sustained, quasi-rhythmic performances, many species have two or more structural classes of such vocalizations, and members of only one class are usually called songs.

Field guides recognize as songs only the most widely accepted cases (herein loosely termed "ordinary" songs). Other vocalizations are listed, if at all, as nonsong "calls." The simpler vocalizations of many species are relegated to nonsong status even if often uttered in the sustained, quasi-rhythmic bouts diagnostic of singing performances. Examples include the *chip* of the Northern and Vermilion Cardinals (*C. cardinalis* and *C. phoeniceus*), the *chrenk* of Rufous-sided Towhees (*Pipilo erythrophthalmus*), the *tih* and *tcheh* of Song Sparrows (*Melospiza melodia*), and the *tp* of Eastern Phoebes. All were used abundantly in approaches to interactive playback, and in some species often replaced ordinary songs. Many species sing these simple vocalizations as the principal advertisements of nonbreeding territories (e.g., the *chek* of the Louisiana Waterthrush and similar vocalizations of most migratory parulines on winter territories; Rappole and Warner 1980).

Along with simple vocalizations, many more complex "calls" such as the *t-slink* and *click-rasp* of Carolina Chickadees (*Parus carolinensis*) are commonly relegated to nonsong status. Apparently, after a particular structural class of songs has been accepted as the ordinary songs of a species (e.g., the trilled songs of Swamp Sparrows and the "whistled" *fee-bee fee-bay* of Carolina Chickadees), all vocalizations of other structural classes are considered calls.

An alternative is to recognize that songs can be of more than one structural class. Swamp Sparrows (see above) have both trilled songs and a more complex song that is sometimes affixed (Nowicki et al. 1991). This species also has brief vocalizations that, like those of the congeneric Song Sparrow, are often rhythmically repeated. These sparrows have a total of three structural classes of songs.

The distinction made herein between singing performances and songs is important because each represents a wholly different signaling repertoire. The distinction between songs and nonsong vocalizations is within the basic "display" repertoire, however, and is less crucial. Songs occur primarily in sustained, quasi-rhythmic bouts. Nonsong vocalizations do so only incidentally, and typically correlate more with momentary actions such as attacking, diving for cover, or alighting from flight (W. J. Smith 1977). Yet many species have some vocalizations that can be repeated (rhythmically or irregularly) or uttered singly. We might ask simply how those figure in particular events rather than categorizing them precisely.

Interchangeable songs. In diverse species the different ordinary song forms of each individual apparently do not correlate with different kinds of behavior. Each male Swamp Sparrow has about two to four ordinary songs (Searcy et al. 1981a, Marler and Sherman 1985), and we detected no consistent selections from among them as subjects approached playback. Most subjects did not change from their preplayback songs. Comparable results were obtained in trials with more than 15 species of emberizids, vireonids, parids, and other oscines. Perhaps some species categorize their song forms in ways we have yet to recognize (e.g., into groups with related intonational patterns; see also Hailman and Ficken, this volume).

Perhaps one reason individuals have interchangeable songs is that singers can use them to "match" songs uttered by opponents or mates. By matching, a singer may be indicating the individual to which it is being attentive (Brémond 1966, p. 49). Further, matching involves familiar song forms and may facilitate estimates of a singer's distance from a listener (Shy and Morton 1986). Perhaps designating other individuals and facilitating estimates of distance can be as useful as providing information about the extent to which interactional initiative is being taken. Matching and distinguishing levels of initiative may be evolutionary alternatives. Matching, which indicates heightened attentiveness to a particular individual, might imply that an individual is seizing considerable social initiative and may seize more. Not matching (in species that use the procedure) might be equivalent to deferring initiative. It may nonetheless be optimal for individuals to have more than one structural class of songs, as do Swamp Sparrows, thus enabling individuals to match and provide information about behavioral initiative with separate classes of songs.

Other ways largely interchangeable song forms can provide information about behavior depend on organizing songs into singing sequences. In some species (e.g., the Northern Mockingbird, *Mimus polyglottos*), all measures of singing versatility peak during courtship (Derrickson 1988). In other species, increases in the transition versatility (Kroodsma 1978b), or rate of switching among song forms (Western Meadowlark, *Sturnella neglecta:* Falls and d'Agincourt 1982, d'Agincourt and Falls 1983), correlate with attentive behavior and even approach toward opponents, mates, or both. Transition versatility also increased with attention and approach in most of the species we initially studied; they shifted the proportions of song forms such that rare ones became increasingly frequent.

The patterned sequences of singing. All procedures that govern how songs are combined into organized performances represent a kind of communicative specialization that is unlike that of the displays themselves. These procedures comprise species-specific repertoires of rules. Using its rules, an individual can produce organized, flexible performances and control how signals occur contextually with one another.

By controlling the durations of intersong intervals and the sequencing of songs, singing provides information not obtainable from a single song unit. For example, many species reduce variation in successive intervals and sing more rapidly with approach to playback. Indeed, rapid, highly regular repetition is perhaps the only distinctive characteristic of approach singing in two furnariid spinetails with which we have done many trials, the lowland *Synallaxis albescens* and the Andean *S. azarae.*

The order in which two or more different songs are sequenced is specialized in many species. Sequences can be regulated by altering transition versatility

(Kroodsma 1978b), or by arranging different song forms into varied string lengths (as described above for *Contopus virens*) or distinctive couplets, triplets, or other combinations (e.g., in *Vireo flavifrons:* W. J. Smith et al. 1978).

When quasi-rhythmic sequencing is imposed on structurally simple songs or songlike vocalizations, perhaps only the rhythm provides information about the extent to which a signaler will seize or defer interactional initiative. The vocalizations themselves may inform about interactional behavior and such transient alternatives as pausing or fleeing.

Modifications of songs. The information provided by songs can be altered or augmented by modifying the songs' physical structures. Each species' procedures for varying songs and other displays constitute a repertoire of signaling specializations that is different in kind from the repertoire of displays themselves (W. J. Smith 1986b, 1990).

Songs from which the terminal components are deleted, for example, are often uttered as birds approach playback. We have recorded this structural modification from several taxonomic families, but particularly commonly from such emberizids as *Cardinalis cardinalis* and *C. phoenicius, Zonotrichia capensis* and *Z. albicollis, Pipilo erythrophthalmus, Spizella pusilla, Passerina cyanea, Sporophila minuta,* and *Piranga olivacea.* Many individuals approached our playback closely and silently, often vigorously and somewhat furtively searching nearby vegetation, and then sang terminally abbreviated songs. Their shortened songs came with cautious behavior that could suddenly lead to close encounters. Subjects took higher perches and resumed full songs when challengers failed to materialize. Occasionally, shortened songs were uttered before subjects first approached and by individuals that did not approach, or that even withdrew. A bird uttering shortened songs may thus be choosing between the alternatives of avoiding and confronting. The probability of avoiding and behaving unstably may be higher than with most singing. Terminally shortened songs may thus approximate nonsong vocalizations in enabling prediction of quick adjustments to ephemeral situations.

Often sung faintly at first, shortened songs get louder as singers take more conspicuous perches. Faint songs must reach fewer recipients. Faintness may be another indicator of cautious behavior, but it and shortening can occur independently: some shortened songs are loud, and full songs were sung faintly by some subjects that had just approached playback. Faintness may indicate that alternatives to interacting limit an individual's approach, and yet provide little information about those alternatives (one, in addition to withdrawal, might be to sing and negotiate without further approach).

Many other structural modifications occur as birds respond to playback. Some are complex (see E. H. Miller 1988 for ways of describing them). Most are unstudied. For example, sometimes the initial (or both initial and terminal) components are deleted. Lengthening songs not by special affixes but by repeating components seems important to some emberizid and formicariid species with which we have begun trials. Various manipulations of amplitude are significant (Dabelsteen and Pedersen 1992). Species such as the Carolina Chickadee sometimes change the relative frequencies of song components as they approach playback. Kentucky Warblers (*Oporornis formosus*) change frequencies to match the energy distribution of the playback stimulus (E. S. Morton and Young 1986).

Joint singing. When two or more birds coordinate their utterances, they produce an example of a formalized interaction. In many species of tyrannids, furnariids, and tropical parulines, mates approach playback together and vocalize both in response to each other and in response to the playback stimuli. Their joint singing is "duetting." Interactive playback enables duetting to be elicited and studied, but to determine whether playback itself elicits particular song forms we have to seek situations in which one mate approaches playback alone.

Some countersinging contests of neighboring males also follow a formalized routine, for example, the remarkable leader-follower sequences of Marsh Wrens (*Cistothorus palustris;* Verner 1976, Kroodsma 1979, 1983b) in which the individual seizing the initiative anticipates each of the other's songs and utters it first. The deferring individual accepts this and adheres to its regular sequence. Their different roles within the encounter provide information about which wren is being the more forceful.

A Phylogenetic Mosaic

Species from diverse lineages use different songs to provide different information about behavior. Oddly, the basic features of these songs are often more similar among distantly than closely related species, and even closely related species with similar songs may differ in how they use them. A phylogenetic mosaic thus exists (see Table 21.1). Individuals of some species distinguish between acts of deferring and seizing initiative in interacting by selecting from a set of discrete song forms in differing proportions. Individuals of other species utter different proportions of a basic song form with or without affixes. In yet other species, each individual has sets of songs so different from one another that they belong to structurally unrelated classes.

Consider an example of divergence. The Ovenbird (*Seiurus aurocapillus;* see Lein 1981) and its congener the Louisiana Waterthrush use different affixing procedures to distinguish B song forms from A forms and have at least a threshold difference for switching between forms. Unlike waterthrushes, Ovenbirds do not sing B song forms from song stations after dawn. They largely restrict B songs to display flights and close, active encounters. Only a few uttered even brief fragments of B on approaching our playback (W. J. Smith and A. M. Smith unpubl. data). Ovenbirds thus sing their primary song both when deferring initiative and with approach. In still other parulines particular song forms have been elicited consistently with approach (*Dendroica nigrescens:* Morrison and Hardy 1983) or during playback (*Vermivora chrysoptera:* Highsmith 1989, approach per se not mentioned), but neither species differentiates song forms by affixes.

More surprising than the divergences among congeners and within subfamilies are the convergences in structure among distantly related species. Affixing differentiates the song forms of Louisiana Waterthrushes, Indigo Buntings, Lesson's Seedeaters, Eastern Kingbirds, and Eurasian Blackbirds, for instance (Table 21.1), but not those of numerous close relatives of these species. A different procedure is to sing two or more distinct song forms in different proportions. Many tyrannids change song proportions as the probability of actively initiating interactions shifts, as does the sylviid Blue-grey Gnatcatcher. The gnatcatcher's tropical congener has, in addition, songs of more than one structural class. The Tropical Gnatcatcher's use of songs of divergent structural classes is somewhat like that of emberizine Swamp Sparrows, which sing ordinary song forms before playback and add quite different song forms (often as prefixes) with approach.

The influence of phylogenetic lineage therefore seems limited. It is as yet unclear what is responsible for the often large differences among species: habitats, social structures, bioacoustic environments (E. H. Miller 1982), seasonal pressures, historical accidents in evolution, or various combinations of these. More species must be studied.

Exploring Differences in What Species Need from Singing

The information individuals need to share, perhaps especially by long-distance signaling, surely differs with the kinds and frequencies of their social encounters. The level of social familiarity and stability in a neighborhood is one major influence; another is the degree of urgency imposed by ecological conditions.

Where arrangements with mates and neighbors are long term and stable, and environments are not sharply seasonal, few subjects may be readily engaged by playback at any one time. Over a longer period, different subjects arrive at crucial stages in their breeding and are highly responsive. In contrast, when new social arrangements must be forged promptly at the start of a short breeding season, neighbors tend to be in phase, and many are highly responsive. Their number falls abruptly as social arrangements stabilize and become less vulnerable to disruption (see also Morton, this volume).

Abrupt escalation of responses to intrusions may occur primarily when both uncertainty and time constraints are maximal. Migrants are pressed for time and must often form new relationships, whereas residents often have reasonably stable social environments. Although temperate residents do breed approximately in concert, they can reduce the urgency of encounters by preparing for nesting over a period of weeks or months. Within those tropical habitats that are not sharply seasonal, both migrants and residents may have relatively loose schedules.

Nonbreeding or winter social arrangements are often simple and place the fewest demands on signaling. The balancing of territorial costs and benefits may be less stringent than during breeding. Clarifying and settling issues may require less elaborate communication and permit relatively leisurely negotiations. Indeed, many species of migrants wintering territorially in the Tropics utter only simple vocalizations (Rappole and Warner 1980), often presented in singing bouts. Similarly, temperate zone residents use simpler vocalizations in winter: Red-bellied Woodpeckers utter primarily bursts of a rolling *kwirr* on breeding territories, and a simpler *kek,* often singly, on winter territories. Northern Cardinals use not their elaborate songs but a simple *chip* in winter.

In general, the evidence suggests that individuals living in unhurried circumstances confront others less, and may signal less about taking strong initiatives, than do those more pressed by urgent events. Perhaps individuals that have stability and time also use relatively subtle differences in signaling to convey changing information. They have both considerable "background" information available and time in which to negotiate.

Exploiting Ways of Actively Interacting with Playback

Basic procedures. In our field trials, the recorded vocalizations used as playback stimuli are undegraded, and we obtain examples of each song form from at least six individuals. Two or more song forms are used (one per subject) to test whether subjects simply match (or avoid matching) playback stimuli. The stimuli are played to sound as loud to us as birds about 20 meters away (86–90 dB at 1 meter from the speaker), although when a subject has approached closely we sometimes mimic the faint vocalizing that is then often elicited. In trials with the pewee, and in our preliminary trials with three other species, we played back with natural intervals between songs before and after attracting subjects. Many subjects fell silent and, if they approached, searched silently. Vocal approaches were more frequent to discontinuous playback.

After recording a preplayback sample, we begin a trial by answering a song with one playback song. If a subject's behavior does not change during its next few songs, we answer up to four more songs. If still no change occurs, playback is moved closer or to the far side of the subject (probing its territory) or laterally. Movements are sometimes constrained by topography, habitat borders, or local cover (e.g., we try to position playback near vegetation that might conceal our simulated intruder; we do not try to conceal ourselves).

The actions of playback and the extent of a subject's approaches, passes, circling, and other behavior are dictated onto the left track of a stereo cassette recorder; the bird's utterances (amplified by a parabolic reflector) are recorded on the right track. Playback (of single songs) is done primarily when a subject stops in one place, is silent for a minute or more, or has passed far beyond our location. Rapidly singing subjects are not interrupted unless their searching behavior suggests attempts to get closer. Once a subject is within 5 meters (or 10 meters for relatively large species such as *Tapera naevia* in open situations), playback moves away to twice that distance or circumferentially to other cover in which the subject might expect to find our simulated intruder.

Any event in which a subject approaches playback and vocalizes is accepted as a trial. Most trials last about 5 minutes and are terminated because the subject's rate of vocalizing has stabilized or declined, or the subject has quit vocalizing or has departed. Trials in which a subject is joined by a mate or neighbor are broken off, because we cannot be sure whether subsequent vocalizations are elicited by playback or by the other bird. When birds approach immediately and are highly vocal and consistent, a trial may be terminated in 3 minutes, although we sometimes continue playback for up to 10 minutes to get a subject to lose interest, asking how it will vocalize as it departs; results from such exploratory procedures are not addressed in this chapter. We never end a trial based on which vocalization is being uttered (e.g., we do not continue until a subject is uttering B songs, then quit). If a subject has approached and is singing only faintly or with abbreviated songs, however, we do continue until its songs are either louder or fuller, or it becomes silent or leaves.

At least four functionally different categories of responses can be defined by behavior. First, some birds become attentively engaged but do not approach. They sing more rapidly or continuously, or answer each playback promptly. Second, subjects may make silent "investigatory" approaches that stop far short of or continue far beyond our playback position. They then scan or more actively search and some come closer, but all remain silent. Third, other subjects approach or pass and, when not close, vocalize. We interpret this as "probing" to elicit responses from an unseen intruder. Probes often begin furtively; then subjects gradually make themselves more conspicuous. Fourth, many subjects quickly or eventually approach closely and vocalize. More than just probing, they apparently attempt to confront. Individuals and species differ in their tendencies to make different kinds of approaches.

Many variables need to be considered in detailed analyses. Measures of the extent to which subjects move about or take conspicuous perches, for instance, reveal that probing and confronting are not fully distinct. Measures of the closeness of approach are problematic and are best assessed during relatively frequent instances of playback. Subjects unable to see an avian intruder that is (temporarily) silent may not themselves feel close.

Simple elaborations. Our lateral moves during a trial simulate jockeying for position or evasive probing. By approaching and withdrawing we escalate or diminish simulated intrusions. Sometimes we withdraw continuously, leading a subject to its territorial boundary. So far, however, our main concern has been to elicit vocal approaches, and we have not exploited these procedures systematically.

Even simple extensions of our basic design may be highly productive. We should be able to make finer distinctions among responses by paying attention to territorial borders and to each subject's phase of the breeding cycle (see Stoddard, this volume). Escalation and de-escalation can be simulated by changing playback stimuli during a trial, although other flexibility must then be constrained to facilitate comparison of responses. Changeable playback from fixed speaker positions (pioneered by Dabelsteen and Pedersen [1990], who established the term *interac-*

tive playback; see McGregor et al. 1992b) can be used, although lack of movement impedes efforts to simulate interacting. If trials using fixed speaker positions are combined with more flexible procedures, each can help calibrate the other.

Interactive playback has limits. It can simulate early phases of territorial intrusions and elicit approach, searching, probing, and attempts to confront, but it cannot confront. In studying signaling that is done primarily by animals that are apart from one another, the deficiency may not be major, because animals use primarily nonsong signals when closely and actively confronting.

Simulating intrusions with nonsong playback. Instead of playing songs to singing subjects, we often play nonsong vocalizations (or vocalizations other than ordinary songs) to birds that were not singing. Any songs then elicited cannot be matched to playback. Under these conditions subjects often uttered both songs and nonsong vocalizations, enabling us to compare the behavior correlated with each.

The usefulness of different kinds of playback stimuli will need to be assessed by studying naturally occurring disputes. Playback of songs from within territories, although it has elicited approach and songs from diverse species, may not always be appropriate. To intrude and continue playing back songs after an owner has approached may be confusing and may violate expectations in species that do not usually utter songs when closely confronting, even with unseen opponents. Playback of nonsong vocalizations or structurally simple vocalizations may elicit more natural responses.

Priming. When a singing individual fails to approach playback, we sometimes try again after 15–60 minutes, or on successive days. In many cases, the subjects then responded strongly. Most such cases took place slightly outside peak seasons of singing by migrants or involved tropical residents not yet into a nesting cycle.

Possibly the initial playback intrusion leads to an increase in circulating testosterone. As little as 10 to 20 minutes of exposure to a challenging stimulus has this effect in various species (Harding and Follet 1979, Wingfield et al. 1987, Wingfield and Wada 1989), although not in nonbreeding seasons (Wingfield and Hahn 1994). Whatever its basis, the effect (tentatively termed "priming") enables comparison of responses by individuals that do and do not need to be primed.

Studying nonsong vocalizations. Some classes of vocalizations may be readily studied with interactive playback because, like songs, they are often uttered by individuals apart from one another. Included are duets (discussed above; duetting vocalizations can be songs or nonsong vocalizations) and so-called contact calls (a poor term because it emphasizes a single function). Interactive playback can test for various information, for example, about an individual's probability of staying at a site or leaving, or of actively seeking to join another individual versus being passively receptive to being joined. We have begun work with the White-breasted Nuthatch (*Sitta carolinensis*), which utters diverse calls as mates associate. Numerous other species that live in social groups are potentially suitable for this type of study.

Conclusions

Songs and singing are surprisingly diversified. Although we find that species of a wide range of lineages provide information about their changing behavior as they sing, we also find great differences among species (comparable with differences already apparent in, for instance, routines of song learning or the numbers of songs in individual repertoires; Kroodsma 1982b). No single explanation exists for why individuals of so many species each have more than one song form: it is not that each form is used as a vehicle for different behavioral information, nor the capacity to "match" songs, nor reduction of monotony; nor is it the effect of sexual selection or the Beau Geste hypothesis. Our knowledge is too fragmentary to explain this diversity. It is particularly puzzling that closely related species often diverge whereas distantly related ones converge, although trends may yet be found that reveal the adaptive significance of this.

Despite the diversity of avian signaling, the provision of information about different behavior is widespread (even though many lineages remain unstudied). Diverse species provide information about the extent to which interactional initiative will be seized or deferred, and about imminent options to interacting that yield behavioral instability. Other information may well emerge as experiments are extended and elaborated. The importance of the urgency of social events in determining the kinds of information provided must also be studied, along with the extent to which urgency affects the subtlety of negotiating moves and the levels of initiative that are commonly taken.

Even though individuals of many species appear to use different songs interchangeably, with no song providing distinctive information about the singer's behavior, it is possible that their singing provides more behavioral information than we realize. Perhaps some species categorize their songs by schemes we have yet to grasp (e.g., by intonational contours, rhyming, or internal cadencing) and select among categories in ways not yet apparent. Behavioral information may be provided with specialized modifications of songs and singing patterns, with songs structurally different from those in the ordinary repertoire, with narrowly predictive vocalizations inserted into singing, or even with special joint singing performances (formalized interactions). When events are less stable than is usual during most singing, a number of effective alternative ways to communicate may exist. Further, different kinds of information might be of comparable value. For instance, is the information provided by matching another individual's songs as useful as the information in songs that distinguish different extents of initiative? Can a signaling procedure that informs about initiative without using ordinary songs free those songs to be used for matching or as a basis for sexual selection?

One useful result of interactive experiments is an enriched understanding of the behavioral messages of signals. *Message* is simply a term for information (as defined above) made available by a signal. It is used in distinguishing among the many sources of information that are relevant in communication, and also to distinguish between social sharing of information and the individual cognitive processing that yields event-specific "meaning." As previously reviewed (W. J. Smith 1977), animals' signals provide information (behavioral selection messages) about the kinds of actions a signaler may perform (its "selections" from its behavioral repertoire) and how it may perform them (e.g., how forcefully, how stably, and with what probability: information that "supplements" the specified selections).

The behavioral selection message termed "interactional behavior" is known from much previous work (W. J. Smith 1977). Broadly predictive, this message does not specify what kinds of interacting to expect (such as attacking or courting). It is characteristic of songs and singing, often along with information about the extent to which (or likelihood that) a signaler will seize or defer initiative. Such signaling may function to solicit responses or may forestall encounters when other individuals accept the status quo. Whether to categorize the activities of seizing and deferring initiative as behavioral selections or to name supplemental messages after them is unclear. Earlier (W. J. Smith 1977) I classed them as "seeking" and "receptive" behavior, each represented by a behavioral selection message. Because each message had to be made available along with information about another selection such as interactional behavior, the messages might better be viewed as supplemental information. Categorization of these messages remains tentative.

In addition, in this chapter I have described another kind of information as being about relatively unstable behavior. A full message analysis will be more complex. The information may actually be about alternatives to interacting (e.g., fleeing, testable by approaching subjects with playback when they are and are not uttering such vocalizations), or the bird may be behaving "indecisively" (i.e., failing to choose between alternatives and hence pausing or vacillating; W. J. Smith 1977), a move useful in negotiating. Both possibilities involve behavioral selections, whereas information about a decrease in stability per se is supplemental to information about selections.

Most results of interactive playback obtained thus far clarify and extend, but are largely compatible with, formulations I proposed earlier (W. J. Smith 1977). The functional differences recognized in investigatory, probing, and confronting approaches, however, may reflect additional information about behavior. Further, dawn singing in some species may provide a message that I recognized only later (W. J. Smith 1986a), information about the behavior of sustaining a formal contest. This message is provided when options such as approach and withdrawal are forgone while vocal probing is continued, the signalers simultaneously providing and eliciting information about commitment to further action.

The future of interactive techniques, both ours and those of others, depends on two distinctive properties: flexibility and a focus on interacting with each subject individually. A subject's responses provide feedback with which an experimenter can adjust to each developing event. By timing playback and moving about, experimenters can respond in varied ways to the subject's behavior, prompting and even directing continued engagement. By choosing different stimuli, even nonsong vocalizations, an experimenter can bias subjects in different ways. Interactions can also be simulated not just with territorial defenders but also with mates, prospective mates, and members of flocks. The relations of different pre-playback behavior of subjects to their subsequent signaling and other actions can be explored with prompt sensitivity to unexpected developments. Interactive playback may even "prime" initially reluctant subjects to respond.

The contributions of songs and singing to interactional behavior cannot be fully revealed by one technique. Observations are difficult and time-consuming, but they remain our touchstone with reality. Fixed stimulus presentations—and the interactional procedures of Dabelsteen and his coworkers, who alter stimuli (from fixed speaker positions) in response to subjects' activities—are basic when comparing the ways animals respond to different stimuli. Our still more actively interactional procedures address another critically important issue. They help us influence singers' activities so that we can learn how their utterances (songs and other signaling) make their behavior predictable. Orderly interactions depend on predictability and can become more complex as the animal provides more information. Interactive playback is beginning to give us glimpses of that information, and we are enthusiastic about the technique's potential.

Acknowledgments

The contributions of Anne Marie Smith have been essential in all phases of this research. I thank the Whitehall Foundation for initial funding, the Smithsonian Tropical Research Institute and the Fundación Branger for support in the Tropics, the University of Pennsylvania Research Foundation for the Kay Elemetrics DSP 5500 Sonagraph, and, for advice on the manuscript, D. Kroodsma, E. Miller, T. Dabelsteen, and D. Spector. The Instituto Nacional de Recursos Naturales Renovables provided permission to work at sites in Panama.

22 Dynamic Acoustic Communication and Interactive Playback

Torben Dabelsteen and Peter K. McGregor

Many animal communication systems are characterized by reciprocity, in the sense that the communicants may alternate between adopting sender and receiver roles, or may perform the two roles more or less simultaneously (see, e.g., Burghardt 1970). Reciprocity implies that communication is a dynamic process involving the interactive exchange of signals. A good example is singing in territorial male birds, in which both the types of song emitted and the timing of song exchange are varied over a short time scale, sometimes song by song (see, e.g., Dabelsteen 1992). Although interactive exchange of songs is usually described as a two-male phenomenon (a dyad), many males are likely to be within singing range of each other and have the potential to interact and form part of a communication network (Dabelsteen 1992, McGregor 1993). The concept of communication networks is discussed more fully by McGregor and Dabelsteen (this volume).

The interactive aspects of communication have been little studied, partly because of the practical difficulties inherent in such studies, and have been more or less ignored by theorists modeling the evolution of animal signals. More rapid progress in our understanding of interactions will result from two developments in equipment and processing. First, digital equipment facilitates interactive experiments in which playback can simulate the presence of a conspecific bird that alters its singing in response to the behavior of the test bird (Dabelsteen and Pedersen 1990, 1991, Bradbury and Vehrencamp 1994, Otter et al. 1994b). Second, a passive acoustic location system (ALS) designed for use with songbirds (McGregor et al. unpubl. ms. a) allows singers to be located in space and simultaneously monitors the timing of their vocal interactions (McGregor and Dabelsteen, this volume). The combination of interactive playback and ALS is a powerful tool for studying communication, because interactive playback can be used to test hypotheses on signal value resulting from the accurate descriptions of the nature of vocal interactions in a communication network derived from the ALS.

Vocal Interactions within Dyads of Males

Vocal interactions between males are a striking behavior, and are probably the most studied feature of acoustic communication in birds. In this section we review the ways that information on vocal interactions has been collected and studied with respect to signal value and outline the potential roles of interactive playback and passive acoustic location techniques in such studies.

Describing vocal interactions. A reasonably comprehensive description is a prerequisite for understanding the nature of vocal interactions. The important information falls into the categories of *who, why,* and *what.* In other words, details are needed on the category of the individual involved (e.g., territorial male, intruder, etc.), the reason for communicating in both the proximate and ultimate senses (e.g., to repel an unknown intruder, to maintain an established territory boundary), and the details of signaling behavior and signal content.

Two aspects of vocal interactions are commonly described: the bird's selection of the vocalization from the repertoire it has available, and the timing of the interaction between birds. A bird can select a number of features of songs during interactions. The simplest level of song selection is to change vocalization when the other individual changes, a behavior usually referred to as synchronization of song switching, as is seen in Western Meadowlarks (*Sturnella neglecta:* Horn 1987). Song switching is combined with song matching in many species. Birds match each other by responding with songs that resemble those of the opponent in some feature (W. J. Smith 1991, Horn and Falls, this volume, McGregor and Dabelsteen, this volume). In some species, however, individuals do the opposite—they respond to matching by singing a different song and therefore avoid being matched (e.g., Western Meadowlarks: Horn and Falls 1986; Winter Wrens, *Troglodytes troglodytes:* X. Whittaker pers. comm.). Song switching interactions in some species are complicated because song types are not functionally equivalent, and features of the song are associated with the level of arousal, intentions, and subsequent behavior. For example, Eurasian Blackbirds (*Turdus merula*) switch between different intensities and types of singing (Dabelsteen 1992).

Several aspects of the timing of singing in interactions can also be varied. The simplest aspect is the degree of synchrony in the interaction. Individuals sometimes attempt to avoid simultaneous vocalization (see, e.g., M. S. Ficken et al. 1974, Wasserman 1977) and sometimes seek to overlap vocalizations (e.g., Dabelsteen et al. in press a; frogs: Tuttle and Ryan 1982, K. D. Wells 1988). On a shorter time scale, individuals may overlap each vocalization during the interaction, or they may coordinate their vocalizations so that they alternate (McGregor et al. 1992b). In a special type of alternating behavior, individuals appear to adopt the role of interaction leader or follower; examples are Marsh Wrens (*Cistothorus palustris:* Kroodsma 1979) and Red-winged Blackbirds (*Agelaius phoeniceus:* D. G. Smith and Norman 1979). Aspects of song selection and timing can be combined in some interactions; Great Tits (*Parus major*), for example, may match song length (strophe length) as well as song type (see, e.g., McGregor and Horn 1992, McGregor et al. 1992b).

Timing the interactions between birds at relatively long distances is complicated by the slow speed of sound transmission in air (Dabelsteen 1992). Timing interactions is further complicated by the relative positions of the birds and the recording microphone (see Dabelsteen 1992, figs. 2, 3). Recordings that use one or two microphones, for example, can never give the precise timing of interactions, even between just two males. A fixed array of microphones (such as an ALS) is needed to overcome the problems of constraints on speed of sound and the relative positions of birds and microphones. An ALS gives information on the position of vocalizing birds and also allows calculation of the timing at the position of each interacting bird. Birds must allow for the delay effect caused by the relatively slow speed of sound and can also use song degradation to perceive distance (Dabelsteen 1992, McGregor and Dabelsteen, this volume). The overall importance of distance effects depends on the biological significance of the precision of timing. In insects, timing can be precise to within tens of milliseconds, and distance-induced delays may therefore be appreciable even at short distances (H. Römer pers. comm.).

Eliciting interactions with loop playback. In most cases, playback using fixed-duration tape loops, or loop playback, cannot simulate interactive features of communication; loop playback can, however, elicit interactive responses. Loop playback has also been valuable in establishing the importance of a number of aspects of vocal behavior in many animal groups, including interactions between songbirds (Falls 1992); for example, loop playback identified the importance of singer identity and song similarity in eliciting song type matching in Great Tits (e.g., Falls et al. 1982). Loop playback also induced song switching by matching song type in Western Meadowlarks (Horn and Falls 1986, see also Stoddard, this volume).

Simulating interactions with digital playback. Playback using digital equipment overcomes the limitations of fixed-duration tape loops. Such limitations are particularly pronounced in interactive playback when interactions involve changes in many song features within a short time.

Digital playback does, however, have some limitations. All digital playback equipment uses the keyboard of a portable computer to control the sound output of the system, and therefore the keyboard speed of the experimenter can be a limiting factor. Digital systems are also limited by the skill of the programmer, but the major limitation of any interactive playback system remains the experimenter's ability to identify changes in test bird vocalizations and to choose the appropriate response (e.g., song type, timing, etc.).

The first system used in experiments with birds (the digital sound emitter, or DSE) was developed in the mid-1980s and therefore had to rely on a portable PC with low processing power and a small RAM (Dabelsteen and Pedersen 1991). The speed of access and memory capacity necessary for interactive playback were achieved by using a customized external digital store. Interactive playback systems are now available in the operating environments of both PCs (Dabelsteen and Pedersen unpubl. ms.) and Macintosh computers (Bradbury and Vehrencamp 1994, Otter et al. 1994b). Because of the high level of performance of current models (in terms of RAM, hard disk size, access time, processing speed, etc.), these systems no longer require an external store, but many still use an additional card or board for greater accuracy and reliability of sound output.

Recent interactive experiments. A number of studies have demonstrated the importance of short-term interactions in signaling behavior. Experiments with interactive playback systems have looked at the signal value of switching between different intensities or types of singing in Eurasian Blackbirds (Dabelsteen and Pedersen 1990, Dabelsteen 1992) and various aspects of matched countersinging in Great Tits (McGregor et al. 1992b). The experiments with blackbirds established that twitter song (also referred to as strangled song) represents the highest degree of arousal of singing males, although the precise meaning depends on twitter song being used interactively (Dabelsteen and Pedersen 1990). The experiments with Great Tits provided support for the idea that successively more detailed levels of matching can indicate the intended receiver more precisely (McGregor and Dabelsteen, this volume). Experiments with Song Sparrows (*Melospiza melodia*) in the early 1990s investigated the combined effects of song type matching and the pattern of switching on both the aggressive response elicited and switching (Nielsen and Vehrencamp 1995). Black-capped Chickadees (*Parus atricapillus*) vary the mean frequency of their fee-bee songs (Ratcliffe and Weisman 1992), and interactive playback has been used to study the effect of matching such frequency shifting behavior by males of known dominance in winter flocks (L. M. Ratcliffe pers. comm.). Experiments in progress are using interactive playback to study the functions of variable call repertoires in the parrots *Amazona auropalliata, A. albifrons, Aratinga canicularis,* and *Brotogeris jugularis* (J. Bradbury pers. comm.). All of the experiments described above dealt mainly with the selection of alternative vocalizations from the repertoire, but interactive playback is also used to study timing aspects of singing interactions. Examples are the cadence of song delivery in the Winter Wren (Brémond and Aubin 1992) and the pattern of overlapping or alternating in Great Tits (Dabelsteen et al. in press a) and European Robins (*Erithacus rubecula;* see below).

Designing and Interpreting Interactive Playback

When designing and interpreting playback experiments, one must consider a number of factors (see, e.g., McGregor et al. 1992a). These factors are especially important in interactive playbacks, which can require rapid responses by the experimenter to changes in the signal of the test animal. Here we consider some of these factors. Our intention is not to deter people from using interactive playback; on the contrary, we heartily encourage the use of this technique and want to pass on experience gained in some of the interactive experiments we have attempted.

Background information about natural interactions. Lack of exact knowledge of natural interactions may, of course, cause difficulties in designing experiments and measuring the resulting responses. For instance, the potential of interactive playback to simulate natural interactions can be realized only if the natural interactions are known (see also Stoddard, this volume). Such information includes the context of interaction (which will influence site of playback speaker and choice of test animal), the frequency and extent of the interaction in this context (e.g., how often matching occurs), and the influence of distance (e.g., at what distance birds will attempt to overlap playback). Measuring the approach response of birds to interactive playback is straightforward; natural approach responses are easy to observe and interpret. More troublesome is the detailed patterning of singing responses song by song. The difficulties in interpretation arise partly because our present knowledge of natural responses is limited, and partly because of the difficulties in interpreting such changes in the patterns of singing. For example, if two different playback stimuli elicit different patterns of approach or aggressive behavior (e.g., posturing), then the stimuli obviously aroused the animals in different ways, leading to an interpretation of them as signaling different levels of threat. A similar straightforward interpretation is not possible for a pattern of singing behavior such as delay to reply or even switching between functionally equivalent song types. Although these changes can be measured easily, they are difficult to interpret. Our understanding and interpretation of responses elicited by interactive playback can be improved by using microphone arrays (see above), which can collect information on natural singing patterns and allow correlation with known approach and aggressive or sexual responses.

Unwanted differences between playback stimuli. A potential problem that is inherent in all interactive playback is that the treatments may differ in features other than those of interest, because the pattern and amount of stimulation are driven by the test bird. An example of this problem occurred during an interactive playback experiment with Great Tits. The experiment investigated the effects of two sorts of playback: overlapping, in which playback began and stopped with the bird; and alternating, in which playback did not start before the bird's song ended but then matched it in length. Overlapping has the potential to result in more playback than alternating, because overlapping playback does not have to wait until the bird has finished singing (Dabelsteen et al. in press a). In alternating treatments the experimental period may end before the experimenter is able to respond to the bird's last song. A similar intrinsic difference in stimuli occurred in interactive playback with Song Sparrows, with a higher rate of switching in lead-switching playback (a type of interactive playback in which the experimenter switches before the bird) than in following switching (i.e., playback in which the experimenter switches after the bird does so; Nielsen and Vehrencamp 1995).

One solution to this problem is to abandon tests in which the difference between treatments becomes too large, or to omit those tests from subsequent analyses. This approach raises the question of how large a difference is acceptable, which can be answered only by playback experiments that directly address this feature. Another solution that may be feasible in many cases is to assume that such unwanted differences are small in relation to the magnitude of differences resulting from the experimental design. For example, in the Great Tit experiment, the duration of songs played in alternating treatments averaged 25% longer than during overlapping; however, overlapping treatments overlapped 100% of songs, whereas alternating treatments overlapped 0% (Dabelsteen et al. in press a).

Individual differences in response. The response of a territory owner to playback of song may vary considerably with different environmental and social factors. These factors should always be considered by the experimenter, who must try to ensure that they are equivalent between playback treatments. Some factors are relatively easy to control, such as time of day, stage of the territory owner's breeding cycle, and location of the speaker (the apparent intruder) in relation to territory boundaries. Other factors, however, will almost certainly vary between males in ways that are difficult or very time-consuming to identify before the start of experiments. For example, one territory owner's competitive ability and affiliation to his territory may be quite different from another's. A result of such variation between playback test subjects could be different kinds of response, particularly when playback is considered to be a serious threat by the territory owner (e.g., Dabelsteen 1985, Stoddard, this volume). Such variation suggests the exciting possibility that the competitive abilities of birds can be examined with this technique. On the other hand, such effects can also obscure differences between the stimuli being compared.

The extent of such differential response effects is illustrated by our interactive experiment with 13 Great Tits (Fig. 22.1). Although the means for measures of approach and amount of song did not differ significantly in response to the two playback treatments, a number of measures of the pattern of singing did differ significantly. These differences are consistent with the interpretation that males seemed to respond in different ways when they were overlapped by playback (Fig. 22.1). Some males increased the variation in strophe length during overlapping treatments without changing the mean strophe length, a pattern of response that could be interpreted as the male's attempt to avoid being overlapped, and hence having his signal masked by the apparent intruder. Other males responded to alternating and overlapping playback with the same variation of the strophe length but with reduced singing during the overlapping treatment. These males seemed unwilling to continue the interaction with the simulated intruder and responded with apparent caution. Others responded by increasing song output, in extreme

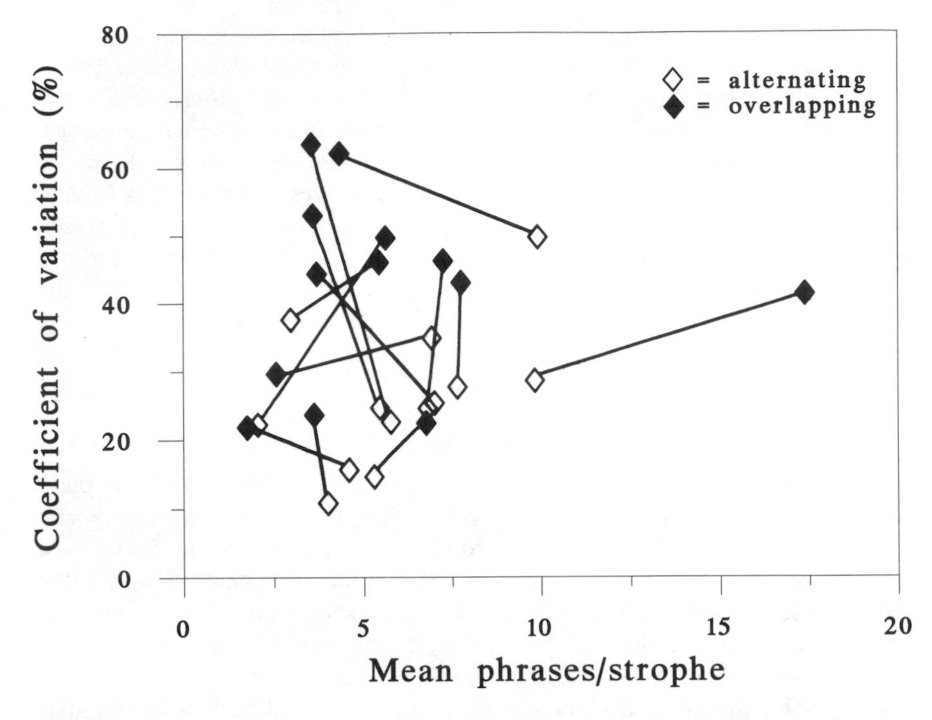

Fig. 22.1. Response of 13 male Great Tits to two forms of interactive playback. A given male's response to alternating playback is represented by an open symbol, which is connected to the solid symbol representing his response to overlapping playback. The symbols represent coefficients of variation of strophe length (number of phrases per strophe; see McGregor and Horn 1992) as a function of average strophe length.

cases to almost continuous singing. We interpret this behavior as a strong response attempting to completely mask the intruder's signal. Such a response may prevent any interaction with the intruder and therefore precludes gathering information on relative abilities through eavesdropping (see McGregor and Dabelsteen, this volume). This example shows that it may be necessary to divide test males into different categories for analyzing response to playback. Of course, responses to treatments may still differ overall, but it is possible that the use of different response strategies may cancel out any overall average difference when males are pooled.

Two further examples illustrate the need to divide males on the basis of their type of response. The first example comes from interactive playback experiments that investigated the signal value of different intensities and types of singing in Eurasian Blackbirds. When males approached to within 10 meters of the loudspeaker playing high-intensity song, playback was changed to twitter song. The test males could be divided into two response groups. The first group approached to within 10 meters only once or twice, but the other group repeatedly approached to within 10 meters, behaving in the same way as males involved in prolonged border disputes (Dabelsteen and Pedersen 1990). Thus, the males that rarely approached the speaker did not increase their aggressive response when playback was switched from high-intensity song to twitter song, but the other males did. Pooling the response of males to playback would not have shown any obvious difference in the effect of playback switching from high-intensity to twitter song. The other example involves overlapping and alternating playback with European Robins. Some males replied to playback from some distance with the usual full song (combinations of low- and high-frequency phrases: Brémond 1968), but other males approached very closely, postured aggressively, and used a twittering song (Brémond 1968) consisting only of high-frequency elements (Dabelsteen et al. in press b). The clear difference between the two groups of males explains the lack of difference in a number of measures of average response if the males are pooled.

If males can be categorized according to their expected response before the playback experiments, then the problems of inappropriately pooling males can be overcome. This approach should be used with caution, however, because males may change response categories in a relatively short time as a result of factors such as mate guarding and the outcome of recent territorial disputes. In principle, a priori categorization of males into subgroups can be avoided if a repeated-measures design is used with a large enough sample of males. In practice, however, rapid changes in response strategy and the logistics of achieving a large sample size may limit the usefulness of this solution.

Of course, the problem of males adopting different response strategies applies to all playback experiments and cannot be avoided by using noninteractive playback. The major advantage of interactive playback is that it can be used to investigate many more aspects of singing behavior than is possible using noninteractive playback.

Future Applications of Interactive Playback

Playback based on digital systems is characterized by its flexibility, ease of control, and potential resulting from a combination of a large storage capacity and flexible programming. These are advantages both for interactive experiments and for traditional noninteractive playback that would previously have used fixed-duration tape loops. The storage, whether it is a hard disk or an external digital store, allows selection from a large collection of artificial sounds and natural vocalizations such as songs and calls from different individuals and populations. The size of the collection is limited only by the storage capacity. Programs controlling the output from the collection can be tailored to almost any task by varying features of the vocalization such as type, interval between, and number, and whether the output is determined by the experimenter or by the program. The flexibility of the system is a consequence of the number of possible combinations of these features. Some idea of the possibilities of digital playback systems can be gained from the examples below.

Patterning of vocalization. A number of species with large song repertoires produce their varied songs by combining selections from a much smaller repertoire of components (syllables, elements, etc.; e.g., Sedge Warblers, *Acrocephalus schoenobaenus:* Catchpole 1976; and European Robins: Brindley 1991). A digital playback system has the flexibility to vary song length and combine units according to a variety of rules, such as random selection of units using the computer's random numbers table as a control. Such a design could be used to test ideas about the rules used for combining components. An interactive component could be added by varying these features with the singing behavior of the test bird during an experiment. In this way it is possible to directly test the effect of the interactive component, as has been done with Great Tits and Eurasian Blackbirds (see above). Many species use calls in association with song, for example, changing from song to calls at close range, as in Great Tits (pers. obs.) and Eurasian Blackbirds. In the blackbird, playback elicits alarm rattles, chooking, and alarm seets (Dabelsteen 1982). Interactive playback can be used to investigate the interrelationships among these vocalizations (Smith, this volume).

Signaling in communication networks. Interactive playback could be used to interfere with a dyad of interacting males as a way to begin to investigate communication within a network. It has been argued that one of the important aspects of communication networks is the potential for eavesdropping (McGregor 1993, McGregor and Dabelsteen, this volume). Interactive playback could be used to simulate an intruder with varying levels of aggression and fighting ability when such information is contained in the signal. Subsequent playback to the neighbors of the test birds could show whether the neighbors' responses were influenced by the interaction with the apparent intruder and therefore whether they were eavesdropping. Pilot experiments have shown that this approach is feasible (McGregor et al. unpubl. ms. b).

ALS studies of communication networks (see above) can provide data to construct models of how individuals behave in a network. The resulting models could be tested by playback from a range of locations in the territory (i.e., a series of loudspeakers arranged to cover the territory).

Interactive speaker replacement experiments. A further development of interactive playback would be to extend the classic "speaker replacement" experiments (Göransson et al. 1974, Krebs et al. 1978, Yasukawa 1981a) by interacting with an intruder in a variety of ways that are thought to target an intruder and signal increased willingness to escalate, depending on distance from the territory border. Such an experiment requires information about the location of the intruder. Simultaneous recording and monitoring using an ALS would help to locate birds and determine the timing and nature of interactions to direct playback; they would also help to interpret the response. Telemetry of the intruder would provide accurate locations in real time, and someday it may be feasible to monitor internal factors such as heart rate and therefore use them as measures of response.

Computer-driven timing and selection of playback stimuli. In theory, the limitations of a human observer in achieving very rapid time synchronization and identification of vocalizations can be overcome by computerized pattern recognition combined with computer-controlled playback; in other words, the computer would decide what and when to play back. Computerized pattern recognition will make song type matching easier in species with large repertoires, but an additional computer may be required. The computer-controlled playback of output could be used to investigate the effect of varying delay, the extent of overlap, and also the level of song matching.

Incorporating other signaling modalities. Interactive playback enables the researcher to study the effects of combining acoustic signals with other signaling modalities. A good example is a study of the waggle dance of honey bees (*Apis mellifera*) that used a robot dancing bee to combine the various signal modalities and movement patterns used in the dance. The robot allowed a direct experimental test of the effect of the dance components on recruiting other foraging bees (Michelsen et al. 1989, 1992). Another system now being developed will similarly allow a direct test of the effects of different combinations of acoustic and visual elements in the territory defense behavior of the Common Coot (*Fulica atra*). The system consists of a remote-controlled stuffed coot that is capable of swimming, can vocalize, and can also perform different postures by combining stretching and bowing of the neck with raising the wings independently (B. B. Andersen, Dabelsteen, and O. N. Larsen unpubl. data).

Conclusions

Digital technology permitting interactive playback now exists for both the PC and Macintosh environments. The first experiments using this technique addressed a number of topics and produced a variety of interesting results, such as the possibility that in some species behavioral or motivation information has signal value only when the signal is presented interactively. Hardly any of these results could have been obtained using noninteractive playback based on fixed-duration tape loops. In this respect digital interactive playback represents a major step forward, and the technique has the potential to be useful in investigations of many other aspects of bird acoustic communication as well.

Interactive playback also presents a number of challenges, not the least of which are the demands placed on the experimenter to identify signals rapidly, select the appropriate stimulus, and play it back, all in the noisy, distracting environment of the field. Other challenges include adequate descriptions of interactive aspects of acoustic communication and measurements of responses to playback.

Interactive playback clearly has the potential to simulate the dynamic and interactive features that characterize bird acoustic communication. Our success in meeting the challenges posed by the technique will to some extent depend on the number of researchers involved. We hope this chapter will inspire others to try interactive playback.

Acknowledgments

The playback technique was developed during collaboration between Simon Boel Pedersen and TD. TD was funded by a Niels Bohr Fellowship from the Royal Danish Academy of Sciences and Letters and the Tuborg Foundation, a Carlsberg Foundation Fellowship, and grant no. 11-5215 from the Danish Natural Research Council. Our collaborative research was funded by the Carlsberg Foundation, a NATO Science Fellowship, and the Danish National Research Foundation (TD), the Royal Society (PKM), the NATO Collaborative Research Grant scheme (PKM and TD), and Jarlfonden (TD). We thank Jack Bradbury, Ken Otter, Laurene Ratcliffe, and Sandy Vehrencamp for permission to cite unpublished work, and the following people for discussion or comments on the manuscript: John Bower, Jack Bradbury, Jo Holland, Ken Otter, Tom Peake, Laurene Ratcliffe, Candy Rowe, José Pedro Tavares, and Sandy Vehrencamp.

23 Communication Networks

Peter K. McGregor and Torben Dabelsteen

The concept of communication networks is based on a simple premise: most of the acoustic signals used in bird communication systems can be transmitted over long distances, so many signalers and receivers are within signaling range of each other and have the potential to interact. For this reason, communication using long-range signals is best considered as occurring within a network of signalers and receivers. Although the concept of communication networks is inherently simple, the idea has a number of important implications for the evolution of signaling systems and the perceptual abilities of receivers. The network concept applies to all communication using long-range signals—acoustic, visual, olfactory, and electric—and may prove to be the most ubiquitous feature of communication systems.

Despite this presumed commonness and importance, two factors have resulted in a dearth of studies of communication networks to date: (1) the theories and models that underlie and stimulate many studies are based on a communication dyad of one receiver and one signaler, for reasons of mathematical tractability; and (2) it is difficult to monitor the vocalizations of several individuals simultaneously. A consideration of communication occurring in networks, however, can generate several testable predictions. A main aim of this chapter is to identify these hypotheses and describe the sort of information that could be used to test them. Although the concept of communication networks is new, recent studies with classically territorial songbirds of the north temperate zone have begun to produce supporting evidence. The potential for extending such studies to other animal groups is exciting.

Background

The idea of communication networks follows directly from consideration of the transmission distance of acoustic signals. Many attributes of birdsong are adapted for long-distance transmission (R. H. Wiley and Richards 1982, R. H. Wiley 1991, Dabelsteen et al. 1993; we use the term *long distance* in the sense of several times

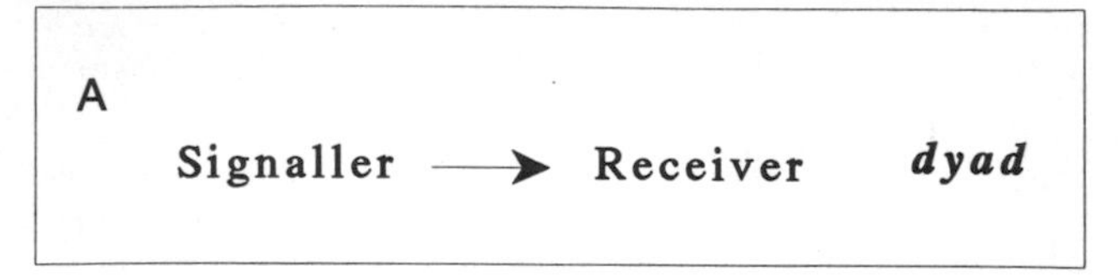

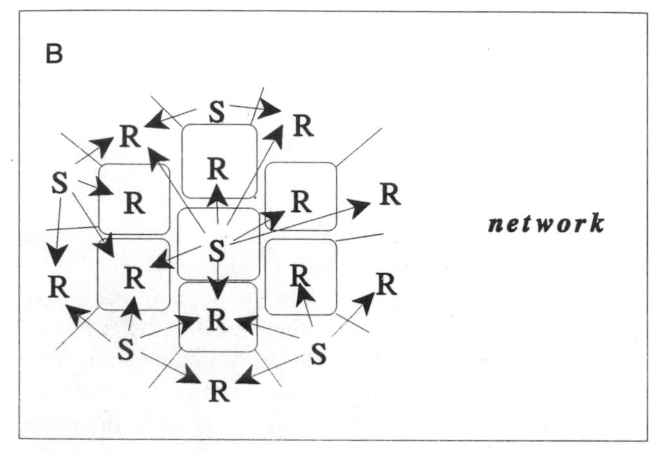

Fig. 23.1. Diagrammatic representations of communication. (A) The signaler-receiver dyad that forms the basis of theoretical models. (B) A communication network consisting of many signalers (S) and receivers (R) at one instant. The figure does not show the dynamic aspects of networks, such as changes in signaler/receiver roles and in the number of participants.

the usual distance separating birds). Furthermore, transmission distance is closely related to the effectiveness of some song functions (e.g., mate attraction). The same is true for the advertising vocalizations of nonpasserines; striking examples are the booming of bitterns and the yodeling of loons. Some songbird calls also seem to have been designed for long-distance transmission, such as distress calls and certain alarm calls (Brémond and Aubin 1990). Given such long transmission distances, it is simplistic to characterize communication as occurring between one signaler and one receiver (Fig. 23.1A). In the real world, many signalers and receivers are within range of each other and can be considered as forming communication networks (Fig. 23.1B; McGregor 1993). A further complication of the real world is the dynamic nature of communication interactions, involving changing signaler/receiver roles. Some aspects of dynamic communication are considered elsewhere in this volume by Dabelsteen and McGregor, and Smith.

Chorusing animals were the first to be considered networks, particularly choruses of insects (Otte 1974) and frogs (see, e.g., Ryan et al. 1981, K. D. Wells 1988). In choruses, individuals are in close proximity and form obvious communication networks. Interestingly, at the high densities of individuals in most choruses, the effective range of a signal can be determined by the level of interference from simultaneously calling individuals rather than by the carrying properties of the vocalization itself (Gerhardt and Klump 1988). Few bird choruses can equal those of insects and frogs in the extent of simultaneous singing (see, e.g., Atwood et al. 1991), although *Acrocephalus* and *Locustella* warblers singing simultaneously in marshes may prove equivalent. We will develop ideas about communication networks by concentrating on species defending contiguous territories, because such networks are as clear as those in choruses and we know far more about territoriality.

Our approach to the topic of communication networks is to identify aspects of acoustic communication behavior that we consider to have evolved in response to the selective challenges and opportunities of networks. To clarify the arguments, we begin by considering signalers and receivers separately, although we recognize that in many instances the signaler and receiver roles are interchangeable. For example, a territorial male songbird may sing at a neighbor (signal) and then listen to the reply (receive).

Implications for Signalers: Directing and Restricting Signals

Directing songs in networks. Signaling in a communication network inevitably entails simultaneously addressing several receivers, but communication must often be directed to one particular individual. In a territorial context, that bird may be a neighbor beginning to encroach on a mutual boundary or a male approaching a mate. If a signal is directed at such a bird as soon as the threat becomes apparent to the signaler, then the receiver usually retreats immediately. If the threat is allowed to continue, intruders will retreat only after the use of dangerous and time-consuming close-range aggression. Rapidly "targeting" an individual bird by directing a signal to it is therefore an advantage.

One way a singer can target a particular receiver at long range is by changing some feature of its singing behavior. At the simplest level, a male may start singing; or if he is already singing, he may switch from one song type to another. Both changes in singing behavior are known to deter intruders (Horn 1987). The idea of selective addressing through matching (also referred to as matched countersinging and song matching) was suggested by Brémond (1968). Most matching studies deal with song type matching—that is, changing the song to sing the same song type as the intended receiver (Lemon 1968, Dixon 1969, Kroodsma and Verner 1978, Krebs et al. 1981b, Slater 1981, Falls et al. 1982, Kramer and Lemon 1983, Schroeder and Wiley 1983b, Falls 1985, Horn and Falls 1986, W. J. Smith 1991, Stoddard et al. 1992a). Matching, however, may occur for any feature of song organization, including song length (see, e.g., McGregor et al. 1992b). Similarly, repertoire matching has been documented in Song Sparrows (*Melospiza melodia*). In this behavior birds reply to a neighbor with a song type known to be in the neighbor's repertoire but avoid the current song type (Beecher et al. in press). Another way of targeting a receiver relies on the shadow effect for high frequencies created by the singer's body: the ratio of high to low frequencies in a wide-band song will give a receiver information on the direction the singer is facing (Larsen and Dabelsteen 1990), and in many species a singing bird turns to face a neighbor that has begun to sing.

The idea of matching as a means of targeting a particular receiver can be tested using interactive playback (Dabelsteen and Pedersen 1990, 1991, Dabelsteen 1992, Dabelsteen and McGregor, this volume, Smith, this volume). Results from interactive playback experiments with Great Tits (*Parus major*) support the idea that matching songs and singing behavior in increasing detail gives a progressively more specific indication that a particular receiver is being addressed (McGregor et al. 1992b, Dabelsteen and McGregor, this volume).

Restricting transmission in a network. On some occasions signalers will gain an advantage by withholding information from members of the communication network, as when, for example, the signals would elicit interference from other individuals (e.g., the response of satellite males and kleptogamists to courtship and mating signals). Also, if signals indicate a high degree of commitment to a behavior such as an aggressive territorial interaction, then other individuals could use the information that the signaler is preoccupied and could intrude into the territory with less likelihood of being noticed. Thus we predict that signals used in such attention-intensive circumstances would not be transmitted to other members of the network, either because of their low amplitude, their high mean frequency, or both. The whispered, or quiet, song of European Robins (*Erithacus rubecula*) and the twitter song of Eurasian Blackbirds (*Turdus merula*) are both produced in very aggressive disputes and therefore fit these predictions (Dabelsteen and Pedersen 1991, Dabelsteen et al. 1993, Dabelsteen and McGregor, this volume). Comparable examples of switches to short-range signals during mating behavior are the copulation trill of female blackbirds (Dabelsteen 1988) and behavior exhibited by male bluehead wrasse (*Thalassoma bifasciatum*), which change to rapid fin movements and alter their body color from yellow-green to more cryptic neutral gray immediately before spawning (M. S. Dawkins and Guilford 1994).

Even if the signal is directed to a particular receiver, such as by song matching, such behavior will not restrict the spread through the communication network of information encoded in a song or a singing interaction, because sound propagates through roughly spherical spreading, and restricting the spread of sound signals is inherently difficult. Some directionality does occur, however, from the combined effects of head and body size, bill gape, and frequency of the vocalization (e.g., M. L. Hunter et al. 1986, Larsen and Dabelsteen 1990).

Another way of minimizing the indiscriminate spread of a long-range signal through the communication network is to switch to, or combine it with, a signaling modality with greater possibilities for "private" communication channels. Visual signals, for example, are effective only if signaler and receiver are in line of sight, and they can be screened relatively easily. There is evidence that visual signals may be used as a more restricted, directed alternative to sound by some birds (e.g., plumage in Western Meadowlarks, *Sturnella neglecta:* Horn 1987) and fish (e.g., pelvic fin flicks in convict cichlids, *Cichlasoma nigrofasciatum:* Shennan et al. 1994). Visual signals are combined with twitter song in blackbirds; indeed, a receiver can determine whether the singer is sexually or aggressively aroused only by the difference in head feather position (raised in sexual arousal, flattened in aggression; Snow 1958, Dabelsteen and Pedersen 1988b, Dabelsteen 1994).

Implications for Receivers: Eavesdropping

Many individuals other than those taking part in an interaction can receive signaling interactions in a communication network, even if the signal is directed toward a particular receiver. A receiver can thus obtain information either as a result of direct involvement in a signaling interaction with another individual or by monitoring the interactions between other individuals. Such eavesdropping behavior (Otte 1974, R. H. Wiley 1983, McGregor 1993, Shennan et al. 1994), which we will define as "extracting information from an interaction between other individuals," must be a universal feature of communication networks. Receivers who eavesdrop have also been referred to as "unintended" (Myrberg 1981) and "illegitimate" (Verrell 1991) receivers, and their behavior described as intercepting, overhearing, and bystanding. In this section we explore the concept of eavesdropping and its consequences.

Eavesdropping and Territory Defense

Monitoring the activity of neighbors through their use of calls and visual displays could provide useful information on the route or position of intruders in territorial systems (e.g., Beletsky et al. 1986). For example, in communities of Red-capped Cardinals (*Paroaria gularis*), information on the route and position of intruders is clearly indicated through interactions with neighbors. This species defends linear territories along lake and river edges; intruders elicit calling and chasing and are accompanied to a territory boundary by owners in highly visible escorting flights. Most intruders (>90%) were detected immediately if they had just been expelled from a neighbor's territory, but they were rarely detected immediately (<15%) if they had previously remained unnoticed on the neighbor's territory. Most intrusions came from areas with no neighbors and had an associated immediate detection rate of about 55%, thus illustrating both the attention territory holders paid to neighbors and the tendency for intruders to avoid defended areas en route to an intrusion (Eason and Stamps 1993).

Receivers can also monitor neighbors through the latter's use of song (e.g., Horn 1987). We recognize this phenomenon in our playback experiments because interference from neighbors may cause trials to be abandoned. The possibility of eavesdropping by neighbors is also implicit in playback protocols that are careful to avoid neighbors as successive subjects in a short time period (McGregor et al. 1992a). A more direct illustration of the monitoring role of song is the effect of song playback on neighbors of territorial male Zitting Cisticola (*Cisticola juncidis*). Males decreased their own vocal activity by 53% during playback to their neighbors and increased it by 254% after playback stopped (S. J. Gray pers.

comm.). This pattern of response is consistent with males attending to the vocal interaction between the neighbor and intruder simulated by playback (i.e., eavesdropping). The information gained by eavesdropping was that the neighbor had apparently successfully defended its territory but that the intruder's position was currently unknown because it was silent (i.e., playback had ceased); eavesdroppers therefore increased their territory proclamations.

Eavesdropping on the interaction between a neighbor and an intruder could provide far more information than simply the presence of the intruder. An eavesdropping territory holder could use its neighbors in a communication network as a yardstick against which to assess other individuals (McGregor 1993). This idea of assessment by proxy follows from the premise that a territory holder will have accurate estimates of the fighting abilities of its neighbors because of the frequency of its own disputes with them and because of interactions between other neighbors. Monitoring neighbors' interactions will allow the relative competitive abilities of the individuals involved to be assessed, provided that such information is contained in the signals and can be extracted by the eavesdropper.

Song contains information vital for assessing relative competitive ability. First, song is individually distinctive, and birds' abilities to discriminate neighbors from strangers and to identify individuals on the basis of song are well established (see the reviews in Falls 1982, Ydenberg et al. 1988, and Stoddard, this volume; also see McGregor and Avery 1986, Weary 1988, 1992, McGregor 1988b, Weary et al. 1990b, Godard 1991, 1992, Weary and Krebs 1992). Second, the nature of singing interactions, particularly detailed timing, contains information on an individual's willingness to escalate an aggressive encounter and also on the relative competitive ability of the interacting individuals (Dabelsteen et al. in press a, b, Dabelsteen and McGregor, this volume). Determining the timing of singing interactions between individuals requires the ability to "range," or estimate the distance to a singer. Ranging ability is required because long-range signaling incurs a time delay as a result of the relatively low speed of sound transmission in air. If the delaying effects of the speed of sound are not taken into account, receivers could mistake an overlapping pattern of singing for alternating, with a consequent misinterpretation of the interaction (Dabelsteen 1992, Dabelsteen and McGregor, this volume). Birds' ability to range using cues from song is well established (see the reviews by McGregor 1991, 1994, and R. H. Wiley 1994; also see Richards 1981, E. S. Morton 1982, 1986, R. H. Wiley and Richards 1982, McGregor et al. 1983, McGregor and Falls 1984, McGregor and Krebs 1984b, R. H. Wiley and Godard 1992, Naguib and Wiley 1994). Birds therefore have the means to assess distance-induced time delays.

Assessment by proxy has four selective advantages for the eavesdropper: (1) information is gathered without the risk of injury inherent in close-range aggressive encounters; (2) because listening at a distance involves only forgoing other behaviors such as feeding, the cost of eavesdropping is low; (3) animals are more likely to win an aggressive encounter if they have advance warning that it will occur (Hollis 1984); and (4) information on the relative fighting ability of a contestant will allow a territory owner to balance the cost and benefits of defending the resource and to choose the best possible response strategy (Dabelsteen 1985). A territory holder might benefit from eavesdropping on a dispute between neighbors by learning about short-term changes in neighbors' fighting ability (caused by disease, feeding success, etc.). Information on current fighting ability could in turn be used to assess the probable success of attempts to usurp resources or gain extra-pair copulations by intrusion. Although these advantages of eavesdropping remain speculative at present, they can be investigated experimentally using interactive playback (Dabelsteen and McGregor, this volume) and a microphone array (see below).

The level of information processing demanded by eavesdropping as discussed above is equivalent to that involved in transitive inference problems studied by experimental psychologists (e.g., Siemann and Delius 1993). Transitive inference is the ability to infer relationships between items using indirect information. For example, if A > B and C < B, what is the relationship between A and C? Such an inference ability has been demonstrated in nonhuman primates (McGonigle and Chalmers 1977, Gillan 1981) and Rock Pigeons (*Columba livia:* Fersen et al. 1991). Speculations about the adaptive significance of transitive inference center on the fitness gains in social interactions (e.g., Fersen and Delius 1992), including "the ability of a social animal to rank a newcomer relative to itself by observing the newcomer's performance in relation to other individuals" (Fersen et al. 1991, p. 345). The only difference of any substance between this quote and our ideas on eavesdropping discussed above is that we have focused on acoustic signals as the medium for information transfer. We conclude that animals have the necessary cognitive potential to use information gathered by eavesdropping in the way we suggest.

Eavesdropping in Other Contexts

Eavesdropping could also be used in contexts other than territory defense. The context of mate choice is likely to yield many examples of eavesdropping by both males and females. Females in territorial systems have the potential to eavesdrop on disputes between males and could use information on the relative fighting abilities of neighbors to select males for extra-pair copulations. In lek systems, a male interacting with a female will be providing information for individuals other than the female being courted, including other lekking males. The female "audience" could gather information on male features pertinent to future choice by observing the interactions, and the risks involved in closely approaching the male, such as forced copulations and interception by peripheral males, could be avoided. This idea of female eavesdropping is related to the idea of "female copying" (Balmford 1991). Other males on the lek could acquire information on, for example, the presence of females and the performance of the male. Both sorts of information would allow the other males to assess the potential gain and likelihood of success in usurping that position in the lek. Such eavesdropping during mate choice is a specific example of what Verrell (1991) considered to be the profound effect that "illegitimate receivers" have had on the evolution of sexual signaling systems.

A Note on Definitions of Eavesdropping

Several authors used the term *eavesdropping* to describe the behavior of predators locating prey (e.g., Barclay 1982, Beletsky et al. 1986, Balcombe and Fenton 1988, Harper 1991, Morris et al. 1994). Examples of such predators include fringe-lipped bats (*Trachops cirrhosus*) preying on chorusing tungara frogs (*Physalaemus pustulosus:* Tuttle and Ryan 1982, Tuttle et al. 1982) and Mediterranean house geckos (*Hemidactylus tursicus*) intercepting female decorated crickets (*Gryllodes supplicans*) as they move toward calling males (Sakaluk and Belwood 1984). Eavesdropping in such contexts is defined as "signals intended for one receiver [that] are intercepted by another" (R. H. Wiley 1983, p. 167); in such circumstances the receiver's fitness increases at the expense of the signaler's fitness (Morris et al. 1994).

Our definition of eavesdropping is more restrictive than that used in the predation context. Although many of the examples we discuss in this chapter may fit the broad definitions of eavesdropping (R. H. Wiley 1983, Morris et al. 1994), we consider it a prerequisite of eavesdropping that a third party (the eavesdropper) gains information from an interaction that could not be gained from a signal alone. We define eavesdropping in this way to clarify the level of information transfer in this behavior. Extracting information from an interaction (i.e., eavesdropping by our definition) is a different level of information transfer from detecting a signal (or any other simple cue) to locate potential prey (i.e., eavesdropping as defined in studies of predator-prey relations). The level of information transfer in communication has important implications for the evolution of signals and signaling behavior (Zahavi 1993) and is a cause of confusion and debate in the communication literature. Further discussion is beyond the scope of this chapter (see M. S. Dawkins 1995), but explicit statements about information transfer should go some way toward clarifying current ideas.

Implications for Signalers and Receivers: Alarm Calls

Many animals produce far-carrying alarm (warning) calls in the presence of conspecific individuals and a predator (e.g., the passerine seet call; Marler 1955). A puzzling feature of alarm calls such as seets is that they contain relatively little information regarding the predator's location (e.g., Charnov and Krebs 1975). The central role of communication networks in the study of alarm-calling behavior in birds is shown by the identification of "audience effects" in such behavior (reviewed by Gyger 1990; see also K. Sullivan 1985, Gyger et al. 1986, Karakashian et al. 1988). In this section we develop the idea that information on predator

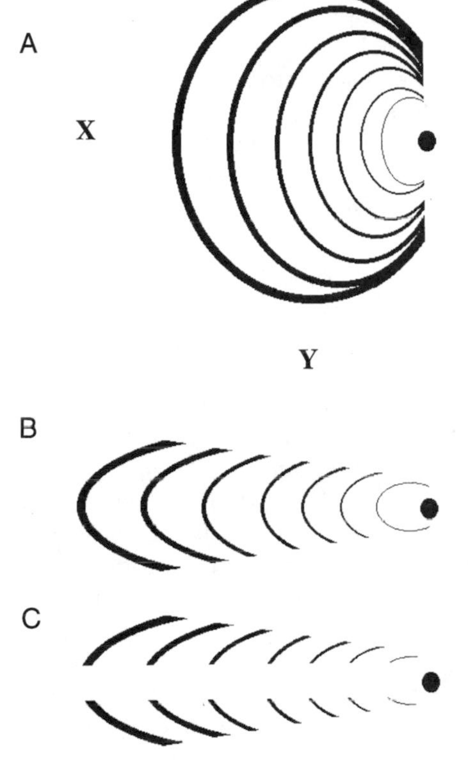

Fig. 23.2. The effect of three assumptions on the propagation pattern of alarm calling through a population. The site of the first alarm call elicited by a predator is marked by the large solid dot. Curves join individuals calling simultaneously, and thickness of lines indicates progression of time (thick = most recent). (A) Propagation pattern generated by individuals calling in response to the presence of a predator and in response to other calling individuals. An individual at the position marked X could deduce that it was in the path of an approaching predator, and an individual at Y could deduce that it was not in the predator's path. (B) As in A, but the pattern is truncated because individuals beyond a certain distance do not call. (C) As in B, but the pattern is further affected by lack of calling in the immediate presence of the predator.

location could be obtained from the pattern of propagation of calls in space and time through a communication network of potential prey. (Beletsky et al. [1986] used "call contagion" to mean propagation pattern.)

A word model based on simple assumptions of bird alarm-calling behavior can generate a number of propagation patterns. The resulting patterns contain varying amounts of information on predator location. Birds generally give alarm calls on detecting a predator, in response to another alarm call, or both, and continue calling for some time (McGregor and Dabelsteen unpubl. data). Initially we will assume that alarm calling costs little, perhaps because predators find the signals relatively difficult to locate (Marler 1955, 1977, C. H. Brown 1982; but see Shalter 1978, Klump and Shalter 1984) or detect (Klump et al. 1986). Alarm calling will therefore spread outward from the first caller. The propagation pattern will be a distorted circle, the distortion being caused by the path of the predator (Fig. 23.2A). Information on predator location is gained by listening to the propagation pattern. For example, a bird listening at position X (Fig. 23.2A) could deduce that it was in the path of an approaching predator, whereas one at Y could deduce that the predator had passed by at some distance. If we incorporate a second assumption into the model by setting a distance beyond which alarm calls (or the predator) do not elicit further calling, then the propagation pattern will be a more distorted circle (Fig. 23.2B). This second assumption seems reasonable, either because calls and predators are unlikely to be detected at some distance or because a distant predator does not constitute a danger. Information on the predator's location is now more detailed because the resulting propagation pattern more accurately indicates the predator's path. If we further assume that calling in the immediate presence of a predator is unlikely because of the danger to the caller, then the propagation pattern acquires a silent path that accurately tracks the route of the predator (Fig. 23.2C).

Propagation patterns can be studied with an acoustic location system (ALS) based on an array of microphones, because such a system records the pattern of calling in space and time. An important aim for studies of propagation patterns is to determine whether birds use the information on predator location that such patterns contain. In two array recordings of an aerial predator (a Cooper's Hawk, *Accipiter cooperii,* moving through an alarm-calling songbird community consisting largely of Red-winged Blackbirds, *Agelaius phoeniceus*), the propagation pattern appeared to have a silent path or zone of about 40 meters around the predator (i.e., it resembled Fig. 23.2C more than Fig. 23.2B; J. L. Bower and C. W. Clark unpubl. data). Also, most calls were given by birds in the path of the predator and about 40–50 meters distant from it, supporting the idea that they were predicting the predator's path (J. L. Bower pers. comm.). A propagation pattern with a silent path suggests that the potential prey (Red-winged Blackbirds) suffer a cost when calling in the immediate presence of a Cooper's Hawk. This cost could indicate the lack of a difference in auditory acuity between hawk and blackbird, information on perceptual abilities that can be derived otherwise only by laboratory psychophysical experiments (e.g., Klump et al. 1986, Klump, this volume).

Methods for Studying Networks

Studying acoustic communication networks requires simultaneous recordings of vocal activity and the location of the individuals forming the network. A number of factors make the collection of such data difficult. First, positional information is difficult to obtain in habitats where direct observation is constrained by vegetation (e.g., forests, reedbeds) or in open habitats (e.g., grasslands) with few landmarks, particularly if a species performs song flights. Second, it is difficult for a single sound recordist equipped with two directional microphones to monitor the locations and sounds of two interacting territorial males (J. P. Tavares pers. comm., Dabelsteen pers. obs.; but see Kramer and Lemon 1983). It is particularly difficult for a recordist to reach a position from which both males can be recorded adequately and at the same time to observe and record the start of an interaction. Furthermore, the two-microphone technique cannot in itself provide the accurate location information necessary to assess the timing of the interactions perceived by the participants (Dabelsteen 1992). Third, the feasibility of radiotelemetry is restricted by the need for high-fidelity transmission of vocalizations. This requirement would result in transmitters with a life of 24 hours weighing about 5 grams and would therefore restrict such telemetry to large birds (>500 grams). Telemetry can help to locate silent individuals, however. For example, Eurasian Blackbirds tend to perch higher when listening, a behavior that minimizes the effects of degradation and facilitates ranging (Dabelsteen et al. 1993). Telemetry could be used to document this response and also to locate eavesdroppers.

Techniques that locate animals by the sounds they produce (see, e.g., Watkins and Schevill 1972, Schleidt 1973b) do not suffer from these problems and in addition have three advantages: (1) acoustic location is passive, and the effect on an animal's behavior will therefore be negligible; (2) the sounds of interest are recorded as a part of the location process; and (3) the technique can remove the possibility of observer effects (Rosenthal 1976) and observer bias (Sharman and Dunbar 1982) inherent in visual distance estimation.

All acoustic location systems localize the sound source by using the differences in the times at which a vocalization arrives at different microphones or hydrophones. Arrival time differences can be determined by cross-correlation (e.g., Helstrom 1975); in theory, this technique can be effective even when the signal is below the level of background noise (Speisberger and Fristrup 1990). Arrival time differences determine a set of hyperbolic cross bearings that localize the sound to a particular region of space. Acoustic location (or direction finding) has been used successfully with whales (R. A. Walker 1963, Watkins and Schevill 1971, Clark 1980) and Northern Bobwhite (*Colinus virginianus:* Magyar et al. 1978).

An ALS has been developed for use with the Canary sound-analysis package (developed by the Bioacoustics Research Program at the Cornell University Laboratory of Ornithology; reviewed by Wilkinson 1994). The Canary-based ALS was first used with hydrophone arrays to locate whales (Clark et al. 1986, Clark 1989, Clark and Ellison 1989, Kiernan 1993), but it has now been modified for use with microphone arrays (McGregor et al. unpubl. ms. a). The microphone array ALS uses recordings of songs or calls with an array of four omnidirectional microphones (Fig. 23.3A) to locate the bird in three-dimensional space. The arrival time differences between all pairs of microphones in the array are calculated using the cross-correlation routine of the Canary software (Fig. 23.3B). These values are then entered into an error minimization algorithm to determine the most likely position of the signaler (Fig. 23.3C).

The usefulness of an ALS in the study of communication networks is related to the accuracy of the locations it generates. An initial study found that playback from a loudspeaker 10 centimeters in diameter could be located to within 10 centimeters of its actual location (C. W. Clark unpubl. data). Such a level of accuracy is more than adequate for studying territorial songbird communication networks in which individuals are many meters apart.

In theory and in practice, ALS accuracy is affected by a number of factors. One

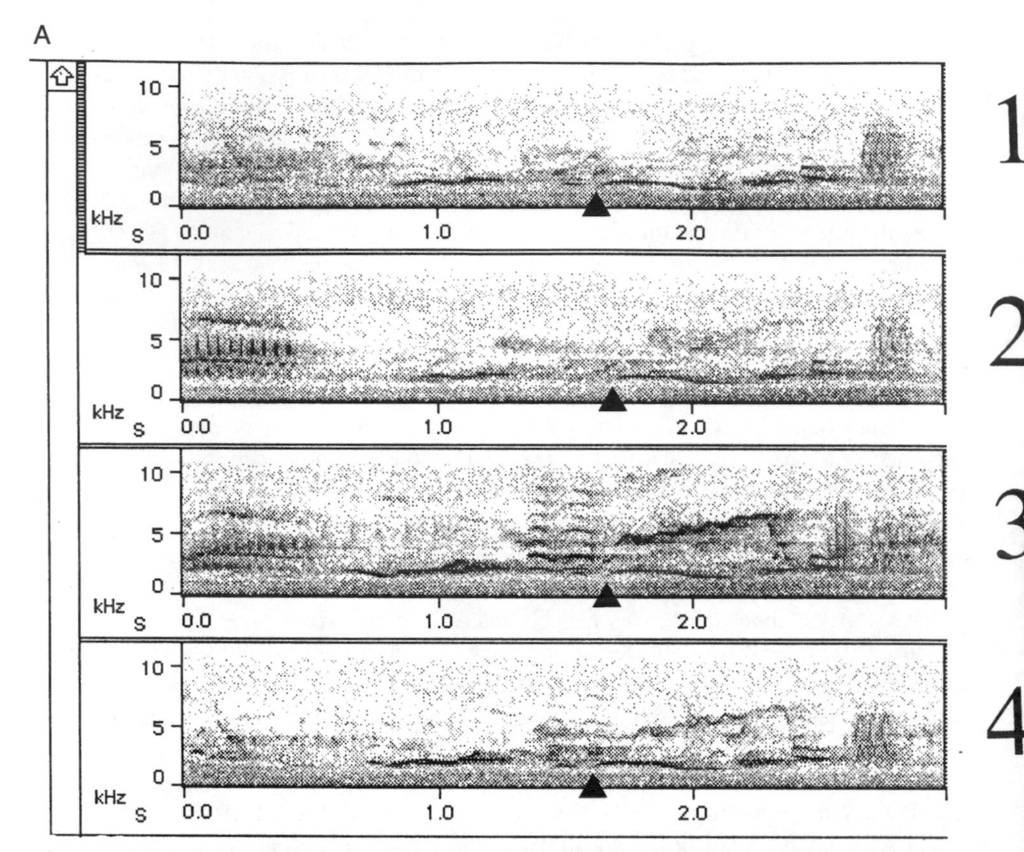

Fig. 23.3. An illustration of the stages involved in an acoustic location system based on Canary 1.1 software. The recording is of Eurasian Blackbird (*Turdus merula*) song made at Strødam Biological Field Station, Denmark, with an array of four omnidirectional microphones (McGregor et al. unpubl. ms. a). (A) Sonograms of a three-second section of simultaneous recordings from the array. Sonograms are from microphones 1 (top) to 4 (bottom), and the approximate position of the start of the same sound is marked in each sonogram with an arrow. (B) Plots of the cross-correlation functions ($\times 10^3$, no dimensions on Y axis) showing time delays for all possible combinations of four microphones. The time of maximum correlation gives the arrival time difference for the microphone pair. (C) The six hyperbolic cross bearings of possible sound source locations resulting from time delays (B) plotted on the location of the array with microphones (∗) 1–4. Axes show distance in meters relative to the position of microphone 1. The estimated position of the sound source (o) is shown by the intersection of the hyperbolic cross bearings.

factor is a consequence of sound morphology (or shape): wide-band sounds are easiest to locate because of effects related to the theoretical relationship between the time-bandwidth product of the sound, the signal-to-noise ratio, and the effectiveness of cross-correlation (Speisberger and Fristrup 1990, pp. 124–128). The major factors influencing ALS accuracy were identified in trials using the ALS and singing male Song Sparrows and playback (Bower and Clark unpubl. ms., McGregor et al. unpubl. ms. a). An important factor is the ability of the observer to unambiguously identify the same signal in each channel. Experience with the vocalizations of both the study species and any other species likely to be vocally

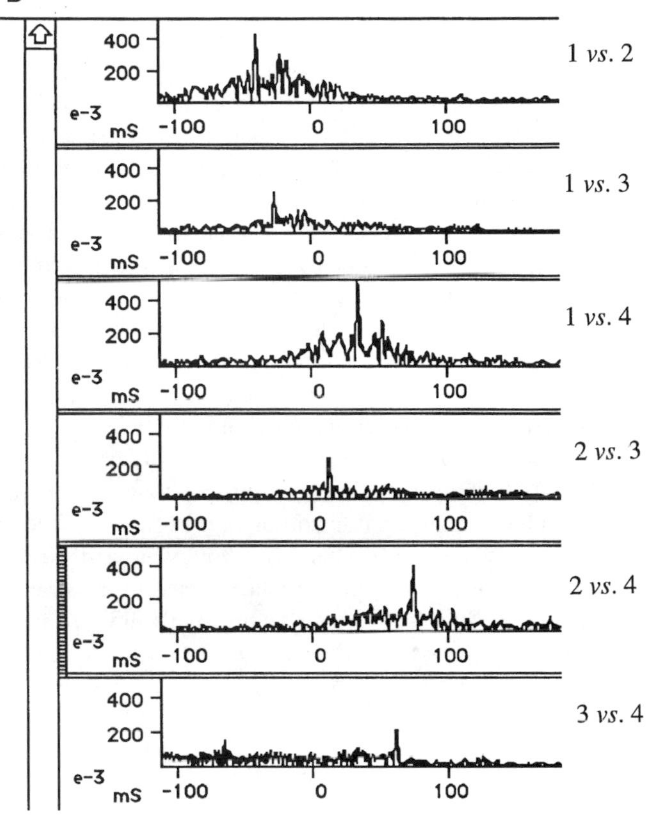

400
200
e-3
mS
-100
0
100
1 vs. 2
1 vs. 3
1 vs. 4
2 vs. 3
2 vs. 4
3 vs. 4

C

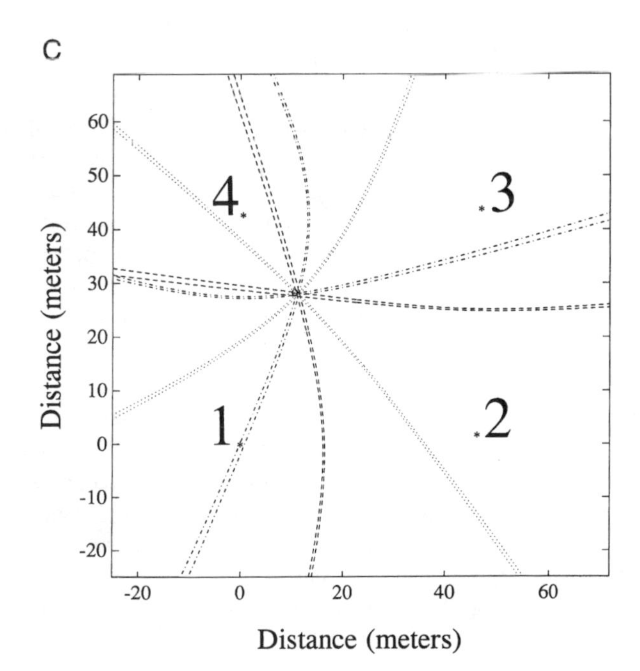

Distance (meters)
Distance (meters)
60
50
40
30
20
10
0
-10
-20
-20
0
20
40
60
1
2
3
4

active at the same time can help to reduce the errors caused by misidentification of the signal. All the other factors found to affect accuracy during field trials of the ALS were consequences of transmission. During transmission, all sounds are attenuated and blurred (degraded) and their signal-to-noise ratio is reduced. Attenuation and blurring are inversely proportional to the height of both microphone and sound source, increase with distance and the sound's degree of modulation, and differ between habitats (e.g., R. H. Wiley and Richards 1978, 1982, Dabelsteen et al. 1993). As attenuation and blurring increase, arrival time differences determined by cross-correlation become less accurate. The accuracy of location achieved by the ALS for male Song Sparrows defending territories in an abandoned field was highest within the array (mean difference between estimated and actual location, ± SE, was 0.83 ± 0.11 meters, $N = 22$) and least accurate outside the array (6.91 ± 1.02 meters, $N = 11$; Bower and Clark unpubl. ms.).

A second system of acoustic location, acoustic tomography, is based on the same principle of arrival time differences at an array of microphones as the ALS. This approach uses tomographic techniques to incorporate the effects of fluctuations in acoustic propagation speed caused by changes in both wind and speed of sound. Acoustic tomography can theoretically produce locations accurate to within 10 centimeters in real time (Speisberger and Fristrup 1990) and is currently being developed to locate Eurasian Skylarks (*Alauda arvensis*) during their aerial singing displays (J. M. C. Hutchinson pers. comm.).

Wider Implications of Networks

We have argued that communication networks will prove to be ubiquitous in animals that use long-range signals. Communication networks are therefore one of the most important aspects of the social environment in which signals have evolved, equal in importance to the transmission characteristics of the physical environment (e.g., E. S. Morton 1975, Michelsen 1978, R. H. Wiley and Richards 1978, 1982, R. H. Wiley 1991, Römer 1992). In this section we explore the wider implications of communication networks by integrating the concept with current ideas on communication and highlighting the disparity between the concept and current models of signaling. These implications are of general interest not only to students of communication behavior but also to those interested in conservation biology.

Networks and communication theories. Communication networks clearly provide new perspectives on signaling and receiving, but if the concept is as important as we claim, then it should also integrate with current ideas. In two areas communication networks provide explanations of behavior that are complementary or alternative to existing ideas. First, in the context of receivers' perceptual abilities, acoustic identification of neighbors and ranging are generally considered to have evolved as means of increasing the efficiency of territory defense (Falls

1982, McGregor 1991, 1993) through the "dear enemy" effect (Fisher 1954, Ydenberg et al. 1988, 1989, Getty 1989, Temeles 1994). We suggest that the advantages gained by having the ability to assess another individual by proxy—with no risk and little cost—through eavesdropping may have been just as important. Second, a consequence of ubiquitous communication networks is that signals will have evolved in a social environment in which eavesdropping occurs (A. Grafen pers. comm.). Zahavi (1979b) suggested that the signals used in aggressive disputes may be directed at individuals other than the opponent, thereby facilitating eavesdropping. Alternatively, behavior and signals intended for private use may have evolved to minimize eavesdropping (Verrell 1991, Endler 1993). We discussed the evolution of unobtrusive signals in this context, but the signals could have evolved for two other reasons, too. First, the mutual benefits to cooperating animals could lead to modifications of the signals involved so that they become "conspiratorial whispers" (Krebs and Dawkins 1984, M. S. Dawkins 1986). This explanation can be distinguished from our idea of restricting transmission by identifying the extent of cooperation involved in the interaction of interest, although the two ideas are not mutually exclusive. Second, unobtrusive signals could result from the selection pressures to remain inconspicuous to predators and parasites; for example, tachinid flies parasitize the cricket *Gryllus integer* (Cade 1975, 1979). Predators and parasites probably constitute "rare enemies" in an evolutionary sense (R. Dawkins and Krebs 1979), and this explanation therefore seems relatively unlikely. Frogs continue to call despite potential predation from bats (Tuttle and Ryan 1982), and crickets continue to call despite the risk of parasitism, although they may modify their behavior (Zuk et al. 1993). Also, signals given only in the presence of a predator (warning calls) may remain unobtrusive by exploiting differences in the hearing abilities of predator and prey (Klump et al. 1986). The idea that unobtrusive signals evolved to reduce transmission through communication networks seems to us to be the most plausible and general explanation.

Modeling. All the current models of communication are based on a single signaler-single receiver dyad (e.g., Grafen and Johnstone 1993). Originally the dyad was a simplifying assumption that rendered models mathematically tractable, but the dyad has become an exclusive modeling paradigm. Alternative approaches such as models based on computing networks (Critchlow 1988) and simulated networks (e.g., Crick 1989) have been ignored. It is possible that the current frustration over the lack of a general theory of signal design (see, e.g., M. S. Dawkins 1993) results from theories and models based exclusively on signaling dyads. We hope this chapter stimulates theoreticians to explore the possibilities of modeling a communication network.

Conservation. The extent of networks (i.e., the number of individuals involved) may be restricted by environmental "noise" of human origin. Such anthropogenic noise could be sound, light, pheromone mimics, or electric fields.

This noise restricts the effective range of signals (Klump, this volume) and therefore the effectiveness and extent of networks. For sound, all other things being equal, the effects will be most severe if the bandwidths of signal and noise overlap, and if the distances over which the signals are transmitted are large. The large baleen whales illustrate the problem well. These animals (e.g., the blue whale, *Balaenoptera musculus*) have vocalizations that potentially could form the basis of oceanwide communication networks (R. Payne and Webb 1971). The current levels of anthropogenic ocean noise probably severely curtail the extent of such networks and therefore restrict the information available through them, exacerbating the problems faced by these already endangered animals.

Conclusions

The study of acoustic communication networks in birds is in its infancy. As such, it is to be expected that currently we have a series of a priori arguments for the importance of networks but little direct experimental evidence to support them. Indeed, some of the best supporting evidence comes from other communication systems such as chorusing in insects and frogs. The technology to collect the necessary information and do the critical experiments now exists and is becoming more widely available. Even without access to such technology, however, it is possible to provide valuable insights into the structure of communication networks by collecting information on distance between song posts, song amplitude, and interactions between groups of neighbors.

We confidently expect that the next few years will see a rapid growth in the number of studies that specifically address communication networks in birds. Colleagues who study animal communication react enthusiastically to these ideas, though all recognize that some of the tests demand excellent experimental skills. We find the opportunities for understanding the evolution of communication systems exciting, and we hope others will join us in meeting these challenges.

Acknowledgments

The ideas expressed in this chapter were developed during a period of collaboration between TD and PKM funded by the Carlsberg Foundation and the Royal Society; we gratefully acknowledge their support and the more recent funding under the NATO Collaborative Research Grant scheme that has allowed Christopher W. Clark to become involved. The ALS he developed (funded by the National Science Foundation, North Slope Borough, and the Office of Naval Research) has the potential to address many of the suggestions we raise, and we have begun to tackle them with funding from NATO, the Danish National Research Foundation, the Natural Environment Research Council, the Junta Nacional de Investigação Científica e Technológica, and the National Science Foundation. We also thank the following people for discussion and comments on various aspects of the manuscript: Mike Beecher, John Bower, Christopher Clark, Juan Delius, Rui De Oliveira, Bruce Falls, James Fotheringham, Alan Grafen, Tim Guilford, Oren Hasson, Jo Holland, Richard Holland, Andy Horn, Vincent Janik, Rufus Johnstone, Glenn Morris, Vince Muehter, Marc Naguib, Ken Otter, Tom Peake, Laurene Ratcliffe, Candy Rowe, Rachel Scudamore, José Pedro Tavares, Pat Weatherhead, Becky Whittam, Haven Wiley, and Amotz Zahavi.

24 The Dawn Chorus and Other Diel Patterns in Acoustic Signaling

Cynthia A. Staicer, David A. Spector, and Andrew G. Horn

Anyone who rises early enough during the songbird nesting season in the north temperate zone will be impressed by the intense vocalizing of birds in what is widely known as the "dawn chorus." Beginning typically 30–60 minutes before daybreak, territorial males of many species sing at rapid rates, more or less simultaneously, then become relatively quiet around sunrise; afterward, their singing tends to be less intense, such that fewer individuals are vocalizing for shorter periods and at lower rates. The acoustic signaling of many other animals—including crickets, frogs, and gibbons—also exhibits strong diel rhythms linked to the light-dark cycle (Tenaza 1976, T. J. Walker 1983, Fritsch et al. 1988); yet the functional significance of diel patterns in acoustic signaling is considerably less obvious than is the significance of patterns in the other behaviors (e.g., movement and feeding) for which most animals show diel cycles (see Aschoff 1981). The dawn chorus of birds is of particular interest because its functions remain poorly understood.

The functions of the dawn chorus have been speculated on extensively, but only recently have they been the subject of synthetic attempts (J. M. C. Hutchinson et al. 1993). The papers on the dawn chorus published to date ignore many widespread patterns in the timing of acoustic signals. One of the most striking features of dawn bouts is the use of different songs and singing behaviors as compared with daytime bouts (Spector 1992). Also, an individual's dawn singing changes in character through the presunrise period as well as through the season (Staicer 1991). Furthermore, diel patterns vary among species in aspects such as the timing of peaks in vocalizing through the day or night and throughout the season, and the kinds of behavior that accompany song. Clearly, if we are to understand the dawn chorus, we need a more comprehensive approach that considers other diel patterns, encompasses a wider range of species, and examines individual variation.

In this chapter we attempt to outline and explain the diversity of diel patterns of acoustic signaling in birds, with special reference to the dawn chorus as a common example of a diel peak. First, we summarize adaptive explanations for dawn singing, including a few new hypotheses. Second, we survey diel patterns in acoustic signaling across diverse species of birds, relate this variation to behavioral and ecological factors, and provide some adaptive interpretations. Third, we examine whether the hypotheses appear to predict the three widespread patterns mentioned above. Throughout we highlight aspects of dawn singing not considered in previous treatments of the dawn chorus. Our intent is to provide a broader context in which to view the various diel patterns of acoustic signaling among birds, and we hope to stimulate thoughts about and further research into the timing of animal signals.

Hypotheses

Several hypotheses have been proposed to explain why male birds sing more intensely at dawn than later in the day. Here we present an overview of the major hypotheses, including some (flagged by an asterisk) that, as far as we know, are developed here for the first time. The 12 hypotheses fall into three rough classes: (1) singing more intensely at dawn could serve a function *intrinsic* to the singer's internal state, e.g., hormonal levels; (2) dawn singing could have a *social* function (inter- or intrasexual) that is best served at dawn; and (3) birds might sing more at dawn because of the daily timing of *environmental* selective pressures. We note, however, that dawn song may have more than one function, even in a single species; consequently, many of the hypotheses are not mutually exclusive and may be difficult to distinguish. The main purpose of this section is to point out aspects of dawn singing that deserve more attention rather than to provide a detailed critique of the hypotheses.

Intrinsic Functions

**Circadian cycles.* A dawn peak in singing may be a nonfunctional consequence of elevated levels of hormones (e.g., testosterone) that influence song. Increased testosterone generally causes increased singing and increased responsiveness to external stimuli, such as the singing of conspecific males, that in turn stimulate singing (Wingfield and Farner 1993). At least for Japanese Quail (*Coturnix japonica*), however, the timing of the dawn bout is independent of testosterone (Wada 1983). Whether changes in hormone levels are correlated with diel patterns in song activity among other birds is presently unclear. Although circadian rhythms (biological cycles of approximately 24 hours) in various hormones have been established for mammals, similar data for birds are scant, though luteinizing hormone is known to peak at night (Wingfield et al. 1981, Rattner et al. 1982, Wingfield and Farner 1993).

If the dawn chorus is attributable to a peak in testosterone before sunrise, then seasonal changes in dawn singing might be expected to parallel seasonal changes in testosterone (though a correlation would not indicate cause and effect). Year-round residents start singing earlier in relation to sunrise, lengthening their dawn

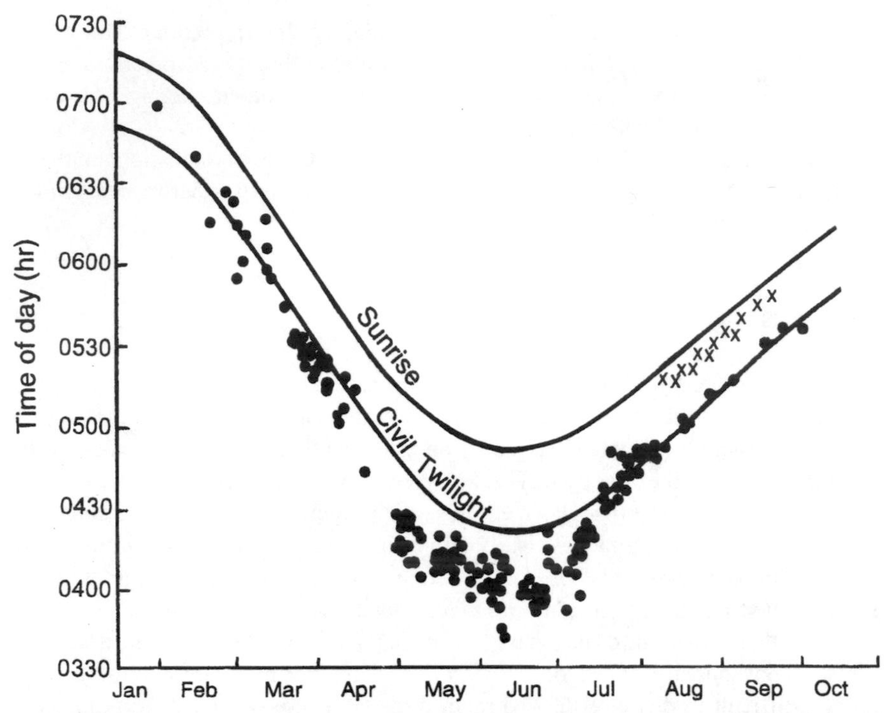

Fig. 24.1. Seasonal trends in the time at which a temperate zone resident, the Rufous-sided Towhee (*Pipilo erythrophthalmus*), sang its first songs of the day. Males begin to sing earlier in relation to civil twilight (when the sun is 6° below the horizon) as the breeding season progresses. Data from clear mornings, January–September 1955–1956, Hastings Reservation, central California (36° N); on five mornings during May and June, birds were already singing when observations began. An X indicates lack of song before sunrise on particular dates. Breeding data showed that testes became active in January and February and most males reached breeding condition around the middle of April; gonads were in regression by 20 July and had become inactive by mid to late September. The shift to earlier song start times in relation to civil twilight is thus correlated with the period of gonad growth. Also note that as daylength increases between the spring equinox and the summer solstice, civil twilight occurs earlier in relation to sunrise, causing the dawn period to lengthen. Modified from Davis (1958).

bouts, as their testes enlarge (Fig. 24.1; see also "Patterns," below). Evidence for other permanently territorial north temperate birds similarly indicates that testosterone is elevated only during the breeding season (Wingfield and Hahn 1994), when the dawn chorus occurs. Yet an individual's song activity appears less closely linked to its testosterone levels than this seasonal correlation might suggest (Nowicki and Ball 1989, Wingfield and Hahn 1994). Much variation in dawn singing, among and within individuals, remains unexplained (see, e.g., Davis 1958). Of interest would be data comparing circadian cycles of testosterone for diurnal species and whether they show a dawn chorus (e.g., the Chaffinch, *Fringilla coelebs,* does not: Hanski and Laurila 1993).

**Self-stimulation.* Dawn singing may stimulate hormone production in singers and prepare them for social interactions (e.g., mating or agonistic encounters).

The songs of conspecific males and involvement in territorial interactions are known to stimulate singing and territorial behavior, and to increase testosterone levels (Wingfield et al. 1990). Evidence is also accumulating that a bird's *own* vocalizations can affect the growth of its gonads and, presumably, the associated hormone levels (Cheng 1992, Wingfield and Farner 1993). Indeed, given that males respond most strongly to playback of their own song, even at the level of the neuron, and that males listen to songs with the same pathways that they use to produce them, a bird's own song may be especially effective at stimulating gonad growth and hormone production (Falls 1985, H. E. Williams and Nottebohm 1985, Margoliash 1986).

The self-stimulation hypothesis suggests that the dawn bout might be a way for birds to regulate the seasonal growth of their gonads, the production of their hormones, or both; the singing of other males would likely add to the effect of the bird's own song. Whereas we see no reason for gonadal stimulation to be regulated specifically at dawn, it could be the best time to prime hormone levels for territorial or mating interactions, which tend to be most frequent just after sunrise. This hypothesis predicts that the songs and singing behaviors used at dawn would be the most effective at producing hormonal states appropriate to territorial defense and mating. Our suggestion that dawn singing regulates readiness for social encounters is akin to applying the challenge hypothesis of Wingfield et al. (1990) to diel signaling patterns.

The self-stimulation hypothesis proposes that singing at dawn changes the hormone levels of the singer, whereas the circadian cycles hypothesis proposes that a circadian increase in hormone levels stimulates singing. If positive feedback occurs between singing and increasing hormone levels (which is likely in view of the evidence reviewed above), the two hypotheses would not be mutually exclusive. Each might predict different patterns in changing hormone levels *within* a given dawn bout. The self-stimulation hypothesis would probably predict an increase in hormone levels during the bout, but whether the circadian cycles hypothesis predicts a decrease is unclear; any such predictions would probably oversimplify the connections between hormones and singing. A detailed understanding of the relationship between dawn song and hormones will have to wait until hormone levels can be measured and manipulated while birds are singing and throughout their diel cycle.

Social Functions

Mate attraction. Males may sing at dawn because that is the best time to attract females. In several species of lekking grouse and shorebirds, females arrive on the leks at particular times and spark a peak of displaying, including vocalizing (e.g., deVos 1983). Even in some nonlekking species (including those with "exploded" leks), the peak of male displaying appears to be timed to coincide with the peak of female foraging or the time when females are more sexually receptive (e.g.,

Eurasian Woodcock, *Scolopax rusticola:* Cramp 1983). Mate attraction might explain the night singing of some birds; only unpaired male Northern Mockingbirds (*Mimus polyglottos*) sing at night (Merritt 1985). Many passerines migrate at night, and thus the number of new, unpaired females might peak in the early morning. If males sing at dawn to attract these females, a more marked dawn chorus should occur after nights favorable for migration. Visits of female European Pied Flycatchers (*Ficedula hypoleuca*) to male territories peak in the early morning, though well after sunrise (Dale et al. 1990). The mate attraction hypothesis might be tested by playing male song at traps at different times of day.

A major prediction of the female attraction hypothesis—that the dawn chorus should be most intense during the period of mate attraction, usually the first week or two after male arrival—has little empirical support. Among a wide array of territorial species that exhibit diel peaks in singing, at different times of day or night, diel patterns are considerably less marked early in the season. For example, the singing of a nocturnal migrant, the Marsh Wren (*Cistothorus palustris*), peaks at night in the middle, rather than at the beginning, of the season (Barclay et al. 1985). Also, the dawn chorus of wood-warblers (Parulinae) is usually absent early in the season (Nolan 1978, Highsmith 1989, Kroodsma et al. 1989, Staicer 1989, 1996, Spector 1991). Unpaired Marsh Warblers (*Acrocephalus palustris*) sing at night early in the season, but their night song rates are similar to those of the morning periods (Kelsey 1989).

The female attraction hypothesis also predicts that males who become unpaired should increase their dawn signaling (e.g., singing more songs per unit time, lengthening bout duration, or alternating song types more frequently). For at least some species, however, pairing status does not affect dawn song. Only after sunrise is there a notable difference between the singing of paired and unpaired male wood-warblers, and removal of their mates does not change the dawn singing of males (Kroodsma et al. 1989, Spector 1991, Staicer 1996). Male Black-capped Chickadees (*Parus atricapillus*) sing longer bouts after mate loss, but this might occur simply because the male is not joined by a female around sunrise (Otter and Ratcliffe 1993). For many species, diel patterns in singing are actually less marked in unpaired than in paired males, and mate loss typically results in greater increases in daytime than dawn song rates (Fig. 24.2).

**Mate stimulation.* Males may sing at dawn to stimulate the reproductive development of their mates. Male vocalizations, especially song, trigger and enhance the nesting behavior of females as well as ovarian follicle growth (Wingfield and Farner 1993). Females might be most affected by male vocalizations heard at dawn. Male Japanese Quail vocalizations that were played back one to two hours before sunrise (the time of the dawn chorus in this species) led to greater follicle growth in females than the same calls played back at other times of the day (Guyomarc'h and Guyomarc'h 1982). Playback of male song to female White-crowned Sparrows (*Zonotrichia leucophrys*) enhanced their gonadal response to changing photoperiod, although changes in photoperiod caused greater

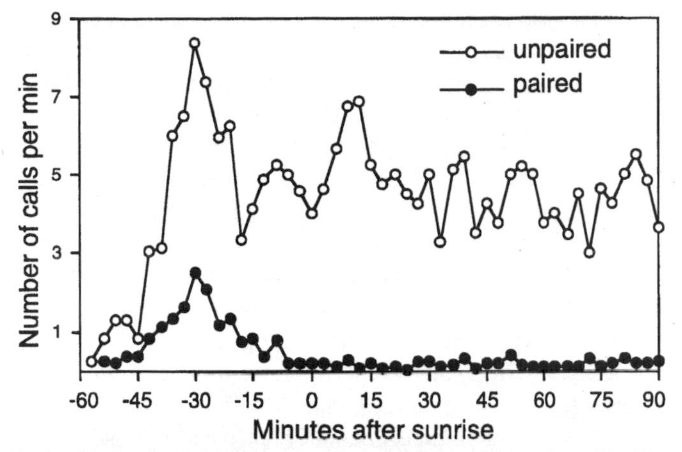

Fig. 24.2. Vocal activity patterns in relation to mating status of a male Mourning Dove (*Zenaida macroura*). The dawn peak in vocal behavior typically is more obvious in paired males than in unpaired males, although unpaired males sing more overall. Data are frequency of cooing for the same captive (penned) male when paired and unpaired (after his mate was experimentally removed), for three-minute periods beginning one hour before sunrise. Modified from Frankel and Baskett (1961).

increases in follicle growth than did playback (M. L. Morton et al. 1985). Furthermore, vocalizations are zeitgebers for daylength in some species (Reebs and Mrosovsky 1990). We speculate that males might be able to increase the responsiveness of females to changing daylength and hasten egg production by singing earlier and longer at dawn.

The mate stimulation hypothesis predicts an association between a male's dawn singing and his mate's reproductive state. Contrary evidence comes from wood-warblers in that the dawn bout of unpaired males is similar to that of paired males (e.g., Staicer 1996). Supporting evidence includes the trend for the dawn bouts of male Great Tits (*Parus major*) and Eurasian Blackbirds (*Turdus merula*) to lengthen over several days until the mate lays her first egg (R. Mace 1987a, Cuthill and Macdonald 1990), although several alternative interpretations are possible. Previously, these data have been used to support the mate-guarding hypothesis.

Mate guarding. Males may sing at dawn to guard their mates from extra-pair copulations (EPCs), which are common and can greatly decrease a male's reproductive success (e.g., Westneat et al. 1990). Many male behaviors, including song, could reduce the risk of EPCs. R. Mace (1986) suggested that EPC risk is highest at dawn, and thus dawn is the most crucial time for males to sing—both to repel their male neighbors and to ensure being heard by their mates. Although the mate guarding hypothesis has been embraced by many authors, it does not provide a robust explanation for the dawn chorus.

Three arguments have been made in support of the mate-guarding hypothesis.

First, dawn bouts of some species may be longest on days that mates lay eggs (e.g., R. Mace 1986). Many species, however, exhibit a dawn chorus throughout the nesting period, and, at least in some (e.g., wood-warblers), a male's dawn bout lengthens after his mate's clutch is laid (Highsmith 1989, Kroodsma et al. 1989, Staicer 1991). Similarly, the duration of the dawn bouts of male Collared Flycatchers (*Ficedula albicollis*) is not longer during their mates' fertile period (Pärt 1991). Second, if females are most fertile immediately after laying an egg, birds that lay at dawn should be more likely to show a dawn chorus (R. Mace 1986). Sheldon and Burke (1994), however, found little evidence for a "fertility window" after egg laying. As well, females may be reluctant to copulate after egg laying commences (Willow Warblers, *Phylloscopus trochilus:* Arvidsson 1992). In several species that show a marked dawn chorus, females actually lay their eggs well after sunrise (e.g., Skutch 1952, Cuthill and Macdonald 1990, Pärt 1991). Third, R. Mace (1986) showed experimentally that a male Great Tit ends his dawn bout when his mate emerges from the nest box and concluded that a male sings at dawn to prevent other males from copulating with his mate (but see below for a different interpretation).

Males of many species end their dawn bout around sunrise, when their mates leave the nest, or when they deliver food to mates on the nest, and often copulate soon afterward (W. P. Smith 1942, R. Mace 1986, Cuthill and Macdonald 1990, Pärt 1991, Otter and Ratcliffe 1993, Horn unpubl. data, W. J. Smith pers. comm., Spector unpubl. data, Staicer unpubl. data). We suggest that males terminate their dawn bout when their mates join them because dawn singing is incompatible with mate guarding. By singing near the female during her fertile period, a male would draw attention to his mate's location at just the time when EPCs would be most detrimental to him. A male should thus be silent near his fertile mate; indeed, roosting with one's mate might be a more effective mate-guarding strategy than singing some distance away from her (e.g., Purple Martins, *Progne subis:* C. R. Brown 1980). Territorial males of many species sing least frequently while guarding a mate. A male wood-warbler, for example, sometimes skips his dawn bout completely and silently accompanies his mate while neighbors sing (Staicer 1991, unpubl. data). Also, females have some control over EPCs; in one population of Great Tits, females successfully rejected EPCs whether or not they were accompanied by their mates (Björkland et al. 1992).

A major difficulty with relating singing behavior to paternity assurance is establishing a cause-and-effect relationship. To date, studies of the dawn chorus have studied its causes, not its effects (R. Mace 1987b). Peaks of singing during the fertile period and during temporary absences of the female (including at dawn) are consistent with a number of alternative hypotheses. A satisfying, albeit difficult, test of the mate-guarding hypothesis might require replacing males with song playback through speakers (analogous to experiments that showed that playback deters intrusions; Falls 1988) and measuring the frequency of their mates' EPCs.

Territory defense. Males may sing at dawn because it is crucial to advertise territory ownership before daybreak. The territory defense hypothesis assumes that a burst of territorial activity is needed after a period of inactivity (e.g., night), and that a male's risk of losing his territory to another male is highest at dawn. The limited evidence for this hypothesis includes a dawn peak in intrusions by floaters in a Rufous-collared Sparrow (*Zonotrichia capensis*) population (S. M. Smith pers. comm.). Intrusions into Great Tit territories, however, peak during the first four hours *after* dawn (Kacelnik and Krebs 1982). The territory defense hypothesis assumes that dawn singing is necessary to repel intruders; tropical wood-warblers and tyrant flycatchers, however, maintain territories during the non-breeding season without having a dawn chorus (Staicer 1991, unpubl. data). The territory defense hypothesis could be tested by replacing males with playback to see if songs broadcast at dawn delay reoccupation longer than songs broadcast later in the morning. Interestingly, males seem to respond more strongly (by singing more, approaching the speaker, or altering their singing pattern) to playback during the day than at dawn (Shy and Morton 1986, Horn unpubl. data, Spector unpubl. data, Staicer unpubl. data; but see Dabelsteen 1984a). More research is needed to determine how males respond to playback at dawn and what this tells us about the role of dawn song in territory defense.

Another version of the territory defense hypothesis attempts to explain the dawn chorus by the conflicting demands of territory defense and mate-guarding activities. Slagvold et al. (1994) suggested that a male Great Tit's dawn singing peaks when his mate is fertile because during the daytime he spends most of his time guarding her and has little time to advertise ownership of his territory by singing. This explanation assumes that dawn singing has a mainly territorial function, regardless of the fertility of the mate or the pairing status of the male.

**Social dynamics.* Dawn singing may play an important role in signaling and adjusting social relationships among territorial neighbors. The dawn chorus should present the best opportunity for both singers and nonsingers to assess the competitiveness of and social relationships between singers, because essentially the dawn bout is a prolonged simultaneous display. Although dominance relationships have long been considered important for birds that forage, travel, or roost in social groups, the social relationships among territorial neighbors have received little attention. Social relationships can be expected to have important effects on a territorial male's reproductive success by influencing both his territory ownership and his mating opportunities, which are often closely connected. This hypothesis emphasizes the complex and dynamic nature of social relations among individuals. We suggest that changes in these relationships are mediated significantly through vocal interactions at dawn; perhaps males who assert their competitiveness at dawn have an advantage in social competition during the day.

The social dynamics hypothesis seeks to explain the characteristics of the dawn chorus, not just the fact that the song occurs at dawn. First, the more lengthy and complex countersinging at dawn, as compared with singing later in the day, is likely to provide information beyond singer identity, because individual recognition can be achieved with just a few songs (Stoddard, this volume). Second, dawn singing appears to be socially contagious. For instance, the dawn chorus of a tropical wood-warbler develops over several weeks in a contagious manner until all the territorial males in the population are singing a dawn bout (i.e., the behavior spreads from neighbor to neighbor, more and more of the residents sing dawn bouts, and these lengthen in duration over a period of weeks; Staicer 1991). Also, male *Tyrannus* kingbirds with conspecific neighbors sing more intensely before sunrise than males without neighbors (W. J. Smith pers. comm.). Third, dawn singing is often directed at particular neighbors. Male wood-warblers and Red-crowned Ant-Tanagers (*Habia rubica*) sing at dawn from particular song posts clustered near certain edges of adjacent territories (Willis 1960, Lemon et al. 1987, Staicer 1989, unpubl. data). During pauses between songs, males turn their heads from side to side as if listening intently to neighbors' songs and appear to respond to particular neighbors (e.g., by matching song types and cadence).

Social interactions might be especially unstable at dawn if males or females are more likely to be discovered to be dead than at other times of the day. Such a pattern could result either from higher predation rates at night or simply from a constant death rate accumulating through the night (Kacelnik and Krebs 1982, Pärt 1991). This factor predicts differences in dawn singing among species that differ in overnight mortality rates. Although such accumulated nocturnal death could contribute to dawn social instability and thus to the need for vocal negotiation of relationships, the simple announcement that an individual had survived the night (Kacelnik and Krebs 1982) would not require an extended song bout.

The social relationships among territorial birds can be visualized as a complex and dynamic web, manifested in territory ownership, pair-bonds, and genetic offspring. Territory edges and contents (e.g., mates, nest sites, and young) may change from day to day (see, e.g., Nero 1956). When a neighbor disappears or a new bird arrives, the remaining birds shift their territory boundaries; a mate's presence, reproductive state, and behavior may also change daily and clearly influence the behavior of a territorial bird (e.g., Staicer 1991). For example, when a female selects a nest site peripheral to her mate's territory, he typically enlarges the territory to encompass the area surrounding the nest. In many species, rates of nest predation and abandonment are high, such that the nesting stage of a pair might change abruptly several times during a season. All these changes likely result in, and partly from, changes among the social relationships of neighbors.

The social dynamics hypothesis predicts that dawn singing will reflect social relationships among neighbors. Preliminary data suggest that male Black-capped Chickadees with higher winter dominance ranks tend to start their dawn bouts earlier, and thus sing longer bouts, than individuals that were subordinate the preceding winter (S. Rice and L. Ratcliffe unpubl. data). This hypothesis might be tested by relating changes in territory borders, as well as changes in mate or nest status, to variation among the dawn bouts of an individual. More dominant males may have territorial and mating advantages. Birds that were dominant in winter flocks tend to obtain higher-quality breeding territories than do subordinates (S. M. Smith 1976, Lambrechts 1992). Social dominance also influences territorial relationships of migratory birds; older male American Redstarts (*Setophaga ruticilla*), for example, prevent younger males from obtaining territories in higher-quality areas, where mates are easier to attract (T. W. Sherry and Holmes 1989).

Social relationships may also influence a male's chances of obtaining EPCs and guarding his mate from them. Recent paternity studies suggest that females tend to engage in EPCs with neighbors rather than with strangers or floaters (Alatalo et al. 1989, Gibbs et al. 1990). Females might select neighbors because they are dominant over nonterritorial males, and thus should have better genes. For instance, female Black-capped Chickadees seek EPCs from males whose dominance rank in winter flocks was higher than their mate's rank (S. M. Smith 1988). Assuming that females pay attention to the dawn singing of neighbors, the social dynamics hypothesis suggests that singing at dawn may attract females that are paired. These females are potential partners for EPCs or potential replacement mates, both in the present and in future breeding seasons. This hypothesis predicts that males should sing at dawn *when* they would have a reasonable chance of mating with presently paired females (which might be more likely than the chance of losing one's mate and needing to attract a new one)—that is, during the pooled fertile periods of females within earshot. The seasonal period in which males engage in a dawn chorus typically coincides with the seasonal period in which females are fertile. For example, both paired and unpaired male Collared Flycatchers cease dawn song after females cease building nests (e.g., Pärt 1991).

The social dynamics hypothesis predicts that variation among species in dawn singing is correlated with the instability of social relationships. Populations in which social relationships are less stable (i.e., in which EPCs are more frequent or territorial boundaries are more dynamic) would thus be expected to show a more lengthy or intense dawn chorus.

Handicap. Dawn singing may be a reliable signal of male quality if singing is more costly at dawn than at other times of day (Montgomerie 1985). If dawn singing were more costly, receivers who needed to assess male quality should attend especially to singing then, which in turn would select for an intense dawn chorus. The underlying assumption is that dawn is a particularly costly time to sing. In diurnal species, energy reserves should be lowest at dawn, either because they have not been built up overnight or because the night might have been cold, or both. This argument does not imply that singing represents an important energetic cost; rather, lacking sufficient energy reserves, a bird may forage instead of sing. Although birds of some species do sing while foraging, these two activities often seem to be mutually exclusive. In addition, the risk of his mate engaging in extra-pair copulation may be higher for a male participating in the dawn chorus, especially if singing is incompatible with mate guarding. The risk of predation might also be higher because dawn straddles the foraging periods of diurnal and nocturnal predators. Evidence that any of these costs is higher at dawn is scant, however. Earlier workers (Kacelnik and Krebs 1982, R. Mace 1987b) rejected this hypothesis because it presented singing as a handicap (Zahavi 1975), but handicap models have since become more respectable (Grafen 1990b).

The evidence that singing is costly is indirect and is based mainly on the result that artificial provisioning of food elevates singing (e.g., Strain and Mumme 1988). Food provisioning also causes dawn bouts to lengthen (Cuthill and Macdonald 1990, K. Otter and L. Ratcliffe pers. comm.), although increased food availability may lead to a general increase in song activity rather than an increase specifically at dawn. A bird provisioned with food might sing more because it spends less time foraging or because its provisioned territory is more valuable; whether singing is energetically costly is presently unclear. Laboratory measurements of oxygen consumption by crowing Red Junglefowl (*Gallus gallus*), which have a marked dawn chorus (Collias and Collias 1967), indicate that crowing is energetically very cheap (Chappell et al. 1995, Horn et al. 1995), but similar measurements on singing Carolina Wrens (*Thryothorus ludovicianus*) suggest a greater metabolic cost (Eberhardt 1994; but see Horn et al. 1995). More studies are needed to determine the costs of singing in relation to time and energy budgets, especially for small birds.

The handicap hypothesis predicts a correlation between dawn song output and some attribute of the singer, for example, parasite load or territory quality. Similarly, a stressor such as a very cold night should increase the interindividual variance in dawn song output. Such observations, though, would not support the handicap hypothesis as an explanation for concentration of song at dawn unless the observed effects were significantly weaker for daytime singing.

Functions Related to Environmental Pressures

Low predation. Dawn may be a less costly time to sing if the risk of predation then is lower than during the daytime, the opposite of what the handicap hypothesis predicts. The chance that an individual will be preyed on may be lower when light levels are low and when many other individuals are also singing, which may dilute a predator's attention (see, e.g., Tenaza 1976). Predation appears to influence the timing of calling in some insects and frogs. For example, crickets start and stop calling more abruptly on islands where a parasitoid is present than elsewhere where it is absent (Zuk et al. 1993), and frogs stop calling on moonlit nights when predatory bats pass overhead (Tuttle et al. 1982). The idea that predation is lower at dawn could be studied experimentally by playing back recorded songs near stuffed birds at different times of day (Gotmark 1992) to test the prediction that fewer would be attacked at dawn than at other times. Also, the presence or absence of primarily auditory hunters, such as diurnally active owls (e.g., *Glaucidium, Asio flammeus*), is expected to have a stronger effect on singing behavior than the presence of primarily visual hunters (e.g., *Falco, Accipiter*). This hypothesis also predicts that males should sing from more exposed perches or locations at dawn than later in the day.

Acoustic transmission. Birds may sing most intensely at dawn because their songs propagate farthest then. Songs should travel farther at dawn than later in the day because the atmospheric conditions (e.g., low wind, high humidity) at dawn are more conducive to sound propagation (Henwood and Fabrick 1979). On the other hand, considerably more background noise occurs at dawn because many more songs are being delivered per unit time by many males. Songs of different species are more likely to overlap at dawn (R. Mace 1987b), though evidence to date does not suggest that masking affects the timing of songs (Brémond 1978). The acoustic transmission hypothesis predicts that birds sing songs with particular acoustic characteristics at times of day when such sounds should travel farther. The idea that songs are more easily detected at dawn could be tested by conducting playback experiments at different times of day and partitioning male response into detectability and response criteria (see also Klump, this volume).

The relative lack of wind and convection at dawn might result in the songs of birds reaching their neighbors relatively undegraded, as compared with later in the day. This effect could have several repercussions. First, the birds might sound closer to each other, depending on the extent to which they use different aspects of degradation to judge distance (Morton 1982). Neighbors' songs might appear to come from within a male's territory rather than outside it, thereby stimulating the male to sing as he does in territorial disputes. Second, it might enable birds to establish relatively accurate baselines for comparison of song degradation later in the day. Third, males may be able to obtain more information from their neighbors' singing if signals are less degraded. If this effect is involved in the proximate control of the dawn bout, it might be detected by comparing dawn bouts in locations where the dawn air is relatively stable with bouts in areas (e.g., coastal or volcanic) where wind or convection is common at dawn.

Inefficient foraging. Males may sing at dawn because foraging at dawn is less efficient than during the day. In an experimental study, male Great Tit foraging success was poorer at dawn relative to later hours, both because the low light levels of dawn interfered with the birds' ability to search for prey and because the prey were less mobile, and thus harder to detect, than later in the day (Kacelnik and Krebs 1982). Yet many noninsectivorous birds also have a dawn chorus, and other studies indicate that invertebrate prey are not necessarily slower in the morning (R. Mace 1987b). As well, birds breeding in the Arctic under continuous daylight show a "dawn" peak at the time of day when light intensity is changing most rapidly (R. G. B. Brown 1963, J. B. Falls pers. comm.).

Unpredictable conditions. Males may sing at dawn because they *must* sing for some time each day, and they usually have energy reserves to do so at dawn. This hypothesis was generated through dynamic programming, which can simulate the optimal daily routine of animals by specifying some desired end state (e.g., a given value of reproductive success) and then calculating the optimal choice of behavior at each previous step. If a bird's daily routine is assumed to consist simply of foraging and singing, the optimal allocation of time (given uncertainty about daily energy requirements because of unpredictable overnight conditions) is to concentrate song activity into the early hours of the day (McNamara et al. 1987, J. M. C. Hutchinson et al. 1993). This dynamic programming model suggests that males gather enough energy to survive the coldest nights, which is more energy than needed for most nights, and thus have an energy surplus at dawn on most days. This model also predicts a dusk chorus, especially if food supplies are uncertain, because again, birds should gather food well before nightfall to meet their needs on the worst nights; if they have gathered enough by dusk, they have extra time to sing.

An underlying assumption of this hypothesis, however, often may not hold, because the degree to which foraging and singing are mutually exclusive varies considerably among species. Eurasian Skylarks (*Alauda arvensis*) sing in flight and feed on the ground, for example, but vireos (*Vireo* species) sing while they forage in the canopy. Also, data for Great Tits suggest that the time available for foraging has little effect on daily singing routines, mainly because birds have more "free time" than is generally assumed; these data come from field studies of populations at different latitudes and from experiments that manipulated the timing of food availability for captive birds (R. Mace 1989a–c).

Furthermore, the available data do not support two obvious predictions of the unpredictable conditions hypothesis. First, the intensity of dawn singing and the length of the dawn chorus would be expected to increase with latitude because overnight energy requirements increase with decreasing and more variable daily temperatures. Birds at higher latitudes thus should have more energy reserves at dawn than birds at lower latitudes. This prediction is confounded, however, by the lengthening of the dawn (and dusk) twilight period with increasing latitude (see any astronomical almanac). Many tropical as well as temperate birds have a dawn chorus as lengthy as the local twilight period (e.g., Skutch 1972, 1981, Sick 1993). Second, the dawn chorus should be most intense early in the season, but the dawn chorus of many species is absent or brief until after nesting has begun and temperatures are warmer (Highsmith 1989, Kroodsma et al. 1989, Staicer 1989). As well, the dawn chorus of tropical residents is strongly seasonal even though temperatures vary little during the year (Staicer 1996).

Nonetheless, dynamic programming provides a way to calculate how conflicting needs can be optimally balanced in daily routines, and, overall, the integration of mathematical modeling and observations of natural singing behavior appears to be a promising approach toward explaining the dawn chorus.

Summary

We summarize by considering how these hypotheses meet the three criteria for a good model: precision, realism, and generality (Levins 1966). As applied to the dawn chorus, these criteria are (1) precision: Do the hypotheses specify the timing of singing? (2) realism: Are the hypotheses testable? and (3) generality: Are the hypotheses applicable to a wide range of taxa?

Precision. Many of the hypotheses are weak on explaining timing. A useful hypothesis must show that its proposed function depends on singing before sunrise rather than address a general function of song that could be served by singing during the first several hours of daylight. Indeed, evidence for diel rhythms in the factors proposed to be driving the dawn chorus are poor in most cases, being based on one or two examples that were used to develop the hypotheses in the first place. This problem can be addressed by collecting data for other species representing a wide range of taxonomic affinities, nesting ecologies, social systems, and life histories. Still, the more precise an explanation of timing, the less likely it will generalize to all species and the more difficult it will be to test. For example, testing the low predation or handicap hypotheses will require detailed measurements of time-dependent predation rates or metabolic costs of singing, the importance of which are likely to vary among species.

Realism. Most of the hypotheses will be difficult to test, especially in the field. Much laboratory work is needed before hypotheses about the intrinsic and mate-stimulating functions of dawn song can be adequately tested in the field. For example, little is known about the relations of circadian changes in testosterone levels and song activity, or how circadian patterns in male song relate to female reproductive state. Hypotheses that suggest social functions for dawn song are testable in the field, although obtaining adequate sample sizes will be difficult. Hypotheses based on functions related to environmental pressures, such as whether acoustic transmission and predation vary during the day, might be easier to test. The many variables involved in models of tradeoffs between singing and foraging appear daunting, but some have been manipulated successfully in the lab. J. M. C. Hutchinson et al. (1993) provided useful suggestions for manipulating variables in the field.

Generality? None of the hypotheses fares very well on the criterion of generality. In nearly every case, diametrically opposed examples can be raised. A separate explanation for every species seems too specific, but the other extreme—attributing all temporal patterns in vocalizing to diel patterns in vague costs and benefits—does not aid our understanding of the dawn chorus. Overall, the more generally applicable a hypothesis is, the less precise will be its predictions for any given species' dawn bout, and the more difficult it will be to test.

Given the diverse ecologies and phylogenies of the species that sing at dawn, we consider it unlikely that any one hypothesis will provide a general explanation of dawn bouts and their variation. The timing, sharpness, intensity, and complexity of signaling during peak periods vary greatly among bird species. The functions of acoustic signaling in general, and of displaying at peak periods in particular, undoubtedly also vary among species. In the next section, we summarize common patterns as we explore this unexplained variety in greater detail.

Patterns

The 12 hypotheses discussed above attempt to explain why certain species sing most intensely at dawn. As we have suggested, however, the focus on dawn is rather narrow considering the sheer diversity of diel patterns of vocalizing among birds. The purpose of this section is to set dawn singing into a broader perspective. Here we survey a broad range of taxa and highlight features of diel patterns in singing that previous treatments of the hypotheses did not consider. This information is organized into five topics: (1) interspecific variety in the timing of diel peaks in acoustic signaling, (2) patterns in the timing of dawn and dusk choruses noted in early studies, (3) use of special signals at dawn that are qualitatively different from those used in daytime singing, (4) dawn choruses in the Tropics, and (5) variation in dawn singing as a function of social system.

A number of problems arise in any such survey. Much of the information is anecdotal, and it is seldom clear how much effort was made to quantify singing activity. Definitions of *dawn* have not always been consistent, and sometimes have included the period after sunrise (e.g., Catchpole 1973). For our modest purposes these limitations are not important; for example, although we point out phylogenetic patterns where obvious, we make no attempt to compare species quantitatively. We hope that further work on the dawn chorus, especially in comparative studies, will incorporate these patterns into broader functional interpretations.

Variation among species. Species vary in the sharpness and timing of the peaks in acoustic signaling that occur during their activity cycles (Fig. 24.3). First, the peak of vocalizing varies in sharpness from species that show little or no variation in song rate through the morning hours (e.g., Western Meadowlark, *Sturnella neglecta,* and *Vireo*) to those that sing much more at dawn than later in the day (e.g., Black-capped Chickadee and Western Bluebird, *Sialia mexicana;* Cosens 1984, Horn et al. 1992, S. L. Hopp pers. comm., W. J. Smith pers. comm., Horn pers. obs.). Second, not all diurnally active birds that show a peak of vocalizing have their peak before sunrise. Manakins (Pipridae), nunbirds (Bucconidae), and ovenbirds (Furnariidae) vocalize most at midday, and barbets (Capitonidae) peak in the late afternoon (Snow 1962, Skutch 1981, Sick 1993). Common Loons

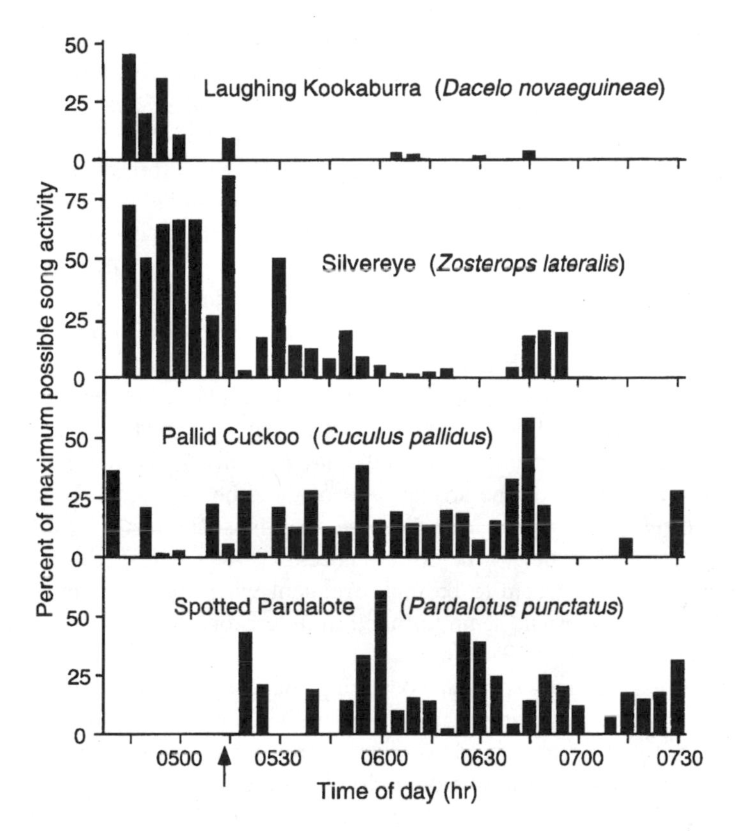

Fig. 24.3. Four species of the eucalypt forest of Australia (Hawkesbury River, New South Wales) exhibit diverse diel patterns in vocal activity. Bars indicate the percentage of each period that each species sang; data are five-day averages (6–15 October 1981, during the breeding season) for the first four hours of song activity per day. Arrow indicates time of sunrise. Daylong patterns (not shown): the Laughing Kookaburra sings mainly before sunrise and after sunset, the Silvereye produces a notable dawn and dusk chorus but also vocalizes at lower rates throughout the day, the Pallid Cuckoo sings at similar rates from dawn through dusk, and the Spotted Pardalote sings only during the daytime (no song at dawn or dusk). Modified from Keast (1985).

(*Gavia immer*) yodel mainly in the middle of the night, when wind is minimal; Great Snipes (*Gallinago media*) peak at midnight; and Marsh Wrens seem to peak near the middle of the night and again at dawn (Cramp 1983, Barclay et al. 1985, L. E. Wentz pers. comm.).

Nonetheless, a common pattern is for birds to vocalize most intensely at the start and end of their diel activity cycle. Many species that call at night tend to vocalize most frequently at dusk, on moonlit nights, and to a lesser extent at dawn, including tinamous (*Tinamus major* and *Crypturellus soui*), potoos (*Nyctibius griseus*), snipe (Scolopacidae), guans (Cracidae), and goatsuckers (Caprimulgidae; Skutch 1983, Sick 1993). Barn Owls (*Tyto alba*) and Boreal Owls (*Aegolius funereus*) also have dusk and dawn choruses (Bunn et al. 1982, Bondrup-Nielsen 1984).

Diurnally active nonpasserines that vocalize most at dawn and dusk include toucans (*Ramphastos swainsonii*), falcons (*Herpetotheres, Micrastur* species), chachalacas (Cracidae), and Whimbrels (*Numenius phaeopus:* Skutch 1972, Skeel 1978, Sick 1993, W. J. Smith pers. comm.). Notably, the dawn and dusk choruses of motmots (Momotidae) dominate certain tropical forests (Skutch 1983, Sick 1993). The Laughing Kookaburra (*Dacelo novaeguineae*), a cooperative breeder, sings mostly in dawn and dusk choruses that appear to function in defending the group territory (Reyer and Schmidl 1988). Among falconiforms, an apparent greeting display between pair members is common early in the morning (Cramp 1980), and male Cooper's Hawks (*Accipiter cooperii*) have a special dawn call (Rosenfield and Bielfeldt 1991).

The signaling peaks of some species seem easily explained by female activity. Lekking species tend to display simultaneously to enhance mating opportunities, and the displays tend to peak when the displaying individuals first gather at the lek, which may depend on the arrival time of the other sex (e.g., deVos 1983). Caciques (*Cacicus*) sing their long-distance song most frequently at dawn, presumably to attract females to the colony (Feekes 1981). The flight displays of various charadriiforms seem to be most frequent when males or females have peaks in other activities and seem to function in territory defense or mate attraction, respectively (Cramp 1983).

In the more gregarious species, vocalizing may peak when the potential for social interaction is highest—when many individuals are in close proximity. Common Swifts (*Apus apus)* and bee-eaters (*Merops*) are most vocal before and after roosting, probably because more individuals are present at communal roosts at those times (Lack 1957, Cramp 1985). The dawn and dusk peaks in vocalizing in the desert-dwelling sandgrouse (Pteroclididae) may be attributable to the timing and nature of their water-gathering behavior; they travel in large flocks to watering holes to transport water to their young in specialized breast feathers, which would dry out more rapidly during the day (Cramp 1983).

Timing of first and last songs. Many early studies of song activity patterns focused on the times of the first and last songs of the day, especially in north-temperate oscines, and noted the timing of songs in relation to environmental factors. Despite the large effort required to establish these patterns, their functional significance has received surprisingly little attention. Although reviewed in detail by Nice (1943) and Armstrong (1963), these patterns were overlooked in previous papers addressing the dawn chorus.

A most intriguing finding, already mentioned (Fig. 24.1), is a pattern that is widespread among birds that exhibit a dawn chorus. As the breeding season progresses, males begin singing earlier and earlier in relation to sunrise, then the pattern reverses after a certain date, which may coincide with the peak of breeding (e.g., Allard 1930, Davis 1958, Nolan 1978, p. 62, Reyer and Schmidl 1988, Staicer unpubl. data). In her studies of the annual cycle of the Song Sparrow

(*Melospiza melodia*), Nice (1943, p. 251) distinguished three stages in the seasonal progression of earlier timing of "awakening song" such that males began the day's singing after civil twilight (stage 1), close to civil twilight (stage 2), and before civil twilight (stage 3). This pattern is important to recognize because it translates into a marked seasonal change in singing at dawn—the lengthening of the dawn bout of an individual. Interestingly, the turning point along the curve is often close to the summer solstice, in both temperate and tropical areas (15° N; Staicer 1991, unpubl. data), suggesting that changing daylength is involved in the proximate generation of this pattern.

Ten other patterns can be gleaned from earlier publications:

1. The starting times for the various species in a community differ, and species typically join the chorus in a predictable order each morning (e.g., Wright 1913, Allard 1930).
2. These species sing their last songs of the day in more or less the reverse order of their first songs, though the timing of last songs is more variable (e.g., Wright 1913, Allard 1930, Leopold and Enyon 1961).
3. Light intensity appears to determine the time at which different species start, but most species start singing at dawn under lower light intensities than those at which they stop at dusk (e.g., Leopold and Enyon 1961).
4. Weather affects starting times, but the underlying factor appears to be light intensity (e.g., Allard 1930).
5. Start times for a species change with latitude, in concert with the time a certain light intensity occurs (W. Craig 1943).
6. Start and end times occur at higher light levels in the Arctic than at temperate and tropical latitudes, but throughout coincide with the time at which light intensity changes most rapidly (e.g., R. G. B. Brown 1963, Falls 1969, pers. comm.).
7. By and large, species within a given taxonomic group have similar singing phenologies; they begin singing at similar light levels (Armstrong 1963).
8. Species differ in their seasonal patterns of song activity (e.g., Cox 1944).
9. Song activity patterns typically change with nesting stage (e.g., Verner 1965, p. 129).
10. As the breeding season progresses, songs delivered at dawn and dusk constitute a larger fraction of the daily total song activity (Falls 1978, p. 67).

Light level is obviously an important proximal cue for the timing of acoustic signaling in birds, but, surprisingly, no one seems to have addressed why different species are cued by different light intensities. For instance, do species that forage in deep shade differ in sensitivity from species that forage in sunny places? In studying sensitivities to light levels, one should seek to distinguish between phylogenetic and ecological effects, as phylogenies may constrain a species' physiology. This behavioral response (song initiation) to light levels is likely mediated by a hormonal response, or at least interacts with levels of hormones (e.g., melatonin). The mechanism that accounts for the shift toward earlier start times as the breeding season progresses may hold clues to understanding why birds exhibit a dawn chorus.

Special dawn signals. Previous reviews of the dawn chorus did not address the fact that many passerines sing qualitatively differently at dawn and dusk in comparison with vocalizations they use in the daytime (Fig. 24.4). Below we summarize some of the obvious patterns and organize the information taxonomically, because notable patterns tend to be shared by related species.

Many tyrant flycatchers (Tyrannidae; including *Elaenia, Pachyramphus, Pitangus, Rhytipterna,* and other genera mentioned below), both temperate and tropical species, sing dawn songs that are distinctive and typically more elaborate than their daytime vocalizations (W. J. Smith 1966, W. E. Lanyon 1978, Skutch 1981, Sick 1993, E. Morton pers. comm.). The dawn song of *Myiarchus* flycatchers, which is delivered in a nearly unbroken sequence for 15–30 minutes, consists of a complex, highly stereotyped, and species-distinctive arrangement of calls, each of which may be used singly during the daytime (W. E. Lanyon 1978). Dawn bouts of Eastern Phoebes (*Sayornis phoebe*) differ from daytime singing in being a particular mix of daytime vocalizations delivered at a higher rate (W. J. Smith 1977, Kroodsma 1985a). The dawn bouts of Eastern Wood-Pewees (*Contopus virens*) are distinctive sequences of vocal units, only some of which are common during the day (W. Craig 1943). The dawn song (regularly repeated vocalization) of the Eastern Kingbird (*Tyrannus tyrannus*) is rarely sung during the day (W. J. Smith 1966).

Many species of thrushes (Muscicapidae) are noted for their dawn singing, but their dawn and daytime songs have rarely been compared, perhaps because both appear very complex. The dawn bouts of Eurasian Blackbirds build in intensity and quickly culminate in song elements that are typical of intense interactions; this singing is accompanied by much calling, chasing of other males, and courting of females (Dabelsteen 1992). At dawn, Common Nightingales (*Luscinia megarhynchos*) sing without obviously interacting, but at other times of day or night neighbors usually countersing with fairly precise timing and switch song types more often (Hultsch and Todt 1982). Western Bluebird dawn song incorporates syllables rarely heard at other times of the day except during close-range intraspecific interactions (Horn unpubl. data).

Black-capped Chickadees sing little during the day and repeat their songs exactly, with little variation in frequency except for occasional shifts of >100 Hz. At dawn, males repeat their songs fewer times between frequency shifts and shift up and down among several different frequencies (Horn et al. 1992). This behavior parallels that of many eventual variety singers, which tend to sing more rapidly and cycle through their song repertoires more quickly at dawn (e.g., Rufous-sided Towhee, *Pipilo erythrophthalmus:* Kroodsma 1971).

Many male wood-warblers use songs and singing behaviors at dawn that are noticeably different, often both qualitatively and quantitatively, from those used

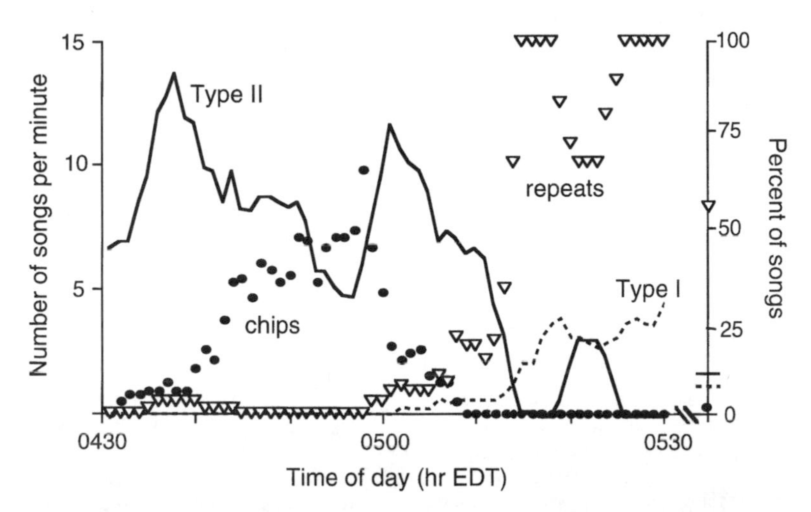

Fig. 24.4. The dawn bout of one male Yellow Warbler (*Dendroica petechia*) recorded on 29 May 1986 in Amherst, Massachusetts, shows that dawn singing can be qualitatively different from daytime singing. Data are five-minute running averages (based on data for each minute) for type I songs (broken line), type II songs (solid line), percentage of songs preceded by chip notes (filled symbols), and percentage of repetitions of the same song type (open symbols). Averages for the 4-hr daytime period (0530–0930; $N = 8$, five-minute samples) were type I and II rates of 0.8 and 1.2 songs per minute, with songs rarely preceded by chips (1%) and many song type repetitions (55%). Dawn singing was characterized by alternation of type II songs interspersed with chips, which were most frequent during the middle of the bout. Daytime singing was characterized by repetitious singing of type I songs and occasional bouts of type II song (but without chips). See Spector (1991, 1992) for further discussion of the differences between type I and II songs and their associated behaviors.

most commonly during the day. The male Ovenbird (*Seiurus aurocapillus*) has an elaborate extended song that combines part of his usual song with call notes and other syllables. This extended song, heard at dawn, dusk, and sometimes in the middle of the night, often accompanies a flight display and also occurs during close encounters with conspecific birds of either sex (Lein 1981).

Male *Dendroica* wood-warblers and closely related genera have two distinct categories of songs (reviewed in Spector 1992). First-category songs dominate daytime singing, are used in communication with females, and predominate in males that have not yet paired or have lost their mates; these songs rarely occur in dawn bouts. Second-category songs are used in territorial encounters between males and for the dawn bout during the breeding season. Second-category singing tends to be more elaborate than first-category singing (see, e.g., MacNally and Lemon 1985, Highsmith 1989, Spector 1991, Staicer 1989, in press), especially at dawn, when second-category songs tend to be delivered at a more rapid rate and with more immediate variety than during the day, and are interspersed with short chip notes rarely used in daytime singing (Staicer 1989, Spector 1992; Fig. 24.4).

The dawn vocalizing of a given male wood-warbler changes through the season and through the dawn period (Fig. 24.4). For example, each male Yellow (*Dendroica petechia*) and Adelaide's Warbler (*D. adelaidae*) has several song

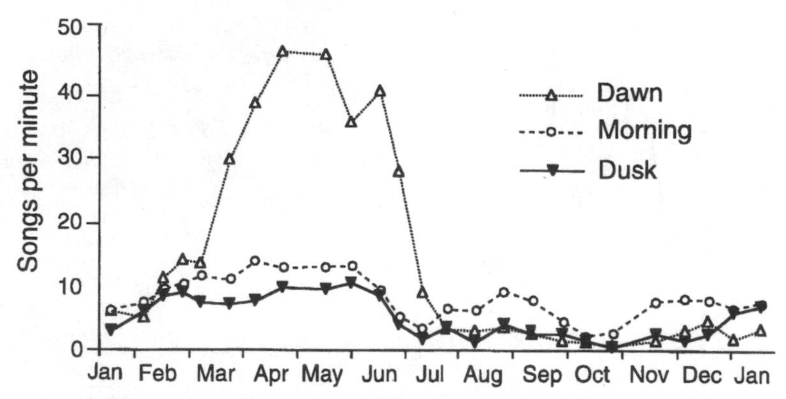

Fig. 24.5. The dawn chorus is restricted to the breeding season of the tropical resident Adelaide's Warbler (*Dendroica adelaidae*). Number of songs per minute given by a group of 10 neighboring males in southwestern Puerto Rico, January 1986–January 1987. Data are averages for each of three census periods: dawn (45 minutes before sunrise), morning (170 minutes after sunrise), and dusk (40 minutes before and 20 minutes after sunset). Note that males sang at considerably higher rates at dawn only during the breeding season (March–June). During the nonbreeding season, males began singing closer to sunrise and stopped singing closer to sunset (thus sang for a smaller fraction of the dawn and dusk census periods), and sang at similar minute-by-minute rates during the three periods; hence nonbreeding season rates averaged lower for dawn and dusk periods than for the morning period. Patterns were similar for individual males. Modified from Staicer (1991).

types in its second-category repertoire (Spector 1991, Staicer in press). On a given morning, song rates and song type switching increase gradually over the first several minutes, build to a peak, level off, and finally decline toward the end of the dawn bout, at around sunrise, when birds typically fall silent (the "sunrise lull") before their first daytime bout. Although this pattern is typical of the nesting period, the dawn bout is essentially absent early in the season. During territory establishment and pairing in the migratory species (Yellow Warbler), and during the nonbreeding season in the resident species (Adelaide's Warbler), males begin singing closer to sunrise and use primarily first-category songs, such that this presunrise singing is similar in character to after-sunrise singing (see also Fig. 24.5). The dawn bout of second-category songs develops as the breeding season progresses and individuals lengthen their bouts (see also Nolan 1978, Highsmith 1989, Kroodsma et al. 1989, Staicer 1989, 1991).

A diversity of special dawn singing behaviors is found among emberizine sparrows. The Field Sparrow (*Spizella pusilla*) has a system similar to that of *Dendroica* wood-warblers, with a distinctive dawn song that also is used in territorial encounters (D. A. Nelson and Croner 1991), and the closely related Brewer's Sparrow (*S. breweri*) sings extremely long, complex songs at dawn but typically gives short, simple songs during the day (D. E. Kroodsma pers. comm.). Male White-throated Sparrows (*Zonotrichia albicollis*) sing longer songs at dawn than during the daytime, and also when unpaired (F. E. Wasserman pers. comm.). Few tropical emberizines or cardinalines appear to sing at dawn; notable exceptions with distinctive dawn songs include *Cyanocompsa parellina, Charitospiza eucosma,* and *Coryphospingus cucullatus* (Sick 1993). Rufous-collared Sparrows infrequently sing unusually elaborate songs at night and when frightened during the day (Lougheed and Handford 1989, Sick 1993).

Many tropical tanagers have distinctive dawn vocalizations (Willis 1960, Moynihan 1962). The dawn songs of Red-crowned Ant-Tanagers, which are sung at a slower cadence than day songs, are sometimes used in territorial encounters during the day (Willis 1960).

Tropical species. The behavior and ecology of tropical birds differ in many ways from those of temperate species (see Morton, this volume). Relatively few tropical species have been studied in any detail, however, and information on their vocal behavior is especially scant (see Kroodsma et al., this volume: Chapter 15). Kunkel's (1974) review of mating systems of tropical birds notes the uncommonness of territoriality, loud song, and sexual dimorphism in vocal behavior. Kunkel also emphasized the commonness of permanent pair-bonds and duetting in tropical birds (Farabaugh 1982). Our tentative conclusions about the influence of duetting and territoriality plus breeding on the behavior of dawn singing are discussed below.

Many tropical species that duet (i.e., pair members closely coordinate their vocalizing) appear to lack a marked dawn chorus (Skutch 1981, Sick 1993). Some duetting tropical passerine species participate in a dawn chorus, though apparently only the males sing at dawn. Examples include *Tyrannus, Myiarchus,* and *Elaenia* flycatchers; a gnatwren (*Ramphocaenus melanurus*); a wren (*Thryothorus thoracicus*); and a sparrow (*Aimophila sumichrasti;* W. J. Smith 1966, Skutch 1968, 1972, Davis 1972, W. E. Lanyon 1978). In duetting Southern Boubou (*Laniarius ferrugineus*), the solo vocalizations given before sunrise (presumably by males) are of the same type as those used later in the day in intense territorial interactions (Harcus 1977). In some tropical nonpasserines that duet, including tinamous (Tinamidae), motmots, chachalacas, and wood-quails (*Odontophorus gujanensis*), both sexes apparently vocalize at dusk or dawn (Skutch 1983, Sick 1993).

The dawn chorus of tropical birds appears largely restricted to the breeding season. Examples include hummingbirds (lekking *Amazilia tzacatl* and *A. amabilis*), flycatchers (*Camptostoma obsoletum* and *Myiarchus tuberculifer*), thrushes (*Turdus grayi,* among others), woodcreepers (Dendrocolaptidae), ant-tanagers, and sparrows (*Aimophila sumichrasti* and *Tiaris bicolor*) (Willis 1960, Davis 1972, Skutch 1972, 1981, Sick 1993, Staicer unpubl. data). Many of these dawn-singing species are territorial; whether individuals maintain territories year-round is unclear.

On the island of Puerto Rico (15° N), tropical residents that maintain territories and pair-bonds throughout the year show a dawn chorus only during the breeding season (Fig. 24.5). The seasonal period during which Adelaide's Warblers, Grey Kingbirds (*Tyrannus dominicensis*), and Puerto Rican Flycatchers (*Myiarchus antillarum*) sing a dawn chorus coincides closely with their nesting period, the timing of which varies among species and years (Staicer 1991, and unpubl. data on color-banded individuals recorded throughout the year). Only males participate in the dawn chorus, and they use songs therein rarely heard outside the breeding season. During the daylight hours throughout the year, both sexes of these species are notably vocal but tend to use simpler or shorter vocalizations, often in duetlike exchanges. Similar observations have been made for Tropical Kingbirds (*T. melancholicus;* W. J. Smith pers. comm.) and Red-crowned Ant-Tanagers on the mainland (Willis 1960).

Social system. Interspecific variation in diel patterns of vocalizing may be attributable to differences in social systems, such as the presence of territoriality and its nature (e.g., exclusiveness, energy expended, and type of territory). Birds that defend nesting plus foraging areas (e.g., wood-warblers and many tyrant flycatchers; see above) often show a marked peak around dawn and dusk but also vocalize regularly during normal hours of activity, day or night, depending on the species. Birds that defend mainly their nest and its immediate vicinity (e.g., *Acrocephalus*) seem to be more likely to vocalize predominantly at dawn or dusk, if at all, after pairing (Catchpole 1973). Highly social species (e.g., some swallows) tend to show early and late peaks in song and contact calls, at times when most individuals are at the colony site. Cardueline finches typically do not defend territories and appear to lack a dawn chorus (Thorpe 1958).

The vireos stand as a counterexample to the many insectivorous territorial species discussed above. Vireos seem not to have distinctive dawn bouts, start singing closer to sunrise than many other sympatric species, and lack a notable diel peak in song activity, although White-eyed Vireos (*Vireo griseus*) sing at elevated rates for the first few minutes of the morning (S. L. Hopp pers. comm., W. J. Smith pers. comm.). Unusual among temperate species, vireos maintain a relatively high rate of singing through the day; perhaps sustained singing at even higher rates at dawn is difficult or somehow less necessary (W. J. Smith pers. comm.).

Although few researchers look specifically at which sex is vocalizing at dawn, the available information suggests that dawn singing is largely a male behavior. At all peak signaling times, males tend to display more frequently than females (which often do not display). Of interest is whether this trend will hold for the Tropics, where females tend to vocalize as much as males during the day (see Morton, this volume). A sexual difference in dawn singing would suggest that competition for the limiting sex (usually females) is an important factor in causing diel patterns of signaling overall, an idea that could be tested with polyandrous species, in which females defend territories and males (the limiting sex). Alternatively, females might be silent at dawn because during the breeding season they would often be sitting on the nest then, and vocalizing would announce its location to predators, brood parasites, and neighbors. Sexual differences in diel signaling patterns and their relationship to social systems clearly deserve more attention.

The effects of social systems on dawn singing could be tested in a number of ways. Diel patterns of signaling could be compared across species that differ in whether they defend any territory and its particular characteristics. Tropical birds would be good subjects because of the large number of species and the great diversity of social systems in the Tropics (see, e.g., Kunkel 1974). Also, many species consist of populations that vary in the degree of territoriality as a result of differences in the environment (Lott 1991). Comparative studies could focus on series of populations of the same or different species arrayed along gradients from multipurpose territoriality to coloniality. In general there is much potential for comparative work relating to various aspects of the diel patterns of acoustic signaling in birds.

Toward a Synthesis

We now attempt a synthesis of hypotheses and patterns by examining whether the hypotheses for functions of dawn singing discussed earlier can predict the patterns in diel signaling identified above. Table 24.1 condenses these signaling patterns into three widespread patterns that have not been recognized previously. The first two, and the conclusions we draw from them, seem to apply mainly to territorial species. We have discussed numerous cases throughout this chapter in which other hypotheses may provide sufficient explanations for diel patterns of particular species that do not defend all-purpose territories (e.g., colonial species), lack pair-bonds (e.g., lekking species), or have an unusual ecology (e.g., sandgrouse). Below we focus on the dawn chorus as an example of a diel peak, but we view the hypotheses broadly with regard to predicting a peak at any time of day or night.

The first pattern includes a seasonal trend widespread among unrelated species: the diel peak in signaling is restricted to and develops through the breeding season (Table 24.1). The dawn chorus of both migratory and resident territorial species develops more or less gradually (often being absent at the start of the breeding season) as the season progresses. In both temperate and tropical species that maintain territories, and often pair-bonds, year-round, the dawn chorus is restricted to the breeding season. These trends are consistent with the circadian cycles and self-stimulation hypotheses because changes in gonad size, testosterone levels, and dawn singing show similar seasonal trends. This seasonal pattern is also consistent with the social dynamics hypothesis because the number and variety of social interactions (territorial and sexual) should increase as the season progresses, and would also be consistent with the territory defense hypothesis provided that a male's territory becomes more valuable as the season progresses (data are needed to verify this assumption).

Not supported by the first pattern are the hypotheses that predict a seasonal timing of the dawn chorus that differs from what is generally observed for territorial species. The mate attraction hypothesis predicts that the dawn chorus is

Table 24.1. Hypotheses for the function of diel peaks in acoustic signaling (e.g., a dawn chorus for diurnal species) and whether they predict three widespread patterns

	Pattern		
Hypothesis	Peak restricted to and develops through the breeding season	Qualitatively different signals used during peak	Diel pattern varies among species in relation to variation in . . .
Circadian cycles	Yes	Yes	—
Self-stimulation	Yes	Yes	Social interactions
Mate attraction	No, predicts different timing	No, signals not used in mate attraction	Mating success
Mate stimulation	No, predicts different timing	Yes, if signals more stimulating to mates	—
Mate guarding	No, predicts different timing	Yes, if signals better at repelling intruders	EPC threat
Territory defense	Yes, if territory value changes	Yes, if signals used in territorial conflicts	Territorial intrusions
Social dynamics	Yes	Yes	Social interactions
Handicap	—	—	—
Low predation	—	—	No, predicts variation with type of predator
Acoustic transmission	—	Yes, if signals travel farther	Transmission properties of signal
Inefficient foraging	—	—	Type of food
Unpredictable conditions	No, predicts different timing	—	Latitude, body size, type of food

Notes: Yes = the hypothesis seems to predict the observed pattern; no = the hypothesis seems to be contradicted by the observed pattern; — = the hypothesis makes no clear prediction relevant to the observed pattern.

best developed (not least developed, as observed) during the time of pairing. The mate stimulation hypothesis predicts that a male's dawn bout should peak earlier in the season than it does (it usually continues to lengthen after egg laying) and that individual males should show increases in their dawn bouts when renesting. The mate-guarding hypothesis predicts that a male's dawn bout will peak during his mate's fertile period (but note that no study has convincingly shown that the timing of a male's dawn bout is not better correlated with the fertile periods of the majority of females in the population). The unpredictable conditions hypothesis predicts a more robust dawn chorus early in the season, when conditions are harsher and more unpredictable (the opposite of the pattern observed). If the handicap, low predation, acoustic transmission, and inefficient-foraging hypotheses predicted seasonal changes in dawn singing, these would probably be the reverse of the patterns observed.

The second widespread pattern includes diel trends we have noted within individual birds: different signals are used during the peak period of signaling (Table 24.1). Dawn signals of many species are characterized by qualitative (different vocalizations, arranged differently in sequence) as well as quantitative differences (more vocalizations per unit time in bouts of longer duration) compared with those used during the daytime. If a hypothesis regarding the function of the dawn chorus is to be generally useful, it must therefore explain why the dawn bout is qualitatively, not just quantitatively, different from daytime singing.

Use of qualitatively different signals at dawn is consistent with predictions of the circadian cycles hypothesis (a dawn peak in hormone levels may result in the use of special vocal signals), the self-stimulation hypothesis (the signals used at dawn may be more effective at raising the singer's hormone levels), and the social dynamics hypothesis (dawn signals seem to carry more information and enable more interactions). This pattern also seems generally consistent with the territory defense hypothesis, as many species use the same special signals at dawn and in territorial conflicts later in the day. Use of special songs at dawn might also be consistent with the mate stimulation, mate-guarding, and acoustic transmission hypotheses, if these signals are more effective for those functions (data are needed to test these assumptions). The mate attraction hypothesis is not supported by the pattern, as the signals used at dawn generally do not seem to be the same signals used for mate attraction during the day. The handicap, low predation, inefficient-foraging, and unpredictable conditions hypotheses do not seem to predict that qualitatively different songs should be used at dawn.

The third major pattern refers to the diversity of diel signaling patterns we have noted among bird species: there is variation in the timing, sharpness, or nature of diel peaks, and social systems and diel signaling patterns may be linked (Table 24.1). Because species differ ecologically in many ways, most of the hypotheses should predict some kind of variation among species; all but three of the hypotheses seem to generate clear predictions about what general factor(s) would be expected to correlate with variation among species in diel signaling. For the most part, however, the data needed to test these predictions are not available. The observed patterns seem inconsistent only with the low predation hypothesis, which fails to predict the variation found among species (e.g., wood-warblers and vireos) preyed on by the same types of predators.

In our view, the self-stimulation hypothesis and the social dynamics hypothesis provide the best explanations for these three widespread patterns. Seasonal trends in dawn singing and the use of different songs and more complex singing behaviors at dawn are consistent with both hypotheses. Both predict that variation in diel peaks among species or populations is related to the intensity with which individuals in a population interact socially. We think these two hypotheses might have complementary functions—self-stimulation at the physiological level and social dynamics at the social level. By participating in the dawn chorus, a bird matches its internal state to that of its neighbors, obtains information about its neighbors, reveals its ability to make and respond to vocal challenges, and adjusts its complex web of social relationships. The information revealed by the singers may influence the behavior of listeners, including nonsinging females and floaters, long after the dawn chorus has ended.

Conclusions

Species differ substantially in their diel patterns of acoustic signaling. Some of this diversity might have relatively simple explanations, such as a tendency to vocalize more at the beginning and end of a diurnal or nocturnal activity period. In other species, peaks in vocal activity are clearly associated with peaks in social interactions. Among lekking species, the peak of male vocalizing tends to coincide with peaks in female presence or activity (though we must further distinguish between male displays prompted by the presence of females and those performed to attract females). In gregarious species that nest or roost in groups, vocal activity often peaks early and late in the day, when the largest number of individuals are present. Why other species show no peak during their diurnal or nocturnal activity period is unclear. Furthermore, such simple explanations do not explain why singing at dawn is so distinctive, complex, and interactive.

Singing at dawn, as in the daytime, might function intrasexually, or intersexually, or both. The dawn chorus appears to be most developed among territorial species, suggesting an intrasexual function related to territory defense. Yet, among tropical as well as temperate species that are territorial, daytime singing coincides seasonally with territory maintenance, whereas the dawn chorus coincides seasonally with breeding. These two seasons overlap extensively in migratory populations but are obviously distinct for permanent residents. The closer temporal association of the dawn chorus with breeding activity rather than territoriality suggests a special intersexual function, but it might also reflect a seasonal change in the value of territories or mates. Two of the hypotheses we have proposed (self-stimulation and social dynamics) attempt to take into account the fact that territorial and sexual functions may be linked through social interactions.

Determining the significance of the various patterns described above will require, above all, detailed studies of variation among and within individuals. Future studies must attend to diel and seasonal patterns *within* the dawn bout of an individual, as these are sources of variation that may greatly influence the data collected for a given individual and population. For example, calidridine sandpipers are often said to give flight displays throughout the nesting season, yet among many species each individual ceases to display immediately on procuring a mate; season-long patterns for the population thus do not mirror those for individuals (E. H. Miller 1979b). Attention to factors that influence individual variation in singing, such as time of day, season, and mating or nesting status, is extremely important in studies that seek to compare male "quality" based on signal characteristics. Experiments could help to explore how various intrinsic, social, and environmental factors influence the dawn signaling of an individual.

Functional explanations need to be extended to encompass the various patterns regarding the dawn chorus and other diel patterns in signaling that we have highlighted. A recent review suggests that "ornithologists should also more thoroughly study when birds sing, total song output, and song bout length (and in particular assess the fitness consequences), rather than solely examining song structure and repertoire size" (J. M. C. Hutchinson et al. 1993, p. 1174). We hope that our review has shown that ornithologists have long been paying attention to the details of the singing behavior of birds, and that much information is already available on dawn singing—as well as other diel patterns—that could be brought to bear on functional questions.

Progress lies in closer examination of diel patterns in signaling. Detailed studies of moment-by-moment changes in the intense vocal interactions between individuals during peak periods of signaling, such as dawn, should add substantially to our knowledge of vocal behavior. Understanding what regulates this peak in signaling is crucial for studies that seek to determine how behavior, hormones, and environmental cues are integrated. Much potential exists for comparative studies to correlate these details of signaling patterns with phylogenetic relationships of signalers—their ecology (e.g., foods, habitats), social system (e.g., mating or pairing systems, pair-bond duration, territoriality), and life history (e.g., sedentary or migratory, tropical or temperate). Until we better understand diel patterns in signaling, such as the dawn chorus, we will not fully understand why birds sing.

Acknowledgments

For suggestions and information we thank M. Dennison, J. B. Falls, C. E. Hill, S. L. Halkin, S. L. Hopp, D. E. Kroodsma, M. L. Leonard, E. H. Miller, E. S. Morton, L. Ratcliffe, G. Ritchison, R. N. Rosenfield, W. J. Smith, F. E. Wasserman, L. Weilgart, L. E. Wentz, and an "energetic" anonymous reviewer.

25 Song and Female Choice

William A. Searcy and Ken Yasukawa

The foremost problem today in the burgeoning field of sexual selection concerns how mate preferences evolve. Research on birdsong can be directly applied to this problem because female attraction is thought to be one of the primary functions of male song. Some progress has been made toward understanding the evolution of female preferences for song, but much more is needed. Our goal in this chapter is to stimulate further research on this issue by framing questions that need to be asked, by reviewing what is known, and by suggesting promising avenues of investigation.

Kirkpatrick and Ryan (1991, p. 33) defined a female preference as "any trait that biases the probabilities that females mate with different kinds of males," pointing out also that a preference "does not necessarily involve any cognitive process." Female preferences, when they exist, often set up sexual selection acting on males. The appropriate female preference is therefore capable of explaining the evolution of almost any discernible trait in males, however bizarre. But why should females prefer males with certain traits, bizarre or not? The easiest answer to this question is that females should not and do not show such preferences, an answer favored by early critics of female choice, such as Wallace (1889). Simple denial of female preferences is no longer tenable, however, in the light of abundant new evidence that such preferences do exist. Our first goal in this chapter is to review evidence for female preferences based on song; evidence for female preferences in general was reviewed by Andersson (1994).

Admitting the existence of female preferences, we are faced with the problem of explaining their origin and maintenance. A large and thorny body of theory has grown up around this problem since the mid-1970s. A second goal of this chapter is to distill from the theoretical literature the lessons appropriate to birdsong research. We seek to identify areas of consensus among the competing hypotheses and to identify models or ideas that are particularly applicable to song preferences in birds. Our third goal is to review the existing evidence bearing on the evolution of song preferences. We attempt throughout to suggest investigations that might lead to a greater understanding of the evolution and maintenance of song-based female preferences, and hence of female mating preferences in general.

Evidence for Song Preferences

We analyze the evidence for song preferences from a hypothetico-deductive viewpoint. The hypothesis that females' preferences are based on males' songs generates a number of testable predictions. Some tests provide indirect (weaker) evidence, and others more direct (stronger) evidence, but in each instance a test should be designed to falsify, if possible, the hypothesis of female preference. Results contrary to the predictions can be interpreted as evidence against song-based female preferences, whereas results consistent with the predictions support the preference hypothesis. In the latter case, we cannot "prove" the preference hypothesis, because other, unstated alternatives may make the same predictions, but repeated support of the predictions should give us considerable confidence that a preference exists. We organize our discussion around types of evidence rather than categories of song features, starting with fairly weak types of evidence and moving on to stronger ones.

Correlations between female behavior and male song. In many species, males sing more before they pair than after; examples include Song Sparrows (*Melospiza melodia:* Nice 1943), European Pied Flycatchers (*Ficedula hypoleuca:* von Haartman 1956), and Sedge Warblers (*Acrocephalus schoenobaenus:* Catchpole 1973; additional examples are given in Møller 1991b). This pattern is often used as evidence that singing functions in mate attraction, but the argument is indirect on two counts. First, the evidence that pairing causes the decline in singing is only correlational; some other event (e.g., a seasonal reduction in trespass rates) may be the real cause. Second, even if pairing is the cause of the drop in singing, it is still not necessarily true that males sing before pairing to attract a mate (e.g., males may forgo singing after pairing in order to bring food to first the female and then the young). In other species, the peak in singing occurs after pair formation, at about the time of fertilization (Møller 1991b). This pattern has been used to support the argument that song functions in part to stimulate females to perform extra-pair copulations (EPCs; Møller 1991b).

Changes in male song have also been associated with shorter-term changes in female behavior. For example, male Northern Mockingbirds (*Mimus polyglottos*) sing more variable songs when associating with a female than when not (Derrickson 1987), and male Red-winged Blackbirds (*Agelaius phoeniceus*) switch between the song types in their repertoires more often with a receptive female present than without (D. G. Smith and Reid 1979). Male Common Starlings (*Sturnus vulgaris*) sing immediately before copulating with mates or other females (Eens and Pinxten 1990). These observations are consistent with female preferences based on song, but they do not constitute strong evidence for such preferences.

Manipulation of female behavior affects male song. This category of evidence differs from the previous one in that the researcher manipulates female behavior rather than simply studying natural variation. Researchers removed mated females and showed that the males then increased song production in Great Tits (*Parus major:* Krebs et al. 1981), Common Starlings (Cuthill and Hindmarsh 1985), Red-winged Blackbirds (Searcy 1988a), Northern Mockingbirds (Logan and Hyatt 1991), and Black-capped Chickadees (*Parus atricapillus:* Otter and Ratcliffe 1993), among others. These studies provide direct evidence that the absence of the female is responsible for the increase in song production, but the evidence for female preferences based on song is still indirect; other explanations for the increase in song production remain possible.

Also included in this category are experiments in which males altered their song production qualitatively, rather than just quantitatively, in response to experimental manipulation of female proximity or behavior. Captive male Cuban Grassquits (*Tiaris canora*) increased their use of a particular song type when presented with a female (Baptista 1978). Free-living male Chestnut-sided Warblers (*Dendroica pensylvanica*) increased use of one of their two song categories when their mates were removed (Kroodsma et al. 1989). Male Red-winged Blackbirds greatly increased the rate of switching between song types when presented with taxidermic mounts of courting females (Searcy and Yasukawa 1990). Captive male Common Starlings quadrupled their song output when confronted with a female (Eens et al. 1993). These results are consistent with female preferences for particular aspects of male song, but the evidence is still indirect.

Correlations between male song and mating success. Natural variation in male singing behavior can be related to some measure of differential mating success. A danger with this type of evidence is that a correlation between song and mating success may not indicate any causal relationship, but may instead be due to the indirect effects of some third variable, which may or may not be identified by the researcher. Howard's (1974) pioneering work on Northern Mockingbirds provides an example. Howard found that males with large repertoires tended to pair early. Pairing date was also correlated with territory quality, and when variation in territory quality was controlled, the correlation between pairing date and repertoire size was no longer significant. Similarly, Yasukawa et al. (1980) showed that repertoire size in Red-winged Blackbirds is correlated with harem size, but also with age; when age was controlled, the correlation between repertoire size and harem size was no longer significant. Catchpole (1986) found a correlation between syllable repertoire size and harem size in the Great Reed-Warbler (*Acrocephalus arundinaceus*), although again, the correlation was not significant when territory quality was controlled.

Correlations between male song and mating success have been found in other species, with no known confounding variables. Examples include the following:

1. A positive correlation between song rate and number of mates in the promiscuous Village Indigobird (*Vidua chalybeata:* R. B. Payne and Payne 1977)

2. Earlier pairing by males with larger syllable repertoires in Sedge Warblers (Catchpole 1980)
3. A correlation between frequency and temporal measures of a whistle display and mating success in male Sage Grouse (*Centrocercus urophasianus:* R. M. Gibson and Bradbury 1985)
4. Earlier pairing by males with higher singing rates in Willow Warblers (*Phylloscopus trochilus:* Radesäter et al. 1987)
5. A negative correlation between relative mating success and the interval between cork notes, holding other measures of male phenotype constant, in Sharp-tailed Grouse (*Tympanuchus phasianellus:* Gratson 1993)
6. Earlier pairing by male Common Starlings with long song bouts and large song repertoires—both in the field, where male arrival date was a confounding factor, and in the laboratory, where male arrival date was controlled (Eens et al. 1991)
7. A preference for approaching and performing EPCs with males exhibiting high song rates in captive Zebra Finches (*Taenopygia guttata*), with beak color controlled (A. E. Houtman 1992, Collins et al. 1994)
8. Larger syllable repertoires among males successful in extra-pair fertilizations than among unsuccessful males in Great Reed-Warblers, with territory attractiveness and various male traits controlled statistically (Hasselquist 1994).

Note that some of these studies controlled other measures of male phenotype and some did not, but even where some were controlled it is still possible that the important confounding variables were missed.

Female response to song playback. This category comprises experiments in which females were exposed to playback of songs and their responses noted. The designs of many of these playback experiments have been criticized, in particular for using too few exemplars of the test stimuli (Kroodsma 1989a, McGregor et al. 1992a). We agree that it would be worthwhile to replicate many of these experiments using more exemplars, but we nevertheless place some confidence in the conclusions of these experiments—more than we do in the results of most purely observational studies. Other methodological issues involving playback experiments were discussed by Kroodsma (1986, 1990) and McGregor et al. (1992a).

Three general types of female responses have been taken as evidence of female preferences: approach, copulation solicitation, and nesting behavior. In one of the first attempts to measure female response to song, Kroodsma (1976a) observed nesting behavior in female Island Canaries (*Serinus canaria*) exposed to playback. Female canaries gathered more string for nest building, and subsequently laid more eggs, when played a large repertoire of syllable types than when played a small repertoire. The conclusion, that females prefer large repertoires, is weakened by the fact that the large repertoire contained certain song types that were not in the small repertoires, so preferences for particular song types can also explain the results.

A few researchers have measured phonotaxis in the field. Eriksson and Wallin (1986) showed that female European Pied Flycatchers and Collared Flycatchers

(*Ficedula albicollis*) approached and entered nest boxes from which conspecific song was played in preference to silent control nest boxes. Mountjoy and Lemon (1991) obtained similar results with Common Starlings. R. M. Gibson (1989) showed that female Sage Grouse were more likely to enter a specific lek territory on days during and immediately after playback of the acoustic display of a successful male than on control days. Female House Wrens (*Troglodytes aedon*) not only approached nest boxes associated with song playback in preference to silent controls, but also competed aggressively for control of speaker-equipped boxes and initiated nest building in them (L. S. Johnson and Searcy, unpubl. data). All the studies in this category of field phonotaxis experiments demonstrate that females prefer song to no song, but they do not demonstrate preferences for certain songs or singing behaviors over others.

Phonotaxis in captive females has also been used as evidence for female preferences. D. B. Miller (1979a, b) demonstrated that female Zebra Finches preferentially approached their mates' songs and their fathers' songs over songs of other males. Similarly, female Zebra Finches preferred to approach songs familiar from tutoring early in life over unfamiliar songs (Clayton 1988) and songs played at high rates over low rates (Collins et al. 1994). A limitation of such evidence is that the function of phonotaxis is not always clear. For example, the preference of female Zebra Finches to approach their fathers' songs may well reflect a social rather than a mating preference. Captive female pied flycatchers previously treated with estradiol preferred to approach nest boxes from which large repertoires were played over nest boxes associated with small repertoires (Eriksson 1991). Approach to a nest box seems more certain to reflect a mating preference than does approach to a loudspeaker.

The most widely used technique for measuring female song preferences is the solicitation display assay. Copulation solicitation is a display given by female birds of many species immediately before and during copulation; it thus seems logical to interpret the display as an indicator of female mating preferences. In some passerines, untreated captive females will give solicitation displays in response to song playback (King and West 1977, West and King, this volume), but females of most species must be treated with estradiol before they will respond (Searcy 1992a). The number, duration, or intensity of displays given by females in response to contrasting playbacks can be used to infer female preferences.

The solicitation display assay has been used to demonstrate preferential response to conspecific over heterospecific songs in a number of species (Searcy 1992a). More interesting in the present context are results on discrimination between songs or singing patterns within the conspecific range of variation. Two discrimination tasks have been studied extensively using the solicitation display assay: response to variation in repertoire size, and response to home versus alien song dialects. Song Sparrows (Searcy and Marler 1981, Searcy 1984), Sedge Warblers (Catchpole et al. 1984), Great Reed-Warblers (Catchpole et al. 1986), Great Tits (M. C. Baker et al. 1986), Yellowhammers (*Emberiza citrinella:* M. C. Baker et al. 1987a), Red-winged Blackbirds (Searcy 1988b), Zebra Finches

(Clayton and Pröve 1989), and White-rumped Munias (*Lonchura striata:* Clayton and Pröve 1989) all preferred larger numbers of song types or song syllables over smaller numbers. Some of these studies controlled for the possibility of female preferences for particular song types and some did not; regardless, the preponderance of evidence indicates that female preferences for more variable song are quite general. Brown-headed Cowbirds (*Molothrus ater:* King et al. 1980), White-crowned Sparrows (*Zonotrichia leucophrys:* M. C. Baker et al. 1981a, 1982, 1987b, M. C. Baker 1983, Casey and Baker 1992), Yellowhammers (M. C. Baker et al. 1987a), Red-winged Blackbirds (Searcy 1990), and Zebra Finches (Clayton and Pröve 1989) all preferred home dialects to alien ones. These experiments are particularly vulnerable to the criticism of having used too few exemplars of the test stimuli, because they thus lose control over whether discrimination is based on dialect rather than on some other difference between the stimuli, but again the overall pattern of results supports a general preference for the home dialect.

Other preferences indicated by the solicitation display assay include (1) a preference for the songs of dominant males over songs of subordinates in Brown-headed Cowbirds (West et al. 1981), (2) a preference for long songs over short songs in Great Reed-Warblers (Catchpole et al. 1986), (3) a preference for more song over less in White-throated Sparrows (*Zonotrichia albicollis:* Wasserman and Cigliano 1991), (4) a preference for foster father's songs over natural father's songs in Zebra Finches (Clayton 1990b), and (5) preferences for songs of particular breeds in canaries (Kreutzer and Vallet 1991, Vallet et al. 1992).

Experimental manipulation of male singing. One of the most direct ways to demonstrate that male song affects female choice is to manipulate how males sing and then measure the mating success of the experimental males relative to controls. The best example of such a study is that by Alatalo et al. (1990) on pied flycatchers. The authors manipulated song rates early in the spring by providing experimental males with food in the form of mealworms. Experimental males sang at higher rates and attracted females significantly earlier than did control males. The results support a female preference for higher rates of song, but alternative explanations exist—for example, that provisioning altered other male behaviors that affect female choice, or that females sampled the extra food directly. We think this type of design should be used more extensively, although it is not clear how certain song attributes, such as repertoire size or dialect, could be manipulated.

Combining evidence. The most convincing cases for song preferences are made when evidence is compiled from more than one of our categories. We give four examples. In the Common Starling, song production decreases after pairing and increases again if the female is removed (Cuthill and Hindmarsh 1985). Males sing before both within-pair and extra-pair copulations (Eens and Pinxten 1990), and captive males sing more songs and more song types when presented with a female than during control periods (Eens et al. 1993). Males with longer song bouts and more song types pair earlier both in the field and in laboratory experiments (Eens et al. 1991). Preferences for long bouts and large repertoires are thus strongly suggested.

In Sedge Warblers, song production decreases dramatically after pairing (Catchpole 1973). Males with large syllable repertoires pair earlier than males with smaller repertoires (Catchpole 1980), and courtship behavior in captive females increases with the number of syllable types played to them (Catchpole et al. 1984). Female preferences for large repertoires are indicated.

In Red-winged Blackbirds, males whose first mates have been removed produce more song than control males (Searcy 1988a). Males increase their rate of switching between song types during natural courtship of females (D. G. Smith and Reid 1979) and when experimentally confronted with a mount of a courting female (Searcy and Yasukawa 1990). Captive females court more when multiple song types are played back than for single song types (Searcy 1988b). These results suggest female preferences for multiple song types, but the preferences may be effective only in within-pair courtship and EPCs, and not in pairing (Yasukawa et al. 1980, Searcy and Yasukawa 1995).

Our fourth example is the Great Reed-Warbler. Captive females of this species court more in response to large syllable repertoires than small ones (Catchpole et al. 1986). In the field, male syllable repertoire size does not affect pairing success (Catchpole 1986, Hasselquist 1994) but does affect success in extra-pair fertilizations; males successful in extra-pair fertilization have larger syllable repertoires than unsuccessful males (Hasselquist 1994). Female preferences for large syllable repertoires are again indicated.

Review of Theory

The first step usually taken in reviewing theories on the evolution of female preferences is to categorize the multitude of mechanisms that have been suggested. A variety of categorization schemes have been used; ours is based on that of Kirkpatrick and Ryan (1991). In our scheme, the major categories are meant to be exhaustive, but the minor ones are not. The major categories are defined by the type of selective forces acting to promote the evolution of the mating preference per se: the preference may be favored by direct selection, by indirect selection, or the preference may be nonselected.

Direct Selection

Direct selection of the female preference means that females possessing the preference have higher survival or fecundity than do females lacking the preference. Preferences could exert direct effects on female fitness in a variety of ways, including (1) facilitating species recognition—maximizing the probability of choosing a male of the correct species; (2) reducing costs of choosing—

minimizing mate choice costs such as energy expended and risks encountered while searching for and evaluating potential mates; and (3) obtaining superior resources—increasing the likelihood of obtaining a male that will provide superior resources to the female and her young. In the latter case, the resources may be provided by the male's territory or by the male himself in paternal care.

A hypothetical example may help to illustrate the operation of direct selection on a female song preference. Suppose that, in a particular species of bird, some females show a mating preference for males with large song repertoires. Direct selection of the preference occurs if females expressing the preference have higher fecundity or survival than females that mate at random. This direct advantage could occur because males with large repertoires are more likely to be of the correct species than are males with small repertoires, because large repertoire size facilitates mate finding and evaluation, or because large repertoire size indicates that a male has a superior territory or will provide superior paternal care.

Theoreticians emphasize indirect selection rather than direct selection in the evolution of female preferences, but only because indirect selection poses the greater theoretical difficulties. This emphasis should not be taken to mean that direct selection of preferences is judged to be less likely than indirect selection. To the contrary, genetic models indicate that female choice for traits increasing female fecundity is to be expected, when such traits exist (Kirkpatrick 1985). In a review of theory, Maynard Smith (1991, p. 146) concluded that "in territorial species, and in species with male parental care, female choice is likely to evolve mainly because of the immediate effects of the male's phenotype on her fecundity, and not because of the effect on the genotypes of her offspring."

Most birds, and especially most passerines, fall into the category of species with territories and male parental care, and so might be expected to obey Maynard Smith's dictum of choice based on traits directly affecting fecundity. Even in species such as these, however, mate choice might respond exclusively to indirect, genetic benefits in one particular context—namely, EPCs. When females choose to copulate with males other than their own mates, neither territorial resources nor paternal care is at stake, so any preferences that are expressed may be for indirect genetic benefits. In this respect, mate choice during extra-pair copulation resembles choice in promiscuous lekking species, another system in which direct benefits are not expected.

A remaining theoretical difficulty concerns whether song can serve as an honest indicator of resource or paternal care quality. Zahavi (1975, 1977) proposed that animal signals could evolve to advertise honestly either genetic or phenotypic quality as long as the signal entailed some cost to the signaler. The cost prevents individuals of low quality from dishonestly signaling high quality. These ideas have prompted considerable debate, especially regarding their application to female choice based on indirect genetic benefits. Grafen (1990a, b) modeled a situation in which the male trait signals phenotypic quality and concluded that honesty of the signal and female preference for males with the signal can be an evolutionarily stable strategy. Grafen (1990a, p. 487) pointed out that under such a handicap model, there should be "rhyme and reason" as to which signals affect female choice, so that "by studying what organisms want to know about their mates, and how signals impose costs, we could in principle explain the diversity of sexual selection." We would therefore expect females to prefer aspects of song that impose costs on males and reveal something about territory or paternal care quality. As an example, high rates of song production might impose severe time and energy costs and reveal either the territory's quality as a food source or the male's foraging ability, and hence his ability to bring food to the young.

Direct selection encompasses preferences minimizing the costs of choice and the risks of hybridization as well as preferences based on resources and paternal care. Choice can be expected to have some cost, such as energy spent and risks incurred while sampling males, the cost of delaying breeding, and the possibility of losing access to certain males while the choice is being made. Preferences selected to minimize time spent and distance traveled during choice are therefore a possibility. Hybridization is currently a risk in certain groups of birds, and must have been a risk in the past in many more, so preferences shaped to minimize hybridization are also possible.

Indirect Selection

Indirect selection of the female preference means that selection does not favor the preference directly, but rather favors some trait with which the preference is genetically correlated (Kirkpatrick and Ryan 1991). As selection exaggerates the correlated trait, the preference evolves also. Another way to view indirect selection is that the fitness effects of the preference are registered in the female's offspring rather than in the female herself.

Two main subcategories of mechanisms for indirect selection have been proposed: the Fisher process and genetic handicaps. In the Fisher process (Fisher 1930), the genetically correlated trait is a male display trait of an essentially arbitrary nature, in that the trait has no fitness benefits other than in mating. The male display trait is favored only because a female preference exists for it, and the preference is favored as a result of the genetic correlation that arises because females with the preference mate with males with the display trait. In Fisher's (1930) original model, the evolution of the male trait could accelerate or "run away" as a result of its genetic correlation with the female preference. In the handicap process (Zahavi 1975, 1977), the genetically correlated trait serves as an "indicator" (Andersson 1994) or an "honest advertisement" (Kodric-Brown and Brown 1984) of male viability. Zahavi originally conceived of such traits as reducing male viability, thus conferring a handicap on males that possess them. In Zahavi's original view, only very fit males could afford such a handicap, so the handicap trait was thought to indicate high viability. Females with the preference choose males with the handicap trait, and males with the handicap also have genes for high viability. The preference thus becomes genetically correlated with viability and is indirectly favored as selection favors high viability.

The role of genetic correlation in indirect selection can be illustrated with the following hypothetical example. Assume that mating preferences vary among females, and that some of the variation is heritable. Also assume, as before, that mating preferences are based on male repertoire size, which we assume is also heritable. Females that prefer males with large repertoires will most likely mate with such males, and the offspring of such matings will therefore inherit both genes for large repertoire preference and genes for large repertoires. The large repertoire preference and the large repertoire trait thus become genetically correlated; that is, the genes for the two occur together in the same individual more often than at random. The Fisher process occurs because when selection favors the large repertoire trait because of its mating advantage, the large repertoire preference also increases because it is genetically correlated with large repertoires. In the handicap process, a genetic correlation is assumed to occur between male repertoire size and male viability; this correlation can arise by a number of different mechanisms (see below). Therefore, when selection favors high viability in males, the large repertoire trait and the large repertoire preference also increase.

Confidence in the likelihood of the Fisher process waxed, waned, and waxed again in the 1980s and 1990s. The original burst of popularity was due to the models of O'Donald (1980), Lande (1981), and Kirkpatrick (1982). Kirkpatrick's (1982) haploid, two-locus model can be used to illustrate the type of result obtained (see Maynard Smith 1991). One locus is assumed to control the male display trait; T males possess the trait, and t males lack it. The second locus controls the female preference; C females prefer T males, and c females mate randomly. In this model, T will go to fixation if C is initially common, but T will be eliminated if C is sufficiently rare. For intermediate initial values of C, a line of equilibria exists with both C and T at intermediate frequencies, up and down which the preference and trait could drift. With the appropriate kind of female preference strategy (e.g., best of *N*, in which a female chooses the best of *N* candidates), this type of model can generate runaway selection, in which both the male trait and the preference become exaggerated (Seger 1985). Quantitative genetics models, in which many loci affect both the preference and the preferred trait, can also generate both the line of equilibria and the possibility of a runaway (Lande 1981).

Confidence in the Fisher process subsequently waned when new models challenged the line-of-equilibria result (Kirkpatrick 1985, Pomiankowski 1987a, Bulmer 1989, Pomiankowski et al. 1991). Bulmer (1989) examined a haploid, two-locus model similar to those of Kirkpatrick (1982) and Seger (1985), and found that direct selection acting on the female preference, however weak, would destroy the line of equilibria, driving the favored preference allele to fixation and dragging the male trait allele along with it. It seems reasonable to assume that preferences have some cost, and Bulmer's results indicate that such a cost would dominate the evolutionary outcome, preventing either the preference or the male trait from evolving.

The Fisher process was resuscitated by Pomiankowski et al. (1991), who used a quantitative genetics model to show that female preferences can evolve despite costs under realistic conditions. The added requirement is that biased mutations act on the male trait—that is, unfavorable mutations will be more likely than favorable ones. Such a bias is reasonable because random changes to a complex trait are more likely to cause deterioration than improvement (Maynard Smith 1991). At equilibrium, several forces balance one another: direct selection acts against the preference because of its cost, indirect selection favors the preference because of its genetic correlation with the male trait, exaggeration of the male trait is favored because of higher mating success, but exaggeration is prevented by biased mutation combined with the effect of selection on the correlated preference.

The other possibility for the evolution of female preferences by indirect selection is through a handicap mechanism. We have already discussed female preferences for handicaps that signal direct benefits, such as a superior territory; here we discuss preferences for handicaps that signal that a male possesses "good genes." This type of handicap mechanism has, like the Fisher process, been in and out of favor with theoreticians. A typical genetic handicap model contains three loci: one locus at which allele A codes for the presence of the male display trait and allele a for its absence; a second locus at which B codes for high viability and b for low; and a third at which C codes for a female preference for A males and c codes for random choice (Andersson 1994). Whether the preference is favored in such a system depends in large part on how expression of the handicap trait is governed, and hence on how a correlation between the handicap and viability arises.

Three possibilities have been envisioned (Maynard Smith 1985, Andersson 1994). (1) In a pure epistatic handicap, all males with allele A develop the handicap, and a correlation between the handicap and viability arises because males with low viability (b) alleles are unlikely to survive with the handicap long enough to be chosen by females. (2) In a condition-dependent handicap, males with the handicap allele (A) produce the handicap trait only if they also possess the allele for high viability (B). The genetic mechanism then directly produces an association between the handicap and viability. (3) In a revealing handicap, all males with the handicap allele (A) produce the handicap trait, but the trait is in poorer condition in males with low viability (b) than in males with high viability (B). Possession of a handicap trait in good condition therefore directly reveals possession of high viability. The essential difference between revealing and condition-dependent handicaps is that all males with revealing handicap alleles produce a handicap and pay the cost of doing so, whereas males with condition-dependent handicap alleles pay the cost of producing the handicap only if they also have high viability.

The consensus among theoreticians seems to be that pure epistatic handicaps will not work, at least not independent of the Fisher process (Maynard Smith 1991, Andersson 1994). The handicap mechanism is more likely to work with condition-dependent and revealing handicaps, though even here it is not clear that female preferences for handicaps will increase from low frequency in simple haploid or diploid models (Pomiankowski 1987a, b, 1988). The best theoretical support for the handicap comes from the quantitative genetics model of Iwasa et al. (1991). Here the handicap trait varies in elaboration, and the preference in intensity. The preference is assumed to have a direct cost to females. The model can encompass any of the three types of handicaps, but preferences evolve only if the degree of elaboration of the handicap increases with viability, which eliminates the pure epistatic handicap. Preferences for revealing and condition-dependent handicaps can evolve, given that mutation on viability is biased toward decreasing viability.

The handicap mechanism for the evolution of female preferences requires that viability be heritable; just how heritable is an open question, although Andersson (1994) proposed that it might be enough if fitness showed just a few percent heritability. Certain forces tend to decrease heritable variation in fitness (e.g., natural selection) and others tend to increase it (e.g., mutation pressure or the host-parasite cycles described in Hamilton and Zuk 1982); what the resultant heritability should be is another much-debated question among theoreticians (Andersson 1994). Fortunately, the heritability of fitness is amenable to empirical measurement, and, fortunately again, some of the best data are from birds. Gustafsson (1986) estimated narrow-sense heritabilities by father-son and mother-daughter regressions for a wild population of Collared Flycatchers. Heritability of lifetime reproductive success was 0.0083 (±0.128 SE) for 108 father-son pairs, and –0.0142 (±0.158) for 105 mother-daughter pairs. Heritability of lifespan was –0.0001 (±0.182) for 111 male-father pairs and –0.0160 (±0.154) for 114 female-mother pairs. These estimates indicate that heritabilities may be too low to sustain a handicap mechanism, but the standard errors of the estimates are so wide that we cannot reject heritabilities in the necessary range. Hasselquist (1994) provided empirical support for the handicap mechanism in birds (see below).

In sum, our sense of the current state of theory is that both the Fisher process and the handicap mechanism may be workable under what *may* be realistic conditions, but neither process should be regarded as inevitable in any way. Theory is perhaps most useful in dismissing certain possibilities, such as pure epistatic handicaps, and in suggesting predictions that can be tested from more likely possibilities. Unfortunately, the predictions made by the indirect selection models are either difficult to test, unable to distinguish between the possibilities, or both.

The clearest prediction made by models of the Fisher process is the existence of a genetic correlation between the male display trait and the female preference. Documenting a genetic correlation for a song preference would necessitate measuring variance in the male trait (easy), measuring variance in the female preference (harder), and then demonstrating a genetic correlation between the two (harder still). Genetic correlations can be measured either by the resemblance of relatives or by an artificial selection experiment (e.g., select for the male trait and measure the response of the preference). As far as we know, variance in female song preferences within a population has not yet been demonstrated in any bird species; such a demonstration would be a useful first step toward testing the Fisher process.

Genetic handicap models predict that females prefer male display traits correlated with above-average viability, and that females that choose such males have offspring of higher than average viability (Andersson 1994). To distinguish indirect from direct handicap mechanisms, one must also show that the offspring acquire high viability genetically (i.e., they inherit "good genes") rather than environmentally (i.e., they benefit from superior resources or paternal care).

Nonselected Preferences

A nonselected preference is one that has not been favored by selection, either directly or indirectly. Within this category, we emphasize the hypothesis that the preference is attributable to a sensory or neural bias that has evolved in females for reasons unrelated to mating and its consequences. Such a female bias may lead to "sensory exploitation," in which males evolve a display trait to exploit the mating preference caused by the bias (Ryan et al. 1990, Ryan and Keddy-Hector 1992). Under this hypothesis, the male trait is favored by sexual selection, but the female mating preference is neutral with respect to selection.

We again use repertoire size in a hypothetical example. Suppose that large repertoires are more effective than small ones in stimulating females to solicit copulations because females are less likely to habituate to a more varied acoustic signal. Habituation can be a peripheral phenomenon in which sensory neurons respond less and less to a repeated stimulus (sensory adaptation), or it can be central, with the reduction in response occurring despite continued activity of the sensory neurons (contingent response decline). Antihabituation therefore can also be either central or peripheral, but in either case the female bias has no direct or indirect fitness consequences in mate choice. In this scenario, males with large repertoires would nevertheless be favored in sexual selection because they are better able to gain copulations than are males with small repertoires.

Empirical support for the sensory bias hypothesis comes from work on the frog *Physalaemus pustulosus* (Ryan et al. 1990, Ryan and Keddy-Hector 1992). Females of this species prefer calls containing low-frequency chucks. This preference is most simply explained by the fact that the female's auditory system is tuned to lower frequencies than those found in the average male's call (Ryan et al. 1990). Tuning of the auditory system is similar in females of a related species (*P. coloradorum*), and females show a preference for conspecific calls with chucks added even though males of their own species do not produce chucks. The evidence is consistent with the hypothesis that the tuning of the auditory system evolved first, in an ancestor of both species, and that the chucks produced by males of *pustulosus* evolved to exploit the already existing sensory bias. Other possible examples of nonselected preferences are female preferences for longer tails in swordtails (*Xiphophorus helleri:* Basolo 1990) and for male visual displays in *Anolis* lizards (Fleishman 1992).

To support the hypothesis that a song preference arose without selection, one must first show that the preference reflects a behavioral bias that functions in other contexts. Stronger evidence would be to show that the preference exists in related species that lack the male trait, as such a result indicates that the preference can evolve without any direct or indirect benefits to females. The strongest evidence would be to map the preference and the preferred trait in a sufficient number of related monophyletic species so that a phylogenetic analysis could determine whether the preference predates the preferred trait in the evolution of the groups within the clade (D. R. Brooks and McLennan 1991).

Evolution of Song Preferences

In this section, we discuss whether the theories describing the evolution of female preferences outlined above can explain song preferences as described in the first section. We also discuss the evidence that would be needed to make a better case for each hypothesis.

Direct Selection

We concluded above that females are likely to prefer male attributes that directly affect female fitness whenever males possess such attributes. We also concluded that displays, such as song, can serve as honest indicators of male attributes important to female fitness if the displays are costly enough so that honesty is enforced. We need to ask, then, which male attributes affect female fitness in birds, and which of these might be signaled, with an appropriate cost, by male song.

The principal male attributes likely to affect female fitness are territory quality, in terms of the availability of food and nest sites, and the quality of paternal care, in terms of ability to provision and defend the young. Both food availability and provisioning ability might be signaled by any display that has a sufficient time or energy cost: the ability of the male to divert time and energy to the display would be a reliable indicator that the territory provides sufficient food, or that the male has sufficient foraging skill to generate surpluses of time and energy that could be devoted to the young.

The song attribute most likely to have the appropriate costs is amount of song produced. Singing must have some energy cost, but the magnitude of this cost is little known. Eberhardt (1994) reported that metabolic rates in Carolina Wrens (*Thryothorus ludovicianus*) are three to five times higher in singing males than in resting males, but Chappell et al. (1995) measured no increase in energy expenditure in crowing Red Junglefowl (*Gallus gallus*). More work on energy expenditure during singing is clearly needed. Regardless of whether singing has a major direct energy cost, we would expect it to have an important indirect effect on the energy budget through its time cost, in that time devoted to singing typically cannot be used in foraging. In a subspecies of Savannah Sparrow (*Passerculus sandwichensis princeps*), for example, singing males forage significantly less than males that do not sing during the same periods (Reid 1987). Males can, therefore, devote time to singing only when their rates of energy intake when foraging are sufficiently high and when other energy expenditures are sufficiently low. Consistent with this logic is the fact that song production is often positively associated with ambient temperatures, meaning that singing increases when the energy costs of self-maintenance are low (e.g., Greig-Smith 1983, Gottlander 1987, Reid 1987). Perhaps the best evidence that song production is limited by time and energy comes from experiments in which males provided with extra food on their territories responded by increasing their song production relative to control males; this result was obtained for Red-winged Blackbirds (Searcy 1979), Carolina Wrens (E. S. Morton 1982), Hedge Accentors (*Prunella modularis:* N. B. Davies and Lundberg 1984), Savannah Sparrows (Reid 1987), and European Pied Flycatchers (Gottlander 1987). Thus, song production must often be a reliable indicator of food availability.

Common Stonechats (*Saxicola torquata*) and Blackcaps (*Sylvia atricapilla*) provide direct evidence that singing rates can signal male parental quality and territory quality. In Common Stonechats, rates of song production early in the breeding season correlate with the male's later participation in both feeding and defending his offspring (Greig-Smith 1982). Such correlations could result from individual differences in foraging skill or energetic efficiency, but they seem more likely to be the indirect consequences of variation in the food resources of different territories. In Blackcaps, male participation in feeding nestlings actually declines with increasing song output, whereas reproductive success is positively correlated with song output and negatively correlated with male provisioning (Hoi-Leitner et al. 1993). The simplest interpretation of the Blackcap results is that food production on the territory is of overriding importance to reproductive success, and that song output acts as a signal of food production because singing is food-limited.

An excellent case can thus be made that female preferences for high song output are favored by direct selection because singing is a signal of a territory's food productivity. The major weakness in this case is in the evidence for the female preferences, which consists mainly of correlations between song production and pairing success (exceptions are described in Alatalo et al. 1990, Wasserman and Cigliano 1991, and Collins et al. 1994). These correlations may be the consequence of female preferences based on song output, but they are also consistent with direct female choice of the amount of available food or of other male behaviors that may be altered by provisioning. To complete the case, we urge that greater attention be given to testing for female preferences for song production using other methods. One possibility would be to measure female attraction to nest sites associated with song playbacks using contrasting rates of song.

Repertoire size is a second song attribute that may signal the quality of direct benefits to the female. The best case can be made for this possibility in a German population of Great Reed-Warblers, in which syllable repertoire size is correlated with measures of territory quality such as the area of reed acceptable for nesting and the length of the interface between reed and open water (Catchpole 1986). Similar correlations arise in Red-winged Blackbirds, apparently because repertoire size is correlated with male age, and male age is correlated with territory quality (Yasukawa et al. 1980). Repertoire size is not correlated with age in the German population of Great Reed-Warblers studied by Catchpole (1986), though repertoire size does increase with age in a Swedish population of the same species (Hasselquist 1994). How correlations between repertoire size and territory quality arise in Catchpole's population is not at all obvious, nor is it obvious how the honesty of such a signal could be maintained. Repertoire size may have a cost in taking up brain space (see Brenowitz and Kroodsma, this volume), but this type of cost would seem unlikely to enforce the honesty of a signal of parental or territory quality. We believe that more attention should be paid to correlations between song attributes and aspects of territory quality and male parental quality, and to the possible costs of such attributes that might maintain the honesty of such signals.

Indirect Selection

Whenever females obtain direct benefits from mate choice, such as territories and paternal care, direct selection should be favored over indirect selection in explaining song preferences. In some species of birds, however, song preferences seem likely even though females do not obtain any obvious direct benefits from choice; for example, Village Indigobirds (R. B. Payne and Payne 1977), Sage Grouse (R. M. Gibson and Bradbury 1985), and Sharp-tailed Grouse (Gratson 1993). Moreover, even in species in which males provide territories or parental care to their mates and offspring, song preferences might influence choice of sire during EPCs without affecting the distribution of such benefits. Indirect selection therefore may well have had some role in the evolution of song preferences.

The central prediction of the Fisher process is a genetic correlation between the elaboration of the male display trait and the strength of the female preference. A genetic correlation could be manifested, for example, by males with large song repertoires having daughters with strong preferences for repertoires, or by females with strong preferences based on song output having sons with high song production. Needless to say, such a genetic correlation has not been demonstrated for any song preference. Below, we outline some steps building up to such a demonstration, each of which would itself represent a considerable advance.

1. A demonstration that the male song attribute under choice is heritable. Hasselquist (1994) measured song-repertoire sizes of fathers and sons in Great Reed-Warblers, using sons whose parentage was confirmed by DNA fingerprinting. Repertoire size of sons was not correlated with repertoire size of fathers, giving no evidence of heritability. Other types of evidence also indicate that heritability of some song attributes is low. Radesäter and Jakobbson (1989) showed that song rates of male Willow Warblers holding the same territory in succession were highly correlated, implying that territory quality has a significant effect on song rate. Song repertoire size in Marsh Wrens is strongly influenced by exposure to song during development (Brenowitz and Kroodsma, this volume), which again points to environmental rather than genetic effects. The one positive result we know regarding heritability of song traits comes from the work of A. E. Houtman (1992), who showed, using parent-offspring and sib-sib resemblances, that song rates in captive Zebra Finches are highly heritable. Heritabilities of more song attributes should be measured, using either correlations between relatives or response to artificial selection in captive populations.

2. *A demonstration of genetic variation in female mating preferences.* The first difficulty lies in measuring individual variation in preferences. As a measure of individual preference, one could observe whether a female chooses phenotype *a* over phenotype *b* in a mating trial or in response to the solicitation display assay. Demonstrating that such preferences are repeatable within females would be a step that has not yet been made with song preferences. Demonstrating genetic heritability would then require correlating preferences in genetic relatives or measuring response to artificial selection (see Majerus 1986 for a nonavian example). In birds, this type of work could be done using captive breeders such as Zebra Finches or canaries.

3. *A demonstration of a genetic correlation between a male song attribute and a female song preference.* Such a correlation could be demonstrated by measuring a song attribute and preferences for that same attribute in male and female relatives. Alternatively, one could show that artificial selection on the song attribute in males causes a response in song preferences in female offspring, or that selection on female preferences causes a response in the male attribute. Again, captive breeding populations could be used for such studies.

A weaker test of the Fisher process would be to look for a correlation between strength of preference and the elaboration of the male trait across different populations. Such correlations have been shown in frogs and fish (Ryan and Wilcynski 1988, Houde and Endler 1990). The closest approach to this kind of evidence for birdsong is a correlation between strength of female preferences for repertoires and male repertoire size across seven species of emberizids (Searcy 1992b). Such correlations are weak evidence because mechanisms other than the Fisher process can explain them (Searcy 1992b).

The best evidence for a genetic handicap mechanism would be to show that females with a certain song preference have offspring of higher viability than females lacking the preference. A recent study closely approaches such a demonstration (Hasselquist 1994). Survival of young Great Reed-Warblers after fledging was positively correlated with the song repertoire size of their fathers. Both laboratory (Catchpole et al. 1986) and field results (Hasselquist 1994) indicated that females prefer males with large repertoires. Females exercising the preference should have offspring of higher fitness than females that do not exercise the preference. Interestingly, in the field the preference is manifested in EPCs but not in pairing, where direct benefits as well as genes are at stake (Catchpole 1986, Hasselquist 1994).

A weaker case for a genetic indicator mechanism can be made for certain other species in which the male song attributes preferred by females have been shown to be correlated with some component of male fitness, without evidence that the fitness component is heritable. In two species (Great Tits and Song Sparrows), males with large song repertoires have higher lifetime reproductive success than males with smaller repertoires (McGregor et al. 1981, Lambrechts and Dhondt 1986, Hiebert et al. 1989). Song repertoire size is also correlated with male age in Red-winged Blackbirds (Yasukawa et al. 1980), Northern Mockingbirds (Derrickson 1987), and Common Starlings (Eens et al. 1992). Why repertoire size is an honest signal of fitness or viability is unclear. One possibility is that older or fitter males are better able to spare the brain space necessary to store a large repertoire. Costs of repertoires deserve more attention, but regardless of whether we can explain the honesty of the signal, it remains true that repertoire size is an honest indicator of age or fitness in some species.

Another song trait that may signal male viability is song output. We have already cited evidence that song output is limited by food availability; song production is therefore a signal of a male's ability to acquire food. That ability may be largely determined by territory quality, but it may be affected by male quality as well. Whether the ability is heritable is unknown. Female preferences for high song output are known to occur in two cases in which females obtain only genes from their mates: during mating in the promiscuous Village Indigobird (R. B. Payne and Payne 1977) and during extra-pair copulations in Zebra Finches (A. E. Houtman 1992). As females obtain no direct benefits from males in these cases, their preferences are unlikely to have evolved through direct selection. The possibility still exists, however, that the preference for high song output is a nonselected one.

Nonselected Preferences

Searcy (1992b) argued that female preferences for repertoires may at least originate as nonselected preferences. In many of the species in which females have been shown to court more in response to multiple song types than single song types, the subjects show a pattern of habituation in their courtship response during repetition of a single song type, and recovery of the response when the song type is switched. Repeated recoveries during the repertoire playback make the overall response higher than for single song types (Searcy et al. 1982, Searcy 1988b, 1992b). Habituation is a nearly universal property of animal response systems (Hinde 1970), found in behaviors ranging from the withdrawal of marine worms in response to shadows (R. B. Clark 1960) to the aggressive reactions of male birds to song (Petrinovich and Patterson 1979). Recovery occurs because the response decrement is specific to a particular stimulus; such stimulus specificity is also very widespread in animal behavior (Hinde 1970). The argument, then, is that the pattern of habituation and recovery, which produces the overall preferential response to repertoires, is probably built into the response system from the beginning, whether or not it is advantageous in mate choice.

Further evidence that repertoire preferences are nonselected is the demonstration of the preference in the Common Grackle (*Quiscalus quiscula*), in which males do not have repertoires. Female Common Grackles court more in response to multiple song types than to single song types (Searcy 1992b). Repertoire size in Common Grackles cannot serve as a signal of paternal care, territory quality, or male genetic quality because no variation in repertoire size exists in this species. Nor can the preference and the male trait have coevolved in a Fisher process, because the male trait does not exist. The most likely explanation of the female preference in Common Grackles is that it is a nonselected consequence of response properties, habituation and stimulus specificity, which are adaptive in other contexts. Female Common Grackles do show a pattern of habituation and recovery in their response to repertoire playbacks. The preference for repertoires is rather weak in Common Grackles, so it is possible that the strength of the preference has been exaggerated in Fisherian coevolution with the male trait in other species in which the preference is stronger and the male trait exists (Searcy 1992b).

It is also possible that preferences based on song output are nonselected, in that females may choose males with high output simply because these males are the first they notice rather than because a preference for males with high output is selectively advantageous. If so, such a preference would be closely akin to "passive attraction," defined by G. A. Parker (1982, p. 182) as occurring when "females are attracted randomly by advertisement" so that a female has the highest probability of choosing "the signal she perceives most intensely." Such preferences contrast with "active choice," in which "a female compares the signals of several males and then chooses a particular phenotype." The difficulty with the passive attraction–active choice dichotomy is that the distinction lies in what the female is thinking, which is unobservable. We believe it is more profitable to concentrate on the question of whether the preference is selected or nonselected. Preferences for higher song output seem likely to be the product of direct selection because high output signals a superior territory (see above), but it is also possible that they are nonselected in cases in which males provide no direct benefits.

Conclusions

Females show mating preferences based on male song attributes in some bird species. The evidence could be strengthened (e.g., by further work in which song or singing behavior is experimentally altered in wild males), but we feel confident in taking the existence of female preferences as given. Female preferences may exist for many song attributes, but the strongest case can be made for a general preference for more variable song over less variable. A general preference for high song rates over low rates is also a good possibility.

Some progress has been made in explaining the evolution of song preferences, but here obviously much work remains in order to test the full variety of theoretical explanations that have been proposed. Three hypotheses have received substantial support so far. First, there is evidence supporting direct selection of female preferences for high song output, with direct selection occurring because song output signals territory quality or paternal quality. Second, evidence exists suggesting that preferences for song variability originate as nonselected response biases that result from the antihabituation effects of variable song. Third, some evidence supports indirect selection favoring preferences for song variability, because variable song acts as an indicator of male genetic quality. Of the major mechanisms for the evolution of female preferences, only the Fisher process gets little or no support from work on birdsong.

We end by reiterating four types of investigations that we think would be most valuable in addressing the problem of evolution of song preferences and would make the greatest contribution to understanding the evolution of mating preferences in general: (1) investigations of correlations between song attributes and the quality of male territories and parental care, and of the costs of the song attributes that may enforce the honesty of such correlations; (2) measurement of the heritability of male song attributes and female song preferences, and of the genetic correlations between the two; (3) comparisons of the viability of the young of females that do or do not express song preferences in mating; and (4) phylogenetic analyses to test whether evolution of the song preference predates the evolution of the preferred song attribute within clades.

Acknowledgments

We thank Malte Andersson for providing us with a prepublication manuscript of his book on sexual selection, and D. E. Kroodsma and E. H. Miller for extensive comments on our manuscript. Financial support was provided by the National Science Foundation (grants IBN-9523635 to WAS and IBN-9306620 to KY).

APPENDIX Natural Sound Archives: Guidance for Recordists and a Request for Cooperation

Donald E. Kroodsma, Gregory F. Budney, Robert W. Grotke, Jacques M. E. Vielliard, Sandra L. L. Gaunt, Richard Ranft, and Olga D. Veprintseva

I urge conservation organizations that fund avifaunal surveys in tropical forests around the world to REQUIRE their participants to use tape recorders systematically. Copies of all recordings should be placed in a professionally maintained sound collection that provides easy access to researchers.

—Theodore Parker III

The conservation of the animal and plant world on this planet is unthinkable without the realization that it is essential to preserve . . . wild animal sound recordings.

—Boris Veprintsev

Editors' note: The final chapter is positioned as an appendix, but it carries an important message that has been endorsed by all the contributors to this volume: tape recordings of bird sounds should be considered just as valuable as such traditional scientific specimens as study skins and skeletons, and should be preserved for posterity. Audio recordings should be documented fully, archived, and curated to ensure proper care and accessibility. As natural populations disappear, sound archives will become increasingly important. To enhance the value of sound archives, we provide advice on how to record animal sounds, what data should accompany recordings, and how private collections and major archives should be maintained. We hope that these suggestions will contribute to a greater effort to archive animal sounds, will enhance understanding of the animals by increasing the accessibility of audio recordings, and will aid in the preservation of the animals themselves. The existing knowledge on acoustic communication in birds is well illustrated by the chapters in this book; properly archiving the information on which this knowledge is based should be an integral part of the work for both scientific and ethical reasons.

The contributors to this volume and the following colleagues associated with major sound collections endorse the goals and primary guidelines of this chapter: Kurt Fristrup, Shiverick House, Woods Hole Oceanographic Institute, Woods Hole, Massachusetts 02543, USA (current address: Cornell Laboratory of Ornithology, 159 Sapsucker Woods Road, Ithaca, NY 14850, USA); Karl-Heinz Frommolt, Tierstimmenarchiv, Humboldt-University, Institute of Biology, Invalidenstrasse 43, D-10115, Berlin, Germany; Peter Fullagar, Sound Library, Australian National Wildlife Collection, CSIRO Division of Wildlife and Ecology, P.O. Box 84, Lyneham, A.C.T. 2602, Australia; J. William Hardy, Bioacoustics Laboratory and Archives, Florida Museum of Natural History, University of Florida, Gainesville, Florida 32611, USA; Ned K. Johnson, Laboratory of Natural Sounds, Museum of Vertebrate Zoology, 3101 Valley Life Sciences Building, University of California, Berkeley, California 94720-3160, USA; Alan Kemp, Fitzpatrick Bird Communication Library, Bird Department, Transvaal Museum, P.O. Box 413, Pretoria 0001, South Africa; Ralph Moldenhauer, Texas Bird Sound Library,

Our goal is to ensure the preservation of recorded animal sounds. If we are to accomplish this goal, we must impress on each individual who records wildlife sounds, whether for research or for any other purpose, that recorded sounds are valuable and should be preserved as scientific specimens in perpetuity. This responsibility requires record keeping, preferably in a standardized format, together with a commitment to deposit the recordings in a publicly accessible archive. We urge each recordist of wildlife sounds to consider the guidelines presented in this appendix, which will facilitate documentation of sound recordings and the integration of the recordings into the collection of an archive. The result of this individual and collective commitment will benefit growing efforts in research, education, and conservation.

The preservation of "bioacoustic specimens" adds to the inventory documenting the biodiversity of this planet, a mission set forth by the scientific community in its Systematics Agenda for the Year 2000, under the auspices of the National Biological Survey ("Systematics Agenda 2000," 1994). Natural habitats on this planet are increasingly under siege, and habitat loss is occurring at a quickening pace. Many recording efforts may in fact be the last opportunity to record at a given location or, even worse, may be the last opportunity to record a given species in nature (e.g., the Little Blue Macaw, *Cyanopsitta spixii,* during the early 1990s). Through research, education, and conservation efforts, sound recordings expand our knowledge of the world. We hope that these efforts will increase awareness of and sensitivity to the richness of communication systems, and will in some measure help to protect our natural world.

A major and significant source of this documentation already resides in the large private collections that often accumulate during the course of a research career in bioacoustics. Not only do these private collections often contain many "last opportunities," they also contain especially valuable series of recordings from individuals, populations, species, or communities. Unlike other scientific specimens such as skins, skeletons, and tissues, the sound recordings on which bioacoustic research is based are rarely deposited in an appropriate archive. Unfortunately, too, private collections are often neglected or lost at the end of a researcher's career. It is this loss of private collections that we wish to reverse by a repeated plea for cooperation among individual researchers and archives (Hardy 1984, E. H. Miller and Nagorsen 1992).

Furthermore, recordings made for one purpose can inevitably be used in unforeseen ways. Surveys, for example, are increasingly crucial for documenting the natural places that remain on the planet, and regional recordings that represent local dialects for each species are needed for these census efforts. Field record-

---

Division of Life Sciences, Sam Houston State University, Huntsville, Texas 77341, USA; I. D. Nikolskii, Phonotek of Animal Voices, Department of Vertebrate Zoology, Biological Science Faculty, Lomonosov State University, 119899 Moscow, Russia; Anna Omedes, Fonoteca Zoologia, Museo de Zoologia, Apartado de Correos 593, 08003 Barcelona, Spain; Roberto J. Stranek, Laboratorio de Sonidos Naturales, Museo Argentino de Ciencias Naturales "Bernardino Rivadavia," Avenida Angel Gallardo 470, 1905 Buenos Aires, Argentina

ings, regardless of their original purpose, are invaluable for addressing questions of how signals vary in space and in time. Archived series of tapes from developmental studies, in which extraordinary effort is involved in raising the subjects and collecting the recordings, can later become a gold mine of data for other research on vocal development. Similarly, archived playback tapes from experimental studies can allow for replication of experiments at later times under similar or different conditions and can also enable the asking of related questions not directly addressed in the original study. Every chapter in this volume is based primarily on private research collections, and these private collections are the primary basis of our knowledge of animal signals. If these private collections are preserved and appropriately archived, their collective use can provide an immense resource for further studies of animals and their signals.

Our basic point, then, is this: with minimal care and documentation, each recorded sound becomes a valuable scientific specimen. When properly preserved in a private collection—and eventually in a publicly accessible sound archive—that specimen will be useful, in perpetuity, in a variety of foreseen and unforeseen ways. We ask for cooperation in a larger effort to study and preserve, both in nature and in the archive, the sounds of the world that we have come to love through our studies of birds.

Current Sound Archives

Some of the earliest recording efforts, using a wax cylinder of an Edison Phonograph and dating back to the 1880s, were documented by Koch (1955) in his autobiography. Early efforts to make field recordings on magnetic tape were chronicled by Kellogg (1962) in his description of the focus on bird sounds at Cornell University's Laboratory of Ornithology. From this focus sprang the first natural sound archive, the Library of Natural Sounds (LNS); numerous other independent archives were subsequently established around the world (see Boswall and Kettle 1979, Kettle 1983, 1989).

One of the first international gatherings of natural sound archivists was organized by James Gulledge, of the LNS, and Jacques Vielliard, of the Arquivo Sonoro Neotropical, at a round-table discussion at the 1986 International Ornithological Congress (IOC) in Ottawa, Canada. A follow-up discussion organized by Ronald Kettle, of the British National Sound Archive, and Vielliard occurred at the 1990 IOC in Christchurch, New Zealand. The participants reached an agreement on which data should be included so as to maximize the usefulness of a given recording. That list of items was subsequently published (Kettle and Vielliard 1991) and is reproduced below.

What to Include in Natural Sound Archives

What follows is a widely accepted list, with brief commentary, of the kinds of information needed for archived tape recordings to have maximal value (see

Kettle and Vielliard 1991). Except for minor modifications, this is the list adopted by participants of the round-table discussion at the 1990 IOC in New Zealand. We urge recordists to adapt this documentation to their needs. How to implement this process will be discussed in the next section.

We recognize that few, if any, private collections meet the standards we advocate here. That doesn't mean those collections are of no value, but poorly documented recordings are, in the long term, not much better than no recordings at all. The value of a recording increases dramatically with added information, and perhaps more than anything else we here encourage a new attitude toward collections and archives, and a greater responsibility in documenting wildlife recordings.

The list of information is punctuated by the following codes:
* Amenable to standardization of categories by different archives
# For completion only where appropriate
[ ] May be completed at the archive

Some information usually applies to all cuts on a tape (a "cut" usually being a continuous recording session of one target animal or a group of animals), and hence the forms chosen for use might have the following information entered only once for each tape:

Basic data
  Recordist: name and address
  Tape number: reels or cassettes should be numbered sequentially, perhaps using the recordist's initials and a year-reel number scheme (e.g., SLLG1994-11 is 11th tape in 1994 recorded by S. L. L. Gaunt). Signify A and B if both sides are used, though use of side B is not recommended—see item 8 in the next section
  Date
Technical data
  Field recorder: make and model
  Microphone(s): make and model
  #Parabola diameter and focal length
  Tape format: cassette, open reel, or R-Dat
  Track format: stereo, mono, 1/2, 1/4, full, etc.
  Speed: ips or cm/s
  Tape type: manufacturer and series number (batch no., if available)
  Equalization standard: either NAB or CCIR for open reel; either high (70 μs) or low (120 μs) for cassettes
  #Noise reduction system, if used (not recommended)
  #Filtering (not recommended)
Geographic data
[Zoogeographic region]
Country

Location (include elevation, [latitude, and longitude], if possible)
*[Climatic zone]

The following information is more likely to change for each cut on a given tape (if the above information changes for the cut, that should be noted, too).

Cut number: clearly announce the beginning and end of each separate recording on a tape
*Habitat
Temporal and meteorological
  Time
  Temperature (where relevant, especially for insects, amphibians)
  Weather
Specific zoological
  [Class of animal]
  [Order]
  [Family]
  Species: state the authority for nomenclature
  Confident identification by sight: yes, no, and degree of certainty (It is important to know if the subject was positively identified by sight regardless of the recordist's confidence in identifying the sound.)
  Number of target individuals (indicate pair, flock, colony, nestling, juvenile, etc., as appropriate)
  #Sex (if known)
  *Age (e.g., nestling, juvenile, adult, and stage or form for insects)
  #Individual (many research collections collect extensive series of recordings from known individuals)
Bioacoustic data
  #*Type of sound (e.g., song, call, drumming, stridulation, duetting, mimicry, chorus, etc.)
  *Behavioral context (e.g., alarm, flight, food begging, mating, etc.) (Some of the most useful information on behavioral context cannot be categorized simply, and archives encourage recordists to include additional written notes for each cut. These notes are often of special significance to researchers.)
Other items
  #In captivity?
  Playback used? (yes or no). Note any perceived response to playback
  [Duration]
  Distance from subject to microphone (in meters)
  [*Quality, include detrimental factors]
  #Background sounds
  #Museum specimen number?
  #Notes, including details of any edits or other special circumstances

Documentation of Recordings, or How to Keep Order

To the average recordist, keeping the above information organized may seem a daunting prospect. In practice, however, the whole process can be relatively simple, and the long-term value of the recordings is increased immeasurably if thorough records are maintained.

At the beginning of each tape, announce the tape number and equipment configuration. Each cut on a given recording is announced, in the field, typically at the end. To ensure that crucial information is not forgotten, the LNS provides tags with a list of items to be announced for each cut (species, date, time, location, behavioral context of sound, habitat description), and those tags can be attached to the recorder as a reminder. Additional critical information should be narrated on the tape, especially behavioral information on little-known species (essentially all Neotropical species). After recording on a tape, the recordist should listen to and create a log for that tape, keeping records of pertinent information that is to be transferred to the data forms. During this listening process, one of us (Kroodsma) uses a form with rows of 60 "periods" broken into 10-second blocks, so that a running account of critical information (e.g., vocalization, voice comment) can be kept on a second-by-second basis for each tape; other recordists adopt different schemes appropriate for their purposes, but essentially the same types of information are documented. Finally, one data form or database entry is completed for each cut on the recording.

Early association with an established archive is especially useful. Because the major archives all have their own methods for documentation, we urge each recordist to contact an archive as soon as practicable. Mutually beneficial mechanisms for data transfer can then be arranged. The Borror Laboratory of Bioacoustics (BLB), the LNS, and the NSA all request much the same information, for example, though in different forms (e.g., see Gulledge 1977, Kettle and Vielliard 1991). The LNS also now provides a FoxPro database system that can be customized for personal use; use of such a standardized database system by LNS associates will greatly facilitate the eventual archiving of their personal collections.

Organizing a personal collection is thus rather straightforward, though admittedly time-consuming. A personal collection has a series of sequentially numbered tapes. Each tape contains recordings of animal sounds and appropriate verbal announcements, especially to signify breaks or cuts. Accompanying each tape is a form or appropriate database entry that provides the information needed to ensure the value of that recording not only to the original recordist but to others as well. Attention to these few small steps will guarantee the permanent value of private collections.

Recommendations for Tape-recording Wildlife Sounds

Many recordists are self-taught individuals who do the best they can with their limited equipment and knowledge. We believe that knowledge is often the more limiting of the factors, however. To improve the recording know-how of recordists, we offer the following advice, which, in our experience, we think is most helpful for field recordists (see also Gulledge 1977, Ranft 1993, Vielliard 1993).

1. One of the most common problems, especially among novice recordists, is distortion of the sound signal because of improper use of the tape recorder's sound-level metering system. The most common cause for this distortion or overloading (tape saturation) is blindly trusting the manufacturer's recommendations for use of the metering system. Recordists must understand the "ballistics" of their record-level meters (see also Wickstrom 1982, pp. 36–38). A VU meter, for example, averages power over time and typically has an integration time of 300 milliseconds, the time it takes for the needle to move from the extreme left of the scale to 0 VU. Birds can produce vocalizations that are only 5–10 milliseconds long, however, so a VU meter has insufficient time to indicate accurately the extent of possible tape saturation during recording. If a recordist follows the owner's manual for the recorder and sets the record level so that the meter just begins to enter the red zone, many bird sounds will be severely distorted. A much faster integration time (10 milliseconds) is available with a Peak Program Meter (PPM), which is therefore more suitable for the rapid attacks and pulses of many bird sounds. Light-emitting diodes (LEDs) are exceedingly fast, equivalent to or better than PPM ballistics, and can be excellent metering devices. An added complication in many systems occurs because most meters are located ahead of the record equalization (preemphasis) circuits, further compounding the distortion problem, especially when recording at slow tape speeds.

As an illustration of these issues, consider the songs of Chipping Sparrows (*Spizella passerina*) and the American Robin (*Turdus migratorius*). The sparrow's song, with its rapid transients, will "beat" a VU meter, but the smooth onset of a robin's song is more similar to music and the human voice, signals for which the VU meter was designed. A good recording of the sparrow's song should not advance the VU meter to more than −10 dB, but that of the robin's could go to −3 or even 0 dB. A recordist using an open-reel recorder can use the highest available tape speeds to help avoid saturating (overloading) the tape with high-frequency, high-energy, brief sounds.

2. The use of headphones can be a great help in making superior recordings. When used with a three-head recorder, headphones allow an instant comparison between what arrives at the recorder (source) and what the recorder is actually storing on the tape (tape). This comparison can detect not only tape saturation (overloading) but other problems as well, such as machine problems (e.g., machine-induced electronic noise) and lack of tape motion (jamming or out of tape). Headphones also help the recordist aim directional microphones or parabolas at subjects and even pinpoint unseen targets. Use of headphones does, however, limit the recordist's ability to monitor other animals in the environment. One compromise used by some recordists is to cover only one ear with the headphones, keeping the other ear free to hear other sound sources.

3. Get the microphone close to the bird with as few intervening obstacles as

possible (all without disturbing the bird, of course). Getting close increases the relative intensity of wanted to unwanted sounds, and a "clear shot" at the bird will reduce degradation of the sound as it travels through vegetation. Also, be aware that the signal is loudest if the bird is facing the microphone (though, of course, birds don't always cooperate).

Playback can be used to bring a bird closer to the microphone, but beware that playback usually alters the bird's behavior, too. Playback usually simulates a natural behavioral context, such as territorial intrusion, and can be useful to stimulate additional singing or to elicit different vocalizations (Dabelsteen and McGregor, this volume, Smith, this volume).

4. Be aware of the tradeoffs between shotgun and parabolic microphones. A shotgun typically has a nearly uniform (flat) frequency response to sounds in front of the microphone, is increasingly directional at higher frequencies, and has a relatively low signal–to–ambient noise ratio. In contrast, parabolas have a low-frequency deficiency at wavelengths below the parabola diameter and an overall high-frequency boost (see Wickstrom 1982, figs. 7, 10, 12). The low-frequency deficiency makes parabolas undesirable for recording low-frequency sounds such as the drumming of Ruffed Grouse (*Bonasa umbellus*).

Differences in the directionality of a parabola and shotgun (see Wickstrom 1982, figs. 3, 6, 9, 11) also influence the amount of environmental signal reverberation recorded on tape. Shotgun microphones are usually less directional than parabolas, and unless the shotgun is relatively close to the subject, signals recorded with the shotgun typically include more reverberation than do signals recorded with highly directional parabolas. On sonograms, the increased reverberation is revealed as a blurring, or echo, of the signal. The shotgun is sometimes said to record a sound more as humans hear it, perhaps because our ears are even less directional than the shotgun. Although sounds recorded with parabolas seem harsher or shriller, those sounds are more likely to reveal the sharply defined details of signals as they left the bird's beak.

5. Know the characteristics and limitations of your chosen tape recorder. Opinions are strong on the merits of digital versus analog and of cassette versus open-reel machines, and decisions are often a personal compromise between technical considerations, cost, and convenience. Here we offer what we believe are "majority opinions," though we acknowledge some differences of opinion.

First, know the limitations of many cassette tape recorders, and know that your machine will do what you need it to do. One popular recorder, for example, will not record much above 9–10 kHz. Cassette recorders also often have difficulty capturing the rapid signal modulations of many bird sounds, mostly because the slow tape speed and narrow width of the recording surface prevent recording of a complex vocalization with high resolution (see Wickstrom 1982). In North America, for example, the songs of many wood-warblers (especially the Blackpoll Warbler, *Dendroica striata*), the Chipping Sparrow, and some thrushes are difficult for most cassette recorders; in the Neotropics many tanagers, hummingbirds, and brush-finches provide special challenges, as do certain finches

(Fringillidae), buntings (Emberizinae), and sparrows (Ploceidae) in the Old World. The safest approach is to record at a relatively low meter level. Distortion will then be minimized, though tape hiss is loud (10–12 dB louder than open-reel machines, for example) relative to the signal at such levels. The performance of a cassette machine can also be improved by calibrating it for a particular tape stock and having it serviced regularly.

Second, digital technology offers the convenience of superb recording machines with long available recording time and no tape hiss (e.g., Ranft 1989). The potential for these small recorders is exciting. The current drawbacks to digital technology, however, should be understood. Digital recorders consume battery power rapidly and are relatively unreliable under extremely high humidity. Affordable machines do not provide for "off-tape" monitoring during recording (they lack discrete playback heads), and they also tend to be less robust than many analog models under field conditions. The current sampling rates of 44.1 kHz and 48 kHz, which enable recording of signals up to 22 kHz, are adequate for most bird sounds, though one manufacturer (Pioneer) is now producing a studio recorder with a 96 kHz sampling rate. At least two of us (GFB and RWG) are convinced that the 96-kHz sampling rate is needed to make digital recordings more lifelike, though the improvement will not affect most uses. (For comments on archiving digital tapes, see item 1 in the next section.)

Third, many recordists still prefer Nagra open-reel machines for use under a variety of field recording conditions and argue that they provide the best combination of reliability and accuracy for field recording. Used Nagras are increasingly available at reasonable prices as recordists switch to other formats (primarily digital).

6. Avoid using recorders or particular settings that intentionally alter signals. The algorithms used for data compression on some digital recorders make assumptions about how humans are believed to hear sounds and are not designed to handle animal sounds. Automatic gain controls (AGCs) and noise reduction (e.g., Dolby) features on cassette recorders also alter signals in ways that are undesirable for research use.

7. Know that all of your equipment will do the job you want it to do. Will the system record the frequencies that you want, for example? Do you have any in-line device (e.g., microphone preamplifier) between the microphone and the tape recorder that distorts signals or limits frequency response, or is in some way a weak link in the record system? Cleaning the recorder regularly ensures reduced maintenance, cleaner tapes, better recordings, and longer life of heads, pinch roller, and capstan. Match the impedance and output level of the microphone to the recorder. Have the machine serviced regularly by a professional (note that in the interest of achieving the highest quality recordings, major archives often provide this service relatively inexpensively).

8. On cassette or open-reel tapes, realize that use of both sides of a tape can result in cross-talk between the tracks. This problem occurs primarily in machines

that have not been serviced regularly and have improper head alignment. We specifically recommend against recording on both sides of a tape.

9. Speak crucial data onto the tape. When possible, review, log, or organize tapes as soon as possible after recording, while the recording conditions are still fresh in your mind.

10. Don't leave a microphone, tape recorder, or tapes in a closed car on a warm day; high temperatures can easily damage them. Insulate your materials in some way, such as by putting them in a cooler or by wrapping them in a blanket or sleeping bag.

11. On analog recordings, record at least 10 seconds from a pitch pipe or tuning fork at the head of each reel. If you suspect machine trouble, record this known tone after each recording, because this technique can help salvage some recordings. (Variation in tape speed on digital machines makes the recordings permanently unretrievable, so this technique is useful only for analog recorders.)

12. Use the proper tape for the recorder. On cassette recorders, use 60-minute (C-60, with 30 minutes/side) or even shorter tapes, because longer tapes are thinner and therefore more subject to problems with pre- and postsignal print-through, stretching, and potential jamming. Also, open-reel tapes of 1.5-mil thickness typically have less print-through than do 1-mil tapes, though, of course, recording time is a tradeoff. Also, calibrate analog recorders for the specific brand and type of tape to be used, and then continue to use that same tape stock once the machine is calibrated.

Recommendations for Storing Tape Collections

Adhering to the following advice will keep a collection in good condition for future use.

1. Be aware of the durability of tape stock. Cassette tapes, for example, are not a good long-term storage medium; print-through can be a serious problem, and high frequencies are lost slowly through intermagnetization of the cramped particles. Thin reel-to-reel tapes (e.g., 0.5 or 1.0 mil thick) have the same potential problems. The archival of digital recordings presents a special concern. One big advantage of digital tapes is that identical copies can be made; the long-term stability of digital recordings is simply unknown, however, because the technology is too new to have a history (Rothenberg 1995). Although both analog and digital tapes degrade in storage, analog tapes can always be stabilized and information can be retrieved; how the electronics or the error correction circuitry of digital systems will respond to tape degradation is unknown (Varela 1994). No long-lasting standards have been set for digital, so formats available today may not be available in the future; indeed, as of early 1994, several major audio manufacturers (Panasonic, Marantz, Casio, and Teac) had discontinued production of portable R-DAT machines. All archivists hope that improvements in digital

recording and stability of the format used will eventually solve these critical problems for long-term storage, but in the meantime it is recommended that digital recordings be backed up in the standard analog format, on ¼-inch tape at 19 or 38 centimeters per second.

2. Avoid temperature and humidity extremes and fluctuations. The ideals for storage are 20°C and 40% relative humidity. High humidity weakens the binder that holds the magnetic coating to the plastic and also promotes fungus growth. Temperature extremes can cause mechanical flexing, which results in permanent deformation of the tape.

3. Deposit copies or originals, or make a commitment to do so, in a long-term, established archive. When you publish, be sure to send voucher specimens to a major archive. Identify the catalog or specimen numbers provided by that archive in your publication so that others can have access to your original data. We encourage journals that publish papers on bioacoustics to require this practice (see also Hardy 1984, E. H. Miller 1985a, 1993, Hunn 1992).

Make the long-term plans for your collection explicit; inform a likely survivor of your intentions and identify them in your will. We plead for planned bequests of collections, either as they are made or at some future time, so that well-documented recordings will be preserved as a resource for others.

4. Protect tapes from magnetic fields. Even the speakers on portable cassette decks can damage tapes. Other sources of high magnetic fields include electric motors, such as those found in vacuum cleaners. To minimize the effects of stray electromagnetic fields, store tapes on earth-grounded nonmagnetizable metal shelving (McKee 1988). Wood shelving is less desirable but can be used provided it is kiln-dried and free of moisture.

5. Clearly label both the tape and the storage box.

6. Keep good records for your collection. Keep the data forms that go with the tapes in a readily accessible place, clearly labeled, in order, and up-to-date. Store a copy of these forms in a safe place, away from the originals. Better yet, maintain an electronic database with copies kept off-site. Such a database facilitates your own use of your collection but can also, if the database matches the format of a chosen major archive, facilitate the incorporation of your private collection into that archive.

7. Back up critical tapes, if they are not already deposited in an archive, and store the backups in a safe place.

8. Avoid splicing original tapes. If you must splice, copy the tapes and then splice the copies. Over the long term, splices can cause severe problems under some conditions.

9. Analog tapes are best stored in the tail-out position after having been played through the recorder (see Wickstrom 1982). This step assures even tape tension and helps avoid tape "staircasing," in which tape wound unevenly from layer to layer can cause tape edge damage and irrevocable tape deformation.

10. Beware of individuals offering to buy your recordings or the rights to your recordings. Commercial entrepreneurs are increasingly trying to buy recordings

and, in the process, also buy the copyright, which may give them exclusive use. If you have questions, consult an archive or a legal expert.

11. To reduce mechanical stress on reel-to-reel tapes, consider storing them on large hub reels. The 2-inch hub diameter of the standard 7-inch reels causes more stress than the 4-inch hubs preferred for archival purposes. (Be aware, of course, that counter numbers based on hub diameter will change from one hub to another; reference numbers based on real time rather than the idiosyncrasies of one recorder's counter are thus more desirable.)

12. When requesting funds for research, include in the budget the cost for safekeeping of the tape collection. Vigorously justify proper archiving practices, because preservation of knowledge in libraries, museums, and archives is likely to be just as important as the findings of the original research for which the recordings were collected.

Obligations of Major Archives

Individual recordists are encouraged to strive for high standards, but major archives have certain obligations, too (see also Lance 1983).

1. Proper archival, safekeeping, and responsible use of bioacoustic specimens have the highest priority. An archive is more than a mere collection of recordings. Ideally, an archive uses an accepted standardized archival format, has climate-controlled storage (manual or automated), duplicates original recordings and stores safety copies off-site, routinely calibrates playback and copy recorders, encourages and facilitates collection of scientifically documented specimens, and provides access to the archive to potential users.

2. A primary obligation of major archives is to use standardized, proven archival methods and technologies while simultaneously investigating new technologies that may provide better long-term and higher-quality storage of acoustic signals (e.g., advances in digital recording technology).

3. Archives are obligated to offer reliable technical advice to recordists on matters of techniques and equipment, and to remain current on new recording technologies for field recording and archives.

4. Electronic databases for each collection should be readily accessible to others and should be kept up-to-date with modern techniques of access. Tracking developments in digital technology, with the goal of making recordings available "on-line" (e.g., via Internet), is important.

5. Broad-scale cooperation must be fostered among archives so that mutual aid and exchange of material to the research community are facilitated.

6. The archive must be used responsibly. The rights of the field recordist—the "owner" of the recorded sound—must be protected, if so requested.

7. Cooperation between the professional and volunteer ornithological communities should be encouraged. This cooperation is facilitated by sharing tapes with each other through a major archive.

8. Archives should make recordings accessible to users by providing listening and copying services.

Advice to the Beginner (and the Virtuoso)

1. Read this appendix.
2. Ask for help. Contact an established archive and ask how to record animal sounds and how to store recordings. Several major archives provide workshops or courses on sound-recording techniques and the principles of bioacoustics. If possible, take one of those courses. At the very least, get advice on equipment and techniques and learn how you can contribute to the larger enterprise.
3. Maintain a well-documented personal collection of wildlife sounds. One of the greatest failings that major archives encounter is the lack of critical voiced information on tape, especially where one recording ends and another begins, and the pertinent information that should accompany each cut. Be consistent in your methods and the way you record data so that someone else could learn your pattern, in your absence, and retrieve important information. Plan so that at any moment the entire collection could be lifted up and deposited, without confusion—and therefore without loss of important data—in an established archive and preserved for future use.
4. Encourage the same action from others who work with wildlife sounds.
5. Remember that current and future use of *your* recordings can benefit others and, perhaps more important, the animals themselves.

Acknowledgments

We thank several deceased colleagues who, through their pioneering recording and archival efforts, helped establish some of the major archives in the world: Arthur A. Allen, Donald J. Borror, Peter Paul Kellogg, Theodore A. Parker III, Paul A. Schwartz, and Boris N. Veprintsev. We thank colleagues who endorsed our primary goals in this chapter, and especially David A. Spector for his editorial help. We also thank the birds themselves, because it is their sounds that we enjoy studying and that have so enriched our lives.

LITERATURE CITED

Numbers in brackets after each reference entry indicate the pages on which the reference is cited in the text.

Adret, P. 1993. Operant conditioning, song learning and imprinting to taped song in the Zebra Finch. Anim. Behav. 46:149–159. [64]

Adret-Hausberger, M. 1983. Variations dialectales des sifflements de l'étourneau sansonnet (*Sturnus vulgaris*) sédentaire en Bretagne. Z. Tierpsychol. 62:55–71. [216]

Adret-Hausberger, M. 1986. Temporal dynamics of dialects in the whistled songs of starlings. Ethology 71:140–152. [215, 216]

Ainley, D. G. 1980. Geographic variation in Leach's Storm Petrel. Auk 97:837–853. [175]

Ainley, D. G., T. J. Lewis, and S. Morrel. 1976. Molt in Leach's and Ashy Storm Petrels. Wilson Bull. 88:76–95. [171]

Alatalo, R. V., A. L. Gustafsson, and A. Lundberg. 1989. Extra-pair paternity and heritability estimates of tarsus length in Pied and Collared Flycatchers. Oikos 56:54–58. [435]

Alatalo, R. V., C. Glynn, and A. Lundberg. 1990. Singing rate and female attraction in the Pied Flycatcher: an experiment. Anim. Behav. 39:601–603. [160, 262, 459, 468]

Alexander, R. D. 1980. Darwinism and human affairs. Pitman, London. [181]

Ali, N. J., S. M. Farabaugh, and R. J. Dooling. 1993. Recognition of contact calls by the Budgerigar (*Melopsittacus undulatus*). Bull. Psychon. Soc. 31:468–470. [108]

Allard, H. A. 1930. The first morning song of some birds of Washington, D.C.: its relation to light. Am. Nat. 64:436–469. [442, 443]

Allee, W. C. 1926. Measurements of environmental factors in the tropical rainforest of Panama. Ecology 7:273–302. [266]

Alvarez-Buylla, A. 1992. Neurogenesis and plasticity in the CNS of adult birds. Exp. Neurol. 115:110–114. [288, 301]

Alvarez-Buylla, A. 1994. Neurogenesis and the unique anatomy of the song control nuclei. J. Ornithol. 135:426. [302]

Alvarez-Buylla, A., M. Theelen, and F. Nottebohm. 1988. Birth of projection neurons in the higher vocal center of the canary forebrain before, during, and after song learning. Proc. Natl. Acad. Sci. USA 85:8722–8726. [288]

Alvarez-Buylla, A., J. Kirn, and F. Nottebohm. 1990. Birth of projection neurons in adult avian brain may be related to perceptual or motor learning. Science 249:1444–1446. [302]

American Ornithologists' Union. 1983. Check-list of North American birds. 6th ed. American Ornithologists' Union, Washington, D.C. [8, 149]

Andersson, M. 1994. Sexual selection. Princeton University Press, Princeton, N.J. [242, 454, 462, 464–466]

Anonymous. 1992. Information for contributors. Condor 94:311–312. [270]

Aoki, K. 1986. A stochastic model of gene-culture coevolution suggested by the "culture historical hypothesis" for the evolution of adult lactose absorption in humans. Proc. Natl. Acad. Sci. USA 83:2929–2933. [217]

Aoki, K. 1987. Gene-culture waves of advance. J. Math. Biol. 25:453–464. [215]

Aoki, K. 1989. A sexual-selection model for the evolution of imitative learning of song in polygynous birds. Am. Nat. 1354:599–612. [217]

Arai, O., I. Taniguchi, and N. Saito. 1989. Correlation between the size of song control nuclei and plumage color change in Orange Bishop Birds. Neurosci. Lett. 98:144–148. [302]

Arcese, P. A. 1987. Age, intrusion pressure and defence against floaters by territorial male Song Sparrows. Anim. Behav. 35:773–784. [65, 69]

Arcese, P. A. 1989a. Intrasexual competition, mating system and natal dispersal in Song Sparrows. Anim. Behav. 37:45–55. [65, 69]

Arcese, P. A. 1989b. Territory acquisition and loss in male Song Sparrows. Anim. Behav. 37:45–55. [362, 367]

Arcese, P. A. 1990. Intrasexual competition, mating system, and natal dispersal in Song Sparrows. Anim. Behav. 38:958–979. [371]

Armstrong, E. A. 1963. A study of bird song. Oxford Univ. Press, London. [214, 216, 442, 443]

Arnold, A. P. 1975. The effects of castration and androgen replacement on song, courtship, and aggression in Zebra Finches. J. Exp. Zool. 191:309–326. [293]

Arnold, A. P. 1992. Developmental plasticity in neural circuits controlling bird song: sexual differentiation and the neural basis of learning. J. Neurobiol. 23:1506–1528. [114]

Arnold, A. P., F. Nottebohm, and D. Pfaff. 1976. Hormone concentrating cells in vocal control and other areas of the brain of the Zebra Finch. J. Comp. Neurol. 165:487–507. [293, 294, 296]

Arnold, A. P., S. W. Bottjer, E. A. Brenowitz, E. J. Nordeen, and K. W. Nordeen. 1986. Sexual dimorphisms in the neural vocal control system in song birds: ontogeny and phylogeny. Brain Behav. Evol. 28:22–31. [114, 291, 296]

Arrowood, P. C. 1988. Duetting, pair bonding and agonistic display in parakeet pairs. Behaviour 106:129–157. [114]

Arvidsson, B. L. 1992. Copulation and mate guarding in the Willow Warbler. Anim. Behav. 43:501–509. [432]

Arvidsson, B. L., and R. Neergaard. 1991. Mate choice in the Willow Warbler—a field experiment. Behav. Ecol. Sociobiol. 29:225–229. [160, 262]

Aschoff, J. 1981. Biological rhythms. Handbook of behavioral neurobiology, no. 4. Plenum Press, New York. [426]

Aschoff, J., and R. Wever. 1962. Beginn und Ende der täglichen Aktivität freilebender Vögel. J. Ornithol. 103:2–27. [333]

Atwood, J. L., V. L. Fitz, and J. E. Bamesberger. 1991. Temporal patterns of singing activity at leks of the White-bellied Emerald. Wilson Bull. 103:373–386. [410]

Austin, J. J., E. Carter, and D. T. Parkin. 1993. Genetic evidence for extra-pair fertilization in socially monogamous Short-tailed Shearwater, *Puffinus tenuirostris* (Procellariiformes, Procellariidae), using DNA-fingerprinting. Aust. J. Zool. 41:1–11. [174]

Austin, J. J., R. W. G. White, and J. R. Ovenden. 1994. Population-genetic structure of a philopatric, colonially-nesting seabird, the Short-tailed Shearwater (*Puffinus tenuirostris*). Auk 111:70–79. [175]

Avise, J. C. 1994. Molecular markers, natural history and evolution. Chapman and Hall, New York. [213]

Avise, J. C., and R. M. Zink. 1988. Molecular genetic divergence between avian sibling species: King and Clapper Rails, Long-billed and Short-billed Dowitchers, Boat-tailed and Great-tailed Grackles, and Tufted and Black-crested Titmice. Auk 105:516–528. [255]

Bain, D. E. 1992. Multi-scale communication by vertebrates. Pages 601–629 *in* Marine mammal sensory systems (J. A. Thomas, R. A. Kastelein, and A. Ya. Supin, Eds.). Plenum Press, New York. [242, 247, 248]

Baker, A. E. M. 1981. Gene flow in house mice: introduction of a new allele into a free-living population. Evolution 35:243–258. [195]

Baker, A. J. 1974. Ecological and behavioural evidence for the systematic status of New Zealand Oystercatchers (Charadriiformes: Haematopodidae). Life Sci. Contr. R. Ont. Mus. 96. [242]

Baker, A. J., and P. A. R. Hockey. 1984. Behavioral and vocal affinities of the African Black Oystercatcher (*Haematopus moquini*). Wilson Bull. 96:656–671. [242]

Baker, A. J., and P. F. Jenkins. 1987. Founder effect and cultural evolution of songs in an isolated population of Chaffinches, *Fringilla coelebs,* in the Chatham Islands. Anim. Behav. 35:1793–1803. [184, 215, 217]

Baker, J. A., and E. D. Bailey. 1987. Sources of phenotypic variation in the separation call of Northern Bobwhite (*Colinus virginianus*). Can. J. Zool. 65:1010–1015. [44, 53, 54]

Baker, J. R. 1938. The relation between latitude and breeding seasons in birds. Proc. Zool. Soc. Lond. A 108:557–582. [259]

Baker, M. C. 1975. Song dialects and genetic differences in White-crowned Sparrows (*Zonotrichia leucophrys*). Evolution 29:226–241. [216]

Baker, M. C. 1982. Genetic population structure and vocal dialects in *Zonotrichia* (Emberizidae). Pages 209–236 *in* Acoustic communication in birds, vol. 2 (D. E. Kroodsma and E. H. Miller, Eds.). Academic Press, New York. [177, 216]

Baker, M. C. 1983. The behavioral response of female Nuttall's White-crowned Sparrows to male song of natal and alien dialects. Behav. Ecol. Sociobiol. 12:309–315. [43, 459]

Baker, M. C. 1987. Intergradation of song between two subspecies of White-crowned Sparrows on the west coast of North America. Ornis Scand. 18:265–268. [213, 215]

Baker, M. C., and M. A. Cunningham. 1985. The biology of bird-song dialects. Behav. Brain Sci. 8:85–133. [4, 11, 15]

Baker, M. C., and L. R. Mewaldt. 1978. Song dialects as barriers to dispersal in White-crowned Sparrows, *Zonotrichia leucophrys nuttalli.* Evolution 32:712–722. [215]

Baker, M. C., and D. B. Thompson. 1985. Song dialects of White-crowned Sparrows: historical processes inferred from patterns of geographic variation. Condor 87:127–141. [213, 215]

Baker, M. C., K. J. Spitler-Nabors, and D. C. Bradley. 1981a. Early experience determines song dialect responsiveness of female sparrows. Science 214:819–821. [290, 459]

Baker, M. C., D. B. Thompson, and G. L. Sherman. 1981b. Neighbor/stranger song discrimination in the White-crowned Sparrow. Condor 83:265–267. [359]

Baker, M. C., K. J. Spitler-Nabors, and D. C. Bradley. 1982. The response of female mountain White-crowned Sparrows to songs from their natal dialect and an alien dialect. Behav. Ecol. Sociobiol. 10:175–179. [459]

Baker, M. C., S. Bottjer, and A. P. Arnold. 1984. Sexual dimorphism and lack of seasonal changes in vocal control regions of the White-crowned Sparrow brain. Brain Res. 295:85–89. [302]

Baker, M. C., T. K. Bjerke, H. Lampe, and Y. O. Espmark. 1986. Sexual response of female Great Tits to variation in size of males' song repertoires. Am. Nat. 128:491–498. [458]

Baker, M. C., T. K. Bjerke, H. Lampe, and Y. O. Espmark. 1987a. Sexual response of female Yellowhammers to differences in regional song dialects and repertoire sizes. Anim. Behav. 35:395–401. [458, 459]

Baker, M. C., K. J. Spitler-Nabors, A. D. Thompson Jr., and M. A. Cunningham. 1987b. Reproductive behaviour of female White-crowned Sparrows: effect of dialects and synthetic hybrid songs. Anim. Behav. 35:1766–1774. [459]

Balaban, E. 1988a. Bird song syntax: learned intraspecific variation is meaningful. Proc. Natl. Acad. Sci. USA 85:3657–3660. [290, 339, 343, 345, 347]

Balaban, E. 1988b. Cultural and genetic variation in Swamp Sparrows (*Melospiza georgiana*). II. Behavioural salience of geographical song variants. Behaviour 105:292–322. [346]

Balcombe, J. P., and M. B. Fenton. 1988. Eavesdropping by bats: the influence of echolocation call design and foraging strategy. Ethology 79:158–166. [416]

Ball, G. F. 1990. Chemical neuroanatomical studies of the steroid-sensitive songbird vocal control system: a comparative approach. Pages 148–167 *in* Hormones, brain and behaviour in vertebrates, vol. 8 (J. Balthazart, Ed.). Karger, Basel. [113, 295, 296]

Ball, G. F. 1994. Neurochemical specializations associated with vocal learning and production in songbirds and Budgerigars. Brain Behav. Evol. 44:234–246. [113]

Balmford, A. 1991. Mate choice on leks. Trends Ecol. Evol. 6:87–92. [415]

Baltz, A. P., and A. B. Clark. 1994. Limited evidence for an audience effect in Budgerigars, *Melopsittacus undulatus.* Anim. Behav. 47:460–462. [98, 104]

Bang, B. G. 1966. The olfactory apparatus of tubenosed birds (Procellariiformes). Acta Nat. 65:391–415. [162]

Baptista, L. F. 1975. Song dialects and demes in sedentary populations of the White-crowned Sparrow (*Zonotrichia leucophrys nuttalli*). Univ. Calif. Publ. Zool. 105:1–52. [6, 13, 15, 50, 213, 215, 216]

Baptista, L. F. 1977. Geographic variation in song and dialects of the Puget Sound White-crowned Sparrow. Condor 79:356–370. [7, 213]

Baptista, L. F. 1978. Territorial, courtship and duet songs of the Cuban Grassquit (*Tiaris canora*). J. Ornithol. 119:91–101. [456]

Baptista, L. F. 1990a. Dialectal variation in the rain call of the Chaffinch. Festschrift honoring Jurgen Nicolai. Vogelwarte 35:249–256. [39, 215]

Baptista, L. F. 1990b. Song learning in White-crowned Sparrows (*Zonotrichia leucophrys*): sensitive phases and stimulus filtering revisited. Pages 143–152 *in* Current topics in avian biology. Proc. Int. 100. DO-G Meeting, Bonn (R. van Elzen, K. L. Schuchmann, and K. Schmidt-Koenig, Eds.). Verlag DO-G, Garmisch-Partenkirchen. [85]

Baptista, L. F. 1993. El estudio de la variación geográfica usando vocalizaciones y las bibliotecas de sonidos de aves neotropicales. Pages 15–30 *in* Curación moderna de colecciones ornitológicas (P. Escalante-Pliego, Ed.). Proc. IV Int. Congr. Neotrop. Birds (Quito, Ecuador, 1991). American Ornithologists' Union, Washington, D.C. [39]

Baptista, L. F., and M. Abs. 1983. Vocalizations. Pages 309–325 *in* Behavior and physiology of the pigeon (M. Abs, Ed.). Academic Press, New York. [44, 46, 47, 50, 52, 53]

Baptista, L. F., and C. K. Catchpole. 1989. Vocal mimicry and interspecific aggression in songbirds: experiments using White-crowned Sparrow imitation of Song Sparrow song. Behaviour 109:247–257. [58]

Baptista, L. F., and S. L. L. Gaunt. 1994. Advances in studies of avian sound communication. Condor 96:817–830. [34, 41, 43, 59, 339]

Baptista, L. F., and H. M. Horblit. 1990. Inheritance and loss of the straw display in estrildid finches. Avic. Mag. 96:141–152. [48, 58]

Baptista, L. F., and R. B. Johnson. 1982. Song variation in insular and mainland California Brown Creepers (*Certhia familiaris*). J. Ornithol. 123:131–144. [215]

Baptista, L. F., and J. R. King. 1980. Geographical variation in song and song dialects of montane White-crowned Sparrows. Condor 82:267–284. [50, 215, 216]

Baptista, L. F., and M. Matsui. 1979. The source of the dive-noise of the Anna Hummingbird. Condor 81:87–89. [54]

Baptista, L. F., and M. L. Morton. 1981. Interspecific song acquisition by a White-crowned Sparrow. Auk 98:383–385. [43]

Baptista, L. F., and M. L. Morton. 1982. Song dialects and mate selection in montane White-crowned Sparrows. Auk 99:537–547. [5, 43, 216]

Baptista, L. F., and M. L. Morton. 1988. Song learning in montane White-crowned Sparrows: from whom and when. Anim. Behav. 36:1753–1764. [5, 6, 43, 63, 71, 215–217]

Baptista, L. F., and L. Petrinovich. 1984. Social interaction, sensitive phases and the song template hypothesis in the White-crowned Sparrow. Anim. Behav. 32:172–181. [43, 44, 58, 61, 63]

Baptista, L. F., and L. Petrinovich. 1986. Song development in the White-crowned Sparrow: social factors and sex differences. Anim. Behav. 34:1359–1371. [12, 43, 44, 61, 63]

Baptista, L. F., and K.-L. Schuchmann. 1990. Song learning in the Anna Hummingbird (*Calypte anna*). Ethology 84:15–26. [44, 59, 97, 277, 294]

Baptista, L. F., and P. W. Trail. 1992. The role of song in the evolution of passerine diversity. Syst. Biol. 41:242–247. [11, 40, 41]

Baptista, L. F., W. I. Boarman, and P. Kandianidis. 1983. Behavior and taxonomic status of Grayson's Dove. Auk 100:907–919. [48, 50]

Baptista, L. F., D. A. Bell, and P. W. Trail. 1993a. Song learning and production in the White-crowned Sparrow: parallels with sexual imprinting. Neth. J. Zool. 43:17–33. [21, 31, 43]

Baptista, L. F., P. W. Trail, B. B. DeWolfe, and M. L. Morton. 1993b. Singing and its functions in female White-crowned Sparrows. Anim. Behav. 46:511–524. [43, 342, 355]

Barclay, R. M. R. 1982. Interindividual use of echolocation calls: eavesdropping by bats. Behav. Ecol. Sociobiol. 10:271–275. [416]

Barclay, R. M. R., M. L. Leonard, and G. Friesan. 1985. Nocturnal singing by Marsh Wrens. Condor 87:418–422. [430, 441]

Barlow, G. W. 1991. Nature-nurture and the debates surrounding ethology and sociobiology. Am. Zool. 286–296. [39]

Barrowclough, G. F. 1980. Gene flow, effective population sizes, and genetic variance components in birds. Evolution 34:789–798. [215]

Basolo, A. L. 1990. Female preference predates the evolution of the sword in swordtail fish. Science 250:808–810. [466]

Bateman, R., I. Goddard, R. O'Grady, V. A. Funk, R. Mooi, W. J. Kress, and P. Cannell. 1990. The feasibility of reconciling human phylogeny and the history of language. Curr. Anthropol. 31:1–24. [213, 214, 219]

Bateson, P. P. G. 1987. Imprinting as a process of competitive exclusion. Pages 151–168 *in* Imprinting and cortical plasticity (J. P. Rauschecker and P. Marler, Eds.). Wiley, New York. [89]

Bauer, H.-G. 1987. Geburtsortstreue und Streuungsverhalten junger Singvögel. Vogelwarte 34:15–32. [215]

Baylis, J. R. 1982. Avian vocal mimicry: its function and evolution. Pages 51–83 *in* Acoustic communication in birds, vol. 2 (D. E. Kroodsma and E. H. Miller, Eds.). Academic Press, New York. [276, 294]

Becker, P. H. 1978. Sumpfmeise lernt künstliche Gesangsstrophe vom Tonband und tradiert sie. Naturwissenschaften 65:338. [236]

Becker, P. H. 1982. The coding of species specific characteristics in bird sounds. Pages 213–252 *in* Acoustic communication in birds, vol. 1 (D. E. Kroodsma and E. H. Miller, Eds.). Academic Press, New York. [160, 175, 339]

Becker, P. H. 1990. Der Gesang des Feldschwirls (*Locustella neavia*) bei Lernentzug. Vogelwarte 35:257–267. [233]

Becker, P. H., G. Thielcke, and K. Wüstenberg. 1980a. Versuche zum angenommenen Kontrastverlust im Gesang der Blaumeise (*Parus caeruleus*) auf Teneriffa. J. Ornithol. 121:81–95. [235]

Becker, P. H., G. Thielcke, and K. Wüstenberg. 1980b. Der Tonhöhenverlauf ist entscheidend für das Gesangserkennen beim mitteleuropäischen Zilpzalp (*Phylloscopus collybita*). J. Ornithol. 121:292–244. [227]

Beecher, M. D. 1989. Signalling systems for individual recognition: an information theory approach. Anim. Behav. 38:248–261. [175]

Beecher, M. D. 1991. Successes and failures of parent-offspring recognition in animals. Pages 94–124 *in* Kin recognition (P. G. Hepper, Ed.). Cambridge Univ. Press, Cambridge. [356, 357]

Beecher, M. D., and P. K. Stoddard. 1990. The role of bird song and calls in individual recognition: contrasting field and laboratory perspectives. Pages 375–408 *in* Comparative perception, vol. 2 (W. C. Stebbins and M. A. Berkeley, Eds.). Wiley, New York. [66, 367]

Beecher, M. D., I. M. Beecher, and S. Lumpkin. 1981. Parent-offspring recognition in Bank Swallows (*Riparia riparia*). I. Natural history. Anim. Behav. 29:86–94. [356]

Beecher, M. D., S. E. Campbell, and J. M. Burt. 1994a. Song perception in the Song Sparrow: birds classify by song type but not by singer. Anim. Behav. 47:1343–1351. [66, 74, 129, 130, 367, 370–373]

Beecher, M. D., S. E. Campbell, and P. K. Stoddard. 1994b. Correlation of song learning and territory establishment strategies in the Song Sparrow. Proc. Natl. Acad. Sci. USA 91:1450–1454. [15, 16, 61, 67, 72, 74, 182, 186, 215, 290, 371, 374]

Beecher, M. D., P. K. Stoddard, S. E. Campbell, and C. L. Horning. In press. Repertoire matching between neighbouring Song Sparrows. Anim. Behav. [66, 78, 372, 411]

Beecher , M. D., S. E. Campbell, and J. M. Burt. Unpubl. ms. A laboratory study of song learning in Song Sparrows. [61, 63, 67, 69]

Beehler, B. M., T. K. Pratt, and D. A. Zimmerman. 1986. Birds of New Guinea. Princeton Univ. Press, Princeton, N.J. [53]

Beer, C. G. 1982. Conceptual issues in the study of communication. Pages 279–310 *in* Acoustic communication in birds, vol. 2 (D. E. Kroodsma and E. H. Miller, Eds.). Academic Press, New York. [25, 28]

Beissinger, S. R., and J. R. Waltman. 1991. Extraordinary clutch size and hatching asynchrony of a Neotropical parrot. Auk 108:863–871. [104]

Bekoff, M. 1988a. Birdsong and the "problem" of nature and nurture: endless chirping about inadequate evidence of merely singing the blues about inevitable biases in, and limitations of, human inference? Behav. Brain Sci. 11:631. [39]

Bekoff, M. 1988b. Motor training and physical fitness: possible short- and long-term influences on the development of individual differences in behavior. Dev. Psychobiol. 21:601–612. [312]

Belcher, J. W., and W. L. Thompson. 1969. Territorial defense and individual recognition in the Indigo Bunting. Jack-Pine Warbler 47:76–83. [359]

Beletsky, L. D. 1983. An investigation of individual recognition by voice in female Red-winged Blackbirds. Anim. Behav. 31:355–362. [359, 373]

Beletsky, L. D., and G. H. Orians. 1989. Familiar neighbors enhance breeding success in birds. Proc. Natl. Acad. Sci. USA 86:7933–7936. [77]

Beletsky, L. D., S. Chao, and D. G. Smith. 1980. An investigation of song-based species recognition in the Red-winged Blackbird (*Agelaius phoeniceus*). Behaviour 73:189–203. [344, 345]

Beletsky, L. D., B. J. Higgins, and G. H. Orians. 1986. Communication by changing signals: call switching in Red-winged Blackbirds. Behav. Ecol. Sociobiol. 18:221–229. [413, 416, 417]

Bennett, P. M., and P. H. Harvey. 1987. Active and resting metabolic rate in birds: allometry, phylogeny and ecology. J. Zool. 213:327–363. [311]

Bereson, R. C., J. Rhymer, and R. Fleischer. In press. Extra-pair fertilizations in Wilson's Warblers and correlates of cuckoldry. Behav. Ecol. Sociobiol. [264]

Bergmann, H. H., and H.-W. Helb. 1982. Stimmen der Vögel Europas. BLV Verlag, Munich. [243, 250, 251, 252, 306]

Bernard, D., and G. F. Ball. 1993. Two independent markers demonstrate an increased volume of HVC in short-day photosensitive relative to long-day photorefractory European Starlings. Soc. Neurosci. Abstr. 19:1017. [302]

Bertram, B. 1970. The vocal behaviour of the Indian Hill Mynah, *Gracula religiosa.* Anim. Behav. Monogr. 3:79–192. [44]

Bierregaard, R. O., Jr., and T. E. Lovejoy. 1989. Effects of forest fragmentation on Amazonian understory bird communities. Acta Amazonica 19:215–242. [269, 273]

Bijnens, L. 1988. Blue Tit, *Parus caeruleus,* song in relation to survival, reproduction and biometry. Bird Study 35:61–67. [308]

Bijnens, L., and A. A. Dhondt. 1984. Vocalizations in a Belgian Blue Tit, *Parus c. caeruleus,* population. Gerfaut 74:243–269. [147, 149, 167]

Birkhead, T. R., and J. D. Biggins. 1987. Reproductive synchrony and extra-pair copulation in birds. Ethology 74:320–334. [263]

Birkhead, T. R., and A. P. Møller. 1992. Sperm competition in birds: evolutionary causes and consequences. Academic Press, London. [263, 264]

Birkhead, T. R., and A. P. Møller. 1993. Female control of paternity. Trends Ecol. Evol. 8:100–104. [341]

Bitterbaum, E., and L. F. Baptista. 1979. Geographical variation in songs of California House Finches (*Carpodacus mexicanus*). Auk 96:462–474. [40]

Bjerke, T. K., and T. H. Bjerke. 1981. Song dialects in the Redwing, *Turdus iliacus*. Ornis Scand. 12:40–50. [216]

Björkland, M., A. P. Møller, J. Sundberg, and B. Westman. 1992. Female Great Tits, *Parus major*, avoid extra-pair copulation attempts. Anim. Behav. 43:691–693. [432]

Blanchard, B. D. 1941. The White-crowned Sparrows (*Zonotrichia leucophrys*) of the Pacific seaboard: environment and annual cycle. Univ. Calif. Publ. Zool. 46:1–178. [215, 216]

Bledsoe, A. H. 1987a. Estimation of phylogeny from molecular distance data: the issue of variable rates. Auk 104:563–565. [214]

Bledsoe, A. H. 1987b. DNA evolutionary rates in nine-primaried passerine birds. Mol. Biol. Evol. 4:559–571. [214]

Bock, W. J. 1958. A generic review of the plovers (Charadriinac, Avcs). Bull. Mus. Comp. Zool. 118:27–97. [241, 252]

Böhner, J. 1990. Early acquisition of song in the Zebra Finch (*Taeniopygia guttata*). Anim. Behav. 39:369–374. [85]

Bondrup-Nielsen, S. 1984. Vocalizations of the Boreal Owl, *Aegolius funereus richardsoni*, in North America. Can. Field-Nat. 98:191–197. [441]

Bonke, B., D. Bonke, and H. Scheich. 1979. Connectivity of the auditory forebrain nuclei in the Guinea Fowl. Cell Tissue Res. 200:101–121. [295]

Bonner, J. T. 1980. The evolution of culture in animals. Princeton Univ. Press, Princeton, N.J. [181]

Borror, D. L. 1959. Songs of the Chipping Sparrow. Ohio J. Sci. 59:347–356. [13, 14, 18]

Borror, D. L. 1967. Song of the Yellowthroat. Living Bird 6:141–161. [13]

Borror, D. L., and W. Gunn. 1965. Variation in White-throated Sparrow songs. Auk 82:26–47. [297]

Boswall, J., and R. Kettle. 1979. A rcviscd world list of wildlife sound libraries. Recorded Sound 74–75:70–72; Biophon 7:3–6. [476]

Bottjcr, S., and S. Hewer. 1992. Castration and antisteroid treatment impair vocal learning in male Zebra Finches. J. Neurobiol. 23:337–353. [293]

Bottjer, S., E. Miesner, and A. P. Arnold. 1984. Forebrain lesions disrupt development but not maintenance of song in passerine birds. Science 224:901–903. [288]

Bottjer, S., K. Halsema, S. Brown, and E. Miesner. 1989. Axonal connections of a forebrain nucleus involved with vocal learning in Zebra Finches. J. Comp. Neurol. 279:312–326. [286]

Boughey, M. J., and N. S. Thompson. 1981. Song variety in the Brown Thrasher (*Toxostoma rufum*). Z. Tierpsychol. 56:47–58. [13]

Bower, G. H. 1970. Organizational factors in memory. Cognit. Psychol. 1:18–46. [95]

Bower, J. L., and C. W. Clark. Unpubl. ms. A test of the accuracy of an acoustic location system based on the singing positions of wild birds. [420, 422]

Bowman, R. I. 1979. Adaptive morphology of song dialects in Darwin's finches. J. Ornithol. 120:353–389. [54]

Boyd, R., and P. J. Richerson. 1985. Culture and the evolutionary process. Univ. Chicago Press, Chicago. [181, 186]

Brackenbury, J. H. 1978. A comparison of the origin and temporal arrangement of pulsed sounds in the songs of the Grasshopper and Sedge Warblers, *Locustella naevia* and *Acrocephalus schoenobaenus*. J. Zool. Lond. 184:187–206. [317]

Brackenbury, J. H. 1979. Power capabilities of the avian sound-producing system. J. Exp. Biol. 78:163–166. [311]

Brackenbury, J. H. 1989. Functions of the syrinx and the control of sound production. Pages

193–220 *in* Form and function in birds (A. S. King and J. McLelland, Eds.). Academic Press, New York. [312, 315, 316]

Bradbury, J., and S. L. Vehrencamp. 1994. SingIt! A program for interactive playback on the Macintosh. Bioacoustics 5:308–310. [398, 401]

Braun, M. J., and T. A. Parker III. 1985. Molecular, morphological, and behavioral evidence concerning the taxonomic relationships of "*Synallaxis*" *gularis* and other synallaxines. Pages 333–346 *in* Neotropical ornithology (P. A. Buckley, M. S. Foster, E. S. Morton, R. S. Ridgely, and F. G. Buckley, Eds.). Ornithological Monographs 36. American Ornithologists' Union, Washington D.C. [269, 272]

Brauth, S. E., and C. M. McHale. 1988. Auditory pathways in the Budgerigar. II. Intratelencephalic pathways. Brain Behav. Evol. 32:193–207. [113]

Brauth, S. E., C. M. McHale, C. A. Brasher, and R. J. Dooling. 1987. Auditory pathways in the Budgerigar. I. Thalamo-telencephalic projections. Brain Behav. Evol. 30:174–199. [113]

Brauth, S. E., J. T. Heaton, S. E. Durand, W. Liang, and W. S. Hall. 1994. Functional anatomy of forebrain auditory pathways in the Budgerigar, *Melopsittacus undulatus.* Brain Behav. Evol. 44:210–233. [113]

Bregman, A. S. 1990. Auditory scene analysis: the perceptual organization of sound. MIT Press, Cambridge, Mass. [125]

Brémond, J.-C. 1966. Recherches sur la sémantique et les éléments physiques déclencheurs de comportements dans les signaux acoustiques du rouge-gorge (*Erithacus rubecula* L.). Thesis, Faculté des Sciences de Paris. [388]

Brémond, J.-C. 1968. Recherches sur la sémantique et les éléments vecteurs d'information dans les signaux acoustiques du rouge-gorge (*Erithacus rubecula* L.). Terre Vie 2:109–220. [359, 369, 405, 411]

Brémond, J.-C. 1978. Acoustic competition between the song of the wren (*Troglodytes troglodytes*) and the songs of other species. Behaviour 65:89–98. [437]

Brémond, J.-C., and T. Aubin. 1990. Responses to distress calls by Black-headed Gulls, *Larus ridibundus:* the role of non-degraded features. Anim. Behav. 39:503–511. [410]

Brémond, J.-C., and T. Aubin. 1992. Cadence d'émission du chant territorial du troglodyte (*Troglodytes troglodytes*): experimentation interactive. C. R. Acad. Sci. Paris Sér. III 314:37–42. [401]

Brenowitz, E. A. 1982a. Aggressive response of Red-winged Blackbirds to mockingbird song imitation. Auk 99:584–586. [346]

Brenowitz, E. A. 1982b. Long-range communication of species identity by song in the Red-winged Blackbird. Behav. Ecol. Sociobiol. 10:29–38. [323, 344, 346]

Brenowitz, E. A. 1982c. The active space of Red-winged Blackbird song. J. Comp. Physiol. A 147:511–522. [321, 323, 324, 330, 336]

Brenowitz, E. A. 1983. The contribution of temporal song cues to species recognition in the Red-winged Blackbird. Anim. Behav. 31:1116–1127. [346]

Brenowitz, E. A. 1991a. Evolution of the vocal control system in the avian brain. Semin. Neurosci. 3:399–407. [113, 290]

Brenowitz, E. A. 1991b. Altered perception of species-specific song by female birds after lesions of a forebrain nucleus. Science 251:303–305. [293, 295, 296, 342, 374]

Brenowitz, E. A. 1994. Flexibility and constraint in the evolution of animal communication. Pages 247–258 *in* Flexibility and constraints in behavioral systems (R. Greenspan and B. Kyriacou, Eds.). Wiley, New York. [301]

Brenowitz, E. A., and A. P. Arnold. 1985. Lack of sexual dimorphism in steroid accumulation in vocal control regions of duetting song birds. Brain Res. 344:172–175. [114, 294, 296, 355]

Brenowitz, E. A., and A. P. Arnold. 1986. Interspecific comparisons of the size of neural song control regions and song complexity in duetting birds: evolutionary implications. J. Neurosci. 6:2875–2879. [291, 296, 298, 299, 342]

Brenowitz, E. A., and A. P. Arnold. 1989. Accumulation of estrogen in a vocal control brain region of a duetting song bird. Brain Res. 480:119–125. [294, 296]

Brenowitz, E. A., and A. P. Arnold. 1992. Hormone accumulation in song regions of the canary brain. J. Neurobiol. 23:871–880. [294, 296]

Brenowitz, E. A., A. P. Arnold, and R. Levin. 1985. Neural correlates of female song in tropical duetting birds. Brain Res. 343:104–112. [114, 291, 296]

Brenowitz, E. A., B. Nalls, J. C. Wingfield, and D. E. Kroodsma. 1991. Seasonal changes in avian song nuclei without seasonal changes in song repertoire. J. Neurosci. 11:1367–1375. [290, 296, 301, 302, 303, 374]

Brenowitz, E. A., B. Nalls, D. E. Kroodsma, and C. Horning. 1994. Female Marsh Wrens do not provide evidence of anatomical specializations of song nuclei for perception of male song. J. Neurobiol. 25:197–208. [298, 299, 374]

Brenowitz, E. A., K. Lent, and D. E. Kroodsma. 1995. Brain space for learned song in birds develops independently of song learning. J. Neurosci. 15:6281–6286. [299]

Brereton, J. L. G. 1971. A self-regulation to density-independent continuum in Australian parrots, and its implication for ecological management. Pages 207–221 *in* The scientific management of animal and plant communities for conservation (E. Duffey and A. S. Watt, Eds.). Blackwell, London. [112]

Brereton, J. L. G., and R. W. Pigeon. 1966. The language of the Eastern Rosella. Aust. Nat. Hist. 15:225–229. [97, 105, 112]

Bretagnolle, V. 1988. Social behavior of the Southern Giant Petrel. Ostrich 59:116–125. [162, 163, 165]

Bretagnolle, V. 1989a. Temporal progression of the Giant Petrel courtship. Ethology 80:245–254. [161, 162, 165, 174]

Bretagnolle, V. 1989b. Calls of Wilson's Storm-Petrel: functions, individual and sexual recognitions and geographic variation. Behaviour 111:98–112. [162, 163, 165, 167, 172–176]

Bretagnolle, V. 1990a. Effet de la lune sur l'activité des pétrels (Aves) aux Îles Salvages (Portugal). Can. J. Zool. 68:1404–1409. [161, 162, 171]

Bretagnolle, V. 1990b. Rôle du comportement dans la formation du couple, l'isolement reproductif, et la spéciation: le cas des pétrels (Aves). Thèse de 3ème cycle, Université de Rennes I. [162]

Bretagnolle, V. 1990c. Behavioral affinities of the Blue Petrel, *Halobaena caerulea.* Ibis 132:102–105. [163, 165, 172, 176]

Bretagnolle, V. 1993. Adaptive significance of seabird coloration: the case of Procellariiformes. Am. Nat. 142:141–173. [162, 177]

Bretagnolle, V. 1995. Systematics of the Soft-plumaged Petrel, *Pterodroma mollis,* complex: new insight from vocalizations. Ibis 137:207–218. [163, 165, 175]

Bretagnolle, V. In press. Information exchange, colonial breeding and long-term monogamy: constraints acting on acoustic communication in waterbirds. *In* Colonial breeding in waterbirds (F. Cézilly, H. Haffner, and D. N. Nettleship, Eds.). Oxford Univ. Press, Oxford. [171]

Bretagnolle, V., and C. Attié. 1991. Status of *Pterodroma baraui:* colony sites, breeding population and taxonomic affinities. Colon. Waterbirds 14:25–33. [165, 167]

Bretagnolle, V., and F. Genevois. Unpubl. ms. a. Geographic variation in the calls of the Blue Petrel, *Halobaena caerulea:* effects of sex and geographic scale. [176, 177]

Bretagnolle, V., and F. Genevois. Unpubl. ms. b. Factors affecting individual stereotypy in the calls of the Blue Petrel, *Halobeana caerulea* (Aves). [176]

Bretagnolle, V., and B. Lequette. 1990. Structural variation in the call of the Cory's Shearwater (*Calonectris diomedea,* Aves, Procellariidae). Ethology 85:313–323. [162, 163, 165, 167, 172–176]

Bretagnolle, V., and P. Robisson. 1991. Species-specific recognition in birds: an experimental investigation in the Wilson's Storm-Petrel (Procellariiformes, Hydrobatidae) using digitalized signals. Can. J. Zool. 69:1669–1673. [172, 176]

Bretagnolle, V., and J.-C. Thibault. In press. A method for sexing fledgings in the Cory's Shearwater, and comments on sex ratio variations over a four year period. Auk. [163]

Bretagnolle, V., R. Zotier, and P. Jouventin. 1990. Population biology of four prions (*Pachyptila*) in the Indian Ocean, and consequences on their taxonomic status. Auk 107:305–316. [165, 175]

Bretagnolle, V., M. Carruthers, M. Cubitt, F. Bioret, and J. P. Cuillandre. 1991. Six captures of a Dark-rumped and Fork-tailed Storm-Petrel in the northeastern Atlantic. Ibis 133:351–356. [175, 176]

Bretagnolle, V., F. Genevois, and F. Mougeot. Unpubl. ms. Intra and intersexual function of the call in a non-passerine bird. [165, 172, 174, 176]

Bril, B., and C. Sabatier. 1986. The cultural context of motor development: postural manipulations in the daily life of Bambara babies (Mali). Int. J. Behav. Dev. 9:439–453. [23]

Brindley, E. L. 1991. Response of European Robins to playback of song: neighbour recognition and overlapping. Anim. Behav. 41:503–512. [359, 369, 406]

Brockway, B. F. 1964a. Ethological studies of the Budgerigar (*Melopsittacus undulatus*): non-reproductive behavior. Behaviour 22:193–222. [97, 98, 100, 104, 105]

Brockway, B. F. 1964b. Ethological studies of the Budgerigar (*Melopsittacus undulatus*): reproductive behavior. Behaviour 23:294–324. [97, 98, 100, 104, 105]

Brockway, B. F. 1965. Stimulation of ovarian development and egg laying by male courtship vocalization in Budgerigars (*Melopsittacus undulatus*). Anim. Behav. 12:493–501. [105, 112]

Brockway, B. F. 1967. The influence of vocal behavior on the performer's testicular activity in Budgerigars (*Melopsittacus undulatus*). Wilson Bull. 79:328–334. [105, 112]

Brockway, B. F. 1969. Roles of Budgerigar vocalization in the integration of breeding behaviour. Pages 131–158 *in* Bird vocalizations (R. A. Hinde, Ed.). Cambridge Univ. Press, Cambridge. [97, 98, 104, 105, 112]

Brooke, M. de L. 1978. Sexual differences in the voice and individual recognition in the Manx Shearwater (*Puffinus puffinus*). Anim. Behav. 26:622–629. [172, 175]

Brooke, M. de L. 1986. The vocal system of two nocturnal burrowing petrels, the White-chinned, *Procellaria aequinoctialis,* and the Grey, *P. cinerea.* Ibis 128:502–512. [163, 165, 168, 172–175, 177]

Brooke, M. de L. 1990. The Manx Shearwater. T. and A. D. Poyser, London. [165, 172, 174, 176]

Brooks, D. R., and D. A. McLennan. 1991. Phylogeny, ecology, and behavior: a research program in comparative biology. Univ. Chicago Press, Chicago. [6, 245, 467]

Brooks, R. J., and J. B. Falls. 1975a. Individual recognition by song in White-throated Sparrows. I. Discrimination of songs of neighbours and strangers. Can. J. Zool. 53:879–888. [345]

Brooks, R. J., and J. B. Falls. 1975b. Individual recognition by song in White-throated Sparrows. III. Song features used in individual recognition. Can. J. Zool. 53:1749–1761. [359]

Brown, C. H. 1982. Ventriloquial and locatable vocalizations in birds. Z. Tierpsychol. 59:338–350. [417]

Brown, C. R. 1980. Sleeping behavior of Purple Martins. Condor 82:170–175. [432]

Brown, E. D. 1985. The role of song and vocal imitation among Common Crows (*Corvus brachyrhynchos*). Z. Tierpsychol. 68:115–136. [114]

Brown, E. D., and S. M. Farabaugh. 1992. Song sharing in a group-living songbird, the Australian Magpie, *Gymnorhina tibicen.* III. Sex specificity and individual specificity of vocal parts in communal chorus and duet songs. Behaviour 118:244–274. [114]

Brown, E. D., S. M. Farabaugh, and C. J. Veltman. 1988. Song sharing in a group-living songbird, the Australian Magpie, *Gymnorhina tibicen.* I. Vocal sharing within and among groups. Behaviour 104:1–28. [109, 114]

Brown, R. G. B. 1963. The behaviour of the Willow Warbler, *Phylloscopus trochilus,* in continuous daylight. Ibis 105:63–75. [437, 443]

Brown, S. D., and R. J. Dooling. 1992. Perception of conspecific faces by Budgerigars (*Melopsittacus undulatus*). I. Natural faces. J. Comp. Psychol. 106:203–216. [98, 110]

Brown, S. D., and R. J. Dooling. 1993. Perception of conspecific faces by Budgerigars (*Melopsittacus undulatus*). II. Synthetic models. J. Comp. Psychol. 107:48–60. [98, 110, 111]

Brown, S. D., R. J. Dooling, and K. O'Grady. 1988. Perceptual organization of acoustic stimuli by Budgerigars (*Melopsittacus undulatus*). III. Contact calls. J. Comp. Psychol. 102:236–247. [98, 108, 115]

Browning, M. R. 1977. Geographic variation in Dunlins, *Calidris alpina,* of North America. Can. Field-Nat. 91:391–393. [244]

Brumfield, R. T., and A. P. Capparella. In press. Historical diversification of birds in northwestern South America: a molecular perspective on the role of vicariant events. Evolution. [275]

Bryant, D. M. 1988. Lifetime reproductive success of House Martins. Pages 173–188 *in* Reproductive success (T. H. Clutton-Brock, Ed.). Univ. Chicago Press, Chicago. [174]

Bucher, T. L., M. J. Ryan, and G. A. Bartholomew. 1982. Oxygen consumption during resting, calling, and nest building in the frog *Physalaemus pustulosus.* Physiol. Zool. 55:10–22. [310]

Buckland, S. T., I. Rowley, and D. Williams. 1983. Survival of juvenile Galahs estimated from resighting of tagged individuals. J. Anim. Ecol. 52:563–573. [98, 101]

Buckley, P. A., M. S. Foster, E. S. Morton, R. S. Ridgely, and F. G. Buckley (Eds.). 1985. Neotropical ornithology. Ornithological Monographs 36. American Ornithologists' Union, Washington, D.C. [269]

Bulmer, M. 1989. Structural instability of models of sexual selection. Theor. Pop. Biol. 35:195–206. [463]

Bunn, D. S., A. B. Warburton, and R. D. S. Wilson. 1982. The Barn Owl. Buteo Books, Vermillion, S.D. [441]

Burghardt, G. M. 1970. Defining "communication." Pages 5–18 *in* Communication by chemical signals (J. W. Johnston, D. G. Moulton, and A. Turk, Eds.). Appleton-Century-Crofts, New York. [398]

Burghardt, G. M. 1977. Ontogeny of communication. Pages 71–97 *in* How animals communicate (T. A. Sebeok, Ed.). Indiana Univ. Press, Bloomington. [30]

Burling, R. 1992. Patterns of language: structure, variation, change. Academic Press, New York. [220]

Burns, J. T. 1982. Nests, territories, and reproduction of Sedge Wrens (*Cistothorus platensis*). Wilson Bull. 94:338–349. [8]

Bursian, S. J. 1971. The structure and function of Killdeer vocalizations. M.S. thesis, Univ. Minnesota, Minneapolis. [252]

Butlin, R. K., and M. G. Ritchie. 1994. Behaviour and speciation. Pages 43–79 *in* Behavior and evolution (P. J. B. Slater and T. R. Halliday, Eds.). Cambridge Univ. Press, Cambridge. [212]

Buus, S. 1985. Release from masking caused by envelope fluctuations. J. Acoust. Soc. Am. 78:1958–1965. [331, 332]

Buus, S. In press. Auditory masking. *In* Handbook of acoustics (M. J. Crocker, Ed.). Wiley, New York. [329]

Byers, B. E. In press. Geographic variation of song form within and between Chestnut-sided Warbler populations. Auk. [16]

Byers, B. E., and D. E. Kroodsma. 1992. Development of two song categories by Chestnut-sided Warblers. Anim. Behav. 44:799–810. [8]

Byrkjedal, I. 1990. Song flight of the Pintail Snipe, *Gallinago stenura,* on the breeding grounds. Ornis Scand. 21:239–247. [242, 252]

Byrkjedal, I. 1996. Plovers. T. and A. D. Poyser, Berkhamsted, England. [244, 255]

Cade, W. H. 1975. Acoustically orienting parasitoids: fly phonotaxis to cricket song. Science 190:1312–1313. [423]

Cade, W. H. 1979. The evolution of alternative male reproductive strategies in field crickets. Pages 343–379 *in* Sexual selection and reproductive competition in insects (M. Blum and N. A. Blum, Eds.). Academic Press, London. [423]

Cameron, E. 1968. Vocal communication in the Red-backed Parrot, *Psephotus haematonotus* (Gould). Honors thesis, Univ. New England, Armidale. [97, 105, 112]

Campbell, S. E., J. C. Nordby, and M. D. Beecher. Unpubl. ms. Song learning and territory establishment in a sedentary population of Song Sparrows. [67, 74, 75]

Canady, R. A., D. E. Kroodsma, and F. Nottebohm. 1984. Population differences in complexity of a learned skill are correlated with the brain space involved. Proc. Natl. Acad. Sci. USA 81:6232–6234. [41, 298, 299]

Cannon, C. E. 1984. Flock size of feeding Eastern and Pale-headed Rosellas. Aust. Wildl. Res. 11:349–355. [102]

Capparella, A. P. 1988. Genetic variation in Neotropical birds: implications for the speciation process. Pages 1658–1664 *in* Acta XIX Congr. Int. Ornithol. (H. Ouellet, Ed.). National Museum of Natural Science, Univ. Ottawa Press, Ottawa. [271, 275]

Capparella, A. P. 1991. Neotropical avian diversity and riverine barriers. Pages 307–316 *in* Acta XX Congr. Int. Ornithol. (D. B. Bell, Ed.). New Zealand Ornithol. Congr. Trust, Christchurch. [271, 275]

Casey, R. M., and M. C. Baker. 1992. Early social tutoring influences female sexual response in White-crowned Sparrows. Anim. Behav. 44:983–986. [459]

Caspers, G. J., J. Wattel, and W. W. DeJong. 1994. Alpha-A-crystallin sequences group tinamou with ratites. Molec. Biol. Evol. 11:711–713. [280]

Castner, J. L. 1990. Rainforests. Feline Press, Gainesville, Fl. [274]

Catchpole, C. K. 1973. The functions of advertising song in the Sedge Warbler (*Acrocephalus schoenobaenus*) and the Reed Warbler (*A. scirpaceus*). Behaviour 46:300–320. [440, 448, 455, 460]

Catchpole, C. K. 1976. Temporal and sequential organisation of song in the Sedge Warbler (*Acrocephalus schoenobaenus*). Behaviour 59:226–246. [406]

Catchpole, C. K. 1980. Sexual selection and the evolution of complex songs among European warblers of the genus *Acrocephalus*. Behaviour 74:149–166. [16, 160, 262, 298, 457, 460]

Catchpole, C. K. 1982. The evolution of bird sounds in relation to mating and spacing behavior. Pages 297–319 *in* Acoustic communication in birds, vol. 1 (D. E. Kroodsma and E. H. Miller, Eds.). Academic Press, New York. [124, 160, 339]

Catchpole, C. K. 1983. Variation in the song of the Great Reed Warbler, *Acrocephalus arundinaceus,* in relation to mate attraction and territorial defence. Anim. Behav. 31:1217–1225. [80]

Catchpole, C. K. 1986. Song repertoires and reproductive success in the Great Reed Warbler, *Acrocephalus arundinaceus.* Behav. Ecol. Sociobiol. 19:439–445. [297, 298, 456, 460, 469, 471]

Catchpole, C. K. 1987. Bird song, sexual selection and female choice. Trends Ecol. Evol. 2:94–97. [16]

Catchpole, C. K. 1989. Pseudoreplication and external validity: playback experiments in avian bioacoustics. Trends Ecol. Evol. 4:286–287. [172]

Catchpole, C. K., and A. Rowell. 1993. Song sharing and local dialects in a population of the European Wren, *Troglodytes troglodytes.* Behaviour 125:67–78. [185, 191]

Catchpole, C. K., J. Dittami, and B. Leisler. 1984. Differential responses to male song repertoires in female songbirds implanted with oestradiol. Nature 312:563–564. [458, 460]

Catchpole, C. K., B. Leisler, and J. Dittami. 1986. Sexual differences in the responses of captive Great Reed Warblers (*Acrocephalus arundinaceus*) to variation in song structure and repertoire size. Ethology 73:69–77. [458, 459, 460, 470]

Cavalli-Sforza, L. L. 1991. Genes, people and languages. Sci. Am. 265(5):104–110. [214]

Cavalli-Sforza, L. L., and M. Feldman. 1981. Cultural transmission and evolution: a quantitative approach. Princeton Univ. Press, Princeton, N.J. [181, 183, 198, 213, 215, 217]

Cavalli-Sforza, L. L., A. Piazza, P. Menozzi, and J. Mountain. 1988. Reconstruction of human evolution: bringing together genetic, archaeological, and linguistic data. Proc. Natl. Acad. Sci. USA 85:6002–6006. [214, 215]

Cavalli-Sforza, L. L., P. Menozzi, and A. Piazza. 1994. The history and geography of human genes. Princeton Univ. Press, Princeton. [215]

Chaiken, M., J. Böhner, and P. Marler. 1993. Song acquisition in the European Starling, *Sturnus vulgaris:* a comparison of the songs of live-tutored, untutored, and wild-caught males. Anim. Behav. 46:1079–1090. [85]

Chapman, C. A., L. J. Chapman, and L. Lefebvre. 1989. Variability in parrot flock size: possible functions of communal roosts. Condor 91:842–847. [102]

Chappell, M. A., M. Zuk, T. H. Kwan, and T. S. Johnson. 1995. Energy cost in an avian display: crowing in the Red Junglefowl. Anim. Behav. 49:255–257. [436, 467]

Charnov, E. R., and J. R. Krebs. 1975. The evolution of alarm calls: altruism or manipulation? Am. Nat. 109:107–112. [416]

Cheng, M.-F. 1992. For whom does the female dove coo: a case for the role of vocal self-stimulation. Anim. Behav. 43:1035–1044. [429]

Chu, P. C. 1994. Historical examination of delayed plumage maturation in the shorebirds (Aves: Charadriiformes). Evolution 48:327–350. [241]

Chu, P. C. 1995. Phylogenetic reanalysis of Strauch's osteological data set for the Charadriiformes. Condor 97:175–196. [241]

Clancey, P. A., R. K. Brooke, and J. C. Sinclair. 1981. Variation in the current nominate subspecies of *Pterodroma mollis* (Gould) (Aves, Procellariiformes). Afr. Mus. Novit. 12:203–213. [175]

Clark, C. W. 1980. A real-time direction finding device for determining the bearing to the underwater sounds of southern right whales, *Eubalaena australis.* J. Acoust. Soc. Am. 68:508–511. [419]

Clark, C. W. 1989. Call tracks of bowhead whales based on call characteristics as an independent means of determining tracking parameters. Report of the Sub-Committee on Protected Species and Aboriginal Subsistence Whaling. Appendix. Rep. Int. Whaling Comm. 39:111–112. [419]

Clark, C. W., and W. T. Ellison. 1989. Numbers and distributions of bowhead whales, *Baleana mysticetus,* based on the 1986 acoustic study off Pt. Barrow, Alaska. Rep. Int. Whaling Comm. 39:297–303. [419]

Clark, C. W., W. T. Ellison, and K. Beeman. 1986. Acoustic tracking and distribution of migrating bowhead whales, *Baleana mysticetus,* off Point Barrow, Alaska in the spring of 1984. Rep. Int. Whaling Comm. 36:502. [419]

Clark, R. B. 1960. Habituation of the polychaete *Nereis* to sudden stimuli. I. General properties of the habituation process. Anim. Behav. 8:82–91. [471]

Clayton, N. S. 1987a. Song tutor choice in Zebra Finches. Anim. Behav. 35:714–722. [63]

Clayton, N. S. 1987b. Song learning in cross-fostered Zebra Finches: a re-examination of the sensitive phase. Behaviour 102:67–81. [85]

Clayton, N. S. 1988. Song discrimination learning in Zebra Finches. Anim. Behav. 36:1016–1024. [458]

Clayton, N. S. 1989. Song, sex and sensitive phases in the behavioural development of birds. Trends Ecol. Evol. 4:82–84. [42, 51]

Clayton, N. S. 1990a. Assortative mating in Zebra Finch subspecies, *Taeniopygia guttata guttata* and *T. g. castonotus.* Philos. Trans. R. Soc. Lond. B 330:351–370. [8, 54]

Clayton, N. S. 1990b. Subspecies recognition and song learning in Zebra Finches. Anim. Behav. 40:1009–1017. [343, 459]

Clayton, N. S., and E. Pröve. 1989. Song discrimination in female Zebra Finches and Bengalese Finches. Anim. Behav. 38:352–354. [459]

Clemmons, J. R. In press. What animals can tell us about linguistic openness: an evolutionary revolution or constraint. *In* Studies in language origins (E. Callary, Ed.). [140]

Clemmons, J. R., and J. L. Howitz. 1990. Development of early vocalizations and the chick-a-dee call in the Black-capped Chickadee, *Parus atricapillus.* Ethology 86:203–223. [145]

Clutton-Brock, T. H., and S. D. Albon. 1979. The roaring of red deer and the evolution of honest advertisement. Behaviour 69:145–169. [384]

Cody, M. L., and J. H. Brown. 1969. Song asynchrony in chapparal birds. Nature 222:778–780. [333]

Colburn, H. S. 1977. Theory of binaural interactions based on auditory-nerve data. II. Detection of tones in noise. J. Acoust. Soc. Am. 61:525–533. [335]

Cole, B. J. 1994. Chaos and behavior: the perspective of nonlinear dynamics. Pages 423–444 *in* Behavioral mechanisms in evolutionary ecology (L. A. Real, Ed.). Univ. Chicago Press, Chicago. [20, 24, 30]

Coles, R. B., and A. Guppy. 1988. Directional hearing in the Barn Owl, *Tyto alba.* J. Comp. Physiol. A 163:117–133. [334]

Colgan, P. 1983. Comparative social recognition. Wiley, New York. [356]

Collar, N. J. 1995. Ted Parker: a personal memoir. Bird Conserv. Int. 5:141–144. [v]

Collar, N. J., L. P. Gonzaga, N. Krabbe, A. Madroño Nieto, T. A. Parker III, and D. C. Wege. 1992. Threatened birds of the Americas: the ICBP/IUCN Red Data book. Smithsonian Institution Press, Washington, D.C. [269, 273]

Collias, N. E., and E. C. Collias. 1967. A field study of the Red Jungle Fowl in north central India. Condor 69:360–386. [436]

Collins, S. A., C. Hubbard, and A. M. Houtman. 1994. Female mate choice in the Zebra Finch—the effect of male beak colour and male song. Behav. Ecol. Sociobiol. 35:21–25. [457, 458, 468]

Connors, P. G., B. J. McCaffery, and J. L. Maron. 1993. Speciation in golden-plovers, *Pluvialis dominica* and *P. fulva:* evidence from the breeding grounds. Auk 110:9–20. [244, 249, 254–256]

Conrads, K. 1966. Der Egge-Dialekt des Buchfinken (*Fringilla coelebs*)—ein Beitrag zur geographischen Gesangsvariation. Vogelwelt 87:176–182. [215]

Conrads, K. 1976. Studien an Fremddialekt-Sängern und Dialekt-Mischsängern des Ortolans (*Emberiza hortulana*). J. Ornithol. 117:438–450. [215]

Conrads, K. 1977. Entwicklung einer Kombinationstrophe des Buchfinken (*Fringilla c. coelebs* L.) aus einer Grünlings-Imitation und arteigenen Elementen im Frieland. Ber. Naturwiss. Ver. Bielefeld 5:91–101. [215]

Conrads, K. 1984. Gesangsdialekte der Goldammer (*Emberiza citrinella*) auf Bornholm. J. Ornithol. 125:241–244. [215]

Conrads, K., and W. Conrads. 1971. Regionaldialekte des Ortolans (*Emberiza hortulana*) in Deutschland. Vogelwelt 92:81–90. [215]

Copi, C., and J. Vielliard. 1990. Phylogeny and biogeography of some Brazilian *Picumnus,* as shown by their bioacoustical characters. Pages 478–479 *in* Acta XX Congr. Int. Ornithol. Suppl. New Zealand Ornithological Congress Trust Board, Wellington. [272]

Cosens, S. E. 1984. Sources of selection on avian songs and singing behaviour in marsh and grassland habitats. Ph.D. dissertation, Univ. Toronto, Toronto. [440]

Cox, P. R. 1944. A statistical investigation into bird song. Br. Birds 38:3–9. [443]

Coyne, J. A. 1994. Ernst Mayr and the origin of species. Evolution 48:19–30. [242]

Coyne, J. A., H. A. Orr, and D. J. Futuyma. 1988. Do we need a new species concept? Syst. Zool. 37:190–200. [242]

Cracraft, J., and R. O. Prum. 1988. Patterns and processes of diversification: speciation and historical congruence in some Neotropical birds. Evolution 42:603–620. [275]

Craig, J. L., and P. F. Jenkins. 1982. The evolution of complexity in broadcast songs of passerines. J. Theor. Biol. 95:415–422. [371]

Craig, W. 1943. The song of the Wood Pewee, *Myiochanes virens,* Linnaeus: a study of bird music. N.Y. State Mus. Bull. 334:1–186. [385, 443, 444]

Cramp, S. (Ed.). 1980. Handbook of the birds of Europe, the Middle East, and North Africa: the birds of the western Palearctic. Vol. 2: Hawks to bustards. Oxford Univ. Press, Oxford. [442]

Cramp, S. (Ed.). 1983. Handbook of the birds of Europe, the Middle East, and North Africa: the birds of the western Palearctic. Vol. 3: Waders to gulls. Oxford Univ. Press, Oxford. [242–244, 250–254, 430, 441, 442]

Cramp, S. (Ed.). 1985. Handbook of the birds of Europe, the Middle East, and North Africa: the birds of the western Palearctic. Vol. 4: Terns to woodpeckers. Oxford Univ. Press, Oxford. [442]

Crawford, R. D. 1977. Polygynous breeding of Short-billed Marsh Wrens. Auk 94:359–362. [17]

Crick, F. 1989. The recent excitement about neural networks. Nature 337:129–132. [423]

Critchlow, J. M. 1988. An introduction to distributed and parallel computing. Prentice-Hall, N.J. [423]

Cronin, H. 1991. The ant and the peacock. Cambridge Univ. Press, Cambridge. [261]

Crow, J. F., and M. Kimura. 1970. An introduction to population genetics theory. Harper and Row, New York. [184, 188, 194]

Crowley, T. 1992. An introduction to historical linguistics. 2nd ed. Oxford Univ. Press, Oxford. [214]

Croxall, J. P. 1984. Seabirds. Pages 533–619 *in* Antarctic ecology, vol. 2 (R. M. Laws, Ed.). Academic Press, London. [161]

Cunningham, M. A., and M. C. Baker. 1983. Vocal learning in White-crowned Sparrows: sensitive phase and song dialects. Behav. Ecol. 13:259–269. [43]

Curio, E. 1994. Causal and functional questions: how are they linked? Anim. Behav. 47:999–1021. [285]

Cuthill, I. C., and A. Hindmarsh. 1985. Increase in starling song activity with removal of mate. Anim. Behav. 33:326–328. [456, 459]

Cuthill, I. C., and W. A. Macdonald. 1990. Experimental manipulation of the dawn and dusk chorus in the blackbird (*Turdus merula*). Behav. Ecol. Sociobiol. 26:209–216. [311, 319, 431, 432, 436]

Cynx, J., and F. Nottebohm. 1992a. Role of gender, season, and familiarity in discrimination of conspecific song by Zebra Finches (*Taeniopygia guttata*). Proc. Natl. Acad. Sci. USA 89:1368–1371. [302, 303, 342, 343, 346–348, 374]

Cynx, J., and F. Nottebohm. 1992b. Testosterone facilitates some conspecific song discriminations in castrated Zebra Finches (*Taeniopygia guttata*). Proc. Natl. Acad. Sci. USA 89:1376–1378. [343]

Dabelsteen, T. 1982. Variation in the response of free-living blackbirds (*Turdus merula*) to playback of song. I. Effect of continuous stimulation and predictability of the response. Z. Tierpsychol. 58:311–328. [406]

Dabelsteen, T. 1984a. Variation in the response of free-living blackbirds (*Turdus merula*) to playback of song. II. Effect of time of day, reproductive status and number of experiments. Z. Tierpsychol. 65:215–227. [433]

Dabelsteen, T. 1984b. An analysis of the full song of the blackbird, *Turdus merula,* with respect to message coding and adaptations for acoustic communication. Ornis Scand. 15:227–239. [329]

Dabelsteen, T. 1985. Messages and meanings of bird song with special reference to the blackbird (*Turdus merula*) and some methodology problems. Biol. Skr. Dan. Vid. Selsk. 25:173–208. [336, 403, 415]

Dabelsteen, T. 1988. The meaning of the full song of the blackbird, *Turdus merula,* to untreated and oestradiol-treated females. Ornis Scand. 19:7–16. [412]

Dabelsteen, T. 1992. Interactive playback: a finely tuned response. Pages 97–110 *in* Playback and studies of animal communication (P. K. McGregor, Ed.). Plenum Press, New York. [82, 398–401, 412, 414, 419, 444]

Dabelsteen, T. 1994. Solsortens sang som signal. Akademisk Forlag, Copenhagen. [413]

Dabelsteen, T., and S. B. Pedersen. 1985. Correspondence between messages in the full song of the blackbird, *Turdus merula,* and meanings to territorial males, as inferred from responses to computerized modifications of natural song. Z. Tierpsychol. 69:149–165. [336, 337, 345]

Dabelsteen, T., and S. B. Pedersen. 1988a. Do female blackbirds, *Turdus merula,* decode song in the same way as males? Anim. Behav. 36:1858–1860. [336, 337, 339, 341, 345, 354]

Dabelsteen, T., and S. B. Pedersen. 1988b. Song parts adapted to function both at long and short ranges may communicate information about the species to female blackbirds, *Turdus merula.* Ornis Scand. 19:195–198. [413]

Dabelsteen, T., and S. B. Pedersen. 1990. Song and information about aggressive responses of blackbirds, *Turdus merula:* evidence from interactive playback experiments with territory owners. Anim. Behav. 40:1158–1168. [381, 383, 393, 398, 401, 404, 412]

Dabelsteen, T., and S. B. Pedersen. 1991. A portable digital sound emitter for interactive playback of animal vocalizations. Bioacoustics 3:193–206. [398, 401, 412]

Dabelsteen, T., and S. B. Pedersen. 1992. Song features essential for species discrimination and behaviour assessment by male blackbirds (*Turdus merula*). Behaviour 121:259–287. [336, 390]

Dabelsteen, T., and S. B. Pedersen. 1993. Song-based species discrimination and behaviour assessment by female blackbirds, *Turdus merula.* Anim. Behav. 45: 759–771. [336, 337, 339–341, 344–347, 354]

Dabelsteen, T., and S. B. Pedersen. Unpubl. ms. DSE-2: a digital sound emitter system for PCs. [401]

Dabelsteen, T., O. Larsen, and S. B. Pedersen. 1993. Habitat-induced degradation of sound signals: quantifying the effects of communication sounds and bird location on blur ratio, excess attenuation, and signal-to-noise ratio in blackbird song. J. Acoust. Soc. Am. 93:2206–2220. [321, 341, 346, 409, 412, 419, 422]

Dabelsteen, T., P. K. McGregor, M. Shepherd, X. Whittaker, and S. B. Pedersen. In press a. Is the signal value of overlapping singing different from that of alternating singing during matching in Great Tits? J. Avian Biol. [399, 401–403, 414]

Dabelsteen, T., P. K. McGregor, J. Holland, J. Tobias, and S. B. Pedersen. In press b. The signal value of overlapping singing in male robins (*Erithacus rubecula*). Anim. Behav. [405, 414]

d'Agincourt, L. G., and J. B. Falls. 1983. Variation of repertoire use in the Eastern Meadowlark, *Sturnella magna.* Can. J. Zool. 61:1086–1093. [388]

Dale, S., T. Amundsen, J. T. Lifjeld, and T. Slagsvold. 1990. Mate sampling behaviour of female Pied Flycatchers: evidence for active mate choice. Behav. Ecol. Sociobiol. 27:87–91. [430]

Darling, F. F. 1938. Bird flocks and the breeding cycle. Cambridge Univ. Press, London. [171]

Darwin, C. 1868. The variation of plants and animals under domestication. J. Murray, London. [46, 50]

Davies, N. B., and A. Lundberg. 1984. Food distribution and a variable mating system in the Dunnock, *Prunella modularis.* J. Anim. Ecol. 53:895–912. [468]

Davies, S. J. J. F. 1970. Patterns of inheritance in the bowing display and associated behaviour of some hybrid *Streptopelia* doves. Behaviour 36:187–214. [48]

Davies, S. J. J. F. 1974. Studies of the three coo-calls of the male Barbary Dove. Emu 74:18–26. [48]

Davis, J. 1958. Singing behavior and the gonad cycle of the Rufous-sided Towhee. Condor 60:308–336. [428, 442]

Davis, L. I. 1964. Biological acoustics and the use of the sound spectrograph. Southwest. Nat. 9:118–145. [56]

Davis, L. I. 1972. A field guide to the birds of Mexico and Central America. Univ. Texas Press, Austin. [447]

Dawkins, M. S. 1986. Unravelling animal behaviour. Wiley, New York. [423]

Dawkins, M. S. 1993. Are there general principles of signal design? Philos. Trans. R. Soc. Lond. B 340:251–255. [423]

Dawkins, M. S. 1995. Unravelling animal behaviour. 2nd ed. Harlow, Longmans, Essex, England. [416]

Dawkins, M. S., and T. Guilford. 1994. Design of an intent signal in the bluehead wrasse (*Thalassoma bifasciatum*). Proc. R. Soc. Lond. B 257:123–128. [412]

Dawkins, R. 1976a. The selfish gene. Oxford Univ. Press, Oxford. [182]

Dawkins, R. 1976b. Hierarchical organization: a candidate principle for ethology. Pages 7–54 *in*

Growing points in ethology (P. P. G. Bateson and R. A. Hinde, Eds.). Cambridge Univ. Press, Cambridge. [128]

Dawkins, R., and J. R. Krebs. 1978. Animal signals: information or manipulation? Pages 282–309 *in* Behavioural ecology: an evolutionary approach (J. R. Krebs and N. B. Davies, Eds.). Blackwell, Oxford. [128, 160]

Dawkins, R., and J. R. Krebs. 1979. Arms races between and within species. Philos. Trans. R. Soc. Lond. B 205:489–511. [423]

Derrickson, K. C. 1987. Yearly and situational changes in the estimate of repertoire size in Northern Mockingbirds (*Mimus polyglottos*). Auk 104:198–207. [13, 455, 471]

Derrickson, K. C. 1988. Variation in repertoire presentation in Northern Mockingbirds. Condor 90:592–606. [388]

Desforges, M. F., and D. G. M. Wood-Gush. 1976. Behavioral comparison of Aylesbury and Mallard Ducks: sexual behaviour. Anim. Behav. 24:391–397. [50]

Deutsch, D. 1982. The processing of pitch combinations. Pages 271–312 *in* The psychology of music (D. Deutsch, Ed.). Academic Press, New York. [125]

DeVoogd, T. J. 1986. Steroid interactions with structure and function of avian song control regions. J. Neurobiol. 17:177–201. [113, 295]

DeVoogd, T. J. 1991. Endocrine modulation of the development and adult function of the avian song system. Psychoneuroendocrinology 16:41–66. [342]

DeVoogd, T. J., B. Nixdorf, and F. Nottebohm. 1985. Formation of new synapses related to acquisition of a new behavior. Brain Res. 329:304–308. [302]

DeVoogd, T. J., J. Krebs, S. Healy, and A. Purvis. 1993. Relations between song repertoire size and the volume of brain nuclei related to song: comparative evolutionary analyses amongst oscine birds. Proc. R. Soc. Lond. B 254:75–82. [296, 298]

Devort, M., M. Trolliet, and J. Veiga. 1990. Les bécassines et leurs chasses. Editions de l'Orée, Bordeaux, France. [242]

deVos, G. J. 1983. Social behaviour of Black Grouse, an observational and experimental field study. Ardea 71:1–103. [429, 442]

DeWolfe, B. B., L. F. Baptista, and L. Petrinovich. 1989. Song development and territory establishment in Nuttall's White-crowned Sparrows. Condor 91:397–407. [5, 15, 21, 31, 63]

Dhondt, A. A., and J. Hublc. 1968. Fledging-date and sex in relation to dispersal in young Great Tits. Bird Study 15:127–134. [5]

Dhondt, A. A., and M. M. Lambrechts. 1991. The many meanings of Great Tit song. Belg. J. Zool. 121:247–256. [312]

Diesselhorst, G., and J. Martens. 1972. Hybriden von *Parus melanolophus* und *P. ater* im Nepal-Himalaya. J. Ornithol. 113:374–390. [223]

Dietrich, K. 1980. Vorbildwahl in der Gesangsentwicklung beim Japanischen Mövchen (*Lonchura striata* var. *domestica,* Estrildidae). Z. Tierpsychol. 52:57–76. [42]

Dilger, W. C. 1960. The comparative ethology of the African parrot genus *Agapornis.* Z. Tierpsychol. 17:649–685. [110]

Dixon, K. L. 1969. Patterns of singing in a population of the Plain Titmouse. Condor 71:94–101. [411]

Dixon, K. L., and R. A. Stefanski. 1970. An appraisal of the song of the Black-capped Chickadee. Wilson Bull. 82:53–62. [140, 154]

Dobson, C. W., and R. E. Lemon. 1977. Markovian versus rhomboidal patterning in the song of Swainson's Thrush. Behaviour 62:277–297. [80, 83]

Dobzhansky, T., and S. Wright. 1947. Genetics of natural populations. XV. Rate of diffusion of a mutant gene through a population of *Drosophila pseudoobscura.* Genetics 32:303–324. [195]

Dooling, R. J. 1979. Temporal summation of pure tones in birds. J. Acoust. Soc. Am. 65:1058–1060. [329]

Dooling, R. J. 1982. Auditory perception in birds. Pages 95–130 *in* Acoustic communication in birds, vol. 1 (D. E. Kroodsma and E. H. Miller, Eds.). Academic Press, New York. [107, 124, 129, 321]

Dooling, R. J. 1986. Perception of vocal signals by the Budgerigar (*Melopsittacus undulatus*). Exp. Biol. 45:195–218. [104, 107]

Dooling, R. J. 1991. Hearing in birds. Pages 545–559 *in* The evolutionary biology of hearing (D. Webster, R. Fay, and A. Popper, Eds.). Springer Verlag, New York. [107]

Dooling, R. J., and K. Okanoya. 1995. The method of constant stimuli in testing auditory sensitivity in small birds. Pages 155–164 *in* Methods in comparative psychoacoustics (G. M. Klump, R. J. Dooling, R. R. Fay, and W. C. Stebbins, Eds.). Birkhäuser, Basel. [326]

Dooling, R. J., and J. C. Saunders. 1975. Hearing in the parakeet (*Melopsittacus undulatus*): absolute thresholds, critical ratios, frequency difference limens, and vocalizations. J. Comp. Physiol. Psychol. 88:1–20. [98, 107, 327, 328]

Dooling, R. J., and M. H. Searcy. 1981. Amplitude modulation thresholds for the parakeet (*Melopsittacus undulatus*). J. Comp. Physiol. A 143:383–388. [323]

Dooling, R. J., and M. H. Searcy. 1985. Temporal integration of acoustic signals by the Budgerigar. J. Acoust. Soc. Am. 77:1917–1920. [329]

Dooling, R. J., J. A. Mulligan, and J. D. Miller. 1971. Auditory sensitivity and song spectrum of the canary (*Serinus canarius*). J. Acoust. Soc. Am. 50:700–709. [330]

Dooling, R. J., S. R. Zoloth, and J. R. Baylis. 1978. Auditory sensitivity, equal loudness, temporal resolving power, and vocalizations in the House Finch (*Carpocadus mexicanus*). J. Comp. Physiol. Psychol. 92:867–876. [330]

Dooling, R. J., S. D. Brown, T. J. Park, S. D. Soli, and K. Okanoya. 1987a. Perceptual organization of acoustic stimuli by Budgerigars (*Melopsittacus undulatus*). I. Pure tones. J. Comp. Psychol. 101:139–149. [97, 98]

Dooling, R. J., B. F. Gephart, P. H. Price, C. McHale, and S. E. Brauth. 1987b. Effects of deafening on the contact call of the Budgerigar (*Melopsittacus undulatus*). Anim. Behav. 35:1264–1266. [97–99, 115, 294]

Dooling, R. J., T. J. Park, S. D. Brown, K. Okanoya, and S. D. Soli. 1987c. Perceptual organization of acoustic stimuli by Budgerigars (*Melopsittacus undulatus*). II. Vocal signals. J. Comp. Psychol. 101:367–381. [97, 98, 104, 108, 115, 121, 131]

Dooling, R. J., T. J. Park, S. D. Brown, and K. Okanoya. 1990. Perception of species-specific vocalizations by isolate-reared Budgerigars (*Melopsittacus undulatus*). Int. J. Comp. Psychol. 4:57–78. [108]

Dooling, R. J., S .D. Brown, G. M. Klump, and K. Okayana. 1992. Auditory perception of conspecific and heterospecific vocalizations in birds: evidence for special processes. J. Comp. Psychol. 106:20–28. [121]

Doupe, A. 1993. A neural circuit specialized for vocal learning. Curr. Opin. Neurobiol. 3:104–111. [94, 286–289, 293]

Doupe, A., and M. Konishi. 1991. Song-selective auditory circuits in the vocal control system of the Zebra Finch. Proc. Natl. Acad. Sci. USA 88:11339–11343. [289]

Drury, W. H., Jr. 1961. Breeding biology of shorebirds on Bylot Island, Northwest Territories. Auk 78:176–219. [244]

Dufty, A. M., Jr. 1985. Song sharing in the Brown-headed Cowbird (*Molothrus ater*). Ethology 69:177–190. [36]

Durham, W. H. 1991. Coevolution: genes, culture and human diversity. Stanford Univ. Press, Palo Alto. [198, 217]

Dykhuizen, D. E., and D. L. Hartl. 1980. Selective neutrality of 6PGD allozymes in *E. coli* and the effects of genetic background. Genetics 96:801–817. [197]

Dyrcz, A. 1983. Breeding ecology of the Clay-colored Robin, *Turdus grayi,* in lowland Panama. Ibis 125:287–304. [265]

Eales, L. A. 1985. Song learning in Zebra Finches: some effects of song model availability on what is learnt and when. Anim. Behav. 33:1293–1300. [63]

Eales, L. A. 1987. Do Zebra Finch males that have been raised by another species still tend to select a conspecific song tutor? Anim. Behav. 35:1347–1355. [42]

Eason, P. K., and J. A. Stamps. 1993. An early-warning system for detecting intruders in a territorial animal. Anim. Behav. 46:1105–1109. [413]

Eberhard, W. G. 1985. Sexual selection and animal genitalia. Harvard Univ. Press, Cambridge, Mass. [242]

Eberhardt, L. S. 1994. Oxygen consumption during singing by male Carolina Wrens (*Thryothorus ludovicianus*). Auk 111:124–130. [311, 436, 467]

Eck, S. 1980. *Parus major*—ein Paradebeispiel der Systematik? Falke 27:385–392. [227]

Eck, S., and R. Piechocki. 1977. Eine Kontaktzone zwischen den *bokharensis*-Subspezies und den *major*-Subspezies der Kohlmeise, *Parus major,* in der Südwest-Mongolei. Mitt. Zool. Mus. Berl. 53 (Suppl.), Ann. Ornithol. 1:127–136. [225]

Eens, M., and R. Pinxten. 1990. Extra-pair courtship in the starling, *Sturnus vulgaris.* Ibis 132:618–619. [455, 459]

Eens, M., R. Pinxten, and R. F. Verheyen. 1991. Male song as a cue for mate choice in the European Starling. Behaviour 116:210–238. [457, 460]

Eens, M., R. Pinxten, and R. F. Verheyen. 1992. No overlap in song repertoire size between yearling and older starlings, *Sturnus vulgaris.* Ibis 134:72–76. [471]

Eens, M., R. Pinxten, and R. F. Verheyen. 1993. Function of the song and song repertoire in the European Starling (*Sturnus vulgaris*): an aviary experiment. Behaviour 125:51–66. [456, 459]

Ehret, G. 1987. Categorical perception of sound signals: facts and hypotheses from animal studies. Pages 301–331 *in* Categorical perception (S. Harnad, Ed.). Cambridge Univ. Press, Cambridge. [122]

Eldredge, N. (Ed.) 1992. Systematics, biogeography, and the biodiversity crisis. Columbia Univ. Press, New York. [273]

Eldredge, N., and J. Cracraft. 1980. Phylogenetic patterns and the evolutionary process. Columbia Univ. Press, New York. [245]

Elgar, M. A. 1989. Predator vigilance and group size in mammals and birds: a critical review of the empirical evidence. Biol. Rev. 64:13–33. [103]

Elowson, A. M., and J. P. Hailman. 1991. Analysis of complex variation: dichotomous sorting of predator-elicited calls of the Florida Scrub Jay. Bioacoustics 3:295–320. [141, 142]

Emlen, S. T. 1971a. Geographic variation in Indigo Bunting song (*Passerina cyanea*). Anim. Behav. 19:407–408. [200]

Emlen, S. T. 1971b. The role of song in individual recognition in the Indigo Bunting. Z. Tierpsychol. 28:241–246. [218, 359]

Emlen, S. T. 1972. An experimental analysis of the parameters of bird song eliciting species recognition. Behaviour 41:130–171. [172, 200]

Emlen, S. T., J. D. Rising, and W. L. Thompson. 1975. A behavioral and morphological study of sympatry in the Indigo and Lazuli Buntings of the Great Plains. Wilson Bull. 87:145–179. [63, 276]

Endler, J. A. 1993. Some general comments on the evolution and design of animal communication systems. Philos. Trans. R. Soc. Lond. B 340:215–225. [423]

Eriksson, D. 1991. The significance of song for species recognition and mate choice in the Pied Flycatcher, *Ficedula hypolecua.* Ph.D. dissertation, Uppsala University, Uppsala, Sweden. [458]

Eriksson, D., and L. Wallin. 1986. Male bird song attracts females—a field experiment. Behav. Ecol. Sociobiol. 19:297–299. [262, 354, 457]

Ewens, W. J. 1972. The sampling theory of selectively neutral alleles. Theor. Pop. Biol. 3:87–112. [188]

Ewert, D. N., and D. E. Kroodsma. 1994. Song sharing and repertoires among migratory and resident Rufous-sided Towhees. Condor 96:190–196. [6]

Eyckerman, R. 1979. Oecologische functies van het koolmeesvocabularium. Ph.D. dissertation, Univ. Ghent, Ghent, Belgium. [313]

Falls, J. B. 1963. Properties of bird song eliciting responses from territorial males. Pages 259–

271 *in* Proc. XIII Int. Ornithol. Congr. (C. G. Sibley, Ed.). Allen Press, Lawrence, Kansas. [344, 345]

Falls, J. B. 1969. Functions of territorial song in the White-throated Sparrow. Pages 207–232 *in* Bird vocalizations (R. A. Hinde, Ed.). Cambridge Univ. Press, New York. [443]

Falls, J. B. 1978. Bird song and territorial behavior. Pages 61–89 *in* Aggression, dominance and individual spacing (L. Krames, P. Pliner, and T. Alloway, Eds.). Plenum Press, New York. [443]

Falls, J. B. 1982. Individual recognition by sound in birds. Pages 237–278 *in* Acoustic communication in birds, vol. 2 (D. E. Kroodsma and E. H. Miller, Eds.). Academic Press, New York. [175, 176, 301, 356, 368–370, 372, 373, 414, 423]

Falls, J. B. 1985. Song matching in Western Meadowlarks. Can. J. Zool. 63:2520–2524. [130, 132, 359, 372, 411, 429]

Falls, J. B. 1988. Does song deter territorial intrusion in White-throated Sparrows (*Zonotrichia albicollis*)? Can. J. Zool. 66:206–211. [171, 432]

Falls, J. B. 1992. Playback: a historical perspective. Pages 11–33 *in* Playback and studies of animal communication (P. K. McGregor, Ed.). Plenum Press, New York. [107, 115, 400]

Falls, J. B., and J. R. Brooks. 1975. Individual recognition by song in White-throated Sparrows. II. Effects of location. Can. J. Zool. 53:412–420. [358–360, 363, 364, 371]

Falls, J. B., and L. G. d'Agincourt. 1981. A comparison of neighbor-stranger discrimination in Eastern and Western Meadowlarks. Can. J. Zool. 59:2380–2385. [132, 359, 362, 365, 367–369]

Falls, J. B., and L. G. d'Agincourt. 1982. Why do Meadowlarks switch song types? Can. J. Zool. 60:3400–3408. [80, 388]

Falls, J. B., J. R. Krebs, and P. K. McGregor. 1982. Song matching in the Great Tit (*Parus major*): the effect of similarity and familiarity. Anim. Behav. 30:977–1009. [130, 131, 359, 371, 372, 400, 411]

Falls, J. B., A. G. Horn, and T. E. Dickinson. 1988. How Western Meadowlarks classify their songs: evidence from song matching. Anim. Behav. 36:579–585. [130, 132, 370]

Falls, J. B., T. E. Dickinson, and J. R. Krebs. 1990. Contrast between successive songs affect the response of Eastern Meadowlarks to playback. Anim. Behav. 39:717–728. [131]

Fandiño-Mariño, H. 1989. A comunicação sonora do Anu branco, *Guira guira:* avaliações eco-etológicas e evolutivas. Editora da UNICAMP, Campinas, Brazil. [278]

Farabaugh, S. M. 1982. The ecological and social significance of duetting. Pages 85–124 *in* Acoustic communication in birds, vol. 2 (D. E. Kroodsma and E. H. Miller, Eds.). Academic Press, New York. [114, 258, 259, 261, 276, 291, 298, 303, 447]

Farabaugh, S. M., E. D. Brown, and C. J. Veltman. 1988. Song sharing in a group-living songbird, the Australian Magpie. II. Vocal sharing between territorial neighbors, within and between geographic regions, and between sexes. Behaviour 104:105–125. [114]

Farabaugh, S. M., E. D. Brown, and R. J. Dooling. 1992. Analysis of warble song of the Budgerigar, *Melopsittacus undulatus.* Bioacoustics 4:111–130. [97–99, 102, 104]

Farabaugh, S. M., A. Linzenbold, and R. J. Dooling. 1994. Vocal plasticity in Budgerigars (*Melopsittacus undulatus*): evidence for social factors in the learning of contact calls. J. Comp. Psychol. 108:81–92. [97, 98, 104–106, 114, 115]

Fastl, H. 1993. A masking noise for speech intelligibility tests. Proc. Acoust. Soc. Japan, Technical Committee Hearing, H93–70:1–6. [332, 333]

Fay, R. R. 1988. Hearing in vertebrates: a psychophysics databook. Hill-Fay, Winnetka, Ill. [337]

Feekes, F. 1981. Biology and colonial organization of two sympatric caciques, *Cacicus c. cela* and *Cacicus h. haemorrhous* (Icteridae, Aves) in Suriname. Ardea 69:83–107. [442]

Fersen, L. von, and J. D. Delius. 1992. Schlussfolgerndes Denken bei Tauben. Spektrum der Wissenschaft, July 1992:18–22. [415]

Fersen, L. von, C. D. L. Wynne, and J. D. Delius. 1991. Transitive inference formation in pigeons. J. Exp. Psych. Anim. Behav. Processes 17:334–341. [415]

Ficken, M. S. 1981. What is the song of the Black-capped Chickadee? Condor 83:384–386. [139, 154]

Ficken, M. S. 1989. Acoustic characteristics of alarm calls associated with predation risk in chickadees. Anim. Behav. 39:400–401. [153]

Ficken, M. S. 1990a. Vocal repertoire of the Mexican Chickadee. I. Calls. J. Field Ornithol. 61:380–387. [142, 148, 149, 153]

Ficken, M. S. 1990b. Vocal repertoire of the Mexican Chickadee. II. Song and song-like vocalizations. J. Field Ornithol. 61:388–395. [148, 149, 155]

Ficken, M. S., and R. W. Ficken. 1973. Effect of number, kind and order of song elements on playback responses of the Golden-winged Warbler. Behaviour 46:114–127. [161]

Ficken, M. S., and J. W. Popp. 1992. Syntactical organization of the gargle vocalization of the Black-capped Chickadee, *Parus atricapillus*. Ethology 91:156–168. [143, 144, 150]

Ficken, M. S., and C. M. Weise. 1984. A complex call of the Black-capped Chickadee (*Parus atricapillus*). I. Microgeographic variation. Auk 101:349–360. [139, 143, 150]

Ficken, M. S., and S. R. Witkin. 1977. Responses of Black-capped Chickadee flocks to predators. Auk 94:56–157. [146]

Ficken, M. S., R. W. Ficken, and S. R. Witkin. 1978. Vocal repertoire of the Black-capped Chickadee. Auk 95:34–48. [107, 114, 137, 139, 145, 147, 148, 348]

Ficken, M. S., R. W. Ficken, and K. M. Apel. 1985. Dialects in a call associated with pair interactions in the Black-capped Chickadee. Auk 102:145–151. [143, 150]

Ficken, M. S., C. M. Weise, and J. A. Reinartz. 1987. A complex vocalization of the Black-capped Chickadee. II. Repertoires, dominance and dialects. Condor 89:500–509. [17, 139, 143, 150]

Ficken, M. S., E. D. Hailman, and J. P. Hailman. 1994. The chick-a-dee call system of the Mexican Chickadee. Condor 96:70–82. [157]

Ficken, R. W., M. S. Ficken, and J. P. Hailman. 1974. Temporal pattern shift to avoid acoustic interference in singing birds. Science 183:762–763. [84, 333, 399]

Fisher, R. A. 1930. The genetical theory of natural selection. Clarendon, Oxford. [462]

Fisher, R. A. 1954. Evolution and bird sociality. Pages 71–83 *in* Evolution as a process (J. Huxley, A. C. Hardy, and E. B. Ford, Eds.). Allen and Unwin, London. [423]

Fitzpatrick, J. W., and J. P. O'Neill. 1986. *Otus petersoni*, a new screech owl from the eastern Andes, with systematic notes on *O. columbianus* and *O. ingens*. Wilson Bull. 98:1–14. [270]

Fitzpatrick, J. W., and D. E. Willard. 1990. *Cercomacra manu*, a new species of antbird from southwestern Amazonia. Auk 107:239–245. [270]

Fleischer, R. C., C. L. Tarr, E. S. Morton, A. Sangmeister, and K. C. Derrickson. In press. Mating system of the Dusky Antbird (*Cercomacro tyrannina*), a tropical passerine, as assessed by DNA fingerprinting. Condor. [263]

Fleishman, L. J. 1992. The influence of the sensory system and the environment on motion patterns in the visual displays of anoline lizards and other vertebrates. Am. Nat. 139:S36-S61. [466]

Fletcher, H. 1940. Auditory patterns. Rev. Mod. Phys. 12:47–65. [327]

Formozov, N. A., A. B. Kerimov, and V. V. Lopatin. 1993. New hybridization zone of the Great Titmouse and *Parus bokharensis* in Kazakhstan and relationships in forms of *Parus major* superspecies. Pages 118–146 *in* Hybridization and the problem of species in vertebrates (O. L. Rossolimo, Ed.). Arch. Zool. Mus. Moscow State Univ. 30. [225, 227]

Forrester, B. C. 1993. Birding Brazil. John Geddes Printers, Irvine, Scotland. [274]

Forshaw, J. W. 1989. Parrots of the world. Landsdowne Editions, Willoughby, New South Wales. [102, 104]

Fortune, E., and D. Margoliash. 1992. Cytoarchitectonic organization and morphology of cells of the Field L complex in male Zebra Finches. J. Comp. Neurol. 324:1–17. [289]

Foster, M. S. 1993. Research, conservation, and collaboration: the role of visiting scientists in developing countries. Auk 110:414–417. [274]

Fowler, J. A., M. E. Hulbert, and G. Smith. 1986. Sex ratio in a sample of tape-lured Storm Petrels, *Hydrobates pelagicus,* from Shetland, Scotland. Seabirds 9:15–19. [174]

Franck, D. 1974. The genetic basis of evolutionary changes in behavior patterns. Pages 119–140 *in* The genetics of behavior (J. H. F. Van Abeleen, Ed.). North-Holland, Amsterdam. [48]

Frankel, A. I., and T. S. Baskett. 1961. The effect of pairing on cooing of penned Mourning Doves. J. Wildl. Manage. 25:372–384. [431]

Freeberg, T. M., A. P. King, and M. J. West. 1995. Social malleability in cowbirds: species and mate recognition in the first two years of life. J. Comp. Psychol. 109:357–367. [29]

Frenzel, B. 1967. Die Klimaschwankungen des Eiszeitalters. Vieweg, Braunschweig. [234]

Frenzel, B. 1968. The Pleistocene vegetation of northern Eurasia. Science 161:637–649. [234]

Freyschmidt, J., M. L. Kopp, and H. Hultsch. 1984. Individuelle Entwicklung von gelernten Gesangsmustern bei Nachtigallen. Verh. Dtsch. Zool. Ges. 77:244. [85, 93]

Frith, H. J. 1982. Pigeons and doves of Australia. Rigby, New York. [48]

Fritsch, B., M. J. Ryan, W. Wilczynski, T. E. Hetherington, and W. Walkowiak. 1988. The evolution of the amphibian auditory system. Wiley, New York. [426]

Furness, R. W., and S. R. Baillie. 1975. Factors affecting capture rate and biometrics of Storm Petrels on St. Kilda. Ringing & Migr. 3:137–148. [171, 174]

Futuyma, D. J. 1986. Evolutionary biology. 2nd ed. Sinauer, Sunderland, Mass. [255]

Gahr, M., H. R. Güttinger, and D. E. Kroodsma. 1993. Estrogen receptors in the avian brain: survey reveals general distribution and forebrain areas unique to songbirds. J. Comp. Neurol. 327:112–122. [10, 113, 293, 295, 296]

Garland, T., and S. C. Adolph. 1994. Why not to do 2-species comparative studies: limitations on inferring adaptation. Physiol. Zool. 67:797–828. [357, 367]

Garson, P. J., and M. L. Hunter. 1979. Effects of temperature and time of year on the singing behaviour of wrens (*Troglodytes troglodytes*) and Great Tits (*Parus major*). Ibis 121:481–487. [311]

Gaunt, A. S. 1983. An hypothesis concerning the relationship of syringeal structure to vocal abilities. Auk 100:853–862. [316]

Gaunt, A. S. 1987. Phonation. Pages 71–94 *in* Bird respiration, vol. 1 (T. J. Seller, Ed.). Academic Press, New York. [311, 312, 315–318]

Gaunt, A. S., and S. L. L. Gaunt. 1985a. Syringeal structure and avian phonation. Pages 213–245 *in* Current Ornithology, vol. 2. (R. F. Johnson, Ed.). Plenum Press, New York. [46]

Gaunt, A. S., and S. L. L. Gaunt. 1985b. Electromyographic studies of the syrinx in parrots (Aves, Psittacidae). Zoomorphology 105:1–11. [316]

Gaunt, S. L. L., L. F. Baptista, J. E. Sanchez, and D. Hernandez. 1994. Song learning as evidenced from song sharing in two hummingbird species (*Colibri coruscans* and *C. thalassinus*). Auk 111:87–103. [53, 59, 277]

Genevois, F., and V. Bretagnolle. 1994. Male Blue Petrels reveal their condition when calling. Ethol. Ecol. Evol. 6:377–383. [162, 165, 174, 176]

Genevois, F., and V. Bretagnolle. 1995. Sexual dimorphism of voice and morphology in Thin-billed Prions (*Pachyptila belcheri*). Notornis 42:1–10. [165]

Gerhardt, H. C. 1992. Conducting playback experiments and interpreting their results. Pages 59–77 *in* Playback and studies of animal communication (P. K. McGregor, Ed.). Plenum Press, New York. [364, 366]

Gerhardt, H. C., and G. M. Klump. 1988. Masking of acoustic signals by the chorus background noise in the green treefrog: a limitation on mate choice. Anim. Behav. 36:1247–1249. [410]

Getty, T. 1989. Are dear enemies in a war of attrition? Anim. Behav. 37:337–339. [423]

Gibbs, H. L. 1990. Cultural evolution in Darwin's finches: frequency, transmission, and demographic success of different song types in Darwin's Medium Ground Finches (*Geospiza fortis*). Anim. Behav. 39:253–263. [218]

Gibbs, H. L., P. J. Weatherhead, P. T. Boag, B. N. White, L. M. Tabak, and D. J. Hoysak. 1990. Realized reproductive success of polygynous Red-winged Blackbirds revealed by DNA markers. Science 250:1394–1397. [435]

Gibson, E. J. 1969. Principles of perceptual learning and development. Appleton-Century-Crofts, New York. [26]
Gibson, E. J. 1991. An odyssey in learning and perception. MIT Press, Cambridge, Mass. [26]
Gibson, J. J. 1966. The senses considered as perceptual systems. Houghton Mifflin, Boston. [26]
Gibson, J. J. 1979. The ecological approach to perception. Houghton Mifflin, Boston. [26]
Gibson, R. M. 1989. Field playback of male display attracts females in lek breeding Sage Grouse. Behav. Ecol. Sociobiol. 24:439–443. [458]
Gibson, R. M., and J. W. Bradbury. 1985. Sexual selection in lekking Sage Grouse: phenotypic correlates of male mating success. Behav. Ecol. Sociobiol. 18:117–123. [457, 469]
Gifford, E. W. 1931. The Galapagos Dove. Avic. Mag. 3:17–19. [50]
Gill, F. B., D. H. Funk, and B. Silverin. 1989. Protein relationships among titmice (*Parus*). Wilson Bull. 101:180–181. [149]
Gill, F. B., A. M. Mostrom, and A. L. Mack. 1993. Speciation in North American chickadees. I. Patterns of mtDNA genetic divergence. Evolution 47:195–212. [149]
Gillan, D. J. 1981. Reasoning in the chimpanzee. II. Transitive inference. J. Exp. Psychol. Anim. Behav. Processes 7:150–164. [415]
Gillespie, J. H. 1989. When not to use diffusion processes in population genetics. Pages 57–70 *in* Mathematical evolutionary theory (M. W. Feldman, Ed.). Princeton Univ. Press, Princeton, N.J. [215]
Gillespie, J. H. 1991. The causes of molecular evolution. Oxford Univ. Press, New York. [213–215]
Gittleman, J. L., and D. M. Decker. 1994. The phylogeny of behaviour. Pages 80–105 *in* The evolution of behaviour (T. R. Halliday and P. J. B. Slater, Eds.). Cambridge Univ. Press, Cambridge. [245]
Glaubrecht, M. 1989. Geographische Variabilität des Gesangs der Goldammer, *Emberiza citrinella,* im norddeutschen Dialekt-Grenzgebiet. J. Ornithol. 130:279–292. [215]
Glaubrecht, M. 1991. Gesangsvariation der Goldammer (*Emberiza citrinella*) in Norddeutschland und auf den dänischen Inseln. J. Ornithol. 132:441–445. [215]
Gleich, O., and P. M. Narins. 1988. The phase response of primary auditory afferents in a songbird (*Sturnus vulgaris* L.). Hear. Res. 32:81–92. [335]
Glutz, U. N., K. M. Bauer, and E. Bezzel (Eds.). 1975. Handbuch der Vögel Mitteleuropas. Band 6. Charadriiformes (1. Teil). Akademische Verlagsgesellschaft, Wiesbaden, Germany. [251, 252]
Glutz, U. N., K. M. Bauer, and E. Bezzel (Eds.). 1977. Handbuch der Vögel Mitteleuropas. Band 7. Charadriiformes (2. Teil). Akademische Verlagsgesellschaft, Wiesbaden, Germany. [242, 243, 250]
Gochfeld, M. 1978. Intraspecific social stimulation and temporal displacement of songs in the Lesser Skylark, *Alaula gulgula.* Z. Tierpsychol. 48:337–344. [84]
Gochfeld, M. 1980. Mechanisms and adaptive value of reproductive synchrony in colonial seabirds. Pages 207–270 *in* Marine birds (J. Burger, B. L. Olla, and H. E. Winn, Eds.). Plenum Press, New York. [171]
Godard, R. 1991. Long-term memory of individual neighbours in a migratory songbird. Nature 350:228–229. [264, 359, 360, 368, 414]
Godard, R. 1992. Tit for tat among neighbouring Hooded Warblers, *Wilsonia citrina.* IVth Int. Behav. Ecol. Congr. Abstr. T17a. [414]
Godard, R. 1993. Red-eyed Vireos have difficulty recognizing individual neighbors' songs. Auk 110:857–862. [359, 360, 366, 368]
Goldman, P. 1973. Song recognition by Field Sparrows. Auk 90:106–113. [359]
Goldman, S., and F. Nottebohm. 1983. Neuronal production, migration and differentiation in a vocal control nucleus of the adult female canary brain. Proc. Natl. Acad. Sci. USA 80:2390–2394. [343]
Goldstein, R. B. 1978. Geographic variation in the "hoy" call of the Bobwhite. Auk 95:85–94. [53, 54]

Gompertz, T. 1961. The vocabulary of the Great Tit. Br. Birds 54:369–394, 409–418. [147, 149, 167, 313]
Goodwin, D. 1964. Observations on the Dark Firefinch, with some comparisons with Jameson's Firefinch. Avic. Mag. 70:80–105. [45]
Goodwin, D. 1965. Instructions to young ornithologists. IV. Domestic birds. London Museum Press, London. [54]
Goodwin, D. 1982. Estrildid finches of the world. Cornell Univ. Press, Ithaca, N.Y. [45]
Goodwin, D. 1983. Pigeons and doves of the world. British Museum of Natural History, London. [48, 52]
Göransson, G., G. Högstadt, J. Karlsson, H. Källander, and S. Ulfstrand. 1974. Sångens roll för revirhållandet hos näktergal *Luscinia luscinia*—några experiment med playback-teknik. Vår Fågelvärld 33:201–209. [406]
Gotmark, F. 1992. Anti-predator effect of conspicuous plumage in a male bird. Anim. Behav. 44:51–55. [436]
Gottlander, K. 1987. Variation in the song rate of the male Pied Flycatcher (*Ficedula hypoleuca*): causes and consequences. Anim. Behav. 35:1037–1043. [262, 311, 468]
Gottlieb, G. 1992. Individual development and evolution: the genesis of novel behavior. Oxford Univ. Press, New York. [25]
Grafen, A. 1990a. Sexual selection unhandicapped by the Fisher process. J. Theor. Biol. 144:473–516. [461]
Grafen, A. 1990b. Biological signals as handicaps. J. Theor. Biol. 144:517–546. [436, 461]
Grafen, A., and R. A. Johnstone. 1993. Why we need ESS signalling theory. Philos. Trans. R. Soc. Lond. B 340:245–250. [423]
Gralla, G. 1993. Modelle zur Beschreibung von Wahrnehmungskriterien bei Mithörschwellenmessungen. Acustica 78:233–245. [332]
Gramza, A. F. 1970. Vocal mimicry in captive Budgerigars (*Melopsittacus undulatus*). Z. Tierpsychol. 27:971–998. [97]
Grant, G. S., J. Warham, T. N. Pettit, and G. C. Whittow. 1983. Reproductive behavior and vocalizations of the Bonin Petrel. Wilson Bull. 95:522–539. [163, 165, 167]
Gratson, M. W. 1993. Sexual selection for increased male courtship and acoustic signals and against large male size at Sharp-tailed Grouse leks. Evolution 47:691–696. [457, 469]
Graves, G. R. , J. P. O'Neill, and T. A. Parker III. 1983. *Grallaricula ochraceifrons,* a new species of antpitta from northern Peru. Wilson Bull. 95:1–6. [270]
Green, A. J. 1991. Competition and energetic constraints in the courting great crested newt, *Triturus cristatus* (Amphibia: Salamandridae). Ethology 87:66–78. [312]
Green, C. 1991. Birding Ecuador. 1208 N. Swan Rd., Tucson, Ariz. [274]
Green, D. M., and J. A. Swets. 1966. Signal detection theory and psychophysics. Krieger, New York. [324, 325]
Green, S., and P. Marler. 1975. Analysis of animal communication. Pages 73–158 *in* Handbook of behavioral neurobiology. Vol. 3: Social behavior and communication (P. Marler and J. G. Vandenbergh, Eds.). Plenum Press, New York. [121]
Greenberg, R., and J. Gradwohl. 1986. Stable territories and constant densities in tropical forest insectivorous birds. Oecologia 69:618–625. [259]
Greenewalt, C. H. 1968. Bird song: acoustics and physiology. Smithsonian Institution Press, Washington, D.C. [244, 253, 312]
Greenough, W. T., J. E. Black, and C. S. Wallace. 1987. Experience and brain development. Child Dev. 58:539–559. [34]
Greenwood, J. G. 1979. Geographical variation in the Dunlin, *Calidris alpina* (L.). Ph.D. dissertation, Liverpool Polytechnic, Liverpool, England. [244]
Greenwood, J. G. 1986. Geographical variation and taxonomy of the Dunlin, *Calidris alpina* (L.). Bull. Br. Ornithol. Club 106:43–56. [244]
Gregorius, H.-R. 1987. The relationship between the concepts of genetic diversity and differentiation. Theor. Appl. Genet. 74:397–401. [187]

Greig-Smith, P. W. 1982. Song-rates and parental care by male Stonechats (*Saxicola torquata*). Anim. Behav. 30:245–252. [310, 468]

Greig-Smith, P. W. 1983. Uses of perches as vantage points during foraging by male and female Stonechats. Behaviour 86:215–235. [468]

Griffin, D. 1981. The question of animal awareness. 2nd ed. Rockefeller Univ. Press, New York. [357]

Groothuis, T. G. G. 1993. A comparison between development of bird song and development of other displays. Neth. J. Zool. 43:172–193. [39, 55, 59]

Groth, J. G. 1993. Call matching and positive assortative mating in Red Crossbills. Auk 110:398–401. [39, 59]

Grubb, T. C. 1974. Olfactory navigation to the nesting burrow in Leach's Storm-Petrel. Anim. Behav. 22:192–202. [162]

Grudzien, T. A. 1976. The breeding behavior of the Common Snipe (*Gallinago gallinago delicata* Ord) in central Michigan. M.S. thesis, Central Michigan Univ., Mount Pleasant. [243, 250]

Guilford, T. C., and M. S. Dawkins. 1991. Receiver psychology and the evolution of animal signals. Anim. Behav. 42:1–14. [121, 127, 339]

Guillotin, M., and P. Jouventin. 1980. Le pétrel des neiges à Pointe Géologie. Gerfaut 70:51–72. [165, 175]

Gulledge, J. L. 1977. Recording bird sounds. Living Bird 15:183–203. [479, 480]

Gurney, M., and M. Konishi. 1980. Hormone-induced sexual differentiation of brain and behavior in Zebra Finches. Science 208:1380–1383. [292]

Gustafsson, L. 1986. Lifetime reproductive success and heritability: empirical support for Fisher's fundamental theorem. Am. Nat. 128:761–764. [465]

Gustafsson, L. 1993. Bad health and Darwinian fitness in the Collared Flycatcher. Abstr. IV Congr. Europ. Soc. Evol. Biol., p. 161. [319]

Güttinger, H.-R. 1970. Zur Evolution von Verhaltensweisen und Lautäusserungen bei Prachtfinken (Estrildidae). Z. Tierpsychol. 27:1011–1075. [50, 56]

Güttinger, H.-R. 1974. Gesang des Grünlings (*Chloris chloris*): Lokale Unterschiede und Entwicklung bei Schallisolation. J. Ornithol. 115:321–337. [53]

Güttinger, H.-R. 1976. Zur systematischen Stellung der Gattungen *Amadina, Lepidopygia* und *Lonchura* (Aves, Estrildidae). Bonn. Zool. Beitr. 27:218–244. [56]

Güttinger, H.-R. 1979. The integration of learnt and genetically programmed behaviour: a study of hierarchial organization in songs of canaries, greenfinches and their hybrids. Z. Tierpsychol. 49:285–303. [53, 54]

Güttinger, H.-R., and J. Nicolai. 1973. Struktur und Funktion der Rufe bei Prachtfinken (Estrildidae). Z. Tierpsychol. 33:319–334. [45, 56]

Guyomarc'h, C., and J. C. Guyomarc'h. 1982. La stimulation du développement sexuel des femelles de caille japonaise, *Coturnix coturnix japonica,* par des chants de mâles: mise en évidence de périodes privilégiées dans le nycthémère. C. R. Acad. Sci. Paris Sér. III 295:37–40. [430]

Guyomarc'h, J. C., Y. A. Hemon, C. Guyomarc'h, and R. Michel. 1984. Le mode de dispersion des mâles de caille des blés, *Coturnix c. coturnix,* en phase de reproduction. C. R. Acad. Sci. Paris, Sér. III 299(19):805–808. [53]

Gyger, M. 1990. Audience effects on alarm calling. Ethol. Ecol. Evol. 2:227–232. [416]

Gyger, M., S. J. Karakashian, and P. Marler. 1986. Avian alarm calling: is there an audience effect? Anim. Behav. 34:1570–1572. [416]

Hackett, S. J. 1993. Phylogenetic and biogeographic relationships in the Neotropical genus *Gymnopithys* (Formicariidae). Wilson Bull. 105:301–315. [271]

Hackett, S. J., and K. V. Rosenberg. 1990. Evolution of South American antwrens (Formicariidae): comparisons of phenotypic and genetic differentiation. Auk 107:473–489. [271]

Haffer, J. 1982. General aspects of the refuge theory. Pages 6–24 *in* Biological diversification in the Tropics (G. T. Prance, Ed.). Columbia Univ. Press, New York. [234]

Haffer, J. 1985. Avian zoogeography of the Neotropical lowlands. Pages 113–146 *in* Neotropical ornithology (P. A. Buckley, M. S. Foster, E. S. Morton, R. S. Ridgely, and F. G. Buckley, Eds.). Ornithological Monographs 36, American Ornithologists' Union, Washington D.C. [275]

Haffer, J. 1990. Avian species richness in South America. Stud. Neotrop. Fauna Environ. 25:157–183. [269, 272]

Haffer, J. 1992. On the "river effect" in some forest birds of southern Amazonia. Bol. Mus. Para. Emilio Goeldi Nova Ser. Zool. 8:217–245. [272, 275]

Haftorn, S. 1973. Lappmeisa, *Parus cinctus,* in hekketiden: forplantning, stemmeregister og hamstring av næring. Sterna 12:91–155. [147, 148, 152, 154, 155]

Haftorn, S. 1990. Social organization of winter flocks of Willow Tits, *Parus montanus,* in a Norwegian subalpine birch forest. Pages 401–413 *in* Population biology of passerine birds (J. Blondel et al., Eds.). Springer Verlag, Berlin. [149]

Haftorn, S. 1993. Ontogeny of the vocal repertoire in the Willow Tit, *Parus montanus.* Ornis Scand. 24:267–289. [148–151, 153, 154]

Hägerstrand, T. 1967. Innovation diffusion as a spatial process. Univ. Chicago Press, Chicago. [195]

Hailman, J. P. 1988. Operationalism, optimality and optimism: suitabilities vs. adaptations of organisms. Pages 85–116 *in* Process and metaphors in the evolutionary paradigm (M.-W. Ho and S. Fox, Eds.). Wiley, Chichester, Sussex, England. [153]

Hailman, J. P. 1989. The organization of major vocalizations in the Paridae. Wilson Bull. 101:305–343. [139, 147, 149, 154, 157, 163, 167]

Hailman, J. P. 1994. Constrained permutation in "chick-a-dee"-like calls of a Black-lored Tit (*Parus xanthogenys*). Bioacoustics 6:33–50. [157]

Hailman, J. P., and S. Haftorn. 1995. Siberian Tit. *In* The birds of North America, no. 196 (A. Poole and F. Gill, Eds.). The Academy of Natural Sciences, Philadelphia; The American Ornithologists' Union, Washington, D.C. [149, 159]

Hailman, J. P., M. S. Ficken, and R. W. Ficken. 1985. The "chick-a-dee" calls of *Parus atricapillus:* a recombinant system of animal communication compared with written English. Semiotica 56:191–224. [107, 114, 137, 140, 143]

Hailman, J. P., M. S. Ficken, and R. W. Ficken. 1987. Constraints on the structure of combinatorial "chick-a-dee" calls. Ethology 75:62–80. [143, 312]

Hailman, J. P., S. Haftorn, and E. D. Hailman. 1994. Male Siberian Tit, *Parus cinctus,* dawn serenades: suggestion for the origin of song. Fauna Norv. Ser. C 17:15–26. [149, 152, 155, 156, 159]

Hale, K., M. Krauss, L. Watahomigie, A. Yamamoto, C. Craig, L. M. Jeanne, and N. England. 1992. Endangered languages. Language 68:1–42. [220]

Hall, B. K. (Ed.). 1994. Homology. The hierarchical basis of comparative biology. Academic Press, San Diego. [245]

Hall, J. W., M. P. Haggard, and M. A. Fernandes. 1984. Detection in noise by spectro-temporal analysis. J. Acoust. Soc. Am. 76:50–56. [331]

Hall, M. F. 1962. Evolutionary aspects of estrildid song. Symp. Zool. Soc. Lond. 8:37–55. [50]

Halliday, T. R. 1987. Physiological constraints on sexual selection. Pages 247–264 *in* Sexual selection: testing the alternatives (J. W. Bradbury and M. B. Andersson, Eds.). Wiley, Chichester, Sussex, England. [305, 310, 311, 316]

Halsema, K., and S. Bottjer. 1992. Chemical lesions of a thalamic nucleus disrupt song development in male Zebra Finches. Soc. Neurosci. Abstr. 18:1052. [288]

Hamilton, W. D. 1990. Mate choice near or far. Am. Zool. 30:341–352. [340]

Hamilton, W. D., and M. Zuk. 1982. Heritable true fitness and bright birds: a role for parasites? Science 218:384–387. [465]

Hansen, P. 1979. Vocal learning: its role in adapting sound structures to long-distance propagation and a hypothesis on its evolution. Anim. Behav. 27:1270–1271. [186]

Hansen, P. 1984. Neighbour-stranger song discrimination in territorial Yellowhammer, *Emberiza citrinella,* males, and a comparison with responses to own and alien song dialects. Ornis Scand. 15:240–247. [359]

Hanski, I. K., and A. Laruila. 1993. Variation in song rate during the breeding cycle of the Chaffinch, *Fringilla coelebs.* Ethology 93:161–169. [428]

Harcus, J. L. 1977. The functions of duetting in some African birds. Z. Tierpsychol. 43:23–45. [447]

Harding, C. F., and B. K. Follett. 1979. Hormone changes triggered by aggression in a natural population of blackbirds. Science 203:918–920. [394]

Harding, C. F., K. Sheridan, and M. Walters. 1983. Hormonal specificity and activation of sexual behavior in male Zebra Finches. Horm. Behav. 17:111–133. [293]

Hardy, J. W. 1963. Epigamic and reproductive behavior of the Orange-fronted Parakeet. Condor 65:169–199. [97, 105, 110]

Hardy, J. W. 1984. Depositing sound specimens. Auk 101:623–624. [475, 484]

Hardy, J. W., and T. A. Parker III (Eds.). 1985. Voices of the New World thrushes. ARA 10, Gainesville, Fla. [276]

Hardy, J. W., J. Vielliard, and R. Straneck. 1993. Voices of the tinamous. ARA 18, Gainesville, Fla. [280]

Harnad, S. 1987a. Categorical perception. Cambridge Univ. Press, Cambridge. [121, 122]

Harnad, S. 1987b. Categorical perception and representation. Pages 535–565 *in* Categorical perception (S. Harnad, Ed.). Cambridge Univ. Press, Cambridge. [125]

Harper, D. G. C. 1991. Communication. Pages 374–397 *in* Behavioural ecology: an evolutionary approach. 3rd ed. (J. R. Krebs and N. B. Davies, Eds.). Blackwell, Oxford. [416]

Harris, M. A., and R. E. Lemon. 1972. Songs of Song Sparrows (*Melospiza melodia*): individual variation and dialects. Can. J. Zool. 50:301–309. [191]

Harris, M. A., and R. E. Lemon. 1976. Response of male Song Sparrows, *Melospiza melodia,* to neighbouring and non-neighbouring individuals. Ibis 118:421–424. [359]

Harrison, C. J. O. 1962. Solitary song and its inhibition in some Estrildidae. J. Ornithol. 4:369–373. [45]

Harrison, C. J. O. 1969. Some comparative notes on the Peaceful and Zebra Doves (*Geopelia striata* ssp.) with reference to their taxonomic status. Emu 69:66–71. [48]

Hartley, R. S. 1990. Expiratory muscle activity during song production in the canary. Respir. Physiol. 81:177–188. [312, 317]

Hartshorne, C. 1973. Born to sing. An interpretation and world survey of bird song. Indiana Univ. Press, Bloomington. [79, 128, 305, 316]

Harvey, P. H., and A. H. Harcourt. 1984. Sperm competition, testes size, and breeding systems in primates. Pages 589–600 *in* Sperm competition and the evolution of animal mating systems (R. L. Smith, Ed.). Academic Press, New York. [263]

Harvey, P. H., and M. D. Pagel. 1991. The comparative method in evolutionary biology. Oxford Univ. Press, Oxford. [6, 177]

Hasselquist, D. 1994. Male attractiveness, mating tactics and realized fitness in the polygynous Great Reed Warbler. Ph.D. dissertation, Lund University, Lund, Sweden. [457, 460, 465, 469–471]

Hausberger, M. 1993. How studies on vocal communication in birds contribute to a comparative approach of cognition. Etología 3:171–185. [44]

Hayman, P., J. Marchant, and T. Prater. 1986. Shorebirds. An identification guide to the waders of the world. Houghton Mifflin, Boston. [241, 242, 251]

Heaton, J. T., S. M. Farabaugh, and S. E. Brauth. 1995. Effect of syringeal denervation in the Budgerigar (*Melopsittacus undulatus*): the role of the syrinx in call production. Neurobiol. Learn. Memory 64:68–82. [113]

Heckershoff, G. 1979. Die Bedeutung von Vererbung und Lernen für die Jugendentwicklung zweier Gesangsformen der Weidenmeise (*Parus montanus*). Diploma thesis, Zool. Inst, Cologne Univ. [235]

Helbig, A. J., M. Salomon, M. Wink, and J. Bried. 1993. Absence de flux génétique mitochondrial entre les pouillots "véloces" médio-européens et ibériques (Aves: *Phylloscopus collybita collybita, P. (c.) brehmii*); implications taxinimiques. Résultats tirés de la PCR et du séquencage d'ADN. C. R. Acad. Sci. Paris Sér. III 316:205–210. [228, 238]

Helbig, A. J., J. Martens, F. Henning, B. Schottler, I. Seibold, and M. Wink. 1996. Phylogeny and species limits in the Palearctic Chiffchaff (*Phylloscopus collybita*) complex: mitochondrial genetic differentiation and bioacoustic evidence. Ibis. In press. [229, 238]

Hellmayr, C. E. 1929. Catalogue of birds of the Americas, part 6. Oxyruncidae-Pipridae-Cotingidae-Rupicolidae-Phytotomidae. Publ. Field Mus. Nat. Hist. 266. [270]

Helstrom, C. W. 1975. Statistical theory of signal detection. Pergamon Press, New York. [419]

Hennessy, D. F., D. H. Owings, M. P. Rowe, R. G. Coss, and D. W. Leger. 1981. The information provided by a variable signal: constraints on snake-elicited tail flagging by California ground squirrels. Behaviour 78:188–226. [266]

Henning, F., B. Schottler, and J. Martens. 1994. Inselspezifische Rufe Kanarischer Zilpzalpe (*Phylloscopus collybita canariensis*). Verh. Dtsch. Zool. Ges. 87.1:43. [239]

Henry, L. 1994. Influences du contexte sur le comportement vocal et socio-sexuel de la femelle étourneau (*Sturnus vulgaris*). Ph.D. dissertation, Univ. Rennes, France. [44]

Henwood, K., and A. Fabrick. 1979. A quantitative analysis of the dawn chorus: temporal selection for communicatory optimization. Am. Nat. 114:260–274. [333, 437]

Herrmann, K., and A. Arnold. 1991. The development of afferent projections to the robust archistriatal nucleus in male Zebra Finches: a quantitative electron microscopic study. J. Neurosci. 11:2063–2074. [289, 293]

Herrnstein, R. J. 1982. Stimuli and the texture of experience. Neurosci. Biobehav. Rev. 6:105–117. [122]

Hiebert, S. M., P. K. Stoddard, and P. Arcese. 1989. Repertoire size, territory acquisition and reproductive success in the Song Sparrow. Anim. Behav. 37:266–273. [471]

Hienz, R. D., and M. B. Sachs. 1987. Effects of noise on pure-tone thresholds in blackbirds (*Agelaius phoeniceus* and *Molothrus ater*) and pigeons (*Columba livia*). J. Comp. Psychol. 101:16–24. [323, 324, 328]

Highsmith, R. T. 1989. The singing behavior of Golden-winged Warblers. Wilson Bull. 101:36–50. [16, 390, 430, 432, 438, 445, 446]

Hill, B. G., and M. R. Lein. 1987. Function of frequency-shifted songs of Black-capped Chickadees. Condor 89:914–915. [13, 126]

Hill, K., and T. DeVoogd. 1991. Altered daylength affects dendritic structure in a song-related brain region in Red-winged Blackbirds. Behav. Neural Biol. 56:240–250. [302]

Hill, W. G. 1979. A note on the effective population size with overlapping generations. Genetics 92:317–322. [184]

Hinde, R. A. 1958. Alternative motor patterns in Chaffinch song. Anim. Behav. 6:211–218. [80, 83]

Hinde, R. A. 1970. Behavioural habituation. Pages 3–40 *in* Short-term changes in neural activity and behaviour (G. Horn and R. A. Hinde, Eds.). Cambridge Univ. Press, Cambridge. [471, 472]

Hinde, R. A. 1985. Expression and negotiation. Pages 103–116 *in* The development of expressive behavior (G. Zivin, Ed.). Academic Press, New York. [379]

Hine, J. E., R. L. Martin, and D. R. Moore. 1994. Free-field binaural unmasking in ferrets. Behav. Neurosci. 108:196–205. [334]

Hoch, H. H. 1991. Principles of historical linguistics. 2nd ed. Mouton de Gruyter, New York. [213, 219]

Hoelzel, A. R. 1986. Song characteristics and response to playback of male and female robins, *Erithacus rubecula*. Ibis 128:115–127. [359]

Hoi-Leitner, M., H. Nechtelberger, and J. Dittami. 1993. The relationship between individual differences in male song frequency and parental care in Blackcaps. Behaviour 126:1–12. [468]

Hollis, K. L. 1984. The biological function of Pavlovian conditioning: the best defense is a good offense. J. Exp. Psychol. Anim. Behav. 10:413–442. [414]

Holm, B. 1982. Lappmesens sång. Vår Fågelvärld 41:106. [154]

Horn, A. G. 1987. Repertoires and song switching in Western Meadowlarks (*Sturnella neglecta*). Ph.D. dissertation, University of Toronto, Toronto. [132, 399, 411–413]

Horn, A. G. 1992. Field experiments on the perception of song types by birds: an overview. Pages 191–200 *in* Playback and studies of animal communication (P. K. McGregor, Ed.). Plenum Press, New York. [129, 130, 132, 134]

Horn, A. G., and J. B. Falls. 1986. Western Meadowlarks switch song types when matched by playback. Anim. Behav. 34:927–929. [399, 400, 411]

Horn, A. G., and J. B. Falls. 1988a. Repertoires and countersinging in Western Meadowlarks (*Sturnella neglecta*). Ethology 77:337–343. [80, 127, 132]

Horn, A. G., and J. B. Falls. 1988b. Responses of Western Meadowlarks, *Sturnella neglecta,* to song repetition and contrast. Anim. Behav. 36:291–293. [131, 132]

Horn, A. G., and J. B. Falls. 1988c. Structure of Western Meadowlark (*Sturnella neglecta*) song repertoires. Can. J. Zool. 66:284–288. [185]

Horn, A. G., and J. B. Falls. 1991. Song switching in mate attraction and territory defense by Western Meadowlarks (*Sturnella neglecta*). Ethology 87:262–268. [125, 132]

Horn, A. G., M. L. Leonard, L. Ratcliffe, S. A. Shackleton, and R. G. Weisman. 1992. Frequency variation in songs of Black-capped Chickadees (*Parus atricapillus*). Auk 109:847–852. [13, 126, 139, 145, 440, 444]

Horn, A. G., M. L. Leonard, and D. M. Weary. 1995. Oxygen consumption during crowing by roosters: talk is cheap. Anim. Behav. 50:1171–1175. [311, 436]

Horning, C. L., M. D. Beecher, P. K. Stoddard, and S. E. Campbell. 1993. Song perception in the Song Sparrow: importance of different parts of the song in song type classification. Ethology 94:46–58. [66, 129, 367]

Houde, A. E., and J. A. Endler. 1990. Correlated evolution of female mating preferences and male color patterns in the guppy, *Poecilia reticulata.* Science 248:1405–1408. [470]

Houtman, A. E. 1992. Female Zebra Finches choose extra-pair copulations with genetically attractive males. Proc. R. Soc. Lond. B 249:3–6. [457, 470, 471]

Houtman, A. M., and J. B. Falls. 1994. Negative assortative mating in the White-throated Sparrow, *Zonotrichia albicollis:* the role of mate choice and intra-sexual competition. Anim. Behav. 48:377–383. [44]

Howard, R. D. 1974. The influence of sexual selection and interspecific competition on mockingbird song (*Mimus polyglottos*). Evolution 28:428–438. [456]

Howard, R., and A. Moore. 1980. A complete checklist of the birds of the world. Academic Press, London. [161]

Howell, S. N. G. 1993. Taxonomy and distribution of the hummingbird genus *Chlorostilbon* in Mexico and northern Central America. Euphonia 2:25–37. [270]

Howell, S. N. G. 1994. The specific status of Black-faced Antthrushes in Middle America. Cotinga 1:20–25. [271]

Hultsch, H. 1980. Beziehungen zwischen Struktur, zeitlicher Variabilität und sozialem Einsatz im Gesang der Nachtigall, *Luscinia megarhynchos.* Ph.D. dissertation, Freie Universität Berlin. [80–82, 84]

Hultsch, H. 1983. Behavioral signficance of duet interactions: cues from antiphonal duetting between males. Behaviour 86:89–99. [291]

Hultsch, H. 1991a. Early experience can modify singing styles—evidence from experiments with nightingales, *Luscinia megarhynchos.* Anim. Behav. 42:883–889. [80, 86]

Hultsch, H. 1991b. Correlates of repertoire constriction in the song ontogeny of nightingales (*Luscinia megarhynchos*). Verh. Dtsch. Zool. Ges. 84:474. [92]

Hultsch, H. 1991c. Song ontogeny in birds: closed or open developmental programs? Page 576 *in* Synapse-transmission, modulation (N. Elsner and H. Penzlin, Eds.). Thieme Verlag, Stuttgart. [93]

Hultsch, H. 1992. Time window and unit capacity: dual constraints on the acquisition of serial information in songbirds. J. Comp. Physiol. A 170:275–280. [89]
Hultsch, H. 1993a. Ecological versus psychobiological aspects of song learning in birds. Etología 3:309–323. [80, 86]
Hultsch, H. 1993b. Tracing the memory mechanisms in the song acquisition of birds. Neth. J. Zool. 43:155–171. [92, 93]
Hultsch, H., and M. L. Kopp. 1989. Early auditory learning and song improvisation in nightingales, *Luscinia megarhynchos*. Anim. Behav. 37:510–512. [85]
Hultsch, H., and D. Todt. 1981. Repertoire sharing and song-post distance in nightingales. Behav. Ecol. Sociobiol. 8:183–188. [93]
Hultsch, H., and D. Todt. 1982. Temporal performance roles during vocal interactions in nightingales (*Luscinia megarhynchos* B.). Behav. Ecol. Sociobiol. 11:253–260. [84, 85, 444]
Hultsch, H., and D. Todt. 1986. Signal matching. Z. Semiotik 8:233–244. [84]
Hultsch, H., and D. Todt. 1989a. Song acquisition and acquisition constraints in the nightingale (*Luscinia megarhynchos*). Naturwissenschaften 76:83–86. [86]
Hultsch, H., and D. Todt. 1989b. Context memorization in the learning of birds. Naturwissenschaften 76:584–586. [12, 13, 87]
Hultsch, H., and D. Todt. 1989c. Memorization and reproduction of songs in nightingales (*Luscinia megarhynchos*): evidence for package formation. J. Comp. Physiol. A 165:197–203. [89]
Hultsch, H., and D. Todt. 1992a. Acquisition of serial patterns in birds: temporal cues are mediators for song organization. Bioacoustics 4:61–62. [88]
Hultsch, H., and D. Todt. 1992b. The serial order effect in the song acquisition of birds. Anim. Behav. 44:590–592. [88]
Hultsch, H., and D. Todt. In press. Discontinuous and incremental processes in the song acquisition of birds—evidence for a primer effect. J. Comp. Physiol. A. [88]
Hultsch, H., R. Lange, and D. Todt. 1984. Pattern-type labeled tutoring: a method for studying song-type memories in repertoire birds. Verh. Dtsch. Zool. Ges. 77:249. [86]
Hunn, E. 1992. The use of sound recordings as voucher specimens and stimulus materials in ethnozoological research. J. Ethnobiol. 12:187–198. [484]
Hunter, F. M., T. Burke, and S. E. Watts. 1992. Frequent copulation as a method of paternity assurance in the Northern Fulmar. Anim. Behav. 44:149–156. [174]
Hunter, M. L., and J. R. Krebs. 1979. Geographical variation in the song of the Great Tit (*Parus major*) in relation to ecological factors. J. Anim. Ecol. 48:759–785. [182, 186, 313]
Hunter, M. L., Jr., A. Kacelnik, J. Roberts, and M. Vuillermoz. 1986. Directionality of avian vocalizations: a laboratory study. Condor 88:371–375. [412]
Hurly, T. A., L. Ratcliffe, and R. Weisman. 1990. Relative pitch recognition in White-throated Sparrows, *Zonotrichia albicollis*. Anim. Behav. 40:176–181. [345]
Hurly, T. A., L. Ratcliffe, D. Weary, and R. Weisman. 1992. White-throated Sparrows (*Zonotrichia albicollis*) can perceive pitch change in conspecific song by using the frequency ratio independent of the frequency difference. J. Comp. Psychol. 106:388–391. [345]
Hutchinson, J. M. C., J. M. McNamara, and I. C. Cuthill. 1993. Song, sexual selection, starvation and strategic handicaps. Anim. Behav. 45:1153–1177. [340, 426, 438, 439, 453]
Hutchison, L. V., and B. M. Wenzel. 1980. Olfactory guidance in foraging by Procellariiformes. Condor 82:314–319. [162]
Immelmann, K., J. Steinbacker, and H. E. Wolters. 1965. Vogel in Käfig und Voliere, Prachtfinken, vol. 1. Verlag Hans Limberg, Aachen, Germany. [45]
Immelmann, K., J. Steinbacker, and H. E. Wolters. 1977. Vogel in Käfig und Voliere, Prachtfinken, vol. 2. Verlag Hans Limberg, Aachen, Germany. [56]
Ince, S. A., and P. J. B. Slater. 1985. Versatility and continuity in the songs of thrushes, *Turdus* spp. Ibis 127:455–364. [79]
Ince, S. A., P. J. B. Slater, and C. Weismann. 1980. Changes with time in the songs of a population of Chaffinches. Condor 82:285–290. [198, 213, 216]

Irwin, R. E. 1988. The evolutionary importance of behavioural development: the ontogeny and phylogeny of bird song. Anim. Behav. 36:814–824. [8]

Isaac, D., and P. Marler. 1963. Ordering of sequences of singing behaviour of Mistle Thrushes in relationship to timing. Anim. Behav. 11:179–187. [80]

Isler, M. L., and P. R. Isler. 1987. The tanagers: natural history, distribution, and identification. Smithsonian Institution Press, Washington, D.C. [274]

Isler, M. L., P. R. Isler, and B. M. Whitney. In press. Biogeography and systematics of the *Thamnophilus punctatus* (Thamnophilidae) complex. *In* Natural history and conservation of Neotropical birds (J. V. Remsen Jr., Ed.). Ornithological Monographs, American Ornithologists' Union, Washington, D.C. [270, 273, 280]

Iwasa, Y., A. Pomiankowski, and S. Nee. 1991. The evolution of costly mate preferences. II. The "handicap" principle. Evolution 45:1431–1442. [465]

Jacobs, L. F., S. J. C. Gaulin, D. F. Sherry, and G. E. Hoffman. 1990. Evolution of spatial cognition: sex-specific patterns of spatial behavior predict hippocampal size. Proc. Natl. Acad. Sci. USA 87:6349–6352. [342]

James, P. C. 1983. Storm-Petrel tape lures: which sex is attracted? Ringing & Migr. 4:249–253. [165, 172, 174]

James, P. C. 1984. Sexual dimorphism in the voice of the British Storm-Petrel, *Hydrobates pelagicus.* Ibis 126:89–92. [165, 172]

James, P. C. 1985a. Geographical and temporal variation in the calls of the Manx Shearwater, *Puffinus puffinus,* and British Storm-Petrel, *Hydrobates pelagicus.* J. Zool. Lond. 207:331–344. [168, 176]

James, P. C. 1985b. The vocal behavior of the Manx Shearwater, *Puffinus puffinus.* Z. Tierpsychol. 67:269–283. [171–173, 176]

James, P. C. 1986. How do Manx Shearwaters, *Puffinus puffinus,* find their burrows? Ethology 71:287–294. [162]

James, P. C., and H. A. Robertson. 1985a. The call of Bulwer's Petrel (*Bulweria bulwerii*), and the relationship between intersexual call divergence and aerial calling in nocturnal Procellariiformes. Auk 102:878–881. [165, 172, 173, 177]

James, P. C., and H. A. Robertson. 1985b. Sexual dimorphism in the voice of the Little Shearwater, *Puffinus assimilis.* Ibis 127:388–390. [165]

James, P. C., and H. A. Robertson. 1985c. The call of male and female Madeiran Storm-Petrel (*Oceanodroma castro*). Auk 102:391–393. [165]

Janzen, D. 1967. Why mountain passes are higher in the tropics. Am. Nat. 101:233–249. [272, 275]

Järvi, T., T. Rädesater, and S. Jakobson. 1977. Individual recognition and variation in the Great Tit (*Parus major*) two-syllable song. Biophonics 5:4–9. [359]

Jehl, J. R., Jr. 1968. Relationships in the Charadrii (shorebirds): a taxonomic study based on color patterns of the downy young. San Diego Soc. Nat. Hist. Mem. 3. [241]

Jellis, R. 1977. Bird sounds and their meaning. Cornell Univ. Press, Ithaca, N.Y. [250]

Jenkins, P. F. 1978. Cultural transmission of song patterns and dialect development in a free-living bird population. Anim. Behav. 26:50–78. [15, 64, 183, 186, 217]

Jenkins, P. F., and A. J. Baker. 1984. Mechanisms of song differentiation in introduced populations of Chaffinches, *Fringilla coelebs,* in New Zealand. Ibis 126:510–524. [182, 215]

Jenkinson, M. A. 1993. The American Ornithologists' Union's support of Latin American ornithology. Auk 110:659–660. [274]

Johansen, H. 1954. Die Vogelfauna Westsibiriens. J. Ornithol. 95:319–353. [231]

Johnson, F., and S. Bottjer. 1992. Growth and regression of thalamic efferents in the song-control system of male Zebra Finches. J. Comp. Neurol. 326:442–450. [288]

Johnson, N. K. 1994. Old-school taxonomy versus modern biosystematics: species-level decisions in *Stelgidopteryx* and *Empidonax.* Auk 111:773–780. [288]

Johnson, N. K., and J. A. Marten. 1988. Evolutionary genetics of flycatchers. II. Differentiation in the *Empidonax difficilis* complex. Auk 105:177–191. [272]

Johnson, O. W., and P. G. Connors. 1996. American Golden-Plover (*Pluvialis dominica*), Pacific Golden-Plover (*Pluvialis fulva*). *In* The birds of North America, no. 201–202 (A. Poole and F. Gill, Eds.). The Academy of Natural Sciences, Philadelphia; The American Ornithologists' Union, Washington, D.C. [244, 256]

Johnsrude, I., D. Weary, L. Ratcliffe, and R. Weisman. 1994. Effect of motivational context on conspecific song discrimination by Brown-headed Cowbirds (*Molothrus ater*). J. Comp. Psychol. 108:172–178. [348]

Johnston, T. D. 1988. Developmental explanation and the ontogeny of birdsong: nature/nurture redux. Behav. Brain Sci. 11:617–630. [39, 41, 59]

Jones, I. L., J. B. Falls, and A. J. Gaston. 1989. The vocal repertoire of the Ancient Murrelet. Condor 91:699–710. [167]

Jouanin, C., and J.-L. Mougin. 1979. Order Procellariiformes. Pages 48–121 *in* Peter's checklist of the birds of the world, vol. 1. 2nd ed. (E. Mayr and G. W. Cotterell, Eds.). Harvard Univ. Press, Cambridge, Mass. [161]

Jouventin, P. 1982. Visual and vocal signals in penguins, their evolution and adaptive characters. Adv. Ethol. 24. [163, 167, 168, 173, 175]

Jouventin, P., and J.-L. Mougin. 1981. Les stratégies adaptatives des oiseaux de mer. Rev. Ecol. Terre Vie 35:217–272. [161]

Kacelnik, A., and J. R. Krebs. 1982. The dawn chorus in the Great Tit (*Parus major*): proximate and ultimate causes. Behaviour 83:287–309. [333, 433, 434, 436, 437]

Kale, H. S. II. 1965. Ecology and bioenergetics of the Long-billed Marsh Wren, *Telmatodytes palustris griseus* (Brewster), in Georgia salt marshes. Publ. Nuttall Ornithol. Club 5. Cambridge, Mass. [8]

Karakashian, S. J., M. Gyger, and P. Marler. 1988. Audience effects on alarm calling in chickens (*Gallus gallus*). J. Comp. Psychol. 102:129–135. [416]

Karten, H., and W. Hodos. 1967. A stereotaxic atlas of the brain of the pigeon. Johns Hopkins Univ. Press, Baltimore. [295]

Kaufmann, K. 1993. Theodore A. Parker III. 1953–1993. Am. Birds, Fall 1993:349–351. [v]

Kavanau, J. L. 1987. Lovebirds, cockatiels, Budgerigars: behavior and evolution. Sciences Software Systems, Los Angeles. [39]

Keast, A. 1985. Springtime song, periodicity and sequencing, a comparison of a southern forest and northern woodland bird community. Pages 119–128 *in* Birds of eucalypt forests and woodlands: ecology, conservation, management (A. Keast, H. F. Recher, H. Ford, and D. Saunders, Eds.). Royal Australian Ornithologists' Union and Surrey Beatty and Sons, Chipping Norton, New South Wales. [441]

Keast, J. A. 1958. Infraspecific variation in the Australian finches. Emu 58:219–246. [8]

Kelley, D., and F. Nottebohm. 1979. Projections of a telencephalic auditory nucleus-field L in the canary. J. Comp. Neurol. 183:455–470. [288]

Kellogg, P. P. 1962. Bird-sound studies at Cornell. Living Bird 1:37–48. [476]

Kelsey, M. G. 1989. A comparison of the song and territorial behaviour of a long-distance migrant, the Marsh Warbler, *Acrocephalus palustris,* in summer and winter. Ibis 131:403–414. [430]

Kempenaers, B. 1993. The use of a breeding synchrony index. Ornis Scand. 24:84. [263]

Kempenaers, B., G. R. Verheyen, M. van den Broeck, T. Burke, C. van Broeckhoven, and A. A. Dhondt. 1992. Extra-pair paternity results from female preferences for high-quality males in the Blue Tit. Nature 357:494–496. [264, 340]

Kendrick, D. F., M. E. Rilling, and M. R. Denny. 1986. Theories of animal memory. Erlbaum, Hillsdale, N.J. [89]

Kern, M. D., and J. R. King. 1972. Testosterone-induced singing in female White-crowned Sparrows. Condor 74:204–209. [43]

Ketterson, E. D., and V. Nolan. 1994. Hormones and life histories: an integrative approach. Pages 327–353 *in* Behavioral mechanisms in evolutionary ecology (L. Real, Ed.). Univ. Chicago Press, Chicago. [27]

Kettle, R. 1983. Natural history. Pages 162–176 *in* Sound archives: a guide to their establishment and development (D. Lance, Ed.). Int. Assoc. Sound Arch. Spec. Publ. 4. [476]

Kettle, R. 1989. Major wildlife sound libraries. Bioacoustics 2:171–176. [476]

Kettle, R. 1991. Selected wildlife sound records and cassettes published in the 1980s. Int. J. Anim. Sound Record. 3:71–76. [274]

Kettle, R., and J. Vielliard. 1991. Documentation standards for wildlife sound recordings. Bioacoustics 3:235–238. [476, 477, 479]

Kiernan, V. 1993. The Pentagon's green flag of convenience. New Sci. 139(1886):12–13. [419]

Kimura, D. 1992. Sex differences in the brain. Sci. Am. 267:119–125. [342]

Kimura, M. 1968. Genetic variability maintained in a finite population due to mutational production of neutral and nearly neutral isoalleles. Genet. Res. 11:247–269. [187, 188]

Kimura, M. 1983. The neutral theory of molecular evolution. Cambridge Univ. Press, Cambridge. [184, 185, 197]

Kimura, M., and J. F. Crow. 1964. The number of alleles that can be maintained in a finite population. Genetics 49:725–738. [187]

Kimura, M., and T. Maruyama. 1971. Pattern of neutral polymorphism in a geographically structured population. Genet. Res. 18:125–131. [190]

King, A. P., and M. J. West. 1977. Species identification in the North American cowbird: appropriate responses to abnormal song. Science 195:1002–1004. [5, 458]

King, A. P., and M. J. West. 1983. Epigenesis of cowbird song—a joint endeavour of males and females. Nature 305:704–706. [12]

King, A. P., and M. J. West. 1987a. Different outcomes of synergy between song production and song perception in the same subspecies (*Molothrus ater ater*). Dev. Psychobiol. 20:177–187. [5]

King, A. P., and M. J. West. 1987b. The experience of experience: an exogenetic program for social competence. Pages 153–182 *in* Perspectives in ethology, vol. 7 (P. P. G. Bateson and P. H. Klopfer, Eds.). Plenum Press, New York. [339]

King, A. P., and M. J. West. 1988. Searching for the functional origins of cowbird song in eastern Brown-headed Cowbirds (*Molothrus ater ater*). Anim. Behav. 36:1575–1588. [27]

King, A. P., and M. J. West. 1989. Presence of female cowbirds (*Molothrus ater ater*) affects vocal improvisation in males. J. Comp. Psychol. 103:39–44. [36]

King, A. P., and M. J. West. 1990. Variation in species-typical behavior: a contemporary theme for comparative psychology. Pages 331–339 *in* Contemporary issues in comparative psychology (D. A. Dewsbury, Ed.). Sinauer, Sunderland, Mass. [20, 22, 28]

King, A. P., M. J. West, and D. H. Eastzer. 1980. Song structure and song development as potential contributors to reproductive isolation in cowbirds (*Molothrus ater*). J. Comp. Physiol. Psychol. 94:1028–1039. [339, 459]

Kirkpatrick, M. 1982. Sexual selection and the evolution of female choice. Evolution 36:1–12. [463]

Kirkpatrick, M. 1985. Evolution of female choice and male parental investment in polygynous species: the demise of the "sexy son." Am. Nat. 125:788–810. [461, 463]

Kirkpatrick, M., and M. J. Ryan. 1991. The evolution of mating preferences and the paradox of the lek. Nature 350:33–38. [454, 460, 462]

Kirn, J., and F. Nottebohm. 1993. Direct evidence for loss and replacement of projection neurons in adult canary brain. J. Neurosci. 13:1654–1663. [301]

Kirn, J., R. Clower, D. E. Kroodsma, and T. DeVoogd. 1989. Song-related brain regions in the Red-winged Blackbird are affected by sex and season but not repertoire size. J. Neurobiol. 20:139–163. [302]

Klump, G. M., and E. Curio. 1983. Warum liegen viele Lautäußerungen von Singvögeln oberhalb des Bereichs besten Hörens? Verh. Dtsch. Zool. Ges. 76:182. [328, 330]

Klump, G. M., and U. Langemann. 1995. Comodulation masking release in a songbird. Hear. Res. 87:157–164. [332, 333]

Klump, G. M., and O. N. Larsen. 1992. Azimuth sound localization in the European Starling (*Sturnus vulgaris*): physical binaural cues. J. Comp. Physiol. A 170:243–251. [334]

Klump, G. M., and E. H. Maier. 1990. Temporal summation in the European Starling (*Sturnus vulgaris*). J. Comp. Psychol. 104:94–100. [329]

Klump, G. M., and K. Okanoya. 1991. Temporal modulation transfer functions in the European Starling (*Sturnus vulgaris*). I. Psychophysical modulation detection thresholds. Hear. Res. 52:1–12. [323]

Klump, G. M., and M. D. Shalter. 1984. Acoustic behaviour of birds and mammals in the predator context. I. Factors affecting the structure of alarm signals. II. The functional significance and evolution of alarm signals. Z. Tierpsychol. 66:189–226. [329, 417]

Klump, G. M., E. Kretzschmar, and E. Curio. 1986. The hearing of an avian predator and its avian prey. Behav. Ecol. Sociobiol. 18:317–323. [321, 417, 418, 423]

Koch, L. 1955. Memoirs of a birdman. Scientific Book Club, London. [476]

Kodric-Brown, A., and J. H. Brown. 1984. Truth in advertising: the kinds of traits favored by sexual selection. Am. Nat. 124:309–323. [462]

König, C., and R. Straneck. 1989. Eine neue Eule (Aves: Strigidae) aus Nordargentinien. Stuttg. Beitr. Naturkd. Ser. A (Biol.) 428:1–20. [270]

Konishi, M. 1963. The role of auditory feedback in the vocal behaviour of the domestic fowl. Z. Tierpsychol. 20:349–367. [44, 46]

Konishi, M. 1965. The role of auditory feedback in the control of vocalization in the White-crowned Sparrow. Z. Tierpsychol. 22:770–783. [10, 115, 288]

Konishi, M. 1970. Comparative neurophysiological studies of hearing and vocalizations in songbirds. Z. Vergl. Physiol. 66:257–272. [330]

Konishi, M. 1973. How the owl tracks its prey. Am. Sci. 61:414–424. [334]

Konishi, M. 1985. Bird song: from behavior to neuron. Annu. Rev. Neurosci. 8:125–170. [25, 39, 41, 58]

Konishi, M. 1989. Birdsong for neurobiologists. Neuron 3:541–549. [3, 10, 94]

Konishi, M., and E. Akutagawa. 1985. Neuronal growth, atrophy and death in a sexually dimorphic song nucleus in the Zebra Finch brain. Nature 315:145–147. [288, 291]

Konishi, M., and E. Akutagawa. 1987. Hormonal control of cell death in a sexually dimorphic song nucleus in the Zebra Finch. Pages 173–185 *in* Selective neuronal death. Ciba Foundation Symposium 126. Wiley, New York. [293]

Konishi, M., and F. Nottebohm. 1969. Experimental studies in the ontogeny of avian vocalizations. Pages 29–48 *in* Bird vocalizations (R. A. Hinde, Ed.). Cambridge Univ. Press, London. [39]

Köppl, C. 1994. Phase locking at high frequencies in the Barn Owl's auditory nerve. Abstr. Midwest Res. Meet. Assoc. Res. Otolaryngol. 18:382. [335]

Kramer, H. G., and R. E. Lemon. 1983. Dynamics of territorial singing between neighbouring Song Sparrows (*Melospiza melodia*). Behaviour 85:198–223. [77, 80, 411, 418]

Krebs, J. R. 1971. Territory and breeding density in the Great Tit, *Parus major* L. Ecology 52:2–22. [359]

Krebs, J. R. 1976. Habituation and song repertoires in the Great Tit. Behav. Ecol. Sociobiol. 1:215–227. [298]

Krebs, J. R. 1991. Animal communication: ideas derived from Tinbergen's activities. Pages 60–74 *in* The Tinbergen legacy (M. S. Dawkins, T. R. Halliday and R. Dawkins, Eds.). Chapman and Hall, London. [160]

Krebs, J. R., and R. Dawkins. 1984. Animal signals: mind-reading and manipulation. Pages 380–402 *in* Behavioural ecology: an evolutionary approach (J. R. Krebs and N. B. Davies, Eds.). Blackwell, Oxford. [423]

Krebs, J. R., and D. E. Kroodsma. 1980. Repertoires and geographical variation in bird song. Pages 134–177 *in* Advances in the study of behavior, vol. 11 (J. S. Rosenblatt, R. A. Hinde, C. Beer, and M. C. Busnel, Eds.). Academic Press, New York. [3, 11, 13, 83, 160, 161, 174, 356, 368, 369, 370]

Krebs, J. R., R. Ashcroft, and M. Webber. 1978. Song repertoires and territory defence in the Great Tit. Nature 271:539–542. [298, 406]

Krebs, J. R., M. Avery, and R. J. Cowie. 1981a. Effect of removal of mate on the singing behaviour of Great Tits. Anim. Behav. 29:635–637. [456]

Krebs, J. R., R. Ashcroft, and K. van Orsdol. 1981b. Song matching in the Great Tit, *Parus major* L. Anim. Behav. 29:918–923. [77, 411]

Kreutzer, M. L. 1972. Les variations dans les chants de *Troglodytes troglodytes* et leurs conséquences comportementales. C. R. Acad. Sci. Paris 275:1423–1424. [216]

Kreutzer, M. L. 1981. Etude du chant le bruant zizi: le répertoire, caractéristiques et distribution. Behaviour 71:291–321. [215]

Kreutzer, M. L. 1985. Le chant du bruant zizi (*Emberiza cirlus*): codage–décodage–évolution. Ph.D. dissertation, Univ. Pierre et Marie Curie, Paris. [308]

Kreutzer, M. L. 1990. La durée du chant des oscines: le rôle des processus cognitifs et moteurs chez le bruant zizi (*Emberiza cirlus*). C. R. Acad. Sci. Paris Sér. III 310:423–428. [312]

Kreutzer, M., and H. R. Güttinger. 1991. Konkurenzbeziehungen und Verhaltensantworten gegenüber dem Gesang: Artnorm und individuelle Variabilität bei der Zaunammer (*Emberiza cirlus*). J. Ornithol. 132:165–177. [233]

Kreutzer, M. L., and E. M. Vallet. 1991. Differences in the responses of captive female canaries to variation in conspecific and heterospecific songs. Behaviour 117:106–116. [459]

Kroodsma, D. E. 1971. Songs and singing behavior in the Rufous-sided Towhee, *Pipilo erythropthalmus oregonus*. Condor 73:303–308. [444]

Kroodsma, D. E. 1974. Song learning, dialects, and dispersal in the Bewick's Wren. Z. Tierpsychol. 35:352–380. [15, 16, 64, 215, 217]

Kroodsma, D. E. 1976a. Reproductive development in a female songbird: differential stimulation by quality of male song. Science 192:574–575. [298, 457]

Kroodsma, D. E. 1976b. The effect of large song repertoires on neighbor "recognition" in male Song Sparrows. Condor 78:97–99. [356, 359, 362, 365, 368]

Kroodsma, D. E. 1977a. A re-evaluation of song development in the Song Sparrow. Anim. Behav. 25:390–399. [50, 53]

Kroodsma, D. E. 1977b. Correlates of song organization among North American wrens. Am. Nat. 111: 995–1008. [8, 16, 17, 79, 80, 357]

Kroodsma, D. E. 1978a. Aspects of learning in the ontogeny of bird song: where, from whom, when, how many, which, and how accurately? Pages 215–230 *in* The development of behavior (G. M. Burghardt and M. Bekoff, Eds.). Garland, New York. [5, 20, 63, 86, 215]

Kroodsma, D. E. 1978b. Continuity and versatility in bird song: support for the monotony-threshold hypothesis. Nature 274:681–683. [388, 389]

Kroodsma, D. E. 1979. Vocal dueling among male Marsh Wrens: evidence for ritualized expression of dominance/subordinance. Auk 96:506–515. [9, 83, 89, 390, 399]

Kroodsma, D. E. 1981a. Ontogeny of bird song. Pages 518–532 *in* Early development in man and animals. The Bielefeld project (K. Immelmann, G. Barlow, L. Petrinovich, and M. Main, Eds.). Cambridge Univ. Press, Cambridge. [11]

Kroodsma, D. E. 1981b. Geographical variation and functions of song types in warblers (Parulidae). Auk 98:743–751. [15, 16]

Kroodsma, D. E. 1981c. Winter Wren singing behavior: a pinnacle of song complexity. Condor 82:357–365. [40]

Kroodsma, D. E. 1982a. Learning and the ontogeny of sound signals in birds. Pages 1–23 *in* Acoustic communication in birds, vol. 2 (D. E. Kroodsma and E. H. Miller, Eds.). Academic Press, New York. [44, 97, 160, 183, 213, 258]

Kroodsma, D. E. 1982b. Song repertoires: problems in their definition and use. Pages 125–146 *in* Acoustic communication in birds, vol. 2 (D. E. Kroodsma and E. H. Miller, Eds.). Academic Press, New York. [79, 128, 294, 297, 317, 395]

Kroodsma, D. E. 1983a. The ecology of avian vocal learning. BioScience 33:165–171. [3, 17, 62, 298]

Kroodsma, D. E. 1983b. Marsh wrenditions. Nat. Hist. 92(9):43–46. [390]

Kroodsma, D. E. 1984. Songs of the Alder Flycatcher (*Empidonax alnorum*) and Willow Flycatcher (*Empidonax traillii*) are innate. Auk 101:13–24. [4, 10, 11, 54]

Kroodsma, D. E. 1985a. Development and use of two song forms by the Eastern Phoebe. Wilson Bull. 97:21–29. [4, 10, 54, 444]

Kroodsma, D. E. 1985b. Limited dispersal between dialects? Hypotheses testable in the field. Behav. Brain Sci. 8:108–109. [4]

Kroodsma, D. E. 1986. Design of song playback experiments. Auk 103:640–642. [360, 457]

Kroodsma, D. E. 1988a. Contrasting styles of song development and their consequences among the Passeriformes. Pages 157–184 *in* Evolution and learning (R. C. Bolles and M. D. Beecher, Eds.). Erlbaum, Hillsdale, N.J. [3, 10, 34, 62, 290]

Kroodsma, D. E. 1988b. Song-types and their use: developmental flexibility of the male Blue-winged Warbler. Ethology 79:235–247. [94]

Kroodsma, D. E. 1989a. Suggested experimental designs for song playbacks. Anim. Behav. 37:600–609. [13, 172, 457]

Kroodsma, D. E. 1989b. Male Eastern Phoebes (Tyrannidae, Passeriformes) fail to imitate songs. J. Comp. Psychol. 103:227–232. [4, 10, 35, 373]

Kroodsma, D. E. 1989c. Inappropriate experimental designs impede progress in bioacoustic research: a reply. Anim. Behav. 38:717–719. [172]

Kroodsma, D. E. 1989d. Two North American song populations of the Marsh Wren reach distributional limits in the central Great Plains. Condor 91:332–340. [34, 271]

Kroodsma, D. E. 1990. Using appropriate experimental designs for intended hypotheses in "song" playbacks, with examples for testing effects of song repertoire sizes. Anim. Behav. 40:1138–1150. [35, 131, 360, 457]

Kroodsma, D. E., and J. R. Baylis. 1982. Appendix: a world survey of evidence for vocal learning in birds. Pages 311–337 *in* Acoustic communication in birds, vol. 2 (D. E. Kroodsma and E. H. Miller, Eds.). Academic Press, New York. [10, 160, 294]

Kroodsma, D. E., and B. E. Byers. 1991. The function(s) of bird song. Am. Zool. 31: 318–328. [41, 61, 160, 298]

Kroodsma, D. E., and R. A. Canady. 1985. Differences in repertoire size, singing behavior, and associated neuroanatomy among Marsh Wren populations have a genetic basis. Auk 102:439–446. [41, 62, 298, 299]

Kroodsma, D. E., and F. C. James. 1994. Song variation among populations of the Red-winged Blackbird. Wilson Bull. 106:156–162. [6]

Kroodsma, D. E., and M. Konishi. 1991. A suboscine bird (Eastern Phoebe, *Sayornis phoebe*) develops normal song without auditory feedback. Anim. Behav. 42:477–487. [3, 4, 10, 44, 113, 272, 295, 296, 373]

Kroodsma, D. E., and E. H. Miller (Eds.). 1982. Acoustic communication in birds. 2 vols. Academic Press, New York. [20, 85, 160, 305, 377]

Kroodsma, D. E., and H. Momose. 1991. Songs of the Japanese population of the Winter Wren (*Troglodytes troglodytes*). Condor 93:424–432. [272]

Kroodsma, D. E., and L. D. Parker. 1977. Vocal virtuosity in the Brown Thrasher. Auk 94:783–785. [13, 297]

Kroodsma, D. E., and R. Pickert. 1980. Environmentally dependent sensitive periods for avian vocal learning. Nature 288:477–479. [5, 9, 32, 33, 63, 85, 287]

Kroodsma, D. E., and R. Pickert. 1984a. Sensitive phases for song learning: effects of social interaction and individual variation. Anim. Behav. 32:389–394. [63]

Kroodsma, D. E., and R. Pickert. 1984b. Repertoire size, auditory templates, and selective vocal learning in songbirds. Anim. Behav. 32:395–399. [63]

Kroodsma, D. E., and J. Verner. 1978. Complex singing behaviours among *Cistothorus* wrens. Auk 95:703–716. [7–9, 13, 15, 62, 411]

Kroodsma, D. E., M. C. Baker, L. F. Baptista, and L. Petrinovich. 1984. Vocal "dialects" in Nuttall's White-crowned Sparrow. Pages 103–133 *in* Current Ornithology, vol. 2 (R. F. Johnston, Ed.). Plenum Press, New York. [63, 160]

Kroodsma, D. E., R. E. Bereson, B. E. Byers, and E. Minear. 1989. Use of song types by the Chestnut-sided Warbler: evidence for both intrasexual and intersexual functions. Can. J. Zool. 67:447–456. [18, 126, 430, 432, 438, 446, 456]

Kroodsma, D. E., D. J. Albano, P. W. Houlihan, and J. A. Wells. 1995. Song development by Black-capped Chickadees (*Parus atricapillus*) and Carolina Chickadees (*P. carolinensis*). Auk 112:29–43. [15–17]

Kroodsma, D. E., P. Fallon, and P. Houlihan. Unpubl. ms. Song development by the Gray Catbird. [4]

Kunkel, P. 1974. Mating systems of tropical birds: the effects of weakness or absence of external reproduction-timing factors, with special reference to prolonged pair bonds. Z. Tierpsychol. 34:265–307. [447, 449]

Labutin, Y. V., V. V. Leonovitch, and B. N. Veprintsev. 1982. The Little Curlew, *Numenius minutus,* in Siberia. Ibis 124:302–319. [252]

Lack, D. 1957. Swifts in a tower. Methuen, London. [442]

Lade, B. I., and W. H. Thorpe. 1964. Dove songs as innately coded patterns of specific behaviour. Nature 202:366–368. [52, 56, 58]

Lambrechts, M. M. 1988. Great Tit song output is determined both by motivation and by constraints in singing ability: a reply to Weary et al. Anim. Behav. 36:1244–1246. [310, 312, 316, 319]

Lambrechts, M. M. 1992. Male quality and playback in the Great Tit. Pages 135–152 *in* Playback and studies of animal communication (P. K. McGregor, Ed.). Plenum Press, New York. [308, 319, 435]

Lambrechts, M. M. Unpubl. ms. Song frequency plasticity and phrase organization in Great Tits. [313]

Lambrechts, M. M., and A. A. Dhondt. 1986. Male quality, reproduction, and survival in the Great Tit (*Parus major*). Behav. Ecol. Sociobiol. 19:57–63. [308, 471]

Lambrechts, M. M., and A. A. Dhondt. 1987. Differences in singing performance between male Great Tits. Ardea 75:43–52. [160, 306, 308, 309]

Lambrechts, M. M., and A. A. Dhondt. 1988. The anti-exhaustion hypothesis: a new hypothesis to explain song performance and song switching in the Great Tit. Anim. Behav. 36:327–334. [306, 310, 312, 316, 317]

Lambrechts, M. M., and A. A. Dhondt. 1990. A relationship between the composition and size of Great Tit song repertoires. Anim. Behav. 39:213–218. [314, 317]

Lambrechts, M. M., and A. A. Dhondt. 1994. Individual voice discrimination in birds. Pages 115–139 *in* Current ornithology, vol. 12 (D. M. Power, Ed.). Plenum Press, New York. [370, 373]

Lance, D. (Ed.). 1983. Sound archives. A guide to their establishment and development. Int. Assoc. Sound Arch. Spec. Publ. 4. [485]

Lande, R. 1981. Models of speciation by sexual selection on polygenic traits. Proc. Natl. Acad. Sci. USA 78:3721–3725. [463]

Langemann, U., G. M. Klump, and R. J. Dooling. 1995. Critical bands and critical-ratio bandwidth in the European Starling. Hear. Res. 84:167–176. [328, 329]

Lanyon, S. M. 1992. Interspecific brood parasitism in blackbirds (Icterinae): a phylogenetic perspective. Science 255:77–79. [7, 8]

Lanyon, S. M., D. F. Stotz, and D. E. Willard. 1990. *Clytoctantes atrogularis,* a new species of antbird from western Brazil. Wilson Bull. 102:571–580. [270]

Lanyon, W. E. 1960. The middle American populations of the Crested Flycatcher, *Myiarchus tyrannulus.* Condor 62:341–350. [54]

Lanyon, W. E. 1967. Revision and probable evolution of the *Myiarchus* flycatchers of the West Indies. Bull. Am. Mus. Nat. Hist. 136:329–370. [271]

Lanyon, W. E. 1969. Vocal characters and avian systematics. Pages 291–310 *in* Bird vocalizations: their relation to current problems in biology and psychology (R. A. Hinde, Ed.). Cambridge Univ. Press, Cambridge. [256]

Lanyon, W. E. 1978. Revision of the *Myiarchus* flycatchers of South America. Bull. Am. Mus. Nat. Hist. 161:427–628. [54, 271, 272, 444, 447]

Larsen, O. N., and T. Dabelsteen. 1990. Directionality of blackbird vocalization. Implications for vocal communication and its further study. Ornis Scand. 21:37–45. [411, 412]

Larsen, O. N., M. L. Dent, and R. J. Dooling. 1994. Free-field release from masking in the Budgerigar (*Melopsittacus undulatus*). Page 370 *in* Sensory transduction, vol. 2 (H. Breer and N. Elsner, Eds.). Thieme Verlag, Stuttgart. [334, 335]

Lassaline, M. E., E. J. Wisniewski, and D. L. Medin. 1992. Basic levels in artificial and natural categories: are all basic levels created equal? Pages 327–378 *in* Percepts, concepts, and categories: the representation and processing of information. Advances in Psychology 93 (B. Burns, Ed.). North-Holland, Amsterdam. [133]

Latter, B. D. H. 1973. The island model of differentiation: a general solution. Genetics 73:147–157. [189]

Layton, L. 1991. Songbirds in Singapore, the growth of a pastime. Oxford Univ. Press, New York. [48]

Lehman, W. P. 1992. Historical linguistics. 2nd ed. Routledge, London. [214, 219]

Lehrman, D. S. 1970. Semantic and conceptual issues in the nature-nurture problem. Pages 17–52 *in* Development and evolution of behavior: essays in memory of T. C. Schneirla (L. R. Aronson, E. Tobach, D. S. Lehrman, and J. S. Rosenblatt, Eds.). W. H. Freeman, San Francisco. [22]

Lehrman, D. S. 1974. Can psychiatrists use ethology? Pages 187–196 *in* Ethology and psychiatry (N. F. White, Ed.). Univ. Toronto Press, Toronto. [22]

Lein, M. R. 1978a. Song variation in a population of Chestnut-sided Warblers (*Dendroica pensylvanica*): its nature and suggested significance. Can. J. Zool. 567:1266–1283. [18]

Lein, M. R. 1978b. Territorial and courtship songs of birds. Nature 237:48–49. [94]

Lein, M. R. 1981. Display behavior of Ovenbirds (*Seiurus aurocapillus*). II. Song variation and singing behavior. Wilson Bull. 93:21–41. [390, 445]

Lemon, R. E. 1966. Geographic variation in the songs of cardinals. Can. J. Zool. 44:413–428. [213]

Lemon, R. E. 1968. Coordinated singing by Black-crested Titmice. Can. J. Zool. 46:1163–1167. [411]

Lemon, R. E. 1975. How birds develop song dialects. Condor 77:385–406. [45, 198, 213, 217, 219, 308]

Lemon, R. E., and C. Chatfield. 1971. Organization of song in cardinals. Anim. Behav. 19:1–17. [80]

Lemon, R. E., and M. Harris. 1974. The question of dialects in the songs of White-throated Sparrows. Can. J. Zool. 52:83–98. [7]

Lemon, R. E., J. Struger, M. J. Lechowicz, and R. F. Norman. 1981. Song features and singing heights of American warblers: maximization or optimization of distance. J. Acoust. Soc. Am. 69:1169–1176. [321]

Lemon, R. E., R. Cotter, R. C. MacNally, and S. Monette. 1985. Song repertoires and song sharing by American Redstarts. Condor 87:457–470. [17, 18, 215]

Lemon, R. E., S. Monette, and D. Roff. 1987. Song repertoires of American warblers (Parulinae): honest advertisement or assessment? Ethology 74:265–284. [17, 262, 264, 434]

Lemon, R. E., S. Perreault, and D. M. Weary. 1994. Dual strategies of song development in American Redstarts. Anim. Behav. 47:317–329. [18, 75, 94, 215]

Lencioni-Neto, F. 1994. Une nouvelle espèce de *Chordeiles* (Aves, Caprimulgidae) de Bahia (Brésil). Alauda 62:241–245. [270]

Leopold, A., and A. E. Enyon. 1961. Avian daybreak and evening song in relation to time and light intensity. Condor 63:269–293. [443]

Levi, W. M. 1965. Encyclopedia of pigeon breeds. T. F. H. Publications, N.J. [46]
Levin, R. N. 1988. The adaptive significance of song in the Bay Wren, *Thryothorus nigricapillus*. Ph.D. dissertation, Cornell University, Ithaca, N.Y. [276, 291, 298]
Levin, R. N., and J. Wingfield. 1992. The hormonal control of territorial aggression in tropical birds. Ornis Scand. 23:284–291. [294]
Levins, R. 1966. The strategy of model building in population biology. Am. Sci. 54:421–431. [439]
Lewald, J. 1990. The directionality of the ear of the pigeon (*Columba livia*). J. Comp. Physiol. A 167:533–543. [334]
Lifjeld, J. T., and R. J. Robertson. 1992. Female control of extra-pair fertilization in Tree Swallows. Behav. Ecol. Sociobiol. 31:89–96. [264, 340]
Lockhead, G. R. 1992. On identifying things: a case for context. Pages 109–143 *in* Percepts, concepts, and categories: the representation and processing of information. Advances in Psychology 93 (B. Burns, Ed.). North-Holland, Amsterdam. [124]
Lockhead, G. R., and J. R. Pomerantz. 1991. The perception of structure. American Psychological Association, Washington, D.C. [125]
Loesche, P., P. K. Stoddard, B. Higgins, and M. D. Beecher. 1991. Adaptations for individual vocal recognition by voice in swallows. Behaviour 118:15–25. [367]
Loffredo, C. A., and G. Borgia. 1986. Male courtship vocalization as cues for mate choice in the Satin Bowerbird, *Ptilonorhynchus violaceus*. Auk 103:189–195. [40]
Logan, C. A., and L. E. Hyatt. 1991. Mate attraction by autumnal song in the Northern Mockingbird (*Mimus polygottos*). Auk 108:429–432. [456]
Löhrl, H. 1967. Zur Verwandtschaft von Fichtenammer (*Emberiza leucocephala*) und Goldammer (*Emberiza citrinella*). Vogelwelt 88:148–152. [231]
Lorenz, K. 1961. Phylogenetische anpassung und adaptive modification des verhaltens. Z. Tierpsychol. 18:139–187. [39]
Lotka, A. J. 1931. Population analysis—the extinction of families, parts 1 and 2. J. Wash. Acad. Sci. 21:377–380, 453–459. [213]
Lott, D. 1991. Intraspecific variation in the social systems of wild vertebrates. Cambridge Univ. Press, Cambridge. [449]
Lougheed, S. C., and P. Handford. 1989. Night songs in the Rufous-collared Sparrow. Condor 91:462–465. [447]
Lowther, J. K., and J. B. Falls. 1968. White-throated Sparrow. Pages 1364–1392 *in* Life histories of North American cardinals, grosbeaks, buntings, towhees, finches, sparrows, and allies (A. C. Bent et al., Eds.). U.S. Natl. Mus. Bull. 237. [44]
Luders, D. J. 1977. Behavior of Antarctic petrels and Antarctic fulmars before laying. Emu 77:208–214. [162, 165]
Ludescher, F. B. 1973. Sumpfmeise (*Parus palustris* L.) und Weidenmeise (*P. montanus salicarius* Br.) als sympatrische Zwillingsarten. J. Ornithol. 114:3–56. [149]
Lynch, A. 1991. Cultural evolution in Chaffinch song: a population memetics approach. Ph.D. dissertation, University of Toronto, Toronto. [190]
Lynch, A., and A. J. Baker. 1986. Congruence of morphometric and cultural evolution in Atlantic island Chaffinch populations. Can. J. Zool. 64:1576–1580. [194]
Lynch, A., and A. J. Baker. 1993. A population memetics approach to Chaffinch song evolution: meme diversity within populations. Am. Nat. 141:597–620. [185–187, 189, 191, 192, 194, 214, 217]
Lynch, A., and A. J. Baker. 1994. A population memetics approach to cultural evolution in Chaffinch song: differentiation among populations. Evolution. 48:351–359. [189, 190]
Lynch, A., G. M. Plunkett, A. J. Baker, and P. F. Jenkins. 1989. A model of cultural evolution of Chaffinch song derived with the meme concept. Am. Nat. 133:634–653. [182, 183, 185, 193, 194, 215]
Mace, R. 1986. The importance of female behaviour in the dawn chorus. Anim. Behav. 34:621–622. [354, 431, 432]

Mace, R. 1987a. The dawn chorus in the Great Tit, *Parus major,* is directly related to female fertility. Nature 330:745–746. [155, 431]

Mace, R. 1987b. Why do birds sing at dawn? Ardea 75:123–132. [432, 436, 437]

Mace, R. 1989a. A comparison of Great Tits' (*Parus major*) use of time in different daylengths at three European sites. J. Anim. Ecol. 58:143–151. [438]

Mace, R. 1989b. Great Tits choose between food and proximity to a mate. The effect of time of day. Behav. Ecol. Sociobiol. 24:285–290. [438]

Mace, R. 1989c. The relationship between daily routines of singing and foraging: an experiment on captive Great Tits, *Parus major.* Ibis 131:415–420. [438]

Mace, W. M. 1977. James J. Gibson's strategy for perceiving: ask not what's inside your head, but what your head's inside of. Pages 43–66 *in* Perceiving, acting, and knowing (R. Shaw and J. Bransford, Eds.). Erlbaum, Hillsdale, N.J. [26]

Mackintosh, N. J. 1965. Selective attention in animal discrimination learning. Psychol. Bull. 64:124–150. [127]

Maclean, G. L. 1969. A study of seedsnipe in southern South America. Living Bird 8:33–80. [247]

Maclean, G. L. 1985. Roberts' birds of southern Africa. Trustees of the John Voelcker Bird Book Fund, Cape Town. [244, 252]

Maclean, S. F., Jr., and R. T. Holmes. 1971. Bill lengths, wintering areas, and taxonomy of North American Dunlins, *Calidris alpina.* Auk 88:893–901. [244]

MacNally, R. C., and R. E. Lemon. 1985. Repeat and serial singing modes in American Redstarts (*Setophaga ruticilla*): a test of functional hypotheses. Z. Tierpsychol. 69:191–202. [445]

MacNeil, R., P. Drapeau, and R. Pierrotti. 1993. Nocturnality in colonial waterbirds: occurrence, special adaptations, and suspected benefits. Pages 187–246 *in* Current ornithology, vol. 10 (D. M. Power, Ed.). Plenum Press, New York. [161, 162, 171]

Magyar, I., W. M. Schleidt, and D. Miller. 1978. Localization of sound producing animals using the arrival time differences of their signals at an array of microphones. Experientia 34:676–677. [419]

Mairy, F. 1976. Vocal repertoire of the Ring Dove (*Streptopelia risoria*). Biophon 4:3–9. [48, 52]

Majerus, M. E. N. 1986. The genetics and evolution of female choice. Trends Ecol. Evol. 1:3–7. [470]

Mammen, D. L., and S. Nowicki. 1981. Individual differences and within-flock convergence in chickadee calls. Behav. Ecol. Sociobiol. 9:179–186. [107, 114, 146]

Manley, G. A. 1990. Peripheral hearing mechanisms in reptiles and birds. Springer Verlag, Heidelberg. [323, 327]

Manley, G. A., G. Schwabedissen, and O. Gleich. 1993. Morphology of the basilar papilla of the Budgerigar, *Melopsittacus undulatus.* J. Morphol. 218:153–165. [328]

Manogue, K. R., and F. Nottebohm. 1982. Relation of medullary motor nuclei to nerves supplying the vocal tract of the Budgerigar (*Melopsittacus undulatus*). J. Comp. Neurol. 204:384–391. [113]

Marantz, C. A. 1992. Evolutionary implications of vocal and morphological variation in the woodcreeper genus *Dendrocolaptes* (Aves: Dendrocolaptidae). M.S. thesis, Louisiana State University, Baton Rouge. [272]

Marchant, S., and P. J. Higgins (Eds.). 1990. Handbook of Australian, New Zealand and Antarctic birds, vol. 1. Oxford Univ. Press, Oxford. [163, 168]

Marchant, S., and P. J. Higgins (Eds.) 1993. Handbook of Australian, New Zealand and Antarctic birds, vol. 2. Oxford Univ. Press, Melbourne. [244]

Margoliash, D. 1983. Acoustic parameters underlying the responses of song-specific neurons in the White-crowned Sparrow. J. Neurosci. 3:1034–1057. [289]

Margoliash, D. 1986. Preference for autogenous song by auditory neurons in a song system nucleus of the White-crowned Sparrow. J. Neurosci. 6:1643–1661. [289, 336, 429]

Margoliash, D. 1987. Neural plasticity in birdsong learning. Pages 23–54 *in* Imprinting and neural plasticity: comparative aspects of sensitive periods (J. P. Rauschecker and P. Marler, Eds.). Wiley, New York. [374]

Margoliash, D., and E. Fortune. 1992. Temporal and harmonic combination-sensitive neurons in the Zebra Finch's HVC. J. Neurosci. 12:4309–4326. [289]

Margoliash, D., C. A. Staicer, and S. A. Inoue. 1991. Stereotyped and plastic song in adult Indigo Buntings, *Passerina cyanea.* Anim. Behav. 42:367–388. [85]

Margoliash, D., E. S. Fortune, M. L. Sutter, A. C. Yu, B. D. Wren-Hardin, and A. Dave. 1994. Distributed representations in the song system of oscines: evolutionary implications and functional consequences. Brain Behav. Evol. 44:247–264. [25]

Marler, P. 1955. Characteristics of some alarm calls. Nature 176:6–8. [141, 321, 416, 417]

Marler, P. 1956. The voice of the Chaffinch and its function as a language. Ibis 98:231–261. [167]

Marler, P. 1960. Bird songs and mate selection. Pages 348–367 *in* Animal sounds and communication (W. E. Lanyon and W. N. Tavolga, Eds.). Publ. 7, Am. Inst. Biol. Sci., Washington, D.C. [20, 186, 356]

Marler, P. 1967. Comparative study of song development in sparrows. Pages 231–244 *in* Proc. XIV Int. Ornithol. Congr. (D. W. Snow, Ed.). Blackwell, Oxford. [4]

Marler, P. 1969. Tonal quality of bird sounds. Pages 5–18 *in* Bird vocalizations, their relation to current problems in biology and psychology (R. A. Hinde, Ed.). Aberdeen Univ. Press, Aberdeen, Scotland. [56]

Marler, P. 1970. A comparative approach to vocal learning: song development in White-crowned Sparrows. J. Comp. Physiol. Psychol. 71:1–25. [15, 41, 43, 44, 62, 115, 216]

Marler, P. 1977. The structure of animal communication sounds. Pages 17–35 *in* Workshop on recognition of complex acoustic signals (T. H. Bullock, Ed.). Abacon, Berlin. [417]

Marler, P. 1981. Birdsong: the acquisition of a learned motor skill. Trends Neurosci. 4:88–94. [12]

Marler, P. 1982. Avian and primate communication: the problem of natural categories. Neurosci. Biobehav. Rev. 6:87–94. [121]

Marler, P. 1987. Sensitive periods and the role of specific and general sensory stimulation in birdsong learning. Pages 99–135 *in* Imprinting and cortical plasticity (J. P. Rauschecker and P. Marler, Eds.). Wiley, New York. [85, 115]

Marler, P. 1990. Song learning: the interface between behaviour and neuroethology. Philos. Trans. R. Soc. Lond. B 329:109–114. [71, 290, 314]

Marler, P. 1991a. The instinct for vocal learning: songbirds. Pages 107–125 *in* Plasticity of development (S. E. Brauth, W. S. Hall, and R. J. Dooling, Eds.). MIT Press, Cambridge, Mass. [3, 10, 12, 85]

Marler, P. 1991b. Differences in behavioural development in closely related species: birdsong. Pages 41–70 *in* The development and integration of behaviour. Essays in honour of Robert Hinde (P. P. G. Bateson, Ed.). Cambridge Univ. Press, Cambridge. [6, 91]

Marler, P. 1991c. Song-learning behavior: the interface with neuroethology. Trends Neurosci. 14:199–206. [21, 94, 288]

Marler, P., and P. Mundinger. 1972. Vocal learning in birds. Pages 389–450 *in* Ontogeny of vertebrate behavior (H. Moltz, Ed.). Academic Press, New York. [43]

Marler, P., and D. Nelson. 1992. Neuroselection and song learning in birds: species universals in a culturally transmitted behavior. Neuroscience 4:415–423. [15]

Marler, P., and S. Peters. 1977. Selective vocal learning in a sparrow. Science (N.Y.) 198:519–521. [41, 374]

Marler, P., and S. Peters. 1981. Birdsong and speech: evidence for special processing. Pages 75–112 *in* Perspectives on the study of speech (P. Eimas and J. Miller, Eds.). Erlbaum, Hillsdale, N.J. [1, 71, 94]

Marler, P., and S. Peters. 1982a. Structural changes in song ontogeny in the Swamp Sparrow, *Melospiza georgiana.* Auk 99:446–458. [91]

Marler, P., and S. Peters. 1982b. Subsong and plastic song: their role in the vocal learning process. Pages 25–50 *in* Acoustic communication in birds, vol. 2 (D. E. Kroodsma and E. H. Miller, Eds.). Academic Press, New York. [12, 182, 183, 314]

Marler, P., and S. Peters. 1982c. Developmental overproduction and selective attrition: new processes in the epigenesis of birdsong. Dev. Psychobiol. 15:369–378. [71, 92, 314]

Marler, P., and S. Peters. 1987. A sensitive period for song acquisition in the Song Sparrow, *Melospiza melodia,* a case of age-limited learning. Ethology 76:89–100. [61, 67, 77, 86, 286]

Marler, P., and S. Peters. 1988. The role of song phonology and syntax in vocal learning preferences in the Song Sparrow. Ethology 77:125–149. [61, 62, 67, 77]

Marler, P., and S. Peters. 1989. Species differences in auditory responsiveness in early vocal learning. Pages 243–273 *in* The comparative psychology of audition: perceiving complex sounds (R. Dooling and S. H. Hulse, Eds.). Erlbaum, Hillsdale, N.J. [85]

Marler, P., and R. Pickert. 1984. Species-universal microstructure in the learned song of the Swamp Sparrow, *Melospiza georgiana.* Anim. Behav. 32:673–689. [45]

Marler, P., and V. Sherman. 1985. Innate differences in singing behaviour of sparrows reared in isolation from adult conspecific song. Anim. Behav. 33:57–71. [40, 42, 50, 387]

Marler, P., and M. Tamura. 1962. Song "dialects" in three populations of White-crowned Sparrows. Condor 64:368–377. [216]

Marler, P., and H. S. Terrace (Eds.). 1984. The biology of learning. Springer Verlag, Berlin. [305]

Marler, P., S. Peters, G. Ball, A. Dufty Jr., and J. Wingfield. 1988. The role of sex steroids in the acquisition and production of birdsong. Nature 336:770–772. [293]

Marova, I. M., and V. V. Leonovitch. 1993. Hybridization between Siberian (*Phylloscopus collybita tristis*) and East European (*Ph. collybita abietinus*) Chiffchaffs in the area of sympatry. Pages 147–163 *in* Hybridization and the problem of species in vertebrates (O. L. Rossolimo, Ed.). Arch. Zool. Mus. Moscow State Univ. 30. [228]

Marshall, J. T., Jr. 1964. Voice in communication and relationships among Brown Towhees. Condor 66:345–356. [46]

Marshall, J. T., Jr. 1978. Systematics of smaller Asian night birds based on voice. Ornithological Monographs 25. American Ornithologists' Union, Washington, D.C. [272]

Marten, K., and P. Marler. 1977. Sound transmission and its significance for animal vocalization. I. Temperate habitats. Behav. Ecol. Sociobiol. 2:271–290. [321]

Martens, J. 1975. Akustische Differenzierung verwandtschaftlicher Beziehungen in der *Parus* (*Periparus*)-Gruppe nach Untersuchungen im Nepal-Himalaya. J. Ornithol. 116:369–433. [223]

Martens, J. 1982. Ringförmige Arealüberschneidung und Artbildung beim Zilpzalp, *Phylloscopus collybita.* Das *lorenzii*-Problem. Z. Zool. Syst. Evolutionsforsch. 20:82–100. [228]

Martens, J. 1989. *Phylloscopus borealoides*—ein verkannter Laubsänger der Ost-Paläarktis. J. Ornithol. 129:343–351. [239]

Martens, J. 1993. Lautäußerungen von Singvögeln und die Entstehung neuer Arten. Forschungsmagazin Univ. Mainz 9:34–44. [222, 223, 239]

Martens, J., and S. Eck. 1995. Towards an ornithology of the Himalayas. Systematics, ecology and vocalizations of Nepal birds. Bonn. Zool. Monogr. 39. [229, 235]

Martens, J., and G. Geduldig. 1990. Acoustic adaptations of birds living close to Himalayan torrents. Pages 123–131 *in* Current topics in avian biology. Proc. Int. 100. DO-G Meeting, Bonn (R. van Elzen, K. L. Schuchmann, and K. Schmidt-Koenig, Eds.). Verlag DO-G, Garmisch-Partenkirchen. [239]

Martens, J., and C. Meincke. 1989. Der sibirische Zilpzalp (*Phylloscopus collybita tristis*): Gesang und Reaktion einer mitteleuropäischen Population im Freilandversuch. J. Ornithol. 130:455–473. [228]

Martens, J., and A. A. Nazarenko. 1993. Microevolution of eastern Palearctic Grey Tits as indicated by their vocalizations (*Parus* [*Poecile*]: Paridae, Aves). I. *Parus montanus.* Z. Zool. Syst. Evolutionsforsch. 31:127–143. [223, 234, 235]

Martens, J., and B. Schottler. 1991. Akustische Barrieren zwischen Blaumeise (*Parus caeruleus*) und Lasurmeise (*Parus cyanus*)? J. Ornithol. 132:61–80. [229]

Martens, J., S. Ernst, and B. Petri. 1995. Reviergesänge ostasiatischer Weidenmeisen *Parus montanus* und ihre mikroevolutive Ableitung. J. Ornithol. 136:367–388. [223]

Martin, A. P., and S. R. Palumbi. 1993. Body size, metabolic rate, generation time, and the molecular clock. Proc. Natl. Acad. Sci. USA 90:4087–4091. [214]

Martindale, S. 1980. On the multivariate analysis of avian vocalizations. J. Theor. Biol. 83:107–110. [182]

Martinez, J. L., and R. P. Kesner. 1986. Learning and memory—a biological view. Academic Press, New York. [89]

Massa, B., and M. L. Valvo. 1986. Biometrical and biological considerations on the Cory's Shearwater, *Calonetris diomedea.* NATO Adv. Ser. Study Inst. 12:293–313. [175]

Massaro, D. 1989. Multiple book review of speech perception by ear and eye: a paradigm for psychological inquiry. Behav. Brain Sci. 12:741–794. [122, 129]

Masure, R. H., and W. C. Allee. 1934. Flocking organization of the Shell Parakeet, *Melopsittacus undulatus* Shaw. Ecology 15:388–398. [104]

Mauersberger, G. 1971. Ist *Emberiza leucocephalos* eine Subspezies von *E. citrinella*? J. Ornithol. 112:232–233. [231]

Maynard Smith, J. 1974. The theory of games and the evolution of animal conflicts. J. Theor. Biol. 47:209–221. [384]

Maynard Smith, J. 1985. Sexual selection, handicaps, and true fitness. J. Theor. Biol. 115:1–8. [464]

Maynard Smith, J. 1991. Theories of sexual selection. Trends Ecol. Evol. 6:146–151. [461, 463, 464]

Mayr, E. 1942. Systematics and the origin of species. Columbia Univ. Press, New York. [216]

Mayr, E. 1963. Animal species and evolution. Belknap Press (Harvard Univ. Press), Cambridge, Mass. [242]

Mayr, E. 1974. Behavioral programs and evolutionary strategies. Am. Sci. 62:650–659. [22]

Mayr, E. 1982. Of what use are subspecies? Auk 99:593–596. [270]

Mayr, E. 1993. Fifty years of progress in research on species and speciation. Proc. Calif. Acad. Sci. 48:131–140. [222]

McArthur, P. D. 1986. Similarity of playback songs to self song as a determinant of response strength in Song Sparrows (*Melospiza melodia*). Anim. Behav. 34:199–207. [130]

McArthur, T. (Ed.) 1992. The Oxford companion to the English language. Oxford Univ. Press, Oxford. [213, 220]

McCasland, J. 1987. Neuronal control of bird song production. J. Neurosci. 7:23–29. [286]

McDonald, D. B., and W. K. Potts. 1994. Cooperative display and relatedness among males in a lek-mating bird. Science 266:1030–1032. [278]

McFarland, D. C. 1991a. The biology of the Ground Parrot, *Pezoporus wallicus,* in Queensland. I. Microhabitat use, activity cycle and diet. Wildl. Res. 18:169–184. [102]

McFarland, D. C. 1991b. The biology of the Ground Parrot, *Pezoporus wallicus,* in Queensland. II. Spacing, calling, and breeding behaviour. Wildl. Res. 18:185–197. [97, 102, 104, 112]

McFarland, D. C. 1991c. The biology of the Ground Parrot, *Pezoporus wallicus,* in Queensland. III. Distribution and abundance. Wildl. Res. 18:199–213. [102]

McGonigle, B. O., and M. Chalmers. 1977. Are monkeys logical? Nature 267:694–696. [415]

McGrath, T. A., M. D. Shalter, and W. M. Schleidt. 1972. Analysis of distress calls of chicken × pheasant hybrids. Nature 237:47–48. [48, 54, 55]

McGregor, P. K. 1980. Song dialects in the Corn Bunting (*Emberiza calandra*). Z. Tierpsychol. 54:285–297. [15, 216]

McGregor, P. K. 1988a. Song length and "male quality" in the Chiffchaff. Anim. Behav. 36:606–608. [308]

McGregor, P. K. 1988b. Pro-active memory interference in neighbour recognition by a songbird. Pages 1391–1397 *in* Proc. XIX Int. Ornithol. Congr. (H. Ouellet, Ed.) National Museum of Natural Science, Univ. Ottawa Press, Ottawa. [414]

McGregor, P. K. 1991. The singer and the song: on the receiving end of bird song. Biol. Rev. 66:57–81. [305, 308, 312, 339, 414, 423]

McGregor, P. K. 1992. Quantifying responses to playback: one, many, or composite multivariate measures? Pages 79–96 *in* Playback and studies of animal communication (P. K. McGregor, Ed.). Plenum Press, New York. [365]

McGregor, P. K. 1993. Signalling in territorial systems: a context for individual identification, ranging and eavesdropping. Philos. Trans. R. Soc. Lond. B 340:237–244. [398, 406, 410, 413, 414, 423]

McGregor, P. K. 1994. Sound cues to distance: the perception of range. Pages 74–94 *in* Perception and motor control in birds (M. N. O. Davies and P. R. Green, Eds.). Springer Verlag, Berlin. [414]

McGregor, P. K., and M. I. Avery. 1986. The unsung songs of Great Tits (*Parus major*): learning neighbors' songs for discrimination. Behav. Ecol. Sociobiol. 18:311–316. [336, 359, 367, 368, 371, 373, 414]

McGregor, P. K., and J. B. Falls. 1984. The response of Western Meadowlarks (*Sturnella neglecta*) to the playback of degraded and undegraded songs. Can. J. Zool. 62:2125–2128. [414]

McGregor, P. K., and A. G. Horn. 1991. Song length and response to playback in Great Tits. Anim. Behav. 43:667–676. [308]

McGregor, P. K., and A. G. Horn. 1992. Song length and response to playback in Great Tits. Anim. Behav. 43:667–676. [400, 404]

McGregor, P. K., and J. R. Krebs. 1982. Song types in a population of Great Tits (*Parus major*): their distribution, abundance and acquisition by individuals. Behaviour 79:126–152. [124, 185, 191, 193, 215, 217, 218, 306, 313]

McGregor, P. K., and J. R. Krebs. 1984a. Song learning and deceptive mimicry. Anim. Behav. 32:280–287. [218]

McGregor, P. K., and J. R. Krebs. 1984b. Sound degradation as a distance cue in Great Tit (*Parus major*) song. Behav. Ecol. Sociobiol. 16:49–56. [414]

McGregor, P. K., and J. R. Krebs. 1989. Song learning in adult Great Tits (*Parus major*): effects of neighbours. Behaviour 108:139–159. [75, 215, 218]

McGregor, P. K., and D. B. A. Thompson. 1988. Constancy and change in local dialects of the Corn Bunting. Ornis Scand. 19:153–159. [216]

McGregor, P. K., and G. W. M. Westby. 1992. Discrimination of individually characteristic electric organ discharges by a weakly electric fish. Anim. Behav. 43:977–986. [359]

McGregor, P. K., J. R. Krebs, and C. M. Perrins. 1981. Song repertoires and lifetime reproductive success in the Great Tit (*Parus major*). Am. Nat. 118:149–159. [218, 471]

McGregor, P. K., J. R. Krebs, and L. M. Ratcliffe. 1983. The response of Great Tits (*Parus major*) to the playback of degraded and undegraded songs: the effect of familiarity with the stimulus song type. Auk 100:898–906. [414]

McGregor, P. K., V. R. Walford, and D. G. C. Harper. 1988. Song inheritance and mating in a songbird with local dialects. Bioacoustics 1:107–129. [215, 217]

McGregor, P. K., C. K. Catchpole, T. Dabelsteen, J. B. Falls, L. Fusani, H. C. Gerhardt, F. Gilbert, A. G. Horn, G. M. Klump, D. E. Kroodsma, M. M. Lambrechts, K. E. McComb, D. A. Nelson, I. M. Pepperburg, L. Ratcliffe, W. A. Searcy, and D. M. Weary. 1992a. Design of playback experiments: the Thornbridge Hall NATO ARW consensus. Pages 1–9 *in* Playback and studies of animal communication (P. K. McGregor, Ed.). Plenum Press, New York. [401, 413, 457]

McGregor, P. K., T. Dabelsteen, M. Shepherd, and S. B. Pedersen. 1992b. The signal value of matched singing in Great Tits: evidence from interactive playback experiments. Anim. Behav. 43:987–998. [394, 399–401, 411, 412]

McGregor, P. K., T. Dabelsteen, C. W. Clark, J. L. Bower, J. P. Tavares, and J. Holland. Unpubl. ms. a. An acoustic location system for the study of communication networks: factors affecting location accuracy. [398, 419, 420]

McGregor, P. K., T. Dabelsteen, and J. Holland. Unpubl. ms. b. Eavesdropping by territorial male Great Tits. [406]

McInnes, R. S., and P. B. Carne. 1978. Predation of cossid moth larvae by Yellow-tailed Black Cockatoos causing losses in plantations of *Eucalyptus grandis* in north coastal New South Wales. Aust. Wildl. Res. 5:101–121. [102]

McKee, E. A. 1988. Decay and degradation of disk and cylinder recordings in storage. Pages 193–199 *in* Audio preservation: a planning study. Associated Audio Archives Committee, Gerald Gibson, chairman. Association of Recorded Sound Collections, Rockville, Md. [484]

McKitrick, M. C. 1993. Phylogenetic constraint in evolutionary theory: has it any explanatory power? Annu. Rev. Ecol. Syst. 24:307–330. [257]

McKitrick, M. C., and R. M. Zink. 1988. Species concepts in ornithology. Condor 90:1–14. [19]

McLaren, M. 1976. Vocalizations of the Boreal Chickadee. Auk 93:451–463. [147, 148, 154, 155]

McNamara, J. M., R. H. Mace, and A. I. Houston. 1987. Optimal daily routines of singing and foraging in a bird singing to attract a mate. Behav. Ecol. Sociobiol. 20:399–405. [438]

McNicol, D. 1972. A primer of signal detection theory. Allen and Unwin, London. [324]

Mebes, H. D. 1978. Pair-specific duetting in the Peach-faced Lovebird, *Agapornis roseicollis*. Naturwissenschaften 65:66–67. [114]

Medvin, M. B., P. K. Stoddard, and M. D. Beecher. 1992. Signals for parent-offspring recognition: strong sib-sib call similarity in Cliff Swallows but not Barn Swallows. Ethology 90:17–28. [357]

Merritt, P. G. 1985. Song function and the evolution of song repertoires in the Northern Mockingbird, *Mimus polyglottos*. Ph.D. dissertation, Univ. Miami, Coral Gables, Fla. [430]

Merton, D. V., R. B. Morris, and I. A. E. Atkinson. 1984. Lek behaviour in a parrot: the Kakapo, *Strigops habroptilus,* of New Zealand. Ibis 126:277–283. [112]

Metz, K. J., and P. Weatherhead. 1991. Color bands function as secondary sexual traits in male Red-winged Blackbirds. Behav. Ecol. Sociobiol. 28:23–27. [160]

Meyer de Schauensee, R. 1966. The species of birds of South America with their distribution. Livingston Press, Narbeth, Pa. [270]

Meyer de Schauensee, R. 1970. A guide to the birds of South America. Academy of Natural Sciences, Philadelphia. [270]

Michelsen, A. 1978. Sound perception in different environments. Pages 345–373 *in* Perspectives in sensory ecology (M. A. Ali, Ed.). Plenum Press, New York. [422]

Michelsen, A., B. B. Andersen, W. H. Kirchner, and M. Lindauer. 1989. Honeybees can be recruited by a mechanical model of a dancing bee. Naturwissenschaften 76:277–280. [407]

Michelsen, A., B. B. Andersen, J. Storm, W. H. Kirchner, and M. Lindauer. 1992. How honey bees perceive communication dances, studied by means of a mechanical model. Behav. Ecol. Sociobiol. 30:143–150. [407]

Miller, A. H. 1934. The vocal apparatus of some North American owls. Condor 36:204–213. [55]

Miller, A. H. 1947. The structural basis of the voice of the Flammulated Owl. Auk 64:133–135. [55]

Miller, D. B. 1979a. The acoustic basis of mate recognition by female Zebra Finches (*Taeniopygia guttata*). Anim. Behav. 27:376–380. [458]

Miller, D. B. 1979b. Long-term recognition of father's song by female Zebra Finches. Nature 280:389–391. [43, 458]

Miller, E. H. 1979a. An approach to the analysis of graded vocalizations of birds. Behav. Neural Biol. 27:25–38. [196]

Miller, E. H. 1979b. Functions of display flights by males of the Least Sandpiper, *Calidris minutilla* (Vieill.), on Sable Island, Nova Scotia. Can. J. Zool. 57:876–893. [452]

Miller, E. H. 1982. Character and variance shift in acoustic signals of birds. Pages 253–295 *in* Acoustic communication in birds, vol. 1 (D. E. Kroodsma and E. H. Miller, Eds.). Academic Press, New York. [176, 391]

Miller, E. H. 1983a. The structure of aerial displays in three species of Calidridinae (Scolopacidae). Auk 100:440–451. [241, 249]

Miller, E. H. 1983b. Structure of display flights in the Least Sandpiper. Condor 85:220–242. [253]

Miller, E. H. 1984. Communication in breeding shorebirds. Pages 169–241 *in* Shorebirds: breeding behavior and populations (J. Burger and B. L. Olla, Eds.). Plenum Press, New York. [241, 244, 247, 253, 256]

Miller, E. H. 1985a. Museum collections and the study of animal social behaviour. Pages 139–162 *in* Museum collections: their roles and future in biological research (E. H. Miller, Ed.). British Columbia Provincial Museum, Occasional Paper 25. [484]

Miller, E. H. 1985b. Parental behavior in the Least Sandpiper (*Calidris minutilla*). Can. J. Zool. 63:1593–1601. [256]

Miller, E. H. 1986. Components of variation in nuptial calls of the Least Sandpiper (*Calidris minutilla;* Aves, Scolopacidae). Syst. Zool. 35:400–413. [241, 249, 255]

Miller, E. H. 1988. Description of bird behavior for comparative purposes. Pages 347–394 *in* Current Ornithology, vol. 5 (R. F. Johnston, Ed.). Plenum Press, New York. [163, 389]

Miller, E. H. 1992. Acoustic signals of shorebirds. A survey and review of published information. Royal British Columbia Museum Tech. Rep. Victoria, B.C. [241, 244, 252, 253, 256, 257]

Miller, E. H. 1993. Biodiversity research in museums: a return to basics. Pages 141–173 *in* Our living legacy: proceeds of a symposium on biological diversity (M. A. Fenger, E. H. Miller, J. A. Johnson, and E. J. R. Williams, Eds.). Royal British Columbia Museum, Victoria, B.C. [484]

Miller, E. H. 1995. Sounds of shorebirds: opportunities for amateurs and an update of published information. Wader Study Group Bull. 78:18–22. [253, 257]

Miller, E. H. 1996. Nuptial vocalizations of male Least Seedsnipe: structure and evolutionary significance. Condor 98. In press. [247]

Miller, E. H., and A. J. Baker. 1980. Displays of the Magellanic Oystercatcher (*Haematopus leucopodus*). Wilson Bull. 92:149–168. [241, 242, 248]

Miller, E. H., and D. W. Nagorsen. 1992. Voucher specimens: an essential component of biological surveys. Pages 11–15 *in* Methodology for monitoring wildlife diversity in B.C. Forests (L. R. Ramsay, Ed.). Ministry of Environment, Lands and Parks, Victoria, B.C. [475]

Miller, E. H., and G. G. E. Scudder. 1994. A rose by any other name . . . Pages 29–43 *in* Biodiversity in British Columbia: our changing environment (L. E. Harding and E. McCullum, Eds.). Environment Canada, Canadian Wildlife Service, Vancouver, British Columbia. [270]

Miller, E. H., W. W. H. Gunn, and R. E. Harris. 1983. Geographic variation in the aerial song of the Short-billed Dowitcher (Aves, Scolopacidae). Can. J. Zool. 61:2191–2198. [241, 248, 253]

Miller, E. H., W. W. H. Gunn, J. P. Myers, and B. N. Veprintsev. 1984. Species-distinctiveness of Long-billed Dowitcher song (Aves: Scolopacidae). Proc. Biol. Soc. Wash. 97:804–811. [241, 253, 255]

Miller, E. H., W. W. H. Gunn, and S. F. Maclean Jr. 1987. Breeding vocalizations of the Surfbird. Condor 89:406–412. [256]

Miller, E. H., W. W. H. Gunn, and B. N. Veprintsev. 1988. Breeding vocalizations of Baird's Sandpiper, *Calidris bairdii,* and related species, with remarks on phylogeny and adaptation. Ornis Scand. 19:257–267. [242, 256]

Millington, S. J., and T. D. Price. 1985. Song inheritance and mating patterns in Darwin's finches. Auk 102:342–346. [218]

Miskelly, C. M. 1987. The identity of the Hakawai. Notornis 34:95–116. [242]

Miskelly, C. M. 1990. Aerial displaying and flying ability of Chatham Island Snipe, *Coenocorypha pusilla,* and New Zealand Snipe, *C. aucklandica.* Emu 90:28–32. [242]

Moiseff, A. 1989. Binaural disparity cues available to the Barn Owl for sound localization. J. Comp. Physiol. A 164:629–636. [334]

Møller, A. P. 1991a. Sperm competition, sperm depletion, paternal care and relative testis size in birds. Am. Nat. 137:882–906. [263]

Møller, A. P. 1991b. Why mated songbirds sing so much: mate guarding and male announcement of mate fertility status. Am. Nat. 138:994–1014. [455]

Møller, A.P. 1991c. Parasite load reduces song output in a passerine bird. Anim. Behav. 41:723–730. [311]

Monroe, B. L., Jr., and C. G. Sibley. 1993. A world checklist of birds. Yale Univ. Press, New Haven, Conn. [xx]

Montagu, A. 1959. Human heredity. World, Cleveland. [25]

Montgomerie, R. D. 1985. Why do birds sing at dawn? Abstracts of spoken papers, XIX Int. Ethol. Congr. 1:242. [435]

Mooney, R. 1992. Synaptic basis for developmental plasticity in a birdsong nucleus. J. Neurosci. 12:2464–2477. [293]

Moore, B. J. C. 1989. An introduction to the psychology of hearing. Academic Press, London. [335]

Moore, B. J. C. 1990. Co-modulation masking release: spectro-temporal pattern analysis in hearing. Br. J. Audiol. 24:131–137. [331]

Moore, B. J. C. 1992. Across-channel processes in auditory masking. J. Acoust. Soc. Jpn. 13:25–37. [331]

Moore, J. H. 1994a. Ethnogenetic theories of human evolution. Res. Exploration 10:10–37. [214, 219]

Moore, J. H. 1994b. Putting anthropology back together again: the ethnogenetic critique of cladistic theory. Am. Anthropol. 96:925–948. [219]

Moreau, R. E., and P. Wayre. 1968. On the Palearctic quails. Ardea 56:209–227. [53, 56]

Morris, G. K., A. C. Mason, P. Wall, and J. J. Belwood. 1994. High ultrasonic and tremulation signals in Neotropical katydids (Orthoptera: Tettigoniidae). J. Zool. Lond. 233:129–163. [416]

Morrison, M. L., and J. W. Hardy. 1983. Vocalizations of the Black-throated Gray Warbler. Wilson Bull. 95:640–643. [381, 390]

Morse, D. H. 1970. Territorial and courtship songs of birds. Nature 226:659–661. [18, 161]

Morse, D. H., and S. W. Kress. 1984. The effect of burrow loss on mate choice in the Leach's Storm-Petrel. Auk 101:158–160. [175]

Morton, E. S. 1971. Nest predation affecting the breeding season of the Clay-colored Robin, a tropical songbird. Science 171:920–921. [265]

Morton, E. S. 1975. Ecological sources of selection on avian sounds. Am. Nat. 108:17–34. [266, 305, 321, 333, 422]

Morton, E. S. 1976. Vocal mimicry in the Thick-billed Euphonia. Wilson Bull. 88:485–487. [276]

Morton, E. S. 1977. On the occurrence and significance of motivation-structural rules in some bird and mammal sounds. Am. Nat. 111:855–869. [279]

Morton, E. S. 1980. The ecological background for the evolution of vocal sounds used at close range. Pages 737–741 *in* Acta XVII Congr. Int. Ornithol. (R. Nöhring, Ed.). Verlag Deutsche Ornithol.-Gesellschaft, Berlin. [259]

Morton, E. S. 1982. Grading, discreteness, redundancy, and motivation-structural rules. Pages 183–213 *in* Acoustic communication in birds, vol. 1 (D. E. Kroodsma and E. H. Miller, Eds.). Academic Press, New York. [262, 264, 266, 267, 336, 414, 437, 468]

Morton, E. S. 1983. Yiquirro, Clay-colored Robin. Pages 610–611 *in* Costa Rican natural history (D. H. Janzen, Ed.). Univ. Chicago Press, Chicago. [265]

Morton, E. S. 1986. Predictions from the ranging hypothesis for the evolution of long distance signals in birds. Behaviour 99:65–86. [17, 262, 266, 414]

Morton, E. S., and K. C. Derrickson. In press. Song ranging by the Dusky Antbird, *Cercomacro tyrannina:* ranging without song learning. Behav. Ecol. Sociobiol. [259, 267]

Morton, E. S., and H. Gonzalez. 1982. The biology of *Torreornis inexpectata.* I. A comparison of vocalizations in *T. i. inexpectata* and *T. i. sigmoni.* Wilson Bull. 94:433–446. [259]

Morton, E. S., and K. Young. 1986. A previously undescribed method of song matching in a species with a single song "type," the Kentucky Warbler (*Oporornis formosus*). Ethology 73:334–342. [13, 390]

Morton, E. S., L. Forman, and M. Braun. 1990. Extrapair fertilizations and the evolution of colonial breeding in Purple Martins. Auk 107:275–283. [264]

Morton, M. L. 1992. Effects of sex and birth date on premigration biology, migration schedules, return rates and natal dispersal in the mountain White-crowned Sparrow. Condor 94:117–133. [64]

Morton, M. L., M. E. Pereyra, and L. F. Baptista. 1985. Photoperiodically induced ovarian growth in the White-crowned Sparrow (*Zonotrichia leucophrys gambelli*) and its augmentation by song. Comp. Biochem. Physiol. A 80:93–97. [431]

Mougeot, F., F. Genevois, and V. Bretagnolle. Unpubl. ms. Skuas and petrels: predation on burrowing petrels by Brown Skuas in a Kerguelen Island. [171]

Mougin, J.-L., B. Despin, C. Jouanin, and F. Roux. 1987. La fidélité au partenaire et au nid chez le puffin cendré, *Calonectris diomedea borealis,* de l'île Selvagen Grande. Gerfaut 77:353–369. [162]

Mougin, J.-L., C. Jouanin, and F. Roux. 1988. Les migrations du puffin cendré, *Calonectris diomedea.* Oiseau Rev. Fr. Ornithol. 58:303–319. [175]

Mountjoy, D. J., and R. E. Lemon. 1991. Song as an attractant for male and female European Starlings, and the influence of song complexity on their response. Behav. Ecol. Sociobiol. 28:97–100. [458]

Moynihan, M. 1962. Display patterns of tropical American "nine-primaried" songbirds. II. Some species of *Ramphocelus.* Auk 79:655–686. [447]

Moynihan, M. 1970. The control, suppression, decay, disappearance and replacement of displays. J. Theor. Biol. 29:85–112. [124, 136]

Müller-Bröse, M., and D. Todt. 1991. Lokomotorische Aktivität von Nachtigallen (*Luscinia megarhynchos*) während auditorischer Stimulation mit Artgesang, präsentiert in ihrer lernsensiblen Altersphase. Verh. Dtsch. Zool. Ges. 84:476–477. [86]

Mulligan, J. A. 1963. A description of Song Sparrow song based on instrumental analysis. Pages 272–284 *in* Proc. XIII Int. Ornithol. Congr. (C. G. Sibley, Ed.). Allen Press, Lawrence, Kansas.[50]

Mulligan, J. A. 1966. Singing behavior and its development in the Song Sparrow, *Melospiza melodia.* Univ. Calif. Publ. Zool. 81:1–76. [40, 43, 50, 53]

Mundinger, P. C. 1970. Vocal imitation and individual recognition of finch calls. Science 168:480–482. [107, 114, 115]

Mundinger, P. C. 1975. Song dialects and colonization in the House Finch, *Carpodacus mexicanus,* on the East Coast. Condor 77:407–422. [40, 215]

Mundinger, P. C. 1979. Call learning in the Carduelinae: ethological and systematic considerations. Syst. Zool. 28:270–283. [107, 114]

Mundinger, P. C. 1980. Animal cultures and a general theory of cultural evolution. Ethol. Sociobiol. 1:183–223. [181, 214, 221]

Mundinger, P. C. 1982. Microgeographic and macrogeographic variation in acquired vocalizations in birds. Pages 147–208 *in* Acoustic communication in birds, vol. 2 (D. E. Kroodsma and E. H. Miller, Eds.). Academic Press, New York. [40, 160, 214, 215, 216]

Mundinger, P. C. 1988. Conceptual errors, different perspectives, and genetic analysis of song ontogeny. Behav. Brain Sci. 11:643–644. [42, 46, 55]

Myers, J. P. 1993. Ted Parker. Am. Birds, Fall 1993:346–347. [v]

Myrberg, A. A. 1981. Sound communication and interception in fishes. Pages 395–425 *in* Hearing and sound communication in fishes (W. N. Tavolga, A. N. Popper, and R. R. Fay, Eds.). Springer Verlag, New York. [413]

Naguib, M., and H. Kolb. 1992. Vergleich des Strophenaufbaus und der Strophenabfolgen an den Gesängen von Sposser (*Luscinia luscinia*) und Blaukehlchen (*Luscinia svecica*). J. Ornithol. 133:133–145. [81]

Naguib, M., and R. H. Wiley. 1994. Perception of auditory distance in song birds: how much information does a listener need? J. Ornithol. 135:167. [414]
Naguib, M., H. Kolb, and H. Hultsch. 1991. Hierarchische Verzeigungsstruktur in den Gesansgstrophen der Vögel. Verh. Dtsch. Zool. Ges. 84:477. [81]
Nakamura, H., and K. Shigemori. 1990. Diurnal change of activity and social behavior of Latham's Snipe, *Gallinago hardwickii,* in the breeding season. J. Yamashina Inst. Ornithol. 22:85–113. (In Japanese). [243, 252]
Naugler, C., and L. Ratcliffe. 1992. A field test of the sound environment hypothesis of conspecific song recognition in American Tree Sparrows (*Spizella arborea*). Behaviour 123:314–324. [336]
Nei, M. 1973. Analysis of gene diversity in subdivided populations. Proc. Natl. Acad. Sci. USA 70:3321–3323. [187]
Nei, M., and A. K. Roychoudhury. 1974. Sampling variances of heterozygosity and genetic distance. Genetics 76:379–390. [187]
Nelson, D. A. 1988. Feature weighting in species song recognition by the Field Sparrow (*Spizella pusilla*). Behaviour 106:158–182. [336, 337]
Nelson, D. A. 1989. Song frequency as a cue for recognition of species and individuals in the Field Sparrow (*Spizella pusilla*). J. Comp. Psychol. 103:171–176. [336, 337]
Nelson, D. A. 1992a. Song overproduction and selective attrition lead to song sharing in the Field Sparrow. Behav. Ecol. Sociobiol. 30:415–424. [65, 71]
Nelson, D. A. 1992b. Song overproduction, song matching and selective attrition during development. Pages 121–133 *in* Playback and studies of animal communication (P. McGregor, Ed.). Plenum Press, New York. [290]
Nelson, D. A., and L. J. Croner. 1991. Song categories and their functions in the Field Sparrow (*Spizella pusilla*). Auk 108:42–52. [446]
Nelson, D. A., and P. Marler. 1989. Categorical perception of a natural stimulus continuum: birdsong. Science 244:976–978. [124, 182]
Nelson, D. A., and P. Marler. 1990. The perception of birdsong and an ecological concept of signal space. Pages 443–478 *in* Comparative perception. Vol. 2: Complex signals (W. C. Stebbins and M. A. Berkley, Eds.). Wiley, New York. [336, 337]
Nelson, D. A., and P. Marler. 1993. Innate recognition of song in White-crowned Sparrows: a role in selective vocal learning? Anim. Behav. 46:806–808. [43]
Nelson, D. A., P. Marler, and A. Palleroni. 1995. A comparative approach to vocal learning. II. Intraspecific variation in the learning process. Anim. Behav. 50:83–97. [5]
Nelson, J. B. 1978. The Sulidae, gannets and boobies. Oxford Univ. Press, Oxford. [163]
Nelson, K. 1973. Does the holistic study of behavior have a future? Pages 281–328 *in* Perspectives in ethology (P. P. G. Bateson and P. H. Klopfer, Eds.). Plenum Press, New York. [80]
Nero, R. W. 1956. Behavior study of the Red-winged Blackbird. Wilson Bull. 68:5–37, 129–150. [434]
Nespor, A. A., M. L. Dent, M. J. Lukaszewicz, R. J. Dooling, and G. F. Ball. 1994. Testosterone induction of male-like vocalizations in female Budgerigars. Soc. Neurosci. Abstr. 20:163. [114]
Nethersole-Thompson, D., and M. Nethersole-Thompson. 1986. Waders: their breeding, haunts and watchers. T. and A. D. Poyser, Calton, England. [242, 250]
Nice, M. M. 1943. Studies in the life history of the Song Sparrow. II. The behavior of the Song Sparrow and other passerines. Trans. Linn. Soc. N.Y. 6:1–328. [356, 442, 443, 455]
Nichols, J. 1992. Linguistic diversity in space and time. Univ. Chicago Press, Chicago. [219]
Nicolai, J. 1959. Familientradition in der Gesangsentwicklung des Gimpels (*Pyrrhula pyrrhula L.*). J. Ornithol. 100:39–46. [41, 85, 236]
Nicolai, J. 1964. Brutparasitismus der Viduinae als ethologische Problem. Z. Tierpsychol. 21:129–204. [45]
Nicolai, J. 1969. Tauben Haltung, Zucht und Arten von Ziertauben. Kosmos, Gesellschaft der Naturfreunde Franckh'sche Verlagshandlung, Stuttgart. [50]

Nielsen, B. M. B., and S. L. Vehrencamp. 1995. Responses of Song Sparrows to song-type matching via interactive playback. Behav. Ecol. Sociobiol. 37:109–117. [401, 403]

Nolan, V., Jr. 1978. The ecology and behavior of the Prairie Warbler, *Dendroica discolor.* Ornithological Monographs 26. American Ornithologists' Union, Washington, D.C. [17, 430, 442, 446]

Nordby, J. C., S. E. Campbell, and M. D. Beecher. Unpubl. ms. Maintenance of song sharing over years in the Song Sparrow, a close-ended learner. [67, 71]

Nordeen, E., and K. Nordeen. 1989. Estrogen stimulates the incorporation of new neurons in avian song nuclei during adolescence. Dev. Brain Res. 49:27–32. [293]

Nordeen, E., A. Grace, M. Burek, and K. Nordeen. 1992. Sex-dependent loss of projection neurons involved in avian song learning. J. Neurobiol. 23:671–679. [288]

Nordeen, K., and E. Nordeen. 1988. Projection neurons within a vocal motor pathway are born during song learning in Zebra Finches. Nature 334:149–151. [288, 302]

Nordeen, K., and E. Nordeen. 1992. Auditory feedback is necessary for the maintenance of stereotyped song in adult Zebra Finches. Behav. Neural Biol. 57:58–66. [288]

Nordeen, K., E. Nordeen, and A. Arnold. 1986. Estrogen establishes sex differences in androgen accumulation in Zebra Finch brain. J. Neurosci. 6:734–738. [292]

Nores, M. 1992. Bird speciation in subtropical South America in relation to forest expansion and retraction. Auk 109:346–357. [275]

Nottebohm, F. 1969a. The "critical period" for song learning. Ibis 111:386–387. [293]

Nottebohm, F. 1969b. The song of the Chingolo (*Zonotrichia capensis*) in Argentina: description and evaluation of a system of dialects. Condor 71:299–315. [177]

Nottebohm, F. 1970. Ontogeny of bird song. Science 167:950–956. [97]

Nottebohm, F. 1972. The origins of vocal learning. Am. Nat. 106:116–140. [11, 46, 97, 276, 294]

Nottebohm, F. 1975. Continental patterns of song variability in *Zonotrichia capensis:* some possible ecological correlates. Am. Nat. 109:35–50. [186]

Nottebohm, F. 1976. Phonation in the Orange-winged Amazon Parrot, *Amazona amazonica.* J. Comp. Physiol. A 108:157–170. [113]

Nottebohm, F. 1980a. Testosterone triggers growth of brain vocal control nuclei in adult female canaries. Brain Res. 189:429–436. [293, 296]

Nottebohm, F. 1980b. Brain pathways for vocal learning in birds: a review of the first 10 years. Pages 85–124 *in* Progress in psychobiology and physiological psychology, vol. 9 (J. M. S. Sprage and A. N. E. Epstein, Eds.). Academic Press, New York. [10, 272, 295]

Nottebohm, F. 1980c. Brain correlates of a learned motor skill. Verh. Dtsch. Zool. Ges. 1980:262–267. [298]

Nottebohm, F. 1981. A brain for all seasons: cyclical anatomical changes in song control nuclei of the canary brain. Science 214:1368–1370. [290, 298, 299, 302, 303]

Nottebohm, F. 1987. Plasticity in adult avian central nervous system: possible relation between hormones, learning, and brain repair. Pages 85–108 *in* Handbook of physiology, sec. 1 (F. Plum, Ed.). Williams and Wilkins, Baltimore. [296, 301]

Nottebohm, F. 1991. Reassessing the mechanisms and origins of vocal learning in birds. Trends Neurosci. 14:206–211. [11]

Nottebohm, F. 1993. The search for neural mechanisms that define the sensitive period for song learning in birds. Neth. J. Zool. 43:193–234. [94, 305]

Nottebohm, F., and A. Arnold. 1976. Sexual dimorphism in vocal control areas of the songbird brain. Science 194:211–213. [290, 291]

Nottebohm, F., and M. E. Nottebohm. 1971. Vocalizations and breeding behaviour of surgically deafened Ring Doves (*Streptopelia risoria*). Anim. Behav. 19:313–327. [44]

Nottebohm, F., and M. E. Nottebohm. 1978. Relationship between song repertoire and age in the canary, *Serinus canarius.* Z. Tierpsychol. 46:298–305. [40]

Nottebohm, F., and R. K. Selander. 1972. Vocal dialects and gene frequencies in the Chingolo Sparrow, *Zonotrichia capensis.* Condor 74:137–143. [279]

Nottebohm, F., T. Stokes, and C. Leonard. 1976. Central control of song in the canary. J. Comp. Neurol. 165:457–486. [286]

Nottebohm, F., A. Alvarez-Buylla, J. Cynx, C.-Y. Ling, M. Nottebohm, R. Suter, A. Tolles, and H. Williams. 1990. Song learning in birds: the relation between perception and production. Philos. Trans. R. Soc. B 329:115–124. [287, 288, 342, 346, 348, 354]

Nowicki, S. 1983. Flock-specific recognition of chickadee calls. Behav. Ecol. Sociobiol. 12:317–320. [107, 114, 146]

Nowicki, S. 1987. Vocal tract resonances in oscine bird sound production: evidence from birdsongs in a helium atmosphere. Nature 325:53–55. [370, 373]

Nowicki, S. 1989. Vocal plasticity in captive Black-capped Chickadees: the acoustic basis and rate of call convergence. Anim. Behav. 37:64–73. [107, 114, 115]

Nowicki, S., and G. F. Ball. 1989. Testosterone induction of song in photosensitive and photorefractory male sparrows. Horm. Behav. 23:514–525. [428]

Nowicki, S., and P. Marler. 1988. How do birds sing? Music Percep. 5:391–426. [79, 286]

Nowicki, S., M. Hughes, and P. Marler. 1991. Flight songs of Swamp Sparrows: alternative phonology of an alternative song category. Condor 93:1–11. [383, 387]

Nowicki, S., J. Podos, and F. Valdes. 1994. Temporal patterning of within-song type and between-song type variation in song repertoires. Behav. Ecol. Sociobiol. 34:329–335. [65, 66]

O'Donald, P. 1980. Genetic models of sexual selection. Cambridge Univ. Press, Cambridge. [463]

Odum, E. P. 1942. Annual cycle of the Black-capped Chickadee, 3. Auk 59:499–531. [145, 147]

Okanoya, K. 1995. Adaptive tracking procedures to measure auditory sensitivity. Pages 143–153 *in* Methods in comparative psychoacoustics (G. M. Klump, R. J. Dooling, R. R. Fay, and W. C. Stebbins, Eds.). Birkhäuser, Basel. [326]

Okanoya, K., and R. J. Dooling. 1987. Hearing in passerine and psittacine birds: a comparative study of absolute and masked auditory thresholds. J. Comp. Psychol. 101:7–15. [98, 107, 327–329]

Okanoya, K., and R. J. Dooling. 1988. Hearing in the Swamp Sparrow, *Melospiza georgiana,* and in the Song Sparrow, *Melospiza melodia.* Anim. Behav. 36:726–732. [328]

Okanoya, K., and R. J. Dooling. 1990. Detection of gaps in noise by Budgerigars (*Melopsittacus undulatus*) and Zebra Finches (*Poephila guttata*). Hear. Res. 50:185–192. [98]

Okanoya, K., and R. J. Dooling. 1991. Detection of species-specific calls in noise by Zebra Finches, *Poephila guttata,* and Budgerigars, *Melopsittacus undulatus:* time of frequency domain? Bioacoustics 3:163–172. [98, 107]

O'Loghlen, A. L., and S. I. Rothstein. 1993. An extreme example of delayed vocal development: song learning in a population of wild Brown-headed Cowbirds. Anim. Behav. 46:293–304. [22, 30, 35]

O'Neill, J. P. 1993. Ted Parker's antwren. Am. Birds, Fall 1993:348–349. [v]

Oppenheim, R. W. 1982. Preformation and epigenesis in the origins of the nervous system and behavior: issues, concepts, and their history. Pages 1–100 *in* Perspectives in ethology, vol. 5 (P. K. Klopfer and P. P. G. Bateson, Eds.). Plenum Press, New York. [25]

Oring, L. W. 1968. Vocalizations of the Green and Solitary Sandpipers. Wilson Bull. 80:395–420. [241]

Osherson, D. N., and H. Lasnik. 1990. Language—an invitation to cognitive science. MIT Press, Cambridge, Mass. [95]

Otte, D. 1974. Effects and functions in the evolution of signalling systems. Annu. Rev. Ecol. Syst. 5:385–417. [410, 413]

Otter, K. 1993. Intersexual selection and song in the Black-capped Chickadee, *Parus atricapillus.* M.S. thesis, Queen's Univ., Kingston, Canada. [350]

Otter, K., and L. Ratcliffe. 1993. Changes in singing behavior of male Black-capped Chickadees (*Parus atricapillus*) following mate removal. Behav. Ecol. Sociobiol. 33:409–414. [18, 307, 348, 350, 430, 432, 456]

Otter, K., L. Ratcliffe, and P. T. Boag. 1994a. Extra-pair paternity in the Black-capped Chickadee. Condor 96:218–222. [340]

Otter, K., M. Njegovan, C. Naugler, J. Fotheringham, and L. Ratcliffe. 1994b. An alternative technique for interactive playback experiments using a Macintosh Powerbook computer. Bioacoustics 5:303–308. [398, 401]

Ovenden, J. R., A. Wust-Saucy, R. Bywater, N. Brothers, and R. W. G. White. 1991. Genetic evidence for philopatry in a colonially nesting seabird, the Fairy Prion (*Pachyptila turtur*). Auk 108:688–694. [175]

Owens, I. P. F., T. Burke, and D. B. A. Thompson. 1994. Extraordinary sex roles in the Eurasian Dotterel: female mating arenas, female-female competition, and female mate choice. Am. Nat. 144:76–100. [242]

Owings, D. H., and D. F. Hennessy. 1984. The importance of variation in sciurid visual and vocal communication. Pages 169–200 *in* The biology of ground-dwelling squirrels (J. O. Murie and G. R. Michener, Eds.). Univ. Nebraska Press, Lincoln. [266]

Owren, M. J. 1990a. Acoustic classification of alarm calls by vervet monkeys (*Cercopithecus aethiops*) and humans (*Homo sapiens*). I. Natural calls. J. Comp. Psychol. 104:20–28. [121, 122]

Owren, M. J. 1990b. Acoustic classification of alarm calls by vervet monkeys (*Cercopithecus aethiops*) and humans (*Homo sapiens*). II. Synthetic calls. J. Comp. Psychol. 104:29–40. [121]

Oyama, S. 1985. The ontogeny of information: developmental systems and evolution. Cambridge Univ. Press, Cambridge. [25]

Oyama, S. 1993. Constraints and development. Neth. J. Zool. 43:6–16. [305]

Panov, E. N. 1989. Natural hybridisation and ethological isolation in birds. Nauka, Moscow. [231]

Park, T. J., and R. J. Dooling. 1985. Perception of species-specific contact calls by the Budgerigar (*Melopsittacus undulatus*). J. Comp. Psychol. 99:391–402. [98, 108]

Park, T. J., and R. J. Dooling. 1986. Perception of degraded vocalizations by Budgerigars (*Melopsittacus undulatus*). Anim. Learn. Behav. 14:359–364. [98, 108]

Parker, G. A. 1974. Assessment strategy and the evolution of animal conflicts. J. Theor. Biol. 47:223–243. [267]

Parker, G. A. 1982. Phenotype limited evolutionary stable strategies. Pages 173–201 *in* Current problems in sociobiology (B. R. Bertram, T. H. Clutton-Brock, R. I. M. Dunbar, D. I. Rubenstein, and R. Wrangham, Eds.). Cambridge Univ. Press, Cambridge. [472]

Parker, T. A., III. 1991. On the use of tape recorders in avifaunal surveys. Auk 108:443–444. [v, 269, 273, 275, 474]

Parker, T. A., III, and J. P. O'Neill. 1985. A new species and a new subspecies of *Thryothorus* wren from Peru. Pages 9–15 *in* Neotropical ornithology (P. A. Buckley, M. S. Foster, E. S. Morton, R. S. Ridgely, and F. G. Buckley, Eds.). Ornithological Monographs 36. American Ornithologists' Union, Washington, D.C. [270]

Parkes, K. C. 1982. Subspecific taxonomy: unfashionable does not mean irrelevant. Auk 99:596–598. [270]

Pärt T. 1991. Is dawn singing related to paternity insurance? The case of the Collared Flycatcher. Anim. Behav. 41:451–456. [432, 434, 435]

Partridge, L., and J. A. Endler. 1987. Life history constraints on sexual selection. Pages 265–277 *in* Sexual selection: testing the alternatives (J. W. Bradbury and M. B. Andersson, Eds.). Wiley, Chichester, Sussex, England. [305, 316]

Paton, J. A., K. R. Manogue, and F. Nottebohm. 1981. Bilateral organization of the vocal control pathway in the Budgerigar (*Melopsittacus undulatus*). J. Neurosci. 1:1279–1288. [113, 294, 295]

Paulson, D. R. 1993. Shorebirds of the Pacific Northwest. UBC Press, Vancouver, B.C. [241, 244, 251]

Payne, R., and D. Webb. 1971. Orientation by means of long range acoustic signalling in baleen whales. Ann. N.Y. Acad. Sci. 188:110–141. [424]

Payne, R. B. 1973a. Behavior, mimetic songs and song dialects, and relationships of the parasitic indigobirds (*Vidua*) of Africa. Ornithological Monographs 11. American Ornithologists' Union, Washington, D.C. [41, 45, 213, 217, 219]

Payne, R. B. 1973b. Vocal mimicry of Paradise Whydahs (*Vidua*) and response of male whydahs to song of their hosts (*Pytilia*) and their mimics. Anim. Behav. 21:762–771. [41]

Payne, R. B. 1978a. Microgeographic variation in songs of Splendid Sunbirds, *Nectarinia coccinigaster:* population phenetics, habitats, and song dialects. Behaviour 65:282–308. [182, 216]

Payne, R. B. 1978b. Local dialects in the wingflaps of Flappet Larks, *Mirafra rufocinnamomea.* Ibis 120:204–207. [216]

Payne, R. B. 1979. Song structure, behavior and sequence of song types in a population of Village Indigobird, *Vidua chalybeata.* Z. Tierpsychol. 45:113–173. [161]

Payne, R. B. 1980. Behavior and songs of hybrid parasitic finches. Auk 97:118–134. [48]

Payne, R. B. 1981a. Song learning and social interaction in Indigo Buntings. Anim. Behav. 29:688–697. [12, 65, 85, 199, 202, 205, 212, 215]

Payne, R. B. 1981b. Population structure and social behavior: models for testing the ecological significance of song dialects in birds. Pages 108–120 *in* Natural selection and social behavior (R. D. Alexander and D. W. Tinkle, Eds.). Chiron Press, New York. [215]

Payne, R. B. 1981c. Persistence of local wingflap dialects in Flappet Larks, *Mirafra rufocinnamomea.* Ibis 123:507–511. [216]

Payne, R. B. 1982. Ecological consequences of song matching: breeding success and intraspecific song mimicry in Indigo Buntings. Ecology 63:401–411. [71, 199, 204, 207, 239, 370, 371]

Payne, R. B. 1983a. The social context of song mimicry: song-matching dialects in Indigo Buntings (*Passerina cyanea*). Anim. Behav. 31:788–805. [199, 218, 371]

Payne, R. B. 1983b. Bird songs, sexual selection, and female mating strategies. Pages 55–90 *in* Social behavior of female vertebrates (S. K. Waser, Ed.). Academic Press, New York. [18, 65, 77, 199]

Payne, R. B. 1985. Behavioral continuity and change in local song populations of Village Indigobirds, *Vidua chalybeata.* Z. Tierpsychol. 70:1–44. [199, 213, 215–217]

Payne, R. B. 1986. Bird songs and avian systematics. Pages 87–126 *in* Current ornithology (R. J. Johnston, Ed.). Plenum Press, New York. [241, 242, 245, 256, 257, 271]

Payne, R. B. 1989. Indigo Bunting. Pages 153–172 *in* Lifetime reproduction in birds (I. Newton, Ed.). Academic Press, London. [199, 200, 212]

Payne, R. B. 1990a. Natal dispersal, area effects, and effective population size. J. Field Ornithol. 61:396–403. [215]

Payne, R. B. 1990b. Song mimicry by the Village Indigobird (*Vidua chalybeata*) of the Red-billed Firefinch (*Lagonosticta senegala*). Vogelwarte 35:321–328. [45]

Payne, R. B. 1991. Natal dispersal and population structure in a migratory songbird, the Indigo Bunting. Evolution 45:49–62. [199, 200, 215]

Payne, R. B. 1992. Indigo Bunting. Pages 1–24 *in* The birds of North America, no. 4 (A. Poole and F. Gill, Eds.). The Academy of Natural Sciences, Philadelphia; The American Ornithologists' Union, Philadelphia. [199, 200, 210, 212]

Payne, R. B., and K. Payne. 1977. Social organization and mating success in local song populations of Village Indigobirds, *Vidua chalybeata.* Z. Tierpsychol. 45:113–173. [216, 456, 469, 471]

Payne, R. B., and L. L. Payne. 1989. Heritability estimates and behaviour observations: extra-pair matings in Indigo Buntings. Anim. Behav. 38:457–467. [199, 200, 212]

Payne, R. B., and L. L. Payne. 1990. Survival estimates of Indigo Buntings: comparison of banding recoveries and local observations. Condor 92:938–946. [199, 200, 204, 210, 212]

Payne, R. B., and L. L. Payne. 1993a. Breeding dispersal in Indigo Buntings: circumstances and consequences for breeding success and population structure. Condor 95:1–24. [199, 212, 215]

Payne, R. B., and L. L. Payne. 1993b. Song copying and cultural transmission in Indigo Buntings. Anim. Behav. 46:1045–1065. [21, 199–201, 204, 212, 217]

Payne, R. B., and L. L. Payne. 1996. Demography, dispersal and song dialects and the persistence of partnerships in Indigo Buntings. Pages 305–320 *in* Partnerships in birds: the study of monogamy (J. M. Black, Ed.). Oxford Univ. Press, Oxford. [199]

Payne, R. B., and L. L. Payne. In press. Social learning of bird song: field studies of Indigo Buntings and Village Indigobirds. *In* Social influences on vocal development (C. Snowdon and M. Hausberger, Eds.). Cambridge Univ. Press, Cambridge. [199, 200]

Payne, R. B., and D. F. Westneat. 1988. A genetic and behavioral analysis of mate choice and song neighborhoods in Indigo Buntings. Evolution 42:935–947. [195, 199, 217, 218]

Payne, R. B., W. L. Thompson, K. L. Fiala, and L. L. Sweany. 1981. Local song traditions in Indigo Buntings: cultural transmission of behavior patterns across generations. Behaviour 77:199–221. [185, 186, 199, 200, 208, 210]

Payne, R. B., L. L. Payne, and S. M. Doehlert. 1987. Song, mate choice and the question of kin recognition in a migratory songbird. Anim. Behav. 35:35–47. [199, 200, 215, 217]

Payne, R. B., L. L. Payne, and S. M. Doehlert. 1988. Biological and cultural success of song memes in Indigo Buntings. Ecology 69:104–117. [13, 195, 199–201, 204, 207–210, 212, 218]

Pepperberg, I. M. 1985. Social modeling theory: a possible framework for understanding avian vocal learning. Auk 102:854–864. [107]

Pepperberg, I. M. 1990. Some cognitive capacities of an African Gray Parrot. Pages 357–409 *in* Adv. Study Behav., vol. 19 (P. J. B. Slater, J. S. Rosenblatt, and C. Beer, Eds.). Academic Press, New York. [97, 277, 280]

Pepperberg, I. M. 1993. A review of the effects of social interaction on vocal learning in African Grey Parrots (*Psittacus erithacus*). Neth. J. Zool. 43:104–124. [94]

Perrins, C. M. 1979. British tits. Collins, London. [193]

Pesch, A., and H.-R. Güttinger. 1985. Der Gesang des weiblichen Kanarienvogels. J. Ornithol. 126:108–110. [291]

Peters, J. L. 1931–1986. Check-list of birds of the world. Museum of Comparative Zoology, Cambridge, Mass. [270]

Peters, S. S., W. A. Searcy, and P. Marler. 1980. Species song discrimination in choice experiments with territorial male Swamp and Song Sparrows. Anim. Behav. 28:393–404. [344]

Petrinovich, L. 1985. Factors influencing song development in the White-crowned Sparrow (*Zonotrichia leucophrys*). J. Comp. Psychol. 99:15–29. [44, 50, 73]

Petrinovich, L. 1988a. Individual stability, local variability and the cultural transmission of song in White-crowned Sparrows (*Zonotrichia leucophrys nuttalli*). Behaviour 107:208–240. [215–217]

Petrinovich, L. 1988b. The role of social factors in White-crowned Sparrow song development. Pages 255–278 *in* Social learning: psychological and biological perspectives (T. R. Zentall and B. G. Galef, Eds.). Erlbaum, Hillsdale, N.J. [107]

Petrinovich, L., and L. F. Baptista. 1987. Song development in the White-crowned Sparrow: modification of learned song. Anim. Behav. 35:961–974. [12]

Petrinovich, L., and T. L. Patterson. 1979. Field studies of habituation. I. Effect of reproductive condition, number of trials, and different delay intervals on responses of the White-crowned Sparrow. J. Comp. Physiol. Psychol. 93:337–350. [471]

Phillips, A. R. 1981. Subspecies versus forgotten species: the case of Grayson's Thrush (*Turdus graysoni*). Wilson Bull. 93:301–309. [270]

Phillips, R. E. 1972. Sexual and agonistic behaviour in the Killdeer (*Charadrius vociferus*). Anim. Behav. 20:1–9. [252]

Piazza, A., S. Rendine, G. Zei, A. Moroni, and L. L. Cavalli-Sforza. 1987. Migration rates of human populations from surname distributions. Nature 329:714–716. [215]

Pickens, A. L. 1928. Auditory protective mimicry of the chickadee. Auk 45:302–304. [153]

Pickstock, J. C., and J. R. Krebs. 1980. Neighbor-stranger discrimination in the Chaffinch (*Fringilla coelebs*). J. Ornithol. 121:105–108. [359]

Pidgeon, R. 1981. Calls of the Galah, *Cacatua roseicapilla,* and some comparisons with four other species of Australian parrot. Emu 81:158–168. [97, 105, 112]

Pitelka, F. A., R. T. Holmes, and S. F. Maclean Jr. 1974. Ecology and evolution of social organization in Arctic sandpipers. Am. Zool. 14:185–204. [244]

Platt, J. R. 1964. Strong inference. Science 146:347–355. [37]

Podolsky, R., and S. W. Kress. 1989. Factors affecting colony formation in Leach's Storm-Petrel to uncolonized islands in Maine. Auk 106:332–336. [171, 174]

Podolsky, R., and S. W. Kress. 1992. Attraction of the endangered Dark-rumped Petrel to recorded vocalizations in the Galápagos Islands. Condor 94:448–453. [171, 172, 174]

Podos, J., S. Peters, T. Rudnicky, P. Marler, and S. Nowicki. 1992. The organization of song repertoires in Song Sparrows: themes and variations. Ethology 90:89–106. [65, 66, 127, 313]

Pomiankowski, A. N. 1987a. The costs of choice in sexual selection. J. Theor. Biol. 128:195–218. [463, 465]

Pomiankowski, A. N. 1987b. Sexual selection: the handicap principle does work—sometimes. Proc. R. Soc. Lond. B. 231:123–145. [465]

Pomiankowski, A. N. 1988. The evolution of female mate preferences for male genetic quality. Oxf. Surv. Evol. Biol. 5:136–184. [465]

Pomiankowski, A., Y. Iwasa, and S. Nee. 1991. The evolution of costly mate preferences. I. Fisher and biased mutation. Evolution 45:1422–1430. [463]

Popp, J. W. 1989. Temporal aspects of singing interactions among territorial Ovenbirds (*Seiurus aurocapillus*). Ethology 82:127–133. [84]

Poulson, H. 1959. Song learning in the domestic canary. Z. Tierpsychol. 16:173–178. [55]

Powell, E. F. 1993. Perception of developing vocalizations in the Budgerigar (*Melopsittacus undulatus*). M.S. thesis, Univ. Maryland, College Park. [99, 115]

Power, D. M. 1966a. Agonistic behavior and vocalizations of Orange-chinned Parakeets in captivity. Condor 8:562–581. [97, 105, 114]

Power, D. M. 1966b. Antiphonal dueting and evidence for the auditory reaction time in the Orange-chinned Parakeet. Auk 83:314–319. [105, 114]

Power, D. M. 1967. Epigamic and reproductive behavior of Orange-chinned Parakeets in captivity. Condor 69:28–41. [110, 114]

Power, D. M., and D. G. Ainley. 1986. Seabird geographic variation: similarity among populations of Leach's Storm-Petrel. Auk 103:575–585. [175]

Prestwich, K. N., K. E. Brugger, and M. Toppling. 1989. Energy and communication in three species of hylid frogs: power input, power output, and efficiency. J. Exp. Biol. 143:53–80. [310]

Price, P. H. 1979. Developmental determinants of structure in Zebra Finch song. J. Comp. Physiol. Psychol. 93:260–277. [63]

Promptoff, A. 1930. Die geographische Variabilität des Buchfinkenschlags (*Fringilla coelebs* L.) in Zusammenhang mit etlichen allgemeinen Fragen der Saisonvögelzuge. Biol. Zentralbl. 50:478–503. [216]

Pröve, E. 1983. Hormonal correlates of behavioral development in male Zebra Finches. Pages 368–374 *in* Hormones and behaviour in higher vertebrates (J. Balthazart and R. Gilles, Eds.). Springer Verlag, Berlin. [293]

Prum, R. O. 1994. Species status of the White-fronted Manakin, *Lepidothrix serena* (Pipridae), with comments on conservation biology. Condor 96:692–702. [269–271]

Rabouam, C., and V. Bretagnolle. Unpubl. ms. Individual recognition in birds with special reference to seabirds. [175, 176]

Radesäter, T. S., and S. Jakobson. 1988. Intra- and intersexual functions of song in the Willow Warbler (*Phylloscopus trochilus*). Pages 1382–1390 *in* Proc. XIX Int. Ornithol. Congr. (H. Ouellet, Ed.). National Museum of Natural Science, Univ. Ottawa Press, Ottawa. [161]

Radesäter, T. S., and S. Jakobsson. 1989. Song rate correlation of replacement territorial Willow Warblers, *Phylloscopus trochilus.* Ornis Scand. 20:71–73. [469]

Radesäter, T. S., S. Jakobsson, N. Andbjer, A. Bylin, and K. Nystrom. 1987. Song rate and pair formation in the Willow Warbler, *Phylloscopus trochilus.* Anim. Behav. 35:1645–1651. [160–262, 310, 457]

Raikow, R. J. 1982. Monophyly of the Passeriformes: test of a phylogenetic hypothesis. Auk 99:431–445. [11]

Randi, E., F. Spina, and B. Massa. 1989. Genetic variability in Cory's Shearwater (*Calonectris diomedea*). Auk 106:411–418. [175]

Ranft, R. 1989. Equipment reviews: Sony TCD-D10 R-DAT recorder. Bioacoustics 1:307–312. [482]

Ranft, R. 1993. Sound recording of Oriental birds. Orient. Bird Club Bull. 17:22–27. [480]

Rappole, J. H., and D. W. Warner. 1980. Ecological aspects of migrant bird behavior in Veracruz, Mexico. Pages 353–394 *in* Migrant birds in the Neotropics (A. Keast and E. S. Morton, Eds.). Smithsonian Institution Press, Washington, D.C. [387, 392]

Ratcliffe, L., and R. G. Weisman. 1985. Frequency shift in the fee-bee song of the Black-capped Chickadee. Condor 87:555–556. [13, 139, 145, 349]

Ratcliffe, L., and R. Weisman. 1992. Pitch processing strategies in birds: a comparison of laboratory and field studies. Pages 211–223 *in* Playback and studies of animal communication (P. K. McGregor, Ed.). Plenum Press, New York. [348, 350, 401]

Rattner, B. A., L. Siles, and C. G. Scanes. 1982. Oviposition and the plasma concentrations of LH, progesterone and corticosterone in bobwhite quail (*Colinus virginianus*). J. Reprod. Fertil. 66:147–155. [427]

Read, A. F., and D. M. Weary. 1992. The evolution of bird song: comparative analyses. Philos. Trans. R. Soc. Lond. B 338:165–187. [305, 311]

Reddig, E. 1978. Der Ausdrucksflug der Bekassine (*Capella gallinago gallinago*). J. Ornithol. 119:357–387. [242, 250]

Reddig, E. 1981. Die Bekassine, *Capella gallinago.* Die neue Brehm-Bücherei, A. Ziemsen Verlag, Wittenberg, Lutherstadt. [242, 250]

Reebs, S. G., and N. Mrosovsky. 1990. Photoperiodism in House Sparrows: testing for induction with nonphototic Zeitgebers. Physiol. Zool. 63:587–599. [431]

Reid, M. L. 1987. Costliness and reliability in the singing vigour of Ipswich Sparrows. Anim. Behav. 35:1735–1743. [262, 311, 468]

Remsen, J. V., Jr. 1976. Observations of vocal mimicry in the Thick-billed Euphonia. Wilson Bull. 88:487–488. [276]

Remsen, J. V., Jr., and T. S. Schulenberg. In press. The pervasive influence of Ted Parker on Neotropical ornithology. *In* Natural history and conservation of Neotropical birds (J. V. Remsen Jr., Ed.). Ornithological Monographs, American Ornithologists' Union, Washington, D.C. [v]

Renfrew, C. 1987. Archaeology and language. The puzzle of Indo-European origins. Cape, London. [213]

Renfrew, C. 1992. Archaeology, genetics and linguistic diversity. Man 27:445–478. [214, 219]

Renfrew, C. 1994. World linguistic diversity. Sci. Am. 270(1):116–123. [214]

Reyer, H.-U., and D. Schmidl. 1988. Helpers have little to laugh about: group structure and vocalization in the Laughing Kookaburra, *Dacelo novaeguineae.* Emu 88:150–160. [442]

Rice, J. O., and W. L. Thompson. 1968. Song development in the Indigo Bunting. Anim. Behav. 16:462–469. [205, 212]

Richard, J.-P. 1991. Sound analysis and synthesis using an Amiga micro-computer. Bioacoustics 3:45–60. [162]

Richards, D. G. 1979. Recognition of neighbors by associative learning in Rufous-sided Towhees. Auk 96:688–693. [359]

Richards, D. G. 1981. Estimation of distance of singing conspecifics by the Carolina Wren. Auk 98:127–133. [266, 414]

Richards, D. G., and R. H. Wiley. 1980. Reverberations and amplitude fluctuations in the propagation of sound in a forest: implications for animal communication. Am. Nat. 115:381–399. [266, 331–333]

Ridgely, R. S., and M. B. Robbins. 1988. *Pyrrhura orcesi,* a new parakeet from southwestern Ecuador, with systematic notes on the *P. melanura* complex. Wilson Bull. 100:173–182. [270]

Ridgely, R. S., and G. Tudor. 1989. The birds of South America. Vol. 1: The oscine passerines. Univ. Texas Press, Austin. [272, 274]

Ridgely, R. S., and G. Tudor. 1994. The birds of South America. Vol. 2: The suboscine passerines. Univ. Texas Press, Austin. [9, 270, 272, 274]

Riede, K. 1993. Monitoring biodiversity: analysis of Amazonian rainforest sounds. Ambio 22:546–548. [269]

Ritchison, G. 1988. Responses of Yellow-breasted Chats to the songs of neighboring and non-neighboring conspecifics. J. Field Ornithol. 59:37–42. [359, 363]

Robb, J. 1991. Random causes with directed results: the Indo-European spread and the stochastic loss of lineages. Antiquity 65:287–291. [220]

Robb, J. 1993. A social prehistory of European languages. Antiquity 67:747–760. [213]

Robbins, M. B., and R. S. Ridgely. 1992. Taxonomy and natural history of *Nyctiphynus rosenbergi* (Caprimulgidae). Condor 94:984–987. [271]

Robbins, M. B., G. H. Rosenberg, and F. S. Molina. 1994. A new species of cotinga (Cotingidae, *Doliornis*) from the Ecuadorian Andes, with comments on plumage sequences in *Doliornis* and *Ampelion.* Auk 111:1–7. [270]

Robbins, M. B., G. R. Graves, and J. V. Remsen Jr. In press. Ted Parker remembered. *In* Natural history and conservation of Neotropical birds (J. V. Remsen Jr., Ed.). Ornithological Monographs, American Ornithologists' Union, Washington, D.C. [v]

Robinson, F. N. 1975. Vocal mimicry and the evolution of bird song. Emu 75:23–27. [59]

Robisson, P. 1991. The broadcast distance of the mutual display call in the Emperor Penguin. Behaviour 119:302–316. [168]

Robisson, P., T. Aubin, and J.-C. Brémond. 1993. Individuality in the voice of the Emperor Penguin, *Aptenodytes forsteri:* adaptation to a noisy environment. Ethology 94:279–290. [171]

Römer, H. 1992. Ecological constraints for the evolution of hearing and sound communication in insects. Pages 79–94 *in* The evolutionary biology of hearing (D. B. Webster, R. R. Fay, and A. N. Popper, Eds.). Springer Verlag, Berlin. [422]

Rosch, E., C. B. Mervis, W. D. Gray, D. M. Johnson, and P. Boyes-Braem. 1976. Basic objects in natural categories. Cognit. Psychol. 8:382–439. [125, 133]

Rosenberg, K. V., and D. A. Wiedenfeld. 1993. Directory of Neotropical ornithology. American Ornithologists' Union, Washington, D.C. [274]

Rosenfield, R. N., and J. Bielfeldt. 1991. Vocalizations of Cooper's Hawks during the pre-incubation stage. Condor 93:659–665. [442]

Rosenthal, R. 1976. Experimenter effects in behavioral research. Irvington, New York. [419]

Rost, R. 1987. Entstehung, Fortbestand und funktionelle Bedeutung von Gesangsdialekten bei der Sumpfmeise, *Parus palustris*—Ein Test von Modellen. Ph.D. dissertation, Univ. Konstanz; Hartung-Gorre Verlag, Konstanz. [43]

Rothenberg, J. 1995. Ensuring the longevity of digital documents. Sci. Am. 272(1):42–47. [483]

Rothstein, S. I. 1990. A model system for evolution of avian brood parasitism. Annu. Rev. Ecol. Syst. 21:481–508. [7]

Rothstein, S. I., and R. C. Fleischer. 1987. Vocal dialects and their possible relation to honest signalling in the Brown-headed Cowbird. Condor 89:1–23. [15, 18, 39]

Rothstein, S. I., D. A. Yokel, and R. C. Fleischer. 1988. The agonistic and sexual functions of vocalizations of male Brown-headed Cowbirds (*Molothrus ater*). Anim. Behav. 36:73–86. [22]

Rothstein, S. I., J. C. Ortega, and A. O'Loghlen. 1989. Cowbird song. Nature 339:21–22. [4]

Rothwell, R., and D. Amadon. 1964. Ecology of the Budgerygah. Auk 81:82. [98]

Rowley, I. 1974. Bird life. Collins, Sydney. [98]

Rowley, I. 1980. Parent-offspring recognition in a cockatoo, the Galah, *Cacatua roseicapilla.* Aust. J. Zool. 28:445–456. [97, 98, 100, 101, 103, 105, 107, 112]

Rowley, I. 1983a. Mortality and dispersal of juvenile Galahs, *Cacatua roseicapilla,* in the western Australian wheatbelt. Aust. Wildl. Res. 10:329–342. [98, 100, 101, 103, 104]

Rowley, I. 1983b. Re-mating in birds. Pages 331–360 *in* Mate choice (P. G. Bateson, Ed.). Cambridge Univ. Press, Cambridge. [104]

Rowley, I. 1990. Behavioural ecology of the Galah, *Eolophus roseicapillus,* in the wheatbelt of Western Australia. Surrey Beatty and Sons, Chipping Norton, Australia. [97–101, 103, 104, 107, 112]

Rowley, I., and G. Chapman. 1986. Cross-fostering, imprinting and learning in two sympatric species of cockatoo. Behaviour 96:1–16. [97–100, 103, 107, 110]

Rowley, I., and G. Chapman. 1991. The breeding biology, food, social organization, demography and conservation of the Major Mitchell or Pink Cockatoo, *Cacatua leadbeateri,* on the margin of the western Australian wheatbelt. Aust. J. Zool. 39:211–261. [98–100, 102–104]

Rucker, M., and V. Cassone. 1991. Song control nuclei of the House Sparrow are photoperiodic and photorefractory. Soc. Neurosci. Abstr. 17:1051. [302]

Ruvolo, M. 1987. Reconstructing genetic and linguistic trees: phenetic and cladistic approaches. Pages 193–216 *in* Biological metaphor and cladistic classification (H. M. Hoenigswald and L. F. Wiener, Eds.). Univ. Pennsylvania Press, Philadelphia. [198, 213, 214]

Ryan, M. J. 1985. The tungara frog. Univ. Chicago Press, Chicago. [310]

Ryan, M. J., and E. A. Brenowitz. 1985. The role of body size, phylogeny, and ambient noise in the evolution of bird song. Am. Nat. 126:87–100. [54, 305, 314, 321, 330]

Ryan, M. J., and A. Keddy-Hector. 1992. Directional patterns of female mate choice and the role of sensory biases. Am. Nat. 139:S4-S35. [466]

Ryan, M. J., and A. S. Rand. 1993. Species recognition and sexual selection as a unitary problem in animal communication. Evolution 47:647–657. [339, 341]

Ryan, M. J., and W. Wilczynski. 1988. Coevolution of sender and receiver: effect on local mate preference in cricket frogs. Science 240:1786–1788. [470]

Ryan, M. J., M. D. Tuttle, and L. K. Taft. 1981. The costs and benefits of frog chorusing behavior. Behav. Ecol. Sociobiol. 8:273–278. [410]

Ryan, M. J., J. H. Fox, W. Wilczynski, and A. S. Rand. 1990. Sexual selection for sensory exploitation in the frog *Physalaemus pustulosus.* Nature 343:66–67. [466]

Saberi, K., L. Dostal, T. Sadralodabai, V. Bull, and D. R. Perrot. 1991. Free-field release from masking. J. Acoust. Soc. Am. 90:1355–1370. [334]

Saether, S. A. 1994. Vocalizations of female Great Snipe, *Gallinago media,* at the lek. Ornis Fenn. 71:11–16. [256]

Sakaluk, S. K., and J. J. Belwood. 1984. Gecko phonotaxis to cricket calling song: a case of satellite predation. Anim. Behav. 32:659–662. [416]

Salomon, M. 1987. Analyse d'une zone de contact entre deux formes parapatriques: le cas des pouillots véloces *Phylloscopus c. collybita* et *P. c. brehmii.* Rev. Ecol. Terre Vie 42:377–420. [228]

Salomon, M. 1989. Song as a possible reproductive isolating mechanism between two parapatric forms. The case of the chiffchaffs *Phylloscopus c. collybita* and *P. c. brehmii* in the western Pyrenees. Behaviour 111:270–290. [228]

Sankoff, D., and G. Sankoff. 1973. Wave versus Stammbaum explanations of lexical similarities. Tech. Rep. 282, Centre de Recherche Mathématique, Montreal. [213]

Sauer, E. G. F. 1962. Ethology and ecology of Golden Plovers on St. Lawrence Island, Bering Sea. Psychol. Forsch. 26:399–470. [244]

Sauer, J. R., and S. Droge. 1992. Geographic patterns in population trends of Neotropical migrants in North America. Pages 26–42 *in* Ecology and conservation of Neotropical migrant landbirds (J. M. Hagen III and D. W. Johnston, Eds.). Smithsonian Institution Press, Washington, D.C. [209]

Saunders, A. A. 1935. A guide to bird songs. Appleton-Century, New York. [145]
Saunders, D. A. 1974. The function of displays in the breeding biology of the White-tailed Black Cockatoo. Emu 74:43–46. [98, 110]
Saunders, D. A. 1982. The breeding behaviour and biology of the short-billed form of the White-tailed Black Cockatoo, *Calyptorhynchus funereus.* Ibis 124:422–455. [98, 100, 103, 104]
Saunders, D. A. 1983. Vocal repertoire and individual vocal recognition in the Short-billed White-tailed Black Cockatoo, *Calyptorhynchus funereus latirostris* Carnaby. Aust. Wildl. Res. 10:527–536. [97, 98, 105, 107, 112]
Saunders, D. A. 1986. Breeding season, nesting success, and nestling growth in Carnaby's Cockatoo, *Calyptorhynchus funereus latirostris,* over 16 years at Coomallo Creek, and a method for assessing the viability of populations in other areas. Aust. Wildl. Res. 13:261–273. [98]
Saunders, J. C., R. M. Denny, and G. R. Bock. 1978. Critical bands in the parakeet (*Melopsittacus undulatus*). J. Comp. Physiol. A 125:359–365. [329]
Saunders, J. C., W. F. Rintelmann, and G. R. Bock. 1979. Frequency selectivity in bird and man: a comparison among critical ratios, critical bands and psychophysical tuning curves. Hear. Res. 1:303–323. [327–329]
Scharf, B. 1970. Critical bands. Pages 159–202 *in* Foundations of modern auditory theory (J. V. Tobias, Ed.). Academic Press, New York. [329]
Scharff, C., and F. Nottebohm. 1991. A comparative study of the behavioral deficits following lesions of various parts of the Zebra Finch song system: implications for vocal learning. J. Neurosci. 11:2896–2913. [288]
Scheer, A. 1991. Vergleichende Verhaltensstudie von Elfen- und Feenastrilden (*Brunhilda erythronotos* und *B. charmosyna*). Tropische Vogel 12:81–87. [45]
Schild, D. 1986. Syringeale Kippschwingungen und Klangerzeugung beim Feldschwirl (*Locustella naevia*). J. Ornithol. 127:331–336. [317]
Schleidt, W. M. 1973a. Tonic communication: continual effects of discrete signs in animal communication systems. J. Theor. Biol. 42:359–386. [247]
Schleidt, W. M. 1973b. The noisy channel. 1:1. Univ. North Dakota, Grand Forks. [419]
Schleidt, W. M. 1974. How "fixed" is the fixed action pattern? Z. Tierpsychol. 36:184–211. [313]
Schleidt, W. M., and M. D. Shalter. 1973. Stereotype of a fixed action pattern during ontogeny in *Coturnix coturnix coturnix.* Z. Tierpsychol. 33:35–37. [44, 54]
Schleidt, W. M., G. Yakalis, M. Donnelly, and J. McGarry. 1984. A proposal for a standard ethogram, exemplified by an ethogram of the Blue-breasted Quail (*Coturnix chinensis*). Z. Tierpsychol. 64:193–220. [163]
Schlinger, B., and A. Arnold. 1992a. Plasma sex steroids and tissue aromatization in hatchling Zebra Finches: implications for the sexual differentiation of singing behavior. Endocrinology 130:289–299. [292]
Schlinger, B., and A. Arnold. 1992b. Circulating estrogens in a male songbird originate in the brain. Proc. Natl. Acad. Sci. USA 89:7650–7653. [292]
Schneirla, T. C. [1953] 1972. The concept of levels in the study of social phenomena. Pages 238–253 *in* Selected writings of T. C. Schneirla (L. R. Aronson, E. Tobach, J. S. Rosenblatt, and D. S. Lehrman, Eds.). W. H. Freeman, San Francisco. [22]
Schooneveldt, G. P., and B. C. J. Moore. 1987. Comodulation masking release (CMR): effects of signal frequency, flanking-band frequency, masker bandwidth, flanking-band level, and monotic versus dichotic presentation of the flanking band. J. Acoust. Soc. Am. 82:1944–1956. [331, 332]
Schooneveldt, G. P., and B. C. J. Moore. 1989. Comodulation masking release (CMR) as a function of masker bandwidth, modulator bandwidth, and signal duration. J. Acoust. Soc. Am. 85:273–281. [332]
Schottler, B. 1993. Die Lautäußerungen der Blaumeisen (*Parus caeruleus*) der Kanarischen

Inseln—Variabilität, geographische Differenzierungen und Besiedlungsgeschichte. Ph.D. dissertation, Univ. Mainz; Hartung-Gorre Verlag, Konstanz. [235]

Schroeder, D. J., and R. H. Wiley. 1983a. Communication with shared song themes in Tufted Titmice. Auk 100:414–424. [359, 368, 371]

Schroeder, D. J., and R. H. Wiley. 1983b. Communication with repertoires of song themes in Tufted Titmice. Anim. Behav. 31:1128–1138. [411]

Schubert, M. 1976. Über die Variabilität von Lockrufen des Gimpels, *Pyrrhula pyrrhula.* Ardea 64:62–71. [54]

Schulenberg, T. S. 1995. A tribute to Ted Parker. Bird Conserv. Int. 5:137–139. [v]

Schulenberg, T. S., and L. C. Binford. 1985. A new species of tanager (Emberizidae: Thraupinae, *Tangara*) from southern Peru. Wilson Bull. 97:413–420. [270]

Schulenberg, T. S., and M. D. Williams. 1982. A new species of antpitta (*Grallaria*) from northern Peru. Wilson Bull. 94:105–113. [270]

Schwabl, H. 1993. Yolk is a source of maternal testosterone for developing birds. Proc. Natl. Acad. Sci. USA 90:11446–11450. [299]

Schwartz, J. J., and H. C. Gerhardt. 1989. Spatially mediated release from auditory masking in an anuran amphibian. J. Comp. Physiol. A 166:37–41. [334]

Schwartzkopff, J. 1952. Untersuchungen über die Arbeitsweise des Mittelohres und das Richtungshören der Singvögel unter Verwendung von Cochlea-Potentialen. Z. Vergl. Physiol. 32:46–68. [334]

Scott, D. K. 1988. Breeding success in Bewick's Swans. Pages 220–236 *in* Reproductive success (T. H. Clutton-Brock, Ed.). Univ. Chicago Press, Chicago. [174]

Searcy, W. A. 1979. Sexual selection and body size in male Red-winged Blackbirds. Evolution 33:649–661. [311, 468]

Searcy, W. A. 1984. Song repertoire size and female preferences in Song Sparrows. Behav. Ecol. Sociobiol. 14:281–286. [458]

Searcy, W. A. 1986. Are female Red-winged Blackbirds territorial? Anim. Behav. 34:1381–1392. [160, 174]

Searcy, W. A. 1988a. Do female Red-winged Blackbirds limit their own breeding densities? Ecology 69:85–95. [456, 460]

Searcy, W. A. 1988b. Dual intersexual and intrasexual functions of song in Red-winged Blackbirds. Pages 1373–1381 *in* Proc. XIX Int. Ornithol. Congr. (H. Ouellet, Ed.) National Museum of Natural Science, Univ. Ottawa Press, Ottawa. [458, 460, 471]

Searcy, W. A. 1988c. Song development from evolutionary and ecological perspectives. Behav. Brain Sci. 11:647–648. [39]

Searcy, W. A. 1989. Pseudoreplication, external validity, and the design of playback experiments. Anim. Behav. 38:715–717. [172]

Searcy, W. A. 1990. Species recognition of song by female Red-winged Blackbirds. Anim. Behav. 40:1119–1127. [339–341, 344–347, 459]

Searcy, W. A. 1992a. Measuring responses of female birds to male song. Pages 175–189 *in* Playback and studies of animal communication (P. K. McGregor, Ed.). Plenum Press, New York. [35, 298, 339, 458]

Searcy, W. A. 1992b. Song repertoire and mate choice in birds. Am. Zool. 32:71–80. [290, 339, 470–472]

Searcy, W. A., and M. Andersson. 1986. Sexual selection and the evolution of song. Annu. Rev. Ecol. Syst. 17:507–533. [160, 171, 298, 357, 366]

Searcy, W. A., and E. A. Brenowitz. 1988. Sexual differences in species recognition of avian song. Nature 332:152–154. [339, 341, 344, 346, 350]

Searcy, W. A., and P. Marler. 1981. A test for responsiveness to song structure and programming in female sparrows. Science 213:926–928. [458]

Searcy, W. A., and K. Yasukawa. 1990. Use of the song repertoire in intersexual and intrasexual contexts by male Red-winged Blackbirds. Behav. Ecol. Sociobiol. 27:123–128. [456, 460]

Searcy, W. A., and K. Yasukawa. 1995. Polygyny and sexual selection in Red-winged Blackbirds. Princeton Univ. Press, Princeton, N.J. [460]

Searcy, W. A., E. Balaban, R. A. Canady, S. J. Clark, S. Runfeldt, and H. Williams. 1981a. Responsiveness of male Swamp Sparrows to temporal organization of song. Auk 98:613–615. [344, 387]

Searcy, W. A., P. Marler, and S. S. Peters. 1981b. Species song discrimination in adult female Song and Swamp Sparrows. Anim. Behav. 29:997–1003. [344]

Searcy, W. A., P. D. McArthur, S. S. Peters, and P. Marler. 1981c. Response of male Song and Swamp Sparrows to neighbour, stranger and self songs. Behaviour 77:152–166. [359, 362, 365, 367]

Searcy, W. A., M. H. Searcy, and P. Marler. 1982. The response of Swamp Sparrows to acoustically distinct song types. Behaviour 80:70–83. [345, 471]

Searcy, W. A., S. Coffman, and D. F. Raikow. 1994. Habituation, recovery, and the similarity of song types within repertoires in Red-winged Blackbirds (*Agelaius phoeniceus*) (Aves, Emberizidae). Ethology 98:38–49. [298, 365]

Searcy, W. A., J. Podos, S. Peters, and S. Nowicki. 1995. Discrimination of song types and variants in Song Sparrows. Anim. Behav. 49:1212–1226. [131]

Seger, J. 1985. Unifying genetic models for the evolution of female choice. Evolution 39:1185–1193. [463]

Serpell, J. 1981. Duets, greetings, and triumph ceremonies: analogous displays in the parrot genus *Trichoglossus*. Z. Tierpsychol. 55:268–283. [110]

Serventy, D. L. 1967. Aspects of the population ecology of the Short-tailed Shearwater, *Puffinus tenuirostris*. Proc. XIV Int. Ornithol. Congr. 165–190. [172]

Serventy, D. L. 1971. Biology of desert birds. Pages 287–331 *in* Avian biology, vol. 1 (D. S. Farner and J. R. King, Eds.). Academic Press, New York. [98]

Seutin, G., J. Brawn, R. E. Ricklefs, and E. Bermingham. 1993. Genetic divergence among populations of a tropical passerine, the Streaked Saltator (*Saltator albicollis*). Auk 110:117–126. [271]

Seyfarth, R. M., D. L. Cheney, and P. Marler. 1980. Monkey responses to three different alarm calls: evidence of predator classification and semantic communication. Science 210:801–803. [279]

Shackell, N. L., R. E. Lemon, and D. Roff. 1988. Song similarity between neighboring American Redstarts (*Setophaga ruticilla*): a statistical analysis. Auk 105:609–615. [17, 215]

Shackleton, S. A., and L. Ratcliffe. 1993. Development of song in hand-reared Black-capped Chickadees. Wilson Bull. 105:637–644. [350]

Shackleton, S. A., and L. Ratcliffe. 1994. Matched counter-singing signals escalation of aggression in Black-capped Chickadees, *Parus atricapillus*. Ethology 97:310–316. [348]

Shackleton, S. A., L. Ratcliffe, and D. M. Weary. 1992. Relative frequency parameters and song recognition in Black-capped Chickadees. Condor 94:782–785. [350, 353]

Shalter, M. D. 1978. Localisation of passerine seet and mobbing calls by Goshawks and Pygmy Owls. Z. Tierpsychol. 46:260–267. [417]

Shannon, C. E., and W. Weaver. 1949. The mathematical theory of communication. Univ. Illinois Press, Urbana. [336]

Sharman, M., and R. I. M. Dunbar. 1982. Observer bias in selection of study group in baboon field studies. Primates 23:567–573. [419]

Sheldon, B. C., and T. Burke. 1994. Copulation behavior and paternity in the Chaffinch. Behav. Ecol. Sociobiol. 34:149–156. [432]

Shennan, M. G. C., J. R. Waas, and R. J. Lavery. 1994. The warning signals of parental convict cichlids are socially facilitated. Anim. Behav. 47:974–976. [412, 413]

Sherry, D. F., M. R. L. Forbes, M. Kurghel, and G. O. Ivy. 1993. Females have a larger hippocampus than males in the brood parasitic Brown-headed Cowbird. Proc. Natl. Acad. Sci. USA 90:7839–7843. [342]

Sherry, T. W., and R. T. Holmes. 1989. Age-specific social dominance affects habitat use by breeding American Redstarts (*Setophaga ruticilla*): a removal experiment. Behav. Ecol. Sociobiol. 25:327–333. [435]
Shields, W. M. 1982. Philopatry, inbreeding, and the evolution of sex. State Univ. New York Press, Albany. [215]
Shils, E. 1981. Tradition. Univ. Chicago Press, Chicago. [210]
Shiovitz, K. A. 1975. The process of species-specific song recognition in the Indigo Bunting (*Passerina cyanea*) and its relationship to the organization of avian acoustic behaviour. Behaviour 55:128–179. [200, 205]
Shiovitz, K. A., and R. E. Lemon. 1980. Species identification of song by Indigo Buntings as determined by responses to computer generated sounds. Behaviour 74:167–199. [341]
Shiovitz, K. A., and W. L. Thompson. 1970. Geographical variation in song composition of the Indigo Bunting, *Passerina cyanea.* Anim. Behav. 18:151–158. [200]
Shy, E., and E. S. Morton. 1986. The role of distance, familiarity, and time of day in Carolina Wren responses to conspecific songs. Behav. Ecol. Sociobiol. 19:393–400. [267, 359, 388, 433]
Shy, E., P. K. McGregor, and J. R. Krebs. 1986. Discrimination of song types by male Great Tits. Behav. Processes 13:1–12. [129]
Sibley, C. G., and J. E. Ahlquist. 1990. Phylogeny and classification of birds: a study in molecular evolution. Yale Univ. Press, New Haven, Conn. [214, 241, 280, 359]
Sibley, C. G., and B. L. Monroe Jr. 1990. Distribution and taxonomy of birds of the world. Yale Univ. Press, New Haven, Conn. [xx, 161, 244, 279]
Sibley, C. G., and B. L. Monroe Jr. 1993. A supplement to distribution and taxonomy of birds of the world. Yale Univ. Press, New Haven, Conn. [xx, 228]
Sibley, C. G., J. E. Ahlquist, and B. L. Monroe. 1988. A classification of the living birds of the world based on DNA-DNA hybridization studies. Auk 105:409–423. [11, 161, 296]
Sibley, C. H. 1955. Behavioral mimicry in the titmice (Paridae) and certain other birds. Wilson Bull. 67:128–132. [153]
Sick, H. 1939. Über die Dialektbildung beim "Regenruf" des Buchfinken. J. Ornithol. 87:568–592. [214]
Sick, H. 1993. Birds in Brazil: a natural history. Princeton Univ. Press, Princeton, N.J. [438, 440, 442, 444, 447]
Siegel, P. B., R. E. Phillips, and E. F. Folsom. 1965. Genetic variations in the crow of adult chickens. Behaviour 24:229–235. [46]
Siemann, M., and J. D. Delius. 1993. Implicit deductive responding in humans. Naturwissenschaften 80:364–366. [415]
Silva, W. R. 1991. Padróes ecológicos, bioacústicos, biogeográficos e filogenéticos do complexo *Basileuterus culicivorus* (Aves, Parulidae). Ph.D. dissertation, Univ. Campinas, S.P., Brazil. [271]
Silva, W. R., and J. Vielliard. 1988. O repertório vocal do Beija-flor-tesoura *Eupetomena macroura* (Trochilidae) e suas funções biológicas. Page 475 *in* Resumos XV Congr. Brasil. Zool., Curitiba. [278]
Simon, H. A. 1974. How big is a chunk? Science 183:482–488. [95]
Simpson, H., and D. Vicario. 1991. Early estrogen treatment of female Zebra Finches masculinizes the brain pathway for learned vocalization. J. Neurobiol. 22:777–793. [292]
Simpson, M. J. A. 1968. The display of the Siamese fighting fish, *Betta splendens.* Anim. Behav. Monogr. 1:1–73. [384]
Skeate, S. T. 1984. Courtship and reproductive behaviour of captive White-fronted Amazon Parrots, *Amazona albifrons.* Bird Behav. 5:103–109. [104, 110]
Skeel, M. A. 1978. Vocalizations of the Whimbrel on its breeding grounds. Condor 80:194–202. [442]
Skutch, A. F. 1952. On the hour of laying and hatching of birds' eggs. Ibis 94:49–61. [432]
Skutch, A. F. 1968. The nesting of some Venezuelan birds. Condor 70:66–82. [447]

Skutch, A. F. 1969. Life histories of Central American birds, III. Cooper Ornithological Society, Berkeley. [259]

Skutch, A. F. 1972. Studies of tropical American birds. Publ. Nuttall Ornithol. Club 10. Cambridge, Mass. [438, 442, 447]

Skutch, A. F. 1981. New studies of tropical American birds. Publ. Nuttall Ornithol. Club 19. Cambridge, Mass. [438, 440, 444, 447]

Skutch, A. F. 1983. Birds of tropical America. Univ. Texas Press, Austin. [441, 442, 447]

Slagsvold, T., S. Dale, and G.-P. Saetre. 1994. Dawn singing in the Great Tit (*Parus major*): mate attracting, mate guarding, or territorial defence? Behaviour 131:115–138. [433]

Slater, P. J. B. 1978. A simple model for competition between behaviour patterns. Behaviour 67:236–257. [83]

Slater, P. J. B. 1981. Chaffinch song repertoires: observations, experiments and a discussion of their significance. Z. Tierpsychol. 56:1–24. [411]

Slater, P. J. B. 1983a. Bird song learning: theme and variations. Pages 475–499 *in* Perspectives in ornithology (A. H. Brush and G. A. Clark Jr., Eds.). Cambridge Univ. Press, Cambridge. [3]

Slater, P. J. B. 1983b. Sequences of songs in Chaffinches. Anim. Behav. 31:272–281. [80]

Slater, P. J. B. 1986. The cultural evolution of bird song. Trends Ecol. Evol. 1:94–97. [181, 199]

Slater, P. J. B. 1989. Bird song learning: causes and consequences. Ethol. Ecol. Evol. 1:19–46. [3, 5, 6, 10, 16, 73, 85, 199, 215, 290, 305]

Slater, P. J. B. 1991. Learned song variations in British Storm-Petrels? Wilson Bull. 103:515–517. [168]

Slater, P. J. B., and S. A. Ince. 1979. Cultural evolution in Chaffinch song. Behaviour 71:146–166. [182, 183, 198, 213, 217, 219]

Slater, P. J. B., S. A. Ince, and P. W. Colgan. 1980. Chaffinch song types: their frequencies in the population and distribution between repertoires of different individuals. Behaviour 75:207–218. [182, 185, 191, 193, 213, 218, 219]

Slater, P. J. B., F. A. Clements, and D. J. Goodfellow. 1984. Local and regional variations in Chaffinch song and the question of dialects. Behaviour 88:76–97. [217]

Slater, P. J. B., L. A. Eales, and N. S. Clayton. 1988. Song learning in Zebra Finches: progress and prospects. Adv. Study Behav. 18:1–34. [5, 20, 43]

Slater, P. J. B., C. Richards, and N. I. Mann. 1991. Song learning in Zebra Finches exposed to a series of tutors during the sensitive phase. Ethology 88:163–171. [63]

Slatkin, M. 1985. Rare alleles as indicators of gene flow. Evolution 39:53–65. [190, 193]

Smith, D. G., and D. O. Norman. 1979. "Leader-follower" singing in Red-winged Blackbirds. Condor 81:83–84. [83, 399]

Smith, D. G., and F. A. Reid. 1979. Roles of the song repertoire in Red-winged Blackbirds. Behav. Ecol. Sociobiol. 5:279–290. [455, 460]

Smith, E. E., and D. L. Medin. 1981. Categories and concepts. Harvard Univ. Press, Cambridge, Mass. [122, 126]

Smith, G. T. 1991. Breeding ecology of the Western Long-billed Corella, *Cacatua pastinator pastinator.* Wildl. Res. 18:91–110. [98, 99, 104]

Smith, G. T., E. A. Brenowitz, M. D. Beecher, S. E. Campbell, and J. C. Wingfield. 1995a. Hormonal and behavioral correlates of seasonal plasticity in the song nuclei of a wild songbird. Soc. Neurosci. Abstr. 21:962. [302]

Smith, G. T., E. A. Brenowitz, J. C. Wingfield, and L. F. Baptista. 1995b. Seasonal changes in song nuclei and song behavior in Gambel's White-crowned Sparrows. J. Neurobiol. 28:114–125. [302]

Smith, L. B., and D. Heise. 1992. Perceptual similarity and conceptual structure. Pages 233–272 *in* Percepts, concepts, and categories: the representation and processing of information. Advances in Psychology 93 (B. Burns, Ed.). North-Holland, Amsterdam. [127, 129]

Smith, N. G. 1969. Polymorphism in Ringed Plovers. Ibis 111:177–188. [243]

Smith, R. L. 1959. The songs of the Grasshopper Sparrow. Wilson Bull. 71:141–152. [18]

Smith, S. M. 1976. Ecological aspects of dominance hierarchies in Black-capped Chickadees. Auk 93:95–107. [435]
Smith, S. M. 1983. The ontogeny of avian behavior. Pages 85–160 *in* Avian biology, vol. 7 (D. S. Farner, J. R. King, and K. C. Parkes, Eds.). Academic Press, New York. [39]
Smith, S. M. 1988. Extra-pair copulations in Black-capped Chickadees: the role of the female. Behaviour 107:15–23. [264, 340, 435]
Smith, S. M. 1991. The Black-capped Chickadee. Cornell Univ. Press, Ithaca, N.Y. [140, 156, 348]
Smith, S. T. 1972. Communication and other social behavior in *Parus carolinensis*. Publ. Nuttall Ornithol. Club 11. Cambridge, Mass. [147, 148, 150]
Smith, W. J. 1966. Communication and relationships in the genus *Tyrannus*. Publ. Nuttall Ornithol. Club 6. Cambridge, Mass. [444, 447]
Smith, W. J. 1969. Displays of *Sayornis phoebe* (Aves, Tyrannidae). Behaviour 33:283–322. [385]
Smith, W. J. 1971. Behavioral characteristics of serpophaginine tyrannids. Condor 73:259–286. [167]
Smith, W. J. 1977. The behavior of communicating: an ethological approach. Harvard Univ. Press, Cambridge, Mass. [25, 28, 121, 140, 144, 160, 167, 175, 377–379, 385–387, 396, 444]
Smith, W. J. 1986a. An "informational" perspective on manipulation. Pages 71–86 *in* Deception (R. W. Mitchell and N. S. Thompson, Eds.). State Univ. New York Press, Albany. [378, 379, 384, 396]
Smith, W. J. 1986b. Signaling behavior: contributions of different repertoires. Pages 315–330 *in* Dolphin cognition and behavior: a comparative approach (R. J. Schusterman, J. A. Thomas, and F. G. Wood, Eds.). Erlbaum, Hillsdale, N.J. [386, 389]
Smith, W. J. 1988. Patterned daytime singing of the Eastern Wood Pewee, *Contopus virens*. Anim. Behav. 36:1111–1123. [380]
Smith, W. J. 1990. Communication and expectations: a social process and the cognitive operations it depends upon and influences. Pages 234–253 *in* Interpretation and explanation in the study of animal behavior, vol. 1 (M. Bekoff and D. Jamieson, Eds.). Westview Press, Boulder, Colo. [386, 389]
Smith, W. J. 1991. Singing is based on two markedly different kinds of signaling. J. Theor. Biol. 152:241–253. [377, 378, 385, 386, 399, 411]
Smith, W. J., and A. M. Smith. 1992. Behavioral information provided by two song forms of the Eastern Kingbird, *T. tyrannus*. Behaviour 120:90–102. [380]
Smith, W. J., and A. M. Smith. In press a. Vocal signaling of the Great Crested Flycatcher, *Myiarchus crinitus*. Ethology [373, 382]
Smith, W. J., and A. M. Smith. In press b. Vocal approaches elicited in playback interactions with Great Crested Flycatchers, *Myiarchus crinitus*. Ethology [373, 382]
Smith, W. J., and A. M. Smith. In press c. Information about behavior provided by Louisiana Waterthrush songs (Parulinae, *Seiurus motacilla*). Anim. Behav. [383]
Smith, W. J., and A. M. Smith. Unpubl. ms. Information about behavior is provided by songs of the Striped Cuckoo (*Tapera naevia*). [383]
Smith, W. J., J. Pawlukiewicz, and S. T. Smith. 1978. Kinds of activities correlated with singing patterns of the Yellow-throated Vireo. Anim. Behav. 26:862–884. [389]
Smith, W. P. 1942. Nesting habits of the Eastern Phoebe. Auk 59:410–417. [432]
Snow, B. K. 1974. Vocal mimicry in the Violaceus Euphonia, *Euphonia violacea*. Wilson Bull. 86:179–180. [276]
Snow, D. W. 1958. A study of blackbirds. Allen and Unwin, London. [413]
Snow, D. W. 1962. A field study of the Black-and-white Manakin, *Manacus manacus,* in Trinidad. Zoologica 47:65–104. [440]
Snow, D. W. 1968. The singing assemblies of Little Hermits. Living Bird 7:47–55. [97, 277]
Snow, D. W. 1979. Tityrinae, Pipridae, Cotingidae. Pages 229–308 *in* Check-list of birds of the

world, vol. 8 (M. A. Traylor Jr., Ed.). Mus. Comp. Zool., Harvard Univ., Cambridge, Mass. [271]

Snow, D. W. 1982. The cotingas: bellbirds, umbrellabirds, and other species. Cornell Univ. Press, Ithaca, N.Y. [277]

Snowdon, C. T. 1987. A naturalistic view of categorical perception. Pages 332–354 *in* Categorical perception (S. Harnad, Ed.). Cambridge Univ. Press, Cambridge. [123]

Snyder, N. F. R., J. W. Wiley, and C. B. Kepler. 1987. The parrots of Luquillo: natural history and conservation of the Puerto Rican Parrot. Western Foundation of Vertebrate Zoology, Los Angeles. [104, 110]

Sohrabji, F., E. Nordeen, and K. Nordeen. 1989. Projections of androgen-accumulating neurons in a nucleus controlling avian song. Brain Res. 488:253–259. [288]

Sokal, R. R., N. L. Oden, and C. Wilson. 1991. Genetic evidence for the spread of agriculture in Europe by demic diffusion. Nature 351:143–145. [215]

Sossinka, R. 1982. Domestication in birds. Pages 373–403 *in* Avian biology, vol. 6 (D. S. Farner, J. R. King, and K. C. Parkes, Eds.). Academic Press, New York. [46, 54]

Spector, D. A. 1991. The singing behaviour of Yellow Warblers. Behaviour 117:29–52. [17, 430, 445, 446]

Spector, D. A. 1992. Wood warbler song systems. A review of paruline singing behaviors. Pages 199–238 *in* Current ornithology, vol. 9 (D. M. Power, Ed.). Plenum Press, New York. [8, 15, 17, 18, 158, 264, 340, 426, 445]

Spector, D. A. 1994. Definition in biology: The case of "bird song." J. Theor. Biol. 168:373–381. [377, 386]

Spector, D. A., L. K. McKim, and D. E. Kroodsma. 1989. Yellow Warblers are able to learn songs and situations in which to use them. Anim. Behav. 38:723–725. [94]

Speisberger, J. L., and K. M. Fristrup. 1990. Passive localization of calling animals and sensing of their acoustic environment using acoustic tomography. Am. Nat. 135:107–153. [419, 420, 422]

Spieth, P. T. 1974. Gene flow and genetic differentiation. Genetics 78:961–965. [185]

Staicer, C. A. 1989. Characteristics, use, and significance of two singing behaviors in Grace's Warbler *(Dendroica graciae).* Auk 106:49–63. [8, 17, 94, 430, 434, 438, 445, 446]

Staicer, C. A. 1991. The role of male song in the socioecology of the tropical resident Adelaide's Warbler (*Dendroica adelaidae*). Ph.D. dissertation, Univ. Massachusetts, Amherst. [426, 432–434, 443, 446, 448]

Staicer, C. A. 1996. Honest advertisement of pairing status: evidence from a tropical resident wood-warbler. Anim. Behav. 51:375–390. [430, 431, 438]

Staicer, C. A. In press. Acoustical structure of two song categories of the Adelaide's Warbler. Auk. [445, 446]

Stamps, J., A. Clark, P. Arrowood, and B. Kus. 1985. Parent-offspring conflict in Budgerigars. Behaviour 94:1–40. [98, 99]

Stamps, J., A. Clark, B. Kus, and P. Arrowood. 1987. The effects of parent and offspring gender on food allocation in Budgerigars. Behaviour 101:177–199. [98, 99]

Stamps, J., A. Clark, P. Arrowood, and B. Kus. 1989. Begging behavior in Budgerigars. Ethology 81:177–192. [98, 99]

Stamps, J., B. Kus, A. Clark, and P. Arrowood. 1990. Social relationships of fledgling Budgerigars, *Melopsittacus undulatus.* Anim. Behav. 40:688–700. [98, 99]

Stap, D. 1990. A parrot without a name. The search for the last unknown birds on earth. Knopf, New York. [v]

Stein, B. E., and M. A. Meredith. 1993. The merging of senses. MIT Press, Cambridge, Mass. [26]

Stein, R. C. 1963. Isolating mechanisms between populations of Traill's Flycatchers. Proc. Am. Philos. Soc. 107:21–50. [271]

Stiles, F. G. 1982. Aggressive and courtship displays of the male Anna's Hummingbird. Condor 84:208–225. [54, 278]

Stiles, F. G. 1983. The taxonomy of *Microcerculus* wrens (Troglodytidae) in Central America. Wilson Bull. 95:169–183. [271, 279]

Stiles, F. G. 1984. The songs of *Microcerculus* wrens in Costa Rica. Wilson Bull. 96:99–103. [271, 272, 279]

Stiles, F. G. 1992. A new species of antpitta (Formicariidae: *Grallaria*) from the eastern Andes of Colombia. Wilson Bull. 104:389–399. [270]

Stiles, F. G. 1996. A new species of Emerald Hummingbird (Trochilidae, *Chlorastilbon*) from the Sierra de Chiribiquete, southeastern Colombia, with a review of the *C. mellisugus* complex. Wilson Bull. 108:1–27. [270]

Stiles, F. G., and A. F. Skutch. 1989. A guide to the birds of Costa Rica. Cornell Univ. Press, Ithaca, N.Y. [277, 279]

Stiles, F. G., and L. L. Wolf. 1979. Ecology and evolution of lek mating behavior in the Long-tailed Hermit hummingbird. Ornithological Monograph 27. American Ornithologists' Union, Washington, D.C. [277]

Stoddard, P. K. 1989. Song repertoire use and perception by male Song Sparrows (*Melospiza melodia*) in the Puget Sound region. Ph.D. dissertation, Univ. Washington. [369]

Stoddard, P. K., M. D. Beecher, and M. Willis. 1988. Response of territorial male Song Sparrows to song types and variations. Behav. Ecol. Sociobiol. 22:125–130. [65, 66, 131, 363]

Stoddard, P. K., M. D. Beecher, C. L. Horning, and M. Willis. 1990. Strong neighbor-stranger discrimination in Song Sparrows. Condor 92:1051–1056. [66, 77, 359, 362]

Stoddard, P. K., M. D. Beecher, C. L. Horning, and S. E. Campbell. 1991. Recognition of individual neighbors by song in the Song Sparrow, a bird with song repertoires. Behav. Ecol. Sociobiol. 29:211–215. [66, 77, 359, 360, 362–364, 371]

Stoddard, P. K., M. D. Beecher, S. E. Campbell, and C. L. Horning. 1992a. Song type matching in the Song Sparrow. Can. J. Zool. 70:1440–1444. [66, 130, 372, 411]

Stoddard, P. K., M. D. Beecher, P. Loesche, and S. E. Campbell. 1992b. Memory does not constrain individual recognition in a bird with song repertoires. Behaviour 122:274–287. [66, 129, 130, 359, 367, 369, 371, 372, 374]

Storey, A. E. 1984. Function of Manx Shearwater calls in mate attraction. Behaviour 89:73–89. [165, 172, 174, 176]

Strahl, S. D. 1992. Furthering avian conservation in Latin America. Auk 109:680–682. [274]

Strain, J. G., and R. L. Mumme. 1988. Effects of food supplementation, song playback, and temperature on vocal territorial behavior of Carolina Wrens. Auk 105:11–16. [267, 436]

Strauch, J. G., Jr. 1978. The phylogeny of the Charadriiformes (Aves): a new estimate using the method of character compatibility analysis. Trans. Zool. Soc. Lond. 34:263–345. [241]

Stresemann, E. 1919. Über die europäischen Baumläufer. Verh. Ornithol. Ges. Bayern 14:39–74. [234]

Stresemann, E. 1947. Baron von Pernau, pioneer student of bird behavior. Auk 64:35–52. [41]

Striedter, G. F. 1994. The vocal control pathways in Budgerigars differ from those in songbirds. J. Comp. Neurol. 343:35–56. [113, 294–296]

Stutchbury, B. J., and E. S. Morton. 1995. The effect of breeding synchrony on the evolution of extra-pair mating strategies in birds. Behaviour 132:675–690. [263, 265, 267]

Stutchbury, B. J., J. M. Rhymer, and E. S. Morton. 1994. Extra-pair paternity in the Hooded Warbler. Behav. Ecol. 5:384–392. [264]

Sullivan, K. 1985. Selective alarm calling by Downy Woodpeckers in mixed species flocks. Auk 102:184–185. [416]

Sullivan, M. S. 1994. Mate choice as an information gathering process under time constraint: implications for behaviour and signal design. Anim. Behav. 47: 141–151. [340]

Sullivan, W. E., and M. Konishi. 1984. Segregation of stimulus phase and intensity coding in the cochlear nucleus of the Barn Owl. J. Neurosci. 4:1787–1799. [335]

Suthers, R. A. 1990. Contributions to birdsong from the left and right sides of the intact syrinx. Nature 347:473–477. [286]

Suthers, R. A. 1994. Variable asymmetry and resonance in the avian vocal tract: a structural basis for individually distinct vocalizations. J. Comp. Physiol. A 175:457–466. [370, 373]

Suthers, R. A., and D. H. Hector. 1985. The physiology of vocalization by the echolocating Oilbird, *Steatornis caripensis.* J. Comp. Physiol. A 156:243–266. [316, 319]

Swadesh, M. 1951. Diffusional cumulation and archaic residue as historical explanations. Southeast. J. Anthropol. 7:1–21. [213]

Swadesh, M. 1952. Lexicostatistic dating of prehistoric ethnic contacts. Proc. Am. Philos. Soc. 96:452–463. [214]

Swatschek, I., D. Ristow, and M. Wink. 1994. Mate fidelity and parentage in Cory's Shearwater, *Calonectris diomedea*—field study and DNA fingerprinting. Mol. Ecol. 3:259–262. [174]

Swets, J. A. 1964. Signal detection and recognition by human observers. Wiley, New York. [326]

Systematics Agenda 2000. 1994. Systematics Agenda 2000: charting the biosphere. Tech. Rep. [475]

Takahashi, T. T., and C. H. Keller. 1992. Simulated motion enhances neuronal selectivity for a sound localization cue in background noise. J. Neurosci. 12:4381–4390. [335]

Taoka, M., and H. Okumura. 1989. Individuality of chatter-calls and selective response to the bird's own call in Leach's Storm-Petrel, *Oceanodroma leucorhoa.* Jpn. Women's Univ. J. 36:107–112. [175]

Taoka, M., T. Sato, T. Kamada, and H. Okumura. 1988. Situation-specificities of vocalizations in Leach's Storm-Petrel, *Oceanodroma leucorhoa.* J. Yamashina Inst. Ornithol. 20:82–90. [163, 165, 167, 172, 174]

Taoka, M., T. Sato, T. Kamada, and H. Okumura. 1989a. Sexual dimorphism of chatter-calls and vocal sex recognition in Leach's Storm-Petrels (*Oceanodroma leucorhoa*). Auk 106:498–500. [165, 172]

Taoka, M., T. Sato, T. Kamada, and H. Okumura. 1989b. Heterosexual response to playback calls of the Leach's Storm-Petrel, *Oceanodroma leucorhoa.* J. Yamashina Inst. Ornithol. 21:84–89. [163, 172, 174]

Taoka, M., W. Pyong-Oh, and H. Okumura. 1989c. Vocal behavior of Swinhoe's Storm-Petrel (*Oceanodroma monorhis*). Auk 106:471–474. [165, 167]

Taylor, R. C. 1978. Geographical variation in wading birds with reference to the Ringed Plover (*Charadrius hiaticula* L.) and related species. Ph.D. dissertation, Liverpool Polytechnic, Liverpool, England. [252]

Tembrock, G. 1984. Verhalten bei Tieren. Neue Brehm-Bücherei 455:139–140. A. Ziemsen Verlag, Wittenberg-Lutherstadt. [221]

Temeles, E. J. 1994. The role of neighbours in territorial systems: when are they "dear enemies"? Anim. Behav 47:339–350. [362, 366, 423]

Tenaza, R. 1976. Songs, choruses and countersinging of Kloss' gibbons (*Hylobates klossi*) in Siberut Island, Indonesia. Z. Tierpsychol. 40:37–52. [426, 436]

ten Cate, C. 1989. Behavioral processes: toward understanding processes. Pages 243–269 *in* Perspectives in ethology, vol. 8: Whither ethology? (P. P. G. Bateson and P. H. Klopfer, Eds.). Plenum Press, New York. [27]

ten Thoren, A., and H.-H. Bergmann. 1987. Die Entwicklung der Lautäusserungen bei der Graugans (*Anser anser*). J. Ornithol. 128:181–207. [54]

Terborgh, J. 1971. Distribution on environmental gradients: theory and a preliminary interpretation of distributional patterns in the avifauna of the Cordillera Vilcabamba, Peru. Ecology 52:23–40. [275]

Terborgh, J., and J. S. Weske. 1972. Rediscovery of the Imperial Snipe in Peru. Auk 89:497–505. [252]

Terborgh, J., S. K. Robinson, T. A. Parker III, C. A. Munn, and N. Pierpont. 1990. Structure and organization of an Amazonian forest bird community. Ecol. Monogr. 60:213–238. [275]

Thelen, E. 1984. Learning to walk: ecological demands and phylogenetic constraints. Pages 113–150 *in* Advances in infancy research (L. P. Lipsitt, Ed.). Ablex, Norwood, N.J. [23]

Thelen, E., and L. B. Smith. 1994. A dynamic systems approach to the development of cognition and action. MIT Press, Cambridge, Mass. [23]

Thelen, E., and B. D. Ulrich. 1991. Hidden skills. Society for Research in Child Development Monogr. 56:1–97. [20, 23, 31, 33]

Thielcke, G. 1961a. Ergebnisse der Vogelstimmen-Analyse. J. Ornithol. 102:285–300. [39, 45]

Thielcke, G. 1961b. Stammesgeschichte und geographische Variation des Gesanges unserer Baumläufer (*Certhia familiaris* L. und *Certhia brachydactyla* Brehm). Z. Tierpsychol. 18:188–204. [216, 234]

Thielcke, G. 1962. Die geographische Variation eines erlernten Elements im Gesang des Buchfinken (*Fringilla coelebs*) und des Waldbaumläufers (*Certhia familiaris*). Vogelwarte 21:199–202. [215, 216]

Thielcke, G. 1965a. Die Ontogenese der Bettellaute von Garten- und Waldbaumläufer (*Certhia brachydactyla* Brehm und *C. familiaris* L.). Zool. Anz. 174:237–241. [45]

Thielcke, G. 1965b. Gesangsgeographische Variation des Gartenbaumläufers (*Certhia brachydactyla*) im Hinblick auf das Artbildungsproblem. Z. Tierpsychol. 22:542–566. [215, 216]

Thielcke, G. 1968. Gemeinsames der Gattung *Parus.* Ein bioakustischer Beitrag zur Systematik. Vogelwelt (Suppl.) 1:147–164. [137, 153, 229]

Thielcke, G. 1969. Die Reaktion von Tannen- und Kohlmeise (*Parus ater, P. major*) auf den Gesang nahverwandter Formen. J. Ornithol. 110:148–157. [223]

Thielcke, G. 1970a. Lernen von Gesang als möglicher Schrittmacher der Evolution. Z. Zool. Syst. Evolutionsforsch. 8:309–320. [39, 63, 233, 236]

Thielcke, G. 1970b. Dei sozialen Funktionen der Vogelstimmen. Vogelwarte 25:204–229. [45]

Thielcke, G. 1971. Versuche zur Kommunikation und Evolution der Angst-, Alarm- und Rivalenlaute des Waldbaumläufers (*Certhia familiaris*). Z. Tierpsychol. 28:505–516. [45, 215]

Thielcke, G. 1973a. On the origin of divergence of learned signals (songs) in isolated populations. Ibis 115:511–516. [45, 222, 234, 236]

Thielcke, G. 1973b. Uniformierung des Gesangs der Tannenmeise (*Parus ater*) durch Lernen. J. Ornithol. 114:443–454. [223]

Thielcke, G. 1974. Stabilität erlernter Singvogel-Gesänge trotz vollständiger geographischer Isolation. Vogelwarte 27:209–215. [215]

Thielcke, G. 1976. Bird sounds. Univ. Michigan Press, Ann Arbor. [250]

Thielcke, G. 1983. Entstanden Dialekte des Zilpzalps (*Phylloscopus collybita*) durch Lernentzug? J. Ornithol. 124:333–368. [215, 236]

Thielcke, G. 1986. Constant proportions of mixed singers in tree-creeper populations (*Certhia familiaris*). Ethology 72:154–164. [215, 216]

Thielcke, G. 1987. Langjährige Dialektkonstanz beim Gartenbaumläufer (*Certhia brachydactyla*). J. Ornithol. 128:171–180. [215, 216]

Thielcke, G. 1988a. Neue Befunde bestätigen Baron Pernaus (1660–1731) Angaben über Lautaüßerungen des Buchfinken (*Fringilla coelebs*). J. Ornithol. 129:55–70. [41, 216]

Thielcke, G. 1988b. Buchfinken (*Fringilla coelebs*) eliminieren erlernte Gesange von Baumpiepern (*Anthus trivialis*). Vogelwarte 34:319–336. [216]

Thielcke, G. 1992. Stabilität und Änderungen von Dialekten und Dialektgrenzen beim Gartenbaumläufer (*Certhia brachydactyla*). J. Ornithol. 133:43–60. [216]

Thielcke, G., and M. Krome. 1989. Experimente über sensible Phasen und Gesangsvariabilität beim Buchfinken (*Fringilla coelebs*). J. Ornithol. 130:435–455. [215]

Thielcke, G., and K. E. Linsenmair. 1963. Zur geographischen Variation des Gesanges des Zilpzalps, *Phylloscopus collybita,* in Mittel- und Südwesteuropa mit einem Vergleich des Gesanges des Fitis, *Phylloscopus trochilus.* J. Ornithol. 104:372–402. [215, 228]

Thielcke, G., K. Wüstenberg, and P. H. Becker. 1978. Reaktionen von Zilpzalp und Fitis (*Phylloscopus collybita, Ph. trochilus*) auf verschiedene Gesangsformen des Zilpzalps. J. Ornithol. 119:213–226. [228]

Thimm, F. 1973. Sequentielle und zeitliche Beziehungen im Reviergesang des Gartenrotschwanzes (*Phoenicurus phoenicurus* L.). J. Comp. Physiol. 84:311–334. [81]

Thimm, F. 1980. The function of feedback-mechanism in bird song. Pages 677–681 *in* Acta

XVII Congr. Int. Ornithol. (R. Nöhring, Ed.). Verlag Deutsche Ornithol.-Gesellschaft, Berlin. [83]

Thimm, F. A., A. Clausen, D. Todt, and J. Wolffgramm. 1974. Zeitabhängigkeit und Verhaltensmusterfolgen. J. Comp. Physiol. 93:55–84. [83]

Thompson, W. L. 1970. Song variation in a population of Indigo Buntings. Auk 87:58–71. [182, 191, 199, 200, 202, 210]

Thompson, W. L. 1972. Singing behaviour of the Indigo Bunting, *Passerina cyanea.* Z. Tierpsychol. 31:39–59. [80, 202]

Thompson, W. L. 1976. Vocalizations of the Lazuli Bunting. Condor 78:195–207. [191, 193]

Thönen, W. 1962. Stimmgeographische und verbreitungsgeschichtliche Studien über die Mönchsmeise (*Parus montanus* Conrad). Ornithol. Beob. 59:101–172. [149, 223, 234, 235]

Thönen, W. 1968. Auffallender Unterschied zwischen den instrumentalen Balzlauten der europaischen und nordamerikaneschen Bekassine, *Gallinago gallinago.* Ornithol. Beob. 65:6–13. [250]

Thorpe, W. H. 1958. The learning of song patterns by birds, with especial reference to the song of the Chaffinch, *Fringilla coelebs.* Ibis 100:535–570. [39, 41, 50, 448]

Thorpe, W. H., and J. Hall-Craggs. 1976. Sound production and perception in birds as related to general principles of pattern perception. Pages 171–190 *in* Growing points in ethology (P. P. G. Bateson and R. A. Hinde, Eds.). Cambridge Univ. Press, Cambridge. [125]

Tikhonov, A. V., and S. Yu. Fokin. 1981. Acoustic signalling and behaviour during the nesting period in shorebirds. Byull. Mosk. Ova. Ispyt. Prir. Otd. Biol. 86:31–42. (In Russian). [244]

Timberlake, W. 1993. Behavior systems and reinforcement: an integrative approach. J. Exp. Anal. Behav. 670:105–128. [20, 24, 30]

Tinbergen, N. 1952. "Derived" activities, their causation, biological significance, origin and emancipation during evolution. Q. Rev. Biol. 27:1–32. [160]

Tinbergen, N. 1953. The herring gull's world. Collins, London. [167]

Tinbergen, N. 1959. Comparative studies of behavior of gulls (Laridae): a progress report. Behaviour 15:1–70. [160, 163, 167]

Todd, P. M., and G. A. Miller. 1993. Parental guidance suggested: how parental imprinting evolves through sexual selection as an adaptive learning mechanism. Adapt. Behav. 2:5–47. [22]

Todd, W. E. C. 1953. A taxonomic study of the American Dunlin (*Erolia alpina* subspp.). J. Wash. Acad. Sci. 43:85–88. [244]

Todt, D. 1968. Zur Steuerung unregelmäßiger Verhaltensabläufe. Pages 465–485 *in* Kybernetik (H. Mittelstaedt, Ed.). Oldenbourg, Munich. [80, 83]

Todt, D. 1969. Interactions between periodicitiy and positive feedback controlling the vocalizations of alternative motor patterns in blackbird song. Pages 168–175 *in* Biokybernetik, vol. 3 (H. Drischel and N. Tiedt, Eds.). Fischer Verlag, Jena. [81, 82]

Todt, D. 1970a. Gesangliche Reaktionen der Amsel auf ihren experimentell reproduzierten Eigengesang. Z. Vergl. Physiol. 66:294–317. [80, 83]

Todt, D. 1970b. Gesang und gesangliche Korrespondenz der Amsel. Naturwissenschaften 57:61–66. [82]

Todt, D. 1971. Äquivlente und konvalente gesangliche Reaktionen einer extrem regelmässig singenden Nachtigall (*Luscinia megarhynchos* B.). Z. Vergl. Physiol. 71:262–285. [81–83]

Todt, D. 1974. Social learning of vocal patterns and modes of their application in Grey Parrots (*Psittacus erithacus*). Z. Tierpsychol. 39:178–188. [94]

Todt, D. 1975a. Spontaneous recombinations of vocal patterns in parrots. Naturwissenschaften 62:399–400. [97]

Todt, D. 1975b. Social learning of vocal patterns and modes of their application in Grey Parrots (*Psittacus erithacus*). Z. Tierpsychol. 39:178–188. [97, 114]

Todt, D. 1975c. Short term inhibition of vocal outputs occurring in the singing behaviour of blackbirds (*Turdus merula*). J. Comp. Physiol. 98:289–306. [81–83]

Todt, D. 1977. Zur infradianen Rhythmik im Verhalten von Vertebraten—Ergebnisse aus Analysen des variablen Verhaltens von Singvögeln. Nova Acta Leopold. 225:607–619. [81, 82]

Todt, D. 1981. On functions of vocal matching: effect of counter-replies on song-post choice and singing. Z. Tierpsychol. 57:73–93. [82, 84]

Todt, D., and J. Böhner. 1994. Former experience can modify social selectivity during song learning in the nightingale (*Luscinia megarhynchos*). Ethology 97:169–176. [86]

Todt, D., and H. Hultsch. 1994. Biologische Grundlagen des Dialogs. Pages 53–76 *in* Kommunikation und Humanontogenese (K. F. Wessel and F. Naumann, Eds.). Kleine Verlag, Bielefeld. [83, 94]

Todt, D., and J. Wolffgramm. 1975. Überprüfung von Steuerungssystemen zur Strophenwahl der Amsel durch digitale Stimulierung. Biol. Cybern. 17:109–127. [83]

Todt, D., H. Hultsch, and D. Heike. 1979. Conditions affecting song acquisition in nightingales (*Luscinia megarhynchos*). Z. Tierpsychol. 51:23–35. [85, 86, 369]

Todt, D., H. Hultsch, and F. P. Duvall. 1981. Behavioural significance and social function of vocal and non-vocal displays in the monogamous duet singer *Cossypha heuglini*. Zool. Beitr. 27:421–448. [80, 291]

Tomback, D. F., and M. C. Baker. 1984. Assortative mating by White-crowned Sparrows at song dialect boundaries. Anim. Behav. 32:465–469. [43]

Tomkins, R. J., and J. R. Milne. 1991. Differences among Dark-rumped Petrel (*Pterodroma phaeopygia*) populations within the Galápagos archipelago. Notornis 38:1–35. [165, 167, 176]

Tosi, O. 1979. Voice identification. Theory and legal applications. Univ. Park Press, Baltimore. [314]

Trail, P. W., and P. Donahue. 1991. Notes on the behavior and ecology of the Red-Cotingas (Cotingidae: *Phoenicircus*). Wilson Bull. 103:539–551. [277]

Trainer, J. M. 1983. Changes in song dialect distributions and microgeographic variation in song of White-crowned Sparrows (*Zonotrichia leucophrys nuttalli*). Auk 100:568–582. [216]

Trainer, J. M. 1989. Cultural evolution in song dialects of Yellow-rumped Caciques in Panama. Ethology 80:190–204. [216]

Trainer, J. M., and D. B. McDonald. 1993. Vocal repertoire of the Long-tailed Manakin and its relation to male-male cooperation. Condor 95:769–781. [278]

Treisman, M. 1978. Bird song dialects, repertoire size, and kin association. Anim. Behav. 26:814–817. [107]

Tretzel, E. 1965. Imitation und Variation von Schäferpfiffen durch Hauberlerchen (*Galerida c. cristata* (L.)). Ein Beispiel für spezielle Spottmotiv-Prädisposition. Z. Tierpsychol. 22:784–809. [215]

Tretzel, E. 1967. Imitation und Transposition menschlicher Pfiffe durch Amseln (*Turdus m. merula* L.). Ein weiterer Nachweis relativen Lernens und akustischer Abstraktion bei Vögeln. Z. Tierpsychol. 24:137–161. [215]

Trillmich, F. 1976a. Spatial proximity and mate-specific behaviour in a flock of Budgerigars (*Melopsittacus undulatus;* Aves, Psittacidae). Z. Tierpsychol. 41:307–331. [110, 116]

Trillmich, F. 1976b. The influence of separation on the pair bond in Budgerigars (*Melopsittacus undulatus;* Aves, Psittacidae). Z. Tierpsychol. 41:396–408. [104]

Trivers, R. L. 1971. The evolution of reciprocal altruism. Q. Rev. Biol. 46:35–57. [356]

Trivers, R. L. 1972. Parental investment and sexual selection. Pages 136–171 *in* Sexual selection and the descent of man (B. G. Campbell, Ed.). Aldine, Chicago. [341]

Tubaro, P. L. 1991. Can *Troglodytes aedon* in Argentina "mimic" the songs of *Thyromanes bewickii*? Condor 93:443–445. [187]

Tubaro, P. L., E. T. Segura, and P. Handford. 1993. Geographic variation in the song of the Rufous-collared Sparrow in eastern Argentina. Condor 95:588–595. [279]

Tuck, L. M. 1972. The snipes: a study of the genus *Capella*. Can. Wildl. Serv. Monogr. 5. [242, 243]

Tuttle, M. D., and M. J. Ryan. 1982. The role of synchronised calling, ambient light and ambient noise in anti-bat predator behavior of a treefrog. Behav. Ecol. Sociobiol. 11:125–131. [399, 416, 423]

Tuttle, M. D., L. K. Taft, and M. J. Ryan. 1982. Evasive behaviour of a frog in response to bat predation. Anim. Behav. 30:393–397. [416, 436]

Vallet, E. M., M. L. Kreutzer, and J.-P. Richard. 1992. Syllable phonology and song segmentation: testing their salience in female canaries. Behaviour 121:155–167. [459]

van Tets, G. F. 1965. A comparative study of some social communication patterns in the Pelecaniformes. Ornithological Monographs 2. American Ornithologists' Union, Washington, D.C. [163]

Varela, A. 1994. Art of audio archiving in the '90s. Pro Sound News 16:1, 40. [483]

Veprintsev, B. N. 1979. Wildlife sound recording in the Soviet Union. Recorded Sound 74–75 (April-July):45–50. [474]

Veprintsev, B. N. 1982. Birds of the Soviet Union: a sound guide. Three long-playing disks. Melodiya, All-Union Studio for Recorded Sound, Moscow. (Narration in Russian). [243, 246, 252]

Verheyden, C., and P. Jouventin. 1994. Olfactory behavior of foraging procellariiforms. Auk 111:285–291. [162]

Verner, J. 1965. Time budget of the male Long-billed Marsh Wren during the breeding season. Condor 67:125–139. [443]

Verner, J. 1971. Survival and dispersal of male Long-billed Marsh Wrens. Bird-Banding 42:92–98. [8]

Verner, J. 1976. Complex song repertoire of male Long-billed Marsh Wrens in eastern Washington. Living Bird 14:263–300. [9, 81, 83, 217, 373, 390]

Verner, J., and G. H. Engelsen. 1970. Territories, multiple nest building, and polygyny in the Long-billed Marsh Wren. Auk 87:557–567. [8, 17]

Verrell, P. A. 1991. Illegitimate exploitation of sexual signalling systems and the origin of species. Ethol. Ecol. Evol. 3:273–283. [413, 416, 423]

Vicario, D. S. 1991a. Neural mechanisms of vocal production in songbirds. Curr. Opin. Neurobiol. 1:595–600. [286]

Vicario, D. S. 1991b. Contributions of syringeal muscles to respiration and vocalization in the Zebra Finch. J. Neurobiol. 22:63–73. [312, 316]

Vicario, D. S., and K. Yohay. 1993. Song-selective auditory input to a forebrain vocal control nucleus in the Zebra Finch. J. Neurobiol. 24:488–505. [289]

Vielliard, J. 1982. Brazilian bird songs. Pages 16–18 *in* Beautiful bird songs of the world (R. Kettle, Comp.). XVIII Int. Ornithol. Congr., The U.S.S.R. Academy of Sciences, Moscow. [276]

Vielliard, J. 1983. Catálogo sonográfico dos cantos e piados dos beija-flores do Brasil, 1. Boletim do Museu de Biologia "Mello Leitão," Serie Biologia 58:1–20. [277]

Vielliard, J. 1989. A new species of *Glaucidium* (Aves, Strigidae) of the Amazon region, Brazil. Rev. Bras. Zool. 6:685–694. [270]

Vielliard, J. 1990a. Estudo bioacústico das aves do Brasil: o gênero *Scytalopus.* Ararajuba 1:5–18. [272]

Vielliard, J. 1990b. Uma nova espécie de *Asthenes* da serra do Cipó, Minais Gerais, Brasil. Ararajuba 1:121–122. [270]

Vielliard, J. 1992. Audio cassette review: voices of the woodcreepers, Dendrocolaptidae. Bioacoustics 4:159–160. [273]

Vielliard, J. 1993. Recording wildlife in tropical rainforest. Bioacoustics 4:305–311. [480]

Vigilant, L., R. Pennington, H. Harpending, T. D. Kocher, and A. C. Wilson. 1989. Mitochondrial DNA sequences from single hairs from a southern African population. Proc. Natl. Acad. Sci. USA 86:9350–9354. [196]

Volman, S. F. 1993. Development of neural selectivity for birdsong during vocal learning. J. Neurosci. 13:4737–4747. [289, 336]

von Haartman, L. 1956. Territory in the Pied Flycatcher. Ibis 98:460–475. [455]

von Helversen, O., and D. von Helversen. 1994. Forces driving coevolution of song and song recognition in grasshoppers. Pages 253–284 *in* Neural basis of behavioural adaptations (K. Schildberger and N. Elsner, Eds.). Gustav Fischer, Jena. [241, 242, 257]

von Uexküll, J. 1934. Streifzüge durch die Umwelten von Tieren und Menschen. Springer Verlag, Berlin. [122]

Vu, E., E. Mazurek, and Y.-C. Kuo. 1994. Identification of a forebrain motor programming network for the learned song of Zebra Finches. J. Neurosci. 14:6924–6934. [286]

Wada, M. 1983. Environmental cycles, circadian clock, and androgen-dependent behavior in birds. Pages 191–200 *in* Avian endocrinology: environmental and ecological perspectives (S. Mikami, K. Homma, and M. Wada, Eds.). Springer Verlag, Berlin. [427]

Wagner, H. O. 1944. Notes on the history of the Emerald Toucanet. Wilson Bull. 56:65–76. [97]

Wagner, R. H. 1991. Evidence that female razorbills control extra-pair copulations. Behaviour 118:157–169. [264]

Wagner, R. H. 1993. The pursuit of extra-pair copulations by female birds: a new hypothesis of colony formation. Theor. Biol. 163:333–346. [264]

Walker, R. A. 1963. Some intense, low-frequency, underwater sounds of wide geographic distribution are apparently of biological origin. J. Acoust. Soc. Am. 35:1816–1824. [419]

Walker, T. J. 1983. Diel patterns of calling in nocturnal Orthoptera. Pages 45–72 *in* Orthopteran mating systems (D. T. Gwynne and G. K. Morris, Eds.). Westview Press, Boulder, Colo. [426]

Wallace, A. R. 1889. Darwinism: an exposition of the theory of natural selection with some of its applications. Macmillan, London. [454]

Wallschläger, D. 1980. Correlation of song frequency and body weight in passerine birds. Experientia 36:412. [229]

Wallschläger, D. 1983. Vergleich von Gesangsstrukturen zentralasiatischer Ammern (*Emberiza*). Mitt. Zool. Mus. Berlin Suppl. Ann. Ornithol. 7:85–116. [231]

Waltman, J. R., and S. R. Beissinger. 1992. Breeding behavior of the Green-rumped Parrotlet. Wilson Bull. 104:65–84. [104, 110]

Ward, D. 1992. The behavioural and morphological affinities of some vanelline plovers (Vanellinae: Charadriiformes: Aves). J. Zool. 228:625–640. [241]

Warham, J. 1979. The voice of the Soft-plumaged Petrel (*Pterodroma mollis*). Notornis 26:357–360. [165]

Warham, J. 1988a. Vocalizations of *Procellaria* petrels. Notornis 35:169–183. [163, 168, 173–175, 177]

Warham, J. 1988b. Responses of *Pterodroma* petrels to man-made sounds. Emu 88:109–111. [165]

Warham, J. 1990. The petrels: their ecology and breeding systems. Academic Press, London. [xx, 161, 162, 173]

Warham, J., B. R. Keeley, and G. J. Wilson. 1977. Breeding of the Mottled Petrel. Auk 94:1–17. [165]

Warren, R. M. 1993. Perception of acoustic sequences: global integration versus temporal resolution. Pages 37–68 *in* Thinking in sound: the cognitive psychology of human audition (S. McAdams and E. Bigand, Eds.). Oxford Univ. Press, Oxford. [127]

Wass, J. R. 1988. Song pitch-habitat relationships in White-throated Sparrows: cracks in acoustic windows? Can. J. Zool. 66:2578–2581. [354]

Wasserman, F. E. 1977. Intraspecific acoustical interference in the White-throated Sparrow (*Zonotrichia albicollis*). Anim. Behav. 25:949–952. [399]

Wasserman, F. E., and J. A. Cigliano. 1991. Song output and stimulation of the female in White-throated Sparrows. Behav. Ecol. Sociobiol. 29:55–59. [344, 459, 468]

Watanuki, Y. 1986. Moonlight avoidance behavior in Leach's Storm-Petrel as a defense against Slaty-backed Gulls. Auk 103:14–22. [162, 171]

Watkins, W. A. 1967. The harmonic interval: fact or artifact in spectral analysis of pulse trains. Pages 15–42 *in* Marine bio-acoustics (W. N. Tavolga, Ed.). Pergamon Press, Oxford. [251]

Watkins, W. A., and W. E. Schevill. 1971. Four hydrophone array for acoustic three-dimensional location. Woods Hole Oceanogr. Inst. Tech. Rep. 71–60. [419]

Watkins, W. A., and W. E. Schevill. 1972. Sound source location by arrival times on a non-rigid three-dimensional hydrophone array. Deep-Sea Res. 19:691–706. [419]

Watson, H. W., and F. Galton. 1875. On the probability of the extinction of families. J. Anthropol. Inst. Great Britain & Ireland 4:138–145. [213]

Watterson, G. A. 1978. The homozygosity test of neutrality. Genetics 88:405–417. [188]

Weary, D. M. 1988. Experimental studies on the song of the Great Tit. D. Phil. dissertation, Univ. Oxford, Oxford. [414]

Weary, D. M. 1989. Categorical perception of bird song: how do Great Tits (*Parus major*) perceive temporal variation in their song? J. Comp. Psychol. 103:320–325. [129]

Weary, D. M. 1990. Categorization of song notes in Great Tits: which acoustic features are used and why? Anim. Behav. 39:450–457. [129]

Weary, D. M. 1991. How Great Tits use song-note and whole-song features to categorize their songs. Auk 108:187–189. [129]

Weary, D. M. 1992. Bird song and operant experiments: a new tool to investigate song perception. Pages 201–210 *in* Playback and studies of animal communication (P. K. McGregor, Ed.). Plenum Press, New York. [129, 347, 414]

Weary, D. M., and J. R. Krebs. 1992. Great Tits classify songs by individual voice characteristics. Anim. Behav. 43:283–287. [370, 373, 414]

Weary, D. M., and R. G. Weisman. 1991. Operant discrimination of frequency and frequency ratio in the Black-capped Chickadee (*Parus atricapillus*). J. Comp. Psychol. 105:253–259. [353]

Weary, D. M., R. E. Lemon, and E. M. Date. 1987. Neighbour-stranger discrimination by song in the Veery, a species with song repertoires. Can. J. Zool. 65:1206–1209. [359]

Weary, D. M., J. R. Krebs, R. Eddyshaw, P. K. McGregor, and A. Horn. 1988. Decline in song output by Great Tits: exhaustion or motivation? Anim. Behav. 36:1242–1244. [307, 312]

Weary, D. M., J. B. Falls, and P. K. McGregor. 1990a. Song matching and the perception of song types in Great Tits, *Parus major.* Behav. Ecol. 1:43–47. [130, 131]

Weary, D. M., K. J. Norris, and J. B. Falls. 1990b. Song features birds use to identify individuals. Auk 107:623–625. [308, 370, 373, 414]

Weary, D. M., M. M. Lambrechts, and J. R. Krebs. 1991. Does singing exhaust male Great Tits? Anim. Behav. 41:540–542. [310, 312]

Weary, D. M., R. E. Lemon, and S. Perreault. 1992. Song repertoires do not hinder neighbor-stranger discrimination. Behav. Ecol. Sociobiol. 31:441–447. [357, 359, 369]

Weeden, J. S., and J. B. Falls. 1959. Differential responses of male Ovenbirds to recorded songs of neighboring and more distant individuals. Auk 76:343–351. [359]

Weimerskirch, H. 1992. Reproductive effort in long-lived birds: age-specific patterns of condition, reproduction and survival in the Wandering Albatross. Oikos 63:464–473. [174]

Weimerskirch, H., P. Jouventin, J.-L. Mougin, J.-C. Stahl, and M. van Beveren. 1985. Banding recoveries and the dispersion of seabirds breeding in the French Austral and Antarctic territories. Emu 85:22–23. [175]

Weinrich, U., W. Labov, and M. Herzog. 1968. Empirical foundations for a theory of language change. Pages 97–105 *in* Directions for historical linguistics (W. P. Lehmann and Y. Malkiel, Eds.). Univ. Texas Press, Austin. [213, 214, 219]

Weisman, R. G., and L. M. Ratcliffe. 1989. Absolute and relative pitch processing in Black-capped Chickadees, *Parus atricapillus.* Anim. Behav. 38:685–692. [348, 350, 353]

Weisman, R. G., and L. M. Ratcliffe. 1992. The perception of pitch constancy in bird song. Pages 243–261 *in* Cognitive aspects of stimulus control (W. K. Honig and J. G. Fetterman, Eds.). Erlbaum, Hillsdale, N.J. [125]

Weisman, R. G., L. Ratcliffe, I. Johnsrude, and T. A. Hurly. 1990. Absolute and relative pitch production in the song of the Black-capped Chickadee. Condor 92:118–124. [15, 349, 350]

Wells, K. D. 1988. The effect of social interactions on anuran vocal behavior. Pages 433–454 *in* The evolution of the amphibian auditory system (B. Fritzsch, M. J. Ryan, W. Wilczynski, T. E. Hetherington, and W. Walkowiak, Eds.). Wiley, New York. [399, 410]

Wells, K. D., and T. L. Taigen. 1986. The effect of social interaction on calling energetics in the gray treefrog, *Hyla versicolor.* Behav. Ecol. Sociobiol. 19:9–18. [310, 317]

Wells, K. D., and T. L. Taigen. 1989. Calling energetics of a Neotropical treefrog, *Hyla microcephala.* Behav. Ecol. Sociobiol. 22:13–22. [310–312]

Wells, S., and L. F. Baptista. 1979. Displays and morphology of an Anna × Allen Hummingbird hybrid. Wilson Bull. 91:524–532. [59]

Wells, S., R. Bradley, and L. F. Baptista. 1978. Hybridization in *Calypte* hummingbirds. Auk 96:537–549. [54, 59]

Welter, W. A. 1935. The natural history of the Long-billed Marsh Wren. Wilson Bull. 47:3–34. [8]

Wenink, P. W., and A. J. Baker. 1996. Mitochondrial DNA lineages in composite flocks of migratory and wintering Dunlins (*Calidris alpina*). Auk 113. In press. [244]

Wenink, P. W., A. J. Baker, and M. G. J. Tilanus. 1993. Hypervariable-control-region sequences reveal global population structuring in a long-distance migrant shorebird, the Dunlin (*Calidris alpina*). Proc. Natl. Acad. Sci. USA 90:94–98. [244]

Wenink, P. W., A. J. Baker, and M. G. J. Tilanus. 1994. Mitochondrial control-region sequences in two shorebird species, the Turnstone and Dunlin, and their utility in population genetic studies. Mol. Biol. Evol. 11:22–31. [244]

Wenink, P. W., A. J. Baker, H.-U. Rösner, and M. G. J. Tilanus. 1996. Global mitochondrial DNA phylogeography of Holarctic breeding Dunlins (*Calidris alpina*). Evolution. In press. [244]

Wenzel, J. W. 1992. Behavioral homology and phylogeny. Annu. Rev. Ecol. Syst. 23:361–381. [245]

West, M. J., and A. P. King. 1980. Enriching cowbird song by social deprivation. J. Comp. Physiol. Psychol. 94:263–270. [28]

West, M. J., and A. P. King. 1987. Settling nature and nurture into an ontogenetic niche. Dev. Psychobiol. 20:549–562. [25, 39]

West, M. J., and A. P. King. 1988a. Ontogenetic programs underlying geographic variation in cowbird song. Pages 1598–1605 *in* Proc. XIX Int. Ornithol. Congr. (H. Ouellet, Ed.). National Museum of Natural Science, Univ. Ottawa Press, Ottawa. [5, 12, 15]

West, M. J., and A. P. King. 1988b. Female visual displays affect the development of male song in the cowbird. Nature 334:244–246. [5, 12, 15, 33]

West, M. J., and H. L. Rheingold. 1978. Infant stimulation of maternal instruction. Infant Behav. Dev. 1:205–215. [23]

West, M. J., A. P. King, and D. H. Eastzer. 1981. Validating the female bioassay of cowbird song: relating differences in song potency to mating success. Anim. Behav. 29:490–501. [29, 31, 459]

West, M. J., A. N. Stroud, and A. P. King. 1983. Mimicry of the human voice by European Starlings: the role of social interaction. Wilson Bull. 95:635–640. [28]

Westcott, D. A., and A. Cockburn. 1988. Flock size and vigilance in parrots. Aust. J. Zool. 36:355–349. [103]

Westcott, D. A., and J. N. M. Smith. 1994. Behavior and social organization during the breeding season in *Mionectes oleagineus,* a lekking flycatcher. Condor 96:672–683. [278]

West-Eberhard, M. J. 1983. Sexual selection, social competition, and speciation. Q. Rev. Biol. 58:155–183. [242, 255]

Westneat, D. F., P. W. Sherman, and M. L. Morton. 1990. The ecology and evolution of extra-pair copulations in birds. Pages 331–369 in Current ornithology, vol. 7 (D. M. Power, Ed.). Plenum Press, New York. [263, 264, 431]

White, H., and R. White. 1980. Physics and music. Saunders, Philadelphia. [55]

Whitman, C. O. 1919. The behavior of pigeons. Posthumous works, vol. 3. Carnegie Institute, Washington, D.C. [48]

Whitney, B. M. 1992. Observations on the systematics, behavior, and vocalizations of "*Thamnomanes" occidentalis* (Formicariidae). Auk 109:302–208. [272]

Whitney, B. M. 1994a. Behavior, vocalizations, and possible relationships of four *Myrmotherula* antwrens (Formicariidae) from eastern Ecuador. Auk 111:469–474. [272]

Whitney, B. M. 1994b. A new *Scytalopus* tapaculo (Rhinocryptidae) from Bolivia, with notes on other Bolivian members of the genus and the *magellanicus* complex. Wilson Bull. 106:585–612. [270]

Whitney, B. M., and J. F. Pacheco. 1994. Behavior and vocalizations of *Gyalophylax* and *Megaxenops* (Furnariidae), two little-known genera endemic to northeastern Brazil. Condor 96:559–565. [272]

Whitney, C. L. 1981. Patterns of singing in the Varied Thrush. II. A model of control. Z. Tierpsychol. 57:141–162. [83]

Whitney, C. L. 1992. Temporal stability of song in a local population of Wood Thrushes. Wilson Bull. 104:516–520. [191, 193, 216]

Whitney, C. L., and J. Miller. 1987. Distribution and variability of song types in the Wood Thrush. Behaviour 103:49–67. [193]

Whittam, T. S., H. Ochman, and R. K. Selander. 1983. Multilocus genetic structure in natural populations of *Escherichia coli.* Proc. Natl. Acad. Sci. USA 80:1751–1755. [189]

Whittingham, L. A., A. Kirkconnell, and L. M. Ratcliffe. 1992. Differences in song and sexual dimorphism between Cuban and North American Red-winged Blackbirds (*Agelaius phoeniceus*). Auk 109:928–933. [259, 267, 279]

Wickler, W. 1986. Dialekte im Tierreich. Ihre Ursachen und Konsequenzen. Schriftenreihe Westfälische Wilhelms-Univ. Münster NF 6. [221]

Wickstrom, D. C. 1982. Factors to consider in recording avian sounds. Pages 1–52 *in* Acoustic communication in birds, vol. 1 (D. E. Kroodsma and E. H. Miller, Eds.). Academic Press, New York. [480, 481, 484]

Wiens, J. A. 1989. The ecology of bird communities. Vol. 1: Foundations and patterns. Cambridge Univ. Press, Cambridge. [7, 8]

Wild, J. M. 1993. Descending projections of the songbird nucleus robustus archistriatalis. J. Comp. Neurol. 338:225–241. [286]

Wild, J. M. 1994a. The auditory-vocal-respiratory axis in birds. Brain Behav. Evol. 44:192–209. [113]

Wild, J. M. 1994b. Visual and somatosensory inputs to the avian song system via nucleus uvaeformis (uva) and a comparison with projections of a similar thalamic nucleus in a nonsongbird, *Columba livia.* J. Comp. Neurol. 349:512–535. [113]

Wiley, E. O. 1981. Phylogenetics: the theory and practice of phylogenetic systematics. Wiley-Interscience, New York. [245]

Wiley, R. H. 1971. Song groups in a singing assembly of Little Hermits. Condor 73:28–35. [97, 277]

Wiley, R. H. 1976. Communication and spatial relationships in a colony of Common Grackles. Anim. Behav. 24:570–584. [168]

Wiley, R. H. 1983. The evolution of communication: information and manipulation. Pages 156–189 *in* Animal behaviour. Vol. 2: Communication (T. R. Halliday and P. J. B. Slater, Eds.). Blackwell, Oxford. [413, 416]

Wiley, R. H. 1991. Associations of song properties with habitats for territorial oscine birds of eastern North America. Am. Nat. 138:973–993. [409, 422]

Wiley, R. H. 1994. Errors, exaggeration and deception in animal communication. Pages 157–189 *in* Behavioral mechanisms in evolutionary ecology (L. Real, Ed.). Univ. Chicago Press, Chicago. [414]

Wiley, R. H., and R. Godard. 1992. Ranging of conspecific songs by Kentucky Warblers, *Oporornis formosus,* reduces the possibilities for interference in territorial interactions. IVth Int. Behav. Ecol. Congr. Abstr. T54c. [414]

Wiley, R. H., and D. G. Richards. 1978. Physical constraints on acoustic communication in the atmosphere: implications for the evolution of animal vocalizations. Behav. Ecol. Sociobiol. 3:69–94. [239, 305, 422]

Wiley, R. H., and D. G. Richards. 1982. Adaptations for acoustic communication in birds: sound transmission and signal detection. Pages 131–181 *in* Acoustic communication in birds, vol. 1 (D. E. Kroodsma and E. H. Miller, Eds.). Academic Press, New York. [121, 124, 171, 186, 265, 321, 324, 331, 336, 338, 409, 414, 422]

Wiley, R. H., and M. S. Wiley. 1977. Recognition of neighbors' duets by Stripe-backed Wrens (*Campylorhynchus nuchalis*). Behaviour 62:10–34. [359, 360]

Wilkinson, G. S. 1994. Canary 1.1: sound analysis software for Macintosh computers. Bioacoustics 5:227–238. [419]

Williams, C. L., A. M. Barnett, and W. H. Meck. 1990. Organizational effects of early gonadal secretions on sexual differentiation in spatial memory. Behav. Neurosci. 104:84–97. [342]

Williams, H. E. 1985. Sexual dimorphism of auditory activity in the Zebra Finch song system. Behav. Neural Biol. 44:470–484. [342]

Williams, H. E., and F. Nottebohm. 1985. Auditory responses in avian vocal motor neurons: a motor theory for song perception. Science 229:279–282. [429]

Williams, H. E., J. Cynx, and F. Nottebohm. 1989. Timbre control in Zebra Finch (*Taeniopygia guttata*) song syllables. J. Comp. Psychol. 103:366–380. [370]

Williams, H. E., K. Kilander, and M. L. Sotanski. 1993. Untutored song, reproductive success and song learning. Anim. Behav. 45:695–705. [343, 354]

Williams, J. M., and P. J. B. Slater. 1990. Modelling bird song dialects: the influence of repertoire size and numbers of neighbours. J. Theor. Biol. 145:487–496. [213]

Williams, J. M., and P. J. B. Slater. 1991. Simulation studies of song learning in birds. Pages 281–287 *in* Simulation of adaptive behavior (J.-A. Meyer and S. Wilson, Eds.). MIT Press, Boston. [13, 15, 16]

Williams, L., and M. H. MacRoberts. 1977. Individual variation in songs of Dark-eyed Juncos. Condor 79:106–112. [191]

Williams, L., and M. H. MacRoberts. 1978. Song variation in Dark-eyed Juncos in Nova Scotia. Condor 80:237–240. [191]

Willis, E. O. 1960. Voice, courtship, and territorial behavior of ant-tanagers in British Honduras. Condor 62:73–87. [434, 447, 448]

Willis, E. O. 1992a. Three *Chamaeza* antthrushes in eastern Brazil (Formicariidae). Condor 94:110–116. [270]

Willis, E. O. 1992b. Comportamento e ecologia do Arapacu-Barrado, *Dendrocolaptes certhia* (Aves, Dendrocolaptidae). Bol. Mus. Para. Emilio Goeldi Ser. Zool. 8:151–216. [272]

Wilson, A. C., S. Carlson, and T. J. White. 1977. Biochemical evolution. Annu. Rev. Biochem. 46:573–639. [213, 214]

Wingfield, J. C., and D. S. Farner. 1993. Endocrinology of reproduction in wild species. Pages 163–327 *in* Avian biology, vol. 9 (D. S. Farner, J. R. King, and K. C. Parkes, Eds.). Academic Press, London. [427, 429, 430]

Wingfield, J. C., and T. P. Hahn. 1994. Testosterone and territorial behaviour in sedentary and migratory sparrows. Anim. Behav. 47:77–89. [294, 394, 428]

Wingfield, J. C., and D. Lewis. 1993. Hormonal and behavioral responses to simulated territorial intrusion in the cooperatively breeding White-browed Sparrow Weaver, *Plocepasser mahali.* Anim. Behav. 45:1–11. [303]

Wingfield, J. C., and M. Moore. 1987. Hormonal, social, and environmental factors in the reproductive biology of free-living male birds. Pages 149–175 *in* Psychobiology of reproductive behavior: an evolutionary perspective (D. Crews, Ed.). Prentice-Hall, Englewood Cliffs, N.J. [297]

Wingfield, J. C., and M. Wada. 1989. Male-male interactions increase both luteinizing hormone and testosterone in the Song Sparrow, *Zonotrichia melodia:* specificity, time course and possible neural pathways. J. Comp. Physiol. A 166:189–194. [394]

Wingfield, J. C., C. M. Vleck, and D. S. Farner. 1981. Effect of day length and reproductive state on diel rhythms of luteinizing hormone levels in the plasma of White-crowned Sparrows, *Zonotrichia leucophrys gambelii.* J. Exp. Zool. 217:261–264. [427]

Wingfield, J. C., G. F. Ball, A. M. Dufty Jr., R. E. Hegner, and M. Ramenofsky. 1987. Testosterone and aggression in birds. Am. Sci. 75:602–608. [394]

Wingfield, J. C., R. E. Hegner, A. M. Dufty Jr., and G. F. Ball. 1990. The "challenge hypothesis": theoretical implications for patterns of testosterone secretion, mating systems, and breeding strategies. Am. Nat. 136:829–846. [429]

Wistel-Wozniak, A., and H. Hultsch. 1992. Song performance in nightingales which had been raised without exposure to acoustic learning programmes. Verh. Dtsch. Zool. Ges. 85:246. [89]

Witkin, S. R. 1977. The importance of directional sound radiation in avian vocalization. Condor 79:490–493. [142]

Wittenberger, J. F., and G. L. Hunt. 1985. The adaptive significance of coloniality in birds. Pages 1–79 *in* Avian biology, vol. 8 (D. S. Farner, J. R. King and K. C. Parkes, Eds.). Academic Press, New York. [171]

Wolffgramm, J. 1973. Lautmustersequenzen und periodische Beziehungen im Rollerkanari-Gesang (*Serinus canaria*). J. Comp. Physiol. 35:65–88. [81]

Wolffgramm, J. 1980. The role of periodicities in avian vocal communication. Pages 671–676 *in* Acta XVII Congr. Int. Ornithol. (R. Nöhring, Ed.). Verlag Deutsche Ornithol.-Gesellschaft, Berlin. [81, 82]

Wolfgramm, J., and D. Todt. 1982. Pattern and time specificity in vocal responses of blackbirds *Turdus merula* L. Behaviour 65:264–287. [84, 130]

Wright, H. W. 1913. Morning awakening and even-song. Auk 30:512–537. [443]

Wright, S. 1931. Evolution in Mendelian populations. Genetics 16:97–159. [184]

Wu, C.-I., and W.-H. Li. 1985. Evidence for higher rates of nucleotide substitution in rodents than in man. Proc. Natl. Acad. Sci. USA 82:1741–1745. [214]

Wunderle, J. M. 1978. Differential response of territorial Yellowthroats to the songs of neighbors and non-neighbors. Auk 95:389–395. [359, 363]

Würdinger, I. 1970. Erzeugung, Ontogenie und Funktion der Lautäußerungen bei vier Gänsearten. Z. Tierpsychol. 27:257–302. [54]

Wyndham, E. 1980a. Environment and food of the Budgerigar, *Melopsittacus undulatus.* Aust. J. Ecol. 5:47–61. [98, 99]

Wyndham, E. 1980b. Diurnal cycle, behaviour and social organization of the Budgerigar, *Melopsittacus undulatus.* Emu 80:25–33. [97, 98, 100–102, 104, 105]

Wyndham, E. 1980c. Total body lipids of the Budgerigar, *Melopsittacus undulatus* (Psittaciformes: Platycercidae), in inland mid-eastern Australia. Aust. J. Zool. 28:239–247. [98]

Wyndham, E. 1980d. Aspects of biorhythms in the Budgerigar, *Melopsittacus undulatus* (Shaw), a parrot of inland Australia. Pages 485–492 *in* Acta XVII Congr. Int. Ornithol. (R. Nöhring, Ed.). Verlag Deutsche Ornithol.-Gesellschaft, Berlin. [98]

Wyndham, E. 1980e. Diurnal changes in crop contents and total body lipid of Budgerigars, *Melopsittacus undulatus.* Ibis 122:229–234. [98]

Wyndham, E. 1981. Breeding and mortality of Budgerigars, *Melopsittacus undulatus.* Emu 81:240–243. [98, 104]

Wyndham, E. 1983. Movements and breeding seasons of the Budgerigar. Emu 82:276–282. [98, 99, 103]

Yasuda, N., L. Cavalli-Sforza, M. Skolnick, and A. Moroni. 1974. The evolution of surnames. An analysis of their distribution and extinction. Theor. Pop. Biol. 5:123–142. [213]

Yasukawa, K. 1981a. Song repertoires in the Red-winged Blackbird (*Agelaius phoeniceus*): a test of the Beau Geste hypothesis. Anim. Behav. 29:114–125. [298, 406]

Yasukawa, K. 1981b. Territory establishment in Red-winged Blackbirds: importance of aggressive behavior and experience. Condor 81:258–264. [160]

Yasukawa, K. 1990. Does the "teer" vocalization deter prospecting female Red-winged Blackbirds? Behav. Ecol. Sociobiol. 26:421–426. [160]

Yasukawa, K., J. L. Blank, and C. B. Patterson. 1980. Song repertoires and sexual selection in the Red-winged Blackbird. Behav. Ecol. Sociobiol. 7:233–238. [456, 460, 469, 471]

Yasukawa, K., E. Bick, D. W. Wagman, and P. Marler. 1982. Playback and speaker-replacement experiments on song-based neighbor, stranger and self discrimination in male Red-winged Blackbirds. Behav. Ecol. Sociobiol. 10:211–215. [359]

Ydenberg, R. C., L.-A. Giraldeau, and J. B. Falls. 1988. Neighbors, strangers and the asymmetric war of attrition. Anim. Behav. 36:343–347. [366, 414, 423]

Ydenberg, R. C., L.-A. Giraldeau, and J. B. Falls. 1989. Remarks on Getty's "fighting to learn" hypothesis. Anim. Behav. 37:336–337. [423]

Yoffee, N. 1990. Before Babel, a review article. Proc. Prehist. Soc. 56:299–313. [213]

Yoneda, T., and K. Okanoya. 1991. Ontogeny of sexually dimorphic distance calls in Bengalese Finches (*Lonchura domestica*). J. Ethol. 9:41–46. [58]

Zahavi, A. 1975. Mate selection—a selection for a handicap. J. Theor. Biol. 53:205–214. [436, 461, 462]

Zahavi, A. 1977. The cost of honesty (further remarks on the handicap principle). J. Theor. Biol. 67:603–605. [461, 462]

Zahavi, A. 1979a. Ritualization and the evolution of movement signals. Behaviour 72:77–81. [160]

Zahavi, A. 1979b. Why shouting? Am. Nat. 113:155–156. [423]

Zahavi, A. 1993. The fallacy of conventional signalling. Philos. Trans. R. Soc. B 340:227–230. [416]

Zann, R. 1965. Behavioural studies of the Quarrion (*Nymphicus hollandicus*). Honours thesis, Univ. New England, Armidale, Australia. [97, 105, 112]

Zann, R. 1985. Ontogeny of the Zebra Finch distance call. I. Effects of crossfostering to Bengalese Finches. Z. Tierpsychol. 68:1–23. [43, 58]

Zann, R. 1990. Song and call learning in wild Zebra Finches in south-east Australia. Anim. Behav. 40:811–828. [41, 43]

Zann, R. 1993. Variation in song structure within and among populations of Australian Zebra Finches. Auk 110:716–726. [41]

Zemlin, W. R. 1988. Speech and hearing science. Prentice-Hall, Englewood Cliffs, N.J. [305, 314, 315]

Zigmond, R., R. Detrick, and D. Pfaff. 1980. An autoradiographic study of the localization of androgen concentrating cells in the Chaffinch. Brain Res. 182:369–381. [296]

Zimmer, J. T. 1926. Catalogue of the Edward E. Ayer ornithological library. Field Mus. Nat. Hist. Publ. Zool. Ser. 16. [270]

Zimmer, K. J. 1993. Ted Parker remembered. Birding 25:377–380. [v]

Zink, R. M. (Convenor). 1993. Species concepts. Special symposium, 111th Annual Meeting, American Ornithologists' Union, Fairbanks, Alaska. [273]

Zink, R. M., and D. L. Dittmann. 1993. Population structure and gene flow in the Chipping Sparrow and a hypothesis for evolution in the genus *Spizella*. Wilson Bull. 105:399–413. [18]

Zuk, M. 1993. Feminism and the study of animal behavior. BioScience 43:774–778. [339]

Zuk, M., L. W. Simmons, and L. Cupp. 1993. Calling characteristics of parasitized and unparasitized populations of the field cricket, *Telogryllus oceanicus*. Behav. Ecol. Sociobiol. 33:339–343. [423, 436]

Zwicker, E., G. Flottrop, and S. S. Stevens. 1957. Critical band width in loudness summation. J. Acoust. Soc. Am. 29:548–557. [328]

SUBJECT INDEX

TAXONOMIC INDEX

Latin or common names preceded by an asterisk (*) are not recognized by Monroe and Sibley (1993; see p. xx).